~~RESTRICTED~~

NAVAL ORDNANCE and GUNNERY

NAVPERS 16116

PREPARED FOR TRAINING IN NAVAL ORDNANCE AND GUNNERY

Bureau of Naval Personnel
Training Division
May, 1944

This text has been digitally watermarked to prevent illegal duplication.

©2013 Periscope Film LLC
All Rights Reserved
ISBN#978-1-937684-22-8
www.PeriscopeFilm.com

TABLE OF CONTENTS

	PAGE
PREFACE	v
CHAPTER 1—Introduction to weapons	1
CHAPTER 2—Explosives	7
A. Explosive reactions	7
B. Service explosives and their uses	9
C. Ammunition	17
CHAPTER 3—Gun ammunition	21
A. General types	21
B. Primers	23
C. Propelling charges	24
D. Projectiles	31
E. Fuzes and tracers	37
CHAPTER 4—Gun assemblies	42
A. Gun barrels	42
B. Breech assemblies	49
C. Mounts	58
D. Sights	69
CHAPTER 5—A major caliber turret installation	78
A. Introduction	78
B. Gun and breech assembly	79
C. Slide assembly	87
D. Elevating, training, and sight gear	88
E. Ammunition handling	92
F. Safety precautions	100
CHAPTER 6—Semi-automatic guns	104
A. Introduction	104
B. Operation of the breech mechanism	104
C. 3"/50 caliber gun and mounts	113
D. 5"/38 caliber gun and mounts	118
CHAPTER 7—Machine guns	149
A. Introduction	149
B. 40 mm. gun assemblies	149
C. 20 mm. gun and mounts	164
D. Caliber .50 machine gun	182
CHAPTER 8—Small arms	201
A. Introduction	201
B. Caliber .30 Springfield rifle	202
C. Caliber .45 automatic pistol	203
D. Thompson submachine gun	204
E. U. S. carbine, caliber .30	206
CHAPTER 9—Bomb-type ammunition and release gear	207
A. Introduction	207
B. Torpedoes	207
C. Mines	220
D. Depth charges	223
CHAPTER 10—Introduction to fire control	229
A. Historical	229
B. Modern armament	231

	PAGE
CHAPTER 11—Range measurement	237
A. Range finders	237
CHAPTER 12—Gun sight principles	251
A. Trajectory analysis	251
B. Range tables	257
C. Gun sights and sight setting	259
D. Gun-sight alignment	261
CHAPTER 13—The surface problem; analytical determination of sight settings	265
A. Range keeping	265
B. The gun ballistic	271
C. Control, arbitrary, total, and initial ballistics	280
D. Other considerations	284
CHAPTER 14—The surface problem; practical determination of sight settings	290
A. Graphic tracking	290
B. Graphic range keeping	292
C. Basic range-keeping mechanisms	296
D. An elementary range keeper	303
E. Basic ballistics computing mechanisms	310
F. An elementary computer	315
CHAPTER 15—The AA problem; analysis of sight and fuze settings	328
A. Positioning computations	328
B. Rate control	332
C. Ballistics computations	336
D. Practical considerations	351
CHAPTER 16—Fundamentals of director control	356
A. Applying sight settings and firing	356
B. Synchros and servos	359
C. An elementary director system	363
D. Director control with measured gun orders	367
E. Director control with computed gun orders	376
CHAPTER 17—Dual-purpose battery fire-control system	387
A. General description	387
B. The Mark 37 director	391
C. The Mark 6 stable element	403
D. The Mark 1 computer	416
E. Gun-control equipment	436
CHAPTER 18—Main battery fire-control system	443
A. General description of the system	443
B. Turret fire-control equipment	456
CHAPTER 19—Machine-gun control	459
A. General	459
B. Ring sight and tracer control	461
C. The Mark 14 sight and Mark 51 director	462
CHAPTER 20—Torpedo control	473
A. Principles of torpedo fire	473
B. Torpedo control problem	476
C. Torpedo director system	479
CHAPTER 21—Spotting	489
A. Salvo analysis and relation to spotting	489
B. The spotter and methods of spotting	494
CHAPTER 22—Organization and interior communications	500
A. Organization	500
B. Communications	505

	PAGE
CHAPTER 23—Suggestions to junior gunnery officers	509
A. The junior officer and the gunnery department	509
B. Sources of information	511
C. Reports and requisitions	516
APPENDIX A—Standard fire-control symbols	518
APPENDIX B—Definitions	525
APPENDIX C—Standard commands	536
APPENDIX D—5-inch range table from O. P. 551	541

NOTICE: This document reproduces the text of a manual first published by the Department of the Army, Washington DC. All source material contained in the reproduced document has been approved for public release and unlimited distribution by an agency of the U.S. Government. Any U.S. Government markings in this reproduction that indicate limited distribution or classified material have been superseded by downgrading instructions that were promulgated by an agency of the U.S. government after the original publication of the document. No U.S. government agency is associated with the publishing of this reproduction.

PREFACE

Improvements in naval ordnance and fire-control methods have been so numerous and comprehensive since the opening of hostilities that the need for a new instruction manual became apparent to officers teaching ordnance and gunnery at Naval Reserve Midshipmen's Schools. Officers in charge of the Standards and Curriculum Section, Training Division, Bureau of Personnel, likewise recognized the need for a manual which would represent the standardized curriculum of these schools. Accordingly, in June, 1943, the preparation of "Naval Ordnance and Gunnery" was authorized, and authors were selected to prepare the manuscript.

The first part of the following manual deals largely with naval weapons, and is intended to provide the student with a basic conception of gun design and construction. Emphasis is given to the specific guns that graduates of Naval Reserve Midshipmen's Schools are most likely to encounter upon joining the Fleet. The second part of the manual is concerned with fire control, including the surface and antiaircraft problems. In this case the emphasis is upon director control principles and the particular systems in common use.

The original manuscript was prepared by the following authors: Lieut. Comdr. Orval H. Polk, U.S.N.R. of U.S.N.R.M.S., Columbia University; Lieut. Ernest H. Dunlap Jr., U.S.N. of U.S.N.R.M.S., Notre Dame University; and Lieut. Frank L. Seyl, U.S.N.R. of U.S.N.R.M.S., Northwestern University. Advantage is taken of this opportunity to acknowledge the fine work of these capable and enthusiastic officers.

The manual was edited by Lieut. Fred L. Fitzpatrick, U.S.N.R., and Lieut. (j.g.) H. E. Shaw, U.S.N.R., Standards and Curriculum Section, Training Division, Bureau of Naval Personnel.

The authors worked at the U. S. Naval Academy. Captain Duncan H. Curry, U.S.N., Head of the Department of Ordnance and Gunnery, made the facilities of his organization freely available. Comdr. H. F. Agnew, U.S.N., was particularly generous in supplying aid and encouragement. Various portions of the manuscript were reviewed and criticised by Comdr. F. V. H. Hilles, U.S.N., and Lieut. Comdr. E. W. Foster, U.S.N. (Ret.).

Acknowledgment is also due to Captain M. R. Kelley, U.S.N., of the Bureau of Ordnance, and to Comdr. Alvin C. Eurich, U.S.N.R., and Comdr. Carlton R. Adams, U.S.N., successively Officers-in-Charge, Standards and Curriculum Section, Training Division, Bureau of Naval Personnel, whose support of the project made the completion of this manual possible.

Recognition of various subject material sources is also in order. Much of the material has come more or less directly from publications of the Bureau of Ordnance. In addition, publications of the General Motors Corporation have been drawn upon in the case of the 20 mm. and caliber .50 guns. Considerable use was made of ordnance department notes compiled at the U. S. Naval Academy and the various Midshipmen's Schools. Illustrations were prepared at the U. S. Naval Training Aids Development Center, New York City, under the immediate supervision of Lieut. (j.g.) Eugene Wasserman, and Lieut. (j.g.) J. S. Virostek.

Suggestions for improvement of the manual are invited. Some of the copies have been bound in loose-leaf form to facilitate revisions that may be made necessary by changes in equipment and design. Although the manual was prepared primarily for use at Naval Reserve Midshipmen's Schools, and represents the ordnance and gunnery curriculum of these institutions, its form permits easy addition or elimination of subject material so as to widen its scope of usefulness.

> EDWIN L. BRASHEARS,
> *Lieutenant Commander, U.S.N.R.*

May, 1944.

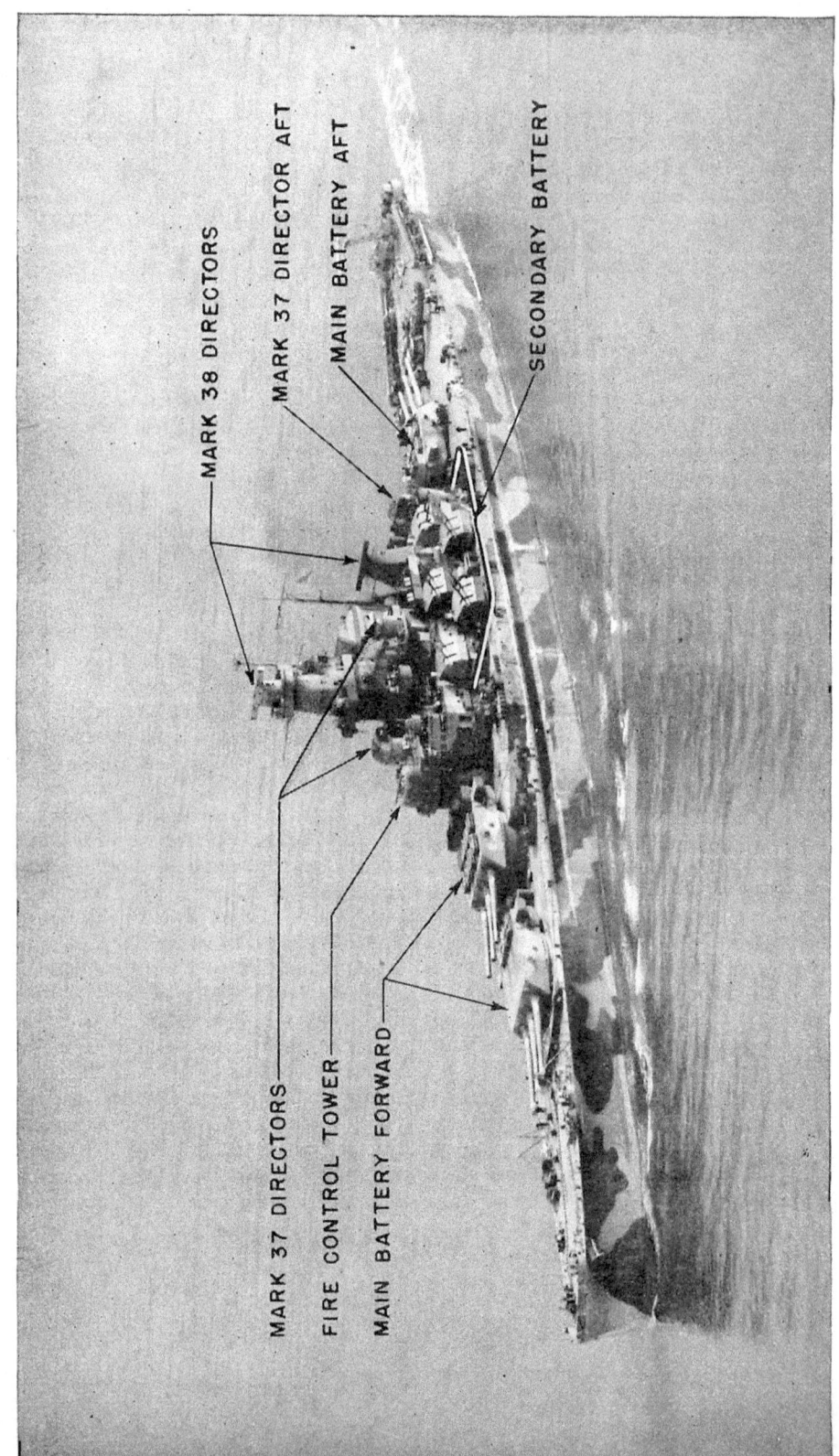

FIGURE 1-1. Arrangement of guns on a battleship.

RESTRICTED

CHAPTER 1

INTRODUCTION TO WEAPONS

1-1. Function of the gunnery department. The effectiveness of a navy must ultimately be gauged by its ability to damage and destroy the enemy. Success in such a venture depends upon various weapons, which aboard ship, are under supervision of the *gunnery department*. Officers and men in this department are primarily concerned with that care and use of armament which will enable the ship to deliver the most effective blows.

1-2. Ordnance. The term ordnance includes weapons and the control equipment used to operate them. The Bureau of Ordnance of the Navy Department supervises the manufacture, installation, upkeep and operation of all naval ordnance.

1-3. Gunnery. Weapons are tools of offense or defense, but are useless without proper application. Strictly speaking, *gunnery* is the art or science of using guns. As the term is employed in this discussion, however, it includes the handling of all armament. An understanding of the weapons and their capabilities is implied, as well as intimate knowledge of the methods and equipment used in controlling them.

1-4. Scope of text. Nearly every naval officer will have some association with guns, while a smaller number will work with torpedoes, mines and depth charges. This manual emphasizes guns and their control, because specialized training usually is provided for those who must have an intimate knowledge of other weapons. The chapters which follow, then, do not represent a book of reference, but rather an introduction to the broader aspects of ordnance and gunnery.

The descriptions of equipment presented therein are primarily designed to show functional operations. They are not intended for use by maintenance personnel, but are to acquaint officers with the operation and use of the equipment they may some day supervise. Discussions pertain strictly to U. S. Navy equipment, although in a few cases they also are applicable to items of Army ordnance.

1-5. Presentation of the subject. Any practical treatment of weapons and their control cannot be sharply divided into sections on ordnance and gunnery. It is practically impossible to study the structure of ordnance mechanisms without giving some thought to their functions; likewise, the successful study of gunnery is dependent upon thorough understanding of the guns and instruments concerned. In this book a compromise has been effected; the first part concentrates upon a study of weapons with minimum reference to control, the second part is largely concerned with fire control.

1-6. Study hints. It is suggested that the student read an assignment rapidly to obtain a general concept of the subject matter. It is better not to linger on a troublesome paragraph during the first reading, for frequently subsequent statements provide the necessary clarification. After the first reading, a thorough study of each article, a paragraph at a time, will usually produce the desired results. The student is warned, however, that ordnance and gunnery cannot be learned successfully by memorizing descriptions. The real key to success is understanding of functions, which in turn depends upon knowledge of the structures which make these functions possible.

Study of actual equipment is also a vital phase of the learning process. Insofar as possible, the inspection and operation of such equipment should be coordinated with study of appropriate sections of the manual.

1-7. Naval weapons. Among the weapons used in naval warfare are (1) guns, (2) torpedoes, (3) mines, (4) depth charges, (5) bombs and (6) chemicals.

INTRODUCTION TO WEAPONS

Guns. Nearly all naval vessels and aircraft carry guns of some type. In time of war even merchant ships are so equipped, their weapons usually being manned by naval crews.

Developments in the manufacture of explosives and guns have greatly increased the destructive power and effective range of naval weapons. The 16-inch guns of some battleships can throw a projectile weighing more than a ton to distances of about 40,000 yards with reasonable accuracy. The general location of guns aboard typical ships is illustrated in figures 1-1 to 1-3.

Torpedoes. Torpedoes are the principal weapons of destroyers, submarines, torpedo planes, and PT boats. The modern version is a development from an early form in which an explosive on the end of a spar was placed against the hull of the target by a small boat or a submarine. Times have changed, however, and today a torpedo carries a very heavy charge of high explosive, and when properly launched, propels itself through the water at high speed and under accurate control to distances of 6 or 7 miles.

Mines. Mines are passive weapons which float in the water or rest on the bottom and explode when a target comes within range. They contain large charges of high explosive, and provide barricades against enemy entrance of harbors and other areas. Mine fields may also be planted so as to surprise the enemy in a place where his future presence may be anticipated, or they may be laid in the path of an enemy force causing it to maneuver to a disadvantage. These weapons are normally placed in position by mine layers, but they may also be planted by destroyers, cruisers, submarines, planes or partol craft.

Depth charges. Depth charges are weapons developed expressly for attacks against submarines. The name comes from the original form of depth charge, which was designed to explode at a definite depth in the water. The destructive effect of such a charge may be considerable even though the explosion occurs several yards from the target. Depth charges are carried by destroyers, patrol craft, destroyer escorts, and some cruisers.

Bombs. The several types of missiles (except torpedoes and mines) dropped from planes are usually classed as bombs. Since this text is designed for deck officers, they are not discussed in detail.

Chemicals. Various chemicals are available for attacks upon personnel, for hindering visibility, and for initiating fires. The specialized subject of *chemical warfare* deals with the use of these chemicals and the protective measures to be taken against them. Most phases of chemical warfare do not lend themselves readily to naval employment, although an obvious exception exists in the case of screening operations.

1-8. **Ballistics.** Ballistics is the science of projectile motion. Volumes have been written on the subject; only an introduction thereto is attempted here. The naval officer necessarily is interested in applying the results of the ballistician's work, just as he employs results obtained by mechanical and electrical engineers, physicists, and chemists.

The general subject of projectile motion may be divided into *interior* and *exterior ballistics*. Interior ballistics considers motion of the projectile within the bore of a gun. The gun designer must understand this subject so that he can build a gun which will expel the projectile at a specified muzzle velocity. Planning for such a result involves not only the interior shape of the gun, but also the type of powder, and many other factors over which the deck officer has no control. The primary interest of the naval officer in interior ballistics is the velocity with which the projectile leaves the bore.

The action of the projectile after it leaves the gun is a matter of exterior ballistics. The projectile is subjected to a number of influences which affect the path it will follow. The nature and magnitude of these influences are important to any officer in gunnery, but the actual analyses of the phenomena are problems for the ballisticians.

1-9. **Fire control.** The practical application of exterior ballistics, and the methods and devices used to control the guns (and other weapons) are included in the subject known as *fire control*. The latter part of the book deals with this subject in some detail, but a few items of fire-control equipment are briefly described here to provide the student with a more inclusive mental picture of the ordnance equipment found aboard most ships.

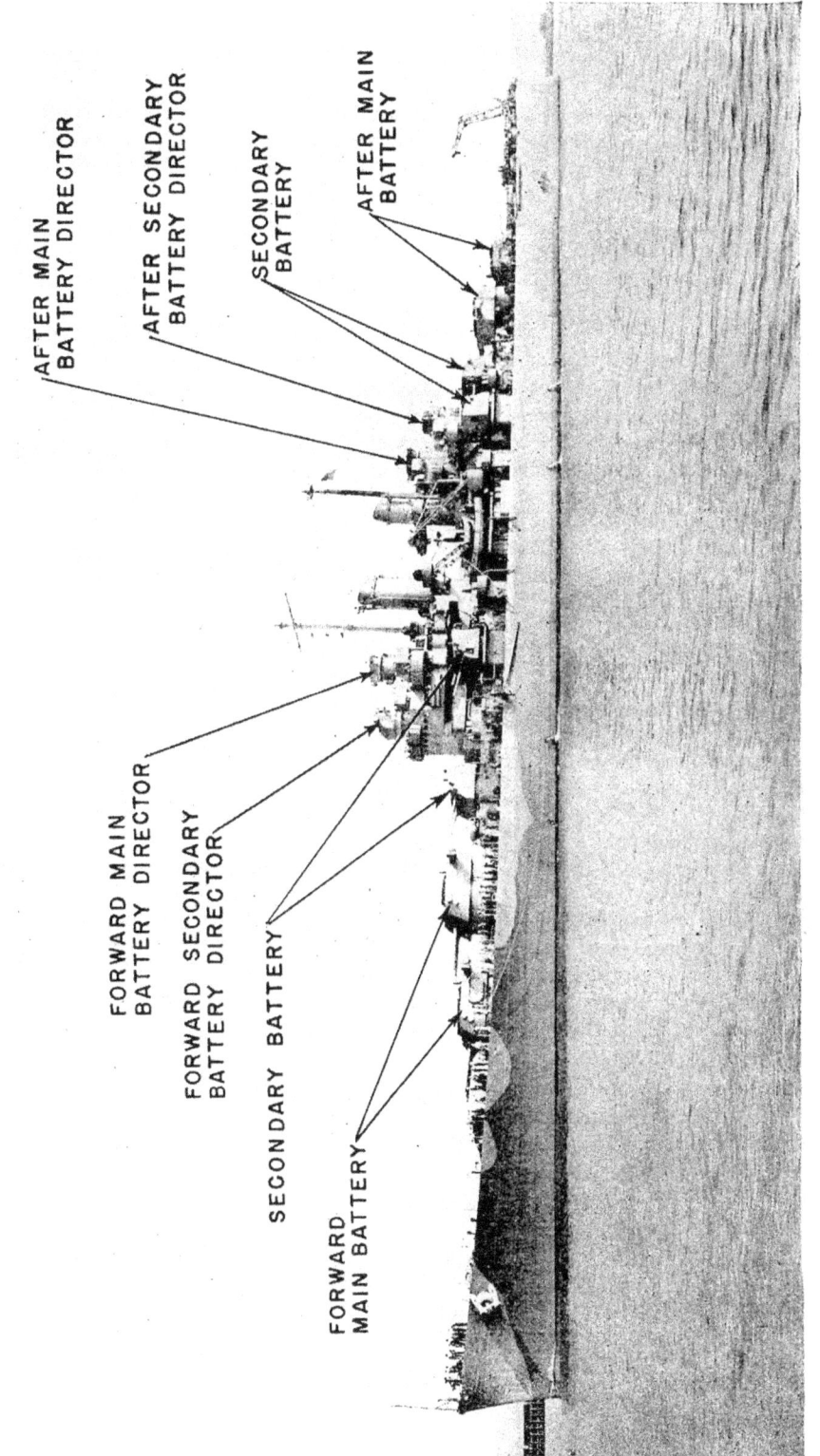

FIGURE 1-2. Arrangement of guns on a cruiser.

INTRODUCTION TO WEAPONS

Range finders are optical instruments used to measure the distance to the target.

Directors are mechanical and electrical instruments which control guns from a distant location. This position is generally located at a higher level than the guns to provide greater visibility.

Range keepers and *computers* are mechanical and electrical instruments which automatically compute information needed to direct gunfire.

Visual indicators are electrical devices which receive certain standard orders and information from a remote control station. The transmitted information appears on dials at the receiving station.

Stable elements are gyroscopically controlled mechanisms which measure movement of the ship with respect to the true horizontal plane, and are employed to increase the accuracy of gunfire.

1–10. Identification of ordnance equipment. Each assembled unit of ordnance equipment is identified by a name, a *mark* number, often a *modification* number, and a *serial* number. This information is stamped either on the equipment itself or on an attached plate. Whenever a basic change in design is made a new mark number is assigned. Modification numbers are added when there has been a minor alteration of design. Individual units of identical design have the same name, mark, and modification numbers, but have different serial numbers.

An example will help to illustrate the use of this identification system: Range keeper, Mark 1, Mod. 0 is the first range keeper designed. Units built to this design are assigned serial numbers. Range keeper, Mark 1, Mod. 1, is similar to the Mark 1, Mod. 0, but has a slight modification. Each range keeper built to this design has a different serial number. From these facts it can be seen that range keeper, Mark 7, Mod. 5 is built according to the fifth modification of the seventh basic design.

In referring to a piece of ordnance, the information required for identification depends upon the circumstances. For example, if reference is made to functions only, the name, mark and modification will be sufficient. If, however, spare parts are requested from the Bureau of Ordnance, the serial number may also be necessary. This number, together with mark and modification numbers, identifies the specific unit involved, and enables the factory to refer to its manufacturing records for precise information.

To consider a further example, a gun mount has a mark and modification number. It also has an *assembly number*. All three of these items should be noted in requesting replacement parts. Within a given mark and modification there may be various assembly numbers, each representing the fact that some component parts differ from components listed under other assembly numbers.

The name of a gun includes its bore diameter. Thus, we find the designation: 5-inch gun, Mark 1, Mod. 2; or 3-inch gun, Mark 21, Mod. 1. There is a separate series of mark numbers for each caliber of gun.

1–11. Knowledge of matériel. Operation of some ordnance equipment calls for relatively detailed knowledge. Such details have been included in this manual but an effort has been made to present them in understandable fashion. An officer must be well acquainted with his equipment, as the following quotation from the *Bureau of Ordnance Manual* testifies:

> "One of the greatest detriments to progress is a tendency on the part of personnel to simply follow general directions and to investigate only where troubles have forced investigation. Results obtained will lead to misdirected effort and erroneous conclusions when the users of matériel fail to equip themselves with detailed and thorough knowledge of the mechanism with which they work. This can only be obtained by hard work, study of pamphlets, blueprints, and ordnance circular letters, etc., without tearing down the material. The starting point, therefore, on the part of officers is a thorough study and understanding of all details of the mechanisms assigned to their charge. Through them the necessary knowledge must reach the enlisted personnel. It is important that no attempt should be made to get results until this knowledge is first obtained."

1–12. Care of matériel. The Bureau of Ordnance issues complete instructions for proper maintenance of ordnance matériel. These instructions should always be consulted and followed in

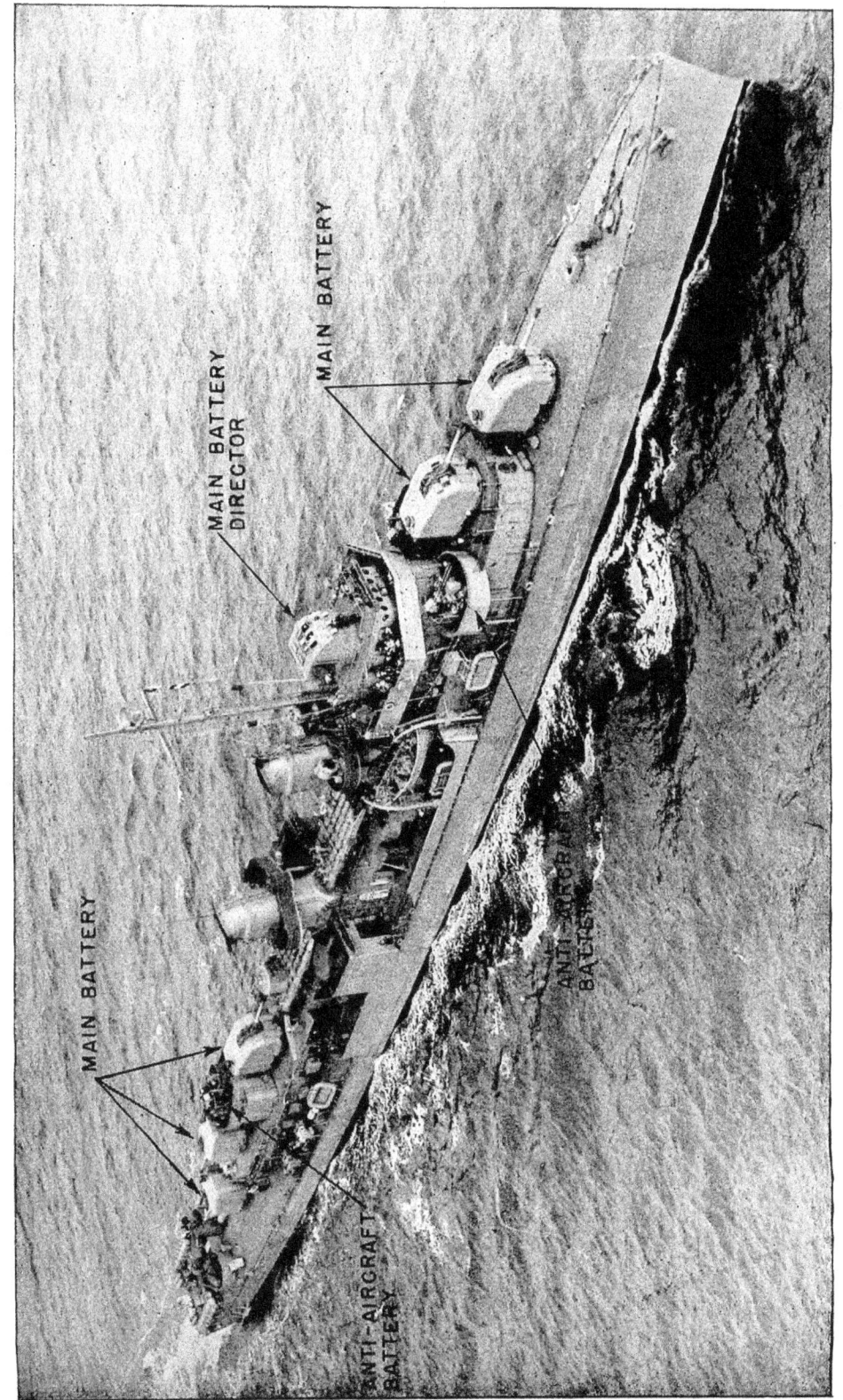

FIGURE 1-3. Arrangement of guns on a destroyer.

detail. Needless to say, they cannot be fully reviewed here, but a few general remarks are appropriate.

All ordnance matériel is carefully manufactured, usually to close tolerances. Any careless treatment is likely to damage seriously a valuable piece of equipment, disabling it when it may be needed most. In using any fine apparatus, it is wise to be governed by common sense. The equipment was built to function. If it does not, something is wrong, and physical forcing will inevitably cause trouble. Levers, knobs, buttons and switches should not be touched by a person who does not know what they will do. The *Bureau of Ordnance Manual* states: "The permanent damage done in a single day of experimentation by inexperienced personnel has frequently exceeded that which, with proper care, might be expected during the entire normal life of the material."

1–13. Safety precautions. Over a period of many years, various rules have been established to prevent casualty to personnel through carelessness. These rules are called *safety precautions*, and are part of *Navy Regulations* (Article 972). Additional precautions, set forth by the Bureau of Ordnance in various publications, have the full force of regulations. They have been built up through actual experience with ordnance during the early part of this century. Selected safety precautions are included at the ends of appropriate sections in this manual; this treatment, however, cannot be complete, and the officer must be familiar with BuOrd listings as well as Article 972. A few general items from this article are included here:

1–14. Safety precautions, selected from Article 972, Navy Regulations.

"1. As familiarity with any work, no matter how dangerous, is apt to lead to carelessness, all persons who may supervise or perform work in connection with the inspection, care, preparation, or handling of ammunition or explosives—

(1) Shall exercise the utmost care that all regulations and instructions are rigidly observed.

(2) Shall carefully supervise those under them and frequently warn them of the necessity of using the utmost precaution in the performance of their work. No relaxation of vigilance shall ever be permitted."

"2. In each part of the ship where ammunition is stored or handled or where gunnery appliances are operated, such safety orders as apply shall be posted in conspicuous places easy of access, and the personnel concerned shall be frequently and thoroughly instructed and drilled in them."

"3. Conditions not covered by these safety orders may arise which, in the opinion of the Commanding Officer, may render firing unsafe. Nothing in these safety orders shall be construed as authorizing firing under such conditions."

"4. The Commanding Officer shall at any time issue such additional safety orders as he may deem necessary, and a report thereof shall be made to his immediate superior and to the Bureau of Ordnance."

"5. When in doubt as to the exact meaning of any safety order, an interpretation should be requested from the Bureau of Ordnance."

"6. The Bureau of Ordnance shall be informed of any circumstances which conflict with these safety orders or which for any other reason require changes in or additions to them."

"7. Helpful suggestions and constructive criticism of these orders are invited. They should be made to the Bureau of Ordnance through official channels."

"8. Changes, modifications in, or additions to ordnance material, or other material used in connection therewith, shall not be made without explicit authority from the bureaus concerned."

"9. Safety devices provided shall always be used to prevent possibility of accident, and shall be kept in good order and operative at all times."

CHAPTER 2

EXPLOSIVES

A. EXPLOSIVE REACTIONS

2A1. Definitions. An *explosion* is a rapid and violent release of energy. An *explosive* is a substance which, when subjected to a suitable initiating impulse, undergoes a sudden chemical change characterized by the liberation of heat and the formation of products which are mainly gaseous. This reaction is accompanied by a rapid rise in pressure caused by heating of the gases.

2A2. Classification by reaction. A hard-and-fast classification of explosives is very difficult to formulate without introducing exceptions. One procedure is to classify explosives according to the *velocity* at which their chemical transformations occur. Thus *low explosives* are relatively slow-acting compared with *high explosives*, which react with extreme rapidity.

2A3. Low explosives. The reaction of a low explosive, such as a propellant, is a burning process. Low explosives usually are ignited by flame. Their reactions occur only on the exposed surfaces of the substance and progress through the mass as each layer is consumed. Smokeless powder is a good example.

2A4. High explosives. High explosives have a nearly instantaneous reaction which takes place throughout the entire substance. Some examples are TNT, fulminate of mercury, explosive D, and tetryl.

2A5. Characteristics of explosive reactions. An explosive reaction proceeds at high velocity, liberates heat, evolves gases, and develops high pressures.

Velocity. The chief characteristic which differentiates an explosive reaction from ordinary combustion, such as the burning of coal, is the high velocity of the explosive reaction. Velocity is a further basis for classification of high and low explosives. A low or propellent explosive reaction proceeds at a rate of about 5 inches per second, whereas a high or disruptive explosive reaction proceeds at a velocity approximating 5 miles per second.

Heat. An explosive reaction is always accompanied by the liberation of heat, which represents the energy of the explosive. This energy is a measure of the explosive's *potentiality* for doing work; for example, in propelling a projectile from a gun, or in causing a projectile to burst. This heat energy is not, however, a measure of an explosive's *capability* to perform such tasks. For example, coal yields about $4\frac{1}{2}$ times as much heat energy per pound as does a propellent explosive. Nevertheless, coal cannot be used as a gunpowder because it does not release its energy with sufficient rapidity. Because of the high temperature at which an explosive reaction takes place, it is nearly always accompanied by flame.

Gases. The principal gaseous products of commonly-used explosives are carbon dioxide (CO_2), carbon monoxide (CO), nitrogen (N_2), nitrogen oxides ($NO_{(n)}$), hydrogen (H_2), methane (CH_4), hydrogen cyanide (HCN), and water (H_2O). Some of the gases are inflammable, or form explosive compounds with air. A secondary explosion of these compounds may occur within a gun as it is opened after firing. The initiation of this secondary explosion may come from the high temperature of the gas or from burning residue in the gun. The resulting blast of flame to the rear of the gun is called *flareback*.

Pressure. The high pressure accompanying an explosive reaction is caused by the heating of the gases. In general, maximum pressure of a low explosion is attained comparatively late in the reaction because of the low velocity at which it proceeds. Maximum pressure of a high explosive reaction, however, is attained almost instantly, and will be much greater than in the case of a low explosive reaction. Some high explosives are *brisant*. A brisant explosive develops maximum pressure so rapidly that the effect is to shatter material that is in contact or nearby.

RESTRICTED

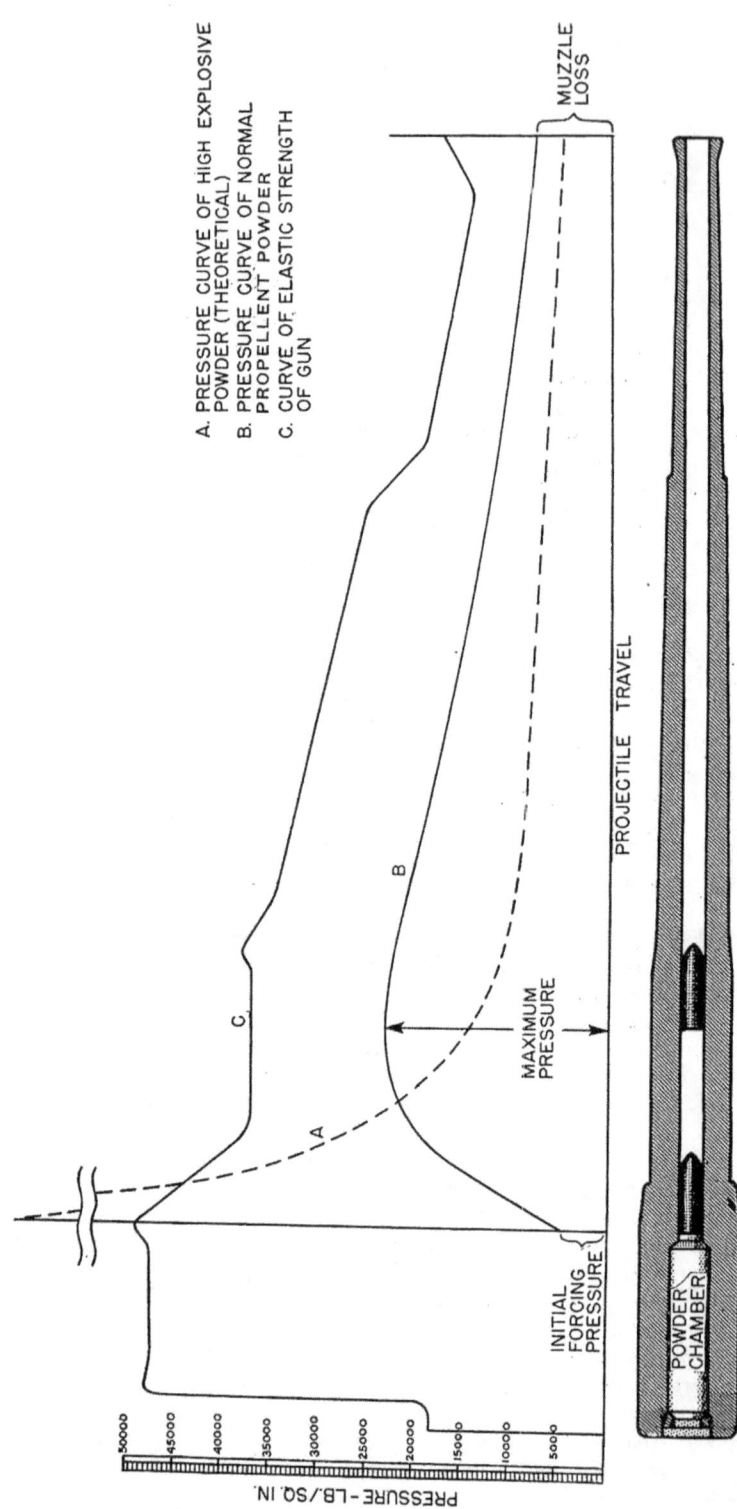

FIGURE 2B1. Gun pressure and strength curves.

CHAPTER 2

2A6. Initiation. To initiate an explosive reaction one of the following causes must exist:
1. Heat, flame, or friction.
2. Percussion, or impact.
3. Detonation, or shock.
4. Influence.

The amount of energy necessary to initiate an explosive reaction is a measure of an explosive's sensitiveness. Thus, when little energy is required, the explosive may be characterized as "sensitive."

Heat, flame, or friction. The low or propellent explosives are commonly ignited by the application of some form of heat; particularly flame. Some high explosives react in a similar manner when sufficient heat is applied, but as a general rule such initiation does not produce the desired high-explosive effect.

Percussion, or impact. Initiation of an explosive reaction by direct impact, or blow, is simply initiation by heat that is derived from the energy of such impact. This method is commonly employed by various mechanisms in which a small amount of a sensitive explosive in a primer cap is struck.

Detonation, or shock. Most of the high or disruptive explosives, such as the main charges of mines and torpedoes, and the fillers of projectiles, require a sudden application of a very strong shock to initiate the explosive reaction. This shock, or detonation, may be obtained by firing a smaller charge of high explosive that is in contact with, or in close proximity to the main charge. This smaller charge may be readily exploded by heat, impact, or detonation.

Influence. Some explosions are said to be initiated by *influence,* which is similar to initiation by shock, except that the means of detonation (another explosive charge) does not have to be in close proximity to the main explosive mass. In this case the energy is transmitted through air or some other medium in the form of a *percussive wave.* When a percussive wave detonates another explosive mass, the result is called a *sympathetic explosion.*

The tremendous energy of a percussive wave in an underwater explosion is evidenced by the immediate upheaval of the water's surface just before the geyser-like disruption occurs when the gases of explosion rise.

B. SERVICE EXPLOSIVES AND THEIR USES

2B1. General. In the preceding section, explosive substances and their reactions were defined. Explosive substances were classified as high and low according to the velocity of their reactions. However, the distinction between the two is not sharply defined. Some low explosives, under proper conditions of fine granulation, confinement, and firing, may have characteristics approximating a high, or detonating, reaction. On the other hand, some high explosives, when unconfined, may simply burn if ignited by flame. Therefore, it is desirable to classify explosives according to their *service use* as follows:
1. Propellent and impulse explosives.
2. Primers and igniters.
3. Disrupting and bursting explosives.
4. Boosters.
5. Detonators.

The number of widely known explosive substances has become so large that it is possible to discuss only those in most common naval use. They are black powder, smokeless powder, TNT and the various mixtures in which it occurs, explosive D, tetryl, and fulminate of mercury.

2B2. Propellent and impulse explosives: characteristics. The primary function of a propellant is to provide pressure which, acting against the base of a projectile, will accelerate the projectile to the required velocity. This pressure must be so controlled that it will never exceed the strength of the gun.

The volume which is occupied by a gas containing a fixed amount of heat determines the pressure of the gas. The gases released by a propellent explosive fill the chamber and that part of

RESTRICTED

the bore behind the projectile. Referring to figure 2B1, it is clear that the volume available for the gases increases as the projectile moves through the bore.

If, on explosion, the explosive is transformed instantaneously and completely to gases, the container of the gases will be the chamber alone, and the maximum pressure will occur before the projectile begins to move. The pressure will thereafter continuously decrease, for no more heat or gases will be liberated as the volume increases. The maximum pressure would have to be very high, as indicated by curve A, to produce the required muzzle velocity. Obviously, the breech end of the gun would have to be excessively heavy to withstand such an explosion.

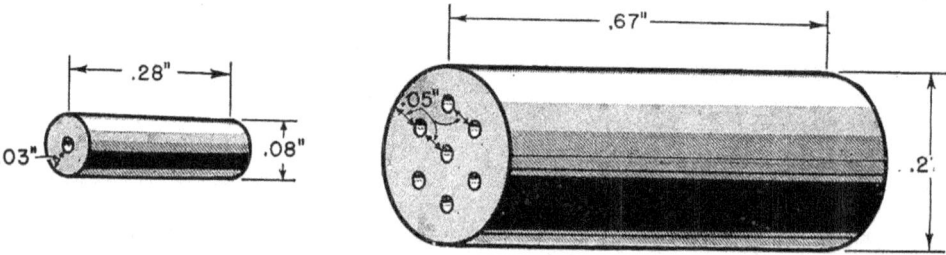

a. SINGLE-PERFORATED GRAIN
FOR A 1.1" 75 CAL. GUN
(Uniform Or Constant Burning Rate)

b. MULTI-PERFORATED GRAIN
FOR A 5"/38 CAL. GUN
(Progressively Increasing Burning Rate)

FIGURE 2B2. Typical smokeless-powder grains.

It is therefore desirable to control the evolution of gases so that the projectile begins to move before the charge has been entirely consumed. Then the space occupied by all of the gases will be greater, and the pressure will be lower than in the case of an equivalent charge having an instantaneous reaction. The pressure of the controlled reaction will follow a curve similar to B in figure 2B1. Consequently, the dimensions of the gun may be reduced as indicated by curve C, with a decided saving in weight.

Only a low or burning explosive can provide a reaction developing pressure at a rate suitable for the propulsion of a projectile from a gun. Since a low explosive is consumed layer after layer, the total burning area, and hence the rate of gas production, will depend on the amount of surface exposed. Actually, it is necessary to provide an *increasing* burning rate if pressure is to be sustained, since the volume available for the gases grows at an increasing rate as the projectile gains velocity.

Any solid powder grain presents a continuously decreasing burning surface. The burning area can, however, be controlled by perforating the grain, as shown in figure 2B2. The small holes, about 1/32-inch in diameter, present an interior burning surface. As burning progresses, the surface around each of these holes *increases* at the same rate as the outer cylindrical area *decreases*. The single-perforated grain maintains a constant area, except for the slight decrease in length as the explosion progresses. This type of grain is used in small guns, since manufacturing difficulties prevent the use of more perforations.

The seven-perforated grain is used in all U. S. naval guns 40 mm. and larger. Since the area surrounding *each* of the seven holes increases as fast as the outside area decreases (excluding the ends), the total burning area, and hence the rate of gas production, *increases* progressively until the holes meet. The burning rate decreases from then on. Seven holes provide the optimum burning rate for propellent powders. It is Navy practice to regulate the burning time of powders for different guns chiefly by varying the size of the grain. Other influencing factors are held within narrow limits.

a. SPHEROHEXAGONAL

b. CANNON POWDER

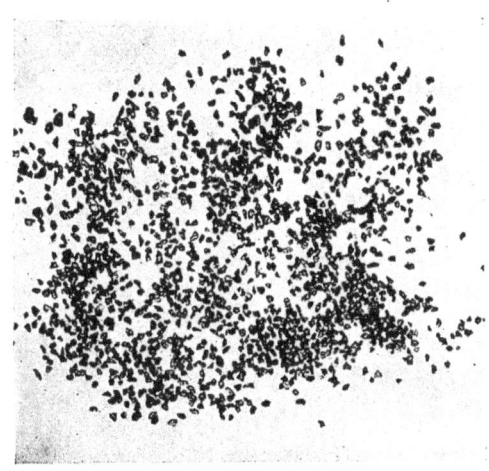
c. SHELL POWDER

d. FINE GRAIN

FIGURE 2B3. Black powder samples.

2B3. Black powder. Black powder is a low explosive, but some of its characteristics are like those of a high explosive. Its grain structure is such that its rapidity of burning produces a pressure curve more like A than B in figure 2B1. In the past many attempts have been made to control its behavior to make it acceptable as a propellent explosive. However, only in the case of certain small arms, depth-charge release gear, and torpedo tubes has black powder been found useful as a propellant. Nevertheless, black powder has a greater variety of service uses than any other explosive, such uses being varied, and of a special nature.

The usual composition of black powder is about 75% potassium nitrate, 15% charcoal, and 10% sulphur. The form of the powder depends upon the use for which it is intended. One variety consists of grains which have a somewhat glazed surface, and are free from dust; in another form, the product is a finely-granulated powder. In general, large grains make for a slower burning rate. The types of black powder used in the naval service, some of which are shown in figure 2B3, are arranged in order of decreasing size of grain as follows:

EXPLOSIVES

TYPE	DESIGNATION	USES
1. Grained (also sodium nitrate black powder)	Spherohexagonal	Impulse charges for torpedo tubes and depth-charge release gear.
2. Granular	Cannon powder	Ignition pads for smokeless powder bag charges.
3. Granular	Shell powder	Burster charges for certain types of projectiles, either alone or in combination with a high explosive.
4. Fine Grain	FFG and FFFG	Primers.
5. Meal	FFFFFFG	Fuze delay trains.

Black powder is one of the most dangerous of explosives, and should always be handled with the greatest of care. A few of the properties of this substance, and precautions that should be observed in handling it are as follows:

1. Possesses practically unlimited stability if stored in tight containers and *kept dry*. Black powder, and assemblies containing it, should be regarded as unserviceable if they have become damp or wet.
2. Deteriorates irregularly when moist.
3. Not affected by moderately high temperatures.
4. Highly inflammable and sensitive to shock, friction, and sparks.
5. Should never be opened in a magazine or in the vicinity of other explosives.
6. Maximum limit of exposed black powder in magazines or turrets is 25 pounds.

2B4. Smokeless powder. Theoretically it would be possible to use any explosive as a propellant if the velocity of its explosion could be controlled. However, there are certain practical considerations, such as the heat of explosion, residue left in the gun, and the volume of smoke produced that must be considered. In the light of such considerations black powder left a good deal to be desired, and this fact gave impetus to the development of smokeless powder.

In the manufacture of smokeless powder a cellulose material (cotton) is nitrated, yielding nitrocellulose—a high explosive that is commonly called guncotton. This nitrocellulose is purified, dissolved in an ether and alcohol mixture (volatiles), dried, and formed into perforated cylindrical grains as shown in figure 2B2. The resultant powder is a low explosive which will burn instead of detonating. A stabilizer, called diphenylamine, is added during manufacture; it opposes any tendency of the powder to decompose chemically and therefore tends to prolong the life of the product.

Such an explosive provides a controlled reaction because the individual grains burn in successive layers from the surface. It may be seen that by altering the shape, size, and perforations of the powder grains, and thus changing the amount of burning surface, the velocity of reaction can be so modified that the same chemically constituted powder may be used in a variety of weapons.

There remains in the finished grain a certain percentage of the ether and alcohol mixture (residual volatiles) which is definitely fixed for each size of granulation, and which should be kept constant as long as the powder is in service. Loss of these volatiles will affect not only the performance of the powder but also its stability. Excessive loss of residual volatiles results in a powder with characteristics of a high explosive; in other words, a "ballistically dangerous" powder which is highly unsuitable for use in a gun. For this reason, it is essential that smokeless powder be kept in air-tight containers, so that loss of volatiles does not occur.

High storage temperatures and moisture also facilitate the decomposition of smokeless powder, and hence the performance, stability, and life thereof. The best storage temperature is 60° F., and it should not be permitted to rise above 90° F. At temperatures below 60° F. stability is not appreciably affected, whereas, above 90° F. deterioration is accelerated. Hence, it is important that precautions be taken to insure the maintenance of a uniformly satisfactory temperature and absence of moisture wherever powder is stored.

CHAPTER 2

A summary of the methods used in controlling the rate of burning and in regulating the performance of smokeless powder includes:

1. In manufacture:
    a. Perforation of grain.
    b. Size of grain, upon which depends the web thickness.
    c. Specific percentage of volatiles for each size of grain.
        (Note: It is present practice to keep the amounts of nitration, stabilizer, and moisture uniform, and to keep the shape of the grain cylindrical.)
2. Aboard ship:
    a. Prevent loss of volatiles by storing powder in air-tight containers.
    b. Prevent access of moisture by storing powder in moisture-proof containers.
    c. Maintain uniformly satisfactory storage temperatures by a daily check of the magazines.
    d. Keep constant check on the stability of powder by the above methods, and by diligent shipboard examinations and tests as specified in the *Bureau of Ordnance Manual*.

The physical properties of smokeless powder are as follows:

1. New powder has an amber color. Some powder found in service has been reworked from old powder; it has a dull brown or black color.
2. Each powder grain has a hard, smooth finish which tends to limit burning to the surface alone.
3. Smokeless powder is difficult to ignite unless a flame of high temperature is supplied for an appreciable time.

2B5. Smokeless powder developments. The introduction of smokeless powder solved many problems created by objectionable features of black powder. It did not, however, provide an ideal explosive, and today efforts are being made to improve smokeless powders. These efforts are directed toward minimizing muzzle flash, reducing the temperature of combustion to lessen erosive action in the gun, and lessening of the effects of ordinary changes in temperature and moisture. Progress has recently been made in the production of a flashless, nonhygroscopic smokeless powder, designated as FNH.

2B6. Primers and igniters. It is necessary to provide some agent which will supply the required hot flame for ignition of the propellant, maintain it long enough to completely inflame the smokeless powder charge, and keep the burning action going. Black powder is most suitable for this purpose. It can be readily ignited by heat, spark, friction, shock, or chemical means. It provides the hot flame necessary to ignite the smokeless powder charge. For this reason, all smokeless-powder charges are ignited by black powder contained in *ignition pads* or *primers*. A primer, which may provide a sufficient initiating impulse, or may be assisted by an additional quantity of black powder, has the function of igniting the main propelling charge. The usual practice is to fire the primer by one or a combination of two methods:

1. By percussion, utilizing a percussion cap containing a very small amount of high explosive which is sensitive to impact and will ignite the black powder.
2. By heating a wire electrically.

The sequence of action is first, firing of the primer cap or the electrical heating of the wire, second, burning of a small primer charge of fine-grain black powder, and finally, burning of the black powder ignition charge. The flame from the ignition charge envelops the propelling charge and initiates combustion of the latter. The various devices for ignition are discussed later in connection with specific types of propelling charges.

2B7. Disrupters, boosters, and detonators. *Disrupting* and *bursting charges* are used as fillers for projectiles of various types, torpedo warheads, mines, depth charges, demolition charges, and other bomb-type ammunition. Filler substances must fulfill different conditions for specific military uses. In general they must:

1. Have proper insensitivity to withstand the shock of handling, loading, gunfire, and impact against armor prior to penetration.

2. Have maximum explosive power.
3. Have stability to withstand adverse storage conditions.
4. Produce proper fragmentation.
5. Be cheap and available, and easy to manufacture.

The substance most nearly fulfilling all of these conditions might be considered the best filler explosive; however, the specific purpose to be served must be considered. For example, a projectile filler must fulfill all of the foregoing conditions, whereas a mine or depth-charge filler need not be so insensitive as to withstand armor impact.

In initiating large disruptive and bursting charges, a small amount of a very sensitive explosive is used to detonate a larger intermediate charge, which, in turn, detonates the main charge. The small initiating charge is termed a *detonator,* and the intermediate charge is called a *booster.* These two terms should be used only in connection with high explosives.

2B8. TNT. *TNT* (trinitrotoluene) is a nitrated derivative of coal tar or petroleum, and is widely used as the main disruptive charge in mines, torpedo warheads, and other bomb-type ammunition. It also has extensive use as a burster charge, either alone or in various combinations with other explosives, in certain projectiles that are not designed to pierce heavy armor. The capacity of TNT to be melted, poured into a container, and cast, makes it particularly adaptable for any of the above-mentioned uses.

TNT is comparatively stable and relatively insensitive; however, it is not so insensitive that it may be handled carelessly with impunity. A dark-brown oily liquid frequently separates out from cast TNT, and may exude from the containers after a period of storage. These *exudates* are of the same composition as TNT, but have different melting points. They are relatively insensitive, but when mixed with a cellulose absorbent, such as a wooden deck, form a low explosive which is easily ignited, burns rapidly, and may even be detonated. An accumulation of exudate is both a fire and explosion hazard. Excessively high temperatures and direct sunlight increase the amount of exudation, and adversely affect the TNT itself. TNT is almost insoluble in water, but experiments have shown that the presence of moisture adds greatly to the difficulty of detonating booster charges of granular TNT.

Grades and uses. At the present time the following grades and forms of TNT are used in the Navy: recrystallized (purified crystals), granular grade A, flaked grade A, and cast grade B. Only grade A is being manufactured at the moment. In pure form, TNT varies from pure white to pale yellow in color. When the proportion of impurities is greater, as in grade B, the color is darker, often being brown.

As mentioned previously, TNT has wide use as a filler explosive in bomb-type ammunition; in certain projectiles not designed to pierce armor it may be employed as a component. Other uses of TNT are in fuzes and booster charges. For these purposes, only the refined, granular, or crystalline form is used.

2B9. RDX. *RDX,* also known as *cyclonite* and *hexogen,* represents a recent development. It is a highly sensitive explosive, and therefore must be mixed with other materials for safe service use. Details concerning the composition of RDX are confidential and only a general statement concerning the explosive can be presented here. It is produced by the nitration of an organic material. Purification is accomplished by crystallization from various substances including acetone or glacial acetic acid. The RDX crystals are then coated with beeswax to reduce sensitivity, and other materials are added to produce the same effect. Various compositions employing RDX as a constituent are:

1. Composition A: a mixture of about 90 per cent RDX and 10 per cent beeswax. It may be used as a projectile filler.

2. Composition B: a mixture of about 60 per cent RDX, 40 per cent TNT, and less than 1 per cent beeswax. In addition to its use as a projectile filler, composition B is employed as a component of torpex which is discussed later.

3. Composition C (also known as P. E.): a plastic mixture of about 90 per cent RDX and

10 per cent emulsifying oil. This composition is used to advantage as a demolition explosive because of its plastic form.

In general, ammunition containing RDX mixtures can be treated in the same manner as that which contains TNT. In the bulk form, however, RDX mixtures must be handled with greater care because of the sensitivity of the RDX ingredient.

2B10. Torpex. Torpex is a mixture of RDX, TNT, aluminum powder, and a fraction of 1 per cent of beeswax. It is a much more powerful explosive than TNT alone. Torpex is cast loaded as an explosive filler in a manner similar to the casting of TNT. On a weight basis, 100 pounds of torpex will produce the same underwater damage as will about 150 pounds of TNT. On a volume basis, 100 volumes of torpex will produce the same underwater damage as 170 volumes of TNT. Therefore, this explosive is especially useful in submarine and aircraft torpedoes.

Torpex does not absorb moisture and has stability comparable to that of TNT. The difference in impact sensitivity between torpex and TNT is slight. There is a critical velocity for a particular fragment above which both TNT and torpex will detonate. There is a lower velocity of the same fragment below which TNT will not detonate but torpex will. There is another lower velocity below which neither TNT nor torpex will detonate. The range of fragment velocities at which torpex will detonate and TNT will not is dependent upon the size of the fragment and probably its shape, but in general this range of velocities decreases as the size of the fragment increases. A point to remember is that a mass of either TNT or torpex will detonate if any part of that mass is caused to detonate.

2B11. Explosive D. The burster charge of an armor-piercing projectile should not explode until the projectile has passed through the side of an armored ship. It has previously been indicated that TNT is unsuitable for such a purpose. On the other hand a bursting charge of ammonium picrate, or *explosive D*, can be used successfully. In strength explosive D is only slightly inferior to TNT. Unlike TNT, however, it must be loaded in a projectile by tamping. The most extensive use of explosive D is as a filler in thick-walled, armor-piercing projectiles because of its ability to withstand severe shock. It is used also as a filler in many antiaircraft projectiles.

Explosive D is a crystalline powder varying in color from yellow to deep red. It has high chemical stability for long periods of time even at temperatures up to 150° F. However, explosive D should not be exposed to moisture as it will absorb up to 5 per cent of water, and thereupon will lose its power and become so insensitive that it is hard to detonate.

2B12. Tetryl. This explosive is more sensitive than either TNT or explosive D, and is utilized almost exclusively as a booster in mines and torpedo warheads, which do not have to undergo the heavy shock of firing. It should be noted, however, that some small projectiles such as the 20 mm. may contain *tetryl* as a burster charge. Sometimes a mixture of tetryl and fulminate of mercury serves as a detonator. Tetryl is also employed in the detonating trains of fuzes. The substance is a yellow, crystalline powder, which is hygroscopic, and is stable at ordinary temperatures.

2B13. Fulminate of mercury. This explosive is the most sensitive of the group in common service. It may be detonated by fire, spark, percussion, or friction, or by contact with certain acids. Its exclusive use is to initiate the action of other explosives, either directly or through the medium of a booster charge. The sensitive nature of *fulminate of mercury* precludes use of large charges, or stowage of even small charges with other explosives. Fulminate ignites, and then rapidly detonates with the powerful disrupting effect that makes it so useful as an initiating agent.

Unlike any other explosive, fulminate becomes less sensitive when stored for long periods of time at high temperatures. It also differs from other explosives in that the larger the fulminate crystals the more sensitive they will be. On the other hand, fulminate is similar to other explosives in that its sensitivity increases with the application of direct heat, and decreases with the addition of moisture, although in a more irregular manner.

A small pellet of fulminate is used in primers and in small-arms cartridges where the flame, produced when the fulminate is set off by percussion, ignites the black powder in the primer or cartridge. In such employment detonation is of no consequence; what is needed is a flame-producer that will ignite the primer charge. Frequently various substances are mixed with the fulminate to vary the degree of sensitiveness, to lengthen the flame produced, and to increase the amount of heat liberated.

2B14. Azides. Certain substances are now used as substitutes for fulminate of mercury. They are *azides* of silver, lead, sodium, and other elements. These substitutes are less sensitive (if the crystals are small), and have more energy than fulminate. They are also more stable and durable in storage. Azides have the property of increased brisance with increased pressure, but are less brisant than fulminate. Brisance may be increased by mixing an azide with fulminate. In this group lead azide is the most suitable compound and is now used as the detonator in major caliber, base-detonating fuzes, because its relatively low sensitivity permits penetration of armor plate before the detonator functions.

2B15. Safety precautions, selected from Article 972, Navy Regulations.

"17. Since the safety in handling and the disposition of ammunition depend upon the correctness of reports and records, care shall be taken not to obliterate identification marks on ammunition or to put it into incorrectly marked containers. When ammunition in other than normal condition is returned to an ammunition depot in compliance with these safety precautions, it shall be marked to indicate its condition and the reason for its return. If smokeless powder is involved the weight of the smokeless powder returned shall also be indicated."

"26. Smokeless powder shall not be exposed to the direct rays of the sun. Powder in bulk, in tanks, cartridge cases, ammunition boxes, ready service boxes, or in any other containers shall be protected against abnormally high temperatures (over 100° F.)."

"27. Whenever smokeless powder has been exposed to adverse storage conditions or treated contrary to the provisions in paragraph 26, it shall be segregated and shall, for purposes of tests, inspections, and reports, be regarded as a separate index; if the ammunition be aboard ship it shall be landed at an ammunition depot at the first opportunity (see par. 17), unless the test unmistakably show normal results."

"28. If any smokeless powder be exposed to temperature higher than 100° F. a special report shall be made to the Bureau of Ordnance immediately, explaining the circumstances in detail and stating the temperature and length of time the powder was so exposed."

"29. Smokeless powder which has been wet from any cause whatever must be regarded as dangerous for dry storage. Such powder shall be completely immersed in fresh water and kept immersed and landed at an ammunition depot at the first opportunity (see par. 17)."

"30. Smokeless powder in leaky containers shall be transferred to airtight containers, and these marked 'Transferred from leaky containers'. If airtight containers are not available or if the container in use cannot be repaired properly, the powder shall be forwarded to an ammunition depot at the first opportunity, the container marked "Leaky container" (see par. 17)."

"31. (a) Powder which shows unmistakable signs of advanced decomposition shall be thrown overboard or destroyed. In the event of such deterioration, every section of the index on board shall be examined, and only such sections as contain decomposing powder shall be destroyed. Decomposition in the sense here used is evidenced by—

(1) Grains being easily crumbled.
(2) Unmistakable odor of nitrous fumes.
(3) Very low violet paper and surveillance tests.

The conditions in (1) and (2) must be confirmed by the tests under (3).

(b) Powder found in a soft or mushy condition shall be thrown overboard or destroyed immediately.

(c) Whenever any powder is landed, thrown overboard, or destroyed because of its unstable condition, 8-ounce samples of each index shall always be preserved in airtight containers and forwarded to the Naval Powder Factory, Indianhead, Md., for examination, and the Bureau of

CHAPTER 2

Ordnance notified of the shipment and the reasons therefor. Such samples shall not be packed in water, but shall be segregated to prevent damage to other materials in case of spontaneous combustion."

"39. Black powder is one of the most dangerous explosives and shall always be kept by itself. Only such quantities as will meet immediate needs shall be taken from the magazine. A container of black powder shall never be opened in a magazine or in the vicinity of a container in which there is any explosive."

"41. Intense heat will detonate cast TNT charges. TNT exudate is explosive and highly inflammable. Therefore the outsides of cast TNT containers shall be kept free from exudate, and it shall not be allowed to accumulate on decks or to come in contact with wood, linoleum, or other materials into which it will soak. TNT exudate shall be removed before it hardens, shall not be scraped off with steel scrapers, and in scrubbing and wiping it up soap or alkaline solution shall not be used. Carbon tetrachloride is a safe and efficient solvent; but if cleared up in time, plain water and a stiff brush will generally remove the exudate from the surfaces to which it adheres."

"103. (a) Bombs containing (1) detonators, or (2) fuzes having detonators or other explosive components shall not be stowed in or near magazines containing explosives.

(b) Bomb fuzes containing integral detonators or other explosive components shall be stored only in specially designated fuze magazines which shall not be located adjacent to magazines containing high explosives.

(c) Detonators for bombs, or other detonators which are not assembled integrally with fuzes, shall be stored only in standard type detonator lockers located in approved places. In large surface craft and submarines, these places shall be below the waterline or protective deck; in small surface craft, these places may be above the weather deck or in the mast.

(d) Electric detonators shall not be located in the same compartment with or near radio apparatus or antenna leads."

C. AMMUNITION

2C1. Definition. *Ammunition* is the complete assemblage of the component parts which, together, form a charge or round for any type of gun, projector, or releasing device. Ammunition thus includes *gun ammunition; bomb-type ammunition,* such as torpedo warheads, aircraft bombs, depth charges, mines, and other thin-walled containers loaded with relatively large bursting charges; *pyrotechnic materials;* all *chemical warfare substances;* and other miscellaneous types.

Ammunition details are the components that form a part of the assembled ammunition. Such details include containers, primers, powder bags, projectiles, fuzes, detonators, boosters, explosives, and other components.

2C2. Classification. For purposes of study, ammunition is here divided into the following types:
1. Gun ammunition:
    a. Bag.
    b. Semi-fixed.
    c. Fixed.
    d. Small-arms cartridges.
2. Bomb type.
3. Pyrotechnic.
4. Other types (chemical, trench warfare, impulse, blank, and dummy drill).

2C3. Bag ammunition. A *bag charge* as used in the U. S. Navy at the present time, is a type of ammunition in which the primer, propelling charge, and projectile are loaded into a gun in separate parts. All major-caliber guns use bag ammunition. The bag, made of silk, contains the propellant, and has a red-dyed ignition pad of black powder quilted into its base. The smokeless-powder grains may be packed into the bags loosely or they may be stacked. In the former case the powder

RESTRICTED

is merely dumped into the bag, and the bag is tightly laced. In making a stacked charge, the powder grains are arranged in successive layers with each grain on end, and the bag is sewed and laced tightly. Stacked charges are more compact and easier to handle; in addition, stacking prevents cutting of the bag due to haphazard grain arrangement.

2C4. Semi-fixed ammunition. *Semi-fixed ammunition* is that in which the primer and propelling charge are firmly fixed in a cartridge case (usually brass), and the projectile is separate, or fits loosely in the case. Projectile and case are loaded separately into the gun. As a rule, the case is not designed to be entirely filled with the powder charge, since this would result in excessive gun pressure. Therefore, a distance piece is inserted between the powder charge and the mouth of the case, and the entire charge held in place by a mouth plug or a sealing wad.

2C5. Fixed ammunition. *Fixed ammunition* is that in which the projectile and primer are firmly secured in a cartridge case containing the propelling charge; thus, the gun can be loaded in one operation. The details of both fixed and semi-fixed ammunition are the same, with the exception that in fixed ammunition the projectile is firmly held in place by the brass case, and the wad and distance piece are similarly secured by the base of the projectile, which has been given a light coat of shellac to insure perfect airtight sealing. The limiting factor on the size of fixed ammunition is the weight that can be handled readily by one man. In general, guns larger than 6 inches use bag charges; 5"/38 cal. and the latest 6-inch guns employ semi-fixed ammunition, and all smaller guns use the fixed type.

2C6. Small-arms cartridges. *Small-arms cartridges* are actually fixed ammunition, differing only in size from types previously discussed. All cartridges up to and including caliber .60 are classified as small-arms ammunition; hence this category includes the calibers .22, .30, .38, .45, and .50. Small-arms ammunition is also specified by class depending upon its use; namely, for hand-carried weapons, heavy machine guns, and aircraft guns.

2C7. Bomb-type ammunition may be enclosed in a thin-walled container which is loaded with a relatively large bursting charge. This ammunition depends for its effect upon the destructive blast of the explosive rather than any penetrative qualities of the container. Included in the group are torpedo warheads, mines, depth charges, and some aircraft bombs. If the contained substance consists of gas, smoke, incendiary, or pyrotechnic materials, the assembled charge is classed as either chemical or pyrotechnic ammunition. Other bombs have rather heavy containers, being designed to penetrate the target before detonation.

2C8. Pyrotechnic ammunition is represented by certain fireworks that have been adapted to military purposes. Four classes, based upon uses to which the ammunition is put, are as follows:

1. Signalling.
2. Illuminating.
3. Screening.
4. Incendiary.

Pyrotechnic materials are mixtures of oxidizing agents and combustibles (powdered metals or non-metals, such as magnesium, and chlorate mixtures) to which may be added compounds designed for a particular purpose, such as to color a flame. *Pyrotechnic ammunition* issued to the service is further classed as ship pyrotechnics or aircraft pyrotechnics, although some items in these groupings are common to both. Ship pyrotechnics include:

1. Ship emergency identification signals.
2. Submarine emergency identification signals.
3. Submarine float signals.
4. Rockets.
5. Grenades (hand, smoke, and gas irritant).
6. Very signalling equipment consisting of pistol and star cartridges.
7. Hand signal flares.
8. Smoke mixtures.

CHAPTER 2

Aircraft pyrotechnics include:
1. Aircraft parachute flares.
2. Bombardment flares.
3. Float lights.
4. Aircraft day distress signals.
5. Very signal apparatus.
6. Aircraft emergency identification signals.
7. Miniature practice bomb signals.

The uses of the different types of pyrotechnic ammunition are prescribed in the various tactical instructions, and detailed descriptions of ship pyrotechnics are found in *Ordnance Pamphlet* No. 725, and of aircraft pyrotechnics in *Ordnance Pamphlet* No. 562. Because of the nature of pyrotechnics, and the unusual hazards involved, special care, tests, and precautions as listed in the *Bureau of Ordnance Manual* must be taken in dealing with such materials.

2C9. Chemical ammunition. The difference between chemical and pyrotechnic ammunition is not sharply defined in many cases. The classification used depends principally upon usage. In general, however, *chemical ammunition* includes any containers of gas, smoke-making substances, and incendiary materials other than pyrotechnic mixtures, or all gas, smoke, and incendiary materials released from generators, aircraft spray tanks, and other projectors. Only four types of chemical ammunition are standard for issue to the naval service. As designated by their code letters, the fillers are:
1. CN—a lachrymator (tear gas).
2. WP—an incendiary smoke.
3. HC—a smoke-making mixture.
4. FM—a smoke.
5. FS—a smoke.

2C10. Trench-warfare ammunition consists of hand grenades, trench-mortar ammunition, and rifle grenades. It is issued to vessels on special service for use by naval landing forces. The details of this ammunition are of a special nature and will not be discussed here.

2C11. Impulse ammunition. This type has already been dealt with under *propellent and impulse charges*. Briefly, impulse ammunition is a specially prepared propelling charge contained in a cartridge case with primer, and assembled as a blank cartridge for the purpose of launching torpedoes, depth charges and line-carrying projectiles, and for use in catapulting aircraft.

2C12. Blank ammunition is that intended to produce a noise, a smoke puff, or both, and is employed in signalling, saluting, and for training purposes. Such ammunition consists of cartridge cases which have been loaded with primers and powder charges only, or with these charges and paper bullets.

2C13. Dummy drill ammunition includes any type of ammunition, or any component of any type, assembled without explosives in imitation of regular ammunition. This type is used for training and testing purposes only, and is carefully marked so that it cannot be confused with service ammunition. To further reduce the possibility of confusion, it should never be stored in magazines.

2C14. Summary. Because of the especially hazardous nature of explosives and ammunition, diligent and careful handling, inspections, examination, and tests are imperative. It is necessary that those responsible for the care and handling of such material carry out the detailed precautions and surveillance routines in connection with all handling and stowage. Such safety precautions and surveillance routines are set forth for strict compliance in *Navy Regulations, Ordnance Instructions*, and the *Bureau of Ordnance Manual*.

2C15. Safety precautions, selected from Article 972, Navy Regulations.

"32. Naked lights, matches, or other flame-producing apparatus shall never be taken into magazines or other spaces used primarily as magazines while these compartments contain explosives".

"33. Before performing any work which may cause either an abnormally high temperature

EXPLOSIVES

or an intense local heat in a magazine or other compartment used primarily as a magazine, all explosives shall be removed to safe storage until normal conditions have been restored".

"34. Magazines shall be kept scrupulously clean and dry at all times. Particular attention shall be paid that no oily rags, waste, or other materials susceptible to spontaneous combustion are stored in them".

"35. Drill charges for bag guns soon become covered with oil and grease, and it is strictly forbidden to store such charges in magazines".

"36. Nothing shall be stored in magazines except explosives, containers and authorized magazine equipment".

"37. During firing no other ammunition than that immediately required shall be permitted to remain outside of the magazines".

"38. During action and during target practice magazine blowers shall be shut down and covers of both supply and exhaust branches to magazines shall be closed".

"40. Pyrotechnic material shall always be kept by itself in regular pyrotechnic storage spaces, if such are provided, or in pyrotechnic lockers on upper decks. In using it only a minimum amount shall be exposed".

"87. The covers of switches, circuit breakers, etc., shall be kept securely closed while powder is exposed in the vicinity".

"107. Smoke boxes which misfire or have been in the water shall not be taken on board ship or inside buildings or structures on shore. Gas masks shall be worn when entering concentrated smoke clouds".

CHAPTER 3

GUN AMMUNITION

A. GENERAL TYPES

3A1. Designation. Ammunition specified as gun ammunition is intended here to mean a complete round or charge for a rifled naval gun. The four types of gun ammunition are discussed by using typical examples for each class; that is, the 16"/45 cal. charge for *bag* ammunition; the 5"/38 cal. charge for *semi-fixed* ammunition; the 3"/50 cal. charge for *fixed* ammunition; and a 20 mm. cartridge for *minor-caliber* ammunition. Figure 3A1 compares these types of ammunition as to their size and general features.

Each of the various guns and its powder charge are designed to impart a particular initial velocity to the projectile for normal service use. However, special or reduced charges may be provided for experimental work, target practice, and training purposes. Such charges produce specific initial velocities which are usually less than service velocities. Some guns are not provided with any but service charges. One-word names are assigned by the Bureau of Ordnance to designate initial velocities and corresponding powder charges in the case of guns for which more than one velocity is provided. The designations are as follows:

1. *Service* designates the normal combat velocity.
2. *Special* designates, for certain guns, velocities only slightly reduced below service velocities.
3. *Target* designates the reduced velocity generally used for gunnery training.
4. *Spotting* designates the very reduced velocity used for certain guns at spotting practices.

Examples of abbreviations used to indicate the type of charge or the proportional part of a charge are:

        Ser. Chg.   —Service charge.
        1/6 Ser. Chg. —One-sixth of service charge.
        1/4 Spec. Chg. —One-fourth of special charge.
        1/2 Tar. Chg. —One-half of target charge.
        1/3 Spot. Chg.—One-third of spotting charge.

The term *initial velocity* means the velocity of the projectile as it leaves the gun muzzle. This velocity is always expressed in *feet per second*. The standard abbreviations for these terms are: I. V. for initial velocity and f.s. for feet per second. Any initial velocity listed in the discussions of various guns is a service velocity. When the term *charge* is used in connection with ammunition for the various guns, a service charge is implied.

3A2. Description. Figure 3A1 indicates the components of a complete round for a 16"/45 cal. gun which are: a combination lock primer, a service charge of six powder bags, and a 2,700-pound projectile. The semi-fixed ammunition, illustrated for the 5"/38 cal. gun, consists of a separately-loaded, 54-pound projectile, and a case assembly weighing about 28 pounds. The brass case contains a case combination ignition primer and the propellent charge. The fixed ammunition shown for the 3"/50 cal. gun has a case percussion ignition primer, the propelling charge, and a 13-pound projectile, all firmly secured in the brass case. The complete assembly, weighing about 30 pounds, is loaded in the gun as a unit. Finally, a 20 mm. cartridge is shown, which is similar to the fixed type of assembly. The details of these typical ammunition assemblies are taken up in the succeeding sections on primers, propelling charges, and projectiles.

3A3. Safety precaution, selected from Article 972, Navy Regulations.

    "10. No ammunition or explosive assembly shall be used in any gun or appliance for which it is not designated."

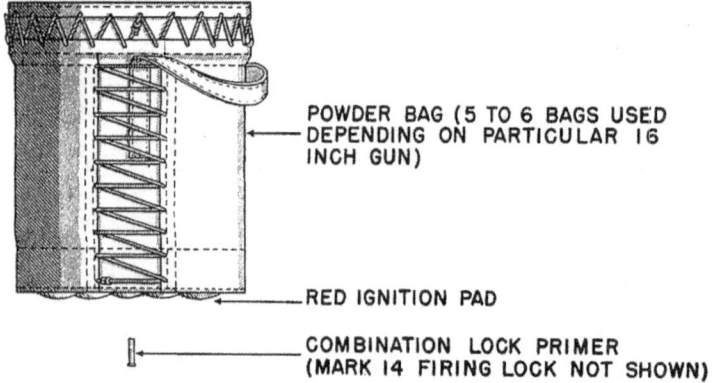

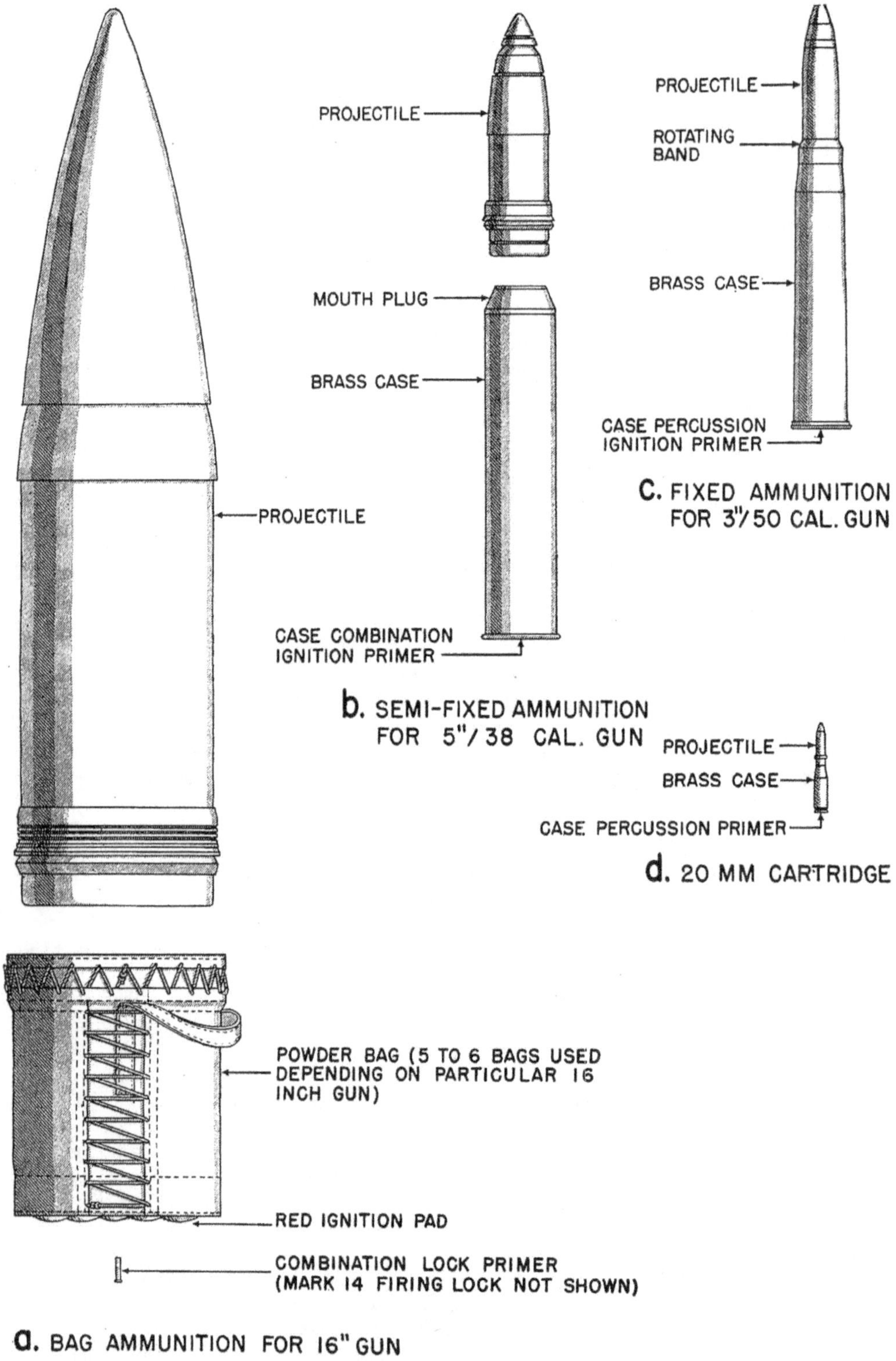

Figure 3A1. Typical gun ammunition.

CHAPTER 3

"12. Service ammunition is supplied to ships for use in battle. It shall not be used for drill, for testing appliances, or for other similar purposes except upon the express authority of the Navy Department. It shall be regarded as a part of the vessel's outfit, shall be kept distinct from the ammunition used for gunnery exercises, and shall never be expended in gunnery exercise unless authorized in the orders for gunnery exercises or special instructions from the Bureau of Ordnance."

"13. Special ammunition is issued for gunnery exercises, except when a part of the ship's allowance of service ammunition is designated for that purpose."

"16. The unexpended portion of such ammunition as may have been issued for a specific gunnery exercise or experimental firing shall be turned in as soon as practicable, after such firing, to an ammunition depot, unless additional firings are immediately authorized by the Navy Department."

B. PRIMERS

3B1. General. A primer is a device which provides a flame for the ultimate ignition of the propelling charge. In case ammunition this flame is applied directly to the propellant, while in bag ammunition it is applied indirectly through an ignition pad. The same size primer is used for all sizes of bag ammunition, but the amount of the black powder in the ignition pad is made proportional to the size of the smokeless-powder charge. A relatively large charge of smokeless powder requires a greater ignition flame, which is obtained by using a large ignition pad. In various types of case ammunition, however, different sizes of primers must be employed in order that each primer may contain in itself the amount of black powder required for the ignition of the propellant.

The two general classes of primers are: (1) the *lock primer* (all bag guns), and (2) the *case primer*, which is fitted into the base of the ammunition case. A primer that can be fired only by impact of a firing pin is called a *percussion primer;* one that can be fired both electrically and by percussion is known as a *combination primer*. Where it is necessary to increase the amount of ignition powder in the primer and to include an ignition tube inside the primer stock, the primer is known as an *ignition primer*. Combinations of these primer designations are employed to describe the types of primers in present use. The four primer types shown in figure 3B1 and examples of their use are as follows:

For case ammunition:
  1. Case percussion primer (small-caliber cartridges).
  2. Case percussion ignition primer (3"/50 cal.)
  3. Case combination ignition primer (5"/38 cal.)

For all bag ammunition:
  4. Lock combination primer.

3B2. Percussion element. The percussion element of a primer consists of the *primer cap*, a *cup* and *anvil*, and a *plunger* and *plunger cup* (except in a simple percussion element). The primer cap usually contains a mixture which has fulminate of mercury as a base. The composition of the mixture varies with the degree of heat, flame, and sensitiveness desired. One or more of the following materials are found in the ordinary mixtures: antimony sulphide, which increases the length of the flame; potassium chlorate, which increases the heat by its oxidizing action, and ground glass, which increases the sensitiveness. The anvil is pierced by holes or vents which allow the flame from the priming mixture to reach the ignition chamber containing black powder. In primers with a plunger, the method of firing is to strike the plunger, which action forces the primer cap against the anvil and explodes the pellet. The simple percussion element of a case percussion primer used in small-caliber ammunition does not have a plunger cup assembly. The primer cap is exploded merely by indenting an inverted cup, and thus driving the pellet against the anvil.

3B3. Electric element. Many primers having an *electric element* are also provided with a percussion element and are therefore called *combination primers*. The construction details of the electric element in combination primers are shown in figure 3B1. This type of primer has an *ignition cup* in which there is a mixture of pulverized guncotton and fine black powder, and a high resistance wire wrapped with a small wisp of guncotton which bridges through the mixture. The wire, attached at one end to the plunger cup, is electrically connected to the plunger through which the current from the firing source travels. The other end of the wire is grounded through the primer stock to the metal of the gun. An electric current heats the bridge wire and ignites the wisp of guncotton and the surrounding mixture in the ignition cup, producing flame. The flame transmitted from the ignition cup fires the primer. Insulation is used to separate the plunger and plunger cup from the ignition cup and primer stock.

The percussion element is a necessary feature of combination primers in case of failure to fire electrically. The normal procedure in firing such primers is to attempt electric firing first. If this fails, an attempt is made to fire by percussion. In firing by percussion, the flame transmitted by the primer cap ignites the mixture in the electric ignition cup, and the flame from both the primer cap and the ignition cup fires the primer.

3B4. Ignition tubes. All sizes of bag guns use the same design of lock combination primer because the amount of ignition powder necessary for the particular size of main charge is contained in the base of each powder bag. However, the ignition charge is combined with the primer in case ammunition, and because the amount of ignition charge varies with the size of main charge, the design of the primer varies with each caliber. For small-caliber case ammunition a primer cap alone suffices to ignite the propellant. The primer for larger-caliber case ammunition requires a greater amount of ignition powder and therefore is fitted with a primer stock extension. The primer assembly for still larger case ammunition contains such a great amount of black powder in the stock extension that a perforated *ignition tube* must be placed inside the outer stock extension. This ignition tube contains a small amount of black powder which makes it possible to readily ignite the larger main-primer charge of black powder in the primer stock extension.

C. PROPELLING CHARGES

3C1. Powder bags. The general features of *powder bags* have been discussed in Chapter 2. The description therein is applicable to all bag charges except in such details as the size and number of bags required for a complete charge. At present, all powder charges are contained in silk bags, but scarcity of silk will undoubtedly lead to the development of a suitable substitute. The material used for the bags is of two weights: light, and heavy. The heavy silk is used for the body of the bag, the handling strap, and the lacing flaps which serve as a means to take up any slack in a loose bag. The light-weight cloth is dyed red and is employed for the ignition pad containing the black powder. An ignition pad is quilted on the base of each powder bag, quilting being necessary to keep the black powder spread evenly throughout the pad.

Two of the powder bags for a 16"/45 cal. gun are shown in figure 3C1. The markings on such a bag should be noted. On the body of the bag these markings indicate the caliber of gun, the index or identification number and the weight of the smokeless powder, the proportion the bag bears to a full service charge, the initial velocity obtained with a complete charge, and the initials of the inspector. On the ignition pad markings indicate the number of grams of black powder contained. Each bag is provided with an identification tag which supplies the above information in addition to other data necessary to make the identity of the charge complete.

The necessity for airtight and watertight storage to maintain standard performance of smokeless powder has been discussed in Chapter 2. The addition of diphenylamine stabilizer only prolongs the life of the powder and does not prevent deterioration under adverse storage condition. It follows that powder tanks are important items of ordnance equipment. If a powder tank is leaky, the ether and alcohol volatiles escape, and air and moisture are admitted. It is important, therefore, to assure the integrity of such tanks at all times. Several types of tanks are to be found in service, but all must fulfill the same basic requirements for the storage of powder. One variety is illustrated in figure 3C1; others have variously constructed top-covers and special handling facilities such as

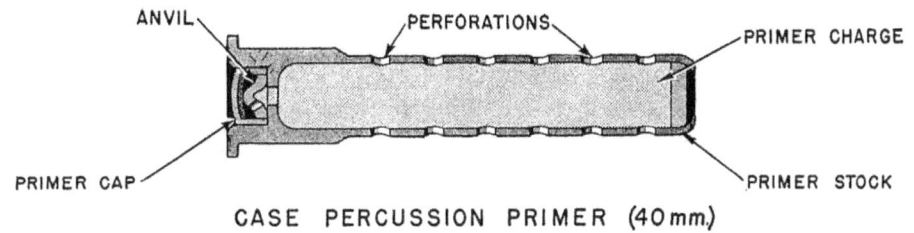
CASE PERCUSSION PRIMER (40 mm.)

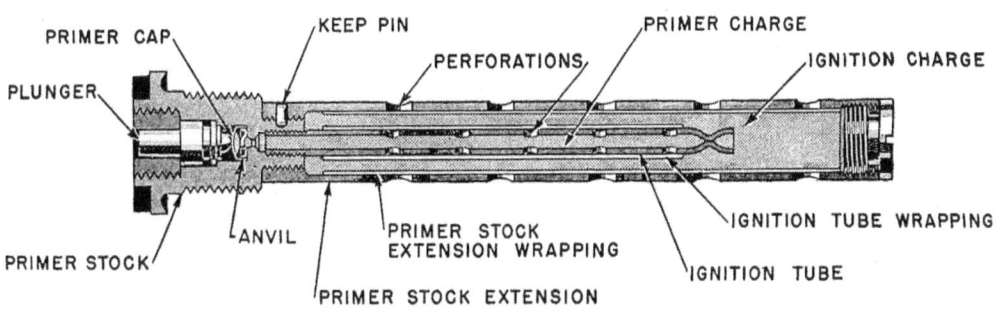
CASE PERCUSSION IGNITION PRIMER (3"/50 cal.)

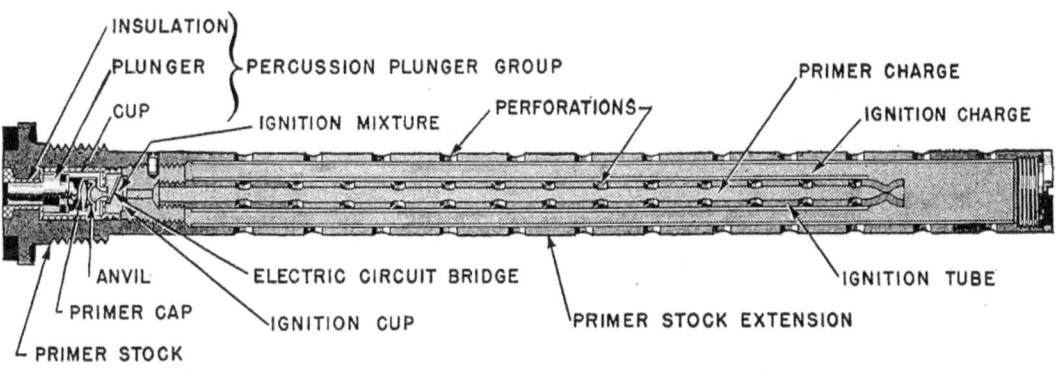
CASE COMBINATION IGNITION PRIMER (5"/38 cal.)

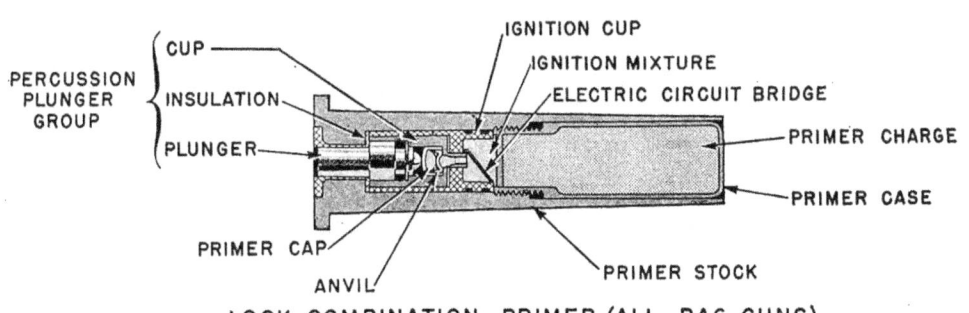
LOCK COMBINATION PRIMER (ALL BAG GUNS)

FIGURE 3B1. Primer types.

CHAPTER 3

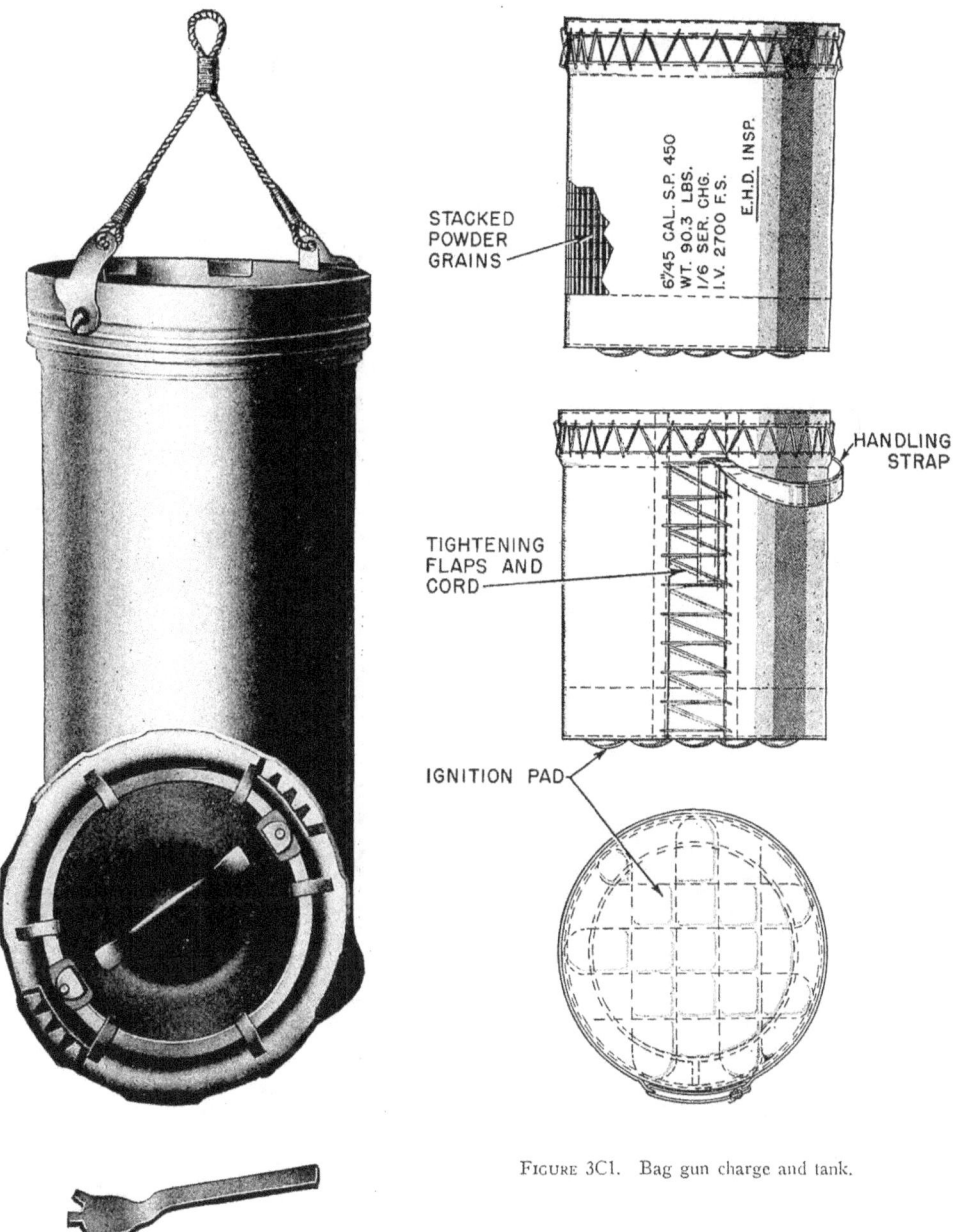

Figure 3C1. Bag gun charge and tank.

lugs (on large tanks) for use with slings, or handles (on smaller tanks). The top-covers are designed to permit quick opening, because the number of loaded tanks allowed open at any one time is limited by safety precautions.

For the purpose of obtaining definite velocities it is important that the powder charge should always be disposed in the gun chamber in the same way. The total length of all the powder bags placed end to end is within two inches of the powder-chamber length whether the charge be ser-

25   RESTRICTED

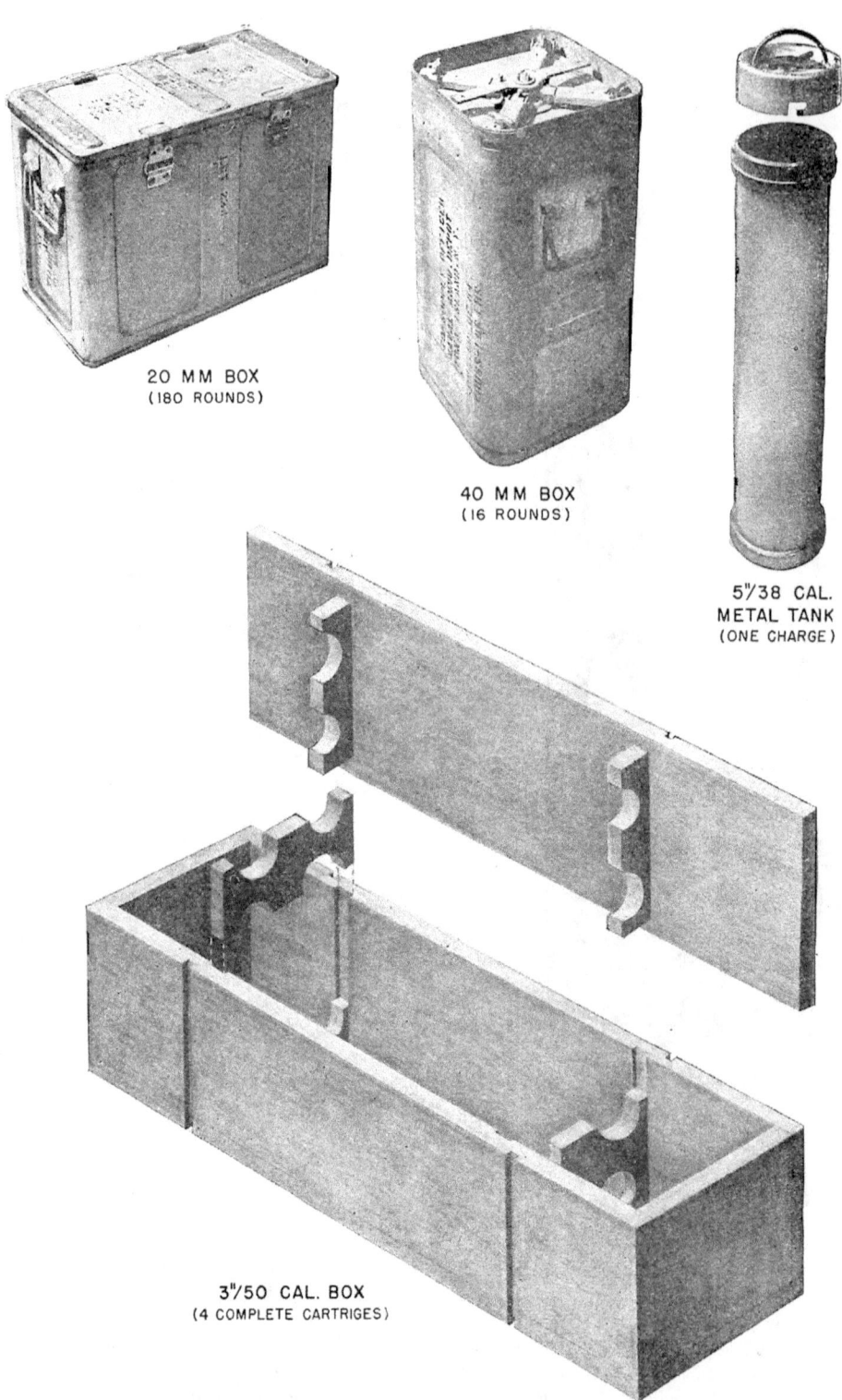

Figure 3C2. Ammunition tanks and boxes.

vice, special, target, or spotting. Thus, the *length* of each powder bag is the same in all cases, and the different effects are produced by varying the *diameter* of the powder bag. Therefore, the same number of bags can be placed in a constant-length powder tank regardless of whether these bags be a part of a service, special, target, or spotting charge. The tank is made large enough to accommodate the service charge, and will consequently accommodate the other types of charges which are of lesser diameter.

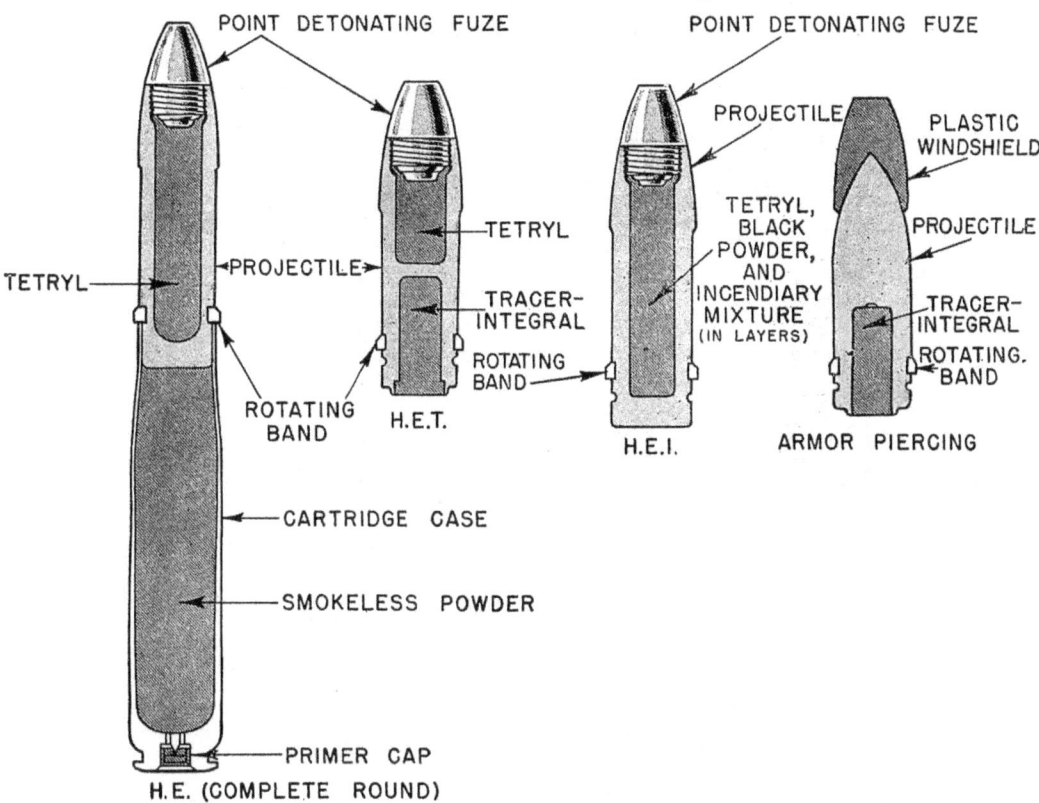

FIGURE 3C3. 20 mm. ammunition.

3C2. **Case ammunition.** The designs of various sizes of case ammunition are similar as may be seen from study of figures 3C3 to 3C6. The preparation of the case assemblies is comparable up to the point at which the mouth of a case is sealed. In fixed ammunition the projectile is the seal; a *mouth plug* is used in semi-fixed charges. There are four steps in the assembly of case ammunition: (1) priming, (2) loading the propellant, (3) fitting a *wad* and *distance piece,* and (4) inserting the *projectile* or mouth plug. In priming, the type of primer used is either screwed (40 mm., and larger) or force-fitted (smaller cartridge ammunition) into the base of the case. The desired weight of smokeless-powder grains is then dumped loosely into the case. In the case of 40 mm. and larger a cardboard disk, or wad, is forced in and a distance piece placed on top. The mouth of a semi-fixed case is then sealed by the insertion of a mouth plug as illustrated by the 5"/38 case in figure 3C6. In fixed ammunition, the mouth of the case is sealed by forcing in the base of the projectile until the rear of the rotating band makes contact with the case.

A distance piece consists of interlocked pieces of cardboard disposed as in an egg carton. The wad and distance piece are for the purpose of filling the entire case so that the propellent charge will not be shaken from around the primer. A mouth plug may be of cork, cardboard, tinfoil, or a combination of these materials, and must be of sufficient strength to keep the contents of the case from spilling out under any conditions of handling or loading. In small-caliber ammunition, the fitting of wads and distance pieces may not be necessary if the propellant fills the case. The brass case itself is a hollow cylinder with either a straight or bottle neck. The base has a rim around its circumference to facilitate extraction of the empty case from the gun. The empty cases can be used again after being reprocessed at an ammunition factory. They are required to stand six service rounds without deterioration, but actually, cases have been found capable of withstanding as many as 30 or 40 rounds before becoming useless.

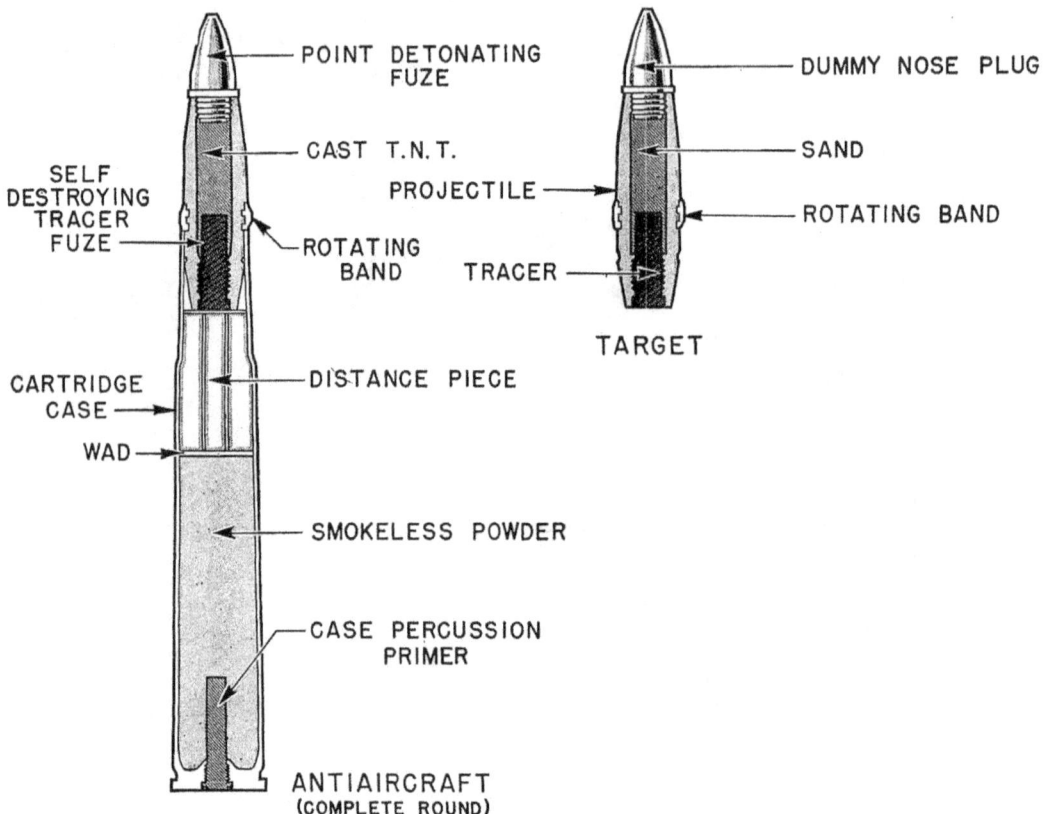

Figure 3C4. 40 mm. ammunition.

3C3. **Ammunition containers.** Since leaks may exist around the primer and the projectile or mouth plug, the case cannot always be relied upon to remain airtight. Therefore, like powder bags, case ammunition is transported and stowed in containers which provide air- and water-tightness. Such containers may be metal tanks or wooden boxes. Metal tanks are employed for case ammunition larger than 3 inches. Either metal tanks or wooden boxes are used for ammunition such as 3"/50 cal. and smaller, although generally wooden boxes which contain several complete rounds are favored.

There are several types of tanks and boxes in service, and no attempt will be made here to describe them in detail. Figure 3C1 shows a metal tank for bag ammunition and figure 3C2 shows ammunition containers for 5"/38 cal., 3"/50 cal., 40 mm., and 20 mm. ammunition. It is sufficient to state that regardless of varying design, the container must, above all, provide proper storage for smokeless powder. In addition, the container should be strong but not unduly heavy, should

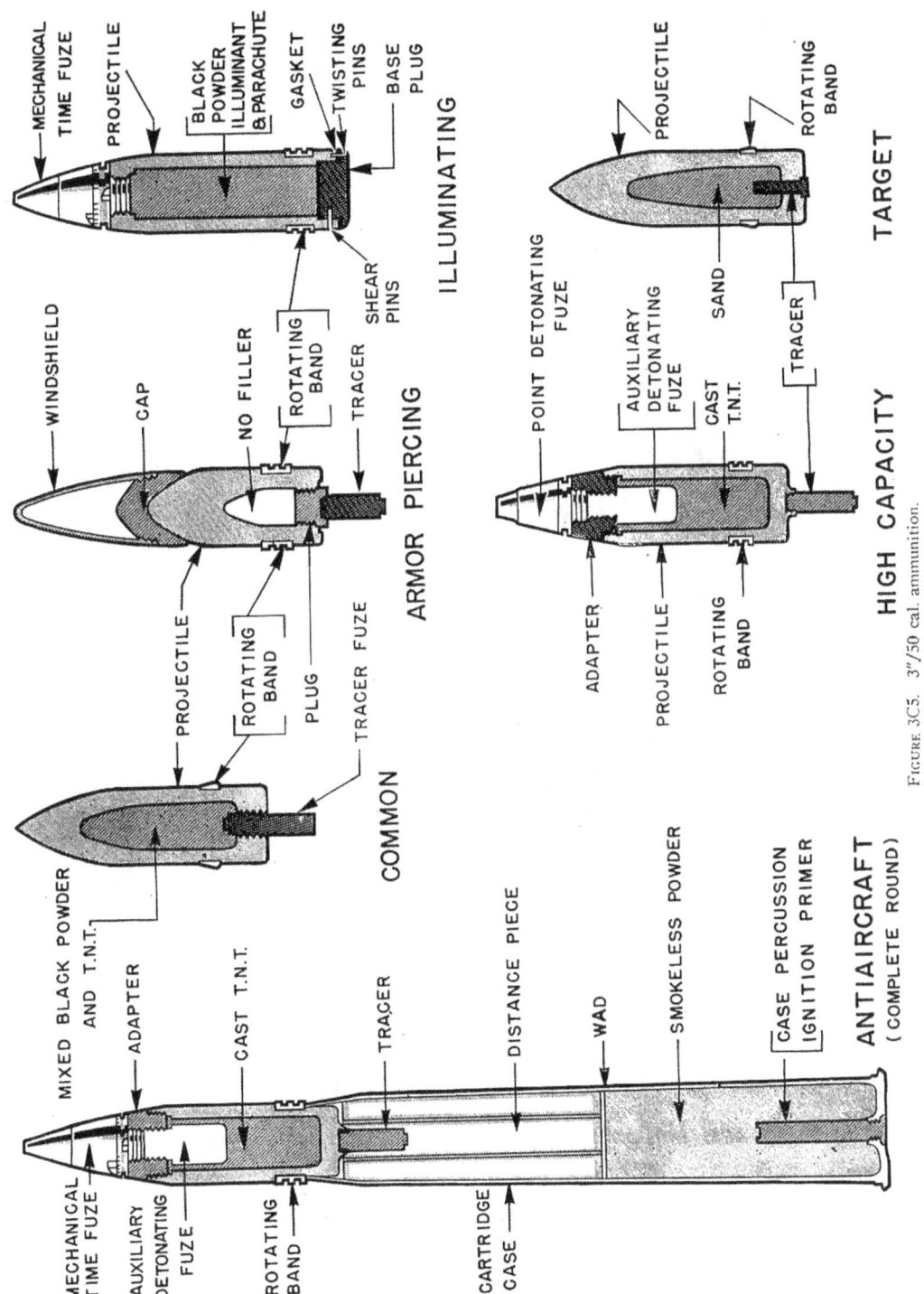

FIGURE 3C5. 3"/50 cal. ammunition.

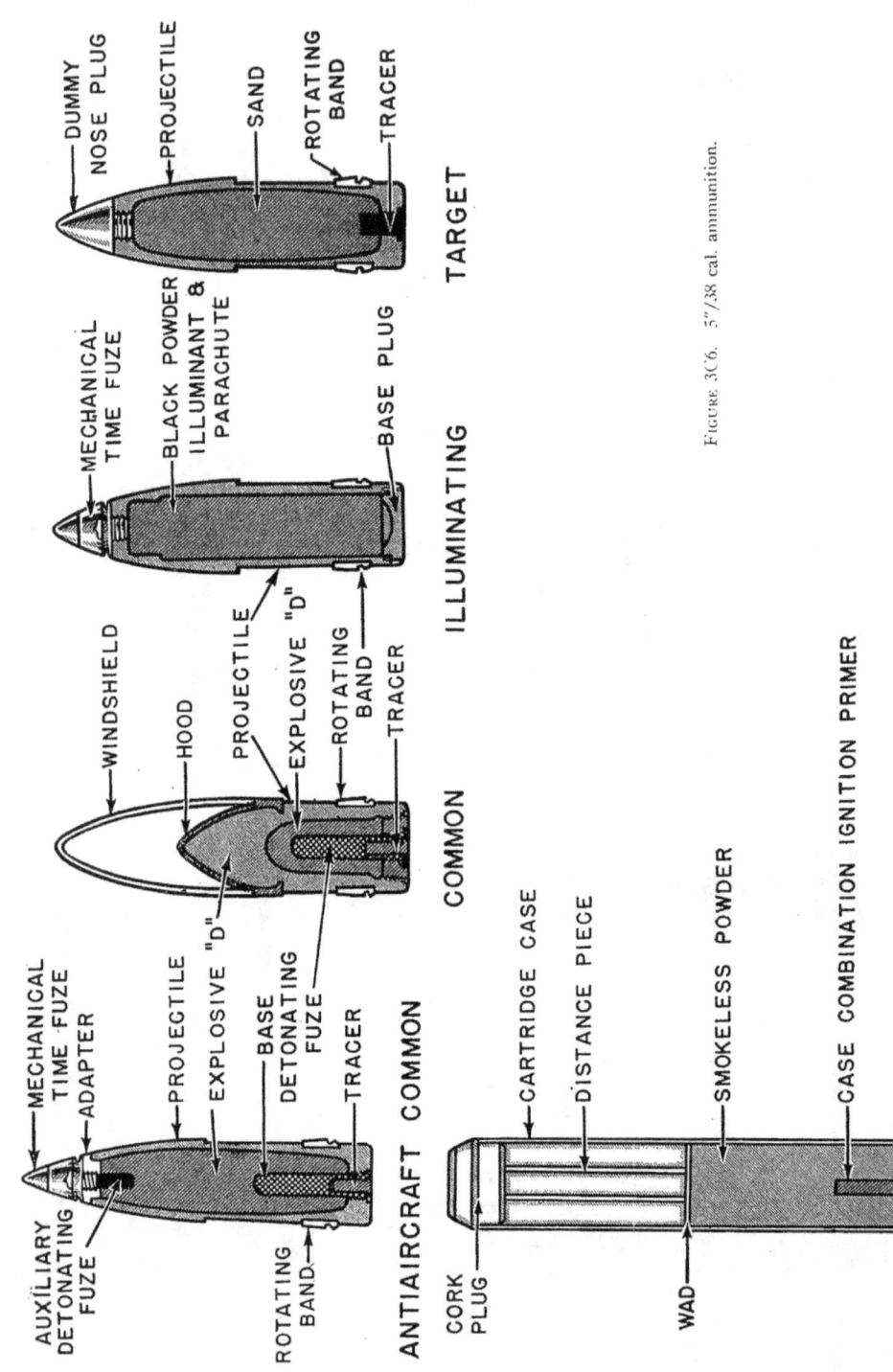

FIGURE 3C6. 5"/38 cal. ammunition.

CHAPTER 3

handle easily, and open quickly. Metal tanks made of aluminum most nearly conform to these requirements, and tanks in current use are of either aluminum or steel construction. Metal tanks are also advantageous in that they provide good storage in ready service racks on deck, or in ammunition-handling rooms not equipped to provide the best storage conditions.

The number of charges packed in a container is determined by the overall size and weight of the filled container so that ease of handling is maintained. The following number of charges per tank or box are standard:

5"/38 cal. (case only)	1 per tank.
3"/50 cal.	1 per tank or 4 per box.
40 mm.	16 per box.
20 mm.	180 per box.

3C4. Safety precautions, selected from Article 972, Navy Regulations.

"11. Handling of ammunition shall be reduced to the minimum to prevent immediate accident and the occurrence of leaky containers, damaged tanks and cartridge cases, loosened projectiles, torn powder bags, etc. Powder stored for a considerable period in a leaky container is likely to deteriorate rapidly, with the attendant danger of spontaneous combustion."

"48. Care shall be exercised to see that all sections of powder charges are entered in the chamber with the ignition ends toward the breech, and that the rear section is so placed as to come within 4 inches of the mushroom when the breech is closed."

"75. Under no circumstances shall the material of powder bags be added to without authority. Should it be necessary to stiffen the charges, additional cloth or tape shall not be used, but the lacings shall be tightened. If the powder bag be badly injured, it should be replaced by a new one from the spares on hand."

"76. When these safety precautions require the removal of smokeless powder from its bag, the ignition charge, together with the bag, shall be thrown overboard."

"79. Fired cartridge cases shall, before storing below, be stood on their bases in the open air for ten minutes in order to avoid danger from inflammable gases."

D. PROJECTILES

3D1. Description. A *projectile* is that component of gun ammunition which is expelled from a gun by the force of the powder-charge explosion. It is an elongated cylinder with a pointed front end and either a cylindrical or tapered after end. The overall length may vary from three to five calibers; the average length of most projectiles being about four calibers. Within reasonable limits, projectiles can be made in various weights for a given gun. In general, the approximate weight of any naval projectile may be found from the formula: weight (lbs.) $= \dfrac{\text{caliber (in.)}^3}{2}$

Standard nomenclature permits use of only the term projectile; the word "shell" is improper.

3D2. Forward end. The front end of a projectile is known as an *ogive*. The name is descriptive of its ogival shape, which is generated by the revolution of the arc of a circle, usually of about seven calibers radius, around its chord which is also in the axis of the projectile. The method is illustrated in figure 3D1. Sometimes, the ogival shape is a combination of several arcs of different radii. Frequently, a conical shape is employed because such a form is easier to manufacture.

A projectile which is designed to pierce heavy armor must have a blunt, solid nose for penetrative ability and must, therefore, have a heavy mass of metal in the forward end. The ogival curvature of this blunt nose is from one and one-half to two calibers. However, it is desirable to taper the front end to obtain longer range and to have the center of gravity slightly aft of the projectile center for better stability. A compromise is made in obtaining these features by fitting a light, streamlined nose-piece over the heavy front end. This addition is called a *false ogive* or *windshield*, and serves to streamline the forward end to an ogival curvature of about seven calibers. An armor-piercing projectile is illustrated in figure 3D2.

RESTRICTED

GUN AMMUNITION

3D3. After end. The after end of nearly all naval projectiles has the edge of the base turned to a small radius, varying from about one-third of an inch for large projectiles down to a mere touch of the file on the corner of small ones. This is called *boat-tailing*. Extreme boat-tailing is not commonly used, although an example is represented by the 40 mm. projectile shown in figure 3C4. The base of the after end may be integral with the body or may consist of a screwed in base plug. A hole may be cut in the base or base plug for a fuze or a tracer.

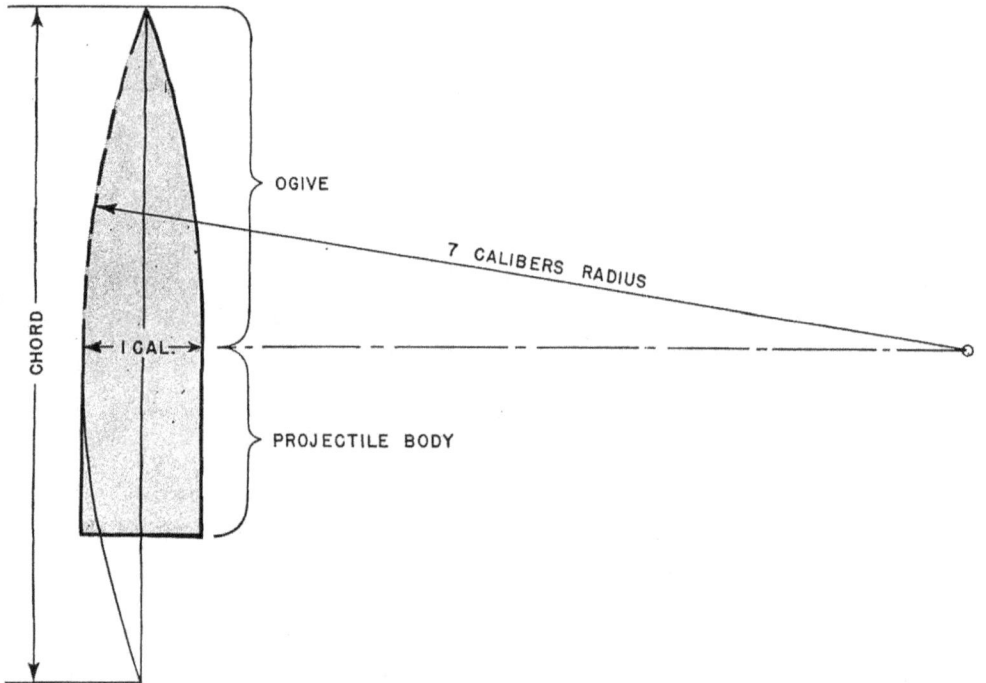

FIGURE 3D1. A projectile ogive.

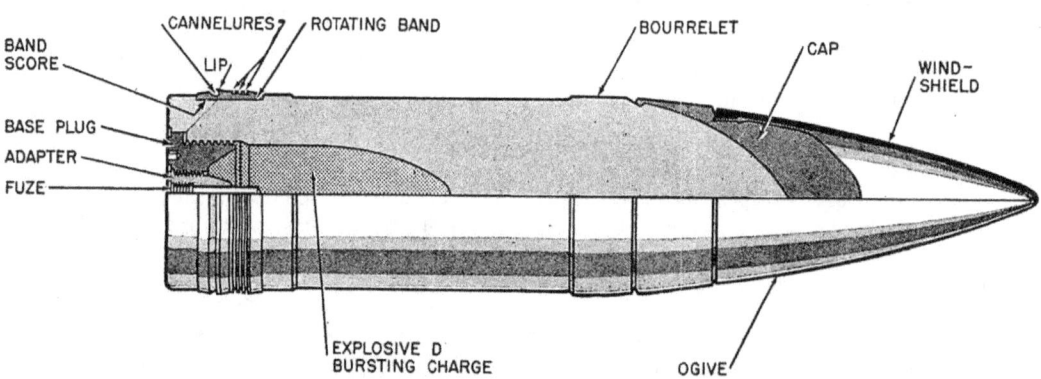

FIGURE 3D2. Armor-piercing projectile.

3D4. Body and exterior finish. Between the front and after ends of a projectile is the cylindrical *body*. Near the after end the body is fitted with a copper *rotating band* which rotates the projectile as it is forced through the rifled barrel. Machined on the forward end of the body is the *bourrelet* which provides a bearing support to steady the projectile and prevent wobbling in its

CHAPTER 3

travel through the gun. Long projectiles may have another bourrelet aft of the rotating band to provide an additional bearing surface. Where there is only one bourrelet, the rotating band provides the support for the after end. Small projectiles usually have no bourrelet; rather, the entire body of the projectile forward of the rotating band functions as a bearing surface.

As a rule only the bourrelet has a smooth-machined finish. The remainder of the body has only a rough-machined finish. The bourrelet is from .015 to .023 of an inch smaller in diameter than the bore of the gun, thus providing clearance in case the gun bore becomes "coppered" with a fine deposit of metal from rotating bands after repeated firing.

3D5. Rotating band. The rotating band serves the following purposes:

1. To rotate the projectile and thus assure stability in flight.
2. To steady the after end of a projectile not fitted with an after bourrelet.
3. To seal the bore, preventing the escape of pressure past the projectile.
4. To prevent overramming of a projectile in a worn or eroded gun.
5. To hold the projectile in its proper position in the gun after ramming, and insure that it does not slip back when the gun is elevated.

The width of a rotating band is about one-third of a caliber. The outer diameter of the band is made slightly larger than the diameter of the gun bore measured between opposite grooves. The rotating band is secured in a score cut in the projectile body, there being a dovetail on each edge of the score to prevent the band from flying off due to centrifugal force. Around the score there are either waved ridges, longitudinal nicks, or knurling, which insure against rotary slipping of the band. A projectile should not be used if the rotating band is loose or badly damaged. A cross-sectional view of the rotating band for a major-caliber projectile is shown in figure 3D2 (various other types may be seen in figs. 3C3 to 3C6). The purpose of the *lip* (see fig. 3D2) is to prevent overramming in an eroded gun. It insures uniform seating of successive charges, thereby contributing to more accurate fire.

The annular grooves of the rotating band, known as *cannelures*, provide a recess for the copper displaced to the rear by the rifling as the projectile travels through the bore. Their purpose is to prevent *fringing*, which occurs if the displaced copper is allowed to be pushed abaft the band. A skirt thus formed is flared outward by the action of the powder gases, and centrifugal force as the projectile leaves the muzzle. Pronounced fringing decreases accuracy, especially in the case of small projectiles.

3D6. Projectile classification. Service projectiles fall in four general classifications: (1) *thick-walled armor piercing;* (2) *common armor piercing;* (3) *high capacity;* and (4) *special purpose.* A projectile in any of the first three groups contains a filler explosive, which damages the target because of detonation alone, high-velocity projectile fragments produced, or both. Special-purpose projectiles are those designed for certain uses such as illuminating, smoke-screening, line-carrying, target practice, and proving-ground work. Under each of these broad classifications there are several types of projectiles, each adapted to a particular use. The number of types which may be provided for a given gun depends upon the functions the gun is designed to serve. The various projectile types and abbreviations are listed under the general classifications in the following table:

1. Thick-walled armor piercing:
    a. Armor piercing (AP).
2. Common armor piercing:
    a. Special-common (Sp Com).
    b. Common (Com).
3. High capacity:
    a. High capacity (HC). This is sometimes called a bombardment projectile (Bbt).
    b. Antiaircraft (AA).
    c. Antiaircraft-common (AA-C).
    d. Shrapnel (Shrap).
    e. Field (Fld).

RESTRICTED

GUN AMMUNITION

4. Special purpose:
    a. Illuminating (SS for star shell).
    b. Smoke (Sm).
    c. Line carrying (LC).
    d. Target (T or Tar).
    e. Proof Shot (PS).

Any of the above-named types may be fitted with tracers which provide means for following the flight of the projectile with the eye. Most of the various classes may be fitted with some type of fuze. Some projectiles have a small amount of dye placed in the nose, which on impact colors the water splash. A definite color of dye assigned to each ship assists in differentiating between salvos fired on a single target from several sources.

3D7. Thick-walled armor piercing projectiles. In general, armor-piercing projectiles are designed to pierce an armor thickness equal to the projectile diameter. Hence, major caliber AP projectiles are primarily designed for use against heavy, face-hardened armor, such as that forming the protective plating of a battleship. Smaller caliber AP projectiles are used to penetrate lighter, face-hardened armor on smaller ships and aircraft. In general, all caliber AP projectiles are constructed with a heavy, hard point, backed up by a softer, more ductile body. A cap of softer steel than the point is attached over the point. Over the cap is fitted the false ogive or windshield which streamlines the front end. A major caliber AP projectile is shown in figure 3D2 and smaller caliber examples in figures 3C3 and 3C5.

Penetration of face-hardened armor tends to shatter, twist, and bend the projectile. It is therefore necessary to incorporate considerable strength in the projectile wall in addition to providing a very hard and strong point. This heavy construction leaves a smaller cavity for the bursting charge than is provided in a projectile not designed to pierce heavy armor. The cap is fitted over the point to assist in penetration, and does this in two ways: (1) It produces an initially-weakened spot in the armor for the point to enter, and (2) being much blunter than the point, the cap increases the angle at which the projectile may strike the armor plate without bouncing off or ricocheting. The cap, being of softer metal than the point, is torn off after performing its functions. The windshield has no part in penetration for it is knocked off at the moment of impact.

The projectile is considered to have served its purpose if it penetrates the armor plate with its bursting-charge cavity still intact. An armor-piercing projectile is equipped with a delay type impact fuze, which detonates the filler explosive after the projectile has pierced the armor. Explosion of the filler disrupts the projectile and produces high-velocity fragments. It is the combination of this confined detonation and the high-velocity fragments within the interior of the target ship which causes the tremendous damage, rather than mere impact of the projectile and penetration of the armor plate.

3D8. Common armor-piercing projectiles. The common type of projectile is used against comparatively lightly armored ships. This type is similar to AP except that the wall of the projectile is lighter, and the point is not usually fitted with a cap or a windshield. The common projectile is designed to pierce armor of only $1/3$ to $1/2$ caliber thickness. This obviates the need for a cap, but a windshield is used in some cases. The lighter wall permits a bursting charge cavity larger than that of AP projectiles. The *special-common projectile* is thinner walled than AP, but is thicker walled than the common projectile. Therefore, the special-common projectile contains a larger bursting charge than that of the AP projectile, but a smaller charge than that of the common projectile.

3D9. High-capacity projectiles. The destructive effect of a high-capacity type of projectile depends almost entirely upon the blast of the very large bursting charge contained therein. The wall is made strong enough to withstand the shock of discharge from a gun, and thick enough to provide destructive projectile fragments. Since high-capacity projectiles do not have to withstand the heavy shock of impact without bursting, various explosives, more sensitive than explosive D, are used as fillers. A high-capacity projectile is shown in figure 3C5.

CHAPTER 3

The high-capacity type of projectiles may be further classified with respect to fuze action as follows:

1. *Impact burst.* These projectiles are fitted with either nose or base detonating fuzes which explode the burster charge immediately upon impact. This type of projectile is sometimes fitted with a self-destroying fuze which bursts the projectile if it does not contact the target before the end of its flight. The 40 mm. AA projectile is an outstanding example.

2. *Time burst.* These projectiles are provided with fuzes which function only after a predetermined length of time.

3. *Combination time and impact.* The majority of high-capacity types such as the AA common, AA, shrapnel, and high-capacity bombardment projectiles are provided with fuzes which will detonate the charge either after a predetermined length of time, or upon impact.

3D10. Special-purpose projectiles. Certain projectiles as defined in Article 3D6 are classed as special-purpose projectiles. The construction of this type incorporates no special strength other than that required to withstand the shock of discharge from the gun. If a filler explosive is contained, as in the case of illuminating and smoke projectiles, the only purpose served is to expel the contents of the projectile. Such projectiles are not intended to produce destructive fragments; therefore, the filler explosive need not produce a high-order detonation, and the burster may consist of a charge of black powder or a relatively small charge of high explosive.

The *illuminating projectile*, commonly called a *star shell*, is used to illuminate a target. Such a projectile has a time fuze which functions so that an illuminant is expelled over and behind the target, producing a silhouette effect. The general construction of two types of SS projectiles may be seen by referring to figures 3C5 and 3C6. The thin-walled body of the projectile contains an illuminating element attached to a small parachute; the parachute lengthens the time of fall. When the time fuze functions, the black powder charge is set off, igniting the illuminant, and blowing out the base plug which has been held in place by *shear pins*. The explosion of the expelling charge drives the illuminant and parachute assembly aft through the base of the projectile. The opening of the parachute is delayed until initial flight velocity is lost, so as to prevent ripping due to rapid motion through the air.

A *smoke projectile* contains a smoke-producing element, and is designed solely to produce a smoke screen. It contains a sufficient charge to split the projectile and thus to produce smoke immediately upon burst. This type of projectile is employed in the 5"/38 cal. gun, but its use is still experimental.

A *line-carrying projectile* is simply a loose-fitting slug which carries a rod extending to the muzzle of the gun. Attached to an eye in the muzzle end of the rod is a light cord which is coiled on the deck. The projectile turns around immediately after firing, and is held to its trajectory by the strain of the trailing cord. After the projectile has passed over the target, successively heavier lines culminating with a hawser are passed.

Target and proof-shot projectiles are used respectively for target practice afloat, and for test work at the proving ground. They are constructed to conform in shape and weight to service projectiles. Unless modified for some particular reason, however, the projectiles are filled with inert materials, and are built at the lowest cost consistent with proper performance and safety in use.

3D11. Projectile markings. Projectiles are painted various colors to designate the kind of projectile filler, the type of projectile, whether a tracer is fitted, and whether the nose is dye-loaded. The method of marking is illustrated in figure 3D3. The kind of projectile filler is indicated by painting the appropriate color on the ogive from the point to a distance of one caliber. Combination fillers call for more than one color. Adjacent to the ogive paint on a projectile fitted with a tracer is a one-inch white band, (black on the all-white shrapnel projectile). A projectile which is dye loaded has a band the same color as the dye adjacent to the tracer band, or adjacent to the ogival paint if not fitted with a tracer. The remainder of the projectile is painted the appropriate color indicating the type. However, a bourrelet is not painted on any projectile, and the portion of the body abaft the rotating band is not painted on a projectile for fixed ammunition. Labeling (stenciled near base), which completes the data on a projectile, is as indicated in figure 3D3.

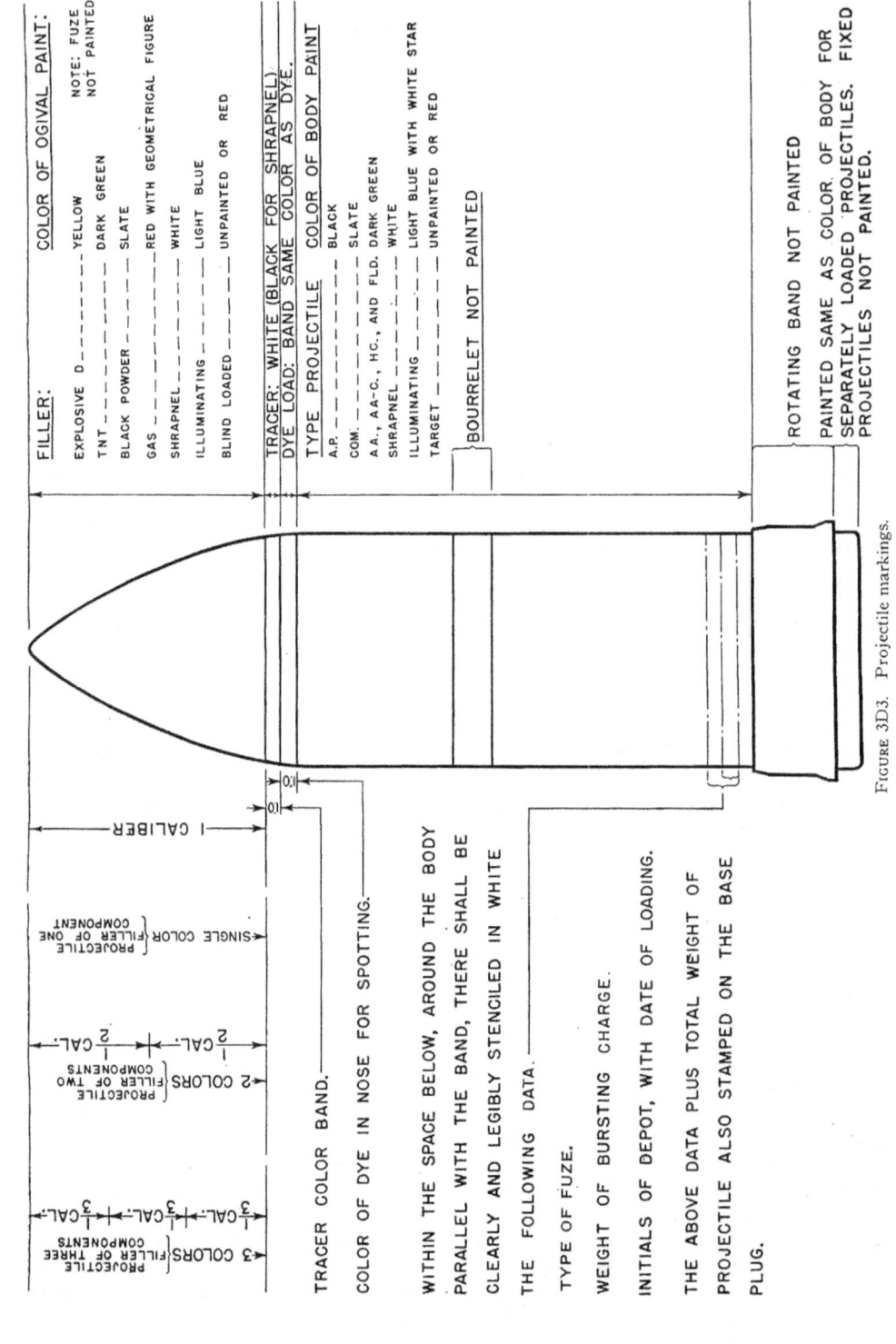

FIGURE 3D3. Projectile markings.

CHAPTER 3

3D12. Safety precautions, selected from Article 972, Navy Regulations.

"18. Projectiles shall not be altered, nor shall fuzes or any other parts be removed or disassembled on board ship without explicit instructions from the Bureau of Ordnance. Projectiles shall not be allowed to rust or to become oversize through paint. Slings and grommets and other similar protective devices shall be removed before loading projectiles into guns. Since the slings are likely to jam the hoists, they shall be removed before sending up the projectile."

"49. Care shall be exercised to insure projectiles from slipping back from their seats, as unseated projectiles may cause abnormally high pressures."

E. FUZES AND TRACERS

3E1. General. A fuze is a device for detonating or igniting the filler explosive of a projectile either on or after impact, or while in flight. Only a few projectiles, such as star shells and shrapnel, may have fillers of black powder, which require an *ignition fuze* for initiation of their low-explosive reaction. Most fillers are high explosives which require a *detonating fuze*.

Fuzes may be grouped in two classes according to their position on the projectile: (1) *nose* or *point fuzes*, and (2) *base fuzes*. A nose fuze is at the extreme forward end of a projectile and is shaped to conform to the curvature of the ogive. A base fuze is located inside the after end of the projectile. It is fitted in a hole in the base or base plug and extends inward to the filler explosive.

Fuzes are further classified according to the methods by which they can be caused to function. A fuze functioning on impact is a *percussion fuze*, one functioning by its own internal action after a definite time in flight is a *time fuze*, while a fuze incorporating both of these features is called a *combination fuze*.

3E2. Fuze design. The designs of current standard fuzes for naval service are maintained in a confidential status; hence, the discussion here is limited to general types of fuzes and principles involved in their functioning. The names and uses of various fuze types may be seen by referring to figures 3C3 to 3C6. Listed by name these fuzes are:

1. *Mechanical time fuze* (time or combination).
2. *Point detonating fuze* (A percussion fuze, usually of a supersensitive type).
3. *Auxiliary detonating fuze* (A supplementary fuze. In the 5"/38 cal. projectile this fuze takes the explosion initiated by the nose fuze and passes it on to the main explosive charge).
4. *Base detonating fuze* (A percussion fuze with or without a delay element).
5. *Self-destroying fuze* (The burning tracer acts as a time fuze and bursts the projectile before the end of its flight).
6. *Ignition fuze* (not shown—A percussion fuze employed to burst black powder fillers).

These fuzes may be grouped into four general classes according to the designed function of the projectile, as follows:

1. *Instantaneous fuzes* on AA-Com and HC projectiles not expected to penetrate more than destroyer plating. There is no time-delay element between the primers of such fuzes and the burster charge.
2. *Supersensitive nose fuzes* on AA projectiles, such as the 20 mm., 40 mm., and the 1.1 inch, whose bursting action must occur immediately after penetration of very light plate or fabric.
3. *Delayed-action fuzes* on AP projectiles which are to penetrate 1½ inches or more of armor plate. There is a time-delay element between the primers of such fuzes and the burster charge.
4. *Time fuzes* on AA and HC projectiles which must function at the desired point in the air. Such fuzes cause the projectiles to explode at some definite time interval after they are fired from the guns.

3E3. Fuze terms. In the design of fuzes the following forces or elements may be utilized so that a fuze can perform its functions:

1. *Setback*. The action caused by the tendency of parts within the fuze body to remain stationary due to inertia as the projectile is accelerated forward.
2. *Creep*. The action caused by the tendency of those parts within the fuze, which are not decelerated by air resistance, to move forward with respect to the fuze body due to inertia.

RESTRICTED

GUN AMMUNITION

3. *Inertia at impact.* The tendency of all parts of the fuze to continue in the line of motion when the projectile strikes an object.

4. *Centrifugal force.* The radial force resulting from the rotation of the projectile.

5. *Friction.* A force tending to slow down moving parts.

6. *Powder trains.* Timing devices operated by the burning of compressed powder pellets or trains.

7. *Clock mechanisms.* Mechanical timing devices operated by springs or centrifugal force.

3E4. **Arming.** A fuze is said to be *armed* when the firing point is in a position to strike a detonating cap. The fuze functions if the firing pin is made to strike this cap. There are three essential requirements for any fuze, as follows:

1. It must be safe to handle; that is, the fuze must not become armed if dropped or joggled.

2. The fuze must be safe within the bore of a gun and for a sufficient distance outside the muzzle to insure security of personnel in the vicinity.

3. The fuze must initiate the explosion of the filler at the proper moment, whether on impact with instantaneous or delayed action, or at the desired time in flight.

The principle employed in arming a fuze in conformance with these requirements, and particularly with regard to safety, is best illustrated by the Semple centrifugal plunger.

3E5. **Semple centrifugal plunger.** The mechanism, shown in figure 3E1, consists of a *firing pin* and *axis*, a *plunger*, *safety pins* and *springs*, and a *creep spring*. In the safe, or unarmed, position, shown in figure 3E1a, the small safety pin springs force the two safety pins inboard to lock the firing pin in the safe position. A side blow would dislodge only one of the pins but not both. Thus, the fuze is safe against side impact. The fuze is also safe if dropped because the firing point is housed in the plunger.

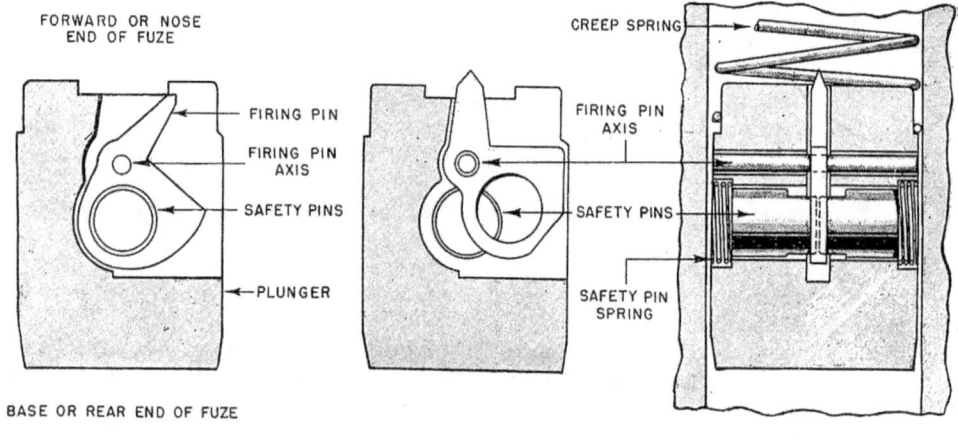

FIGURE 3E1. Semple centrifugal plunger.

The fuze is armed by centrifugal force at a predetermined rotational velocity of the projectile varying from about 1,500 to 3,000 r.p.m. When the projectile is rotated by the rifling, the safety pins are thrown out against the action of their springs by centrifugal force. The same force tends to rotate the heavier part of the firing pin about the firing pin axis to bring the firing point into the armed position. However, the inertia of the heavier part of the firing pin causes the movement of the pin to lag during acceleration of the projectile in the bore. Thus the firing pin is not rotated to the armed position until the projectile leaves the bore; then the firing pin assumes the position shown in figures 3E1b and 3E1c.

CHAPTER 3

There is a tendency for the plunger to creep during flight, as it is not subject to the same retardation due to air resistance as is the projectile. If creep were permitted the plunger would move forward, bringing the firing pin in contact with the primer cap. However, the creep spring placed forward of the plunger prevents this motion (see fig. 3E1c). The spring is only strong enough to prevent creep, but naturally does not prevent the normal firing action of the plunger flying forward due to inertia when the projectile strikes an object. The impact of the firing pin against the cap (not shown) explodes the filler by the detonation of the cap, by way of an explosive train involving a detonator and a booster. A delay element, which may be a pellet that burns a predetermined time before exploding the detonator, also is provided in a fuze for a projectile designed to pierce armor before bursting.

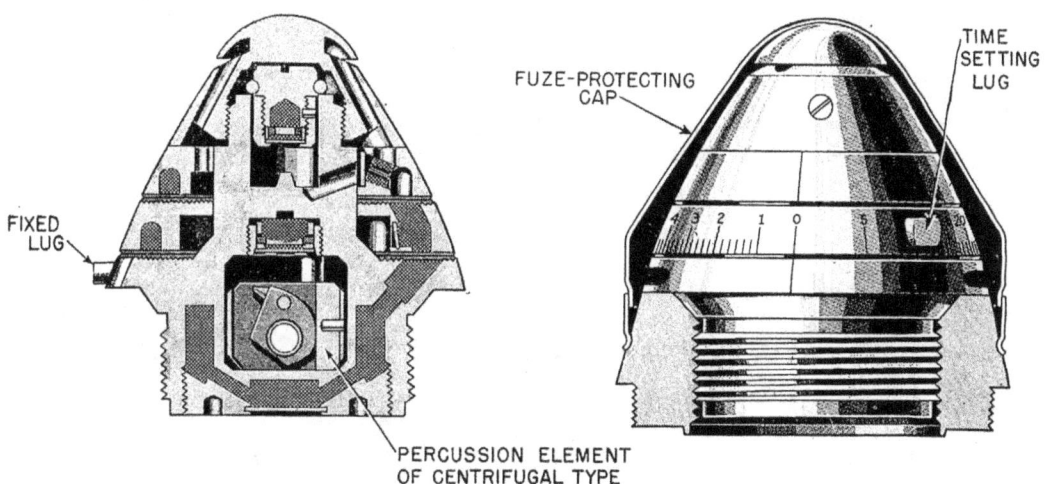

FIGURE 3E2. Frankford Arsenal 21-second combination fuze.

3E6. Time fuzes. For many years the 21-second time fuzes used for 3-inch projectiles and the 45-second time fuze used for 5-inch and larger projectiles were of the powder-train type as shown in figure 3E2. In these, setback causes a small *plunger* carrying a cap to be released from a *split ring* and to impinge on a fixed firing pin which explodes the cap. The flame from the cap ignites one end of the upper powder ring of compressed black powder. The flame progresses from the upper ring to the lower powder ring, the total length of train to be burned being determined by the time setting of the fuze. When the set length of train has burned, the flame flashes through to the detonator which bursts the filler explosive.

Setting of the fuze is accomplished by removing the fuze-protecting cap, and then rotating the movable lower ring on the outside of which is engraved the time scale, so that the desired time is set opposite the index mark of the fixed upper ring (see fig. 3E3). The total length of the two trains to be burned is determined by the position of the end port of the lower train relative to the end port of the upper train.

A *safe* setting is provided so that the fuze cannot accidentally be caused to function by dropping the projectile. The fuze is also designed so that it will not function at a setting below 1.8 seconds. This safety feature is provided so that the projectile will not burst too near the gun crew. The primary disadvantage of the powder-train time fuze is that the burning rate of the rings is affected by change in atmospheric pressure during the projectile's flight, thus causing variation in the time of burst. In addition, the aging and hygroscopic characteristics of black powder make this type of fuze unreliable and inaccurate.

Modern time fuzes are of a mechanical type not subject to the disadvantages of the powder train. There are two kinds of mechanical time fuzes; one which depends solely upon the centrifugal force

RESTRICTED

resulting from rotation of the projectile, and the other a spring-driven variety operating on a clockwork principle. Obviously, mechanical fuzes of the centrifugal type are less affected by long periods of stowage than those of the spring-driven type. However, the spring-driven fuze is more satisfactory for use on larger projectiles where the speed of rotation is slower, and hence, the centrifugal force is less.

The external contour of a mechanical time fuze is almost identical with that of the powder-train time fuze shown in figure 3E2. The method of rotating a movable ring with reference to a fixed index mark is also very similar. A safe setting is provided, as shown, in addition to a minimum setting that prevents functioning too soon after discharge.

3E7. **Tracers.** A *tracer* is a device fitted to the base of a projectile to make it possible for the eye to follow the flight thereof. A tracer alone may be fitted to the base of a projectile, or, when a base fuze is employed, the tracer may be fitted inside the base fuze stock—the actions of fuze and tracer being independent. Another type of tracer is designed to act as a kind of time fuze which explodes the filler when the tracer burns through to the explosive. The time interval is determined by the design of the tracer, and is of such duration as to cause destruction of the projectile in flight. The mechanism in question is called a *self-destroying tracer fuze* and is employed on impact-fuzed AA projectiles to prevent them from landing and bursting on own ships or installations.

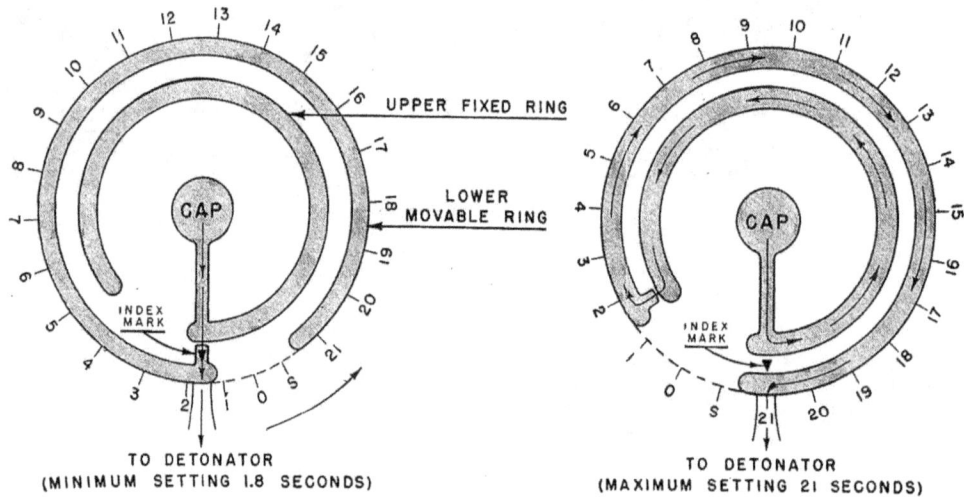

FIGURE 3E3. Schematic of the powder rings of a powder-train time fuze.

3E8. **Tracer description.** Essentially, a tracer is a column of highly compressed pyrotechnic mixture designed to burn slowly with a definite color during enough of the projectile's flight to provide for visibility. Pressed on the outer end of the pyrotechnic column is a small quantity of starter mixture which is capable of easy ignition. Ignition may be accomplished by the flame from the propellent charge, or by the pressure of the propellent explosion which actuates a friction or percussion element at the base of the tracer. The entire column is loaded in a tracer body which is fitted internally or externally to the base of the projectile. Some projectiles, such as the 20 mm., have the tracer mixture loaded directly into the tracer extension of the projectile proper. The tracer colors now standard in the Navy are red or white for AA projectiles, and orange for night tracers in AP and common projectiles.

3E9. **Safety precautions, selected from Article 972, Navy Regulations.**

"19. A fuzed projectile, or a cartridge case, whether in a container or not, if dropped from a height exceeding 5 feet shall be set aside and turned into a naval ammunition depot at the first opportunity. (See par. **17**.) Such ammunition shall be handled with the greatest care. However, if a shrapnel or an illuminating projectile with a 21-second fuze not set at 'safety' is dropped or struck so as to deform the fuze, the complete cartridge, if it is fixed ammunition (otherwise the fuzed projectile only), shall at once be thrown overboard or immersed in water until this can be done."

"21. Certain minor caliber and time fuzes are armed by setback instead of centrifugal action. Care must be used to avoid tapping or otherwise striking projectiles so fuzed. This precaution is particularly applicable to attempts to loosen such a projectile in the cartridge case by repeated light blows of a hammer or mallet. It also applies to unloading such a projectile wedged in the bore of a gun."

"22. Nose fuzes being sensitive, care shall be taken to prevent them from being struck as by the gun in recoil, by ejected cases, dropping, etc."

"23. Time fuzes which have been set shall be reset on 'safety' before sending them below."

"24. In handling projectiles fitted with tracers, care shall be taken not to strike the tracer, as such a blow involves danger of igniting it."

CHAPTER 4

GUN ASSEMBLIES

A. GUN BARRELS

4A1. Definition. A gun is a tube capable of containing and so controlling the explosion of a propelling charge as to discharge a projectile at high velocity. The term *gun* properly designates the tube or barrel, but it is commonly used to refer to the entire assembly of which the barrel is but one part. The terms *gun* and *barrel* are used synonymously in this section. Thus, a *gun assembly*, such as the one shown in figure 4A1, is composed of a number of components which can be grouped under these headings:

1. The gun, or barrel.
2. The breech assembly.
3. The mount.
4. The sight.

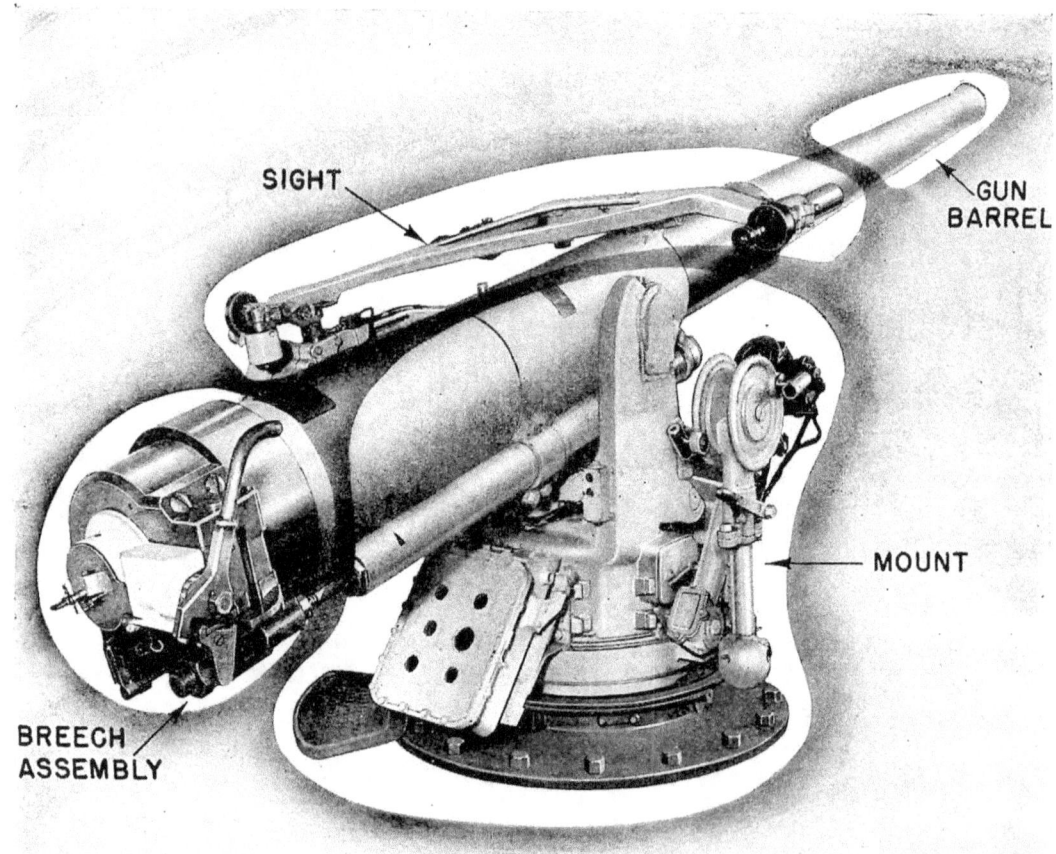

FIGURE 4A1. A typical gun assembly.

4A2. Designation by size. The diameter of a gun's bore is its *caliber*, usually expressed in inches. The length of the gun from the forward face of the breech plug to the muzzle is measured in calibers. A 16″/45 *caliber* gun has a 16-inch caliber, or bore diameter, and an overall length slightly greater than 45 calibers, or 720 inches (16″x45).

Some guns, such as a 1-pounder, may be designated by the weight of their projectiles. Others,

CHAPTER 4

usually of foreign origin, like the 20 mm. and 40 mm., may be referred to by the diameter of the bore expressed in millimeters. Grouped according to bore diameter guns are known as:

1. *Major caliber* if 8 inches or larger.
2. *Intermediate caliber* if between 4 inches and 8 inches.
3. *Minor caliber* if larger than 0.60 inches but not more than .4 inches.
4. *Small arms* if 0.60 inches or smaller.

Designation by size is not a complete description of a gun barrel. There are numerous variations among guns, requiring the use of the mark and modification numbers for identification. Variations among mounts may make it necessary to identify a gun assembly by designating the barrel and the mount.

4A3. Classification by use. Gun assemblies are classified by their use aboard ship as:

1. *Turret*, when mounted in a heavily armored revolving structure. These are the largest guns, and they are mounted on cruisers and battleships. Some intermediate-caliber guns are mounted in gun houses closely resembling turrets, but since the structures are not heavily armored, they are not true turrets.
2. *Antiaircraft*, when designed primarily to fire against aircraft.
3. *Antiaircraft machine guns*, when of the machine gun type and designed to fire against aircraft.
4. *Dual purpose*, when designed for use against both surface and air targets.
5. *Broadside*, when designed for use against surface targets and not installed in turrets. These guns were formerly common, but have been almost entirely replaced by the dual-purpose guns.

4A4. Guns in service. The guns most likely to be found on ships of the Navy at the present time are:

GUNS	CARRIED ON
14 inches and 16 inches	Battleships
12″/50 cal.	Large cruisers
8″/55 cal.	Heavy cruisers
6″/47 cal.	Light cruisers
5″/38 cal.	Battleships, cruisers, destroyers, carriers and auxiliaries.
3″/50 cal.	Submarines, destroyer escorts, subchasers, patrol craft.
40 mm., 20 mm., and caliber .50 machine guns	Any ship from patrol craft to battleships.

4A5. Gun construction. Naval guns in common use today are built in one of the four ways described in succeeding paragraphs. The method of construction is determined chiefly by the size of the gun, but the basic principle governing the design is common to all. This principle is best illustrated in a built-up gun.

4A6. Built-up guns. A gun composed of several concentric cylindrical pieces shrunk over each other is a *built-up gun*, such as that shown in figure 4A2. The outer cylinders, or hoops, are heated and assembled one at a time on the A tube. As the hoops cool, they shrink, and tightly grip the cylinders within them. The locking rings are then added to prevent longitudinal movement of the hoops. After the hoops and locking rings are assembled on the tube, the entire assembly is heated and shrunk on a liner. The 14-inch and 16-inch guns are examples of this construction.

The assembled barrel forms a cylinder within which high pressure is developed as the charge explodes. The effect of the pressure is greatest on the inner cylinder, and diminishes rapidly as it proceeds outward. If the outer hoops were assembled over the tube without shrinkage, they would be subjected to less strain than the tube, and the strength of the gun would be little more than the strength of the A tube. However, the shrinkage of the hoops squeezes the tube, at the same time stretching the hoops. The pressure of explosion can then be increased, for it must overcome the squeeze before it can stretch the tube. The shrinkage is calculated so that each hoop carries a larger share of the strain than it would with no shrinkage.

RESTRICTED

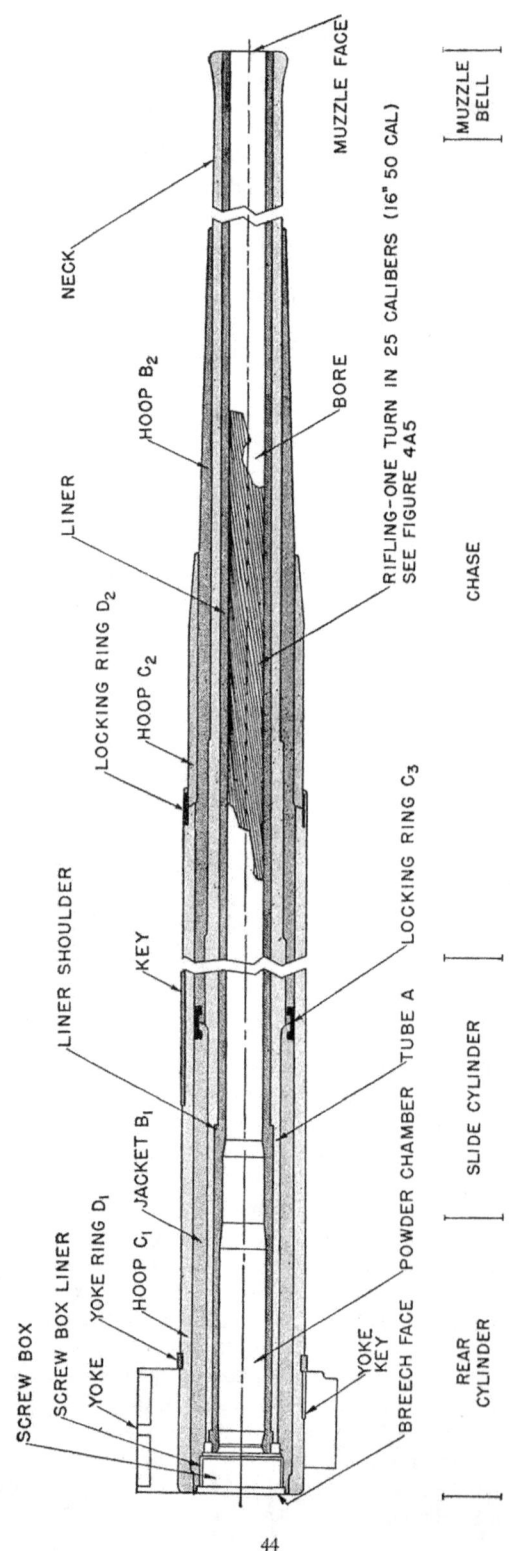

FIGURE 4A2. A built-up gun.

CHAPTER 4

It is apparent that a built-up gun can withstand higher pressures than a gun made of a single untreated cylinder. This characteristic of built-up construction makes possible the large 14-inch and 16-inch rifles used by the Navy. An ordinary single cylinder, even if many times as thick as a built-up cylinder, could not withstand the high pressures developed in modern high-powered guns.

4A7. Radially-expanded guns. A gun made from a single cylinder which has been subjected to a radial-expansion process is called a *radially-expanded monoblock gun*. The process consists of applying high hydraulic pressure to the inside of the cylinder. The pressure, higher than that developed during firing, expands the entire tube radially. After the pressure is removed, each imaginary layer of the tube attempts to return to its original size. However, the inner layers, subjected to greater force and stretched proportionately more than the outer layers, cannot fully return to their original size. As a result, the outer layers grip the inner layers just as the hoops grip the tube in a built-up gun. The monoblock is then in a condition to withstand modern firing pressures without damage to its inner surface.

The radial-expansion process results in a cylinder in which the change from squeeze on the inner layer to stretch of the outer layer is uniform. The change in a built-up gun is in abrupt steps from hoop to hoop, and the strength of the metal is not fully utilized. The metal in a radially-expanded gun is used much more efficiently, hence strength for strength, radially-expanded guns weigh less than built-up guns. The decreased weight of the barrel permits a decrease in the weight of the entire gun assembly.

The radial-expansion process produces guns faster and cheaper than the building-up process. This fact, plus the saving in weight, make radially expanded guns preferable to others. However, the process is limited at present to moderate size guns because of the difficulty of obtaining a single forging large enough for those of major caliber. Typical guns having monoblock barrels are the 3"/50 cal. and 5"/38 cal.

4A8. Combination guns. The built-up and radially expanded features may also be incorporated in a single gun. Thus, the difficulty of obtaining a single forging big enough for the larger guns can be overcome. The 8"/55 cal. gun, for example, has a jacket shrunk on a radially expanded tube.

4A9. Simple one-piece guns. Many small guns such as the 40 mm. and 20 mm. are made from a single steel forging which requires neither the radial expansion nor hoops. The pressures developed per square inch in small guns may be higher than those in large guns, but this may be compensated by increasing the size of the forging which is not excessively large in any event. This type of construction is found in machine guns and small arms.

4A10. Nomenclature. The commonly used names of the various parts of a gun, generally applying to any one of the four types of construction, are labeled in figure 4A2 and defined as follows:

External:

    1. The *breech* is the rear end.

    2. The *yoke* is the large ring surrounding the breech end of the barrel, and is considered part of the gun. It provides a connection between the barrel and the mount. Shoulders on the gun prevent movement between the barrel and yoke.

    3. The *rear cylinder* is the portion which extends from the breech to a point forward of the chamber. The yoke is attached to this part.

    4. The *slide cylinder* is the portion of uniform diameter that supports the gun in a slide, through which the gun moves when it is fired. One or more longitudinal keys are inserted between the slide and slide cylinder to prevent rotation of the gun due to the reaction of the projectile on the rifling. The slide cylinder is sometimes made the same diameter as the rear cylinder to permit removal of the gun from the front of the slide.

    5. The *chase* is the sloping portion forward of the slide cylinder extending to the neck.

    6. The *neck* is the part of least outside diameter.

    7. The *muzzle bell* is the part forward of the neck which flares outward to provide greater

RESTRICTED

GUN ASSEMBLIES

strength and prevent enlargement of the bore at the muzzle. In some guns, especially of smaller sizes, the flare may extend past the end of the rifling to provide a shield which will minimize flash and muzzle blast.

Internal:

1. The *screw-box liner* is screwed into threads cut in one of the hoops. This liner contains the threads which engage the threads on the breech plug described later. A liner is used to facilitate replacement of the threads when they become worn.

2. The *chamber* is the space into which the powder charge is loaded. It is made larger in diameter than the bore in order to shorten its length and to facilitate loading. The details vary among guns, but slopes are provided to guide the projectile into the bore, and to provide a seat for the rotating band to insure uniform loading.

3. The *bore* is that interior part of the tube which is of uniform diameter and extends from the chamber to the muzzle. Its diameter is measured between diametrically opposite lands.

4. The *rifling* is the set of helical grooves cut in the bore. The raised spaces left between the grooves are the *lands*.

5. The *muzzle* is the forward end.

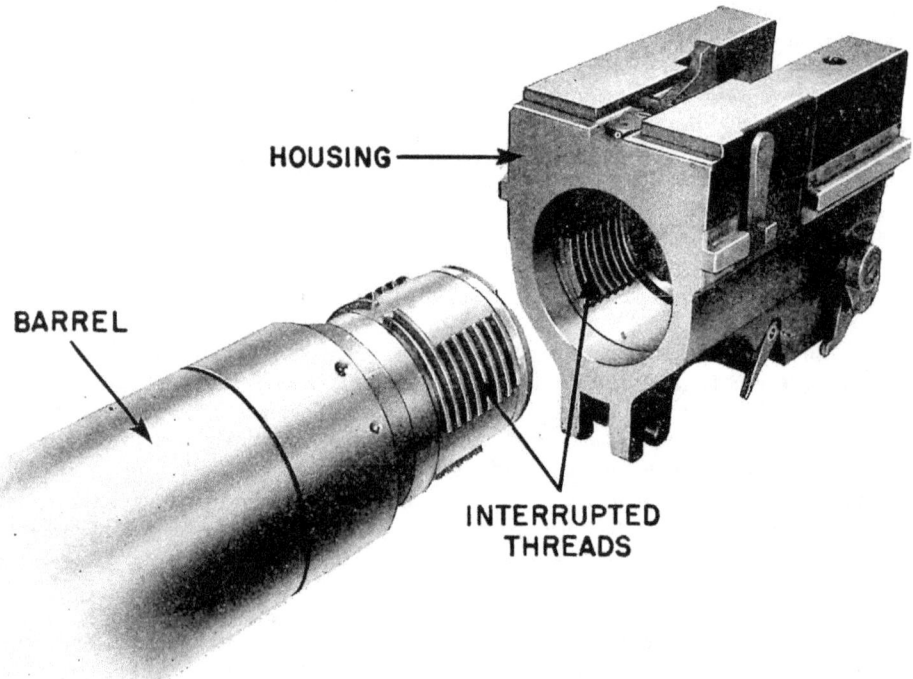

FIGURE 4A3. A typical bayonet-type joint.

4A11. Bayonet-type joint. Many of the smaller guns* now in use have a *housing* which takes the place of the yoke and screw-box liner. This construction, illustrated in figure 4A3, sometimes also permits the elimination of the slide cylinder, as will be seen when individual guns are studied. Such an arrangement permits the use of a *bayonet-type joint,* which makes it possible to remove the barrel without disassembling the entire gun.

The joint is similar to that formed between a screw and a nut. Threads on the barrel engage with threads in the housing, but they are interrupted (have sections cut away) in such a manner that the barrel can be inserted or removed after a partial revolution. This type of joint is employed

* A housing is used in guns through the 6"/47 cal., and this use is now being extended to the 8"/55 cal.

in numerous pieces of ordnance because of its simplicity, and the speed with which the barrel may be replaced.

4A12. Rifling. The rifling, whose purpose is to impart rotation to the projectile, is formed by cutting grooves in the bore after the barrel has been machined. The origin of rifling is in the band front slope, which receives the front edge of the rotating band as shown in figure 4A4.

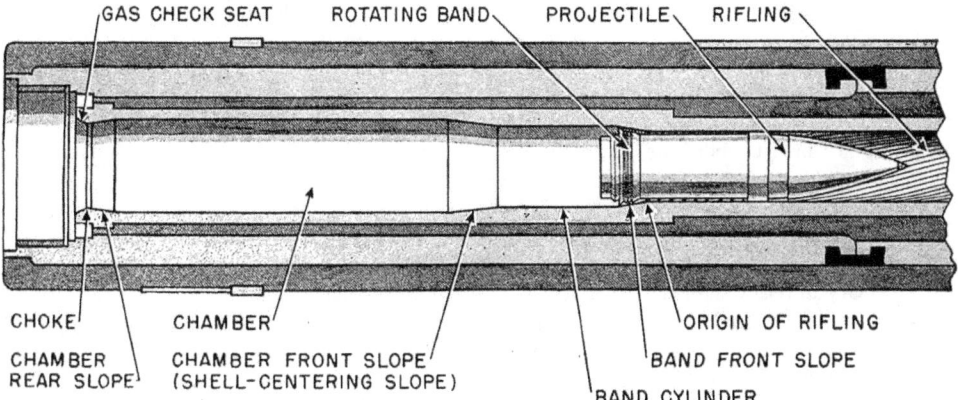

FIGURE 4A4. A projectile seated in the bore.

Generally, the rifling at this point actually starts to engage or engrave the rotating band when the projectile is loaded. Any forward movement after the propellent charge begins to burn will cause rotation of the projectile.

There are two classes of rifling, one *uniform twist,* the other *increasing* twist. Uniform twist causes the same rotation for each foot traveled in the bore, while increasing twist causes greater rotation during the last foot traveled than during the first foot. Both types are in use, although most of the recent designs call for uniform twist. All naval guns have right-hand twist, which produces clockwise rotation of the projectile when viewed from the base. This rotation causes the projectile to drift to the right after it leaves the gun.

Twist is usually expressed by the number of calibers which must be traveled to produce one revolution of the projectile. A twist of 1 turn in 30 calibers means that the projectile turns once as it travels a distance of 30 calibers. The rotational velocity of a projectile depends in part on the twist and the muzzle velocity, and varies considerably between guns. In the 16"/45 cal. gun it is about 4,000 r.p.m., while in the 40 mm. it is about 40,000 r.p.m.

The width of the lands in uniform-twist rifling is made slightly greater at the muzzle than at the origin of rifling in some guns. This makes up for the wearing away of copper from the rotating band, and thereby helps to maintain an effective gas seal.

4A13. Cause and effect of erosion. *Erosion* in a gun is the wearing away of the bore surface. It is caused by hot gases rushing over that surface, and by the rubbing of the projectile. The result is enlargement of the bore, mostly at the origin of the rifling. It would ultimately cause the projectile to be seated farther forward for each successive round, if it were not for the lips provided on rotating bands.

Another effect produced on the bore by repeated firing appears in the form of *heat cracks.* The tendency of the bore to expand with the great increase in temperature at the instant of firing is prevented by the outer layers of cooler metal. The bore surface is crushed by this action, and when the metal cools and contracts, minute fissures appear. They work deeper with repeated firings, but do not reduce the strength of a gun during its useful life. They do provide minute passages for the flow of gas around the rotating band, and in this respect are undesirable. Gas escaping by these paths moves at a much higher velocity than that of the main body of gas behind the projectile, and has a greater erosive action.

The effect of erosion, as it influences the control of gunfire, is a reduction in muzzle velocity. The enlarged bore offers less resistance to projectile movement; consequently, travel begins at a lower gas pressure, hence earlier in the reaction than in a new gun. This means that before maximum pressure is attained, the projectile travels farther through a worn gun bore than through a new gun bore. This lowers the maximum pressure because of the greater volume occupied by the gases, and decreases the average effective pressure. The result is reduced muzzle velocity.

4A14. **Control of erosion.** No way to eliminate erosion has been found, although some case ammunition is provided with a wad or mouth plug containing a lubricating substance which tends to retard this action.

A gun is useful until its rifling becomes so worn that accuracy is lost. If it were not for this, the life of a gun would be indefinite, as the other parts can be kept in working condition by adequate maintenance. The effect of erosion is considerably greater in large guns than in small guns. The life of a major-caliber gun may be about 300 equivalent service rounds, while some minor-caliber guns may successfully fire more than 1,500 equivalent service rounds.

The cost of large guns is so great that they are equipped with liners which can be replaced when it is necessary to renew the rifling. A few built-up guns have no liners, but they are re-rifled by replacing the tube, or by boring out the tube and inserting a liner. The cost of re-rifling many of the smaller guns is not justified, and their barrels are replaced when necessary.

4A15. **Compensation for erosion.** Loss in muzzle velocity reduces the distance a projectile travels from a gun under a given set of conditions. This lost distance can be regained by pointing the gun a little higher. Knowing the velocity loss, the required additional elevation can be readily determined.

Velocity loss can be obtained in two ways from data prepared by the Naval Proving Ground. One may find the velocity loss corresponding to (1) the number of equivalent service rounds previously fired, or (2) the amount of bore enlargement. A record of rounds fired is kept aboard ship so that the velocity loss can be established.

Velocity loss corresponding to bore enlargement is determined by measuring the bore with a star gauge, like the one in figure 4A5. The three points of the star-shaped head are moved outward by

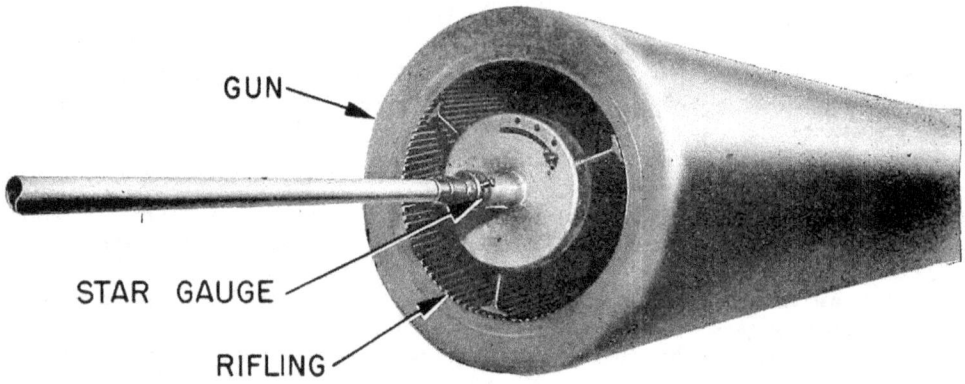

FIGURE 4A5. A star gauge.

a conical bar pushed between them at the center. The diameter of the bore is read from a scale on the bar, and the enlargement is then computed. This method is usually used to check the rounds fired method at appropriate intervals. The necessary star-gauging equipment is not available on all ships, hence the measurements are often made alongside tenders or repair ships, or in navy yards.

4A16. **Care of bore.** Guns require the best of care. Instructions regarding them are found in the *Bureau of Ordnance Manual*, Ordnance Pamphlets and other sources of information. A few of the more important hints are mentioned here, but must not be considered complete.

Dirt in the bore of a gun is not only an invitation to corrosion, but it may offer resistance to the passage of the projectile sufficient to produce excessive pressures with consequent rupture of the gun. Hence, after each firing the bore must be thoroughly washed and dried to remove any residue and then oiled lightly.

The muzzle of a gun is usually protected against the weather by a solid plug, known as a *tompion* (pronounced tomkin). It must be removed before firing to prevent excessive pressures. However, during wartime the guns must always be ready for action, and the plugs are sometimes replaced by canvas muzzle covers through which a projectile can be fired if necessary, provided it is not fitted with a supersensitive nose fuze.

The liner in a gun is usually elongated after repeated firing by the action of the rotating band tending to squeeze the metal of the liner before it. This in itself is no cause for alarm, and the metal extending from the muzzle is simply cut off when the extension exceeds ½ inch.

The projectile has the additional tendency to pull the entire liner forward with it. The conical shape of the liner and the shoulders on it resist this movement, but occasions may arise when the liner will creep during firing. The shoulders on the liner, shown in figure 4A2, may partially override the corresponding shoulders in the A tube. This causes the metal of the liner to be forced inward, reducing the diameter of the bore. Any constriction is dangerous and must be removed to avoid excessive pressures within the gun.

Constrictions are detected with a gun-bore plug gauge, shown in figure 4A6, which is a cylindrical steel plug accurately machined to the diameter of the bore. The gauge must be passed through the bore before and after each firing. If it will not pass without undue forcing, the cause should be determined.

PLUG GAUGE   **LAPPING HEAD**

FIGURE 4A6. Plug gauge and lapping head.

Constrictions may be removed by polishing or lapping the bore with a lapping head like that in figure 4A6. The head is covered with a fine abrasive material and then drawn back and forth at the location of the constriction until the plug gauge will pass without forcing.

4A17. Safety precautions, selected from Article 972, Navy Regulations.

"94. Steel constrictions of the bore, usually caused by the gun liner overriding the retaining shoulders in the tube, are a source of possible danger in firing. It is not always possible to distinguish copper constrictions from steel constrictions. Therefore no gun shall be fired in target practice unless the bore gauge will pass through the entire bore without undue forcing. After target practice the gauge shall be tried in each gun and the bore enlarged, if necessary, until the gauge will pass."

B. BREECH ASSEMBLIES

4B1. General. The breech assembly of a gun includes the mechanisms necessary to close or stop the breech and fire the primer, the attachments required to initiate the firing, automatic and power loading-equipment if used, and the various other devices intimately associated with the breech end of the gun. The assembly must be capable of withstanding the high chamber pressures with no leakage of gas to the rear. It must also provide dependable, rapid means of firing the charge at the proper instant, and be safe in operation.

4B2. Breech mechanisms. The major portion of the breech assembly is the *breech mechanism*. The two general types currently used on nearly all U. S. naval guns, excluding small arms, are the *interrupted screw* and the *sliding wedge*. An interrupted-screw mechanism is recognized by

GUN ASSEMBLIES

a cylindrically shaped, partially threaded plug which screws into and *plugs* the breech. The distinctive part of a sliding-wedge mechanism is a grooved, oblong block which slides across the face of and *blocks* the breech. The breech mechanism in either case includes the block or plug, the system of operating parts required to close the breech, and the devices actually carried on the block or plug.

4B3. Sliding-wedge block. The sliding wedge is used on case guns such as the 3″/50 cal., the 5″/38 cal., the 6″/47 cal., and the 40 mm. The block *slides* vertically in grooves contained in a housing, similar to that in figure 4B1. The sloping surface on the forward face of the block *wedges* the cartridge case home in the chamber. Since this type of block moves vertically, it is often referred to as a *vertical sliding-wedge* to distinguish it from the horizontal type used in some older guns.

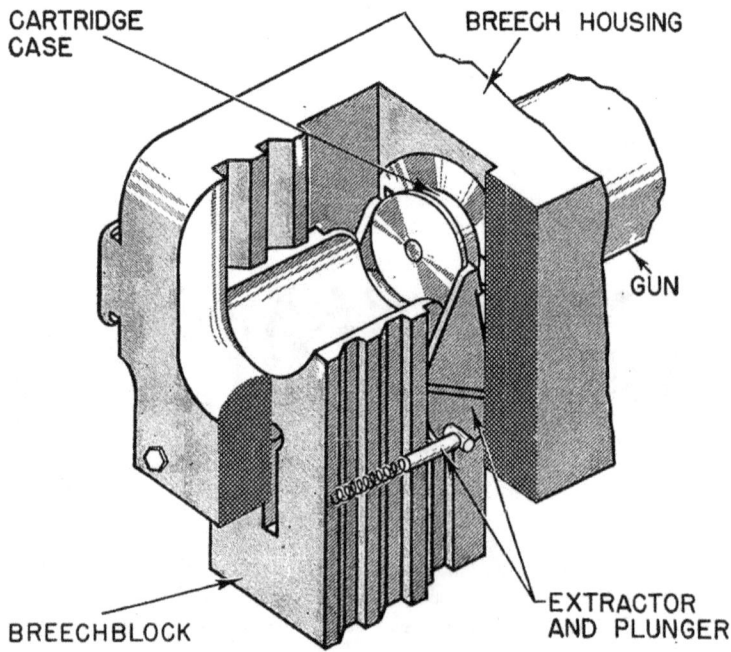

RIGHT SIDE VIEW
FIGURE 4B1. Sliding-wedge breechblock.

The problem of extraction is present with all case guns since the empty case must be removed before the new charge can be loaded. Devices called *extractors* fit under the rim of the case, and operate to extract and eject it when the block is dropped. Generally, the extractors are intimately associated with the operation of the mechanism, and their details will be explained more fully in the study of individual guns.

4B4. Interrupted-screw plug. All modern bag guns and a few old case guns still in service have the Welin type of interrupted-screw breech plug. The plug carries on its periphery a set of interrupted threads as shown in figure 4B2.

There is some similarity between this construction and the bayonet-type joint, but it is to be noted that the threads are stepped. The number of threads on plugs varies among guns, but it is clear that if the threads were not interrupted, the plug would have to be revolved about a dozen times to free it. This would increase the loading time, causing an undesirable decrease in the maximum rate of fire.

CHAPTER 4

It is of interest to note that the simple bayonet-type joint described in the section on guns has interrupted threads on approximately 50 per cent of its circumference. The stepped arrangement makes it possible to have threads on from 60 to 75 per cent of the circumference, resulting in a stronger joint for a given number of threads, or permitting the use of fewer threads. The number of threaded and blank sectors varies among different guns, but the general characteristics are common to all breech mechanisms using the interrupted-screw type plug.

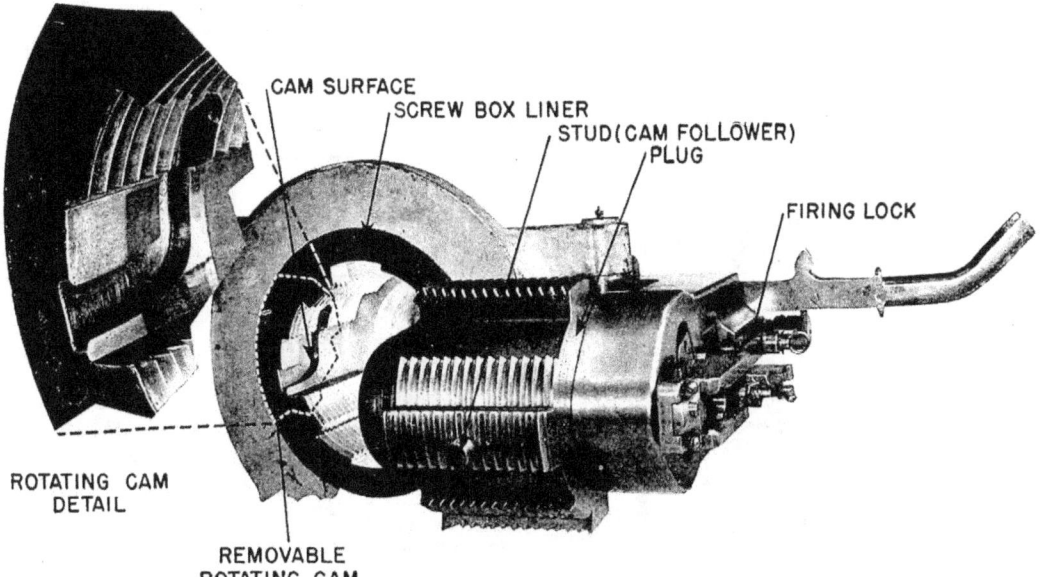

FIGURE 4B2. Welin interrupted-screw breech plug.

During closing, the plug must be swung into place and then rotated. Attached to the *screw-box liner* is a *rotating cam* which engages a *stud* (or sometimes a roller) on the plug and smoothly changes the swinging to rotating motion. While the transition is taking place, the plug is swinging inward and rotating at the same time. The threads can not be engaged until the swinging motion stops. Consequently, the blank sectors must be slightly wider than the threaded steps to allow about 5° of plug rotation and permit the rotating cam to function before the threads engage.

4B5. Gas checks. No breechblock or plug can be so carefully fitted to the breech that it will completely prevent the escape of gas from the chamber to the rear. Gas escaping in this manner is known as *blowback*. While blowback may be injurious to personnel, it must be prevented for another reason as well. If it occurs, the hot gases will erode the accurately machined surfaces, making them still less gas-tight. Any breech mechanism must be supplemented with an effective *gas check*. There are two automatic checks employed, both of which use the pressure of explosion to tightly seal the breech.

4B6. Sealing by expansion. The cartridge case provides the gas check in all case guns. The pressure develops within the case and *expands* it outward against the surface of the chamber. Theoretically, the surfaces of the case and chamber must be perfectly smooth to provide a tight seal. Actually, there are irregularities, but the total area of surface in contact is so large that the seal is effective.

4B7. Sealing by compression. A separate gas check must be provided to form the gas seal in bag guns, since there is no case to perform that function. A device known as the *De Bange gas check* is employed in all U. S. naval bag guns. Its principal parts are indicated in figure 4B3.

The *pad* is firmly pressed against the *gas-check seat* as the breech is closed. The pressure of explosion pushes against the *mushroom,* forcing it slightly to the rear. The curvature on the after side of the mushroom *compresses* the pad, causing it to spread radially against the gas-check seat. The greater the chamber pressure, the harder the mushroom pushes against the pad. The seal is automatic in the sense that it adjusts itself to the pressure within the gun.

The circumference of the gas-check seat and the surfaces between the plug and mushroom are sealed by the pad. The primer vent is sealed by the expansion of the primer case against the *primer chamber walls* in the outer end of the mushroom stem.

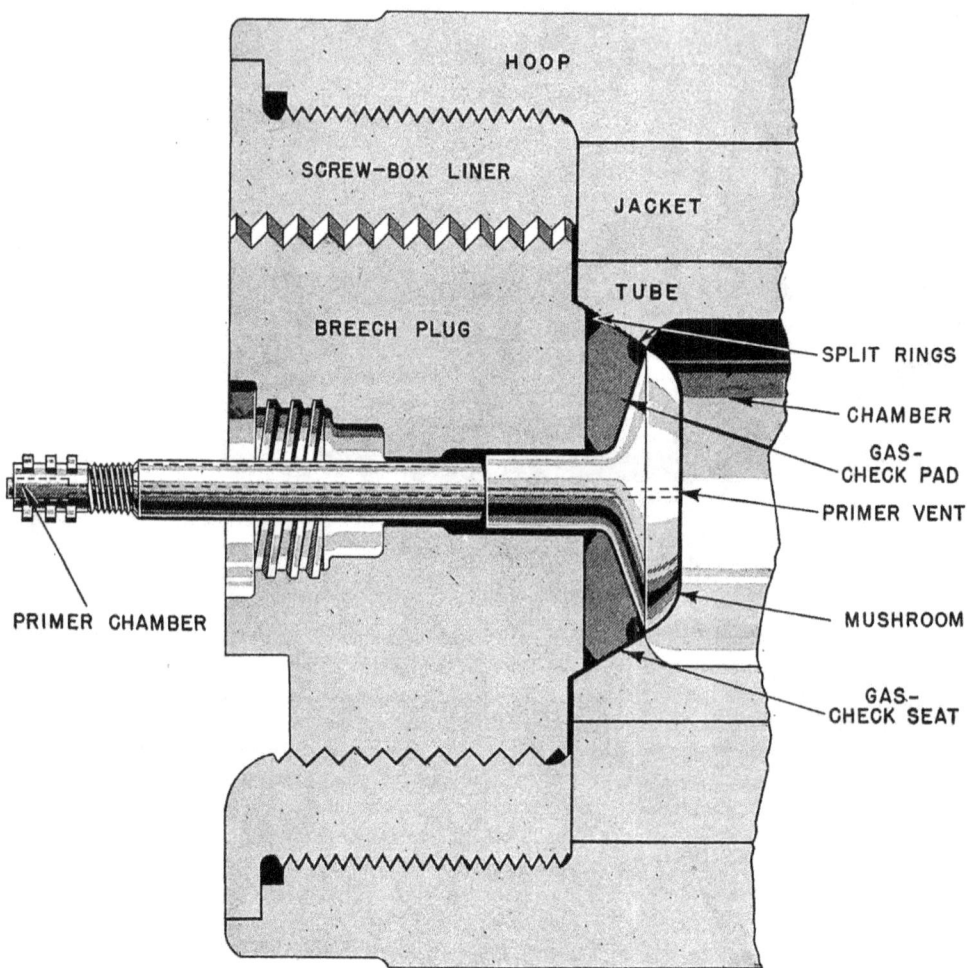

FIGURE 4B3. The De Bange gas check.

Although the pad is plastic and capable of flowing under high pressure, it feels hard to the touch. The material is a compressed mixture of tallow and asbestos, enclosed in a fabric cover. The metal *rings*, protecting the edges of the pad, are split to allow them to expand when the pad is compressed.

After the gun has fired, the pad adheres to the surfaces which it seals. The pad should be pulled to the rear without rotation to prevent damage to that part of its surface which is in contact with the gas-check seat. This is accomplished by preventing rotation of the mushroom while the breech plug rotates around it and moves to the rear. The rearward motion of the plug pulls the mushroom aft, freeing the pad from the gas-check seat.

4B8. Operating mechanism. Any system of breech closure requires a mechanism to operate it. The plug and its attached parts on a 16"/45 cal. gun weigh over 2,000 pounds, while the block on a 40 mm. weighs little more than 10 pounds. Such variations in weight alone introduce differences between mechanisms. Basically, there must be a system of cranks or levers arranged to rotate a plug and swing it out of the way, or to raise and lower a block. Gun size largely determines the additional features.

The three principal classes of breech-operating mechanisms are automatic, semi-automatic, and non-automatic. Guns using these mechanisms are classified by the same names, as shown below. Examples of each of these types of guns will be studied in more detail later.

1. *Automatic* guns are case guns in which part of the energy of explosion is utilized to eject the empty cartridge case, and to energize a device which automatically loads another cartridge and fires the gun continuously as long as the trigger is operated and the ammunition supply is maintained. Such guns are sometimes called *machine guns*. Examples are the 20 mm. and the 40 mm.

2. *Semi-automatic* guns are case guns in which part of the energy of explosion is stored, then later used to automatically open the breech, eject the empty cartridge case, and then close the breech after another round is loaded either by hand or auxiliary power. Examples are the 6"/47 cal., the 5"/38 cal., and the 3"/50 cal.

3. *Non-automatic* guns are in neither the automatic nor semi-automatic classes. All bag guns and some case guns are of this type. The term is not in general use, and usually a gun is considered non-automatic unless it is described as automatic or semi-automatic. Guns of this class have been designated as rapid fire or quick fire.

4B9. Loading equipment. The weight of projectiles and powder for major caliber guns necessitates power-operated handling and ramming equipment. The weight of the charge and the desired high rate of fire for intermediate caliber guns such as the 5"/38 cal. and 6"/47 cal. require power rammers. Small guns such as the 40 mm. and 20 mm. have such high rates of fire that they must be provided with fully automatic loading mechanisms. Such equipment, not strictly part of the breech mechanism, is so intimately associated with it that it is properly considered part of the breech assembly.

4B10. Misfires and hangfires. There is always the possibility of some defect in ammunition or error in loading which might delay the ignition of the charge. As an example, if a powder bag is loaded with the ignition pad forward, the flame from the primer may not be sufficient to cause immediate ignition. The silk bag *may* have been ignited, and may smoulder slowly until the glowing area has spread to the ignition pad, whereupon the gun will fire. Such a delay in discharge is termed a *hangfire*, while a complete failure to fire is called a *misfire*. Common sense dictates the regulation that a failure to fire shall be treated as though a hangfire were in progress. Except in action, the breech shall not be opened before 30 minutes have elapsed from the last attempt to fire the gun.

There are several ways of making safe, repeated efforts to fire the gun if the first attempt fails. Primers can be replaced in bag guns without opening the breech, hence it is possible to examine the primer to see if it has fired. If this or additional primers do not fire, there is still the possibility that a burning ember from the last charge fired has initiated a hangfire, and the gun must not be unloaded for 30 minutes from the last firing. The waiting time for hangfires may be changed during action at the discretion of the Commanding Officer.

4B11. Salvo latches. A *salvo latch* is a device which prevents the unintentional opening of the breech after a gun is loaded, but before it has been fired. Usually more than one gun will be fired at once, forming a salvo, and a gun captain may not be sure that his own gun has fired. The addition of salvo latches to guns was the result of a serious accident in a turret in which the breech of one of the guns, presumed to have fired, was opened while a hangfire was in progress. The charge exploded, wiping out the crew of the turret. Hence, the primary function of a salvo latch is to make it impossible to open the breech of a loaded gun unless done deliberately, with full knowledge of the potential danger. A secondary function is to keep the operating mechanism locked during the shock of firing.

There are two basic types of salvo latches: one *inertia*, the other *positive*. The inertia type employs a pendulous part which carries a catch that engages or locks the breech-operating lever. When the gun is fired and moves to the rear, the inertia of the part causes it to swing on its axis and release the lever. The positive type also engages or locks the operating lever, but requires a positive mechanical movement to operate it. The arrangement of parts causes the latch to be released during the rearward movement of the gun. The positive type is standard now, but there are some older guns which still may be equipped with inertia-operated latches.

Drills at guns require the opening and closing of the breech without actual firing. Salvo latches, unless made inoperative, lock each time the plug is closed, hence a drill with the latch operative will develop the bad habit of releasing it by hand. It is therefore a regulation that the salvo latch be made inoperative at loading drills.

4B12. Firing mechanisms. A *firing mechanism* is that part of a breech mechanism which directly explodes the primer, which in turn fires the gun. The firing mechanism must be capable of delivering the percussive blow necessary to initiate the reaction in the primer cap, and for combination primers, it must also provide for transmission of the necessary electric current.

4B13. Bag-gun firing mechanisms. All bag guns can be fired both by electricity and percussion, electrical firing being the primary method. The primers must be inserted in the primer chamber in the mushroom stem. A firing mechanism known as a *firing lock* (See fig. 4B2) must be provided to control the loading and firing of the primer. Since the primer is essentially the same for all bag guns, regardless of size, firing locks are all quite similar. Most locks in use are various modifications of the Mark 14 firing lock, which is described in Article 5B3.

4B14. Case-gun firing mechanisms. The primer in the base of a cartridge case must be fired by means of a firing pin extending through the face of the breechblock. The mechanism usually consists of a firing pin, suitable springs, levers to withdraw the pin, and a *sear* to hold it in a cocked position until the block is closed and the trigger is pressed. The sear, holding the pin in the retracted position against the compression of the spring, is released by trigger action, allowing the spring to drive the pin into the primer. Case guns capable of being fired electrically require a slightly different arrangement since the pin must be in contact with the primer to permit passage of current. In case of electrical failure, the percussive blow is delivered by a hammer or striker, driven against the firing pin by a spring. These mechanisms are more fully explained in the descriptions of the guns on which they are used.

4B15. Firing attachments. The firing mechanism alone is not capable of firing the gun. Appliances, termed *firing attachments*, must be attached to the mechanism to make it operative. These attachments include batteries, wires, keys, lanyards, and other devices, some of which are attached to the mount rather than the gun. Since they are essential to the firing of the gun, they are here considered as part of the breech assembly, but it must be understood that they are not part of the breech mechanism as previously defined.

4B16. Percussion firing attachments. Usually the pointer, who moves the gun in elevation, has control of the firing. He is provided with either a foot-firing pedal or a hand-firing lever which, when operated, causes a linkage to release the sear, permitting the firing pin to strike the primer. Some guns are also equipped with solenoids (electromagnets) which release the sear when a firing key is closed. The actual firing is by percussion, but the release of the firing pin is accomplished electrically. Many firing mechanisms are arranged so that they can be fired by the pull of a lanyard connected to the mechanism. Firing by lanyard is used to fire the gun if other methods fail.

4B17. Electrical firing attachments. The wires, switches, firing keys, transformers and batteries mounted on a gun or its mount for the purpose of supplying and controlling current to the primer or a solenoid are firing attachments. The arrangements of these parts vary considerably among guns, but the circuits used generally are capable of accomplishing the same results. A representative wiring diagram is illustrated in figure 4B4.

CHAPTER 4

4B18. Firing circuits. One end of the bridge wire in the primer is connected to the primer case which makes contact to ground through the metal of the gun. The other end is connected to the primer plunger through the plunger cup. Contact between the plunger and the external wiring is made by the firing pin in the firing mechanism. Firing mechanisms are generally equipped with a *safety contact,* automatically held open until the breech is fully closed and locked to prevent premature firing. The circuit then passes through the *pointer's firing key* which must be closed before the primer will fire.

The firing key is connected to the *firing snap switch* mounted near the pointer. This switch has two positions marked *battery* and *motor generator.* When the snap switch is set on *battery,* one end of the circuit from the primer is connected to one side of a 6-volt storage battery which is similar to an automobile battery. The circuit is completed from the grounded side of the battery through the metal of the gun to the case of the primer. If the safety contact is closed electric current will pass through the primer when the pointer presses his key.

If the snap switch is set on *motor generator,* the current is supplied from the secondary of the *firing transformer* at 20 volts, ground again being used to complete the circuit. Although 6 volts are normally sufficient to force the necessary current through the circuit, the 20 volts supplied by the transformer provides a reserve which will overcome some extra resistance that might be caused by poor contacts. In addition, the primer will fire quicker when a higher voltage is applied.

A transformer is used to change voltage from one level to another. It consists of two separate coils of wire wound on an iron core, and is represented schematically as shown in the diagram. The coil connected to the source of power is the *primary,* and the side which delivers the power to the firing circuit is the *secondary.* The voltage available to supply the transformers is usually 115. While 115 volts could be used to fire the guns, the operating conditions at a gun make it desirable to use a lower voltage for the safety of the personnel. Hence, a transformer is used.

The power comes to the firing transformer through a *remote firing key.* This remote key can be used to fire the gun, for unless it is closed no current will flow in the firing transformer or in the firing circuit. Obviously this key will not fire the gun unless the other breaks in the firing circuit are closed first. The use of the remote key will be more apparent when the subject of fire control is studied.

4B19. Illuminating circuits. Lamps are provided to illuminate the crosswires in the telescopes, and various dials which members of the crew must read. The bulbs are similar to automobile lamps, and operate on 6 to 8 volts. They receive their power from the battery or from the 6-volt secondary of the illumination transformer, depending on the position of the illumination snap switch.

4B20. Source of power. The power supplied to the transformers must ultimately be obtained from the ship's generators. A generator is a rotating machine, like a motor in appearance, which converts mechanical into electrical energy. The mechanical energy which drives the generator may come from a turbine, an internal-combustion engine, or an electric motor. On a steam-driven ship, the auxiliary generators which supply electrical power to the ship's auxiliaries are usually driven by steam turbines, as indicated in figure 4B4. These generators supply power throughout the ship, and in addition may be used to charge large storage batteries which can supply electrical energy in the event that the generators fail temporarily.

Storage batteries are capable of delivering only direct current, which flows continuously in one direction. Generators can be built to supply either direct or alternating current. Alternating current continually reverses its direction of flow very rapidly, and cannot be used to charge a storage battery. (The reversals occur 120 times per second in the common 60 cycle domestic power systems.) Since batteries must be used if the generators fail, and the auxiliary generators must charge the batteries, the auxiliary generators are usually the direct current type supplying 125 volts. The transfer switch in the diagram is so arranged that 125 volts D.C. may be obtained from either the batteries or the generators. Actually the electrical connections are more complex than indicated, but they are simplified here for illustrative purposes.

GUN ASSEMBLIES

Alternating current has a number of advantages over direct current which make it more useful in certain fire-control equipment to be studied later. Transformers can change the voltage of alternating current, but not of direct current. Hence, it becomes necessary to change direct current to alternating current for the guns and other fire-control equipment. This is done in a motor-generator set consisting of a direct-current motor driving an alternating-current generator. The motor receives its power from the ship's auxiliary generators or stand-by batteries. The A.C. generator supplies 115 volts to the transformers shown in the diagram.

The system shown in figure 4B4 provides a fair degree of flexibility. There are usually at least two motor-generator sets, either one of which can supply all the alternating current needed by the guns and associated equipment. Even if the ship's auxiliary generators fail, the stand-by batteries will keep the motor-generator sets running. If alternating current cannot be obtained because of casualty, the pointer shifts to local power from the standby battery at the gun. If all electrical attempts fail, the gun still may be fired by percussion.

4B21. Care of electrical circuits. One of the rigid requirements for electrical circuits is that they be carefully checked at frequent intervals. There are many electrical connections which must be kept clean and tight, otherwise current will be limited by the extra resistance offered by loose connections. Insulation must be protected against oil and moisture to avoid short circuits which will prevent the current from flowing where it is wanted. There are a number of standard tests, all outlined by the Bureau of Ordnance, but obviously the surest test is the actual firing of a primer.

4B22. Gas-expelling apparatus. Flarebacks are caused when combustible gases remaining in the bore from the preceding explosion come in contact with oxygen, ignite, and explode to the rear when the breech is opened. They are not only directly dangerous to personnel but they may ignite the next charge of powder during the loading of bag guns. A *gas-ejector* is mounted on all bag guns to guard against flarebacks, and in addition is mounted on case guns enclosed in gun houses or turrets.

The device is essentially a system of nozzles directed into the bore in such a manner that when compressed air is admitted it blows the residual gases out of the muzzle. The air is controlled by a gas-ejector valve which is automatically operated when the breech begins to open. Air at a pressure between 50 and 200 pounds p.s.i., depending on the gun, is supplied from the ship's air accumulators through a system of pipes and valves arranged to permit recoil and changes of gun elevation. The valves are capable of being turned off by hand, and on bag guns usually are as soon as a member of the crew has announced "bore clear". The valves will close automatically when the breech is closed, if they were not closed by hand.

The "bore clear" signal indicates that there are no flaming gases, powder fragments or pieces of bag remaining in the bore, chamber, screw box, or on the face of the breech plug. If this condition does not exist, the signal is "foul bore", and steps are taken to clear the chamber with a wet sponge before the bags are loaded.

It is to be noted that a gas ejector does not guarantee a clear bore. Since the system is mechanical, it is not infallible. For emergency use, an auxiliary gas ejector is provided. It consists of a special quick-acting nozzle on the end of a hose located close to the breech of the gun.

4B23. Safety precautions, selected from Article 972, Navy Regulations.

"56. The salvo latch shall be removed or made inoperative during any exercise which requires opening the breech, except when firing a loaded gun."

"56½. The loading of certain guns does not necessarily insure the functioning of their salvo latches. Care shall therefore be taken in loading to open the plug sufficiently far to insure that the salvo latch trips into position to make it operative on the succeeding closure of the breech."

"57. Effective measures shall be taken to guard against prematurely opening the breech of a loaded gun, whether or not the gun is fitted with a salvo latch."

"67. The possibility of a serious accident due to opening the breech of a gun too soon after a misfire demands the constant exercise of the utmost prudence and caution. After an unsuc-

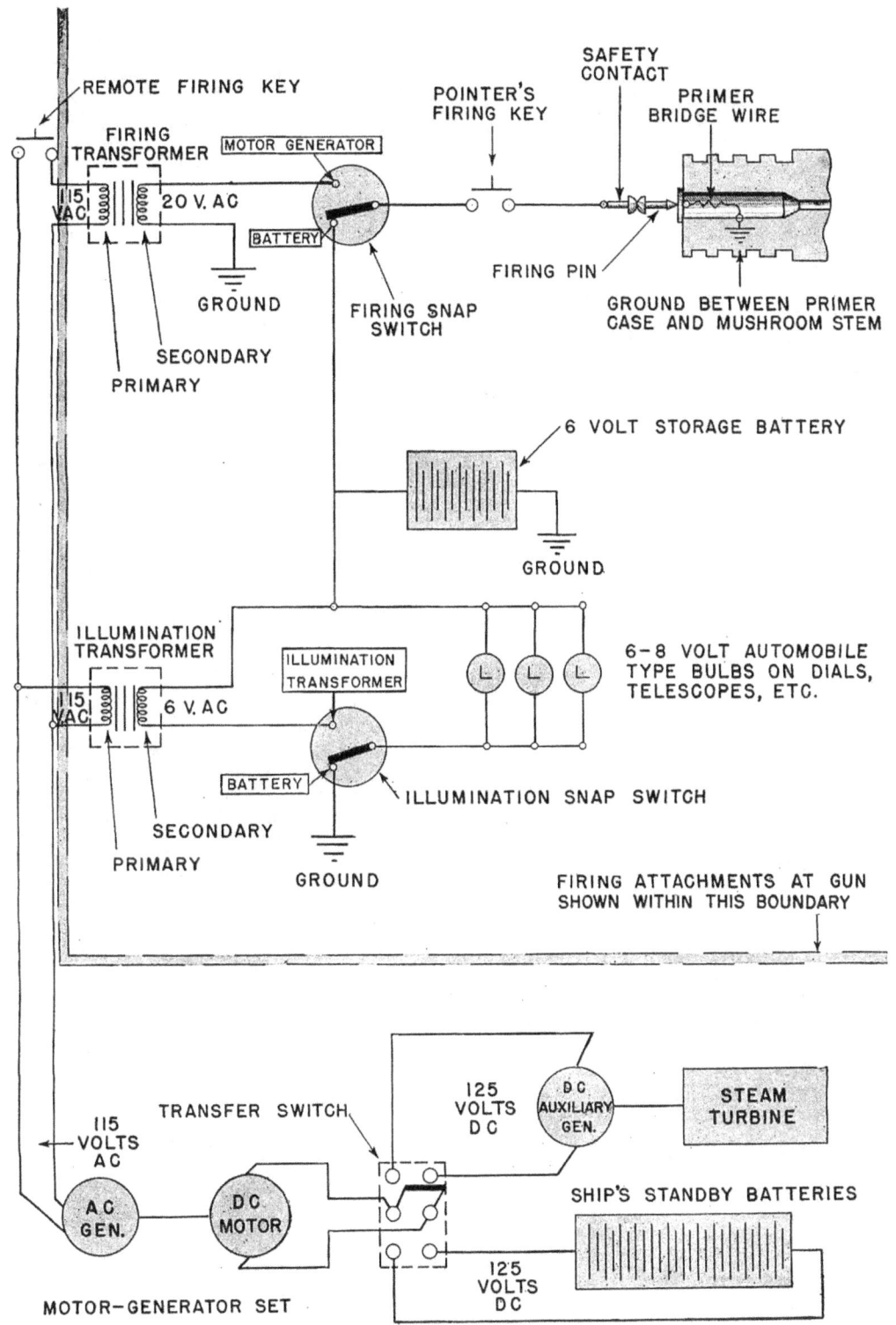

FIGURE 4B4. A typical firing and lighting circuit.

GUN ASSEMBLIES

cessful attempt to fire a gun, it shall be assumed that a hangfire is under way, and the procedure outlined below shall be followed:

(1) Keep the gun pointed and trained in a safe direction.

(2) Continue attempts to fire, if desired, provided such efforts do not involve any movement tending to open the breech.

(3) Do not open the breech for 30 minutes (10 minutes for field and landing guns on shore) after the last attempt to fire. This, at the discretion of the Commanding Officer, is not obligatory in time of action.

When the fire of a bag gun is interrupted otherwise than by a misfire and the gun remains loaded, a hangfire from an undetected ember from the last load shall be assumed to be in progress and the procedure above for hangfires shall apply.

In a bag gun the primer may be extracted, using the priming tools supplied for this purpose, taking care to avoid danger from recoil or blowback. For this purpose, or for shifting primers, the firing lock should not be left open longer than necessary. If examination shows that primers have not fired, and the gun has previously not fired within 30 minutes of the time of the unsuccessful attempt to fire or from the time of interruption, the gun may be unloaded if desired."

C. MOUNTS

4C1. General. Guns must be supported aboard ship in such a manner that they can propel a projectile without transmitting excessive shock to the ship. Proper positioning of a gun requires that it be capable of being elevated above the horizontal, and trained in azimuth. These two motions make it possible to point a gun at any position in space, within structural limits.

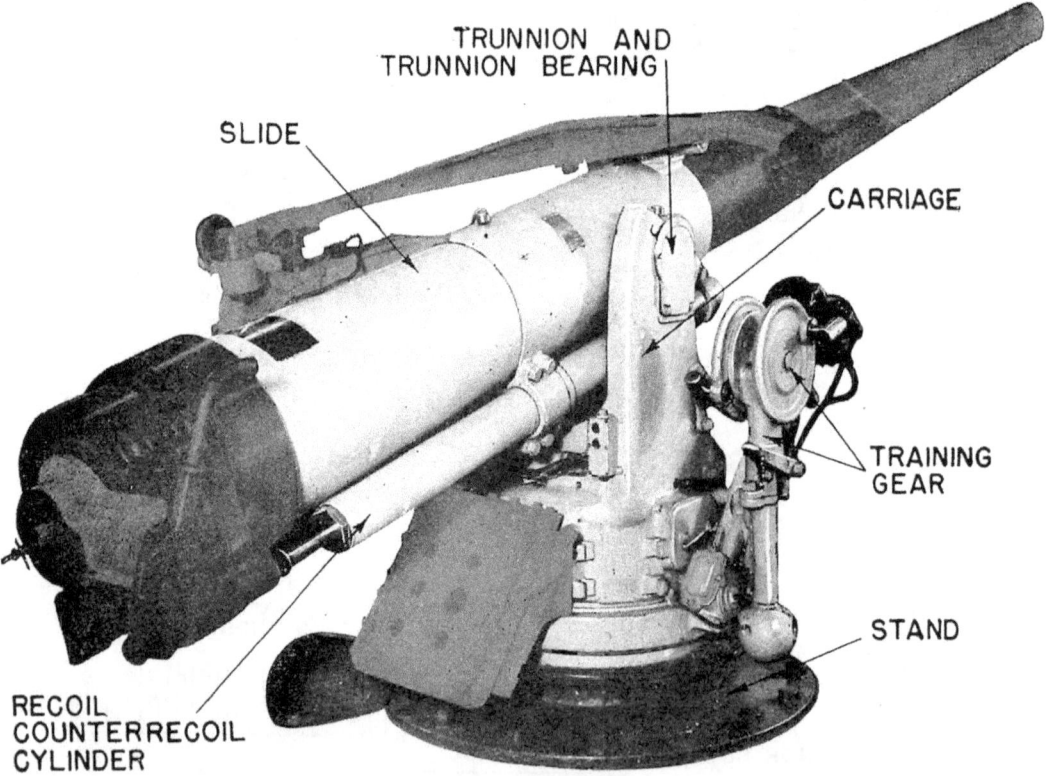

FIGURE 4C1. A pedestal-mounted gun.

CHAPTER 4

The force of explosion pushing against the base of the projectile has an equal and opposite reaction tending to drive the gun to the rear. If the gun were rigidly supported, the forces transmitted to the ship would be very large and the parts supporting the gun would have to be unnecessarily strong and heavy. The forces can be reduced by allowing the gun to recoil or move to the rear a controlled amount. The gun must counterrecoil, or return to the *in-battery* position after each recoil before it can be fired again.

4C2. Types of mounts. A gun *mount* is the entire system of gun-supporting parts, elevating and training mechanisms, and recoil and counterrecoil equipment. There are several types of mounts, but all of them must accomplish essentially the functions outlined above. Most bag guns are supported in heavily armored, rotatable structures known as *turrets*. Minor caliber case guns are supported on *pedestal* mounts. Guns such as the 5"/38 cal.* are sometimes enclosed in lightly armored *gun houses* which outwardly resemble turrets. Small caliber guns are supported by miscellaneous types of mounts. Examples are given in the chapters on specific guns.

4C3. Pedestal mounts. The principal supporting parts of a typical pedestal mount are the *slide*, the *carriage*, and the *stand*, as indicated in figure 4C1.

The *slide* is a hollow support in which the slide cylinder of the gun is housed. During recoil and counterrecoil, the gun rides back and forth in this part. Two horizontal cylindrical projections on the slide, called *trunnions*, provide the axis about which the gun is free to move in the vertical plane. The trunnions are suitably supported on bearings in the arms of the carriage.

The *carriage* supports the slide and rests on roller bearings. It is free to rotate in a horizontal plane, and provides means for training the gun. It is held to the stand by the bearings or suitable holding-down clips.

The *stand* is the stationary part of the mount bolted to the ship. It supports the combined gun, slide and carriage.

4C4. Turret mounts. The elaborate turret mounts used with major-caliber guns perform the same basic functions as pedestal mounts, but they house and carry either two or three identical guns. The size of the equipment makes it necessary to introduce many additional devices which will be considered more thoroughly in the next chapter. A turret is arranged as shown schematically in figure 4C2, and consists of these four structural divisions:

1. The *turret proper*, which is the heavily armored structure containing the guns; it may be seen to revolve from the outside.

2. The *turret revolving structure*, which is the system of platforms suspended from the turret proper.

3. The *turret foundation*, which is the structural system that supports the turret revolving structure on its roller path.

4. The *barbette*, which is the thick armor-plate cylinder surrounding the turret revolving structure and foundation. It extends from the lowest armored deck up to the turret proper.

The *slides* in a turret perform the same functions as the slide in a pedestal mount. Some installations have a single slide which supports all three guns on one pair of bearings. Other installations have three separate slides which can be elevated independently.

The *deck lugs* and *gun girders* are structural parts of the turret. The deck lugs support the bearings which carry the slide trunnions. The gun girders, longitudinally assembled in the turret, support the deck lugs. The turret, performing the functions of a carriage, rotates on the roller path and in addition houses much of the equipment required to operate the guns.

4C5. Elevating and training gear. The equipment provided on a mount to position the gun in space is classed as *elevating and training gear*. The complexity of the gear depends on the type of control required, and on the size of the gun. Provision is made on each kind of mount for one or more of the following methods of moving the gun.

---

* The latest 5"/38 cal. mounts are called "base ring" rather than "pedestal." The majority of 5"/38 cal. gun mounts currently in service, and all having an upper handling room with hoists leading to the loading stations, are of the base ring type.

GUN ASSEMBLIES

4C6. Free swinging. Small guns, such as the 20 mm. and caliber .50 machine guns, are supported on free-swinging mounts. The operator, holding hand grips or strapped to the gun, simply swings the gun to the required position.

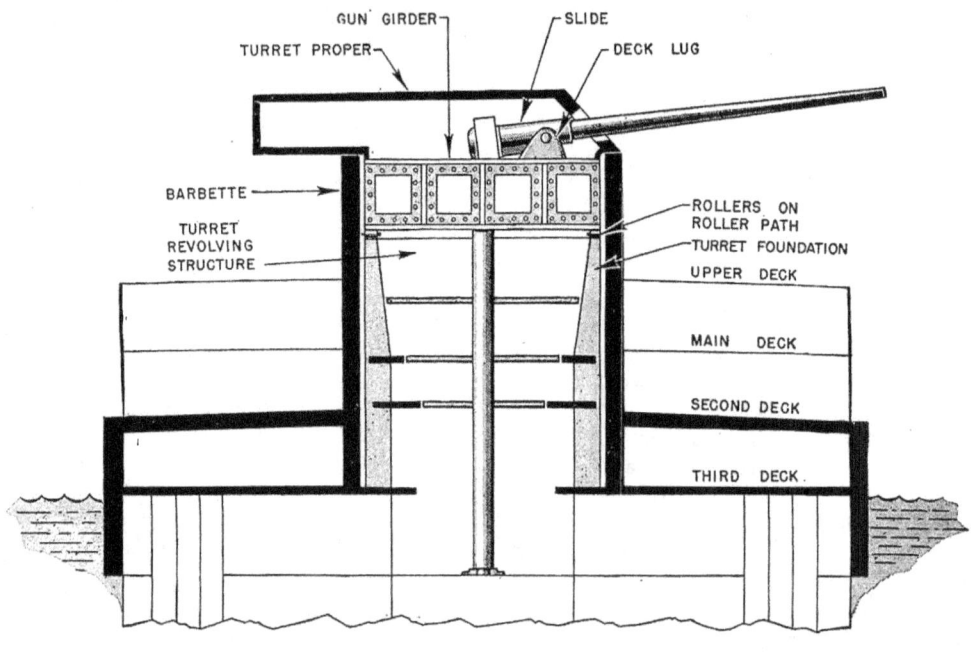

ATHWARTSHIP SECTION - TURRET TRAINED ABEAM

FIGURE 4C2. Schematic arrangement of a typical turret mount.

4C7. Manual drive. Nearly all mounts (including turrets), except the free-swinging type, are provided with a system of handwheels or cranks, shafts, and gears which enable members of the gun crew to move the gun by manual effort. The heavier the mount, the less effective is such equipment, but it is provided for stand-by use if the power drives fail.

The elevating and training gear of a pedestal mount are illustrated in figure 4C3. The pointer's handwheels are geared to a vertical shaft which carries the *elevating worm*. The worm drives the *elevating worm wheel*, which in turn drives the *elevating pinion shaft*. The *elevating pinion* on the elevating shaft meshes with the *elevating arc* attached to the gun slide. Rotation of the handwheels revolves the pinion, which moves the arc and thus elevates or depresses the gun.

The trainer's handwheels are geared to the *training worm* which meshes with the *training circle* attached to the stand. Rotation of the handwheels revolves the worm, which "pulls" the gun around the circle.

4C8. Power drive; local control. Power apparatus used to position a gun may be controlled locally at the gun. A handwheel, or a lever, operated by one of the gun crew, controls the power. The equipment may be *electric* or *electric-hydraulic*.

4C9. Power drive; automatic control. Most of the power drives are capable of being controlled from a remote station. The remote station transmits electrical impulses or signals to control automatically the power equipment that positions the gun. The gun crew only has to maintain the ammunition supply and keep the firing circuit closed.

4C10. Power-drive equipment. There are a number of different power drives in use. Any one of them is complicated to the extent that it cannot be treated fully in this text. For example, the description of the training and elevating gear, including the power drive, covers over 100 pages

CHAPTER 4

in the ordnance pamphlet describing the 5"/38 cal. gun and mount. The details of such installations are of such a nature as to require specialized study. However, a general acquaintance with the principle of power drives is desirable at this time.

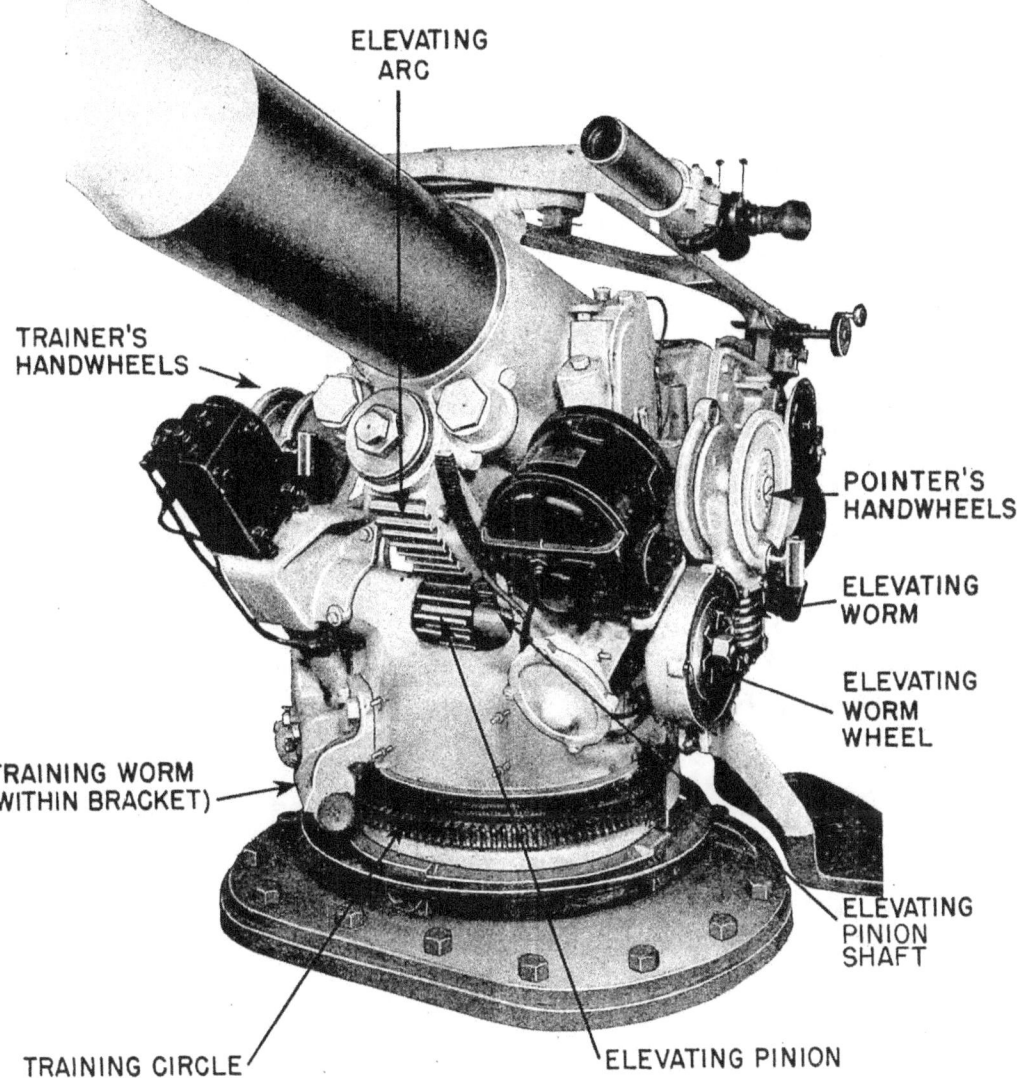

FIGURE 4C3. Training and elevating gear of a pedestal mount.

Since the electric-hydraulic drive is used on a greater variety of guns than the all-electric drive, a brief description of the functions of the principle components of an electric-hydraulic drive is presented. This system, schematically represented in figure 4C4, illustrates the principles of a power drive, but is not necessarily typical of any system in use.

4C11. **Power transmission.** An electric motor drives a hydraulic pump (fig. 4C4) which delivers oil under pressure to a hydraulic motor. The shaft of the hydraulic motor drives the gun through gearing. The hydraulic motor and pump together form a *hydraulic speed gear* which is capable of transmitting power.

RESTRICTED

GUN ASSEMBLIES

The hydraulic motor is called the *B end* of the hydraulic speed gear. It must be reversible, and capable of variable speed, whether its function is to train or to elevate the gun. The B end contains a number of reciprocating pistons which drive the output shaft. These pistons have a fixed length of stroke, hence they always require the same volume of oil for each stroke. They are forced to move at a speed determined by the rate at which oil is delivered to the motor. The direction of rotation of the B-end shaft can be changed by reversing the direction of oil flow, and the speed of rotation can be varied by changing the rate of oil flow.

The hydraulic pump is called the *A end* of the hydraulic speed gear. It is capable of delivering oil in either direction through the lines to the B end, and at variable rates of flow. The A end, like the B end, contains a number of reciprocating pistons. However, the length of the piston stroke in the A end can be varied smoothly from zero to the maximum permitted by the construction. The length of this stroke determines the rate of oil flow. If the stroke is zero, no oil will flow, and the B-end shaft will stand still.

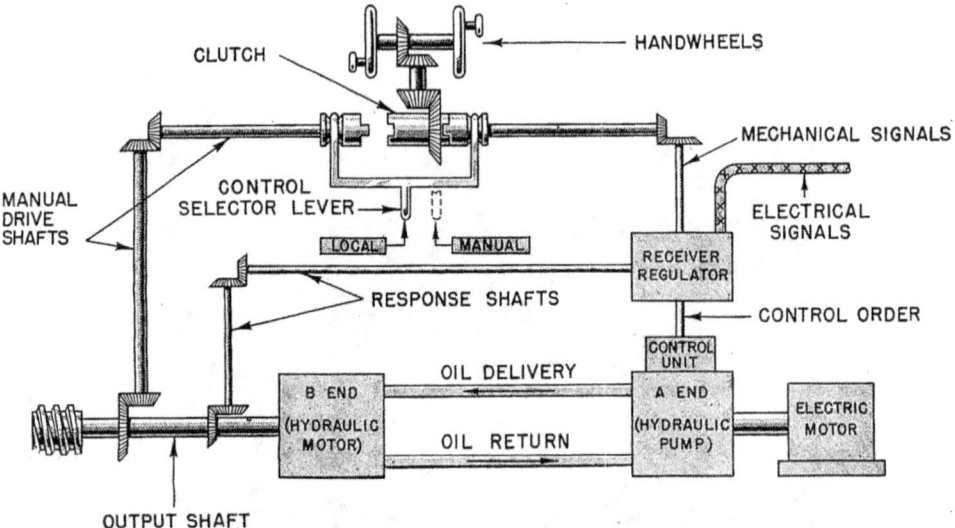

FIGURE 4C4. Schematic diagram of an electric-hydraulic power drive.

At any given instant, half of the pistons in the A end are pumping, or discharging oil to the B end. The other half are sucking, or receiving oil from the B end. The A end is arranged so that this condition can be reversed, making the first half suck, the second half discharge. Thus the direction of flow to the B end can be reversed.

The electric motor drives the A end continuously in one direction at a constant speed. By adjustment of the piston stroke in the A end, the direction and speed of rotation of the B end can be controlled. Thus, a variable speed, reversible drive is available to position the gun.

4C12. Control of power. Most of the intricate components of an electric-hydraulic drive are in the assemblies whose functions are to control the piston stroke in the A end of the speed gear. A *control unit,* containing several valves and linkages to the A end (fig. 4C4), adjusts the piston stroke in response to orders from the *receiver-regulator.*

The receiver-regulator *receives* either mechanical or electrical signals designating the *required* gun position, and mechanical signals designating the *actual* gun position. It *regulates* the gun movement to make the actual position equal the required position.

The *actual* gun position is transmitted mechanically from the B-end driving shaft to the receiver-regulator. The *required* position is transmitted either electrically from a remote station or mechanically from handwheels at the gun. The control of the gun movement is *automatic* (i. e., the

CHAPTER 4

handwheels are not operated) when the remote electrical signals are provided. In the event that electrical signals from the remote station fail, the operator at the gun sets his selector lever on *local* and transmits the required gun position by turning his handwheels.

As soon as a required position signal is received, the receiver-regulator operates the control unit on the A end. This unit in turn sets the piston stroke, and the gun starts to move in the required direction. At the same time actual gun movement is transmitted by *response shafts* to the receiver-regulator. When the total movement of the response shafts equals the requirements of the position signals, the receiver-regulator causes the control unit to set the A-end piston stroke to zero.

In actual practice the response of the power drive to the signals is so rapid that the gun moves as though it were geared directly to the handwheels. The pointer and trainer do not have to allow for lag between handwheel movement and gun movement.

A direct mechanical drive from the handwheels to the B-end drive shaft provides manual control of the gun movement if the power drive fails. The operator simply shifts the selector lever to *manual,* disengaging the power-drive equipment, and operates his handwheels to move the gun. Some guns are too heavy to be moved by hand, and in such cases provision may be made to drive the A end of the speed gear by hand cranks if the electric motor fails. The hydraulic speed gear provides the leverage to enable a few men to move such a heavy installation.

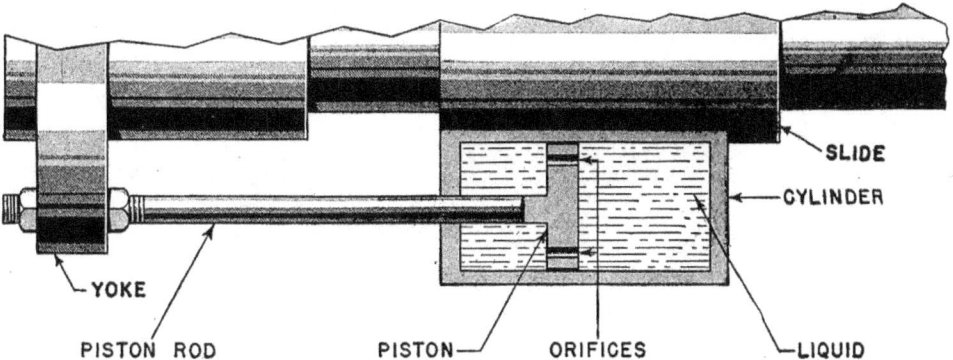

FIGURE 4C5. Schematic arrangement of a recoil brake.

4C13. Recoil mechanism; characteristics. The force tending to push the gun to the rear as the projectile is discharged must be disposed of. If the mount and ship were made strong enough, the shock could be absorbed by the resistance of the supporting structure alone, and the gun could be rigidly attached to its mount. However, the construction required would be too heavy for practical utility.

The force of recoil transmitted to the mount is reduced appreciably if the gun recoils a distance of several calibers. The force of explosion causes the gun to recoil, and the energy stored in the moving parts is dissipated by resisting the movement with a moderate force applied during the time of recoil. The mechanisms used by the Navy to resist recoil are *hydraulic brakes,* often called *recoil cylinders* or *recoil brakes.*

Actually, the recoil mechanism does not absorb all the energy stored in the gun. Part of the energy is transferred to the counterrecoil mechanism, part is used in overcoming friction in the slide, and in some cases, part is stored in the devices which accomplish automatic functions, such as loading mechanisms. The remaining energy is dissipated in the form of heat generated in the hydraulic brake and in the ship's structure which "gives" under the force transmitted to it.

The principal components of a hydraulic brake are a *cylinder,* a *piston* and *rod,* a *liquid,* and a system of orifices, arranged in the manner indicated in figure 4C5. The piston is attached through the rod to the yoke of the gun, and the liquid-filled cylinder is attached to the slide. The orifices in the piston permit the liquid to flow from one side to the other as the piston moves during recoil.

GUN ASSEMBLIES

The force required to pull the piston through the liquid depends on the size of the orifices, the area of the piston, the velocity of the piston, and the viscosity, or thickness, of the liquid. The viscosity factor is held nearly constant by using a standard, non-freezing liquid. It usually consists of a mixture of 80 per cent glycerine and 20 per cent water. In any given brake, the piston area is fixed. Thus the size of the orifices and the velocity of the piston control the resistance to recoil.

The velocity of recoil obviously cannot be constant, since it is zero at its beginning and end, and maximum at some intermediate point. If the orifices are constant in size, they will offer a high resistance at high velocity and low resistance at low velocity, and the force on the mount will have an unnecessarily high peak at the instant of maximum velocity. The weight of the mount can be kept to a minimum if the resistance is kept constant during the entire length of recoil. It is thus apparent that it is desirable to provide orifices which enlarge as the recoil velocity increases, taking into account the additional resistance being offered by the counterrecoil mechanism as recoil progresses. The manner of varying the area of the orifices is the principal difference between the two common types of recoil brakes used by the Navy.

4C14. Recoil brake; throttling-rod type. The recoil brake shown in figure 4C6 is typical of the units used on guns of 6-inch caliber and larger. Tapered *throttling rods* secured at both ends of the cylinder pass through the orifices in the piston. The effective area for the passage of liquid is the difference between the cross-sectional areas of the holes and the parts of the rods within them. The rods are cut so that the smallest section is at the point where the piston speed is greatest. An outstanding advantage of the removable rods is that they can be changed if it becomes necessary to alter the recoil characteristics of the mount.

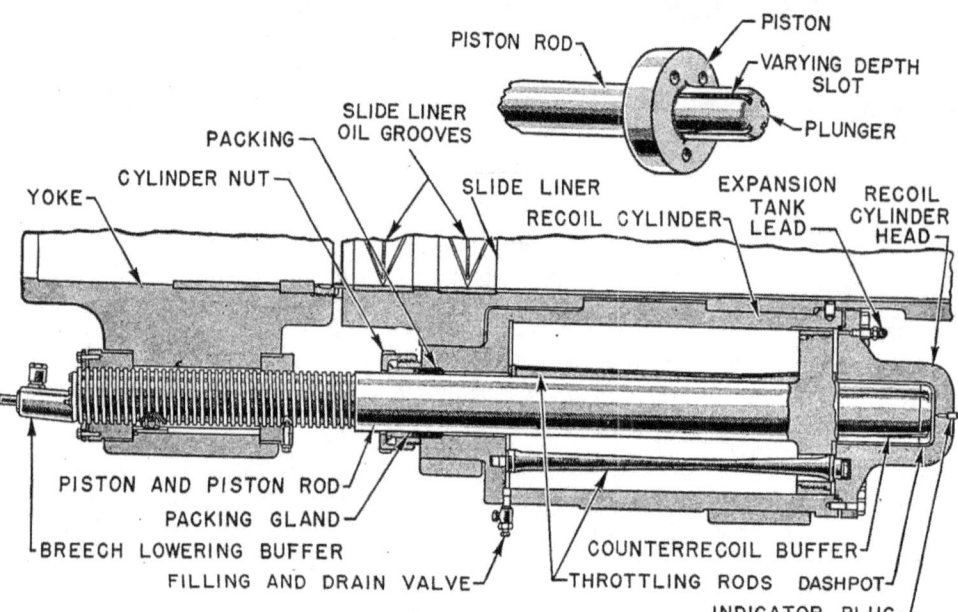

FIGURE 4C6. Recoil brake; throttling-rod type.

A *dashpot* is built into recoil cylinders to terminate the counterrecoil stroke without excessive shock or slam. Counterrecoil mechanisms have reserve force to insure proper return to battery. The recoil cylinder itself exercises some control of the return stroke, but the dashpot is needed to avoid mechanical damage to the mount. The *plunger* in figure 4C6 is provided with four slots whose depth decreases from the plunger end toward the piston. The liquid trapped in the dashpot when the plunger enters must escape through these four slots, which are nothing more than variable orifices. The gradual decrease in speed produced by the resistance of the liquid passing through the slots results in the desired cushioning. The same effect is obtained in some recoil cylinders by slightly tapering the plunger. The escape orifice is then the gradually diminishing space between the dashpot wall and plunger.

The heat generated in the hydraulic brake causes the liquid to expand. The *expansion tank* attached to the cylinder accommodates any overflow. Air trapped in the tank when the cylinder is filled prevents liquid from flowing out of the cylinder into the tank unless the liquid expands.

4C15. Recoil brake; grooved-wall type. Recoil brakes used on minor- and some intermediate-caliber guns are equipped with solid pistons, and have orifices formed by grooves cut in the cylinder walls. A brake representative of this type is shown in figure 4C7. The variation in size of orifice is obtained by cutting the grooves either with constant depth and varying width or varying depth and constant width. Often this type of recoil cylinder contains within itself the counterrecoil springs, and the mechanism controls completely both recoil and counterrecoil.

The dashpot, operating on the same principle as the type previously described, is carried in the end of the piston rod, while the plunger is attached to the cylinder head. The trapped liquid escapes through the longitudinal *inlet hole* in the plunger, past the adjustable *needle valve,* and out the transverse plunger *outlet hole.* Near the end of the counterrecoil stroke, the piston rod partially blocks the escape from the plunger outlet hole. The liquid then must flow through the annular orifice between the plunger and dashpot. The tapered base of the plunger throttles this passage, producing the final cushioning effect. The needle valve is used to vary the amount of cushioning.

No expansion tanks are used with this type of recoil brake. After the cylinders are filled, a small amount of liquid is drained out to make room for expansion. Frequently two or more cylinders of this type are used on one gun mount, and they are interconnected by pipes which equalize the pressure between them.

4C16. Counterrecoil mechanism; characteristics. The assembly of parts whose function it is to return a gun to battery after recoil is known as a *counterrecoil mechanism* or as a *recuperator.* It has an additional function of *holding* the gun in battery after counterrecoil. It must be realized that if a gun does not completely return to battery, it cannot recoil the proper amount at the next firing. The resulting shock might wreck the mount. Suitable marks are painted on the gun and slide to warn the crew when a gun fails to return to battery.

The types of counterrecoil mechanisms used by the U. S. Navy are spring, spring-pneumatic, and pneumatic. The spring type may be a separate cylinder containing springs, a piston, and rod; or it may be combined with the recoil cylinder as shown in figure 4C7. The spring-pneumatic type, in use on some older guns, is similar to the separate-spring type, except that the cylinder is filled with air at about 50 pounds p.s.i. pressure. The springs and air, compressed during recoil, expand and return the gun to battery.

It is difficult to obtain satisfactory springs capable of returning large guns to battery, especially when they are elevated to high angles at which the pull of gravity greatly increases the load. Springs are likely to become permanently set or broken, requiring replacement which consumes considerable time and work. Smaller guns are still equipped with springs, but nearly all guns 5 inches and larger are fitted with pneumatic recuperators having no springs.

4C17. Hydropneumatic recuperators. Pneumatic counterrecoil systems use air to return the gun to battery, but require a liquid-filled auxiliary called a *differential cyplinder*. The function of the latter is to provide a liquid seal and prevent a loss of air pressure around moving parts. These systems are generally described as being *hydropneumatic*. The arrangement shown in figure 4C8 is typical of the installations on some of the larger guns. Secured to the slide are two pneumatic *counterrecoil cylinders* served by a single *differential cylinder*. The *plungers* of the main cylinders are attached to a *plunger yoke* which rides on guide rails on the slide during recoil. This yoke is in turn attached to the gun yoke by *yoke rods.* Recoil of the gun pulls the plungers into the pneumatic cylinders, compressing air which later expands and forces the plungers out, returning the gun to battery.

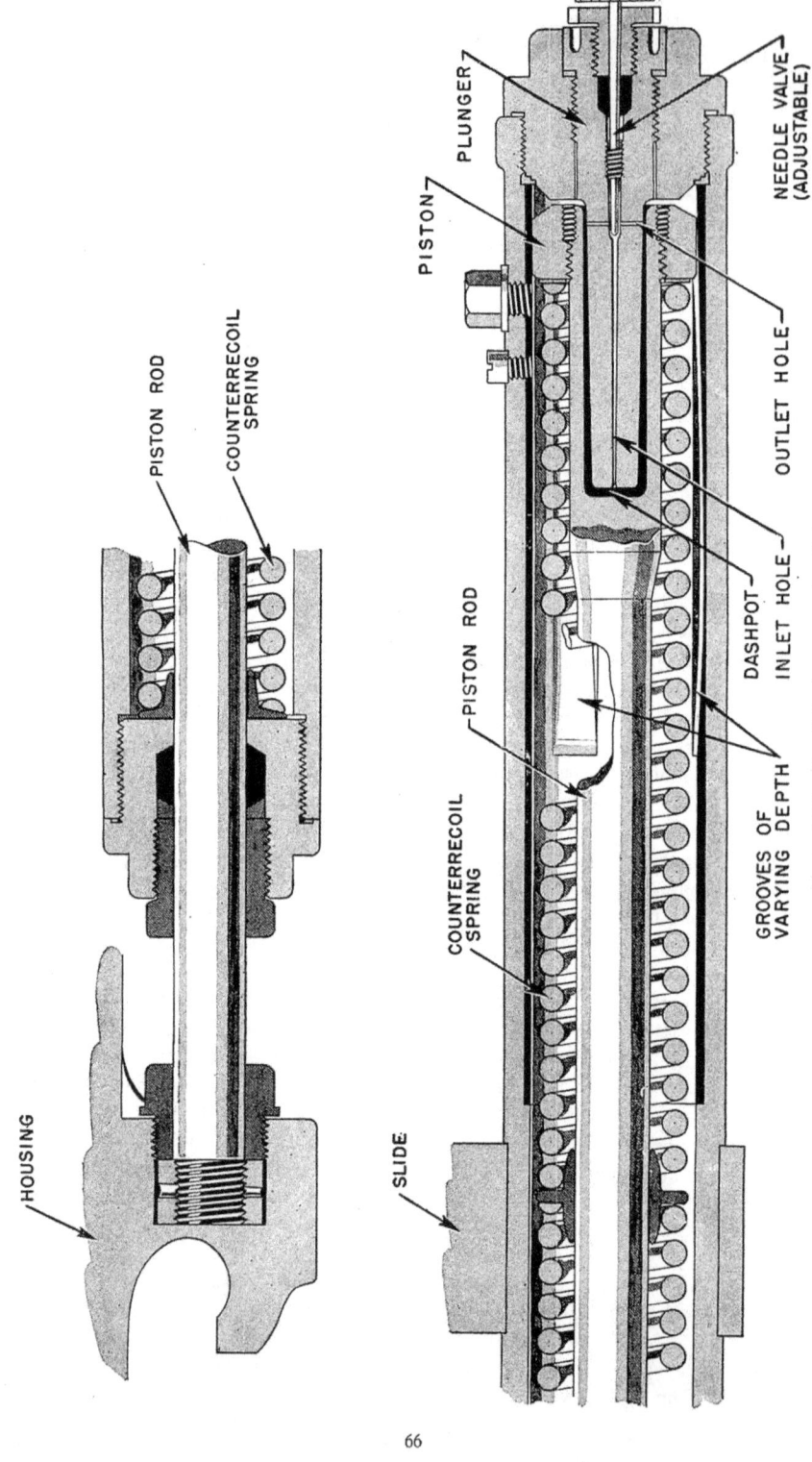

Figure 4C7. Recoil brake; grooved-wall type.

Some of the smaller guns use essentially the same type of counterrecoil system, but they generally have only one main cylinder. In some cases, the differential cylinder is built into the main cylinder forging to eliminate some of the piping. There are some installations, such as the 5"/38 cal., in which the cylinder moves back and forth on a stationary plunger, but the operating principle is the same.

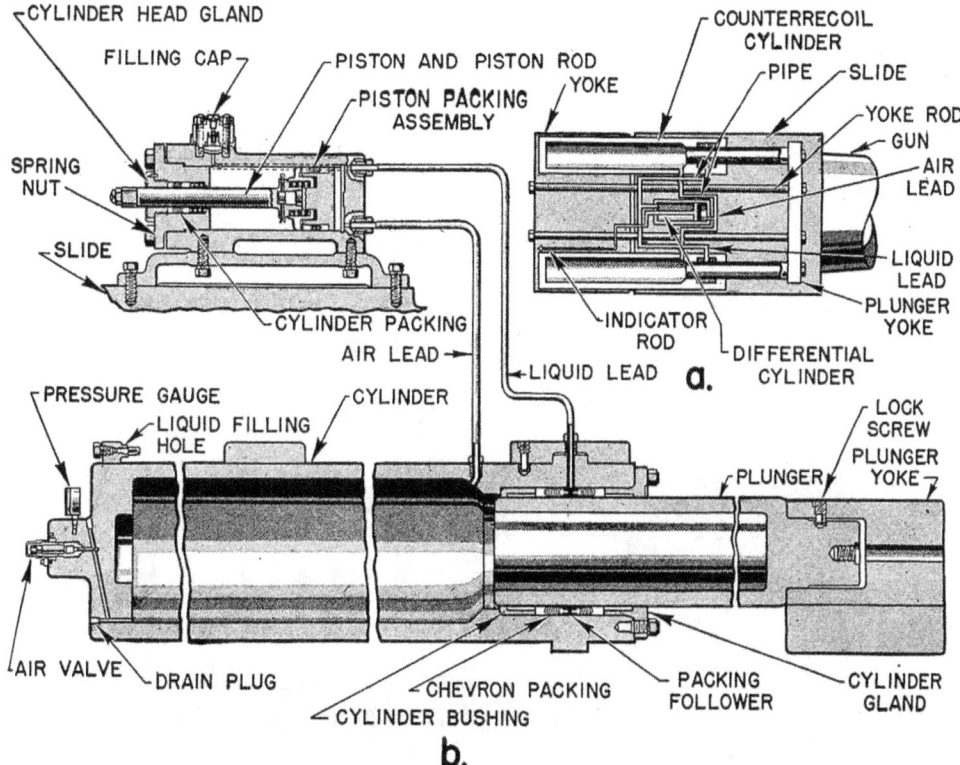

FIGURE 4C8. Arrangement of hydropneumatic recuperator: a shows complete assembly; b represents relationship of one counterrecoil cylinder to a differential cylinder.

4C18. Main cylinders. The main cylinders shown in figure 4C8 are charged initially with air from the high-pressure air system of the ship at pressures ranging from 1,000 to 1,800 pounds p.s.i., depending on the installation. Recoil of the gun pulls the hollow plungers into the cylinders, decreasing the cylinder volume and further compressing the air. The air expands after recoil, forcing the plungers out and returning the gun to battery. The success of this system is dependent on the ability of the packing to prevent leakage of air around the plunger.

The packings used around the plunger of the main cylinders are an assembly of individual composition rings, each having a V-shaped cross section, whence comes the name *chevron-type packing*. They are arranged in two groups, with the open ends of the V's of each group facing each other, as shown in figure 4C9. Between the groups is a packing follower which is grooved in such a manner that liquid under pressure can enter between the packings. The presence of the liquid keeps the packings soft and pliable, and the pressure of the liquid forces the packing against the plunger and the packing-box wall. The pressure of the liquid between the packings is always higher than the pressure of the air, hence the packings form an airtight seal. Any leakage which does occur will be liquid forcing its way into the air cylinder or out along the plunger.

4C19. Differential cylinder. The device used to maintain the liquid on the packings under pressure greater than that of the air is the differential cylinder. It consists essentially of *a cylinder*,

GUN ASSEMBLIES

a piston, and *a piston rod* extending outside the cylinder, as shown in figure 4C8. An air line connects the air in the rear end of the differential cylinder to the air in the main cylinders. A liquid line connects the liquid in the front end to the spaces between the main-cylinder plunger packings.

The rod, having a blank end, is free to move in and out of the cylinder. Hence, the total force of the air on the piston must equal the total force of the liquid. Since the effective area of the liquid side of the piston is less than the effective area of the air side by the cross-sectional area of the piston rod, the pressure of the liquid must always be higher than the pressure of the air.

The pressure differential varies between 15 per cent and 35 per cent, depending on the installation. This differential is sufficient to maintain an effective seal of the packings in the main cylinders. Any leakage in the differential cylinder will be liquid either entering the air side or escaping from the system completely. As liquid is lost, the piston rod of the differential cylinder gradually moves out, giving a visible indication of the loss of liquid.

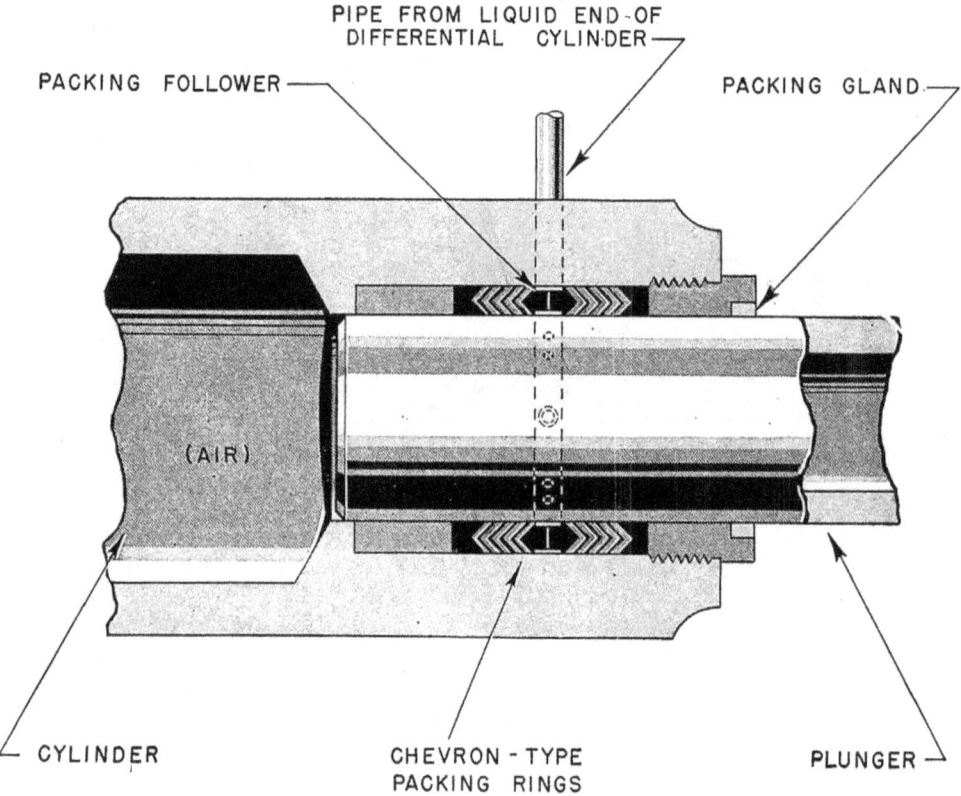

FIGURE 4C9. Main cylinder packing detail.

4C20. Slide-securing device. Loss of air pressure in the recuperator would permit the gun to drop to the rear if no positive restraint were provided. A locking device is employed to guard against such a casualty. It consists of a *safety link* strong enough to hold the gun in battery at elevated positions, but weak enough to part without seriously damaging the gun or mount if the gun is fired with it in place.

The safety link may consist of a metal pin or a metal strip designed to break under a force slightly greater than that required to hold the gun in battery. Since a broken pin or strip must be replaced by a new one, the safety link should be removed before firing.

CHAPTER 4

4C21. Safety precautions, selected from Article 972, Navy Regulations.

"71. A circle shall be marked on the deck to indicate the limiting position of the breech of the gun on recoil, and the gun crew shall be instructed to keep clear."

"72. Marks or indicators shall be provided to indicate whether or not the gun returns to battery, and a member of the gun crew shall be detailed to observe these marks or indicators after each shot. The service of the gun shall be stopped should the gun fail to return to battery."

"73. On guns equipped with hydropneumatic counterrecoil systems, the safety link, locking the gun to the slide, shall be connected up at all times except when firing or when testing and overhauling the counterrecoil systems or when the battery is in a condition of readiness for action. These safety links shall be disconnected after checking the pressure on counterrecoil system and prior to firing."

"82. Whenever any motion of a power-driven unit is capable of inflicting injury on personnel or material not continuously visible to the person controlling such motion, the officer or petty officer who authorizes the unit to be moved by power shall, except at general quarters, insure that a safety watch is maintained in areas where such injury is possible both outside and inside the unit, and shall have telephone or other effective voice communication established and maintained between the station controlling the unit and the safety watch. These precautions are applicable to turrets, gun mounts, guns, directors, range finders, searchlights, torpedo tubes and similar units. Under the conditions stated above, the station controlling shall obtain a report 'all clear' from each safety watch before starting the unit. Each safety watch shall keep his assigned area clear and if unable to do so shall immediately report this unit fouled, and the controlling station shall promptly stop the unit until again clear."

"83. In turrets and enclosed mounts of 5-inch and larger guns, a warning signal shall be installed outside the turret or mount and whenever power train is used, except at general quarters, the officer or petty officer in charge of the turret or mount shall cause warning signals to be sounded before using power and at intervals during its use."

"84. When using director train while firing at gunnery exercises, an observer from the firing vessel for each gun or turret shall cause the firing circuit to be broken whenever the gun or turret is trained dangerously near any object other than the designated target."

"91. (a) Before firing any gun, other than a saluting gun, in time of peace, the recoil cylinders shall be inspected and filled in the presence of the gunnery officer or assistant gunnery officers, and such officer shall check the pressure being carried by the pneumatic counterrecoil cylinders and verify that the air systems are properly charged and that the valves of the gas-ejector system operate freely; and a report thereof shall be made to the Commanding Officer.

(b) Whenever there is a possibility of action, the Commanding Officer shall require all recoil and counterrecoil systems to be kept ready for immediate use and inspected as frequently as safety demands.

(c) These provisions do not apply for firing blank charges."

"92. After filling recoil cylinders not fitted with expansion tanks, the prescribed amount of liquid necessary to allow for the expansion of the liquid due to heat, and no more, shall be withdrawn."

D. SIGHTS

4D1. Introductory. The line of sight from a gun to a target is used to position the gun so that a projectile fired from it will hit the target. Since the projectile will not travel in a straight line, the bore of the gun must be offset from the line of sight by a vertical angle and a horizontal angle. The mechanism mounted on a gun to provide these offsets is called a *gun sight*. Its functions are to provide (1) a sighting axis, (2) means for introducing the vertical angle, and (3) means for introducing the horizontal angle.

4D2. The sighting axis. There are a number of devices used to establish the sighting axis. Each depends on the fundamental fact that two points determine a straight line.

4D3. The open sight. The simplest arrangement for sighting purposes is the open sight

GUN ASSEMBLIES

consisting of a rear peep and a front ring. A common variety of this type is shown in figure 4D1. It is employed as an auxiliary device on some guns to aid the pointer and trainer in "finding" the target quickly before they start using their more accurate sight telescopes. The open sight has little value for aiming a gun accurately, for it is hard on the eyes and obscures a large portion of the target. It also requires a constant change of focus as the eye shifts between the rear and front points and the target. In addition, if the peep is made small enough for accurate alignment, it is smaller than the pupil of the eye and thus reduces the amount of light reaching the eye.

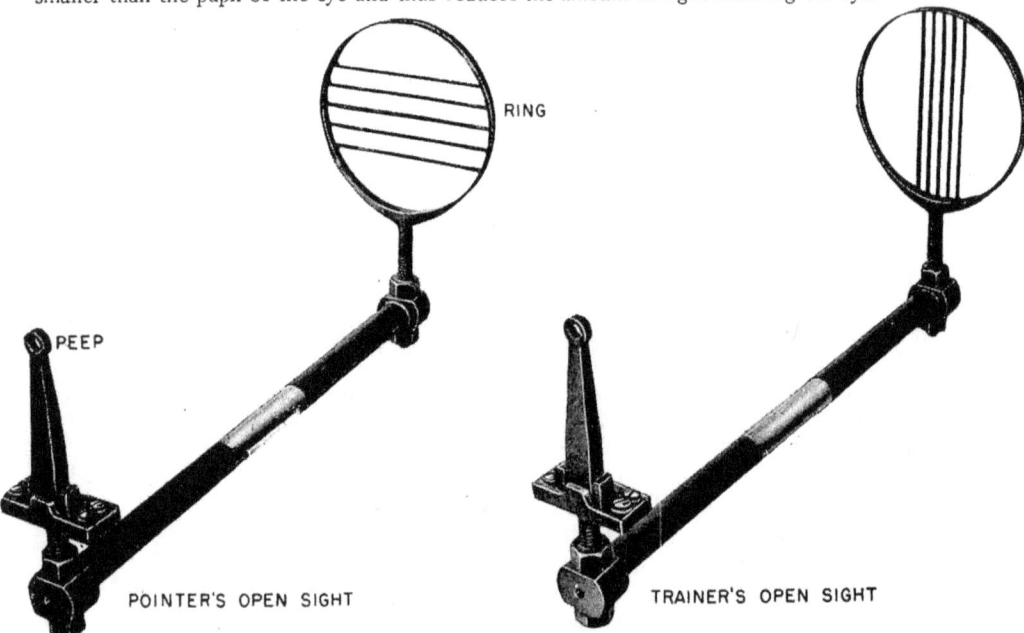

FIGURE 4D1. A pair of open sights.

4D4. The sight telescope. A simple gun-sight telescope is like an ordinary small telescope except that it has within it a pair of crosshairs or lines. The point of intersection of these lines marks the sighting axis. The objective lens (the one nearest the target) forms an image on the plane containing the crosslines. The image and the crosslines are viewed through an eyepiece which presents to the eye a magnified image of the target and crosslines. The optical arrangement permits the eye to see the target and the crosslines distinctly while focused for one distance only. In addition, the eyepiece delivers the light to the eye in such a manner that the eye muscles are permitted to rest. The eye need not be exactly on the axis of a telescope to sight accurately, whereas any movement of the eye in front of a peep sight will change the aim.

The magnification of the target image by the telescope makes possible more-accurate aiming of the gun. Smaller errors of alignment of the sighting axis with the point of aim can be detected more readily with a telescope than with an open sight. The optical characteristics of modern gun-sight telescopes actually make it easier to see a target at night through a telescope than with the unaided eye. Most of the telescopes in service may be classified in one of the following groups.

4D5. Straight telescopes. The simplest and most rugged telescope is a straight tube containing all the optical parts in a straight line, as shown in figure 4D2a. This type is in use on many older guns and directors in service. .

4D6. Fixed-prism telescope. Some telescopes contain fixed prisms, whose function it is to turn the line of sight through one or more angles. These telescopes have a particular application in turrets, where it is desirable to have the telescopes protrude through the side or top armor to avoid unnecessary openings in the front. An example is shown in figure 4D2b.

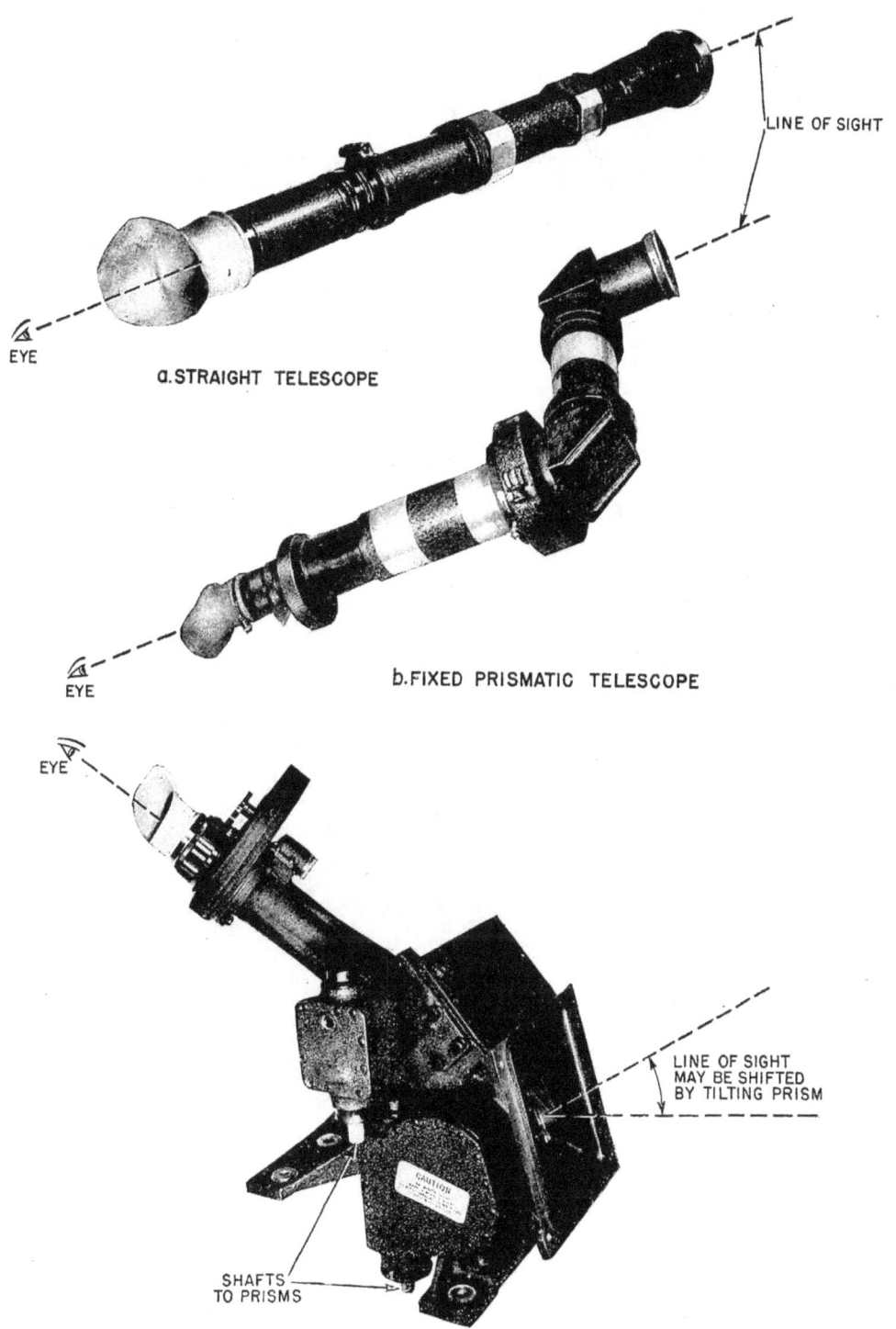

C. MOVABLE PRISM TELESCOPE
FIGURE 4D2. Typical types of sight telescopes.

4D7. Movable prism telescopes. A telescope having fixed optical parts must be moved in its entirety to follow a target. If the target is an airplane nearly overhead, such a telescope puts the operator's head in an uncomfortable position. Telescopes with a movable prism arrangement are used on many AA guns and directors to overcome this difficulty. The telescope body maintains a fixed position with respect to the deck of the ship, while the movable prism shifts the line of sight (see fig. 4D2c).

Telescopes of each of these classes may have a variety of extra features. Some have a choice of two magnifying powers, permitting the operator to select high or low magnification. Some are equipped with an additional eyepiece, or check sight, which permits an observer to check the accuracy of the sighting. Filters are often built in to permit the operator to reduce the intensity of the light. The tubes or housings of most of the latest telescopes are filled with an inert gas which improves optical characteristics and reduces the aging of the optical parts.

4D8. Sight mounts. Any sight must be mounted so that the sighting axis can be offset from the bore of the gun. The mechanism which provides this movement and supports the sight on the gun mount is the *sight mount*.

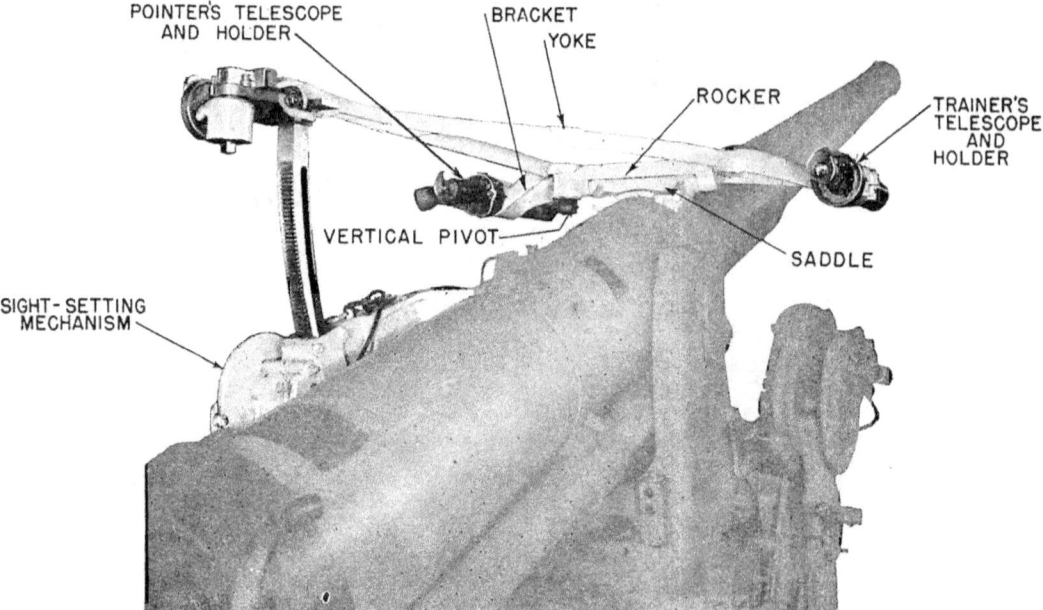

FIGURE 4D3. A yoke-type sight.

4D9. The yoke-type sight. The yoke-type sight derives its name from the *yoke* which carries the telescopes. Figure 4D3 shows a typical example. Separate telescopes are provided for the pointer and trainer since they must both sight on the target at the same time. Each telescope is mounted within a holder which protects it from the weather when not in use. The holders are clamped in brackets secured to the yoke.

The yoke is free to move in a horizontal plane about a *vertical pivot*. The vertical pivot is supported on the *rocker* which rotates about a *horizontal pivot*. Thus, the vertical pivot and the yoke are free to move in a vertical plane. The horizontal pivot is supported by the *saddle* which is secured to the slide of the gun mount. At the rear end of the yoke are the mechanisms which are used to set the angles between the sighting axis and the bore. An enlargement of these mechanisms is shown in figure 4D4.

CHAPTER 4

4D10. Range setting. Attached to the rear end of the yoke is a *sight bar* supported in a bracket. Gear teeth on the sight bar form a rack which meshes with gears driven by the *range-setting crank*. Rotation of this crank moves the yoke up or down, introducing the vertical angle, known as *sight angle*, between the sighting axis and the bore of the gun. Since this angle determines the distance or *range* the projectile will travel, the operation is called *range setting*.

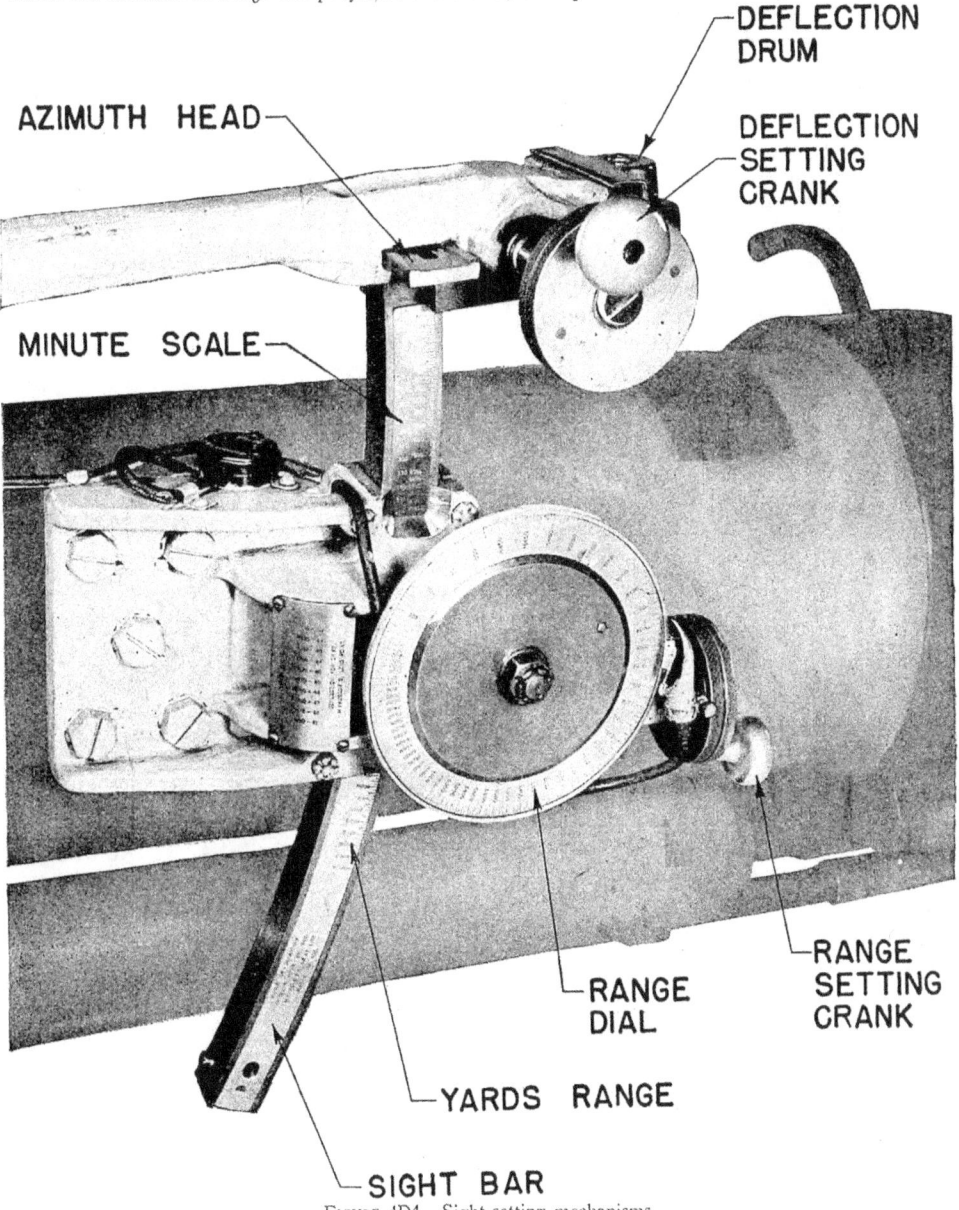

FIGURE 4D4. Sight-setting mechanisms.

The angle is measured by two scales on the *range strip* mounted on the sight bar. One of these scales indicates the angle in minutes of arc, the other indicates the yards of range the projectile will travel under standard conditions.

A *range dial* graduated in yards of range is geared to the range crank and sight bar. Its graduations are farther apart than those on the range strip, hence they are easier to read and permit more-accurate setting of the range.

4D11. Deflection setting. The horizontal angle between the line of sight and the bore of the gun is introduced by the *deflection setting crank*. This crank and its gearing are carried on the *azimuth head* which is mounted on the upper end of the sight bar. The rear of the yoke is arranged to move laterally in the azimuth head, permitting the setting of the horizontal angle, known as *sight deflection*.

A *deflection drum* geared to the deflection crank indicates the horizontal angular offset between the line of sight and the bore. The scale on this drum measures the offset in terms of an angular unit known as the *mil*. A mil is an angle subtended at the center of a circle by an arc whose length is one one-thousandth of the radius of the circle. For practical purposes it is sufficient to consider a mil as the angle subtended by one yard at a distance of 1,000 yards.

The deflection scale is calibrated to read some positive number such as 100 or 500 when the sight deflection is zero. This avoids the use of negative numbers which are likely to confuse a sight setter. The reading on the deflection scale is called simply the *scale*. Thus, an order *scale 510* means set the deflection scale at 510.

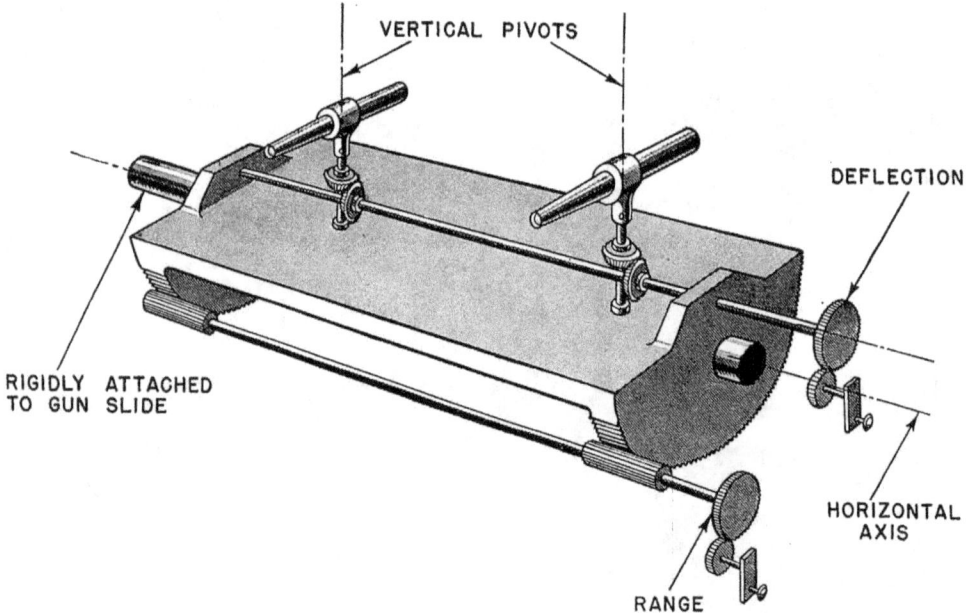

FIGURE 4D5. Parallel-motion sight mount; schematic diagram.

4D12. Parallel motion sight mounts. It can be seen from an examination of a yoke sight mount that if the gun elevates to high angles, the telescopes will move down appreciably. Further, large deflection settings will shift the telescopes sideways. These motions are objectionable from the standpoint of convenience in use, for they cause the pointer and trainer to shift their heads an uncomfortable amount. Such motion of the telescopes on a turret mount would require excessively large openings in the armor.

The displacement of the telescopes can be reduced by mounting them on a *parallel-motion sight mount*. Such a mount supports the telescopes on two individual pivots, in a manner schematically indicated in figure 4D5. These pivots are supported on a base which rotates about a horizontal axis close to the vertical pivots.

CHAPTER 4

Parallel motion mechanisms vary considerably in details. The systems used in modern turrets are intimately associated with the control of the elevating and training gear, and are consequently somewhat involved. The telescopes on some guns and directors have movable prisms which introduce the required angular offsets, eliminating all displacement of the telescopes as the line of sight moves.

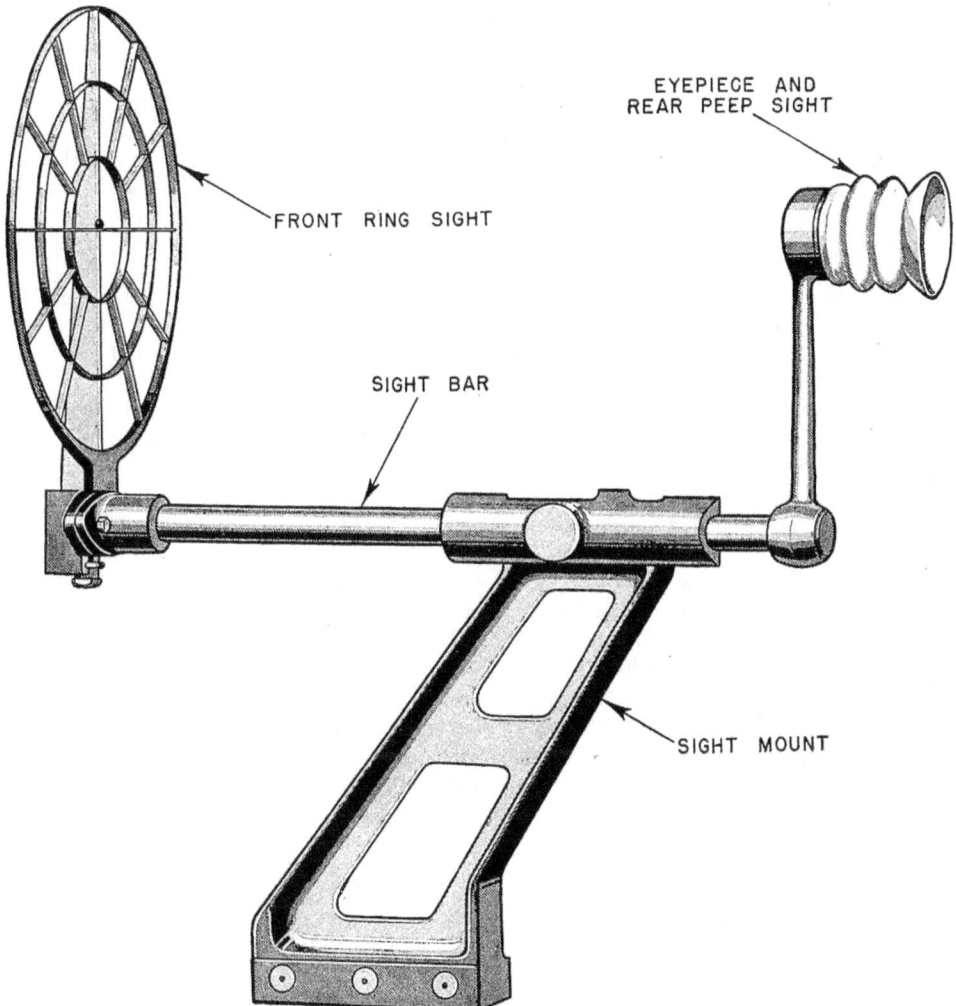

FIGURE 4D6. A peep and ring sight.

4D13. Other sights. Small guns like the 20 mm. and 40 mm. are equipped with rear peep and front ring sights as shown in figure 4D6. Offset of the sighting axis is obtained by selecting the appropriate section of the front ring, and aligning it with the target by means of the peep. Consequently, such a sight requires no movable mount, since the shift of the sighting axis requires nothing more than a shift of the eye.

There is another type of sighting arrangement, such as that used in the Mark 14 sight. The principles involved are discussed in Section 19C.

4D14. Some general safety precautions, selected from Article 972, Navy Regulations.

"14. Only such of the ammunition issued for gunnery exercises as does not contain a

GUN ASSEMBLIES

primer, fuze, or detonator may (at the discretion of the Commanding Officer) be used for testing the fit in hoists, guns, and appliances. Paragraph 69 (a) (2) shall be complied with. Case ammunition may be fitted in the guns as prescribed in paragraph 80."

"15. No other than drill ammunition shall be used for drill."

"20. A loaded and fuzed projectile, seated in the bore of a gun that is hot from previous firing, presents a hazard since detonation of the projectile is probable as a result of being heated. Whenever practicable, such projectiles should be disposed of promptly by firing the round. Whether a gun is hot or cold, the risks attendant upon removing a loaded and fuzed projectile seated in the bore, by backing out, are considered unwarranted except in the case of guns for which existing instructions specifically prescribe this procedure."

"45. During gunnery exercises, charges in excess of the amount required to be available for one run shall not be assembled in the vicinity of guns mounted outside of turrets. No charge for a bag gun shall be removed from its tank, nor shall the tops of tanks be removed or so loosened that the bags may be exposed to flame until immediately before the charge is required for loading."

"51. Except when using a power rammer, no force greater than that which can be applied by the hand alone shall be used in loading a live cartridge into a gun. Any cartridge which does not freely and fully enter the chamber of the gun shall be carefully extracted and put aside, and in peace time no further attempt shall be made to fire such a cartridge."

"63. The utmost care shall be taken to insure that the firing pin and other parts of the firing mechanism of a case gun are in good condition and properly assembled in order to prevent premature discharge."

"66. If a gun is loaded at the order 'Cease firing—'
    (1) The gun shall remain loaded and shall be pointed and trained in a safe direction;
    (2) The breech mechanism shall be kept fully closed;
    (3) The firing key shall be opened and the firing circuit broken elsewhere;
    (4) The firing lanyard, if detachable, shall be unhooked;
    (5) The primer shall be removed from the lock of a bag gun.
The crew shall never leave a loaded gun until these precautions have been carried out."

"68. When the primer of a bag gun misfires during gunnery exercises and it cannot later be fired by electricity, it shall be carefully preserved, distinctly marked, and turned in to an ammunition depot for examination. When a case gun misfires the primer and the other ammunition details shall be disposed of as directed in paragraph 69 (b)."

"69. Ammunition unloaded from a gun may be reloaded if the service of the gun is resumed within a reasonable time. When it is apparent that the service of the gun will not be resumed within a reasonable time, the powder unloaded from a gun shall be disposed of as follows:

    (a) Powder unloaded from a bag gun.

        (1) If the gun was warm from firing when loaded, the powder shall be emptied from its bag into fresh water and in that condition turned in to an ammunition depot at the first opportunity.

        (2) If the gun was not warm from firing when loaded, the charge shall be carefully examined. If found dry, free from grease, and in good condition, it shall be sent back to its magazine. If injured or slightly greasy, an unstacked charge shall be rebagged and then sent back to its magazine; a stacked charge shall be emptied from its bags into containers, marked and turned in to an ammunition depot at the first opportunity. If grease or moisture has in any way gotten to the powder, the powder shall be emptied from its bag into fresh water and in this condition turned in to an ammunition depot at the first opportunity.

    (b) Powder in cartridges unloaded from a case gun.

        The cartridge shall be turned in to an ammunition depot at the first opportunity if—
        (1) The gun was warm when loaded;
        (2) An attempt was made to fire the gun;
        (3) After careful examination the cartridge is found injured or out of alignment.

CHAPTER 4

Crimped cartridges shall not be broken down before being turned in. Uncrimped cartridges shall be broken down and the powder immersed in fresh water before being turned in.

(c) When ammunition is returned to a depot in accordance with the above, paragraph 17 (See Article 2B15) shall be complied with. When cartridges are broken down in accordance with subparagraph (b) above, all the ammunition details composing it, including primers, shall be similarly marked and turned in."

"70. When a gun is being unloaded, all personnel not required for the unloading operation shall be kept at a safe distance from the gun. The division officer shall supervise the unloading."

"80. Fitting fixed ammunition in guns by hand prior to firing may defeat its purpose by canting or loosening the projectile in its case. Such fitting shall not be done except by order of the Commanding Officer, and then not until the firing pins have been removed from the breechblocks and the firing circuits have been disconnected."

"81. In testing primers outside of closed firing locks, no magneto or other device which can possibly supply current sufficient to fire the primer shall be used."

"85. Except in action or when specifically authorized, antiaircraft guns shall not be fired at elevations greater than those prescribed in the current Orders for Gunnery Exercises. When firing antiaircraft guns as such, all personnel not required to be exposed shall be kept under cover."

"86. Except in action, whenever a circuit breaker becomes so sensitive as to function due to the shock of firing, the circuit breaker shall be either overhauled or replaced and shall not be tied or fixed in position so as to be inoperative for the purpose for which designed."

"88. Whenever the guns of a vessel are fired, the fire hose shall be connected and pressure shall be maintained on the fire main. This does not require water to be running through the hose."

RESTRICTED

CHAPTER 5

A MAJOR CALIBER TURRET INSTALLATION

A. INTRODUCTION

5A1. General. Major caliber turret installations include all those with 8-inch or larger guns. Without exception at the present time these are bag guns which conform to common principles of operation and have similar component parts. Turret arrangements and equipment differ considerably in mechanical details, but in general the installations perform the same basic functions. Therefore, if the operation of one of these assemblies is thoroughly understood, it is a relatively simple matter to learn the special features of others as the occasion demands. It is the purpose of this chapter to illustrate modern bag-gun and turret construction, and to give general functional descriptions of the component parts of the gun assembly. The associated ammunition-handling equipment is also described briefly. The 16"/45 cal. triple-gun turret assembly is used as an example of modern gun and turret design in the discussion which follows.

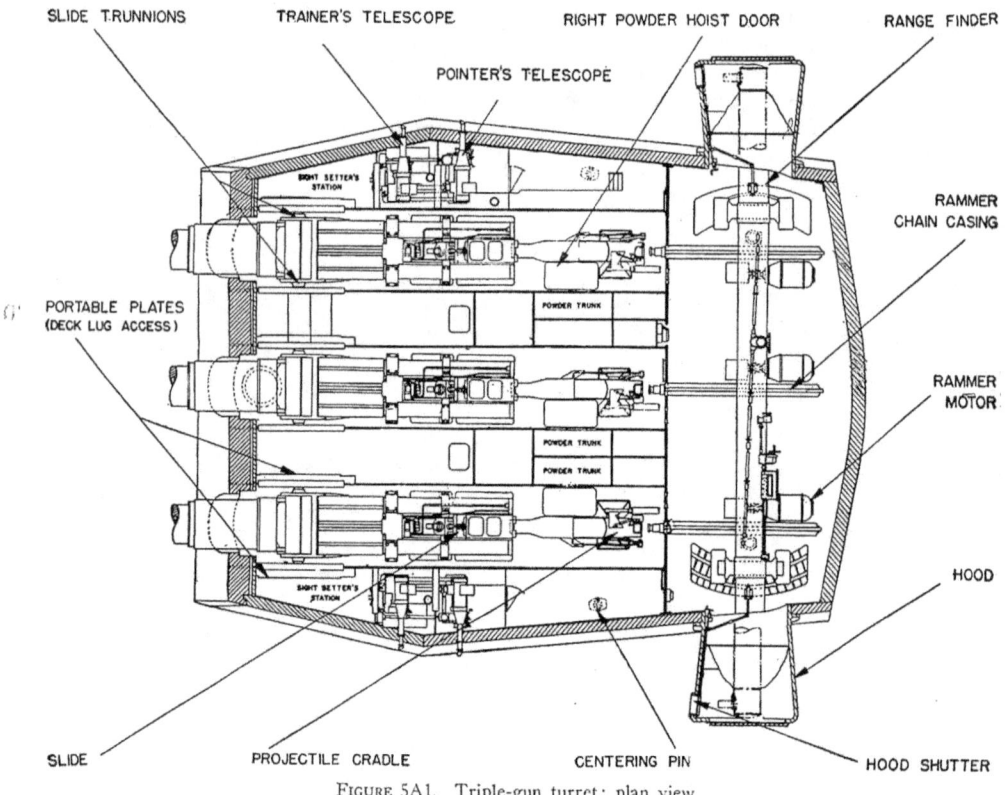

FIGURE 5A1. Triple-gun turret; plan view.

5A2. Turret installations. The main battery on the latest battleships consists of nine 16-inch guns, of either 45 or 50 calibers. These are emplaced in three triple-mount armored turrets located on the ship's centerline, with two forward and one aft. Each turret consists of the conventional subdivisions described in Article 4C4, namely: (1) *turret proper,* (2) *turret revolving structure,* (3) *turret foundation,* and (4) *barbette.* The turret proper is heavily armored and the barbette encloses the turret revolving structure down to the second deck. The lower spaces containing the magazines and handling rooms are protected by an armor belt, protective deck plate and heavy turret foundation plates. Each turret and its associated handling rooms are water, gas, and weather sealed. Ventilation systems circulate air through all compartments and maintain an air pressure somewhat higher than that on the outside. Semi-automatic sprinkling systems are installed in gun compartments, hoist trunks, and projectile stowage spaces within the turret rotating structure.

The turret proper is arranged so that each gun, both sight stations, and the turret officer's booth are in separate compartments, as shown in figure 5A1. The gun compartments occupy the forward part of the turret and the sight stations are on the outboard sides of the wing gun compartments. The turret-officer's booth, abaft the gun compartments, has a range finder mounted transversely in its aft end.

The turret proper and the turret revolving structure are supported by a roller-bearing assembly consisting of 60 large roller bearings placed on a roller path of approximately 34 feet outside diameter. The lower roller path and the training circle are secured to the turret foundation. This supporting structure within the barbette transfers the weight of the guns and the rotating turret structure, and the forces set up by firing to the ship's framework.

The turret revolving structure includes: (1) the *pan floor* below the guns, (2) a *machinery floor* below the pan floor, and (3) either two or three projectile stowage compartments, called *projectile flats,* below the machinery floor. The principal units of the training-gear assembly are located on the machinery floor and beneath the guns in the middle and forward half of the pan floor. The gun-elevating power drives, and the elevating and training control stations are on the machinery floor. Projectile and powder hoists pass up through the turret revolving structure. The powder handling rooms and the magazines are on the second platform deck which is below the lowest projectile flat.

B. GUN AND BREECH ASSEMBLY

5B1. Gun. The 16"/45 cal. guns are built up of six parts. The external profile shows a straight slide cylinder, a stepped and tapered chase, and a straight muzzle cylinder. The maximum diameter is 46 inches (slide cylinder), and the total weight is approximately 96 tons. Figure 5B1 shows the arrangement of these guns in the two forward triple-gun turrets of a modern battleship.

The rifling has a uniform twist of one turn in 25 calibers. The bore is chromium plated from the band slope to the muzzle for the purpose of retarding erosion. The powder chamber is slightly less than 18.5 inches in diameter and about 91 inches long. The choke diameter is 17.5 inches.

The *screw-box liner* is screwed into the breech and its inner surface provides the stepped threads which engage those on the Welin-type plug. Ducts between the gun and this liner, and holes through the latter provide the air leads required for the gas-ejector system.

The *yoke* is mounted around the breech end of the gun to provide the means for securing the recoil piston rod and the two recuperator yoke rods to the gun. The yoke also serves to help counterbalance the gun. The yoke is secured to the gun by a key, a fixed ring, and a locking ring. A shoulder on the yoke butts against the rear face or breech end of the fixed ring secured to the gun in an annular groove. The locking ring, slipped over the gun from the muzzle end, screws into the yoke and holds it in place against the fixed ring. The key prevents rotation about the gun axis.

5B2. Breech mechanism. The breech mechanism is shown in figures 5B2 and 5B3. Its principal parts include: (1) a Welin-type plug, (2) a mushroom and gas check pad, (3) a carrier, (4) counterbalance springs and closing cylinder, (5) rotating cams, (6) an operating lever and connecting rod, and (7) a salvo latch.

A MAJOR CALIBER TURRET INSTALLATION

FIGURE 5B1. 16″/45 cal. turret installation.

Welin plug and gas seal. The *plug* is locked to the screw-box liner by rotating it 29°. The blank sectors permit the two *plug rotating cams* to rotate the plug slightly over 6° before the threads engage as the breech is closed. The cam action assists in obtaining a smooth transition between plug translation and rotation.

The *mushroom* and *gas check pad* are the conventional De Bange type described in Article 4B7. An indication of the mushroom size is its weight of over 220 pounds.

CHAPTER 5

The plug is mounted on the *carrier* as shown in Figure 5B3. The threads on the *carrier journal* and in the plug recess have a pitch of 0.9 inches. This means that the plug will move aft 0.9 inches per counterclockwise revolution. Since the plug is unlocked by a rotation of 29° it is evident that it will move aft on the carrier journal about 0.1 inch when the breech is opened. If the gas check pad is stuck to the seating surface the plug will pull away from the back of the pad and the compressed *spring* around the mushroom stem will tend to force the *collar* and hence the pad aft. This force ordinarily breaks the seal, but if it does not, the continued rearward movement of the plug separates the pad from its seat.

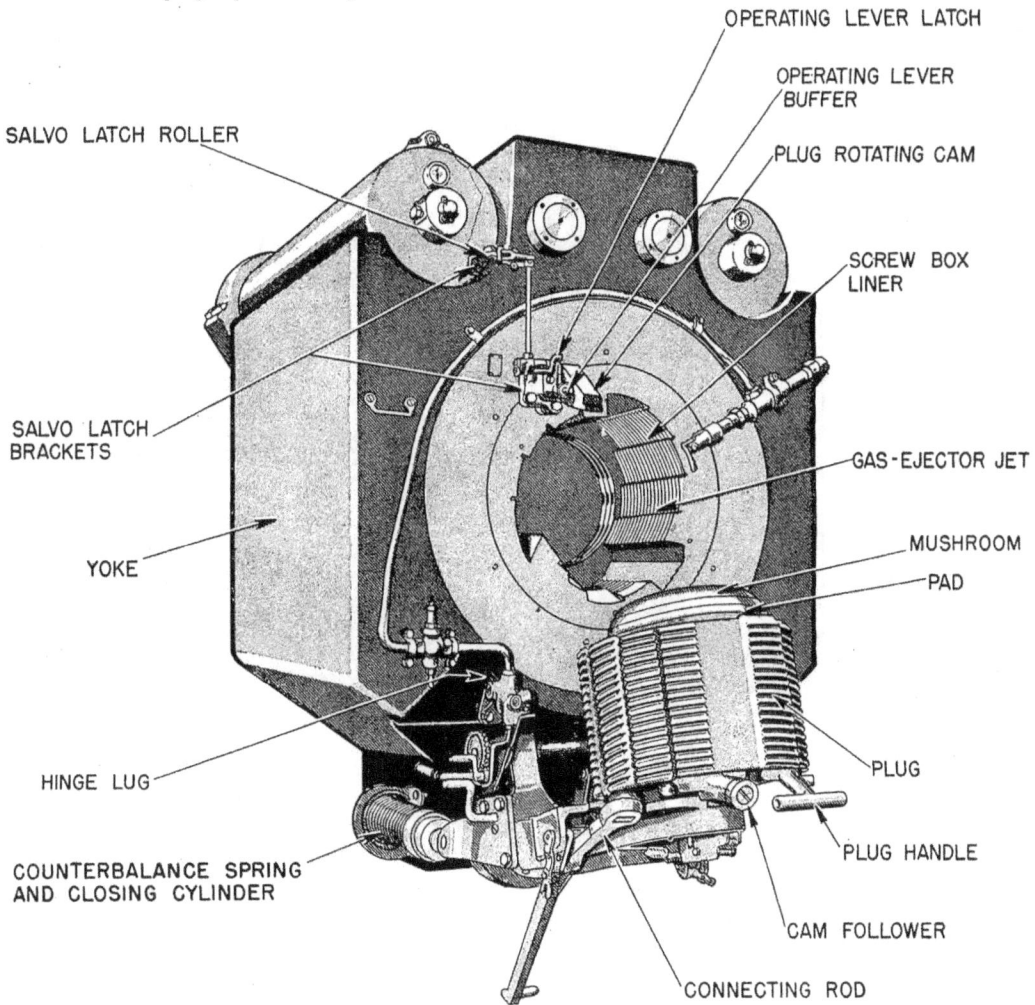

FIGURE 5B2. 16-inch breech mechanism; breech open.

Carrier and associated equipment. The plug and carrier, which have a combined weight of about 2,000 pounds swing about a pivot supported by the *hinge lug* secured to the gun (refer to fig. 5B2). The plug is locked in the open position by a *holding-down latch* until it is released by a foot crank near the loading platform.

The *opening buffer* functions to bring the carrier downward movement to a stop without shock. In construction and operation it is similar to a grooved-wall type of recoil brake. The *counter-*

A MAJOR CALIBER TURRET INSTALLATION

balances and *closing cylinders* are a combination of springs and pneumatic cylinders which function to balance the weight of the breech assembly in opening and to lift it when closing the breech. The cylinders receive their air supply from the gas ejector system after the pressure has been reduced from 200 to about 40 pounds p.s.i. The counterbalance springs are so adjusted that the breech mechanism will swing down and latch without a jarring stop or rebound. The air pressure in the cylinders is set so that the breech will close and swing the operating lever nearly home. The lever is always finally closed by hand.

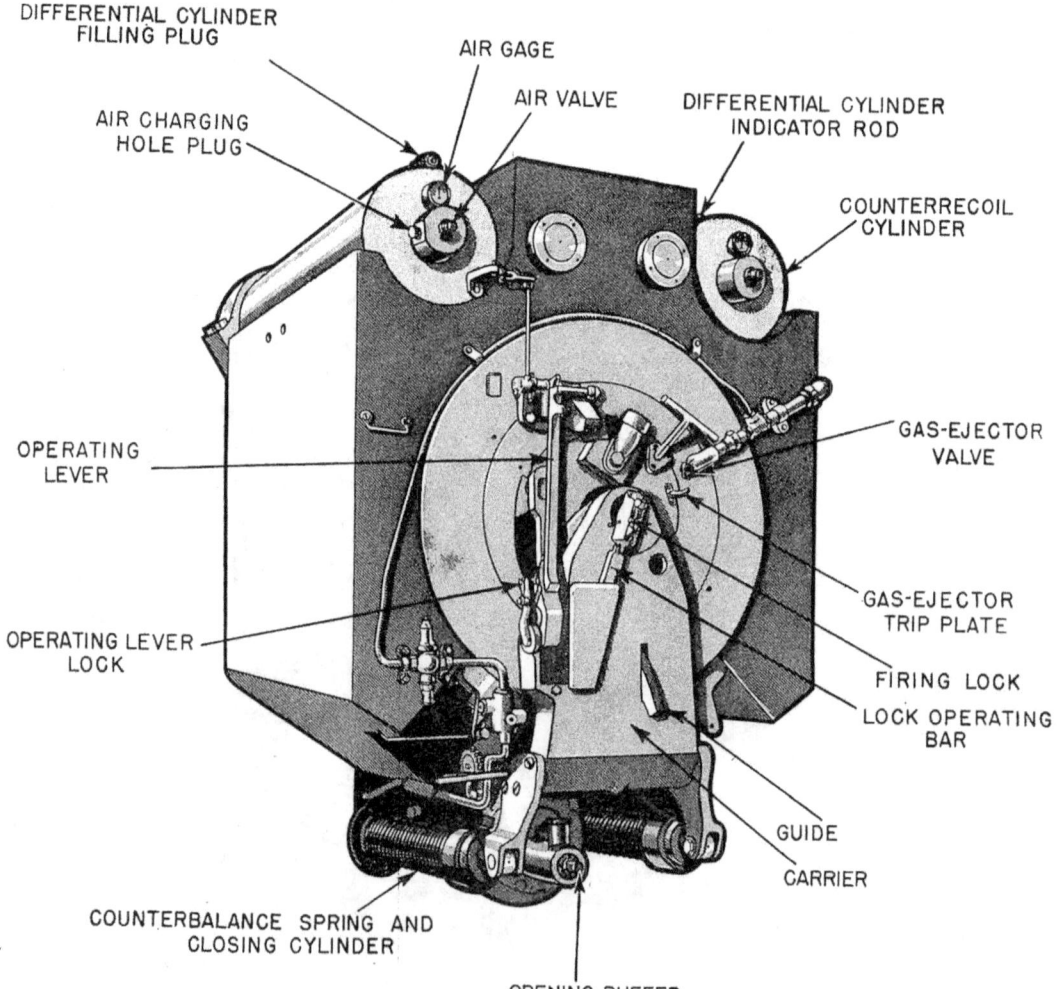

FIGURE 5B3. 16-inch breech mechanism; breech closed.

Operating lever and salvo latch. The *operating lever* is pivoted on the carrier and is connected to the plug by means of the *connecting rod* which extends from a pivot on the lever to a pin on the breech plug rear face. Motion of the operating lever about its fulcrum as it is pulled aft causes the connecting rod to rotate the plug and unlock it from the screw-box liner. A beveled, spring-loaded plunger in the end of the lever engages the *operating lever lock* when the breech is fully closed. The swing of the operating lever is stopped by the *operating lever buffer* which is similar to the opening buffer.

CHAPTER 5

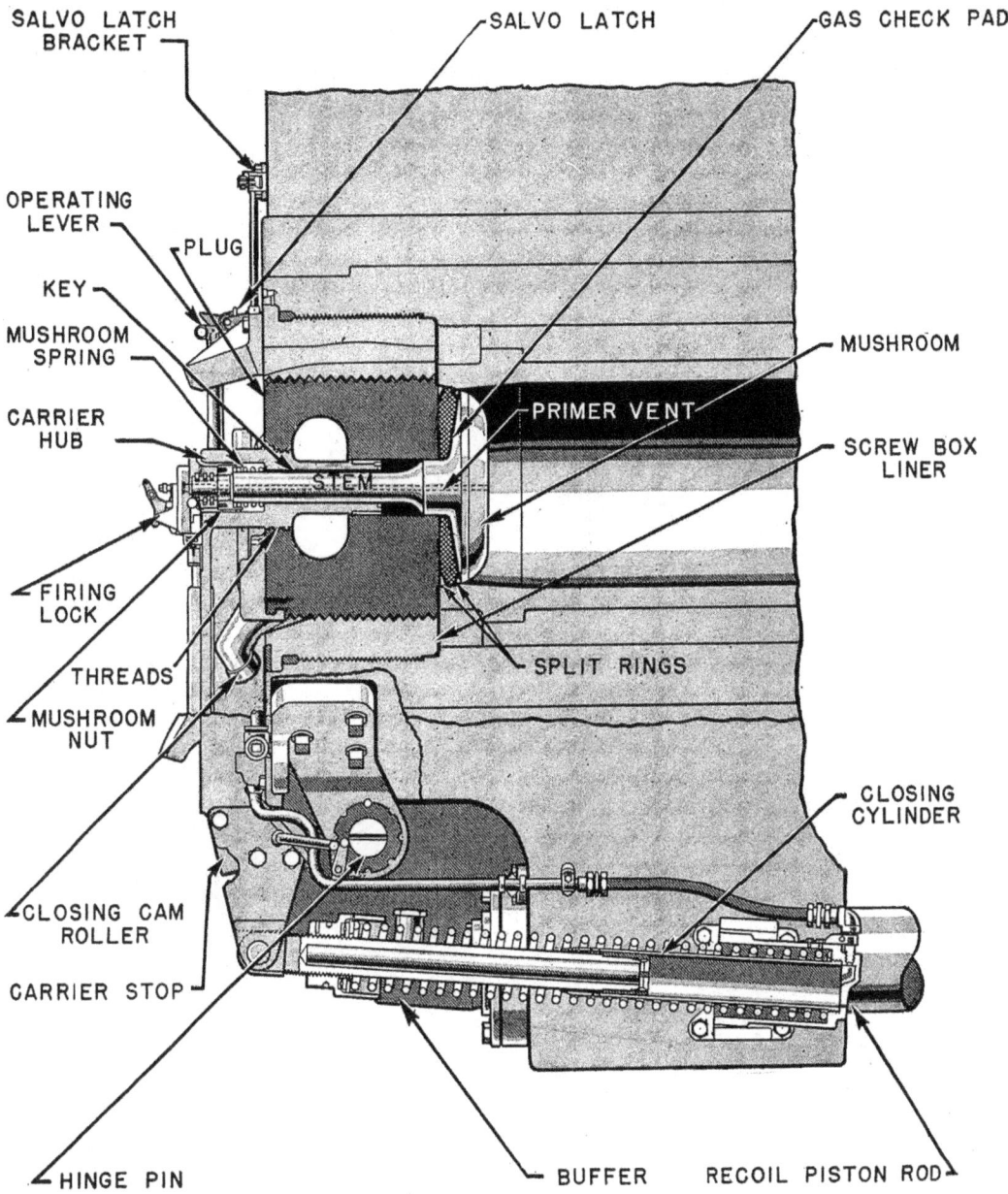

FIGURE 5B4. Breech plug and carrier.

The *salvo latch*, shown in figure 5B5, is a positive type and functions to prevent lifting the operating-lever latch after the breech is closed until the gun recoils. The *salvo latch lever* is positioned in the path of the roller mounted on the *roller support bracket*. The latter is secured to one of the counterrecoil cylinders and hence does not move in recoil. During recoil the salvo latch lever moves over the roller and rotates the connected shaft in the direction shown by the arrow. This moves the *salvo latch locking arm* off the *latch sleeve* and the sleeve rotates enough to prevent the locking arm from returning to its original position. When the operating-lever latch is raised, the latch sleeve is turned sufficiently to allow the locking arm to return to the position shown in figure 5B5. For drills the latch-locking arm can be secured in its unlocked position by screwing the latch-locking pin into hole B. The pin is placed in hole A for normal operation.

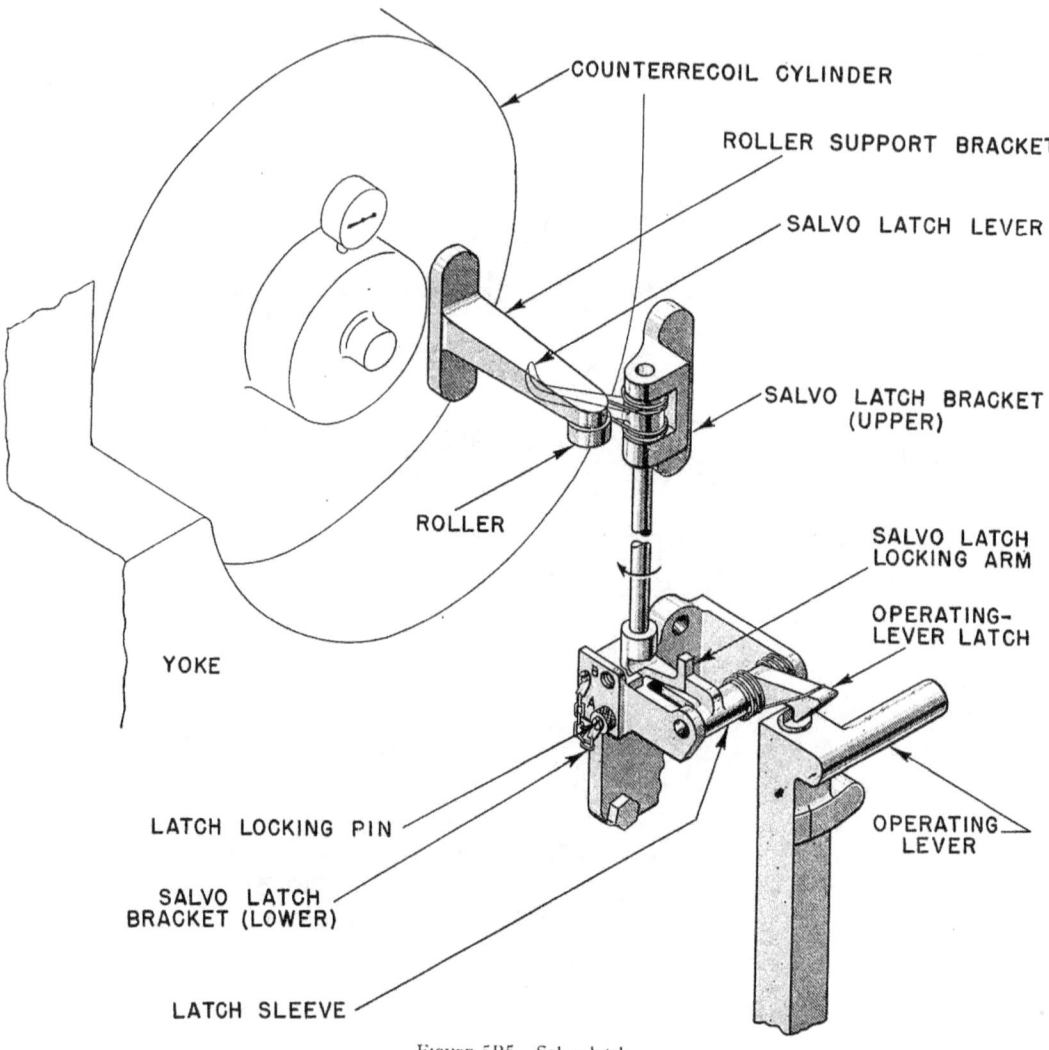

FIGURE 5B5. Salvo latch.

5B3. Firing mechanism. The *firing mechanism* consists of a Mark 14, Mod. 5 *firing lock* and the associated operating devices which synchronize its action with that of the operating lever. The firing lock, shown in figure 5B6, consists essentially of a *receiver* which is attached to the end of the mushroom stem and a *wedge* which slides back and forth in the receiver to open and close the primer chamber in the mushroom stem. The *primer retaining catch* helps to hold the primer in place until the wedge closes; the *extractor* ejects the primer case as the wedge opens. The wedge contains an insulated *firing pin* which carries the current to the primer bridge for electrical

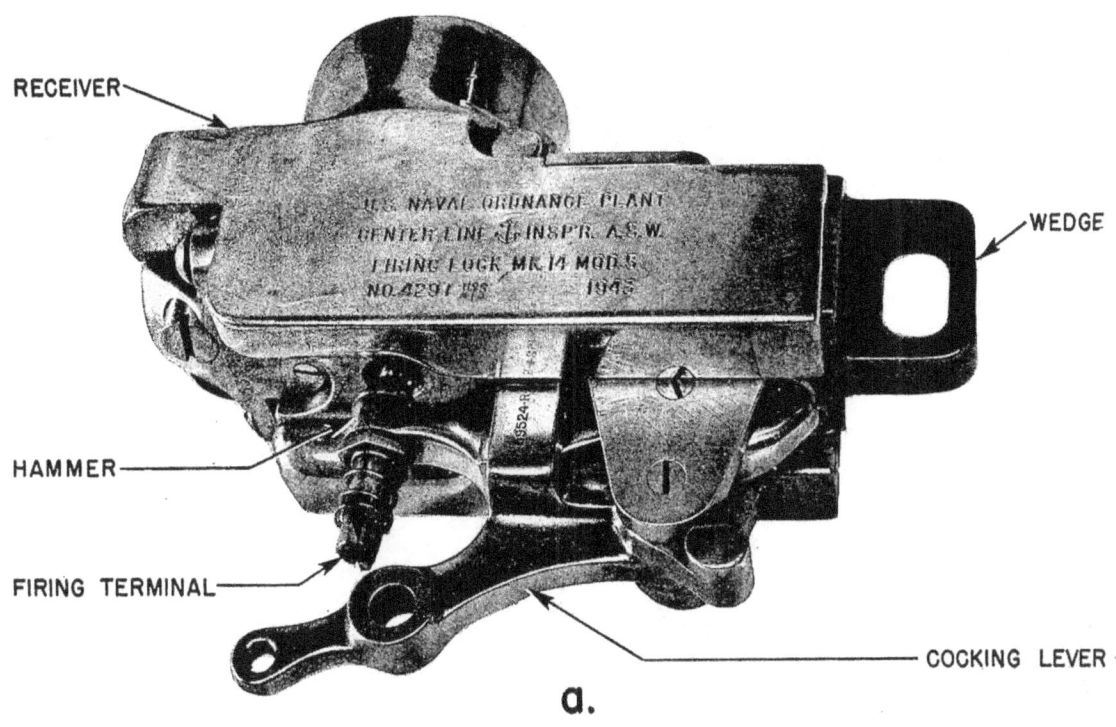
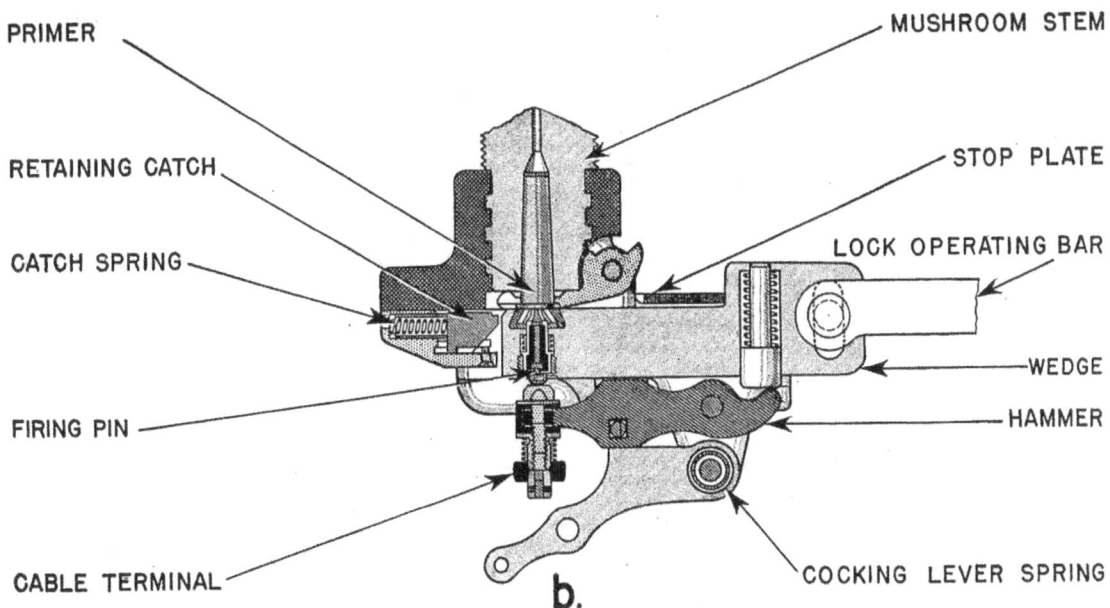

FIGURE 5B6. Mark 14 firing lock: a, photograph with wedge closed; b, electrical firing position.

A MAJOR CALIBER TURRET INSTALLATION

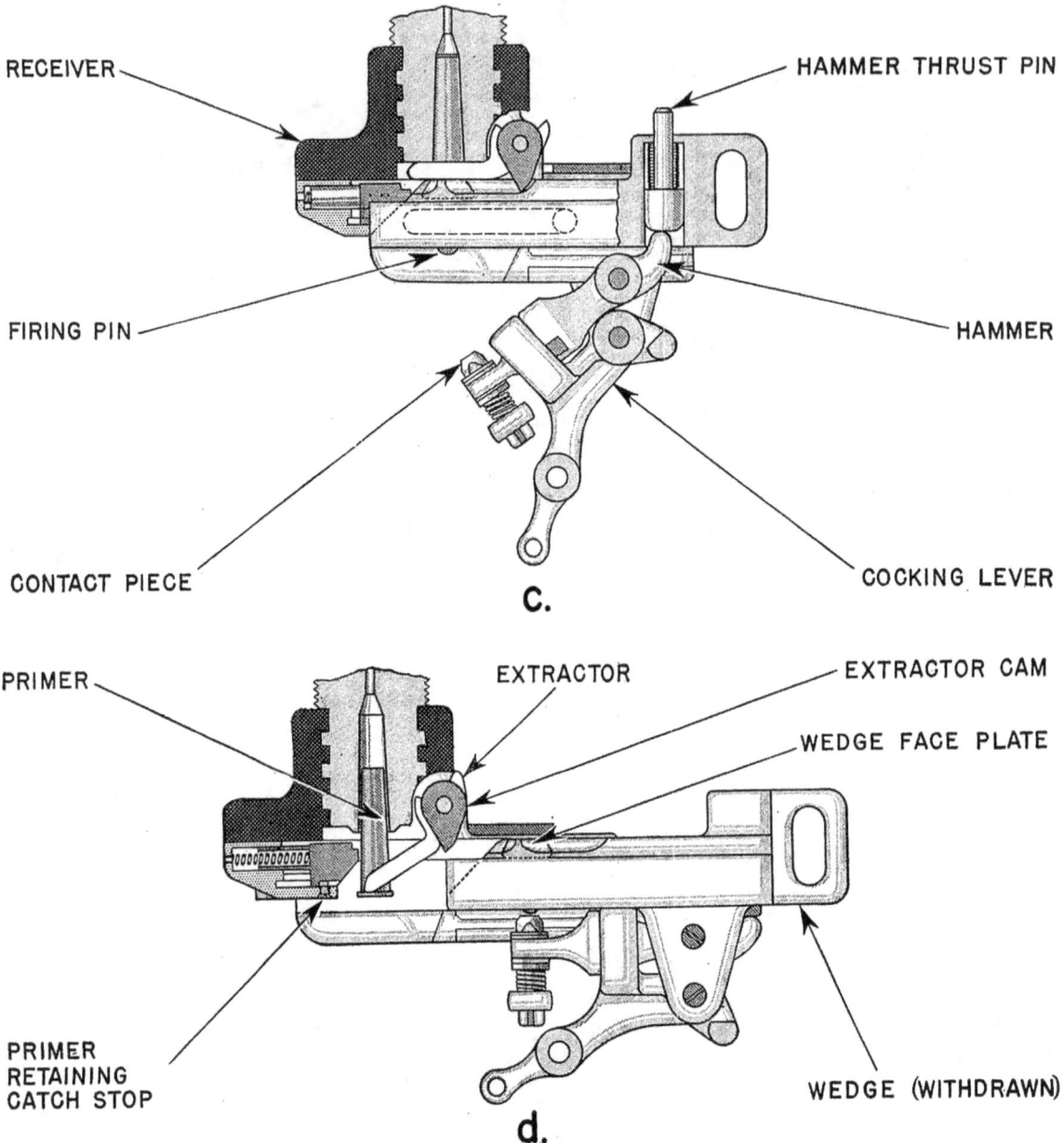

FIGURE 5B6. Mark 14 firing lock: c, percussion firing position; d, ejecting primer after firing.

firing. For percussion firing, a *hammer, contact piece,* and *cocking lever* attached to the wedge function to deliver a blow to the firing pin. This is accomplished by drawing back on the cocking lever which pulls the hammer with it until a catch on the lever releases the hammer. A spring then drives the hammer forward, making the contact piece strike the firing pin, which transmits the blow to the primer.

CHAPTER 5

Safety features incorporated in the lock prevent firing until the gun breech and the firing lock are completely closed. These features include:

 1. A lug on the hammer slides in a groove and lifts the hammer enough to break electrical contact between the firing pin and primer until the wedge is in the closed position.

 2. The same lug also prevents the hammer from being drawn back for percussion firing until the wedge is in the closed position.

 3. The *hammer thrust pin* must line up with a hole in the carrier before the cocking lever can be pulled back for percussion firing. This alignment occurs only when the wedge is fully closed.

 4. The cocking lever is so constructed that it transmits any accidental blow directly to the wedge instead of to the hammer and firing pin. The hammer placed between the cocking lever and the wedge is amply protected from exterior blows.

The wedge is connected to the breech-plug operating lever by means of a *lock-operating bar* (fig. 5B3) and a mechanical linkage which converts the operating lever shaft rotation into linear motion of the wedge. This motion is such that the wedge is withdrawn and the primer extracted when the operating lever is pulled aft and down in opening the breech (fig. 5B2). In case of misfire the wedge can be retracted without opening the breech by unlatching the *operating lever lock* (fig. 5B3) and rotating it without moving the operating lever. The construction is such that the wedge cannot be fully closed unless the breech is closed and locked.

The wedge is secured to the operating bar by means of a pin. The hole in the wedge through which the pin passes is slotted to permit movement of the mushroom stem during firing. The firing lock can be removed by taking out the pin and turning the receiver 90° to unfasten it from the mushroom stem.

5B4. Gas ejector system. The system includes a *reducer valve*, a series of *storage tanks*, piping with the necessary swivel and expansion joints, the *gas-ejector valve* (fig. 5B3), the *gas-ejector trip plate*, passages through the screw-box lining, and nozzles which direct the air into the bore. High-pressure air from the compressors is brought into the turret through the central column for the counterrecoil recuperators, the main gas ejectors, the auxiliary gas ejectors, and the pneumatic breech-closing cylinders. The gas-ejector reducer valve drops the pressure to about 200 pounds p.s.i. before it passes into the storage tanks located in the projectile flats. From the tanks the air is piped to the slide and thence through an expansion joint to the breech. The gas-ejector valve controls the flow of air in the ejector system. It is automatically opened by the gas-ejector trip plate when the plug is turned for opening the breech and is closed manually when the bore is clear. When the valve is opened air is admitted to an annular space between the screw-box liner and the gun. Three equally spaced holes extend through the liner from this space and nozzles are fitted at their ends. The location of one of these jets is shown in figure 5B2. The air jets direct their streams forward toward the bore center.

An auxiliary gas ejector consisting of a hose with a quick-acting valve and nozzle is stowed overhead near the breech. This is a standby device which can be directed into the gun to clear the bore in case normal gas ejection by the primary means fails.

C. SLIDE ASSEMBLY

5C1. General. Each gun has a separate slide which is individually mounted in trunnion bearings and arranged for independent elevating movement. The slide assembly, shown in figure 5C1, includes: (1) the *trunnions*, (2) the *rear end brackets*, (3) the *loader's platform bracket*, (4) the *hydraulic recoil system*, (5) the *counterrecoil recuperator*, (6) the *slide-securing mechanism*, (7) *upper* and *lower shield plates*, and the *gun cover*. The assembly is over 31 feet long and 8 feet high and is 78.5 inches wide across the trunnions. It weighs from 40 to 42 tons.

The slide bore has bearing liners on which the gun slides during recoil and counterrecoil. The *gun cover* is a cylinder with ½-inch wall thickness whose inside diameter is slightly larger than the slide cylinder. It extends through the gun port about 5.5 feet and serves only to provide an

RESTRICTED

oil and weather seal for the gun. A fabric cover clamped on the outer end of the gun cover and attached to the sloping turret front (fig. 5B1) completely encloses the gun port.

The left rear end bracket houses the 4-inch diameter *slide-securing pin* which seats in an aligned tapered socket in an adjacent girder when the gun is at 0° elevation. The pin is forced into the socket by means of a handwheel-operated screw. This arrangement serves to relieve the elevating screw and drive gear from the dead weight load when the gun is not being used. Another socket permits locking the gun at 20° elevation while the recuperator air cylinders are being charged. The right-rear bracket on the slide (not shown in the figure) mounts the large pin about which the elevating screw pivots.

5C2. Recoil mechanism. The recoil mechanism is the conventional hydraulic type and has a single cylinder of 26 inches bore diameter and an overall length of about 5 feet. It is similar to the one shown in figure 4C6. Three equally spaced throttling rods provide the required orifice size-control. The recoil, limited to 48 inches (3 calibers), occurs in approximately 1/3 second.

Counterrecoil buffing action is provided by a plunger-dashpot arrangement at the forward end of the cylinder. Liquid displacement from the dashpot occurs through four grooves of variable depth and the small clearance between the plunger and the dashpot entrance. Buffing action takes place during the last 16 inches of counterrecoil. Two expansion tanks are mounted forward of the recoil cylinders.

5C3. Recuperator. The hydro-pneumatic counterrecoil mechanism consists of two air cylinders mounted on top of the slide and a differential cylinder between them. Its operating principle and arrangement of components are discussed in Article 4C17. The *counterrecoil cylinders*, shown in figure 5C1, are about 10 feet long and have an inside diameter of 12 inches. The plungers are hollow cylinders closed at the forward end, and are about 6 feet long and 9 inches outside diameter. They are connected to the gun yoke by means of a plunger yoke and two yoke rods which are covered by the *plunger cover* (fig. 5C1).

Each plunger packing assembly is about 14 inches long. Normal air pressure is 1,600 pounds p.s.i., increasing to a maximum of approximately 2,400 pounds p.s.i. during recoil. The differential cylinder keeps the packings under a liquid pressure about 15 per cent higher than the cylinder air pressure. The amount of liquid in the system is shown by the *differential cylinder indicator rod* (fig. 5B3) which extends to the aft end of the right-hand counterrecoil cylinder where it can be seen from the loader's platform.

The *yoke-locking device*, which holds the gun in battery when it is not in use, is on top of the slide, at the rear between the counterrecoil cylinders. A safety link, in the form of an eye bolt, is pinned to the slide. When locked, the safety link rests in a notched bracket on top of the yoke, and a nut on the link is tight against the bracket, holding the yoke and gun securely to the slide. The link is stowed by loosening the nut, swinging the pin up and forward, and clamping it out of the way. The link, or eye bolt, is necked so that it will part without damage to the gun if it is not removed before firing.

5C4. Trunnion bearings. Each gun is separately pivoted on 18.5-inch diameter trunnions integral with the slide. The trunnion bearing assembly is made up of a set of bearing retainers, and roller bearings which carry the weight of the gun and take the shock of firing. Each bearing assembly includes 24 cylindrical rollers 2.5 inches in diameter and 3.5 inches long.

The six bearing assemblies are supported by deck lugs on a transverse gun girder at the front of the turret. The bearing seats are precisely machined so that all of the bearings are on a common trunnion axis which is parallel to the turret roller path. This axis is about 15 feet above the roller path and about 11 feet forward of the turret transverse centerline.

D. ELEVATING, TRAINING, AND SIGHT GEAR

5D1. Elevating gear. Each gun has a separate electric-hydraulic elevation power drive and control. The drive, shown schematically in figure 5D1, consists of an electric motor (60 h.p., with an overload rating of 108 h.p.), a reduction gear, and a hydraulic speed gear. The *B-end* output shaft drives a nut on the *elevating screw* which is attached to the gun by means of a pin

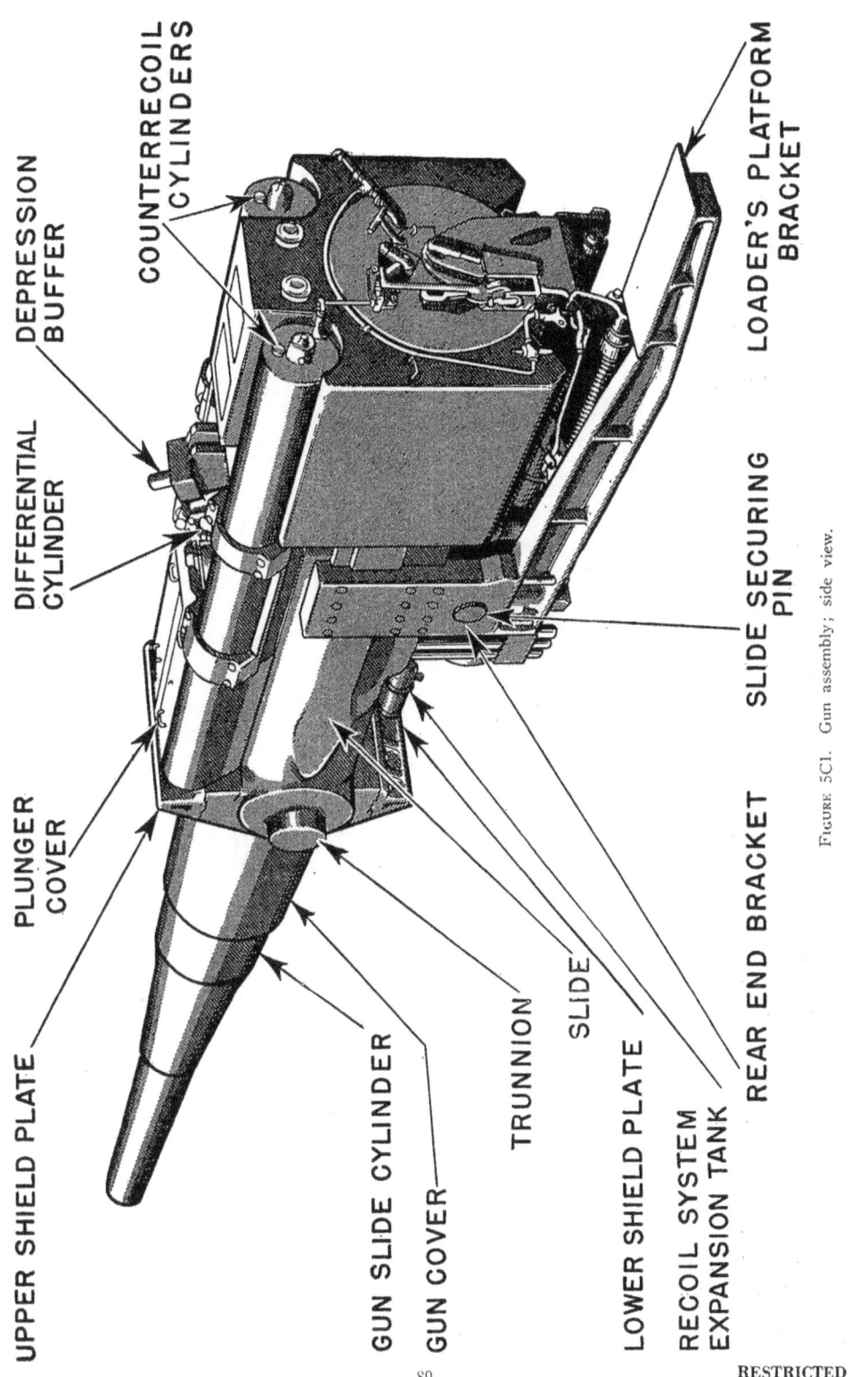

FIGURE 5C1. Gun assembly; side view.

in the rear-end bracket secured to the slide. This screw is 7.75 inches in diameter and almost 12.5 feet long. The power drives and control units are located below the guns on the machinery and pan floors as shown in figures 5D2 and 5D3.

Gun-elevating movement is controlled automatically by means of a receiver-regulator or by rotation of the *gun-layer's handwheels*. In the latter method the gun-laying personnel (one gun layer for each gun) controls the power drive in accordance with dial-indicated gun-elevation orders emanating from either the director system or one of the two sighting stations within the turret. Elevation stops are installed to limit elevation and depression to 45° and 2° respectively. The maximum elevating speed is approximately 12° of arc per second. In some installations the gun may be automatically lowered after firing to the loading angle of 5° elevation by operating a control lever.

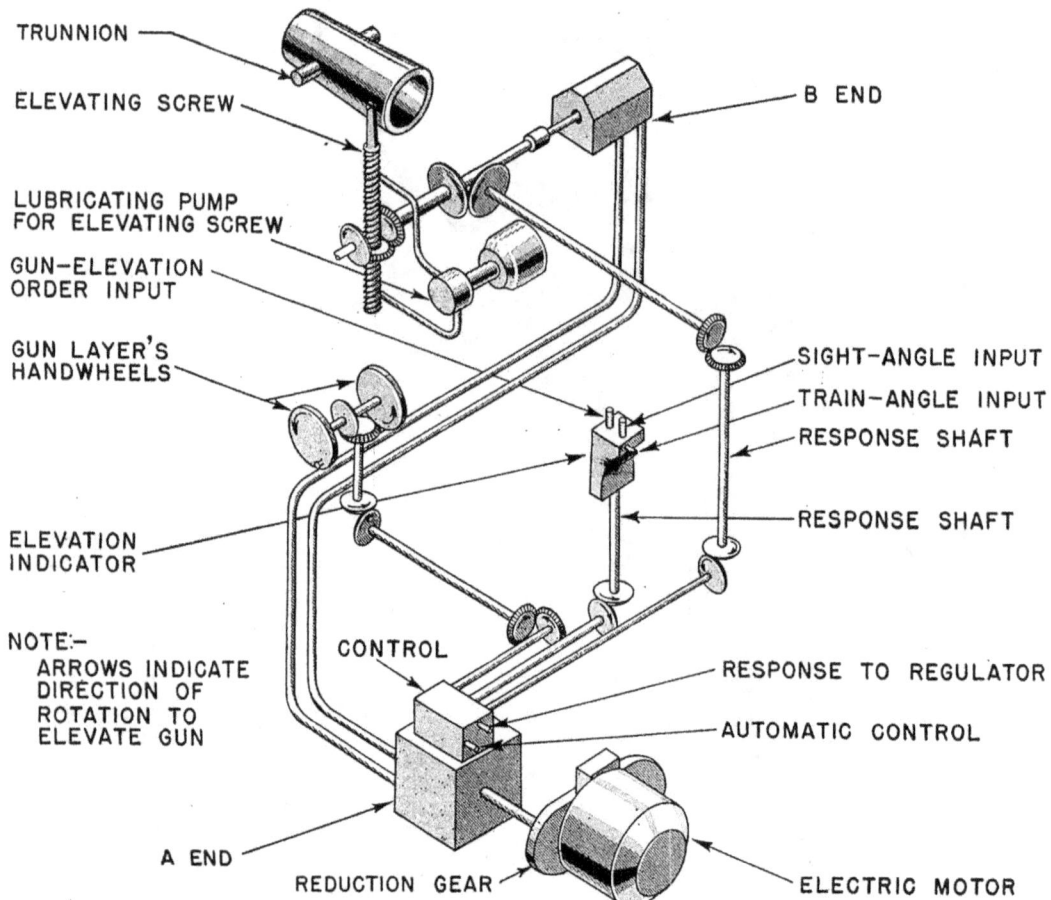

FIGURE 5D1. Elevating gear; schematic.

5D2. Training gear. The training gear and its control equipment, shown schematically in figure 5D4, also consists of an electric-hydraulic power drive which can be automatically or manually controlled. The electric motor (300 h.p., with an overload rating of 540 h.p.) drives the *A end* of the hydraulic transmission unit through a *reduction gear*. Two B-end units are driven by the single A end.

CHAPTER 5

The B ends drive the *training pinions* which mesh with the *training circle* secured to the turret foundation. The *brakes* between the B ends and the pinions function to prevent the turret from turning when the training power is off. The brakes are set by springs and released by *hydraulic cylinders*. When power is off no pressure is on the hydraulic cylinders and the springs take charge. During the training operation a piston within each cylinder is forced upward by oil under pressure, and the brakes are released against the action of the springs.

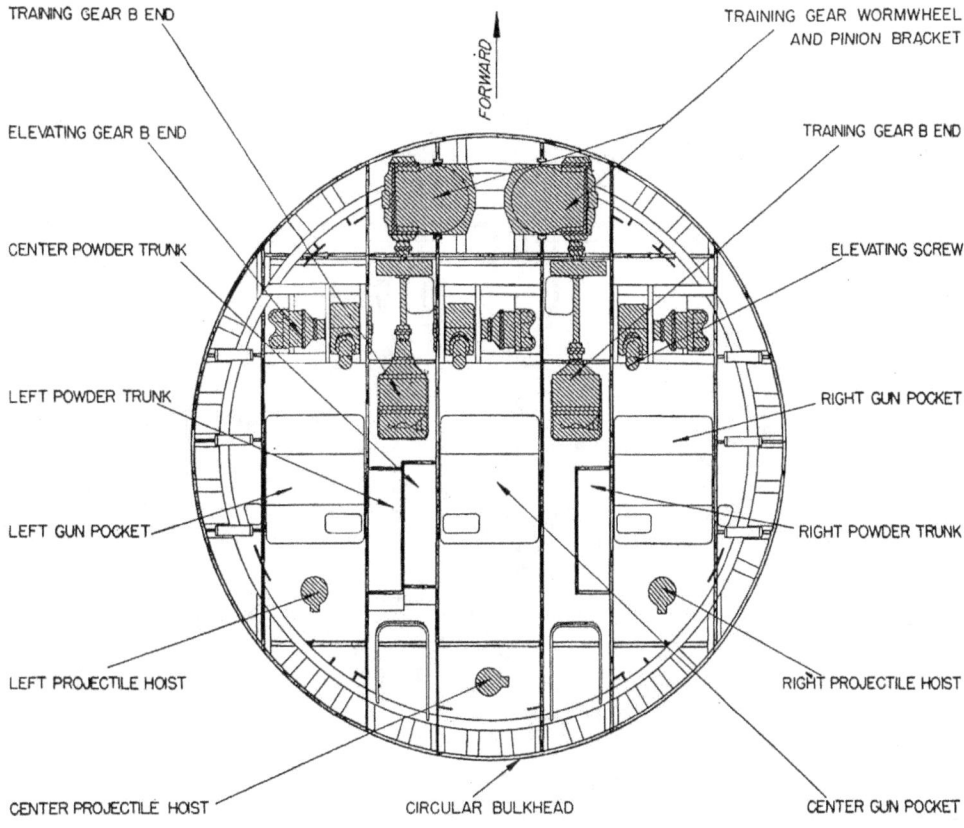

FIGURE 5D2. Pan floor; plan view.

Training may be automatically controlled from the director system through a receiver-regulator, or manually controlled by means of handwheel drives. In handwheel control, either the *train operator* on the machinery floor trains the turret in accordance with dial-indicated train-order signals received from the director system or turret sighting stations, or one of the trainers controls the drive by keeping his telescope on the target.

5D3. **Sighting gear.** Each turret has duplicate sight stations in enclosed compartments located outboard from the wing guns and approximately on the transverse centerline of the turret. The arrangement of telescopes, sight-setter's indicator, and the associated control and transmission equipment shown in figure 5D5 is called a *director-type telescope sight*. It is so designated because it transmits gun-elevation and gun-train orders to the indicators and control gear on the machinery floor from which the power is actually applied to the turret and guns. The pointer's and trainer's telescopes are so mounted that they have parallel motions when the sight settings are made.

Corresponding elements of the two stations are interconnected so that either station can take over sighting operations. Rotation of the pointer's handwheels operates only indicators at the gun-

RESTRICTED

A MAJOR CALIBER TURRET INSTALLATION

layer's station. Rotation of the trainer's handwheels may control the turret-train drive, or actuate indicators at the train-operator's station. The guns may be indirectly positioned from a sight station by first setting the sights, and then turning the pointer's and trainer's handwheels to produce turret and gun movement that will place the telescopes on the target.

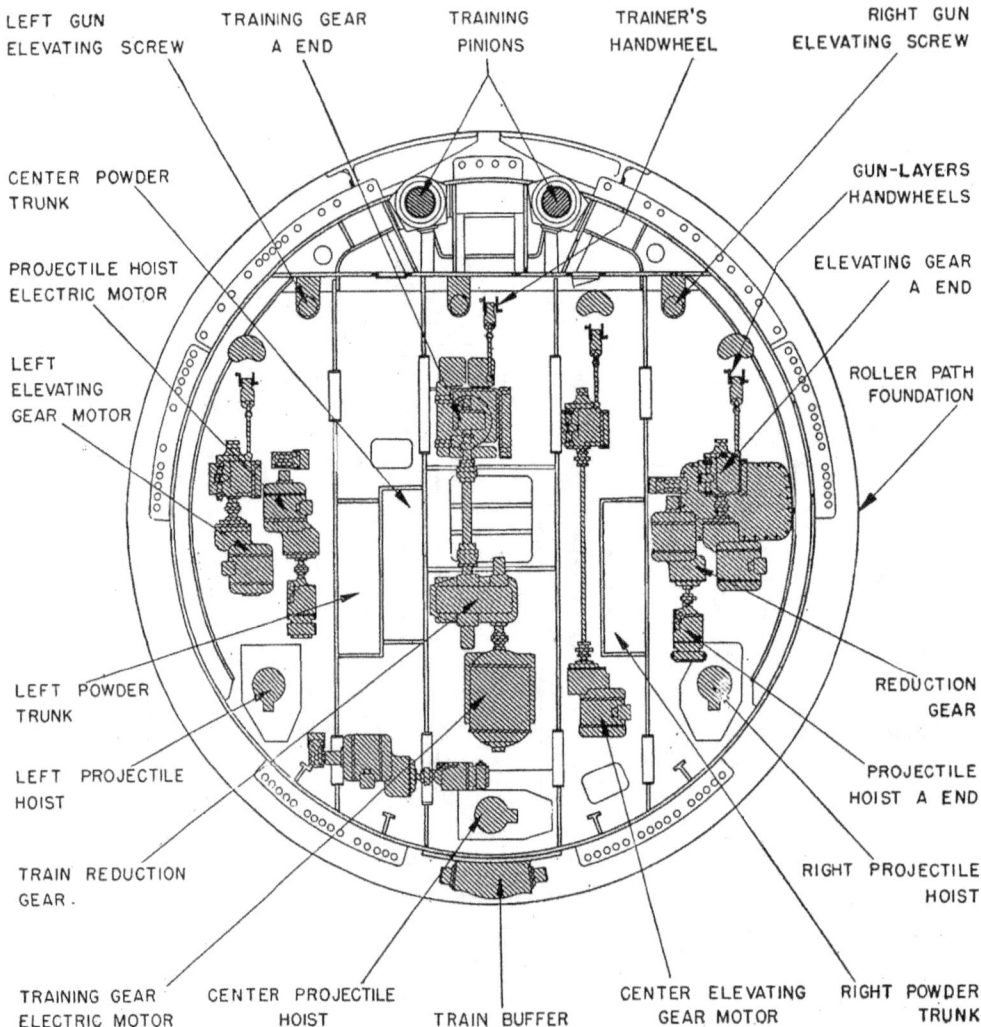

FIGURE 5D3. Machinery floor; plan view.

E. AMMUNITION HANDLING

5E1. Projectile stowage. Projectiles are stowed in two circular compartments, called upper and lower projectile flats, located in the turret revolving structure about 25 feet and 33 feet respectively below the turret proper. Turret 2 has an additional fixed-stowage compartment immediately beneath the lower projectile flat. Figure 5E1 is a plan view showing the arrangements in one of these flats.

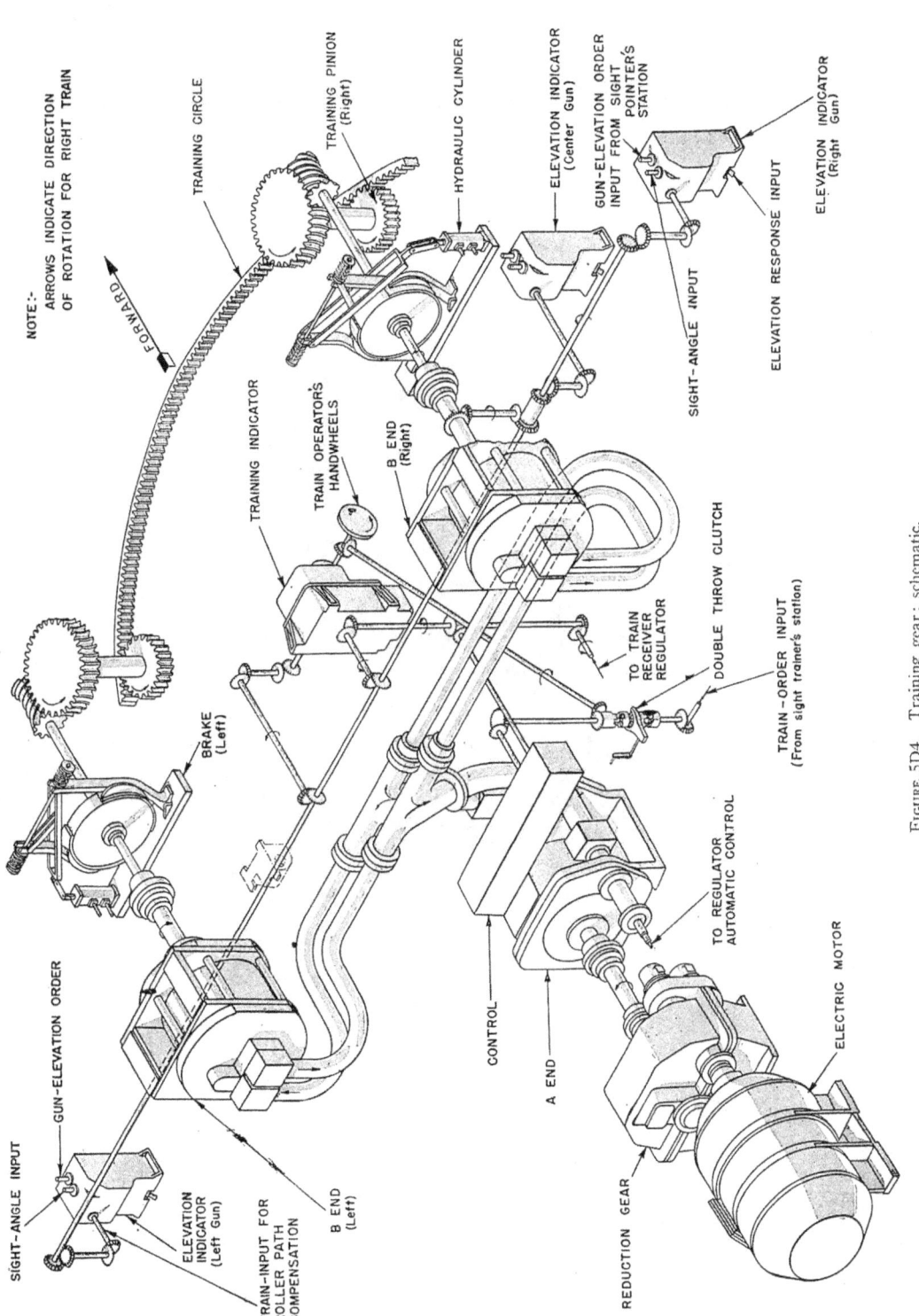

Figure 5D4. Training gear; schematic.

A MAJOR CALIBER TURRET INSTALLATION

The projectile storage spaces are between the roller-path foundation (about 35 feet inside diameter) and a concentric circular bulkhead which encloses a machinery space of 17 feet diameter. In the projectile flats the deck is subdivided into three concentric ring-shaped platforms. The outer ring is a non-rotating shelf attached to the cylindrical turret foundation and provides fixed stowage. The intermediate ring is part of the rotating structure and is the projectile-handling platform. The three projectile hoists pass through this section. The inner ring is mounted on

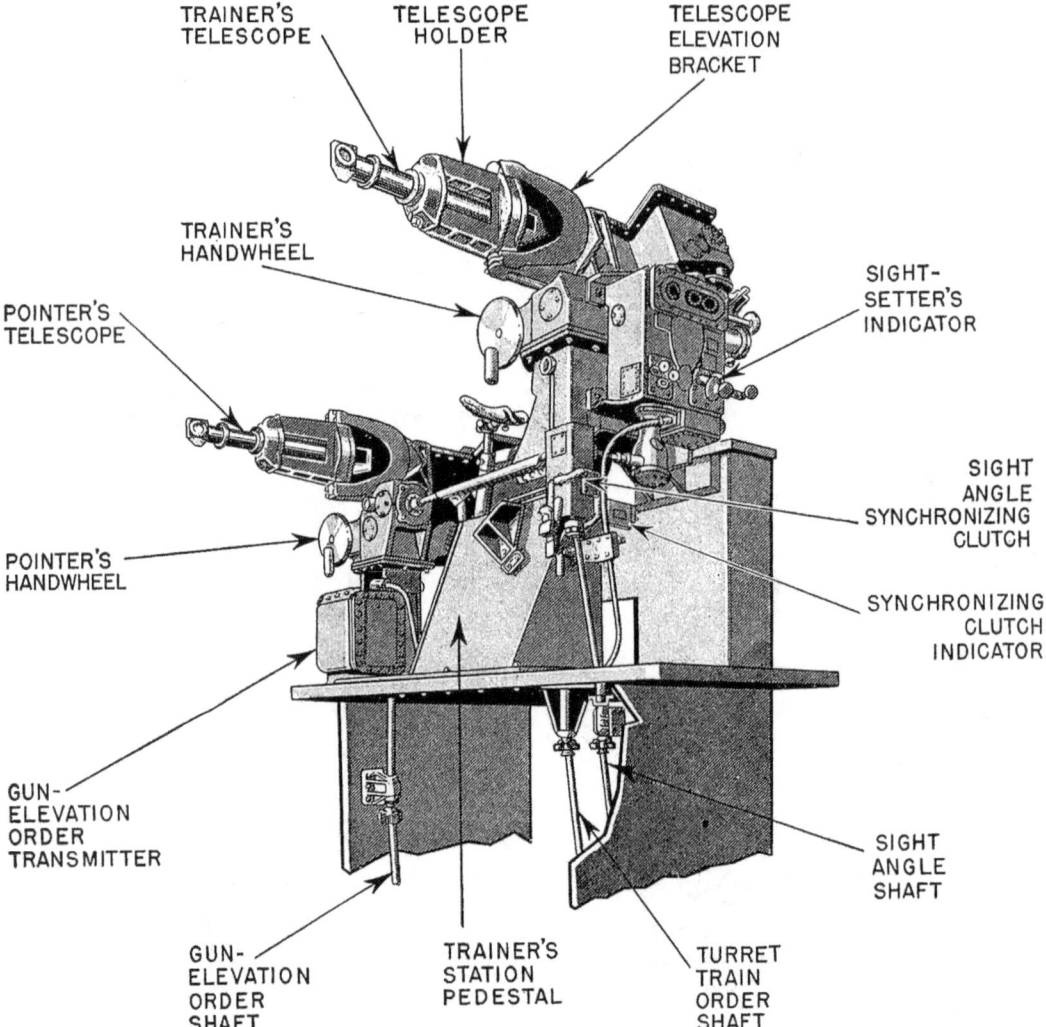

FIGURE 5D5. 16-inch sight gear.

rollers and has a power drive which can rotate the ring in either direction with respect to the intermediate ring. This power drive (located in the machinery compartment) consists of an electric motor and a hydraulic speed gear.

All projectiles are stowed erect, and each is separately lashed to the adjacent bulkhead. The stowage capacity of the inner ring is 72, and that of the fixed platform 130. The additional stowage compartment for turret 2 accommodates 121 more. Hence the number of projectiles that can be

CHAPTER 5

stored in each turret, exclusive of hoist capacity, is 404 in turrets 1 and 3, and 525 in turret 2. The projectiles are brought to the projectile-stowage compartments through the powder-handling rooms below, and hoisted to the second projectile-flat level where they are taken through water-tight doors in the circular foundation bulkhead to the storage space within. From this point they are distributed to the other stowage spaces.

The inner projectile rings function to move projectiles to positions near the hoists. Handling is by means of parbuckling gear which includes motor-driven capstans. The power drive on the projectile ring is semi-automatically controlled so that the ring turns through an arc of 30° and then stops each time the control lever is manually operated. The control station is adjacent to the center hoist.

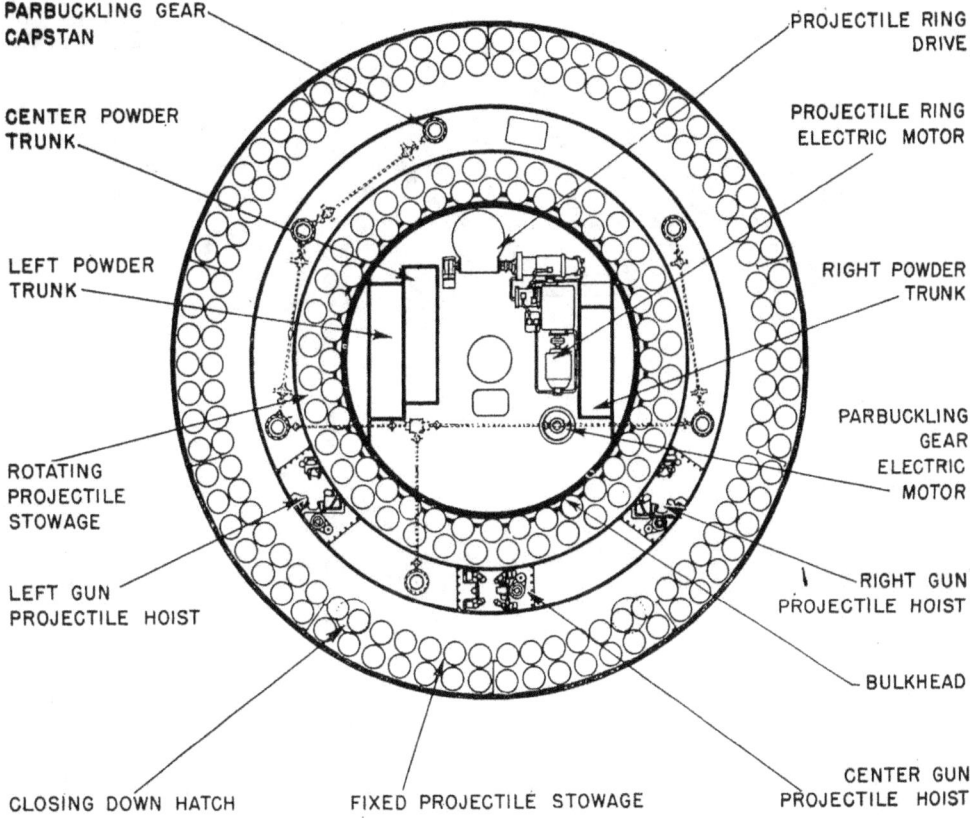

FIGURE 5E1. Projectile flat; plan view.

5E2. Projectile hoist. Each gun has a projectile hoist which raises the projectiles from the stowage compartments in the turret-revolving structure to the breech end of the gun. The hoist (see fig. 5E2) consists of a tube through which the projectile passes (base downward), a hydraulic-ram type lift, and a power drive with its associated control. The three tubes pass through the middle rings of the projectile flats and are almost vertical up to the pan floor. From this point the two outboard tubes slope rearward at an angle of approximately 17° to the cradles (shown in fig. 5E5) at the upper ends of the tubes. The center tube continues its vertical course so that at the guns all cradles are equi-distant from the breech.

The hydraulic lift consists of a hydraulic cylinder vertically mounted between the decks of the

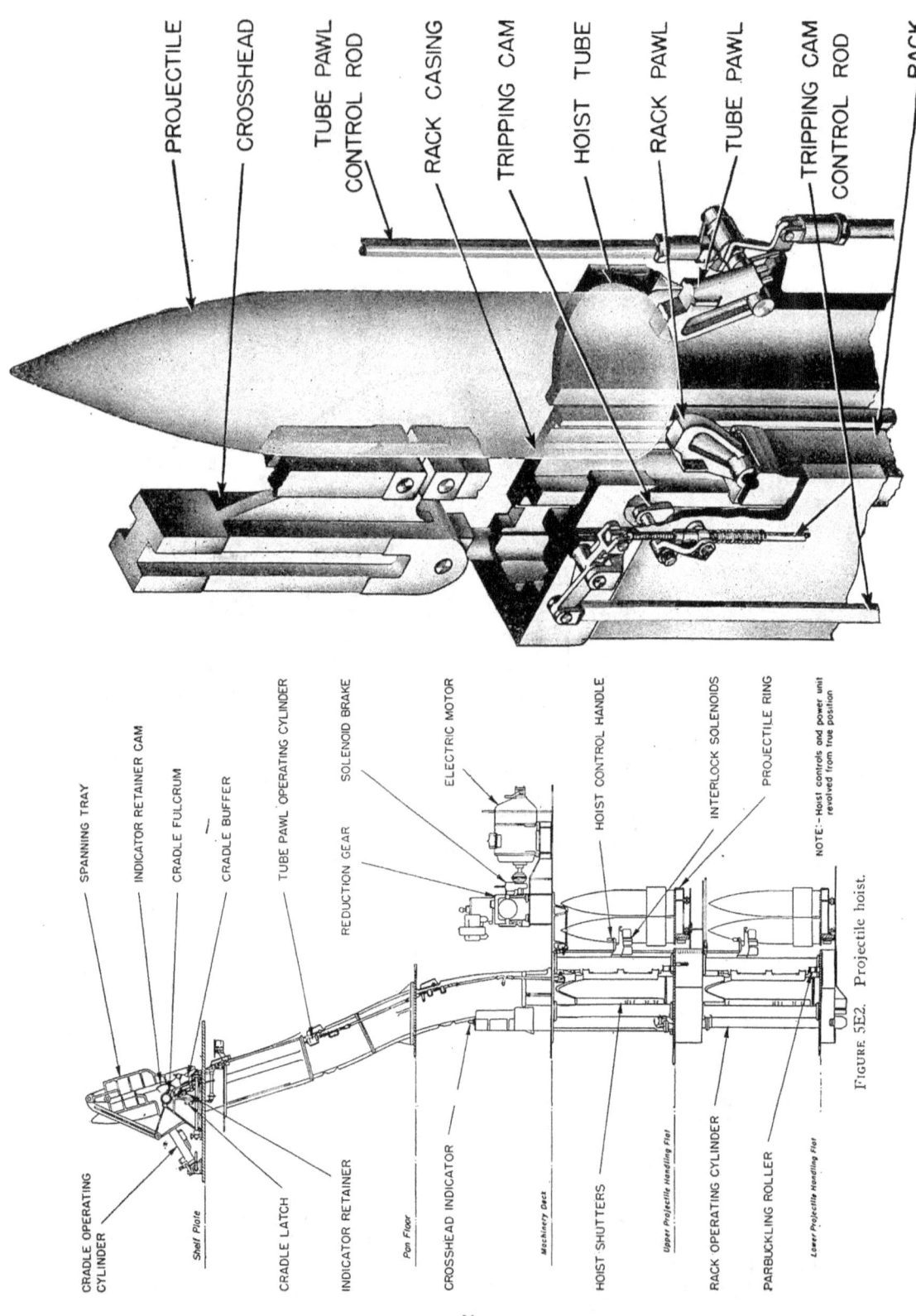

Figure 5E2. Projectile hoist.

CHAPTER 5

projectile flats, a piston, and a piston rod which connects to a rack placed within the tube casing. The piston diameter is about 6 inches, and the piston moves approximately 8 feet during the hoist stroke. With the aid of pawls pivoted in the hoist tube, the hydraulic ram operates to raise or lower the projectiles in equal stages. Each stage is equal to a full stroke of about 8 feet and the distance to the cradle is four stages from the lower projectile flat and three from the upper.

The projectiles are supported at the end of each stroke by the tube pawls which are spring actuated to move beneath the base of the projectile as the projectile is lifted on the rack pawls. The rack pawls are also spring loaded so that they are forced into the hoistway, and when the rack descends after a lifting stroke, the pawls are depressed as they pass by any projectiles in the stages below. A pawl-operating mechanism retracts the pawls in the reverse order when projectiles are lowered.

The power drive for each hoist is located on the machinery floor and consists of an electric motor (60 h.p.), a reduction gear, and the A end of a hydraulic speed gear. The A end is directly connected to the hoist cylinder and develops the 700 pounds p.s.i. pressure required to lift a full load of four projectiles. A solenoid brake on the electric-motor shaft operates to prevent motion of the hoist when there is a power failure.

The main hoist-control system includes duplicate installations of control handles at stations near the hoist-loading apertures in the upper and lower projectile flats, mechanical and electrical interlocks, and audible and visual signals at these stations.

Mechanical and electrical interlocks prevent operation of control handles when projectiles are being loaded on the hoist, when there is a projectile in the cradle, or during the ramming cycle. The signal devices at each control station include:

1. A red signal light which flashes when the rammer reaches the top of its upward stroke to indicate that the projectiles have moved above the tube pawls.

2. A visual indicator which is positioned by motion of the cradle. When the cradle is full or open, *danger* is displayed, and when it is empty and closed *hoist* appears and a gong is sounded.

5E3. Cradle. The cradle assembly shown in figures 5E3 and 5E5 includes the *cradle fulcrum, cradle,* and *spanning tray.* During hoisting the rammer chain is in its retracted position, the cradle is upright so that its axis lines up with that of the hoist tube, and the spanning tray is folded against the cradle. A projectile hoisted into the cradle is supported there by a projectile latch within.

The cradle and spanning tray are operated by a hydraulic cylinder not shown in the figures. The control lever for this unit is located at the aft end of the gun compartment at the side of the rammer and cradle. When the control lever is operated, the cradle tips forward and the spanning tray moves forward and up, forming a smooth tray from the hoist to the chamber. The gun must be elevated to 5° during loading so that the tray lines up with the gun bore.

5E4. Rammer. The rammer consists essentially of a special roller-type chain (shown in fig. 5E3) which meshes with a power-driven sprocket. The arrangement of parts is shown in figure 5E5. The rammer assembly is located in the turret-officer's booth just abaft the gun compartment.

The power drive includes an electric motor (60 h.p. with an overload rating of 108 h.p.) and a conventional hydraulic speed gear whose B end drives the sprocket wheel. The control lever is to the rear and one side of the cradle. The control arrangements provide full-power projectile ramming to a jammed stop. This is necessary since the rotating band must be forced into the band slope so that the projectile will not move to the rear when the gun is elevated. The maximum velocity during the ramming cycle is slightly less than 14 feet per second and the time required for ramming is 1.6 seconds. The powder bags are rammed into the chamber at a slower speed, ramming being stopped by manually operating the controls.

The rammer chain forms a rigid column when it is extended as shown in figure 5E3. The chain assembly including the head link buffer is over 22 feet long. When in the "ready to ram" position

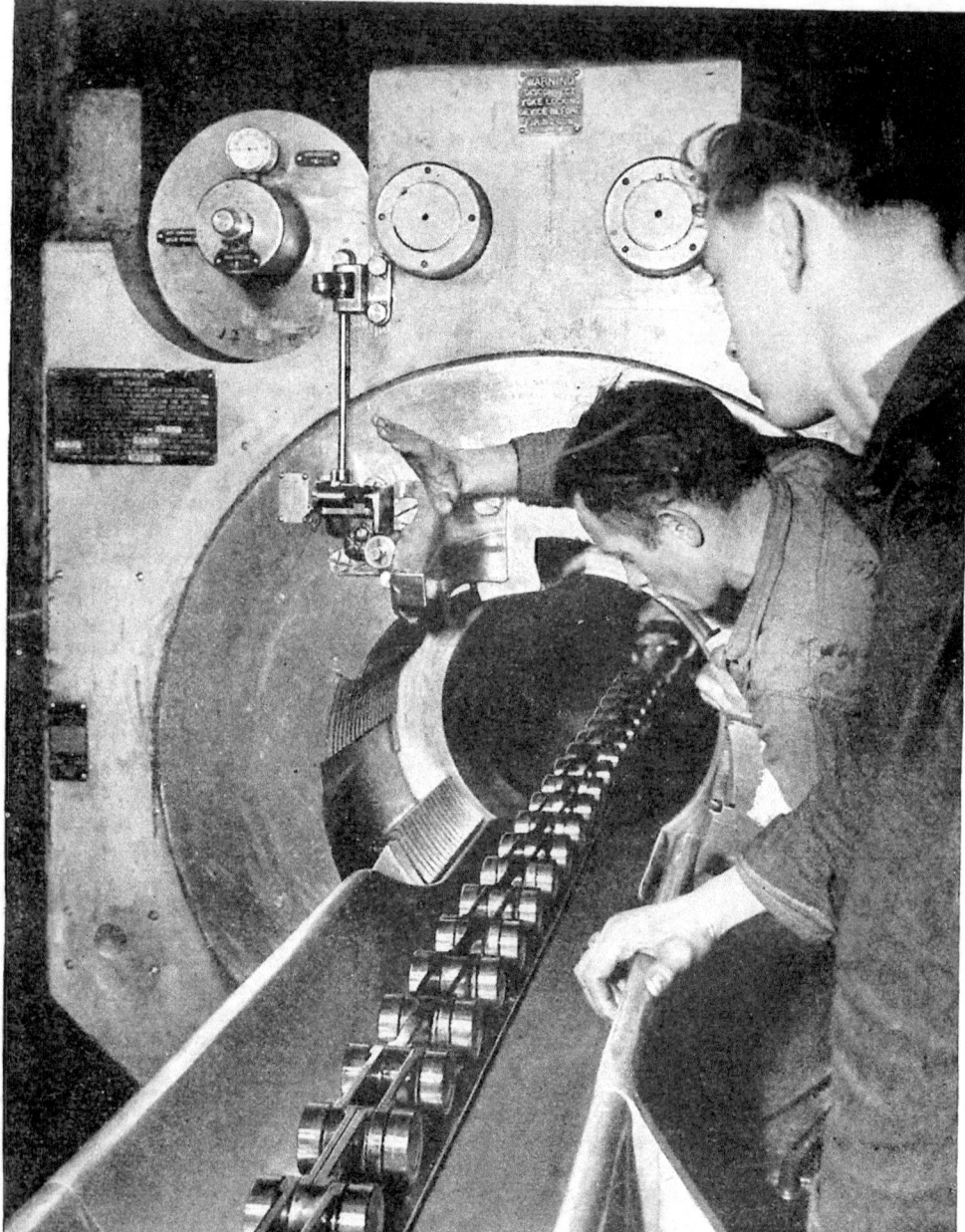

FIGURE 5E3. Cradle and rammer chain.

the ramming head is slightly over 3 inches behind the projectile base and 8½ feet from the breech face. The projectile travel to full seat during ramming is about 18 feet.

5E5. Powder hoist. The powder magazines are outside the turret foundation on the platform deck beneath the lower projectile flat. The powder is passed manually through self-closing scuttles in the powder-handling room bulkheads and the lower foundation, into the hoist-loading space around the lower end of the turret revolving structure, and delivered by hoists to the guns.

CHAPTER 5

FIGURE 5E4. Projectile about to be seated.

An electric-hydraulic powered lift is provided for each gun. Each hoist contains a powder car which rides up and down like an elevator in a closed, inclined trunk which extends through the turret structure from the hoist-loading space to the turret roof. The car is a large, box-shaped structure closed on three sides. Four guide wheels, mounted on the rear corners, ride along the after side of the trunk. The bags are carried on two vertically-spaced trays which can be separately tilted by means of hand levers on one side of the car. A loaded car delivers a full service charge of six bags to the gun without intermediate handling.

The car is loaded in the hoist-loading space beneath the projectile flats through an aperture in the trunk. This opening is sealed by a manually operated vertically-sliding door which must be closed before the car can be hoisted or lowered. It cannot be opened unless the car is in its loading position.

Powder is delivered to the gun through a door located just abaft and to one side of the breech mechanism. The door swings outward and down, providing a shelf across which the powder bags are rolled from the car to the loading tray. The door is operated by a hydraulic power device or an auxiliary hand drive. When the delivery door is open, the car can be lowered, but not raised. The interlocking features make it impossible to have both the loading and delivery doors open at the same time.

The hoist power unit consists of an electric motor (75 h.p., with a 135 h.p. overload rating) and a conventional hydraulic speed gear connected to a hoisting drum. This unit is located in the gun compartment next to the trunk. The car is lowered by gravity and hoisted by means of a $5/8$-inch diameter wire rope which passes over a sheave under the turret top plate and thence to the hoisting drum.

The hoist control is located at the hoist-operator's station near the top of and abaft the trunk. Control arrangements include devices which limit acceleration and deceleration rates, door-operated interlocks, and a power-failure brake on the hoisting drum. The latter prevents the car from falling in case of power failure or rupture of hydraulic leads.

F. Safety precautions, selected from Article 972, Navy Regulations.

"42. When either cartridges or bag charges are outside the magazines, wherever practicable, each flame-proof compartment or space which forms a stage of the ammunition train, including the magazines and gun compartments (in or out of turrets), shall be kept closed from all other compartments or spaces, except when the actual passage of ammunition requires it to be open. Where practicable, no flame-proof stage of the ammunition train shall be open to both the preceding and the following stages at the same time."

"43. (a) In a magazine or handling room in which powder is removed from tanks to be sent to the guns in bags, not more than one charge per gun, for the guns being served by that magazine or handling room, shall be exposed by removal from tanks, by removal of tank tops, or by so loosening the tank tops that the bags may be exposed to flame.

(b) In each subsequent flame-proof stage of the ammunition train, not more than one charge per gun, for the guns being supplied through that stage, shall be allowed to accumulate. For this purpose, the spaces or handling rooms at the tops and bottoms of continuous-chain powder hoists will be considered separate stages (whether or not separated from the hoists by flame-proof doors, flaps, or shutters).

(c) In addition to the above, continuous-chain powder hoists may be kept filled; or if hand passing is used, there may be one bag of powder at the station of each man in the train.

(d) It is the intent of this article to permit sufficient powder to be exposed to provide an adequate supply for the guns being served. The maximum amount specified above should be exposed only if a smaller amount will not assure an adequate supply."

"44. If flame seals be damaged during firing, except in action, so that they cannot fulfill their purpose, the gun or guns concerned shall cease firing until the flame seals are again effective."

"46. As there is an inflammable gas present in the chamber of a gun after firing which, under certain conditions, may constitute a danger by igniting the powder charge which is to be used for the next round, and as smouldering remnants of powder bag may also be present, the following precautions shall be observed:

(1) Bag guns shall not be loaded until a member of the crew has assured himself that the bore is clear of powder gases and remnants and has announced 'bore clear' either by

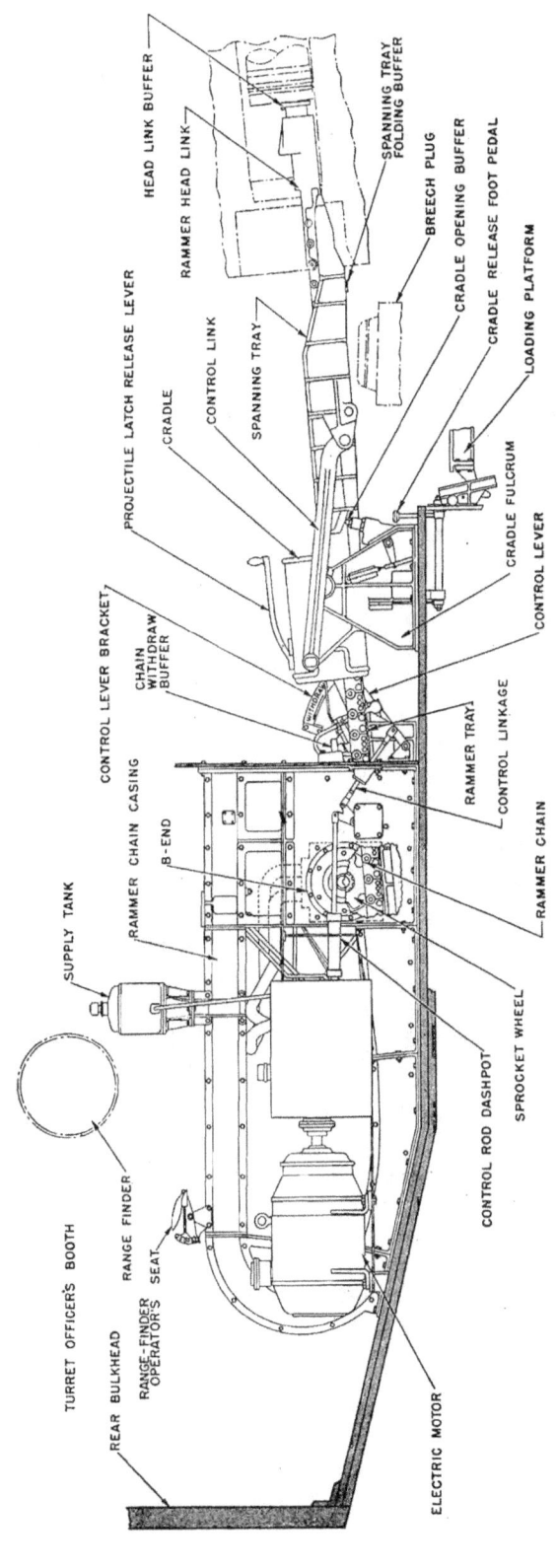

FIGURE 5E5. Projectile-loading mechanism.

voice or by approved signal, such as a whistle, gong, or horn, except that, when the gas-ejector system does not readily clear the bore, the combined sponge and rammer (where provided) may be used. It shall be dipped in water for each load.

(2) Until the 'bore clear' signal above described is given, or the projectile is rammed home with the wet combined sponge and rammer, powder shall not be exposed closer than 4 feet to a gun not mounted in a turret.

(3) In turrets fitted with ammunition cars the car and the center of an open breech shall not be allowed within 6 feet of one another until the 'bore clear' signal has been given. In turrets fitted with continuous-chain powder hoists or for hand passing, the powder shall not be exposed in the turret chamber, nor shall the flame seal, shutter, or flap between the turret chamber and the next stage in the powder train be opened or unlocked until the 'bore clear' signal is given."

"47. In turrets not fitted with bulkheads between guns, the 'bore clear' signal to the turret crew shall not be given until the guns which have been fired and whose breech plugs have been opened are reported clear, when one signal to the entire turret crew shall be given".

"50. The ramming of projectiles in bag guns by interposing one or more sections of a powder charge between the head of the rammer and base of the projectile is prohibited."

"53. As soon as a gun is loaded the breech shall be closed without delay."

"54. The breech plug of a gun shall never be unlocked or opened while there is a live primer in the lock."

"55. A firing lock into which a live primer has been inserted shall never be opened, either independently or by operation of the breech mechanism, unless the firing circuit is broken externally of the lock or breech mechanism (for example, at local-pointer's key or gun-captain's ready switch), except when it is known that the loaded gun has fired. This applies to the firing of primers at drill, to the operation of loaded guns, and the examination of primers referred to in paragraph 67." (See Article 4B23.)

"59. When priming a lock of the sliding-wedge type, care shall be taken to insure the primer being pushed in beyond the primer catch to prevent the primer coming out or being crushed by the operation of the wedge in closing."

"61. The mushroom of every bag gun shall be wiped after each shot with a sponge or cloth dampened with fresh water."

"64. To guard against blowing out primers which may fire at the instant of closure, care shall be taken whenever the breech of a bag gun with a live primer in the lock is being closed, that the operating lever is followed through during the last part of its travel, to prevent any opening of the lock due to rebound."

"65. In loading a gun, neither the gun ready-light switch nor the gun firing circuit cut-out switch (which are combined in some installations) shall be in the closed position until the breech is fully closed and all personnel are clear of the recoil."

"74. If a powder bag is broken to the extent of allowing powder to fall out, the command 'silence' shall be given and the loose powder shall be gathered up. If it is impracticable to utilize this section of the charge satisfactorily in loading, it shall be secured in a flame-proof container or immersed in water."

"77. It shall be the duty of one man of each gun crew of a turret to insure that the loading or spanning tray is properly seated before a projectile is rammed."

"78. A trunked-in ammunition car, with automatic flame seal, shall be made inoperative by opening its switch or locking its control lever in 'off' position before anyone shall lean or reach into the car."

"90. All immersion tanks shall be filled to proper level before firing. Tanks into which turret drains lead shall be filled with water to cover the ends of the drain pipes before firing."

CHAPTER 5

"89. Turret and handling-room sprinkling systems shall be tested and all tanks of these systems filled before firing."

"93. Before firing primers, the division officer will see that gun tompions are removed and mushroom vents clear. In preparing the battery for firing he shall, in addition, see that the gas-ejector system, and the turret blower system are working satisfactorily and that the bore of the gun is in satisfactory condition."

CHAPTER 6

SEMI-AUTOMATIC GUNS

A. INTRODUCTION

6A1. General. Semi-automatic guns are those in which some of the energy of counterrecoil is used to open the breech, eject the empty case, close the breech, and cock the firing mechanism. Loading, ramming, and firing are accomplished by other means. The cycle of operation is as follows: (1) the gun recoils after firing and energy is stored in the counterrecoil system, (2) during counterrecoil the breechblock is automatically lowered and the empty case is extracted, (3) after the next case is rammed into the chamber, the block is automatically raised to close the breech, and (4) the firing mechanism is cocked, either when the block is lowered or when it is raised, depending on the design of the gun, and the gun is then ready for another operating cycle. The firing rate of semi-automatic guns depends largely upon the time required to load.

All semi-automatic guns use fixed or semi-fixed ammunition depending upon their size. The sliding-wedge breechblock back of the cartridge case prevents the case from being blown to the rear when the gun is fired. A firing mechanism is fitted in the breechblock to fire the primer by percussion, and also electrically in most guns.

Semi-automatic guns are used extensively by the U. S. Navy on all classes of ships from submarine chasers to battleships. Typical examples of this type of gun are the 3"/50 cal. and 5"/38 cal. Although these guns differ in mechanical details they both use the vertical sliding-wedge semi-automatic breech mechanism. The operation of this breech mechanism, as outlined in the next section, should be thoroughly understood before the individual guns are studied.

B. OPERATION OF THE BREECH MECHANISM

Figure 6C2 shows the breech mechanism assembled with the component parts in their correct relative positions. The operating principle of the mechanism will be more easily understood if this figure is used along with the sketches in the following section.

6B1. Breech closure. Figure 6B1 shows the gun barrel secured to the forward part of the *housing*. The *breechblock*, or sliding wedge, fits into a rectangular hole (vertical well) cut through the housing from top to bottom. The *ribs* on the wedge mate with grooves in the vertical well and serve to guide the block as it moves up and down to close and open the breech. The force exerted on the block when the gun is fired is transmitted to the housing through these ribs. The upper portion of the forward face of the block is beveled, and the guides slope slightly forward to wedge the cartridge case home as the breech is closed.

6B2. Operating shaft. The vertical motion of the breechblock is controlled by an *operating shaft* which is supported in bearings that are secured to the under side of the housing. The *breechblock crank* rides in a sloping T-slot in the under side of the breechblock. Figure 6B1 shows the relative position of breechblock and crank when the breech is closed. It is apparent that rotation of the operating shaft in the direction indicated by the arrow forces the block downward. Operating-shaft rotation in the opposite direction, back to the original position, raises the block. The operating shaft is locked by a positive type salvo latch when the breech is closed.

6B3. Cam plate. The breechblock is automatically lowered when the *crank arm* strikes a *cam plate* pivoted on the slide. The action of this cam during recoil and counterrecoil is shown schematically in figure 6B2. The numbers alongside the operating shaft indicate successive positions of the shaft and cranks during the operating cycle. The cam plate has a vertical pivot at its for-

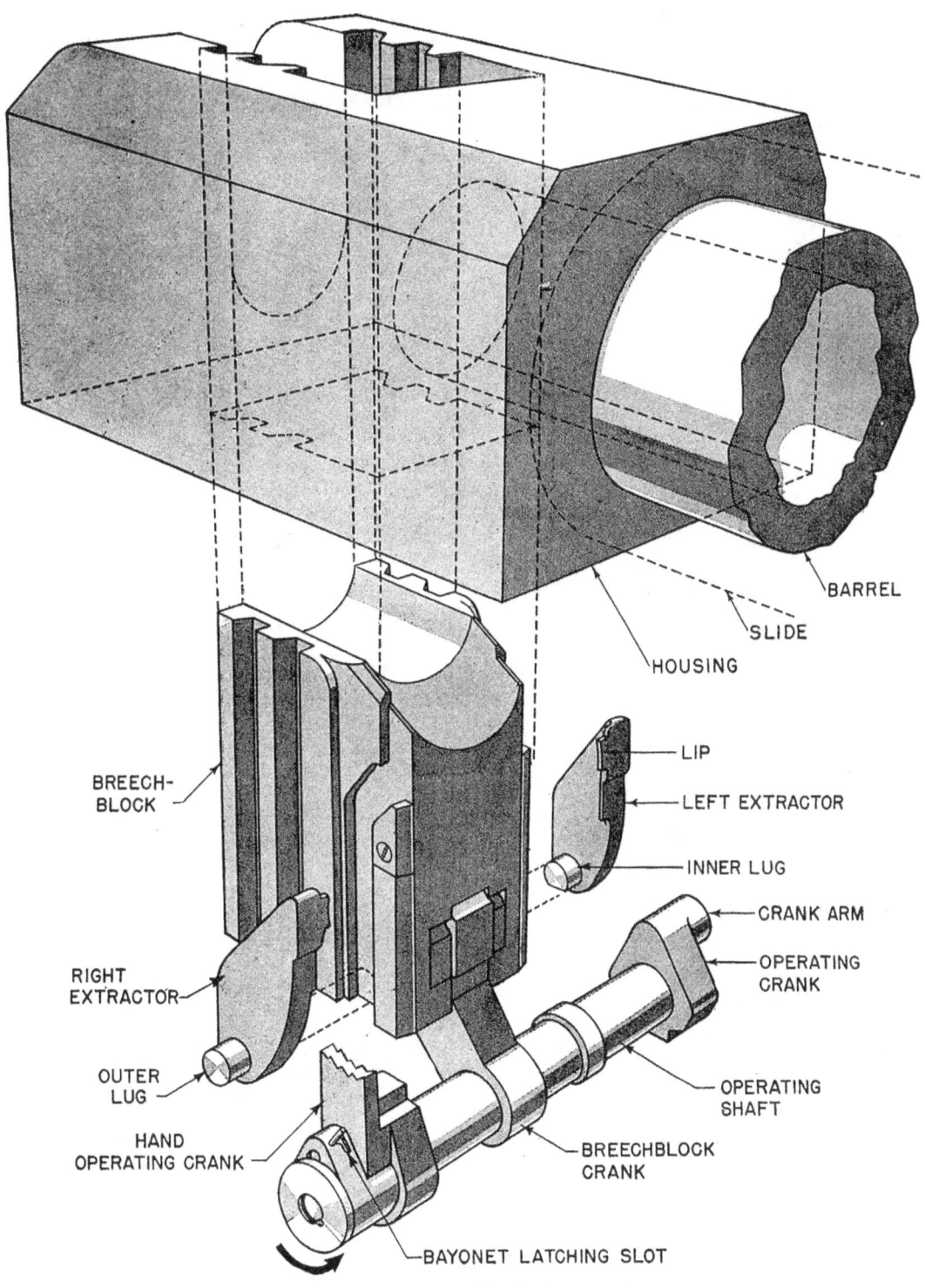

Figure 6B1. Breech closure.

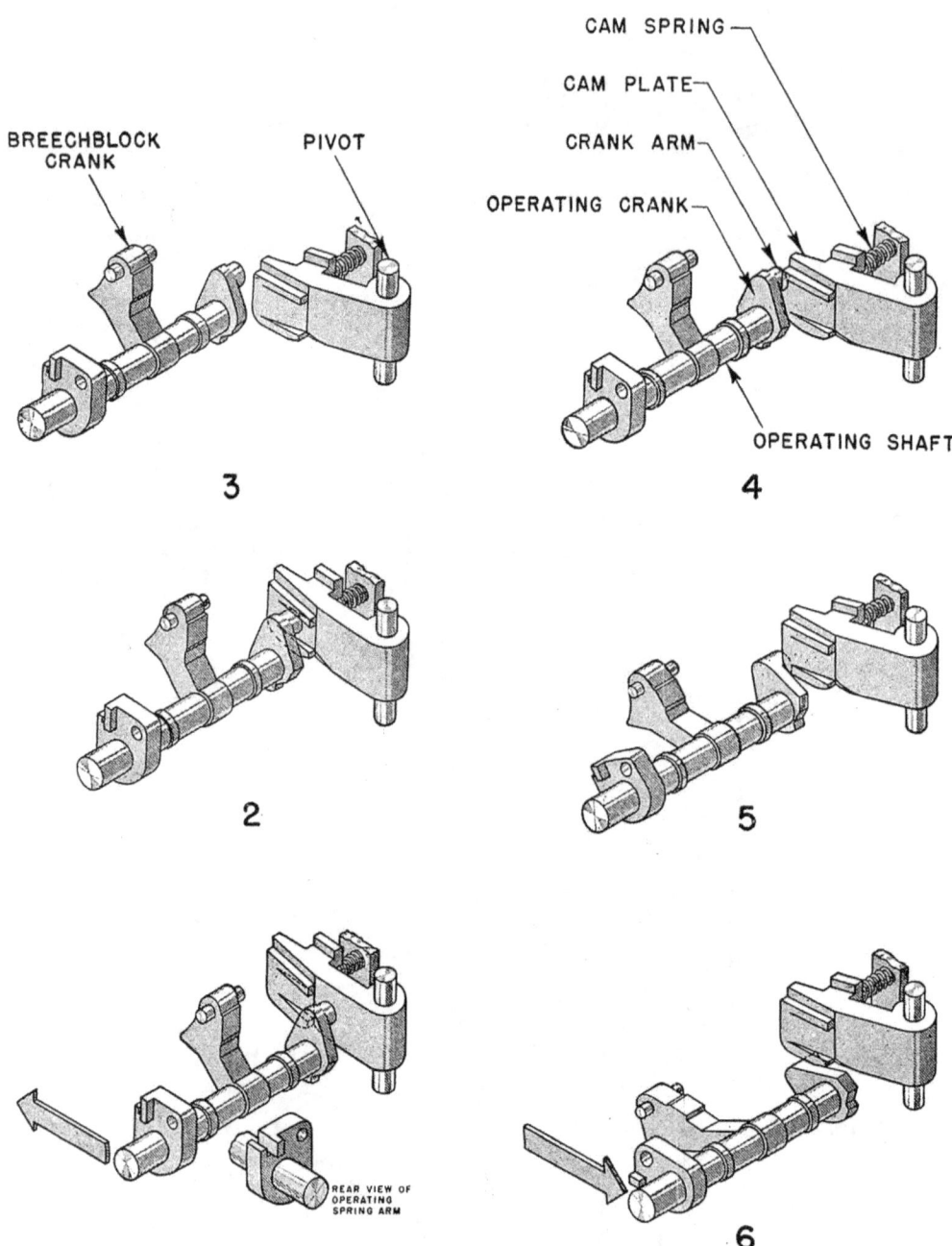

Figure 6B2. Operation of the cam.

ward end. During recoil the crank arm pushes the cam plate outward toward the slide about this pivot. This movement compresses a spring placed between the cam plate and the slide (fig. 6B2—1 and 2). The operating shaft does not rotate at this time. At the instant the crank arm moves abaft the cam plate, the spring snaps it back into position (fig. 6B2—3). During counterrecoil, the crank arm is forced to move down and pass under the curved rear surface of the cam plate (fig. 6B2—4, 5, and 6). Motion of the operating crank, controlled by the cam, rotates the operating shaft and lowers the breechblock. The gun can be operated in single fire (non-automatic) by holding the cam plate in a retracted position, in which case the breechblock must be lowered by hand.

6B4. Operating spring. The force necessary to raise the breechblock is provided by a compressed spring called the *operating spring*. This spring is placed in a cylinder mounted on the housing. The spring is so connected to the operating shaft that it is compressed as the block is lowered. This is shown in figure 6B3. The spring automatically raises the block when it is free to move. The block is held down after the crank arm passes the cam plate in counterrecoil, to prevent the breech from closing when the gun returns to battery. The method used in doing this is discussed in Article 6B6.

6B5. Extractors. Extraction of the fired case is a mechanical operation which is accomplished when the breechblock is dropped. The two separate *extractors*, shown in figure 6B4, are rocker arms pivoted at their bases about *outer lugs*. The outer lugs are placed in slots in the housing on each side of the breechblock. In these slots, the lugs are free to move in the fore-and-aft direction, but not up and down. The extractors have *inner lugs* near their bases, which fit into slots machined in each side of the breechblock. As the block drops, the inner lugs must ride in their slots to the position shown in figure 6B5. The curved portion of the slots has forced the inner lugs forward. Since the extractors are rocker arms, rocking on the breech face, the upper portions of both extractors are forced aft when the inner lugs are forced forward. The *lips* on the extractors engage the rim of the cartridge case. The quick motion of the extractor bases forward causes an accelerated flip of the extractor lips to the rear, thus extracting the case. Spring-loaded *extractor plungers* are mounted horizontally in the sides of the housing. These plungers, also shown in figure 6C2, push forward against the outer extractor lugs, and assist in extraction. The primary function of these extractor plungers is discussed in the next article.

6B6. Pallets. The breechblock is held down against the action of the operating spring by the inner lugs bearing on *pallets*. This is shown in figure 6B5. When the block is in its lowest position the extractor inner lugs are in the curved portion of their slots and are held there by the extractor plungers which are pushing forward on the outer lugs. The lower surfaces of the lugs and the upper surfaces of the pallets are flattened to provide good bearing surfaces.

The breech cannot be closed as long as the lugs bear on the pallets. However, when the gun is loaded the rim of the case strikes the extractor lips and forces them forward. The extractor rocker action moves the inner lugs aft against the action of the extractor plunger springs, until they clear the pallets and are in the vertical portion of the slots. The operating spring can then force the breechblock upward to close the breech.

6B7. Hand operating lever. A *hand operating lever* is provided so that the breech can be opened and closed by hand. This lever is connected to the operating shaft through a mechanical *clutch*. In semi-automatic operation, the clutch is disengaged and the operating lever is latched in an upright position alongside the housing. The clutch is so designed that even though it is disengaged, it is still possible to open, but not close, the breech. With the operating lever properly latched to the housing and the clutch disengaged, the operating shaft can rotate without moving the lever. When the clutch is engaged, the lever and operating shaft work as an integral unit. As a safety precaution, the clutch should never be engaged when the gun is set for semi-automatic operation. It is obvious that if the operating handle is attached to the shaft, it will fly back and forth as the gun operates, and thus endanger the gun crew.

6B8. Operating cycle. Starting with the gun unloaded and the breech closed, the first

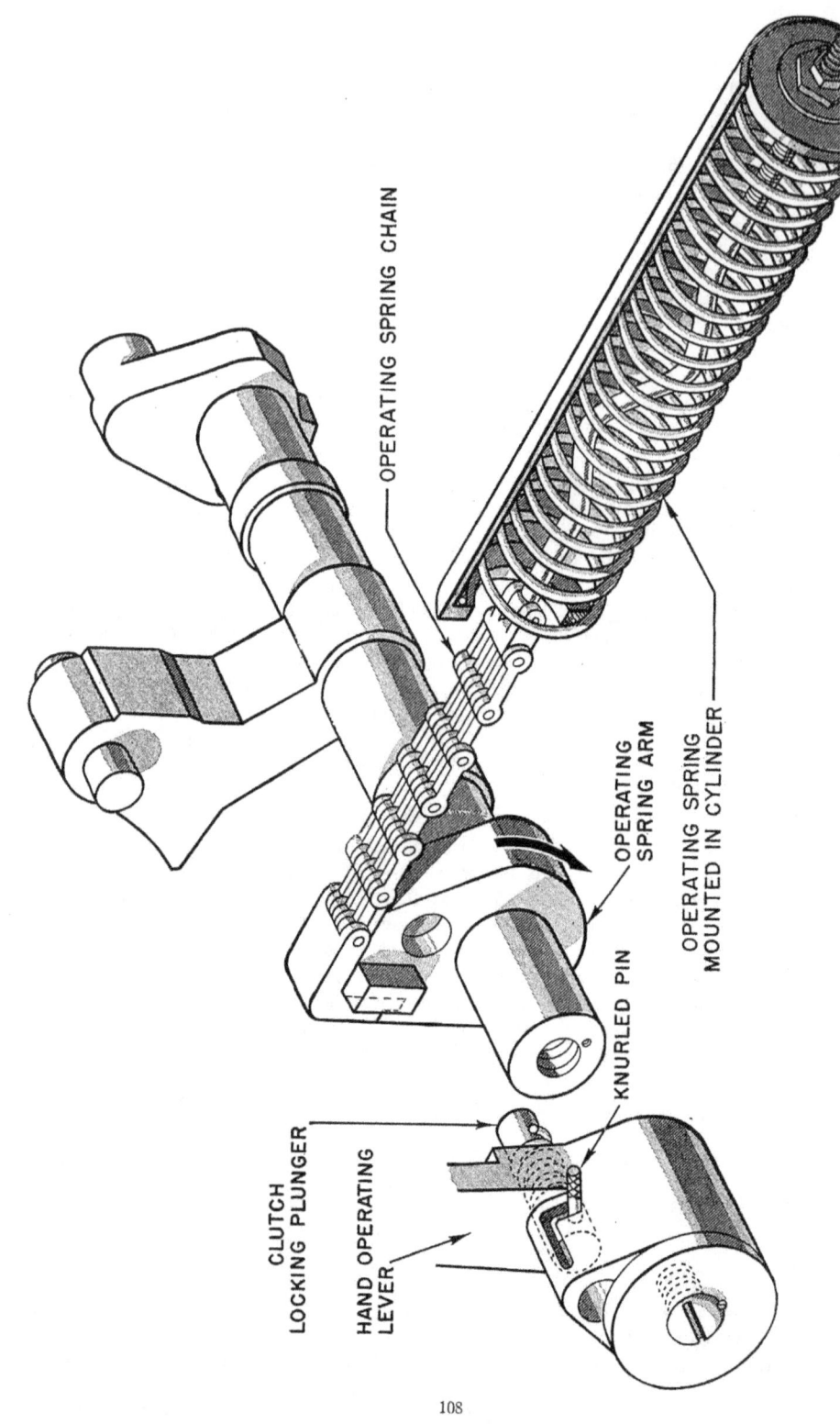

Figure 6B3. Operating spring.

step in the operation of the gun is to trip the positive-type salvo latch by hand. Next the breechblock is lowered by using the hand operating lever. The block is held in the open position by the inner extractor lugs bearing on the pallets. The operating spring is compressed when the breech is opened. The hand operating lever is then returned to its original position, and the gun is ready for loading. As the cartridge case is rammed into the chamber, the rim on the case forces the extractor lips forward, and moves the inner extractor lugs off the pallets. The operating spring then

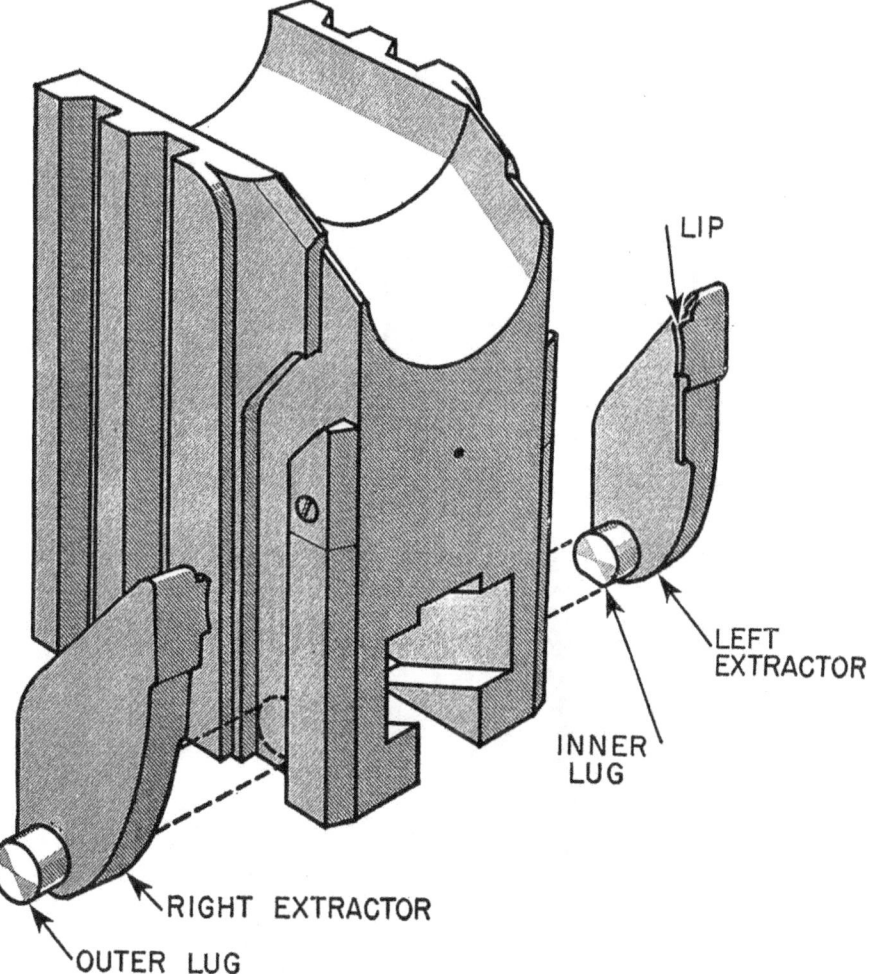

FIGURE 6B4. Extractors.

turns the operating shaft to close the breech. When the breech is fully closed, the salvo latch again engages a lug on the shaft. On some guns the firing mechanism is cocked when the block is lowered, on others, when it is raised. The gun can now be fired. After firing, the gun recoils, the salvo latch is tripped, and the cam plate is pushed aside by the crank arm. On counterrecoil the crank arm, guided by the cam, rotates the operating shaft. This rotation lowers the breechblock, and compresses the operating spring. When the block is lowered, the extractor lugs are forced forward, thus moving the extractor lips aft to extract the empty case. The block is held down by the inner extractor lugs bearing on the pallets. The cycle is now complete and operation may be continued by loading another round.

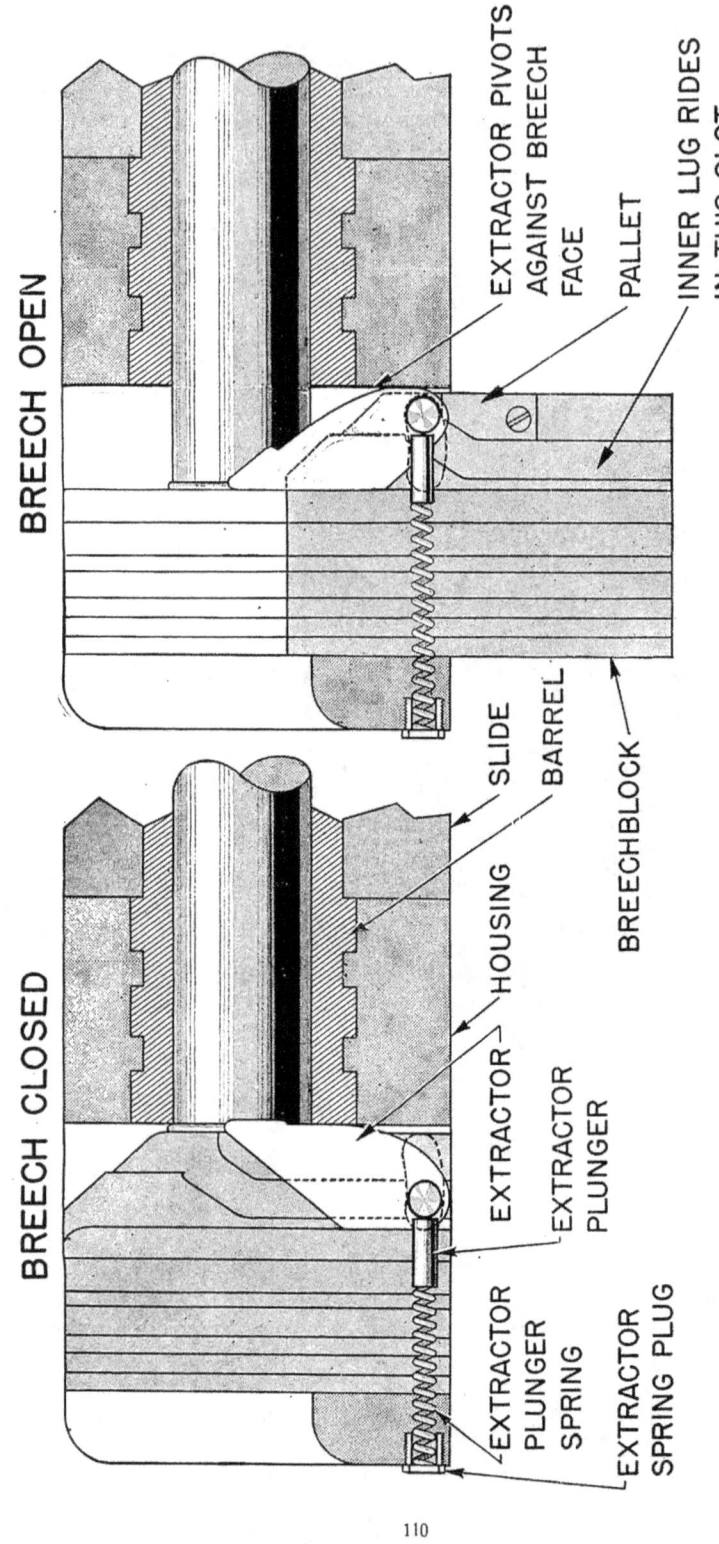

Figure 6B5. Extraction.

FIGURE 6C1. Personnel arrangement; 3"/50 cal. gun.

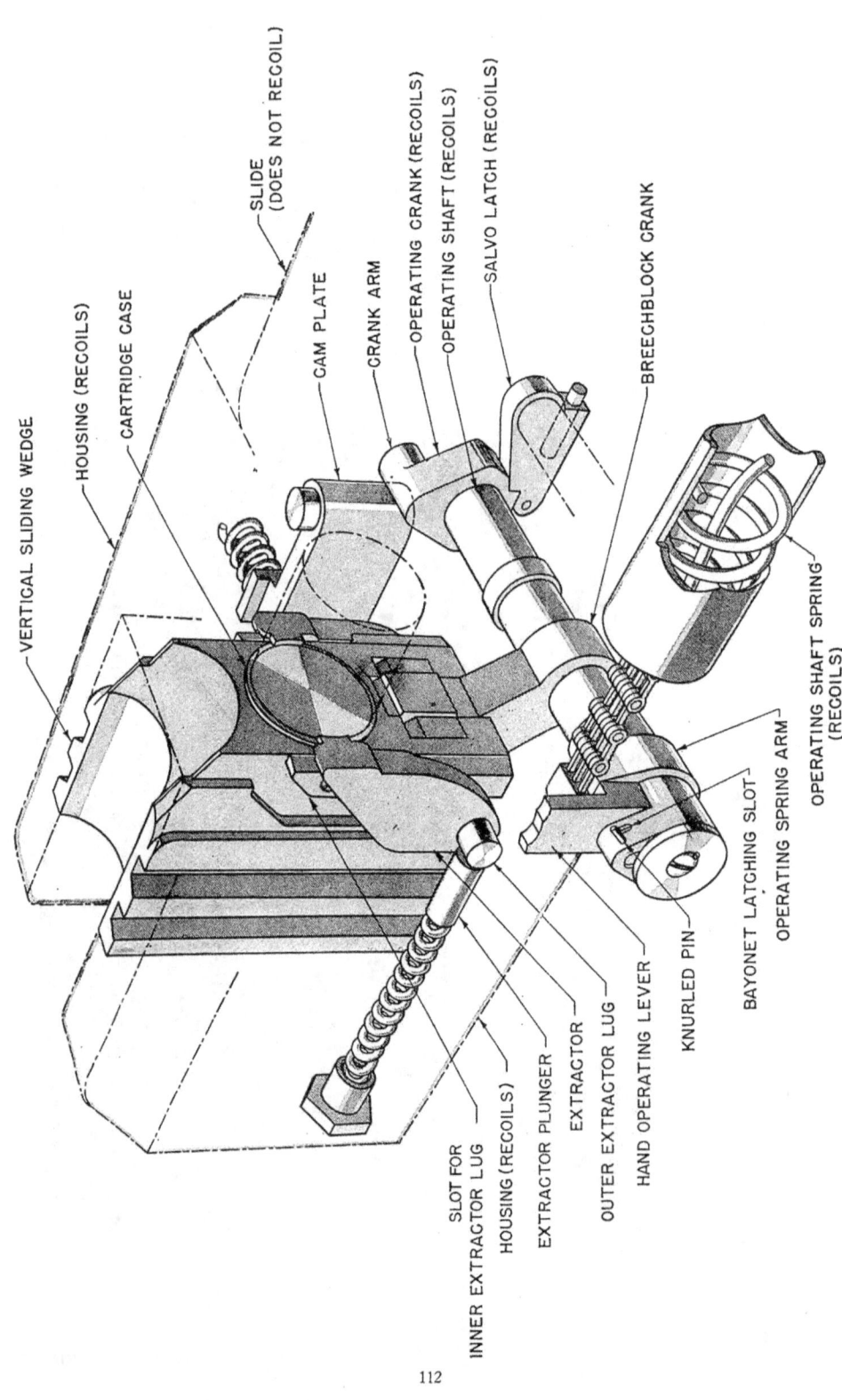

Figure 6C2. Breech mechanism of the 3"/50 cal. gun.

CHAPTER 6

C. 3″/50 CALIBER GUN AND MOUNTS

6C1. General. Figures 6C5 and 6C6 show a 3″/50 cal. gun on an open, pedestal-type, dual-purpose mount. This assembly is used for deck emplacement on many of the smaller ships including the new destroyer escorts, auxiliaries, freighters, and transports. A wet-type, open mount is used on a number of submarines.

The gun is positioned by manually operated training and elevating gear. The elevation limits

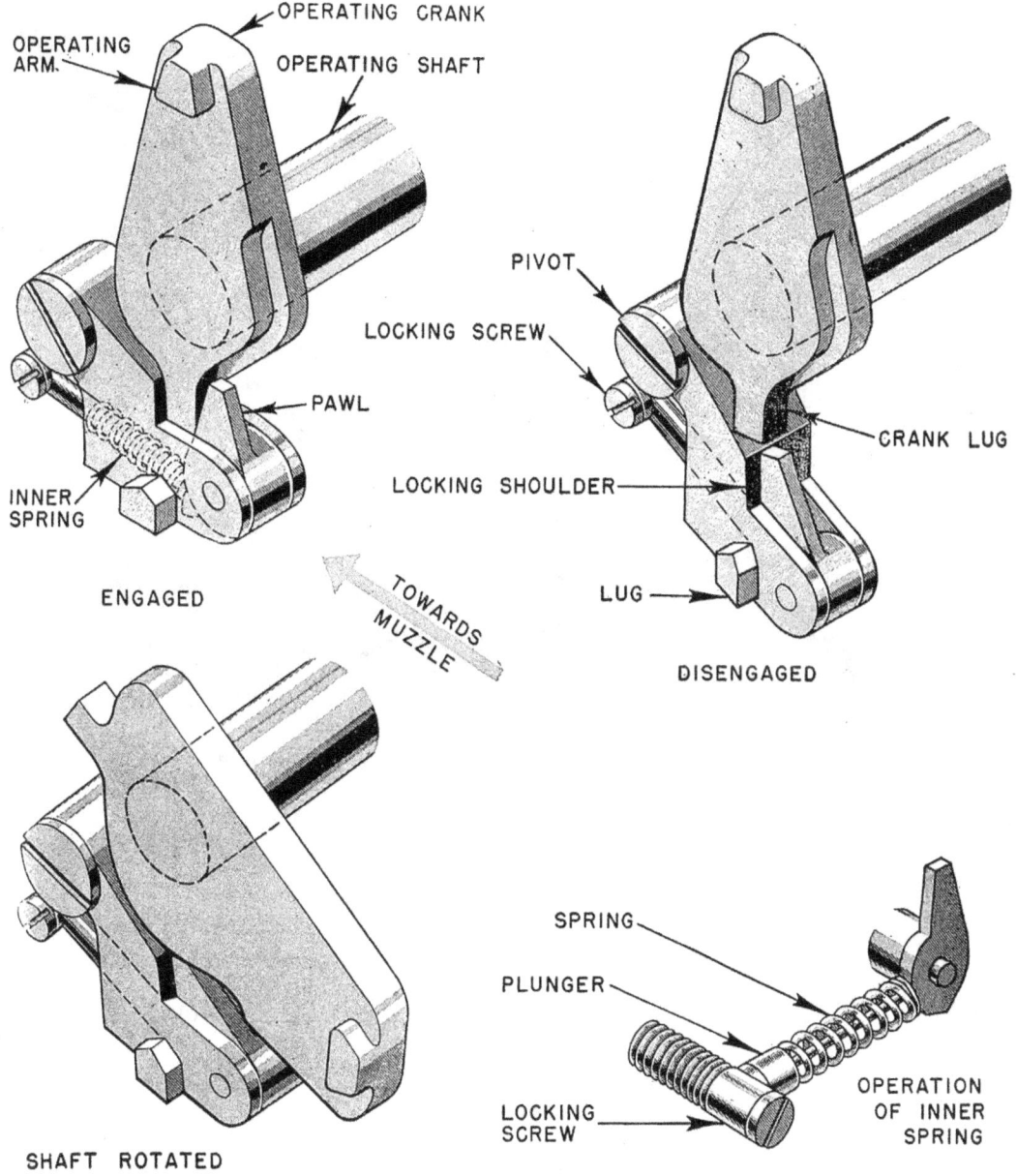

FIGURE 6C3. Operation of the salvo latch; 3″/50 cal. gun.

RESTRICTED

SEMI-AUTOMATIC GUNS

on the gun are approximately 13° depression and 85° elevation. Training stops are installed to prevent training the line of fire into the ship's superstructure. After the sights are set, the gun is positioned by the pointer and trainer bringing their telescopes on the target. The ammunition is hand loaded and hand rammed. New developments probably will include a power rammer to increase the firing rate. A number of installations use director control for more accurate firing.

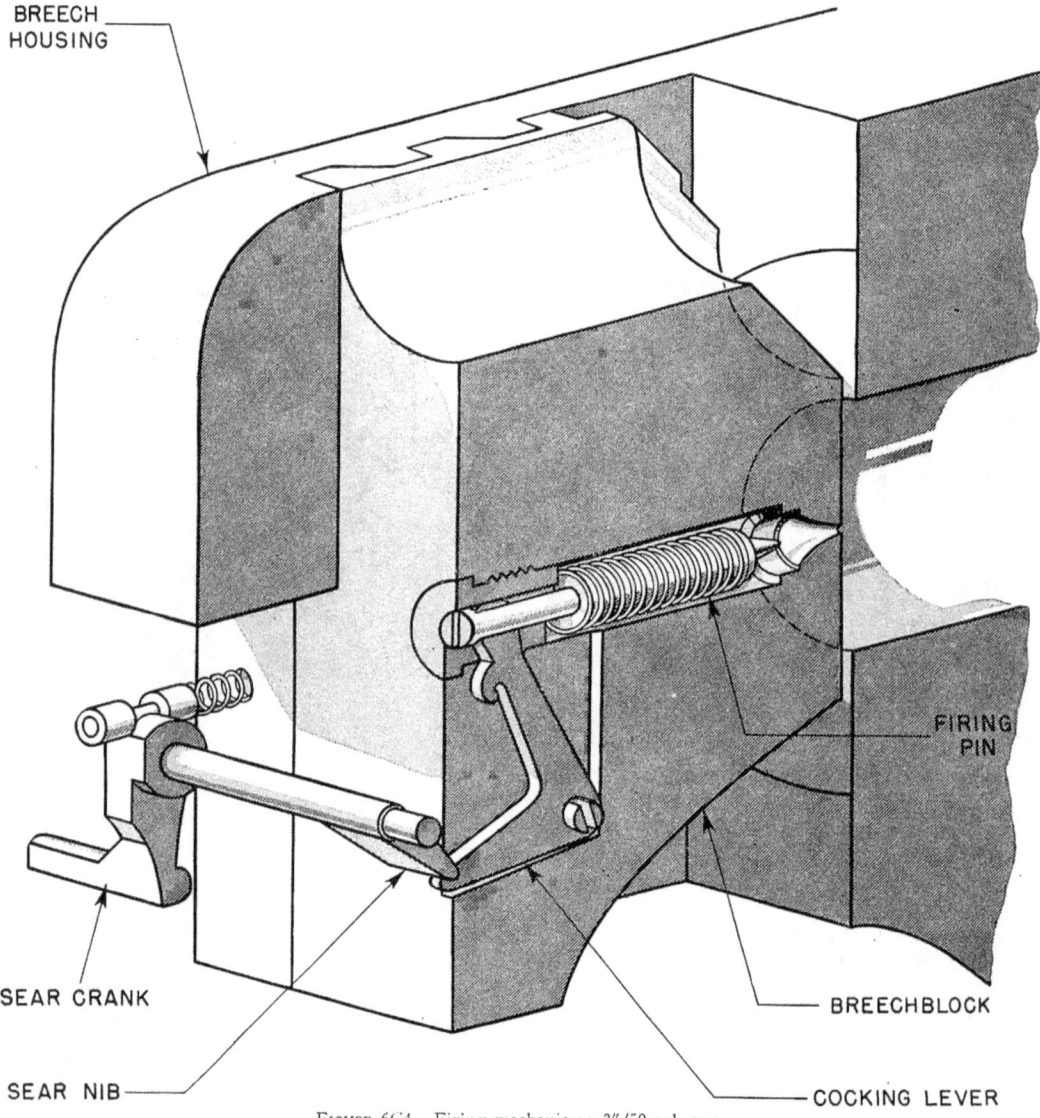

FIGURE 6C4. Firing mechanism; 3"/50 cal. gun.

The ammunition used in the 3"/50 cal. gun is the fixed type designed for an initial velocity of 2,700 f.s. The maximum horizontal range is approximately 12,000 yards, and the maximum ceiling range about 21,000 feet. The projectiles used in the 3"/50 cal. gun include the following types: (1) antiaircraft, (2) common, (3) armor piercing, and (4) illuminating. These projectiles all weigh approximately 13 pounds, and the different types vary in length from about 9 to 13 inches.

CHAPTER 6

Figure 6C1 shows the arrangement of personnel for a typical 3"/50 cal. gun crew. In some cases the gun captain also serves as plugman, and some crews have a sight checker for drills.

6C2. Guns. The three guns in general use are:

1. Mark 20, a built-up gun with barrel and housing an integral unit. It has rifling which increases in twist from zero to one turn in 25 calibers.

2. Mark 21, a radially-expanded gun with a separate housing to which the gun is secured

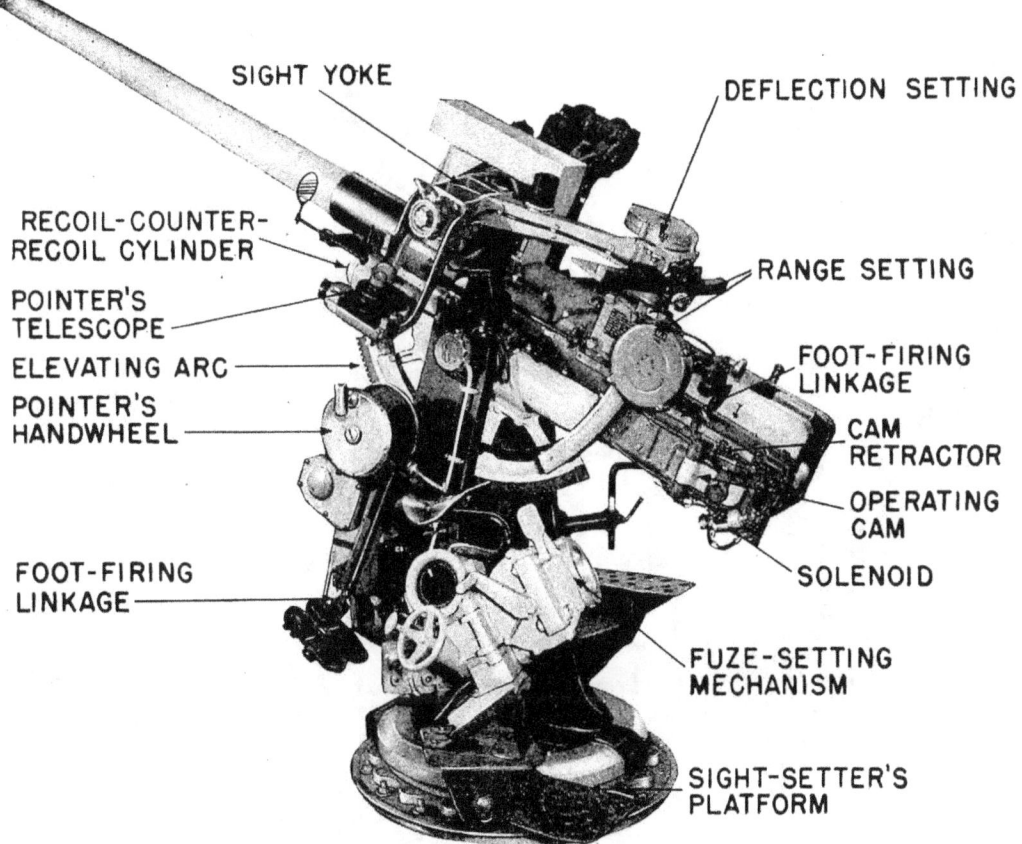

FIGURE 6C5. The 3"/50 cal. gun and mount (Mark 21); left-side view.

by means of a bayonet-type joint. The bore is chromium plated and the rifling has a uniform twist of one turn in 32 calibers.

3. Mark 21, Mod. 1, is of the same construction as the Mark 21 except for additional chromium plating in the chamber and on some exterior surfaces. This gun is provided with watertight tompions for breech and muzzle. These features make it suitable for wet-type mounts.

6C3. Breech mechanism. The vertical sliding-wedge semi-automatic breech mechanism used on the 3"/50 cal. gun is shown in figure 6C2. Its principle of operation is discussed in section 6B. The top surface of the breechblock and the aft end of the housing form a U-shaped trough which serves as a loading tray.

The clutch used to connect the hand operating lever to the operating shaft is shown in figures 6B3 and 6C2. A *knurled pin* in a *bayonet latching slot* positions the *plunger*. This plunger is forced into a hole in the operating shaft to lock the lever and shaft together for hand operation.

SEMI-AUTOMATIC GUNS

6C4. Salvo latch. The *salvo latch* shown in figure 6C2 is mounted on a horizontal pin which is secured to the left side of the housing. In figure 6C3 it is seen that the latch is pivoted on this pin and rotates against the action of a restraining *spring*. This spring tends to keep the latch in contact with the operating crank. A *crank lug* bears on a locking shoulder when the latch engages the shaft. This prevents rotation of the shaft and the opening of the breech.

During recoil a lug on the salvo latch passes under an unlatching cam secured to the slide. This cam forces the salvo latch downward until the crank lug clears the locking shoulder. At this

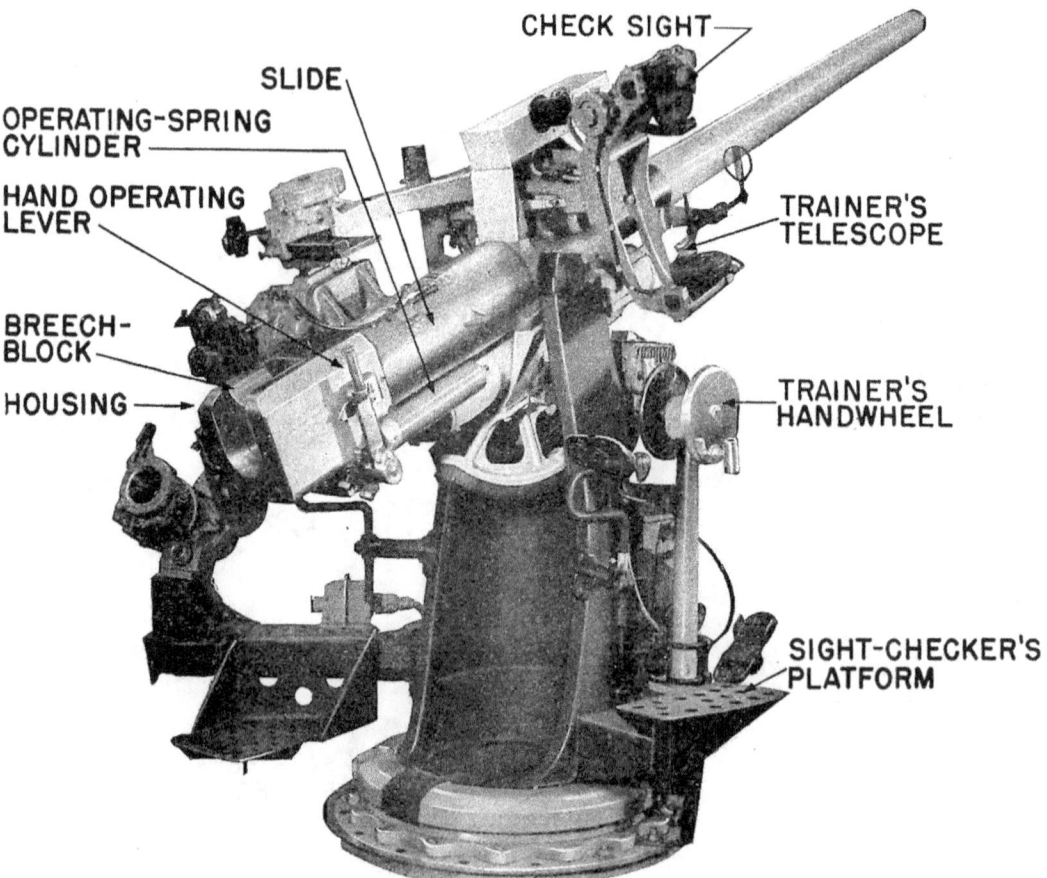

FIGURE 6C6. The 3"/50 cal. gun and mount (Mark 21); right-side view.

instant, a *pawl* mounted on the aft end of the latch snaps forward, and prevents the re-engagement of latch and shaft. The shaft is now free to turn, and the breech will be opened during counterrecoil. When the breech is closed again, rotation of the crank lug forces the pawl back until the lug passes the locking shoulder. At this point the salvo latch again snaps up to engage the lug on the crank.

6C5. Firing mechanism. The *firing pin, cocking lever,* and *firing spring* are mounted in the breechblock as shown in figure 6C4. The cocking lever is placed in a vertical slot below and slightly to the rear of the *firing pin*. The upper end of the vertical arm on the cocking lever fits in a slot through the firing pin. The cocking lever pivots on a transverse shaft in the breechblock. The horizontal arm extends slightly to the rear of the breechblock in a position to contact a sear nib as the block rises.

116

CHAPTER 6

In general, the function of the sear on any gun is to hold the firing mechanism in a cocked position. On the 3"/50 cal. gun, however, the sear is also used to cock the firing mechanism. The *sear* shown in figure 6C4 is mounted in a transverse, horizontal hole near the aft end of the housing, so that the sear crank projects from the left side. As the block rises, the horizontal arm of the cocking lever strikes the *sear nib*. As the block continues to rise, the cocking lever is rotated, the firing pin is drawn aft, and the firing spring is compressed. The sear nib then holds the firing mechanism in a cocked position. When the foot-firing pedal is depressed or the firing key

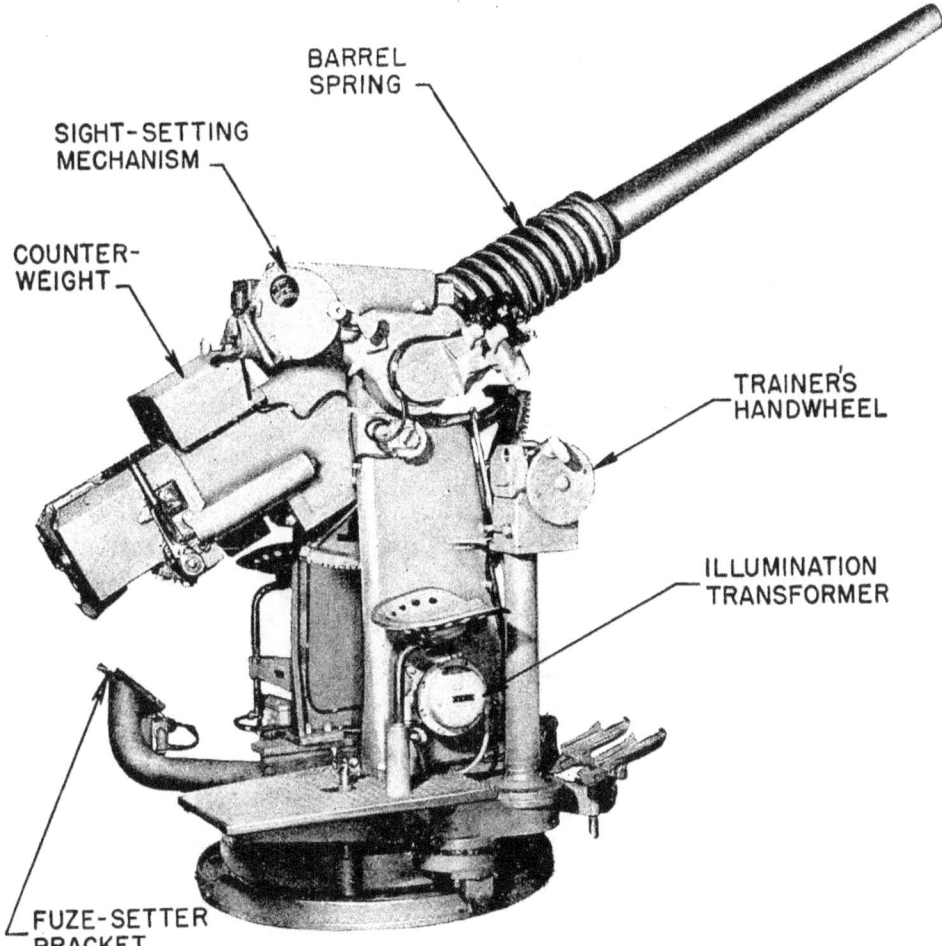

FIGURE 6C7. 3"/50 cal. gun and Mark 24 mount.

is closed, the sear is rotated against the action of a *sear spring*. The sear nib is turned to clear the cocking lever, and the firing pin is driven forward by the firing spring until it strikes the primer in the case.

The primer is fired by percussion only. The power to rotate the sear may come from the foot-firing pedal through a mechanical linkage or from a solenoid controlled by the firing key on the pointer's handwheel. If the foot-firing pedal is held down or the firing key is kept closed, the gun will fire automatically whenever the breech is closed after loading. The firing mechanism is cocked as the breechblock rises but the cocking lever slides off the sear nib and the gun is fired when the block is fully raised.

SEMI-AUTOMATIC GUNS

6C6. Mounts. A typical 3"/50 cal. mount is shown in figures 6C5 and 6C6. The stand is bolted to the deck and supports the roller-bearing paths and the training rack. The carriage rests on ball or roller bearings and another bearing assembly is used to hold the carriage on the stand. These holding-down bearings are designed to take the radial and upward forces which are present when the gun is fired while depressed.

The elevating and training handwheel gear, seats for the pointer and trainer, and platforms for the sight setter and sight checker are attached to the carriage. The trunnions are supported in sleeve-type bearings. The slide assembly provides the cylindrical bearing for the gun, and mounts the sight yoke and recoil-counterrecoil mechanism. The recoil and counterrecoil mechanisms are the combined type and are similar to that illustrated in figure 4C7. Two of these units are used on the built-up gun and only one on the lighter radially-expanded gun. In both cases the cylinders are mounted on the underside of the slide. The piston rod is secured to the gun housing. Hydraulic throttling action is obtained by using three constant-width, variable-depth grooves in the cylinder lining, and counterrecoil action is controlled by springs. The gun recoils about 11 inches. The piston rod and piston are integral, and the dashpot is formed by boring out the forward end of the piston rod. Buffer action takes place during the last few inches of counterrecoil and is regulated by a needle valve.

Figure 6C7 shows a Mark 24 mount which has radical changes in the recoil mechanism and sight gear. This mount was introduced late in 1943 and is being installed on some merchant ships at the present time.

6C7. Sights. Mark 21 mounts have fixed-prism telescopes for the pointer, trainer, and sight checker. These, along with auxiliary peep-and-ring-type open sights, are mounted on a sight yoke supported by the slide. The position of the sights and their associated equipment is shown in figures 6C5 and 6C6. The sight-setting gear is on the left side of the gun abaft the pointer's station and the sight checker is stationed on the right side near the trainer.

D. 5"/38 CALIBER GUN AND MOUNTS

6D1. General. The 5"/38 cal. gun is a semi-automatic, dual-purpose, pedestal or base ring-mounted gun which uses semi-fixed ammunition. The outstanding features found on nearly all 5"/38 cal. gun assemblies are:
1. Vertical sliding-wedge breech mechanism.
2. Hydraulic recoil and hydropneumatic recuperator systems.
3. Power-operated rammer.
4. Power-operated elevating and training gear.
5. Movable-prism telescopes.
6. Power-operated fuze-setting projectile hoist (except on mounts of class 4 listed below).

The operating principle of the 5"/38 cal. gun is the same for all of the installations found on naval vessels. Minor variations in mechanical features represent improvements in design, and the special requirements of different installations, and are largely associated with the mounts. Each of the various assemblies, typified by figures 6D1 to 6D4, belongs in one of the following classes:
1. Enclosed twin mount with its own handling room below the gun platform.
2. Enclosed single mount with handling room as in figure 6D2.
3. Open single mount with handling room.
4. Open single mount without individual handling room.

Twin mounts are carried by the majority of battleships, heavy cruisers, and carriers. Single mounts are found on destroyers, small carriers, light cruisers, auxiliaries, and merchantmen.

6D2. General. The gun and breech assemblies, the slides, and their associated parts, are almost identical in all mounts. The greatest variation in these parts occurs in the twin mount, where one of the breech and slide assemblies is left handed and the other right handed. This difference changes the appearance of the gun assembly but does not modify the mechanical operating principles.

CHAPTER 6

Since the purpose of this section is to describe basic features of the construction and operation of 5"/38 cal. gun assemblies, and because all assemblies are so similar, a minimum number of structural variations are mentioned. In general, the discussion pertains to a *single enclosed mount*. More detailed information on a particular mount can be obtained from the appropriate Ordnance Pamphlet.

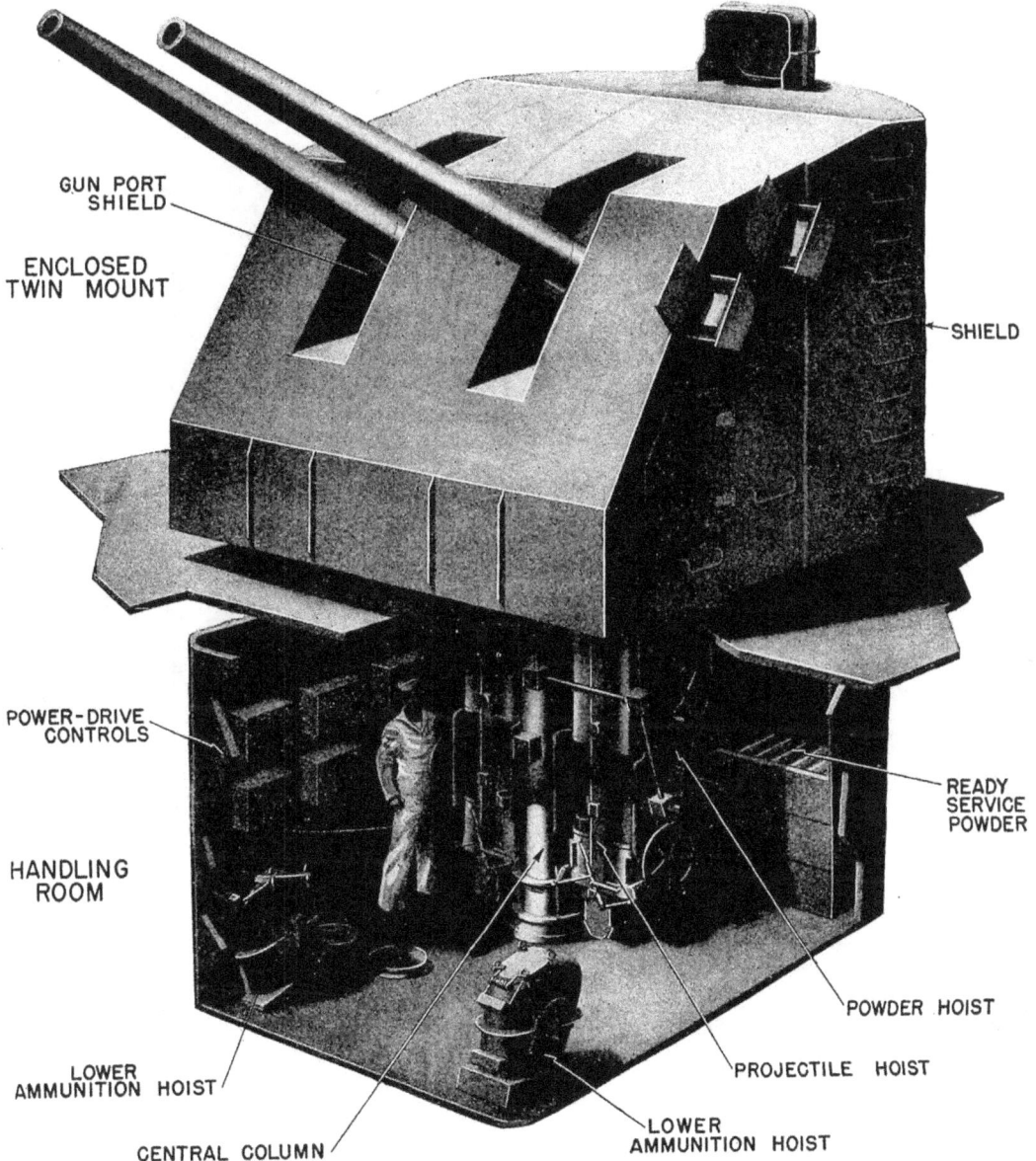

FIGURE 6D1. Enclosed twin mount and handling room; 5"/38 cal. gun.

6D3. Ammunition. The gun uses semi-fixed ammunition consisting of a 54 pound projectile and a case assembly weighing about 28 pounds, which includes a 15 pound powder charge. The projectiles used are: antiaircraft common, common, illuminating, and WP smoke. The bal-

SEMI-AUTOMATIC GUNS

listic performance obtained with a 15 pound service charge is as follows: I.V. 2,600 f.s.; maximum horizontal range, 18,000 yards; maximum vertical range, 37,300 feet. The gun is capable of a sustained firing rate of 15 rounds per minute, but a short period rate of 22 rounds per minute has been reported.

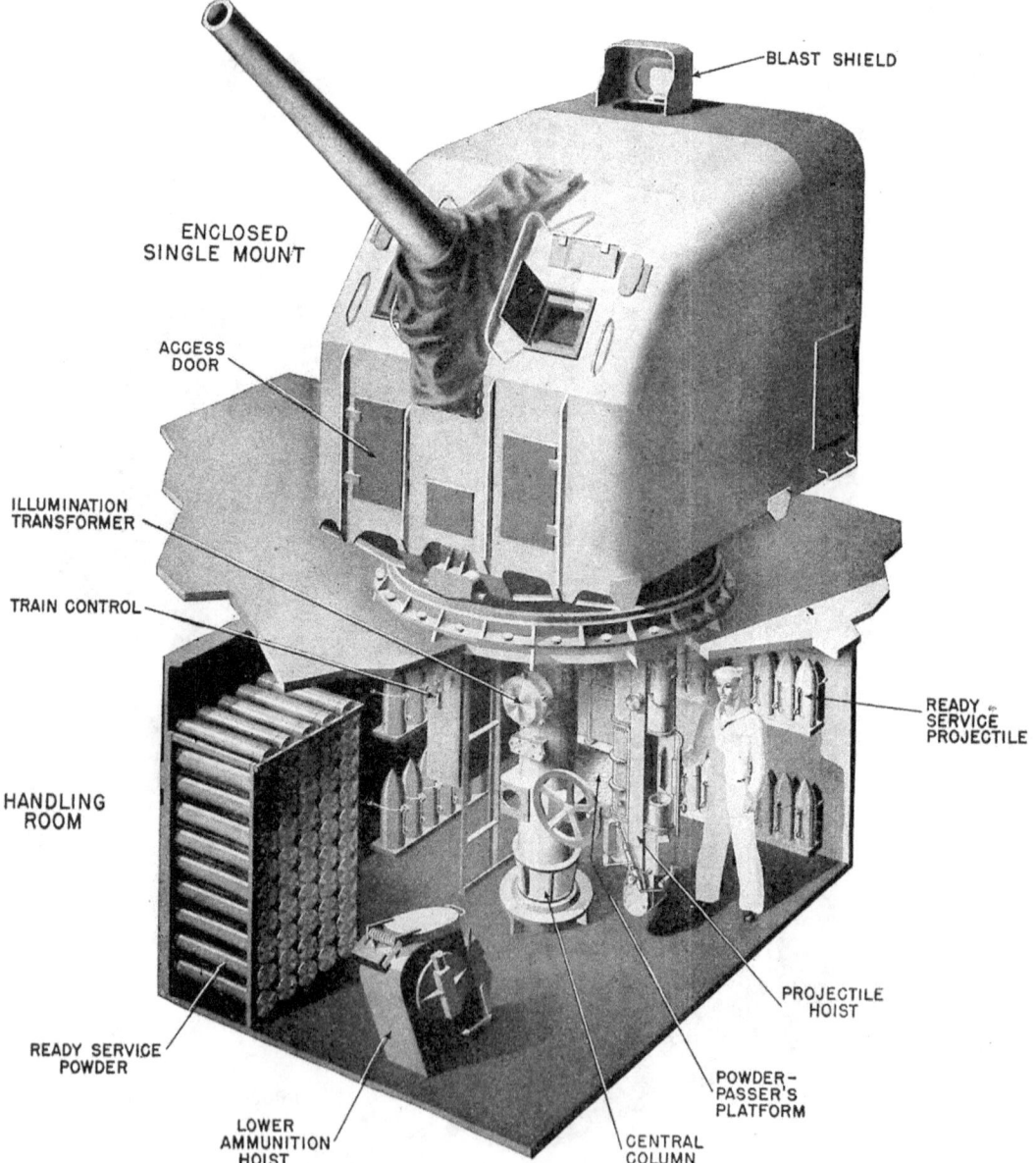

FIGURE 6D2. Enclosed single mount and handling room; 5"/38 cal. gun.

The ammunition is manually served to the loading tray in all assemblies by a projectile man and a powder man. On mounts with handling rooms below the gun platform, the projectiles are delivered to the projectile man by a fuze-setting projectile hoist, and the cases are delivered to the

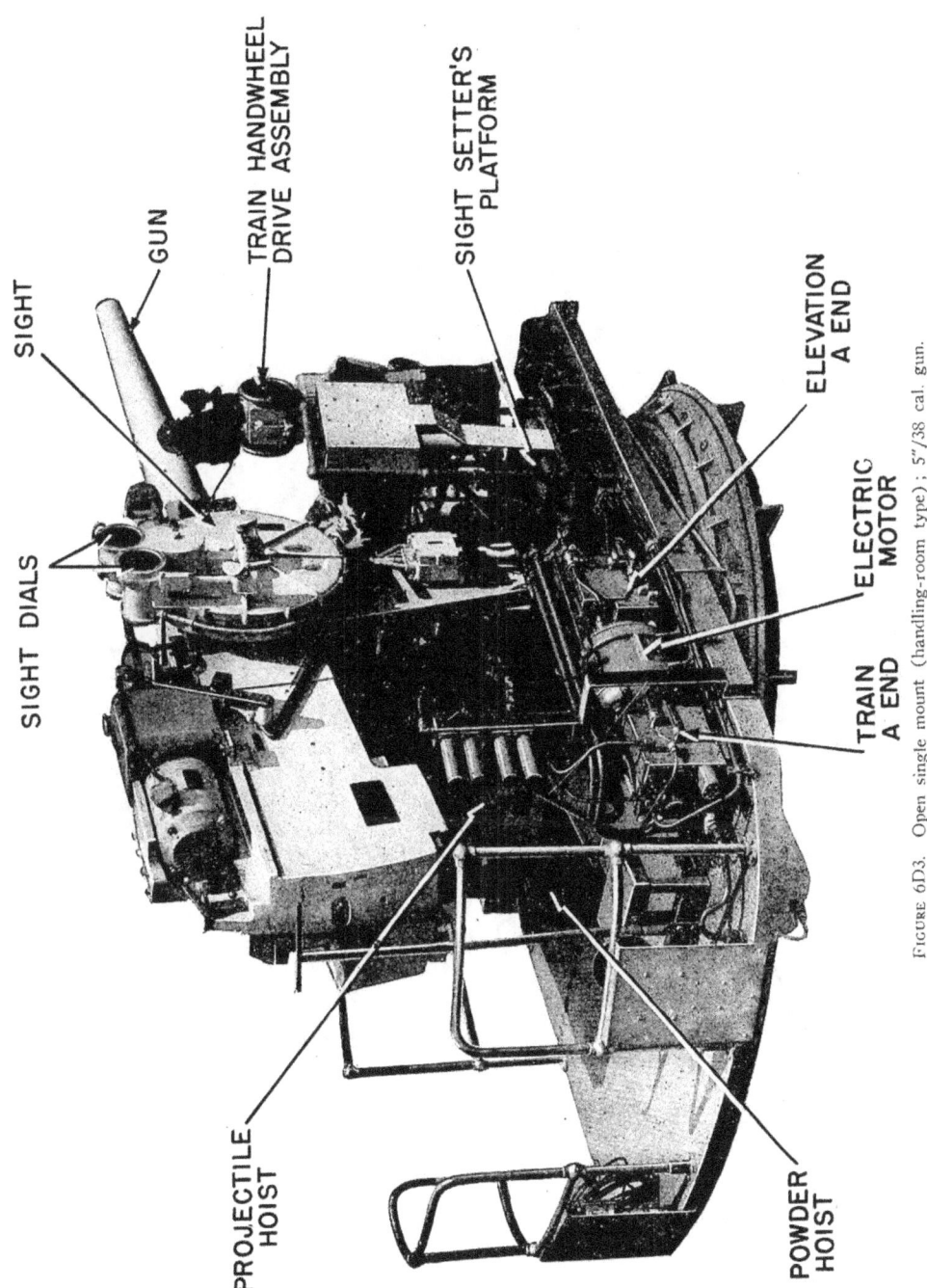

Figure 6D3. Open single mount (handling-room type); 5"/38 cal. gun.

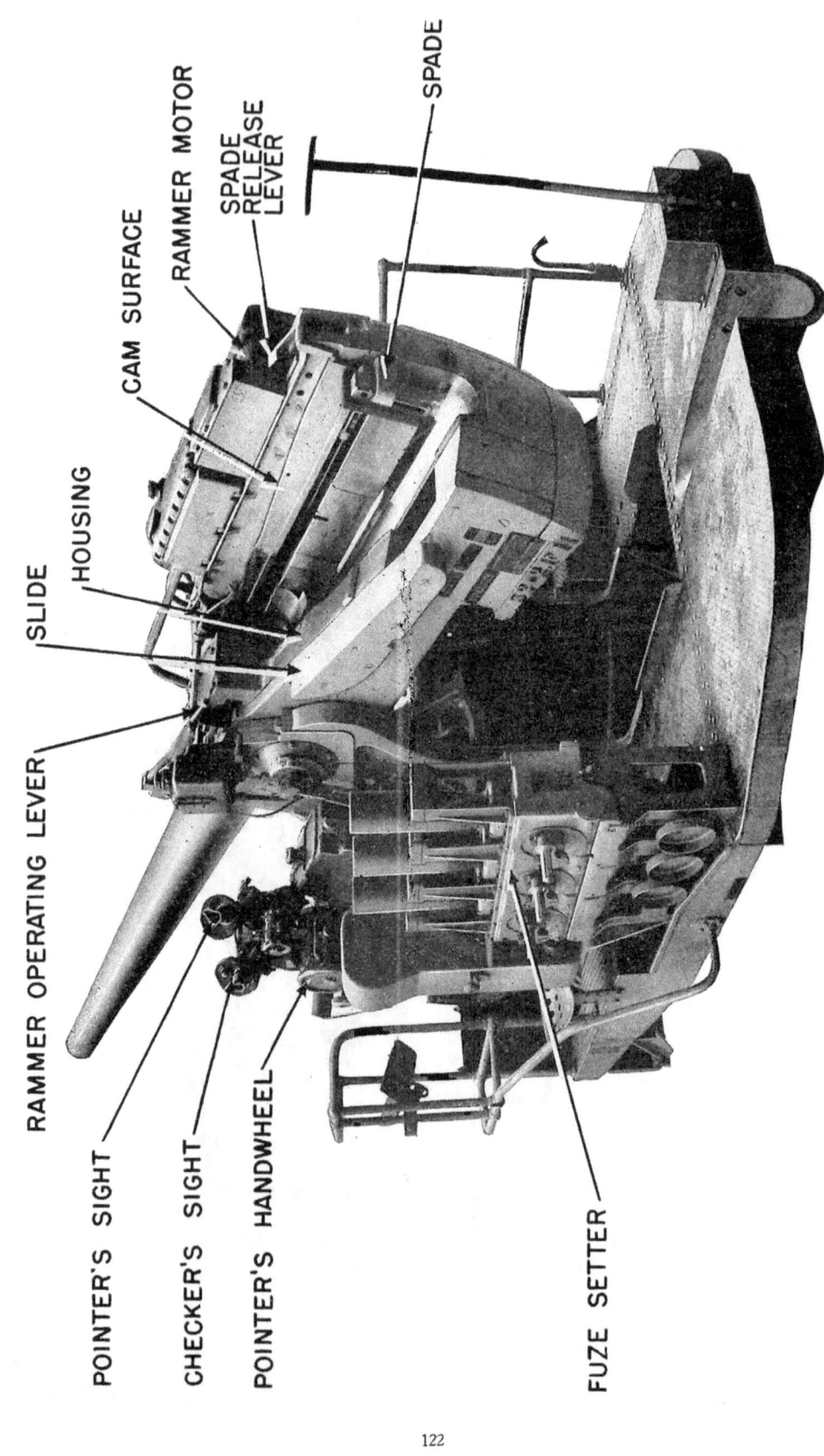

Figure 6D4. Open single mount (without handling room); 5"/38 cal. gun.

powder man, either through a hand passing scuttle, or by a powder hoist. In the case of open mounts without handling rooms below, the ammunition is delivered by passers from hoists a short distance away, and fuzes are set in a fuze setter on the gun platform.

6D4. Gun. The barrel is a radially-expanded monoblock which weighs about two tons. It is secured to the housing by means of a bayonet-type joint which makes it possible to change the gun barrel without dismantling the breech mechanism. The bore is chromium plated from the forward portion of the powder chamber to the muzzle. The rifling has a uniform twist of one turn in 30 calibers.

6D5. Housing. The housing, shown in figure 6D5, is approximately 5 feet long and slightly less than 2 feet across the side faces. Longitudinal *slots* on the side faces ride on guide strips attached to the slide. The forward portion of the housing receives the rear cylinder of the

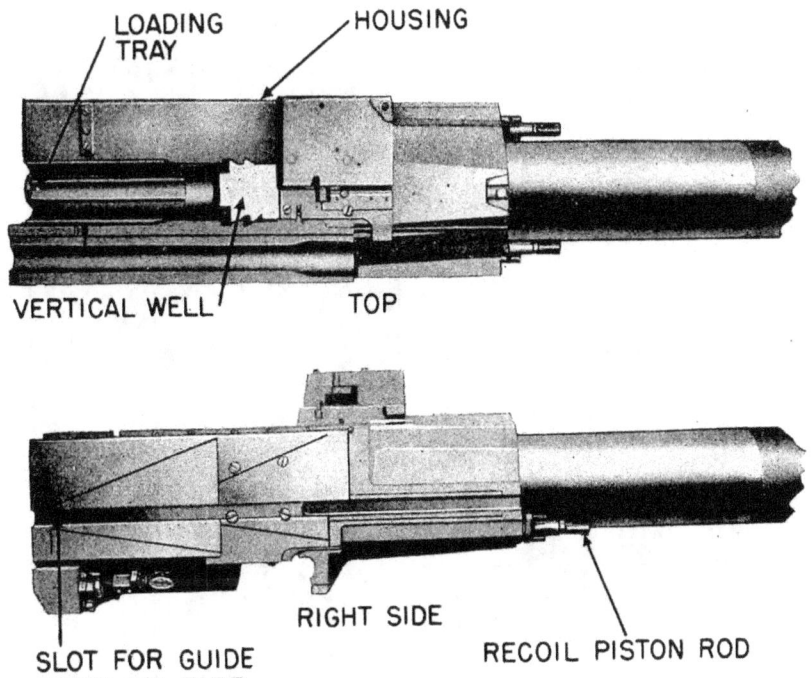

FIGURE 6D5. Housing; 5"/38 cal. gun.

gun barrel and the *vertical well* in the center accommodates the breechblock. To the rear of the well and aligned with the gun bore is a trough-like *loading tray*.

Two longitudinally-bored holes in the under side of the housing contain the recoil cylinders. Figure 6D13 is a cross-sectional view of the housing showing the recoil mechanism in place. The recuperator is mounted between these cylinders and near the aft end of the housing. The broken lines in the figure indicate the position of the recuperator plunger and packing unit.

6D6. Breech mechanism. All 5"/38 cal. guns have a vertical sliding wedge semi-automatic breech mechanism similar to that described in Section 6B. The breech mechanisms on the 5"/38 cal. and 3"/50 cal. guns are very much alike, and the differences which do exist are principally due to variations in arrangements for cocking the firing mechanisms, and connecting the hand operating levers to the operating shafts.

The 5"/38 cal. breechblock and extractors are shown in figure 6D6, and the schematic draw-

SEMI-AUTOMATIC GUNS

ings of the breech mechanism are in figures 6D11 and 6D12. Reference should be made to these latter two figures as the following description of the breech mechanism is studied.

6D7. Hand operating lever. Since the hand operating lever on the 5"/38 cal. gun is not a recoiling part, it is necessary to provide a mechanical linkage which can both engage the operat-

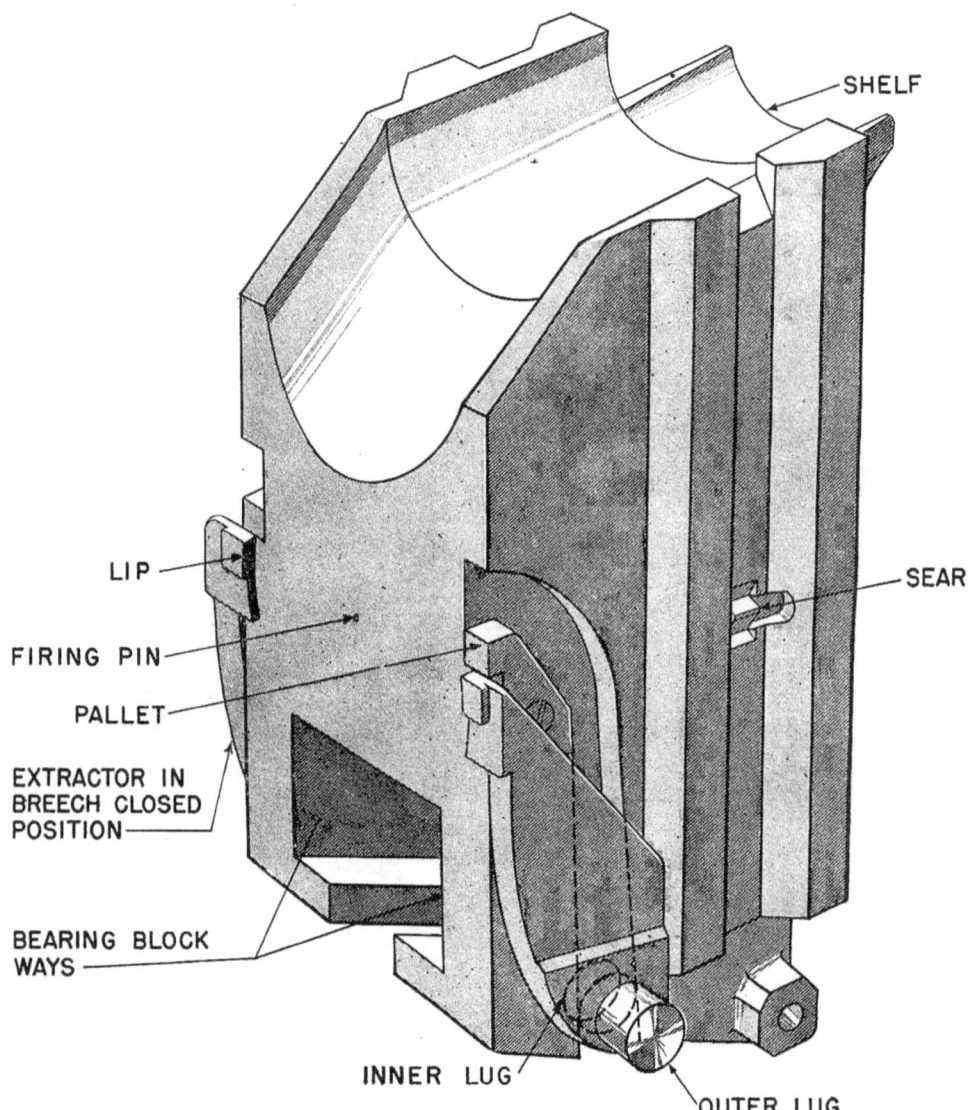

FIGURE 6D6. Breechblock and extractors; 5"/38 cal. gun.

ing shaft when the gun is in battery and permit the shaft to move freely in recoil. This linkage is shown in figure 6D12. It consists of a *latch bell crank* attached to a shaft on which is mounted an *engaging lug*, and a *hand-closing latch* which can be used to lock the bell crank and operating shaft together. The bell crank assembly is mounted in the slide.

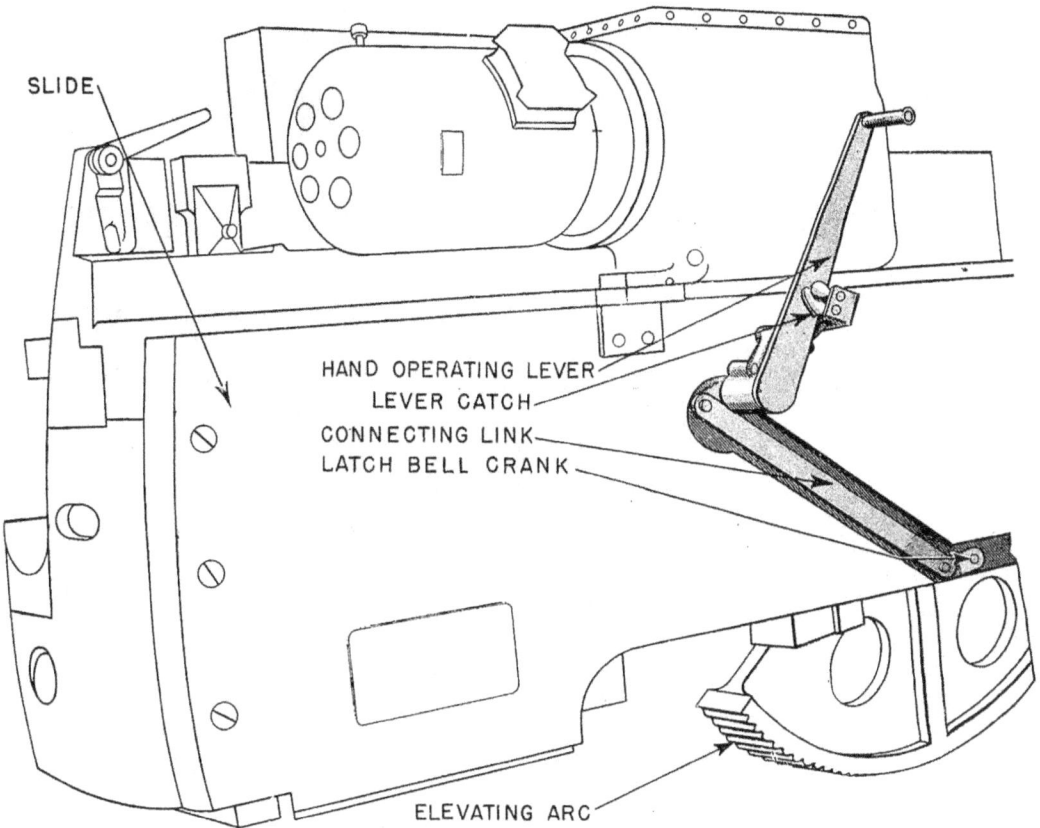

FIGURE 6D7. Breech hand operating lever; 5"/38 cal. gun.

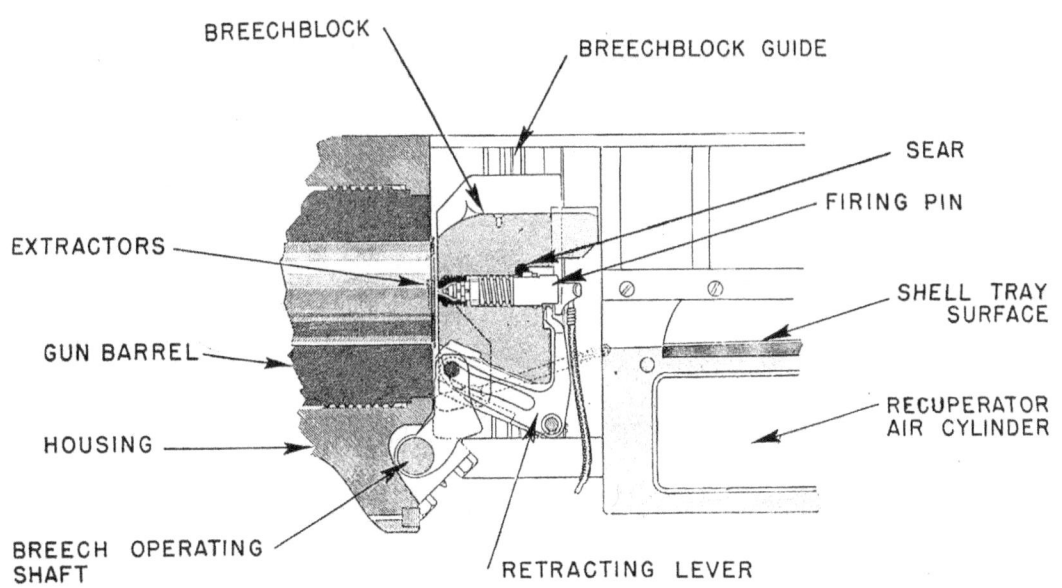

FIGURE 6D8. Breech assembly; sectional view.

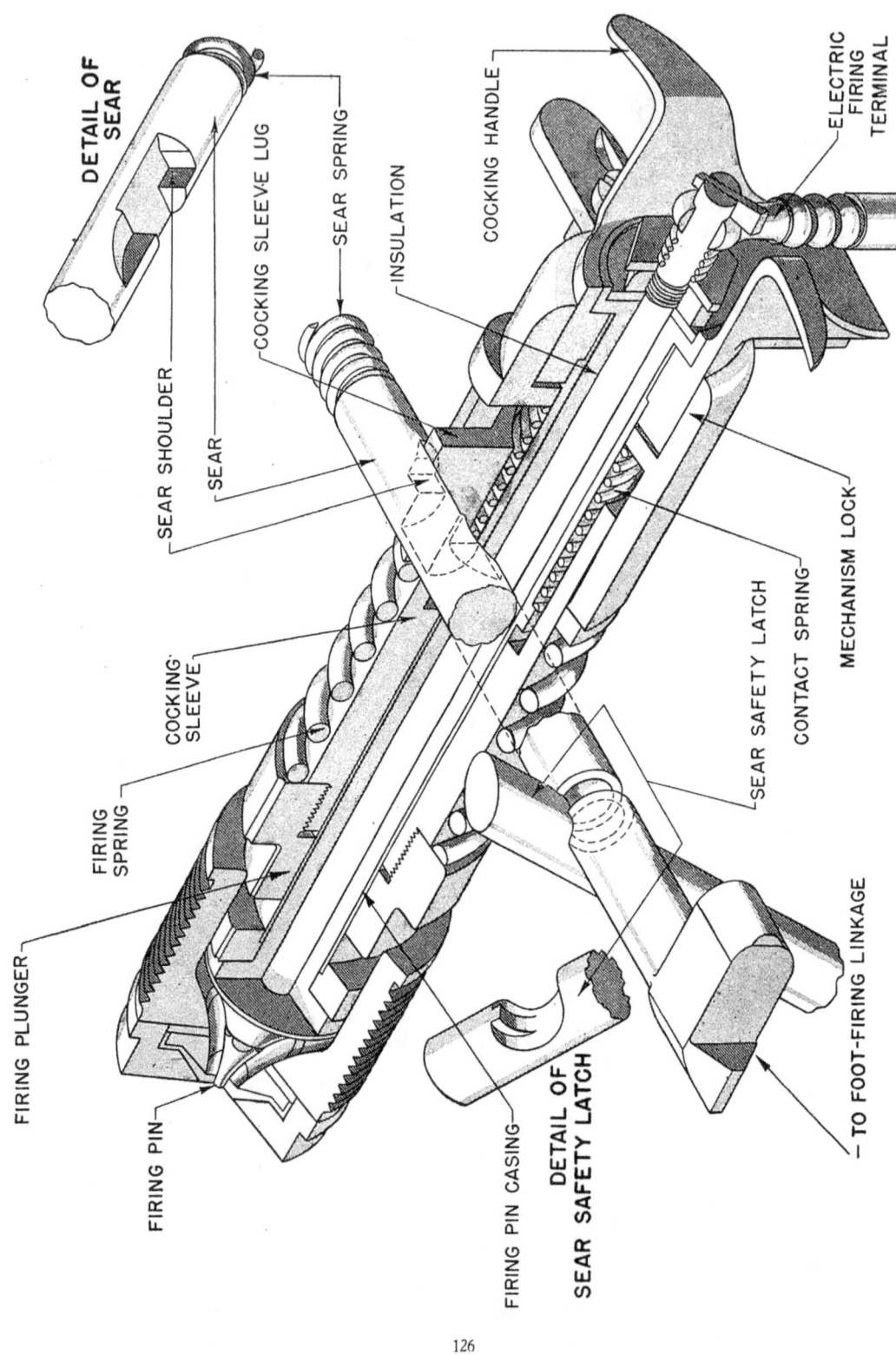

Figure 619. Firing-pin unit.

CHAPTER 6

The hand operating lever is connected to the bell crank by means of a *connecting link* shown in figure 6D7. The *lever catch* holds the operating lever when it is not in use. When the gun is in battery, the engaging lug and *hand operating crank* are engaged in such a way that the hand operating lever can be pulled to the rear and downward to open the breech. The breechblock

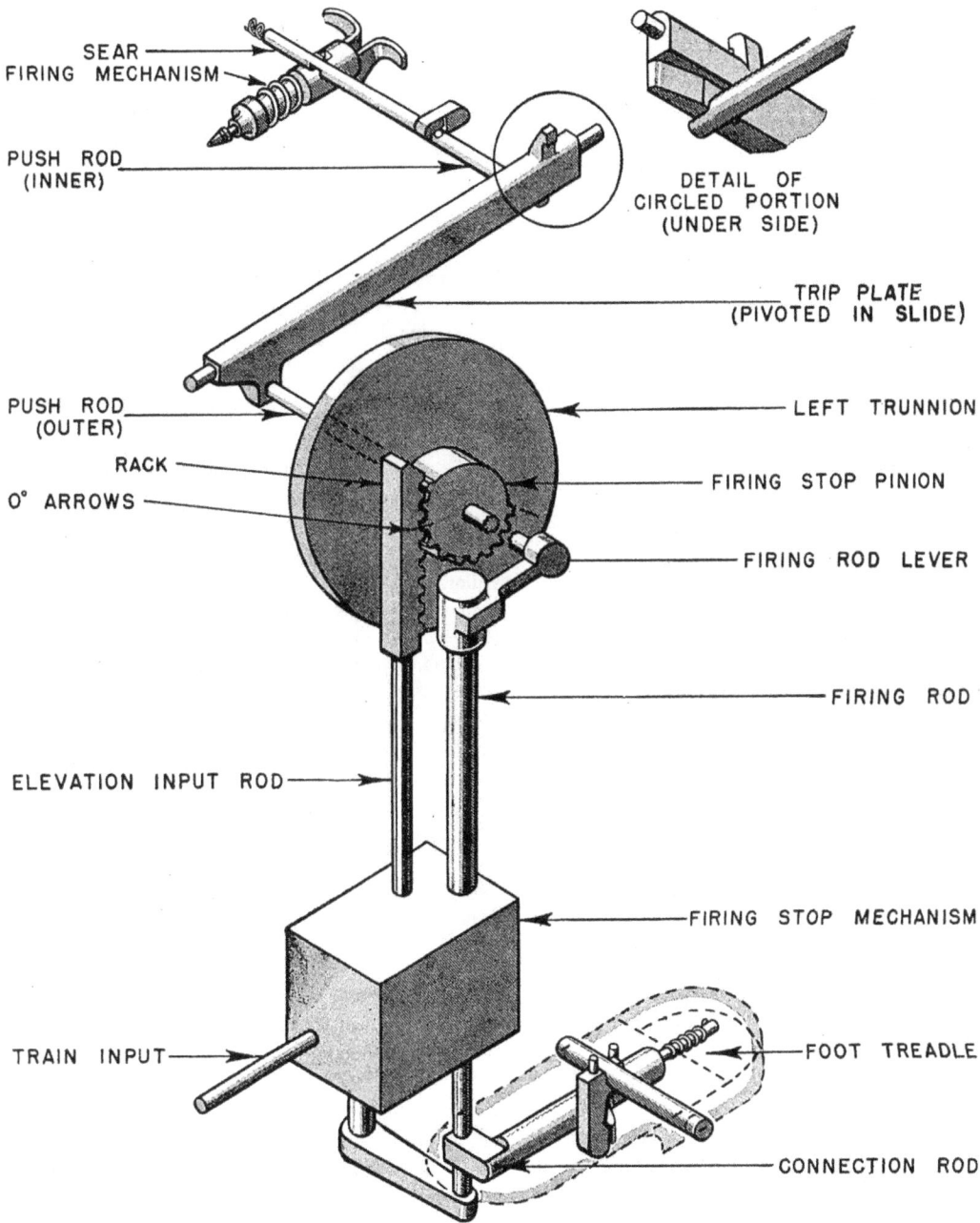

FIGURE 6D10. Foot-firing mechanism; 5"/38 cal. gun.

can be raised by the hand operating lever if the hand closing latch is so positioned that the bell crank and operating shaft are locked together. These relationships are shown in figure 6D12. The latch is automatically unlocked, or pushed in, as the hand-operating crank strikes it when the gun counterrecoils.

6D8. Salvo latch. The positive type *salvo latch*, shown in figures 6D11 and 6D12, engages the *latching lug* on the operating shaft crank when the breech is closed, and prevents rotation of the shaft until the gun has recoiled. During recoil, the *latch operating lug* (fig. 6D12) passes beneath the *salvo latch cam plate* secured to the slide to disengage the latch. The *pawl* prevents the latch from reengaging the latching lug until the operating shaft has turned.

6D9. Firing mechanism. The firing mechanism provides for percussion and electrical firing. Electrical firing requires that the firing pin be insulated from the gun, but in contact with the primer when the breechblock is closed. Percussion firing with the pin resting on the primer is obtained by striking the pin with a plunger. The arrangement of the firing mechanism in the cocked position is shown in figure 6D8.

The *retracting lever* is pivoted in the *breechblock extensions*. The vertical arm of the retracting lever engages a flange on the *firing-pin unit*. A pin carried in the operating shaft central arm passes through the slot in the lower arm of the retracting lever. When the breechblock crank rotates to lower the breechblock, the pin rides in the slot and forces the retracting lever to rotate and cock the firing-pin unit. It is to be noted that cocking is accomplished when the block drops.

The firing-pin unit, shown in figure 6D9, is held in the breechblock by the *mechanism lock*. In the fired position, the plunger rests against the *shoulder* on the *firing pin*. During cocking, the retracting lever pulls the *cocking handle* and firing pin aft, thereby drawing the *plunger* to the rear and compressing the *firing spring*. The *cocking sleeve*, attached to the plunger, moves aft until the *cocking sleeve lug* just passes the *sear shoulder*. At the same time, the *contact spring* is compressed.

As the breechblock closes, the retracting lever allows the contact spring to move the firing pin forward until it gently but firmly contacts the primer *after* the block is fully closed. During the movement of the firing pin, the sear shoulder engages the cocking sleeve lug, holding the cocking sleeve and plunger in the cocked position. The pin is then in position to pass the current for electrical firing, and ready to transmit the blow of the plunger for percussion firing.

Percussion firing is initiated by the pointer who presses a foot treadle connected to the *sear* by a foot-firing linkage (see fig. 6D10). When the treadle is pressed, the sear moves against the *sear spring*, releasing the cocking sleeve lug through the *sear notch*. The firing spring drives the cocking sleeve and plunger forward, and the blow of the plunger drives the firing pin into the primer, firing the gun.

The gun cannot be fired until the breech is *fully* closed, because until then, (1) the firing pin does not touch the primer, (2) the sear is locked by the *sear safety latch* (operated by the central arm—see fig. 6D12), and (3) the sear is out of alignment with the *inner push rod* of the foot-firing linkage.

6D10. Firing attachments. The foot-firing linkage, shown in figure 6D10, functions to actuate the firing mechanism for percussion firing. The *firing rod* is connected to the *foot treadle* through a clutch which is controlled by the firing stop mechanism. The *outer push rod* which passes through the left-side trunnion actuates the *trip plate, inner push rod*, and the *sear* when the breech is closed and the foot treadle is depressed.

The electrical firing circuit, similar to that shown in figure 4B4, provides a means for firing either from a remote director, or a local firing key at the pointer's station. The pointer's key is usually locked closed when firing from the director. A firing snap switch, located near the pointer, functions to connect either a 20-volt transformer or a 6-volt storage battery into the firing circuit.

The firing stop mechanism functions to disconnect the clutch in the foot-firing linkage, and to open a switch in the electrical circuit when the gun is elevated or trained to fire into own-ship

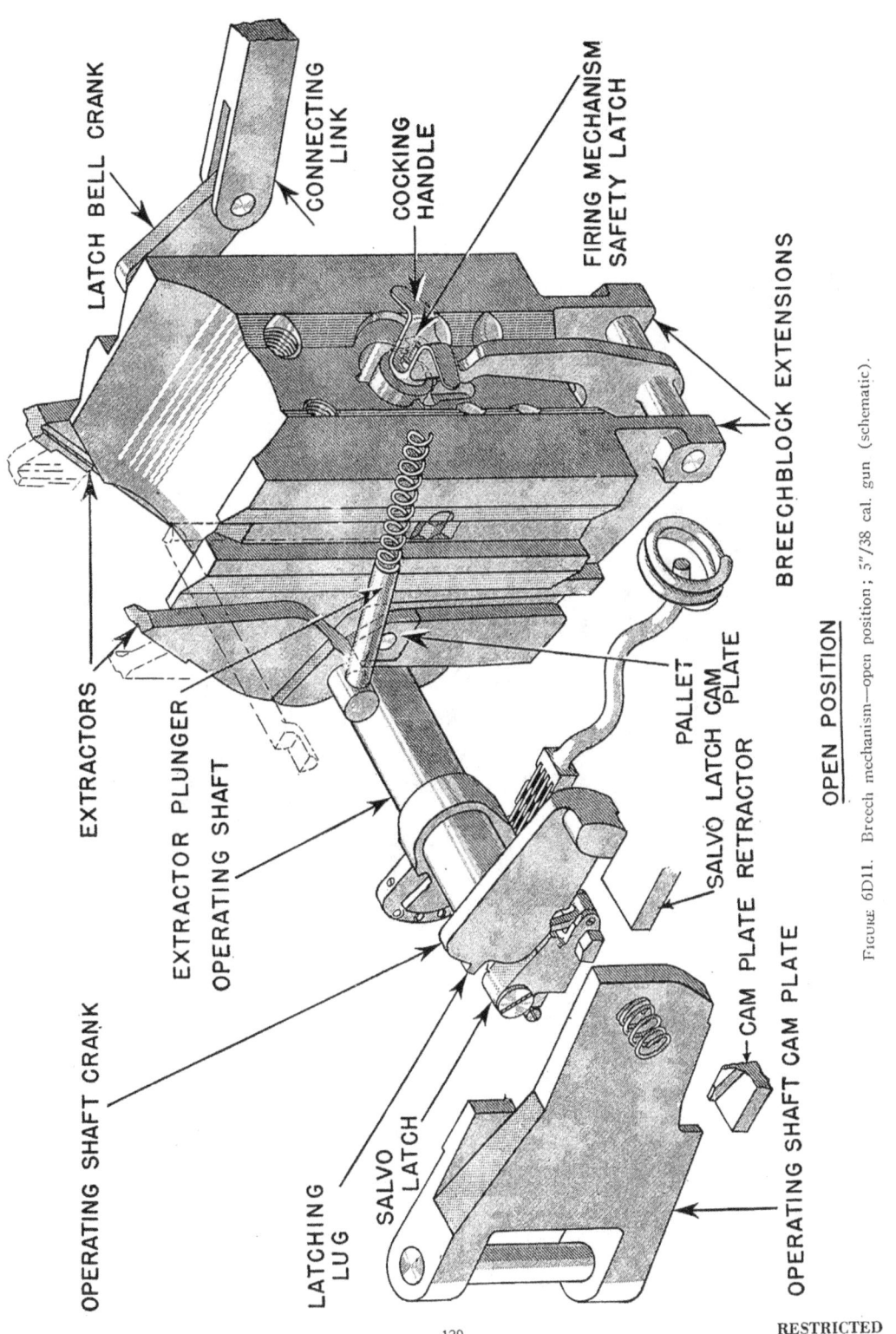

Figure 6D11. Breech mechanism—open position; 5"/38 cal. gun (schematic).

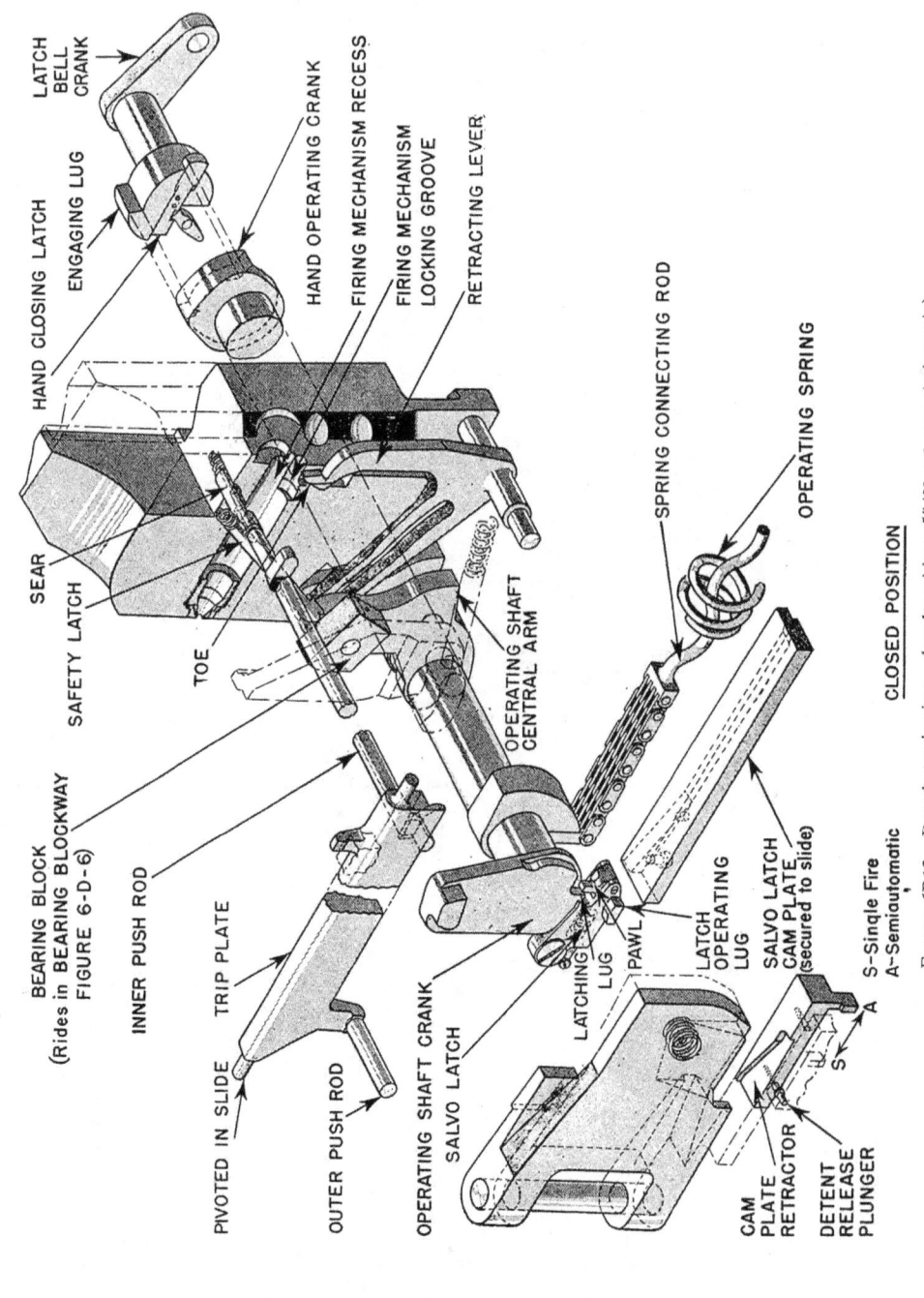

FIGURE 6D12. Breech mechanism—closed position; 5"/38 cal. gun (schematic).

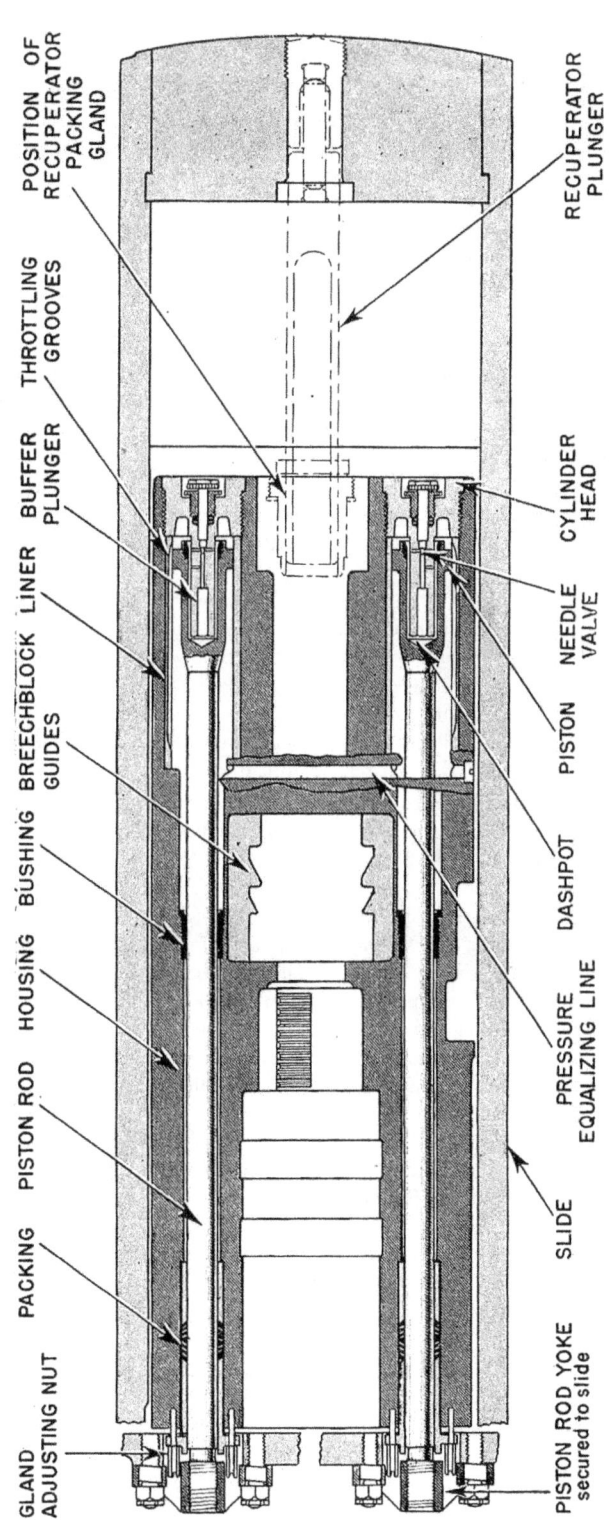

Figure 6D13. Recoil mechanism; 5"/38 cal. gun.

structure. Thus, it is impossible to fire the gun either electrically or by percussion when it is so positioned.

6D11. Gas ejector. Gas ejectors are on all guns that are mounted in enclosures, to prevent the entry of powder gases into the gun houses. Air under approximately 55 pounds p.s.i. pressure is supplied to an orifice in each of the breechblock guides. These orifices direct air toward the powder chamber from the moment a case is extracted until the next case is seated. The air-control valve is automatically opened when the gun counterrecoils and closed by the rammer mechanism at the end of the ram stroke. A hand-control lever is also provided for manual operation of the valve.

6D12. Hydraulic recoil system. The details of the recoil mechanism are shown in figure 6D13. The two cylinders bored in the under side of the housing are each lined with a tube about $18\frac{1}{2}$ inches long and 4 inches inside diameter. Three equi-spaced *throttling grooves* of constant width and variable depth are used to control the recoil-brake action. The cylinders are interconnected by a *pressure-equalizing line*.

The *piston rods* are secured to the forward part of the slide and hence do not move in recoil. The *piston* and rod are integral, and a *dashpot* is machined in the piston end of the rod. This dashpot provides buffer action as the gun returns to battery. The normal recoil is approximately 15 inches.

The *buffer plunger* is a part of the cylinder head which closes the aft end of the cylinder. The dashpot unit functions during a little more than half of the counterrecoil stroke. The buffing action is obtained by forcing the liquid through small clearances around the slightly-tapered plunger and through two orifices, one of which is varied in size by an adjustable *needle valve*. This needle valve is used to regulate the final buffing action near the end of the counterrecoil stroke. The valve is usually set before the gun is installed on a ship, but adjustments can be made whenever it is desired to regulate the rate at which the gun returns to battery.

6D13. Recuperator. The counterrecoil system, shown in figure 6D14, is the hydropneumatic type of recuperator. The operation and the principal component parts of the system are discussed in Articles 4C17 to 4C19 inclusive.

A single air cylinder is fitted into a recess on the under side and at the aft end of the housing. The plunger is a hollow, highly polished cylinder approximately 2 feet long and $3\frac{1}{2}$ inches outside diameter. The *support bar housing* and the *support bar* and *block* provide a coupling between the plunger and the rear plate of the slide. Enough freedom of motion is present in the coupling to allow the plunger to align itself with the packing gland in the air cylinder. This prevents distortion of the packing and consequent leakage of air through the packing gland.

The differential cylinder is a transverse bore in the lower, aft end of the air cylinder forging just below the plunger packing. Short pressure leads drilled in the forging connect the differential cylinder to the air cylinder and packing.

Air in the main cylinder is kept at about 1,500 pounds p.s.i. pressure with the gun in battery. When the gun recoils, the air is compressed further and the pressure rises to nearly 2,000 pounds p.s.i. After recoil, the compressed air forces the air cylinder, and hence the gun and housing, forward into the battery position.

The safety link is a metal strip which secures the housing to the slide when the gun is not being used. It should always be removed before firing, but neither gun or mount will be damaged if this is not done.

6D14. Rammer mechanism. Figure 6D15 shows a rammer assembly. It is a semi-automatic, electric-hydraulic unit mounted alongside the loading tray on the upper rear portion of the slide. It is designed to function properly for any elevation of the gun. The power unit consists of a *hydraulic pump* (mounted within the *oil tank*) driven by an electric motor (about $7\frac{1}{2}$ h.p.) The rammer proper includes: (1) *hydraulic cylinder*, (2) piston and piston rod connected to a *crosshead*, (3) a *ramming head*, called a *spade* or *shell guard*, mounted on the crosshead, and (4) *control valves* and *operating levers*.

CHAPTER 6

Oil, under pressure built up in the pump, is introduced into the cylinder to drive the piston (and hence the spade) forward for ramming and aft for retracting. The rammer stroke is about 4½ feet. Control valves, operated partly by hand and partly automatically, control the motion of the piston. The face of the spade is padded with rubber backed with canvas to prevent damage to the base of the case during the ramming operation. The spade rams the projectile and case into the chamber at the same time. The spade is mounted on the crosshead in such a manner that it can be moved

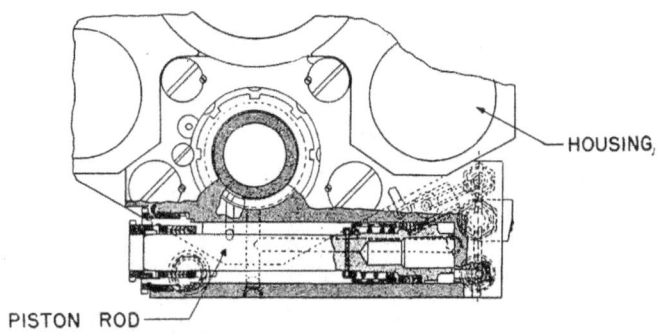

DIFFERENTIAL CYLINDER SECTION A–A

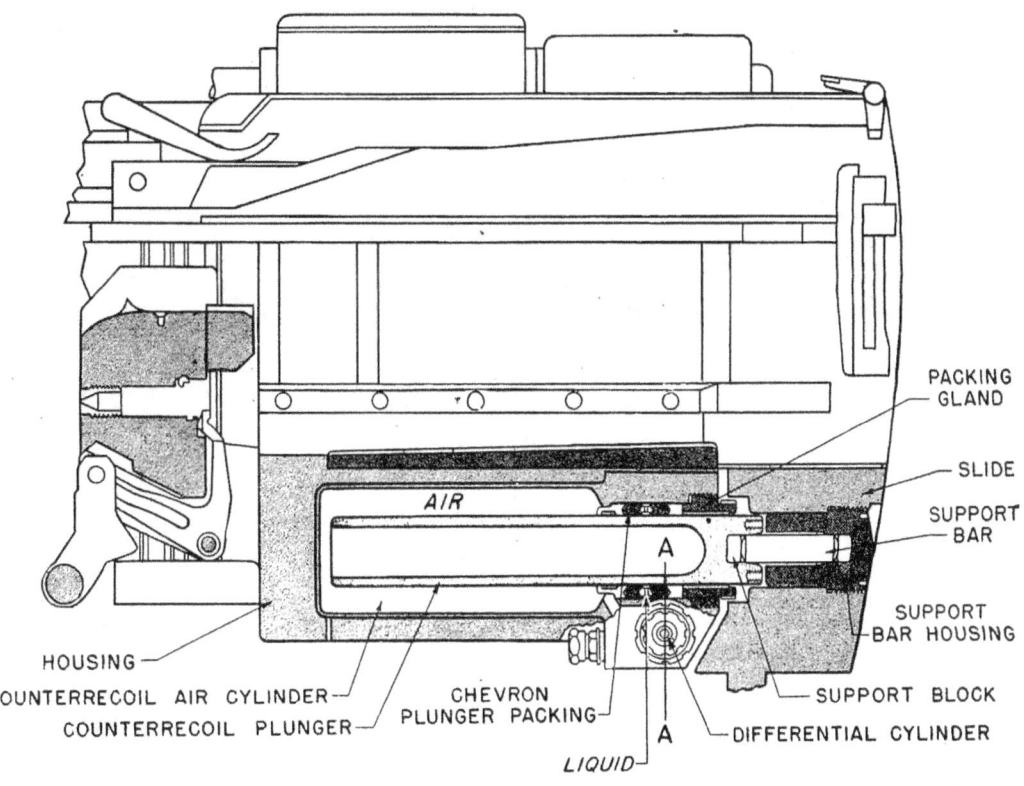

FIGURE 6D14. Recuperator; 5″/38 cal. gun.

SEMI-AUTOMATIC GUNS

upward about 5 inches against the action of a vertical spring placed between the spade and crosshead. A spring-loaded plunger (plunger latch) projects from the inboard side of the spade.

The *crosshead guide and cam plate* is mounted along one side of the loading tray. Its purpose is to guide and support the crosshead, and to provide cam surfaces upon which the plunger latch rides and hence controls the motion of the spade. After a round is rammed, the breechblock rises and forces the spade upward about 3 inches. As the spade and crosshead move aft in retraction, the plunger latch rides up an inclined cam surface on the cam plate and the spade is raised an additional 2 inches. The upward movement of the spade, totalling about 5 inches, is sufficient to permit the ejected case to pass under the spade.

The *rammer control shaft* and control cams are shown in figure 6D16. An *operating lever* on the end of this shaft, the *retraction cam,* and the *latch pin lever* are used to position the valves

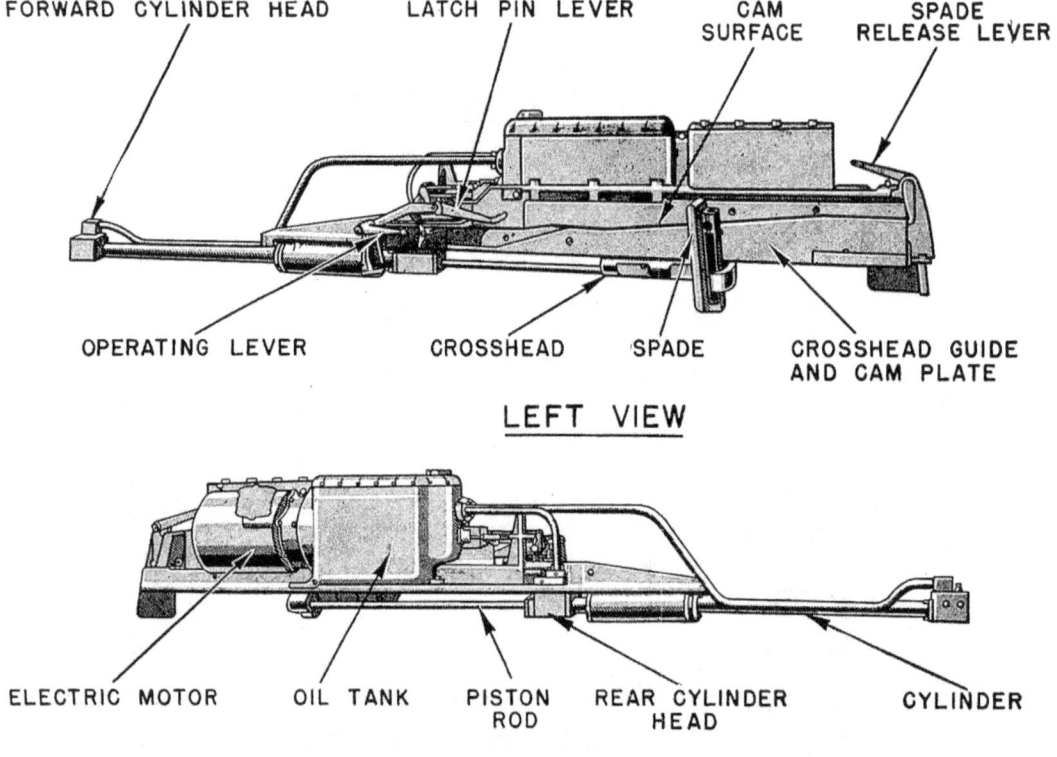

FIGURE 6D15. Hydraulic rammer; 5"/38 cal. gun.

which control the operation of the rammer. The three positions for the operating lever are *ram, neutral,* and *retract.*

6D15. Ramming cycle. Assume that the cartridge case and projectile are in the loading tray and that the spade is behind the case in position for ramming. Refer to figure 6D16. The operating lever is thrown into the ram position, the spade forces the ammunition into the gun, and the crosshead is locked to the housing by the *crosshead interlock latch.* When the rim on the cartridge case strikes the extractors, the breechblock rises and forces the spade upward. As the spade rises the plunger latch forces the *latch pin lever* upward. This action returns the control shaft to the neutral position.

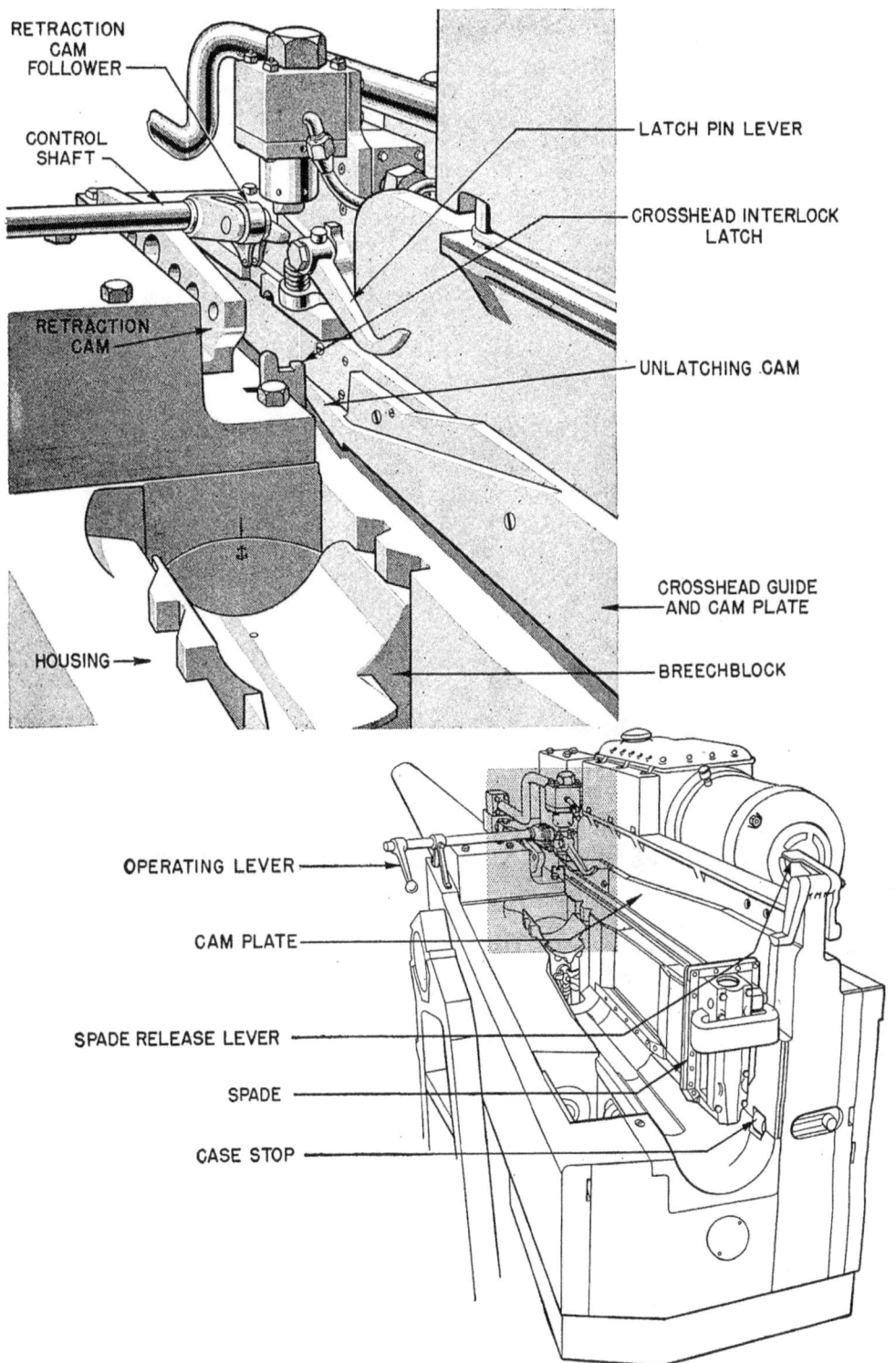

Figure 6D16. Rammer controls; 5″/38 cal. gun.

SEMI-AUTOMATIC GUNS

After the gun is fired the following actions take place during the first part of recoil: (1) the crosshead rides back with the recoiling gun and housing, (2) the *crosshead unlatching cam* pushes against the crosshead interlock latch, and (3) the rammer control shaft is positioned for retraction by the retraction cam, and the rammer continues to retract under its own power.

During retraction the plunger latch rides over the cam surface on the cam plate, and the spade is raised to provide the necessary clearance for the ejected case. Near the end of the retract stroke the crosshead engages a retract-control rod connected to the control shaft. When the rammer forces the retract-control rod a short distance aft, the operating shaft is returned to the neutral position and retraction is stopped. The spade is held in its upper position by a latch which must be released manually by means of the *spade-release lever* before the spade can drop into position to start another ramming cycle.

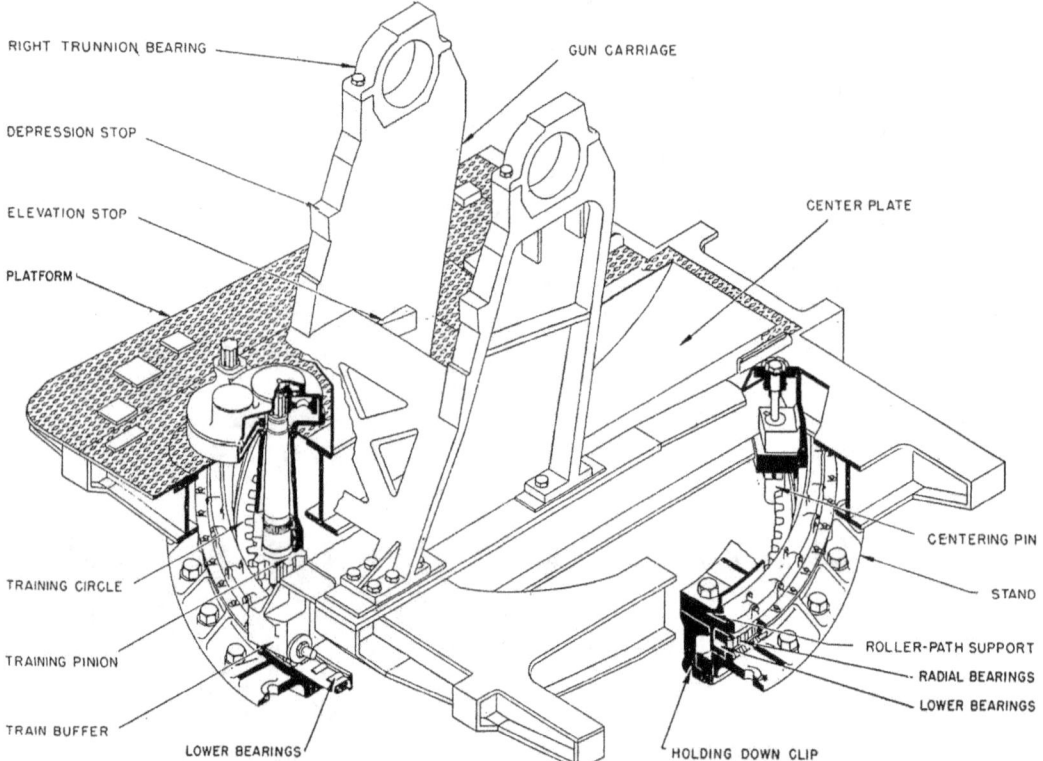

FIGURE 6D17. Stand and carriage; 5"/38 cal. single mount.

The rammer can be controlled manually at any point in the cycle by setting the operating handle in any one of the three positions. Two other cam surfaces on the cam plate are not used in normal operation. One is used when exercising the rammer with the breech mechanism idle. The other prevents damage if the ram stroke is initiated before the spade has been dropped.

6D16. **Loading operation.** The disposition of personnel in a gun compartment is shown in figure 6D23. After the ejected case has cleared the loading tray, the *spade man* depresses the *spade release lever* to drop the spade. The *powder man* takes a powder case out of the hand passing scuttle, removes the primer-protecting cap and places the case on the loading tray. The *projectile man* removes a projectile from the projectile hoist and places it on the loading tray just forward of the case. The projectile man then pulls the rammer lever to ram and the charge is rammed home.

CHAPTER 6

After the gun fires, the *hot case man* disposes of the ejected case through a case hatch in the aft end of the gun compartment. For low angles of fire the cases are allowed to pass back over the loading tray and are caught by the hot case man. For high-angle fire the case stop shown in figure 6D16 is set to block the aft end of the loading tray and the cases must be lifted from the tray.

6D17. Stand, carriage, and slide. The general arrangement of the stand and carriage is shown in figure 6D17. The stand is a ring-shaped casting slightly over 8 feet in diameter. It is machined to provide for two roller-bearing paths, *lower* and *radial,* and the *training circle.*

The entire carriage assembly, including the platform, shield, and suspended projectile hoist, is supported on the roller-bearing assembly placed on the lower bearing path. Radial thrusts are taken on the roller bearings in the upper bearing circle, and the carriage is held on the stand by the *holding down clips.*

The slide, similar to that shown in figure 6D4, is a large box-shaped structure within which the housing moves in recoil and counterrecoil. Its aft end is shaped to form part of the loading tray. A semi-cylindrical bearing is mounted in the front end to provide a forward guide bearing for the gun. The gun housing is supported and guided by two bearing strips bolted to the side plates. The elevating arc, gun-port shield and rammer mechanism are secured to the slide, and the slide assembly is supported by *trunnions* on roller bearings.

6D18. Elevating and training gear. All mounts are equipped with electric-hydraulic power drives for training and elevating the gun. Figure 6D3 shows the location of the power-drive equipment. Training and elevating gear are similar assemblies driven by a single electric motor (about 10 h.p.) through separate hydraulic transmission units.

Each transmission unit consists of a pump connected by pressure lines to a hydraulic motor. A control unit, called an indicator-regulator, contains valves and control devices which regulate the output of the pump to the motor. Through each transmission unit the constant speed, uni-directional input from the electric motor is converted into a reversible, variable-speed output to the respective elevating and training racks. The following three methods of control are available for both elevating and training gear:

    1. *Automatic,* or remote control from a director system by means of electrical signals transmitted to the indicator-regulator, which in turn controls the output of the hydraulic pump.

    2. *Local,* or control of the power drive by means of pointer's and trainer's handwheels. This control is similar to automatic, except that rotation of the handwheels takes the place of the electrical signals from the remote point. A choice of *high* or *low* speed may be selected by means of levers at the pointer's and trainer's stations.

    3. *Manual,* in which the handwheels are geared directly to the training and elevating racks. The indicator-regulator and hydraulic transmission are inoperative. Only low speed should be used for manual drive.

The gun elevation limits are 15° depression and 85° elevation. The train limit varies with the emplacement. Typical elevating and training rates for the three methods of control are tabulated below:

CONTROL	MAXIMUM TRAINING RATE DEGREES PER SEC.	MAXIMUM ELEVATING RATE DEGREES PER SEC.
Automatic	34	18
Local—high speed*	29	15
Local—low speed*	4	4½
Manual*	4	4½

6D19. Projectile hoist. The mount is equipped with the dual-tube projectile hoist shown in figure 6D18. The assembly consists of a hoisting unit in two enclosed tubes, an electric-hydraulic power drive, and semi-automatic control equipment. An auxiliary manual drive is also provided for emergency use. The hoist is equipped with a fuze-setting device which automatically sets the fuzes as the projectiles are lifted from the handling room to the gun compartment.

    * For 180 r.p.m. of the handwheels.

SEMI-AUTOMATIC GUNS

Two *projectile flights* are attached to the endless chain. The chain oscillates and each tube alternately delivers a projectile. Figure 6D19 is a schematic drawing of the hoisting and fuze-setting drives. The projectiles are hoisted nose downward so that the fuzes can be set during the hoisting operation. The flights are so spaced that one is at the upper end of one tube when the other is at the lower end of the opposite tube. These are the limiting points of chain movement, as the drive automatically stops at the positions indicated. When a projectile is placed on the lower flight and the upper one is empty, the drive will automatically hoist. An inter-locking device prevents hoisting when the upper flight is loaded. The projectile men in the upper handling room, shown in figure 6D24, load the hoist with projectiles taken from the ready service racks or from the lower ammunition hoist.

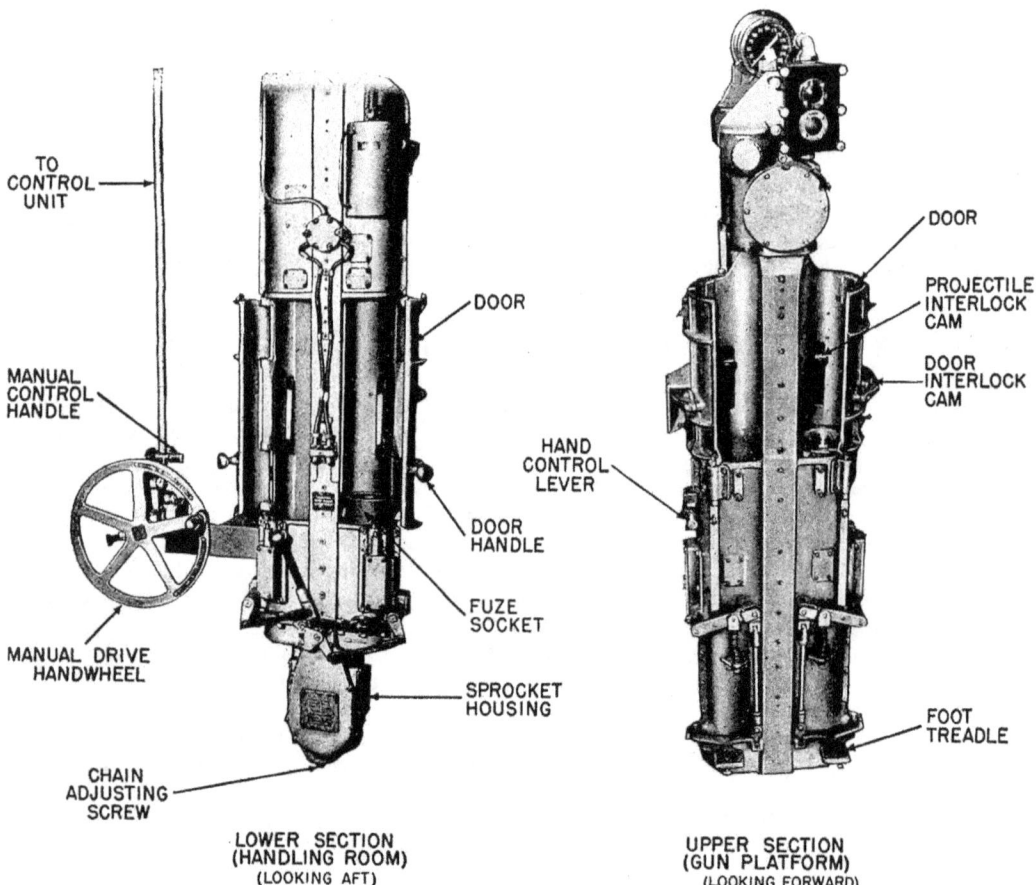

FIGURE 6D18. Projectile hoist; 5"/38 cal. gun.

The hoist power drive consists of an electric motor (about 10 h.p.) and a hydraulic pump and motor unit. The hydraulic pump is connected to the hydraulic motor by pressure lines which incorporate a system of control valves. The hydraulic motor is geared to the chain-driving sprocket. The valves are initially positioned by hand, and from then on loading and unloading the projectile automatically controls the operation of the hoist. A handwheel, geared to the power-drive output shaft through a clutch, provides an auxiliary drive to be used in case of power failure.

6D20. **Fuze-setting mechanism.** The fuze-setting mechanism consists of: (1) a *fuze-setting indicator-regulator*, (2) *fuze-setter chains*, and (3) *projectile flights*. These three components, shown schematically in figure 6D19, are called regulator, fuze chains, and flights in the following discussion.

The regulator is a device which determines fuze setting by controlling the positioning of the fuze chains. It is mounted on the forward face of the upper end of the projectile hoist, below the

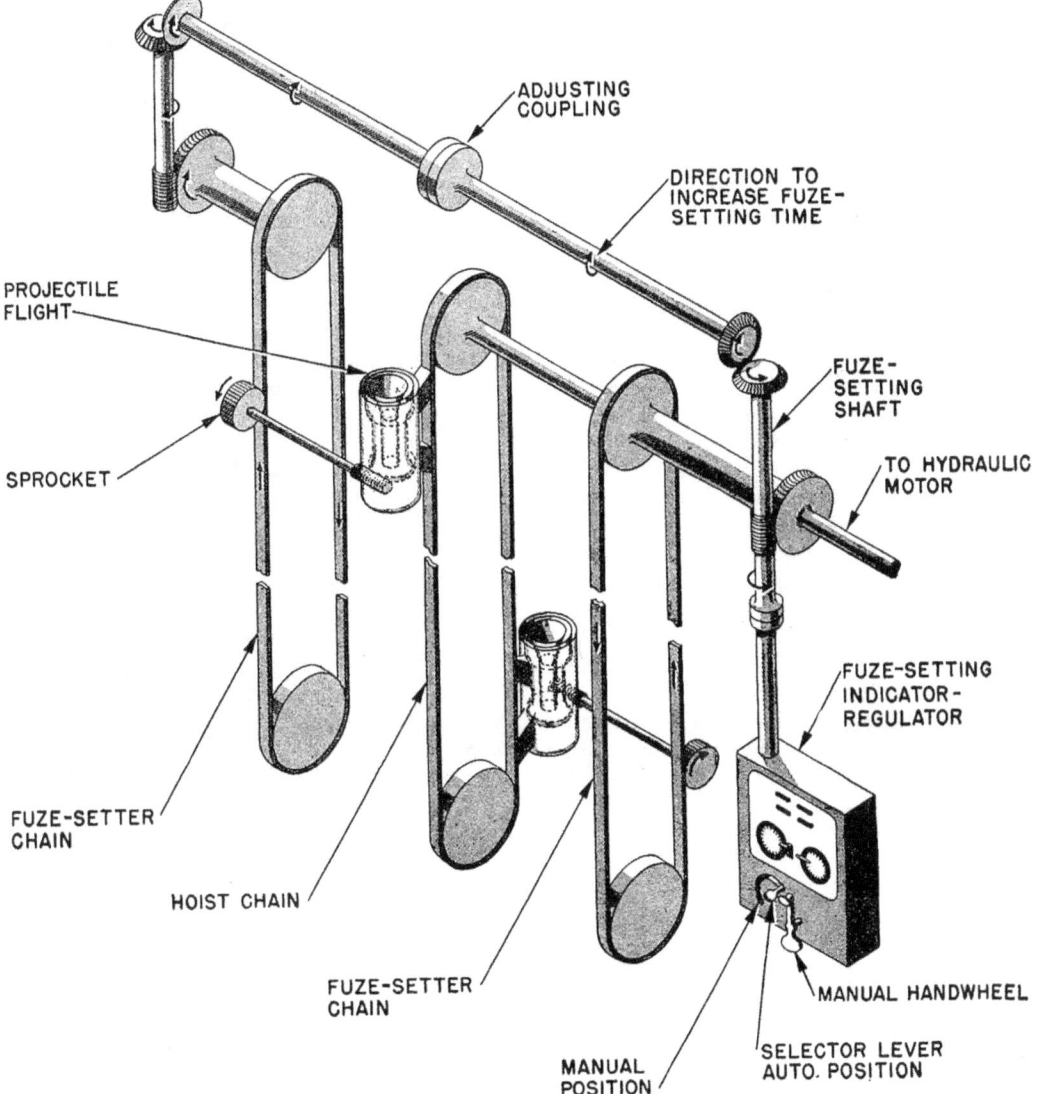

FIGURE 6D19. Hoisting and fuze-setting drives—schematic; 5"/38 cal. gun.

pointer's seat. The regulator can be set for either *manual* or *automatic* control by means of a *selector lever*.

In *manual* control the gun-crew fuze setter turns the *manual handwheel* to actuate the chain-setting drive, and the time setting is effected by one of the following methods:

1. By matching pointers, in which case the fuze setter matches pointers on two sets of

regulator dials, one set of which is actuated by electrical signals from the computer in the director system, and the other set by rotation of the manual handwheel.

2. By setting the desired fuze time on the regulator dials, in which case the manual handwheel is turned until the dials indicate the fuze time opposite fixed index marks. This method is used when the computer is disconnected from the regulator.

In *automatic* control the manual handwheel is disengaged and the fuze chains are automatically positioned by an electric power drive in the regulator. The power drive is controlled by electrical signals from the computer. The gun-crew fuze setter has no control over the regulator in *automatic*.

Figure 6D20a shows the details of a flight. It consists essentially of an *outer socket* attached to the hoist chain, and an *inner socket* so mounted on ball bearings that it can rotate within the outer socket. The inner socket is geared to a *sprocket* which engages the fuze chain. The fuze chains are endless chains mounted on the outboard sides of the hoist chain.

A fuze is set by rotating the *time ring lug* shown in figure 6D20b. The projectile must be placed on the hoist with the fuze *fixed lug* in the *V slot* to obtain correct settings. The initial fuze setting is *safe*, and during the fuze-setting operation the time ring is rotated from the maximum time setting down to the setting required.

The *pawls* on the inner socket will engage the time ring lug when these parts are in alignment. Further rotation of the socket, after engagement, sets the fuze. During hoisting, the point in the flight travel at which engagement does take place depends upon the fuze setting desired.

The sprocket transmits fuze-setting adjustment from the fuze chain to the time fuze in the following ways:

1. By movement of the flight with respect to the fuze chain when the projectile is hoisted. The gearing is so arranged that movement of the flight from the lower to upper positions causes the inner socket to rotate almost exactly one revolution in a direction to *reduce* the time setting on the fuze. A rotation of 315° is equivalent to 45 seconds fuze-setting time.

2. By movement of the fuze chain. This is controlled by the regulator.

From the above discussion it is apparent that if a high fuze setting is desired, the regulator must so position the fuze chain that the pawls will engage the time ring lug late in the hoisting cycle. The time ring will then be turned a small amount during the remainder of the flight travel and the setting will be high. On the other hand, the pawls are positioned for early engagement in making low fuze settings. Thus, in obtaining a given fuze setting it is only necessary to make the setting on the regulator which positions the fuze chain, and then hoist the projectile.

Changes in fuze setting may be introduced at any time while the projectile is in the hoist by changing the setting on the regulator. The fuze will be set for the time indicated by the regulator when it is lifted out of the flight even though it has remained at the delivery end of the hoist for some time.

6D21. Powder scuttle. The powder scuttle, shown in figure 6D21, is a tubular hand-passing unit. Its lower end is accessible from the powder-passer's platform in the upper handling room and its upper end projects a short distance above the gun platform on the left side of the gun. Figure 6D24 shows the arrangement of personnel in the upper handling room. The *first powder man* places the cases in the case holder on the handling platform. The *powder passer* manually inserts the cases, base up, into the scuttle. The *pawl* forces the case over onto the scuttle step where it rests until it is removed by the powder man in the gun house above. An *indicator* in the handling room, actuated by the *indicator cam*, shows *full* when a case is in the scuttle and *empty* when it has been removed.

6D22. Sights. Movable prism-type telescopes are mounted at the pointer's, checker's, and trainer's stations. Figure 6D22 shows a schematic layout of the movable prisms in the telescopes, and the mechanical sight-setting mechanism. The elevating prisms in the three telescopes are all moved simultaneously in response to rotation of the *sight-angle handwheel*, and similarly all the deflection prisms are positioned by the *deflection handwheel*. The sight-setting mechanism is mounted on the right-hand side of the slide with its center over the trunnion.

Both the *sight-angle dial* and *deflection dial* groups, shown in figure 6D22, consist of an inner disk

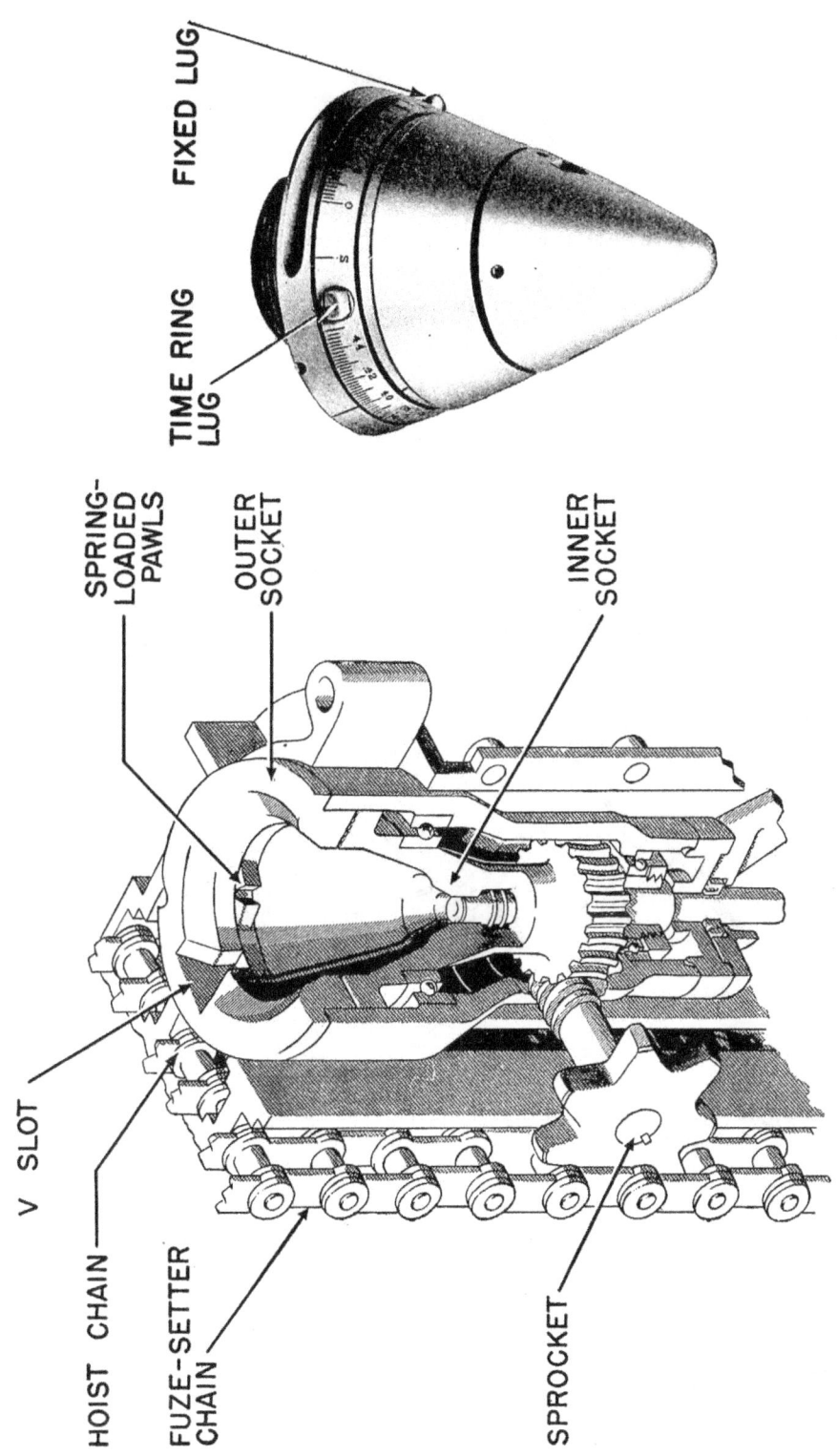

FIGURE 6D20. a: projectile flight (schematic) and, b: fuze; 5"/38 cal. gun.

SEMI-AUTOMATIC GUNS

dial with an index mark, outer ring dial with suitable graduations and an index mark, and a fixed index outside the outer ring dial. The sight-angle outer ring dial is inscribed in minutes of arc, and the deflection outer ring dial in mils. The range dial is geared to the sight-angle handwheel and its scale in yards corresponds to the elevation in minutes indicated on the sight-angle dials. Both inner dials are positioned by electrical impulses from the computer in the director system. The sight-angle and deflection outer ring dials and the range dial are positioned by turning the handwheels.

Sights are set by the sight setter in one of two ways. If the computer is in operation, the proper setting is obtained by training the handwheels until index marks on the outer ring dials match with

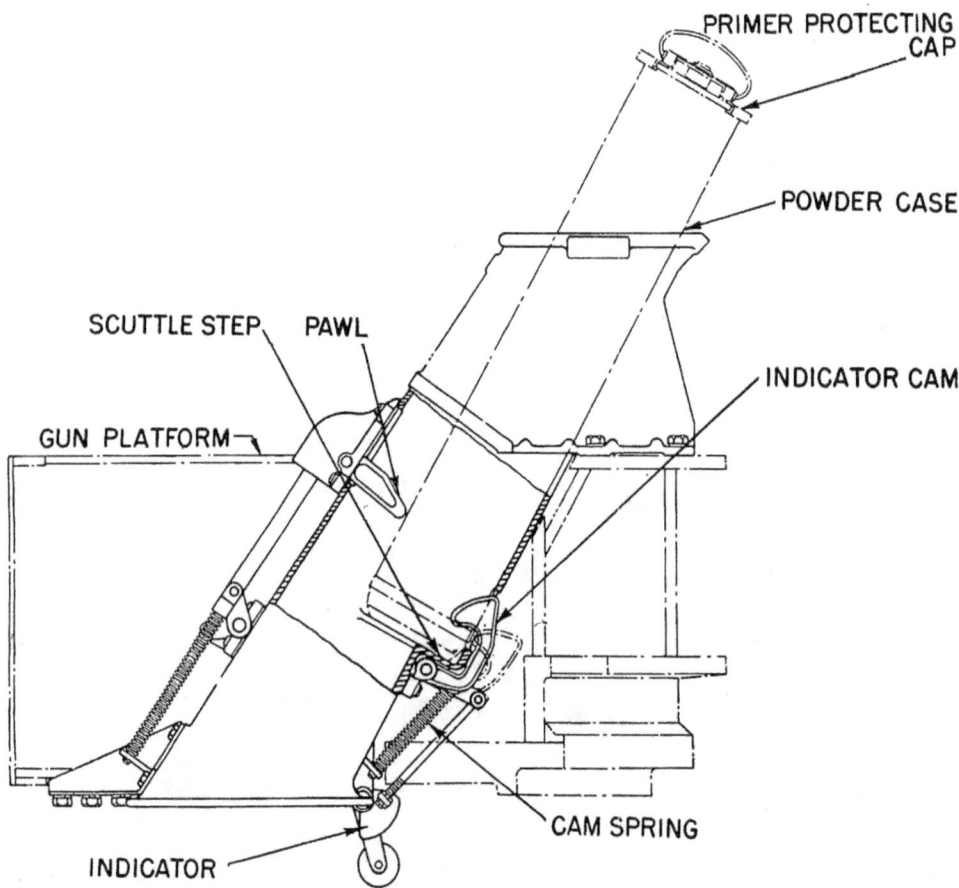

FIGURE 6D21. Powder scuttle; 5"/38 cal. gun (side elevation).

the index marks on the inner disk dials. The other method is to turn the handwheels until the ring dials show the desired settings opposite the fixed pointers. In the latter method the sight-angle setting desired would probably be expressed in yards, and this setting would be made on the range dial.

Since the telescopes are mounted on the carriage instead of the slide, the elevating prisms are rotated as the gun elevates in order to maintain the vertical angle set between the line of sight and the axis of the bore. The sight mechanism automatically moves the prisms to keep this vertical angle equal to that set on the sight-angle dials.

On some enclosed mounts an open sight is installed on top of the shield at the mount captain's hatch. This sight is used by the mount captain to assist the pointer and trainer to get on the target

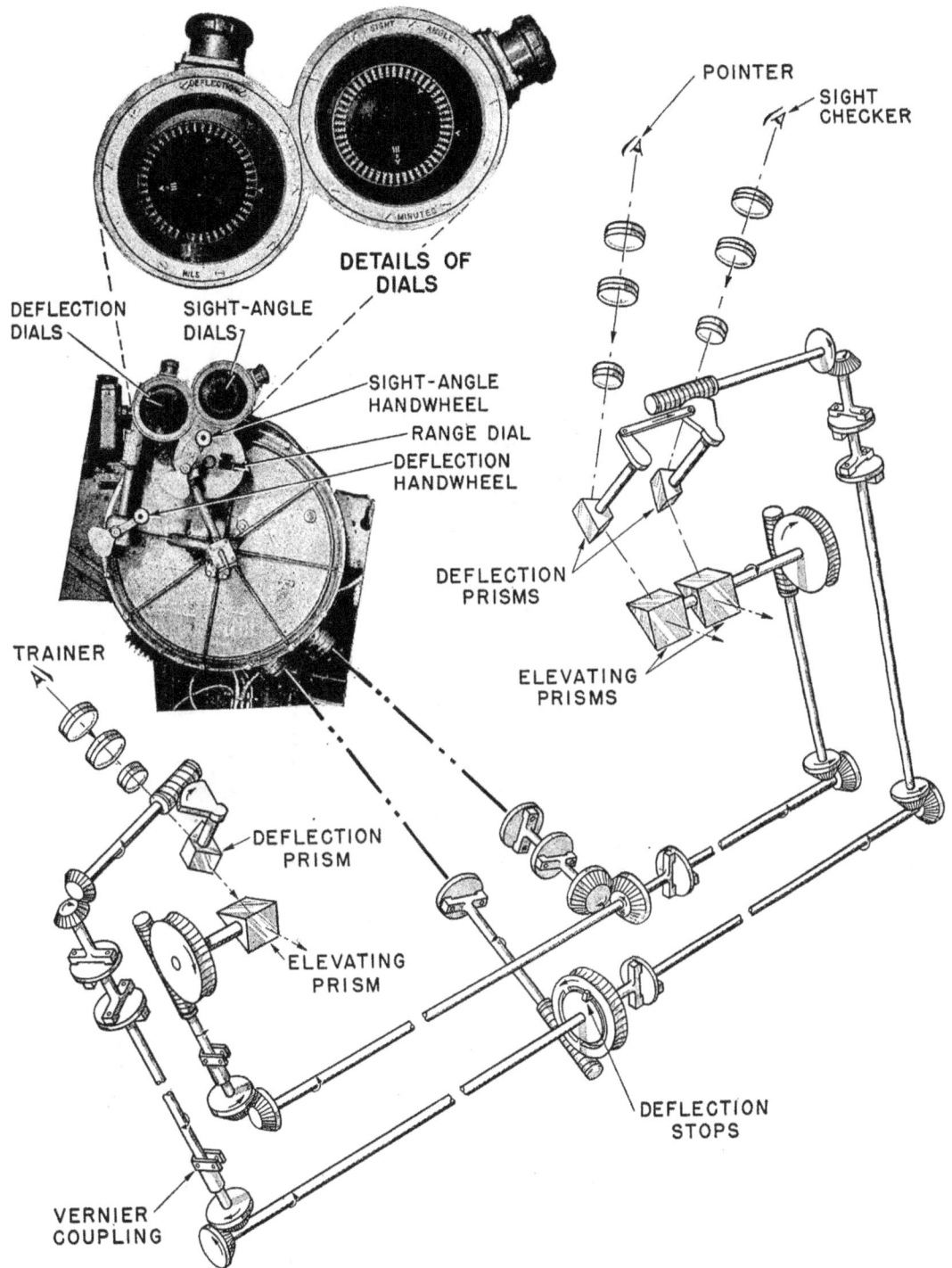

FIGURE 6D22. Sight assembly; 5"/38 cal. gun.

SEMI-AUTOMATIC GUNS

rapidly. The sight is connected to indicators at pointer's and trainer's stations by means of a mechanical linkage. The pointer and trainer turn their handwheels until the indicators show that they are on the designated target.

6D23. Shields. These enclosures are box-like structures of steel plate which provide weather, blast, and splinter protection for the gun crew. Figure 6D2 shows the exterior appearance of a typical enclosed mount. On single mounts the plate thickness varies from $\frac{1}{8}$ inch to $\frac{1}{2}$ inch, but the recent tendency is to standardize on $\frac{1}{4}$ inch of ballistic steel plate.

Spotting windows and telescope ports are fitted with hinged shutters and handcranks so that they can be opened and closed from the inside. There are doors on both right and left sides near the aft end, through which the operating personnel enters or leaves the gun house. Other doors are provided to make the operating mechanisms more accessible for inspection and repair.

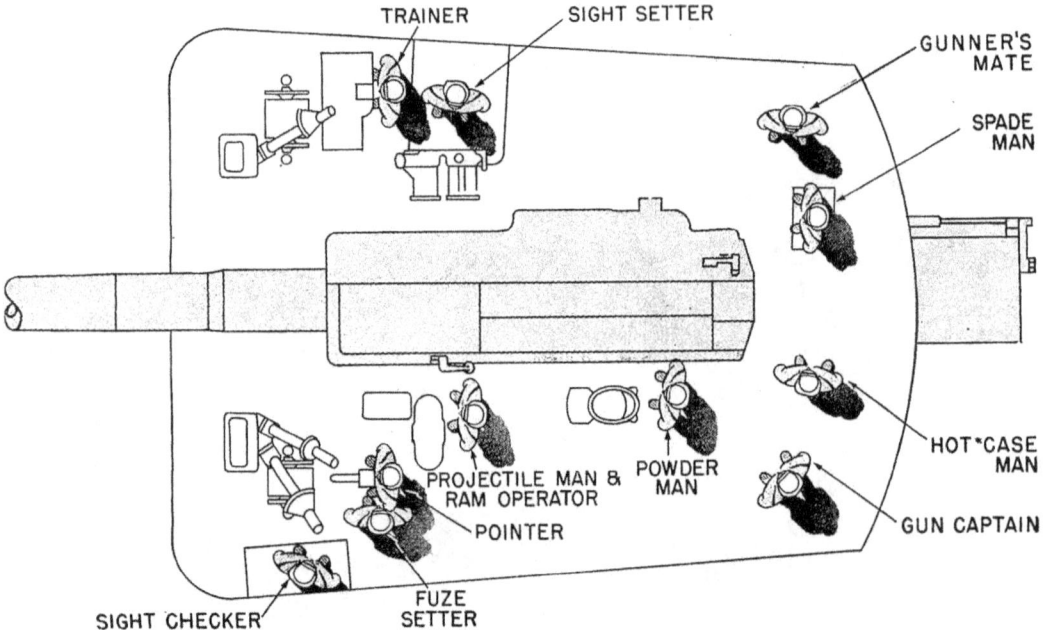

FIGURE 6D23. Personnel arrangement—enclosed mount; 5"/38 cal. gun.

A case-hatch and deflecting hood, placed at gun-platform level in the center of the rear plate, provides an opening through which the empty cases may be passed to the deck outside. A ceiling hatch for the mount captain is located near the aft end of the gun house. On some installations this hatch is equipped with a blast hood or shield.

6D24. Personnel. The minimum crew to man the gun house and handling rooms of an enclosed mount numbers 17 men. Figures 6D23 and 6D24 show the disposition of operating personnel.

6D25. Twin mounts. The twin mount, shown in figures 6D1, 6D25, and 6D26, is an enclosed assembly of two 5"/38 cal. guns mounted on a single carriage and stand similar to that used for the single gun. It is equipped with its own ammunition hoists and handling room beneath the gun platform. The major differences between the single and twin mounts are:

Gun and breech assembly. The two guns and breech assemblies are the same except for changes to make right-hand and left-hand assemblies which can be loaded and operated from the inboard sides of the guns. The operating principles are not changed by these modifications.

CHAPTER 6

Shield. All twin mounts are totally enclosed in a shield of armor plate which varies in thickness from ¾ inch on light cruisers to 2½ inches on battleships. The box-like structure is about 15½ feet long, 15 feet wide, and 10 feet high.

Training and elevating gear. Each of the power drives for elevating and training has its own electric motor. The horse-power ratings are approximately 40 for train and 7.5 for elevation. Both guns are elevated simultaneously by the elevating gear. No provision is made to elevate the guns separately, but clutches are provided so that either gun may be disconnected from the elevating drive. Figure 6D26 shows the location of elevation and training drives and the control gear.

In addition to the three methods of control described in Article 6D18, the twin mount has *hand control*. In this method the handwheels are directly connected to the control mechanism in the hydraulic pump to provide control of the power drive in the event that the indicator-regulator

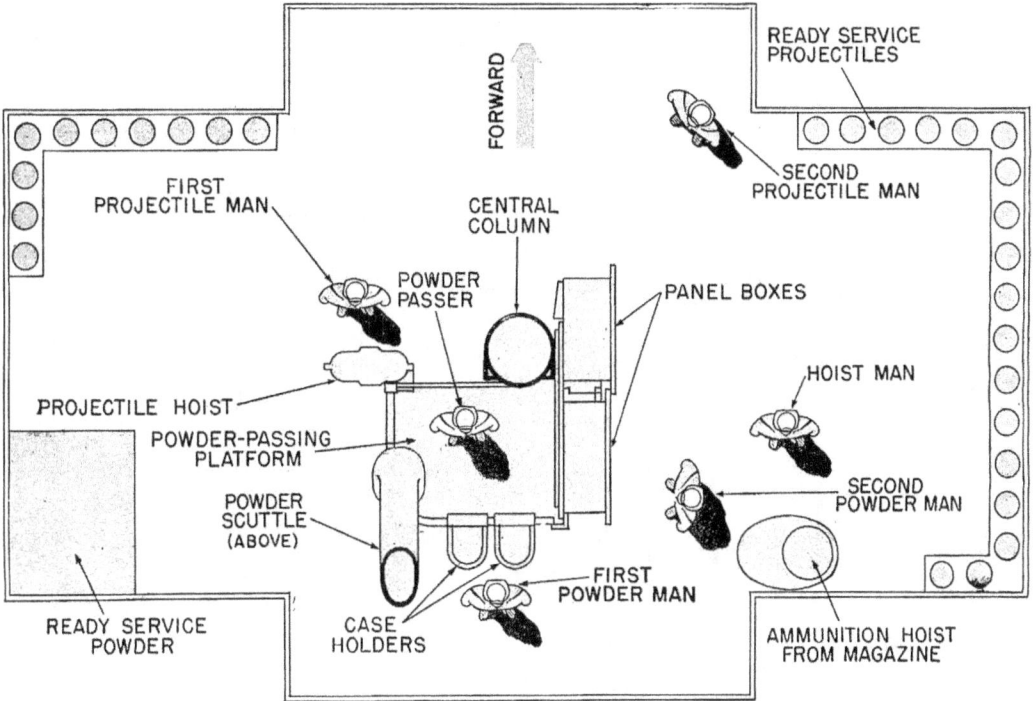

FIGURE 6D24. Personnel arrangement—ammunition-handling room; 5"/38 cal. gun.

fails. High and low speeds in ratios of 8⅓ to 1 for training and 5 to 1 for elevating are available for both *local control* and *hand control*.

Rammer mechanism. The twin mount uses a rammer mechanism which is very much like that described in Article 6D14. The major differences are in the length of rammer cylinder and the method of connecting the piston rod to the crosshead.

For the arrangement shown in figure 6D15, the piston must move in the rammer cylinder a distance equal to a spade travel of approximately 4½ feet. For twin-mount installations the piston travel, and hence the length of cylinder, are shortened by gearing the piston rod to the crosshead. The piston rod mounts a gear which meshes with a rack secured to the crosshead and another fixed rack on the slide. When the gear is moved by the piston it is forced to rotate as it passes over the fixed rack. The combined rotation and travel of the gear drives the rack on the crosshead (and hence the spade) twice as far as the piston moves. The shorter cylinder used with

RESTRICTED

SEMI-AUTOMATIC GUNS

this assembly is mounted entirely within the shield for greater protection. This rammer arrangement can also be used on single enclosed mounts.

Projectile hoists. Each gun has a projectile hoist similar to that described in Article 6D19. These are located inboard between the two guns as shown in figure 6D25.

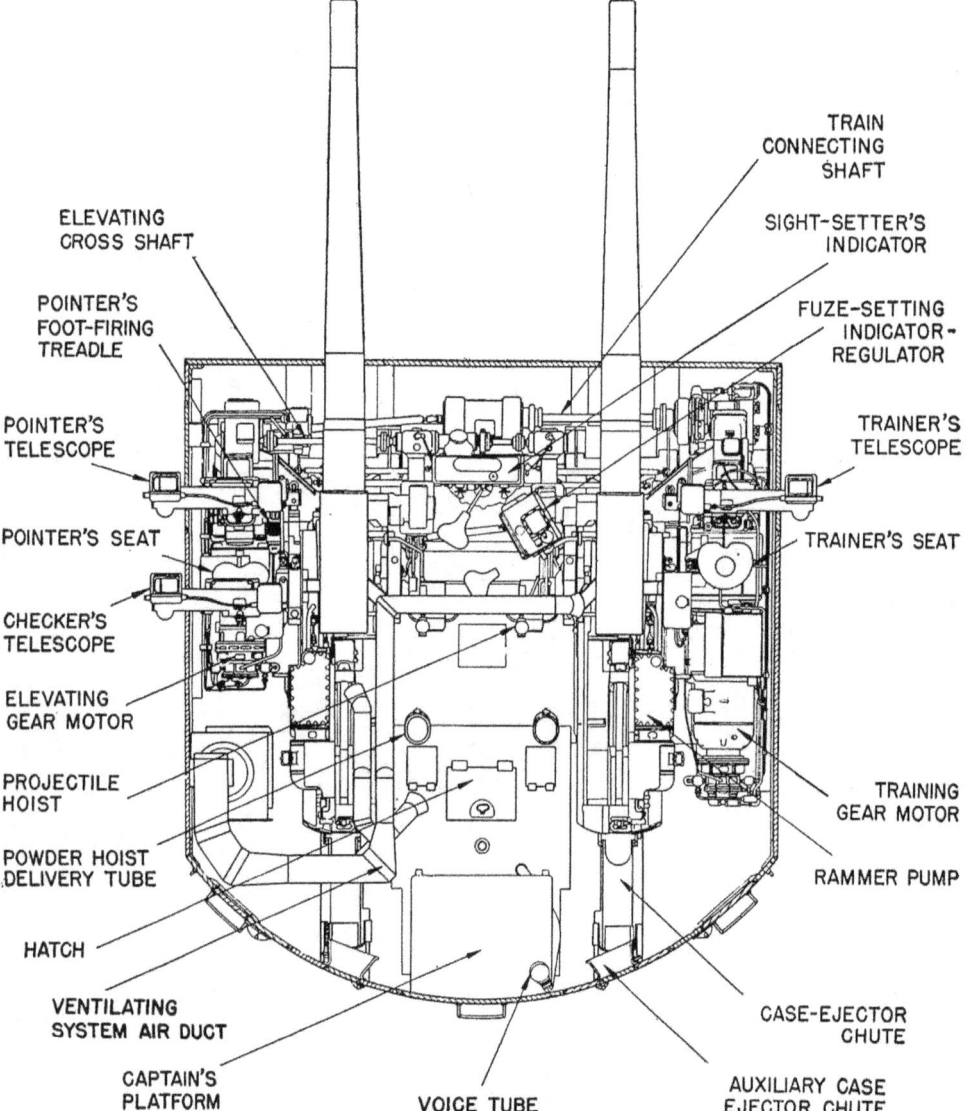

FIGURE 6D25. Twin mount—plan view; 5"/38 cal. gun.

Powder hoists. Each gun is served by a separate powder hoist which delivers the powder cases through delivery tubes located inboard between the guns, and abaft the projectile hoists. These powder hoists are endless chains enclosed in tubes driven by an electric-hydraulic drive which is semi-automatically controlled.

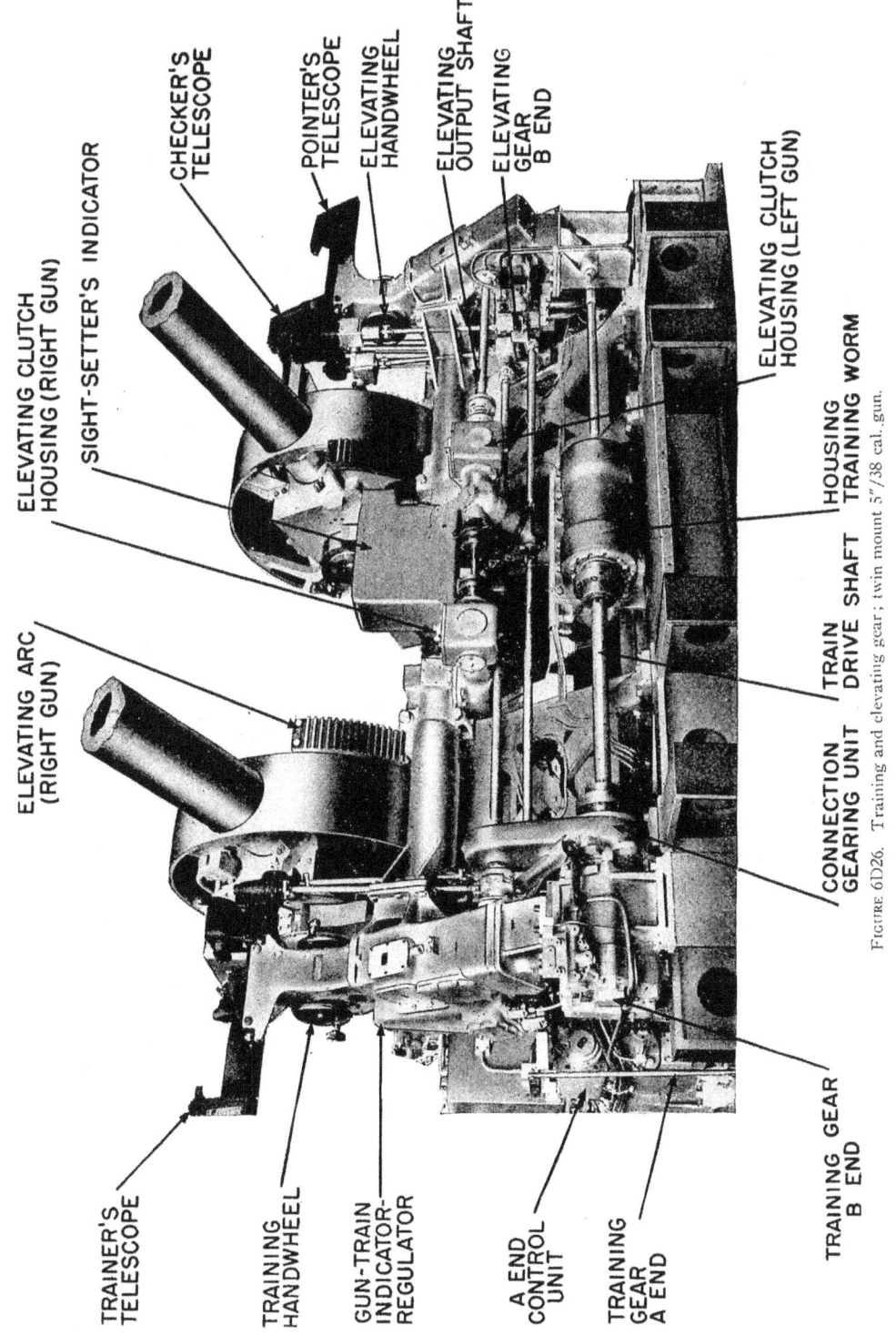

FIGURE 6D26. Training and elevating gear; twin mount 5"/38 cal. gun.

SEMI-AUTOMATIC GUNS

For hoisting, the chain is driven continuously in one direction and the motion is so controlled that one powder case is delivered per cycle of operation. Interlocks are provided which function to prevent operation when there is a case on the hoist at the delivery end. The rotation of the chain may be reversed if it is desired to return ammunition to the handling room. At the lower end is a handwheel which is geared to the power-drive output shaft through a clutch. This provides an auxiliary manual drive for emergency use.

Sights. Figure 6D25 shows the location of the pointer's, trainer's, and checker's telescopes, and the *sight-setting indicator* or sight-setting gear. The telescopes used and the methods of setting sights are similar to those discussed in Article 6D22. The major difference is in the location of the sight-setting gear.

Personnel. A typical crew to man the twin-gun compartment and the handling rooms is at least 27 men. The assignment and location of these men is as follows:

NO. OF MEN AND ASSIGNMENT	LOCATION
1. Gun compartment.	
1 Mount captain	rear center
1 Trainer	right forward
1 Pointer	left forward
1 Checker	left forward
1 Sight setter	center forward
1 Fuze setter	center
2 Projectile men (right and left guns)	center
2 Powder men (right and left guns)	center
2 Hot-case men (right and left guns)	rear center
2 Spade men (right and left guns)	rear (right and left)
2. Upper handing room.	
1 Handling room captain	
2 (or more) Projectile-hoist loader (left and right guns)	
2 (or more) Powder-hoist loader (left and right guns)	
3. Lower handling room.	
4 (or more) Lower-hoist projectile men	
4 (or more) Lower-hoist powder men	

CHAPTER 7

MACHINE GUNS

A. INTRODUCTION

7A1. General. As previously defined in this text, machine guns are case guns in which part of the energy of explosion is utilized to eject the empty cases and to energize a device which automatically loads the cartridges and fires the gun continuously as long as the trigger is operated and the ammunition supply is maintained. All short range (2,000 yards) and medium range (4,000 yards) AA guns aboard ship are machine guns. The larger-sized machine guns such as the 40 mm. and the 1.10 inch, are sometimes called heavy machine guns or simply AA guns.

The heavy machine guns used by the Navy are the water-cooled, pedestal-mounted 40 mm. and 1.10 inch. The 1.10-inch quad mounts were formerly used on the larger ships for medium range AA fire. The 40 mm. twin and quad mounts have replaced the 1.10-inch guns on the heavier ships and are now used extensively on many of the smaller ships as well. The 1.10-inch gun is installed on a number of destroyer escorts, and may still be found on some ships of other classes. Both guns have about the same firing rate and may be controlled either locally or by director.

The 20 mm. and caliber .50 machine guns are generally classed as light machine guns, and they are used on ships and at shore stations for short-range AA fire. The 20 mm. with its heavier, fuzed projectile and bursting charge is more effective than, and has almost entirely replaced, the caliber .50 gun on the larger ships. However, the caliber .50 is still used on a number of small vessels such as P.T., S.C., and landing barges. Both of these lighter machine guns are mounted on free-swinging pedestal-type mounts. The 20 mm. guns are all air cooled, while both air- and water-cooled designs are used for the caliber .50. The firing rate for the caliber .50 varies considerably with design, but it may be more than twice as fast as that of the 20 mm. Both use tracers to control firing.

The operating principles and some characteristics of the 40 mm., 20 mm, and caliber .50 machine guns are described in this chapter. No attempt has been made to describe these weapons completely, and more detailed information can be obtained from appropriate Ordnance Pamphlets.

B. 40MM. GUN ASSEMBLIES

7B1. General. The 40 mm. gun is a recoil-operated, heavy machine gun designed primarily for AA fire. Its distinctive features include: (1) liquid-cooled *barrel*, (2) vertical sliding-wedge *breech mechanism*, (3) hand-fed *automatic loader*, and (4) spring-operated *rammer*. Another notable characteristic of this gun is that the trigger mechanism controls the rammer operation only, and once the ramming cycle is started the round will be loaded and automatically fired without further control. The gun can be operated either fully automatic or single fire. In automatic the firing rate is about 120 rounds per minute.

For naval use the guns are assembled in pairs, and one or two of the pairs is used to make up the twin and quad assemblies shown in figures 7B1 and 7B2. The individual gun mechanisms are alike except for the changes necessary to make them right and left guns in the pairs. Both twin and quad gun assemblies are used on destroyer escorts, destroyers and all classes of larger naval ships.

The conventional pedestal-type mounts, for both twin and quad assemblies, have power-operated elevating and training gear which position all guns as a unit both in elevation and in train. The power drives are controlled either at the mount by means of the pointer's and trainer's handwheels, or from a director through an electrical-control system. Some of the later mounts are provided with a single control lever, or joystick, at the pointer's station which can be used to control both

MACHINE GUNS

elevation and train drives. This type of control provides a unified and effective means for controlling gunfire particularly when employing tracers.

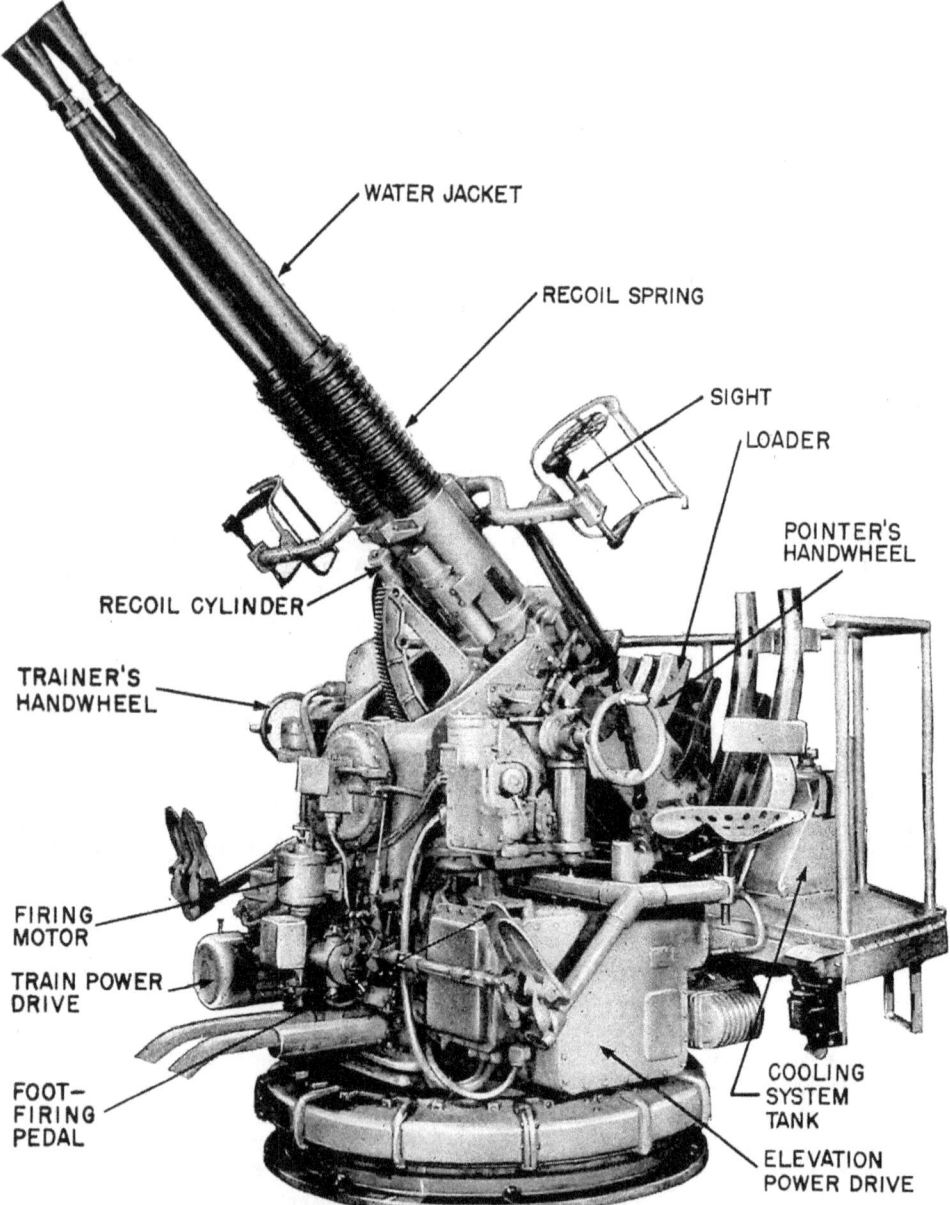

FIGURE 7B1. 40 mm. twin gun assembly.

The ammunition is of the fixed type, loaded into clips having a capacity of 4 rounds each. Each cartridge weighs 4.8 pounds and the projectile alone weighs slightly less than 2 pounds. The projectile is provided with a tracer, and a nose-type impact fuze which is armed by the rotation of the

CHAPTER 7

projectile in flight. The tracer is effective for about 8 or 11 (new Mark) seconds. When it is burned out, a self-destructive fuze is ignited and the projectile burster charge is detonated. The service charge produces an initial velocity of about 2,800 f.s. The maximum horizontal and vertical ranges obtained before self-destructive action occurs are 4,000 to 5,000 yards. The maximum horizontal range without the self-destructive action is approximately 11,000 yards.

The gun mechanism, shown in figure 7B3, a and b, consists of the following five major components: (1) *slide assembly*, (2) *barrel assembly*, (3) *breech mechanism*, (4) *loader assembly*, and (5) *recoil-counterrecoil system*.

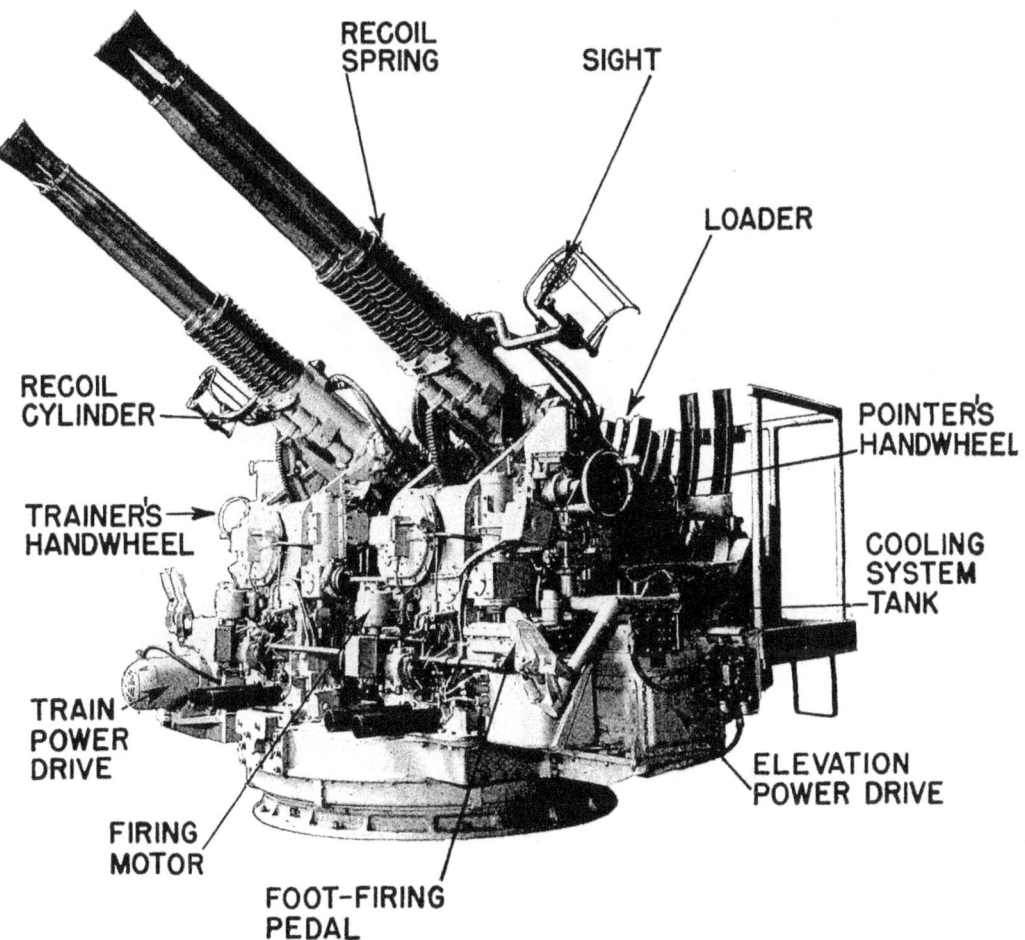

FIGURE 7B2. 40 mm. quad gun assembly.

7B2. **Slide assembly.** The slide assembly is shown in figure 7B6. It consists of the box-shaped rear section and a cylindrical forward section which serve to support and provide working surfaces for the other parts of the gun mechanism. Two slides, arranged for right and left-hand mounting are bolted together. The trunnion on the outboard side of each slide is supported on roller bearings mounted on the carriage. A single elevating arc is attached to the bottom surface of the slide assembly. The cylindrical front portion of the slide provides a bearing surface for a bearing sleeve which surrounds the aft end of the barrel assembly.

151   **RESTRICTED**

MACHINE GUNS

7B3. Barrel assembly. The barrel assembly consists of the gun *barrel*, its *water jacket*, and the *recoil spring* (see fig. 7B3a and b). The barrel passes through the cylindrical fore part of the slide and is attached to the housing within by means of a bayonet joint.

The barrel is a single-piece forging about 60 calibers long. The twist of the rifling varies uniformly from one turn in 45 calibers at the breech to one turn in 30 calibers at the muzzle. The muzzle is provided with a *flash guard* or *flash hider* to shield the eyes of the gun crew.

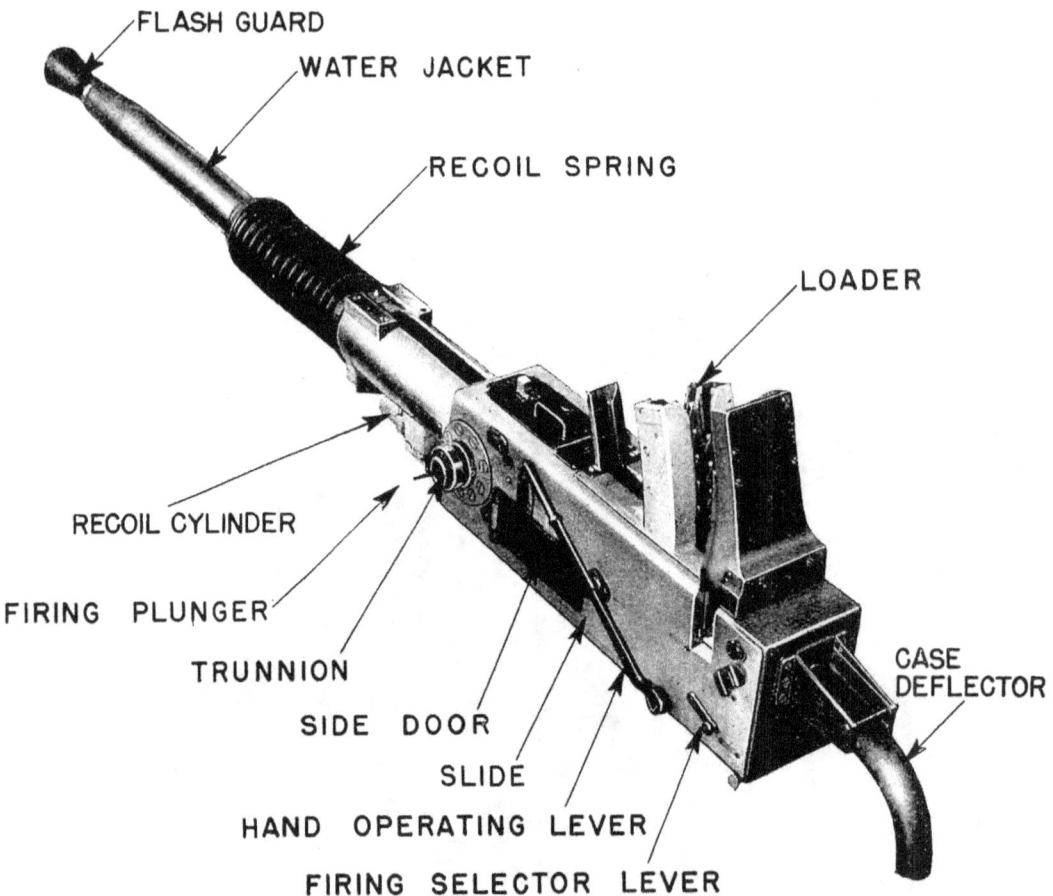

a. LEFT HAND GUN – EXTERIOR VIEW

FIGURE 7B3b. 40 mm. gun mechanism.

The barrel is covered by a water jacket in which a cooling liquid is circulated. Each gun has a complete forced-circulation system consisting of a tank mounted on the aft end of the carriage platform and an electric motor-driven circulating pump underneath the tank. Pipes and flexible tubing carry the cooling liquid to the water jacket. The tank holds about 10 gallons and the entire system about 14 gallons.

7B4. Breech mechanism. The breech mechanism, shown in figure 7B4, consists of *a housing*, *a breechblock*, and associated operating parts. The housing is mounted on bearing guides within

CHAPTER 7

the slide. It has interrupted threads which engage similar threads on the breech end of the gun barrel. The vertical recess accommodates the vertical sliding breechblock. The recoil piston rod is secured to lugs on the underside near the forward end and the rammer tray is attached to the after end.

Although the breech mechanism for the 40 mm. is similar to that used on semi-automatic guns, there are a number of differences in the operation of the two mechanisms. Figure 7B5 is a schematic drawing showing the relative positions of breech-mechanism components in a *left-hand gun*. The

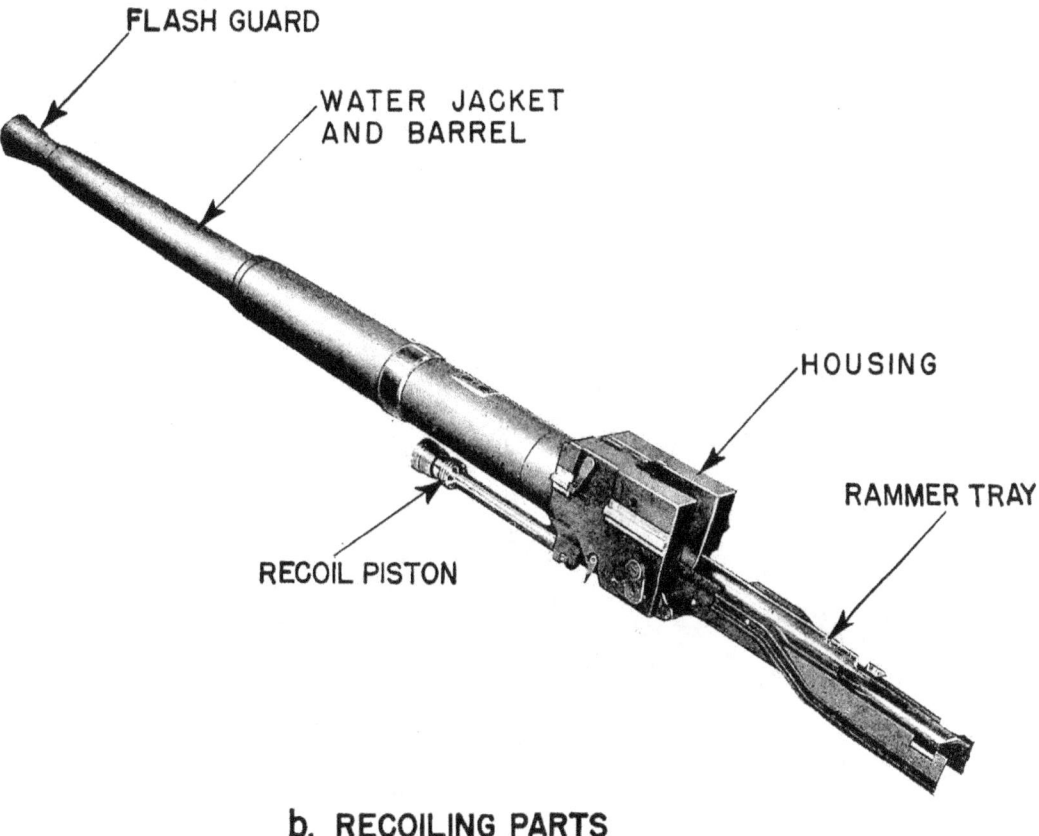

b. RECOILING PARTS

Figure 7B3b. 40 mm. gun mechanism.

arrangement for the right-hand gun is similar; the major difference being the interchange of positions for spring and operating crank so that the operating cam can be placed in an outboard door.

The *crankshaft*, which controls the motion of the breechblock, passes through the housing abaft the breechblock. Instead of a single breechblock crank at the center of the shaft working in a T-slot in the under side of the block, the operating shaft on the 40 mm. carries two cranks called the *right and left inner cranks*. These extend along the sides of the breechblock, and *crank lugs* on the ends fit into horizontal slots in the breechblock.

The spirally-wound *operating spring* has one end attached to the shaft and the other end to the enclosing cylindrical case secured to the housing. The spring is wound when the breechblock is lowered, and provides the force necessary to close the breech.

The *outer crank* used for the 40 mm. gun has a counterpart in the operating crank on the semi-au-

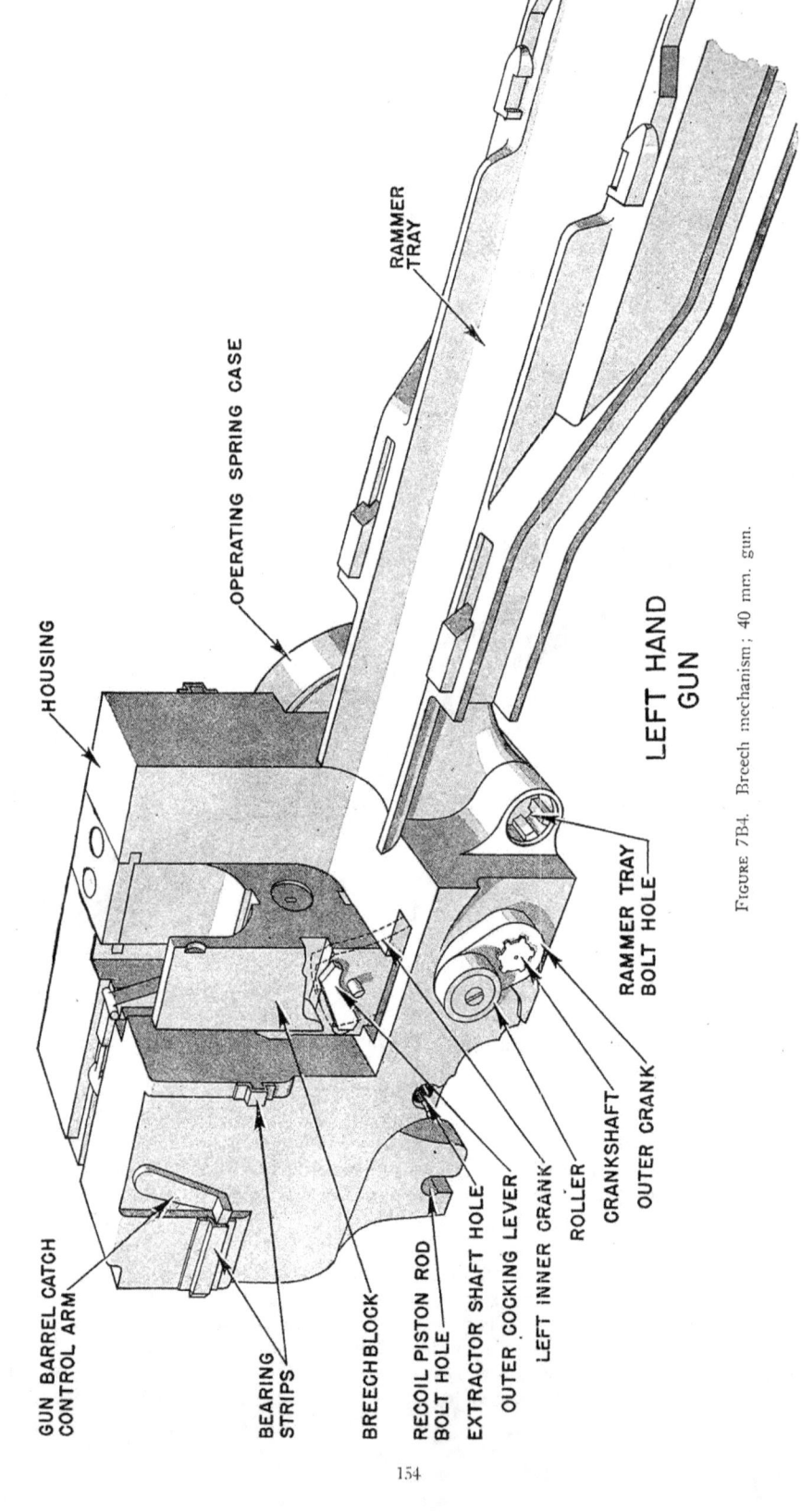

FIGURE 7B4. Breech mechanism; 40 mm. gun.

CHAPTER 7

tomatic guns. When the gun moves in recoil, a *roller* on the outer crank rides on the *operating cam* cut in the side door of the slide. This cam serves the same purpose as the operating cam in semi-automatic guns. At the moment of firing, the roller is at point A on the cam. During recoil the roller is forced downward as it follows the cam to point B. This motion turns the outer crank and the crankshaft so that the breechblock is dropped. During counterrecoil the breechblock is locked down by the extractors, as described below, so the roller follows the lower edge of the side door as the gun returns to battery. As the next round is rammed, the extractors release the breechblock and the operating spring turns the crankshaft to close the breech. This returns the crank

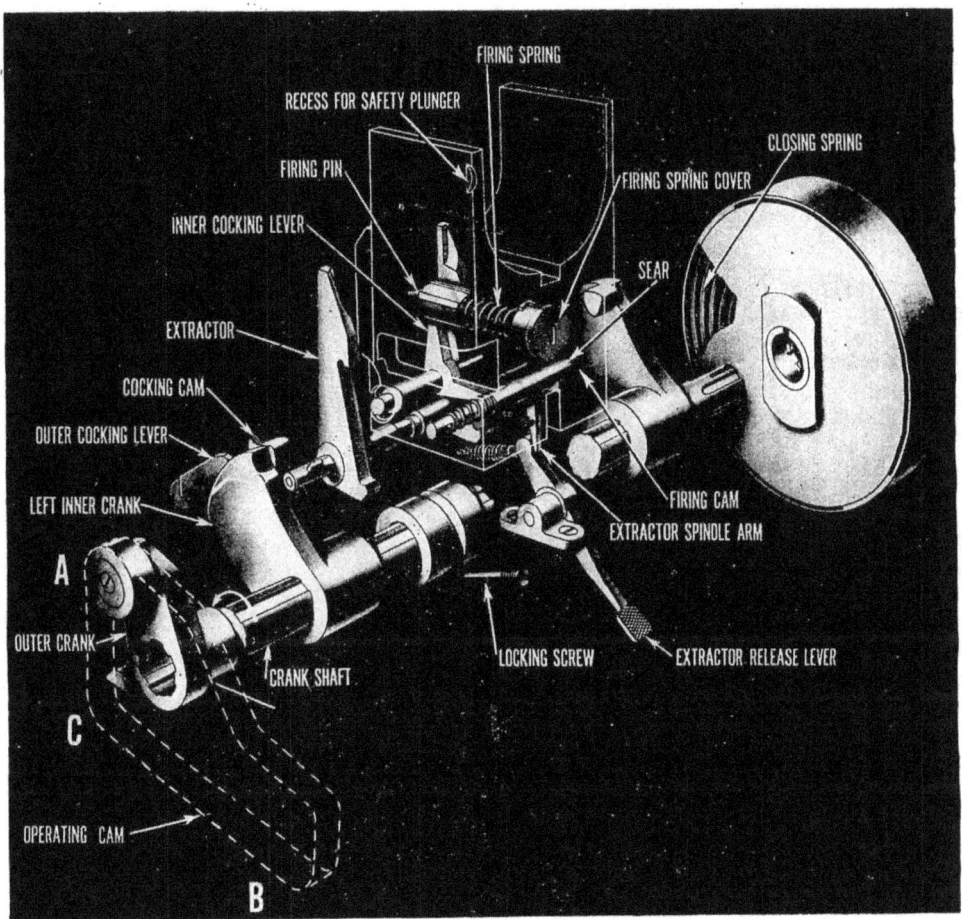

FIGURE 7B5. 40 mm. breech and firing mechanisms.

roller to position A so that it is ready to begin a new cycle. It is to be noted that the 40 mm. breechblock is lowered during recoil.

Figure 7B5 also shows that the *extractors* are not connected to the breechblock by means of slots and inner lugs as in the semi-automatic guns. The two extractors are pivoted on a shaft which is mounted transversely in the housing just forward of the breechblock. As the breechblock drops, the lower ends of the *extractor catch ribs* (on forward face of breechblock) strike the *extractor-operating lugs* and this forces the upper ends of the extractors aft. This action ejects the empty case and positions the *hold down hooks* so that they engage the catch ribs on the breechblock.

155

RESTRICTED

MACHINE GUNS

The breechblock is then held down until the upper ends of the extractors are moved enough to disengage the hooks. This is done when a case is rammed into the chamber.

7B5. Firing mechanism. The percussion firing mechanism is mounted within the breechblock. Figure 7B5 shows the relative location of parts. The mechanism is cocked and fired automatically by cams on the inner cranks. As the block is lowered, the *cocking cam* on the left inner crank bears on the *outer cocking lever* extending from the left side of the block. This action rotates the *inner cocking lever* enough to cock the mechanism. As the block rises, the *sear firing cam* on the end of the *sear plunger* bears on the *firing cam* which projects from the right inner crank. When the breech is closed the sear plunger is forced inboard far enough to release the inner cocking lever and the firing pin. It is to be noted that the firing mechanism is cocked when the breech-

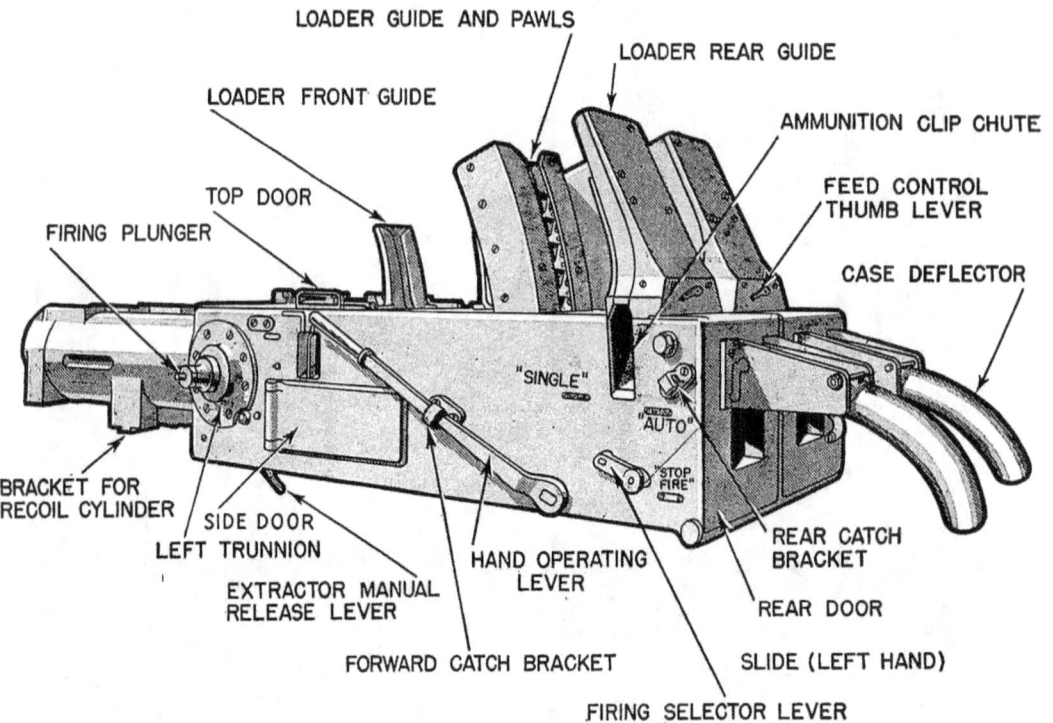

FIGURE 7B6. Arrangement of slide and loader; 40 mm. gun.

block is lowered and automatically fired when the breechblock is raised, and that the trigger does not actuate any part of the firing mechanism.

7B6. Loader assembly. The loader assembly, shown in figures 7B6 and 7B7, automatically feeds cartridges to the rammer tray and then catapults them into the firing chamber. The main operating parts are the: (1) *loader*, (2) *rammer tray*, (3) *rammer*, and (4) *rammer catch levers*.

The principal parts of the loader consist of the *front, center,* and *rear guides,* the *stop* and *feed pawls,* and the *star wheels.* These are mounted above the rammer tray. The three guides position the clips of ammunition and guide the motion of the cartridges as they are fed into the loader. The stop pawls are all pivoted in *fixed pawl holders* while the feed pawls are pivoted in *feed pawl holders* which are free to move up and down with respect to their guides. The pawls are spring actuated and so pivoted that a cartridge can be forced downward into the loader but is restrained from moving in the opposite direction.

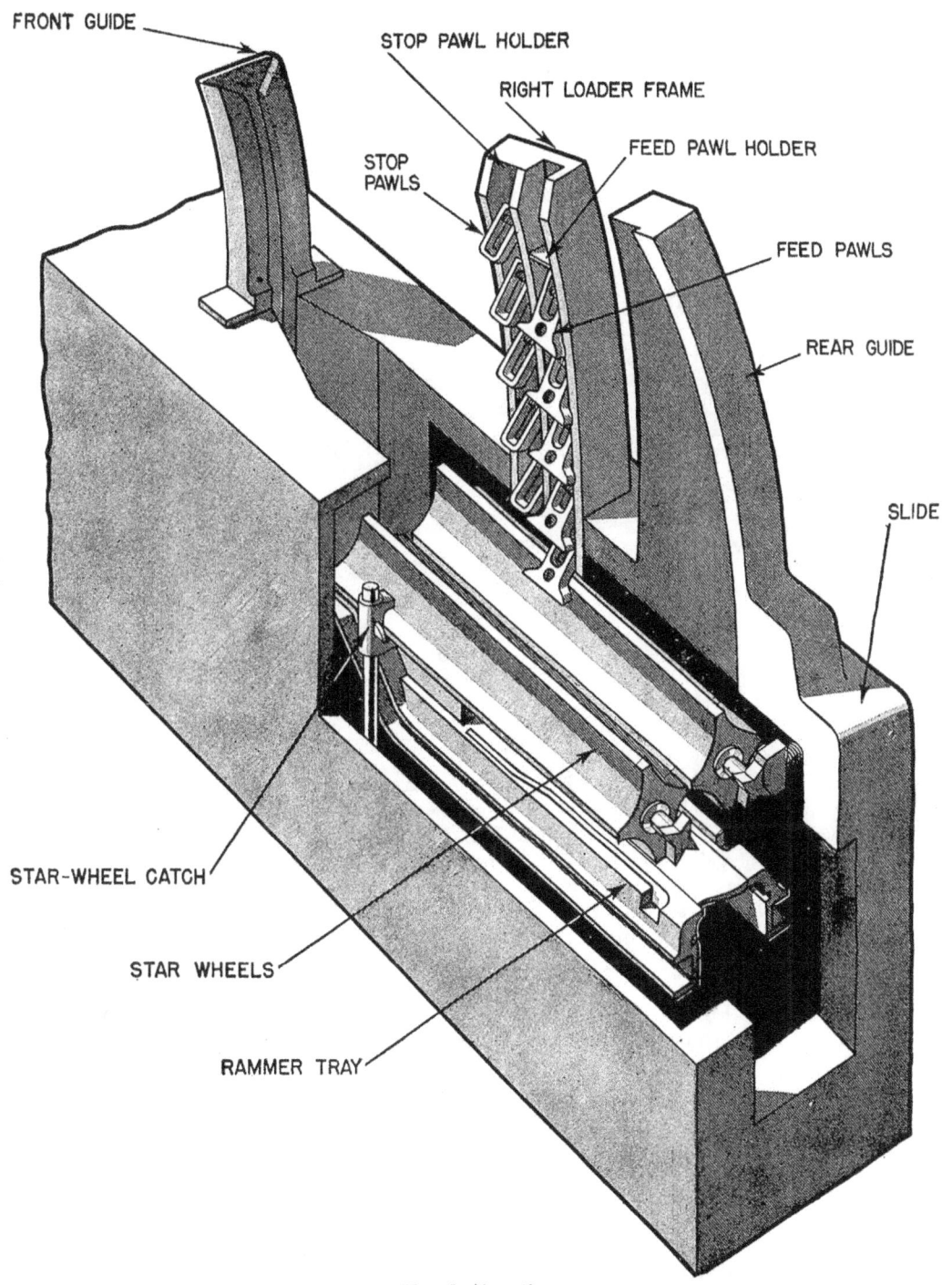

FIGURE 7B7. Loader; 40 mm. gun.

The feed-pawl holders are connected to rods through a spring device which permits relative motion between the rod and holder in case of a jam. Rollers at the lower ends of the rods ride in *guides,* or cam grooves, on the sides of the rammer tray, shown in figure 7B8. The guides force the feed pawls upward on recoil and back down again on counterrecoil. The feed pawls ride over the cartridges as they move upward and engage and force the cartridges downward as they return.

The two *star wheels* (feed cylinders), mounted below the stop and feed pawls, form a trough in which the bottom cartridge rests. The star wheels are locked by *star-wheel catches* (see fig. 7B7) which are tripped by the rammer-tray pawls during counterrecoil. When these star-wheel catches are released, and downward pressure is applied to the lower cartridge, both star wheels will turn inward one-quarter of a turn and the bottom cartridge will pass through to the rammer tray below. The downward pressure required to turn the star wheels is provided by the feed pawls.

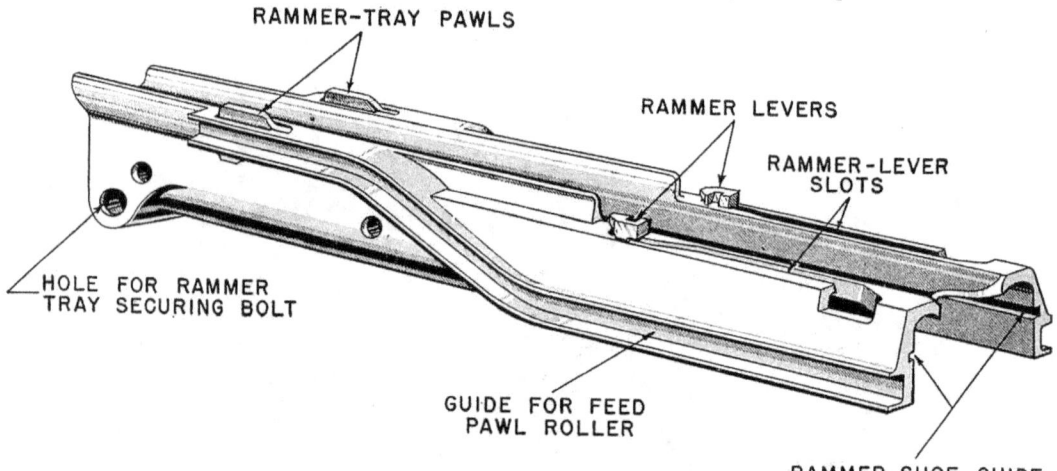

FIGURE 7B8. Rammer tray; 40 mm. gun.

The rammer tray, shown in figure 7B8, is attached to the aft end of the housing and hence is a recoiling part. It contains the spring-operated rammer, supports the cartridge before it is rammed, and provides a guide for the ejected case. The *guides* in the sides of the tray actuate the feed pawls during recoil and counterrecoil and rammer-tray pawls on the top face of the tray unlock the star-wheel catches just before the cartridges are forced downward. The *rammer levers* extend upward through *rammer-lever slots* and engage the rim of the cartridge case during the ramming operation. The upper surface of the tray is trough shaped, and this trough lines up with those in the housing and breechblock when the breechblock is down. These form a continuous tray across which the cartridges and empty cases travel to and from the firing chamber.

7B7. Rammer. Figure 7B9 is a schematic diagram of the loader, rammer tray, and breech mechanism. This diagram, while not correct in mechanical details, illustrates the operating principles.

The rammer is built into the under side of the rammer tray. Its principal parts are: (1) *rammer spring,* (2) *rammer rod,* (3) *rammer shoes,* and (4) *rammer-catch levers.* The rammer spring is placed in a hollow cylinder bored in the underside of the rammer tray near the forward end. The rammer rod is connected to the rammer shoe which rides back and forth in guide slots cut in the downward-projecting sides of the rammer tray shown in figure 7B8. When the rammer shoe is drawn to the rear the spring is compressed and provides the force necessary for the ramming action which is described below. The rammer levers are attached to the rammer shoe and extend upward through slots in the rammer tray. These slots are spread apart at the forward end so that the rammer levers will be disengaged from the rim of the cartridge at the end of rammer travel.

CHAPTER 7

7B8. Check levers. Three levers mounted side by side on a common shaft secured to the floor near the aft end of the slide control the rammer action. These are shown in figure 7B9. Whenever the aft end of any of these levers is elevated, it will engage the rammer shoe as it starts forward in counterrecoil, and will hold the shoe to the rear as the rammer tray returns to battery. This action compresses the spring and cocks the rammer. The catch levers are separately controlled and are designated as follows: (1) *loader catch lever*, (2) *trigger catch lever*, and (3) *tray catch lever*.

The catch levers are of varying length so that they do not latch the rammer shoe in the same position. The levers release the rammer shoe in the following order: tray catch lever, loader catch lever, trigger catch lever.

The loader catch lever is maintained in the releasing position through the action of the *feed-control lever* in the rear guide, whenever the loader contains more than one round above

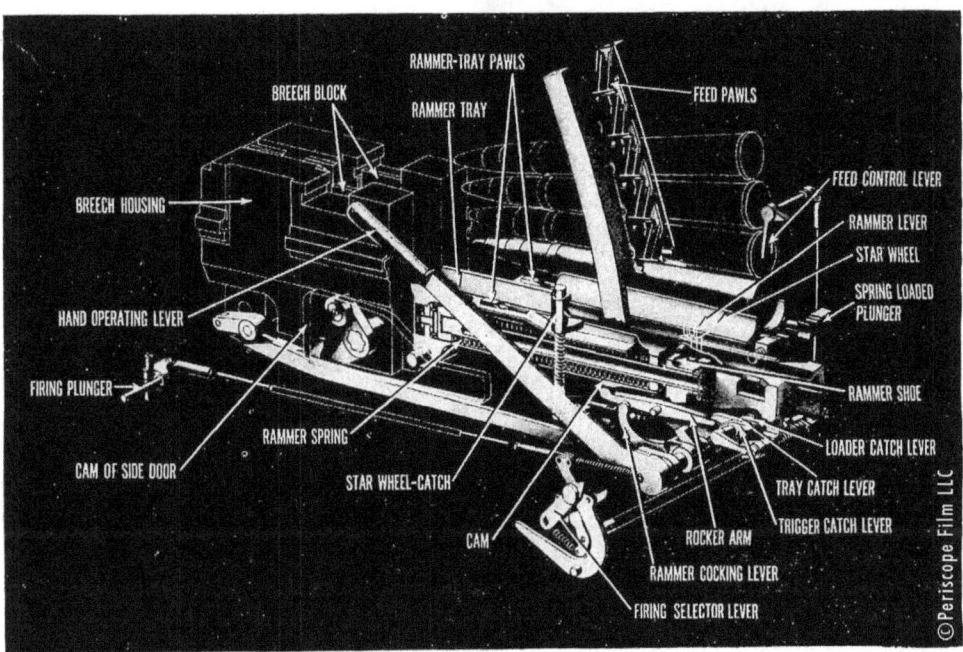

FIGURE 7B9. Loader and breech assemblies—40 mm. gun; schematic diagram.

the star wheels. This lever is provided to prevent the gun from losing its automatic action when ammunition is supplied at a rate less than the rate of fire. This lever releases the rammer shoe automatically when additional ammunition is loaded. The *feed-control thumb lever* (on older models) can be set to keep the loader catch lever continually depressed.

The trigger catch lever is operated by the *trigger* and is maintained in the releasing position whenever the trigger is in the firing position.

The tray catch lever is operated through the *rammer rocker arm* by the *cam* on the bottom of the rammer tray, and is in the releasing position only when the tray is within about one inch of battery position. In automatic operation the tray catch lever engages the rammer shoe at the beginning of counterrecoil, holds it while the rammer tray returns to battery, and then releases it as the tray nears the end of counterrecoil.

7B9. Ramming cycle. The rammer cycle in automatic fire is illustrated in figure 7B10. It is assumed that the gun is fully loaded and is to be fired in *automatic*. The trigger and loader catch levers are both depressed when the firing key is pressed and the loader is filled with ammunition. Hence, the tray catch lever is the only one that can engage the rammer shoe during the

RESTRICTED

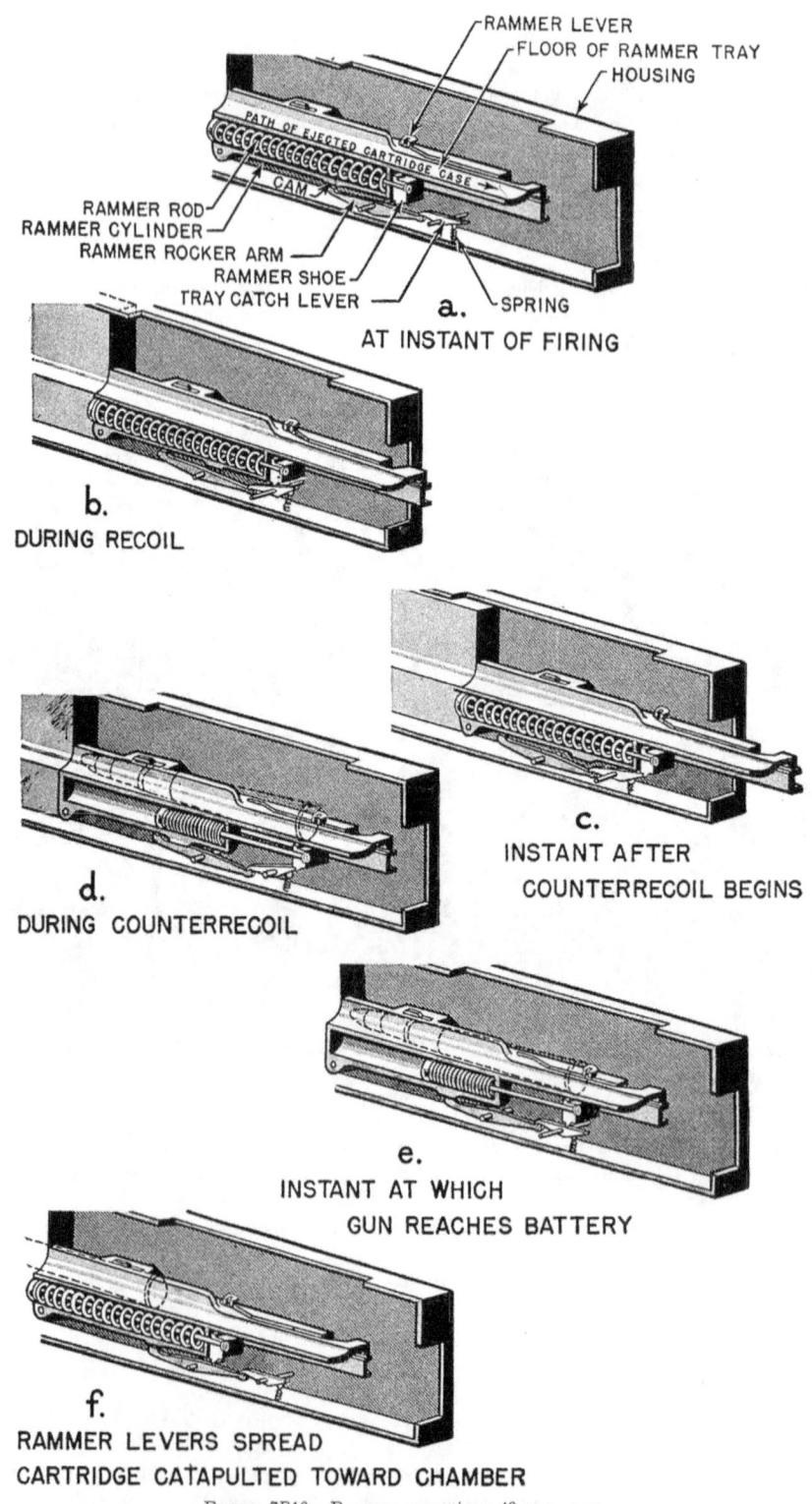

FIGURE 7B10. Rammer operation; 40 mm. gun.

ramming cycle. At the time of firing the rammer shoe is forward, where it has stopped after feeding a cartridge into the gun. The rammer tray is in battery and the rammer rocker arm holds the tray catch lever depressed. After firing, the gun recoils, the tray catch lever springs up, and the rammer shoe is caught and held as the gun starts to counterrecoil. The rammer spring is compressed as the loading tray returns to battery. The tray catch lever is depressed near the end of counterrecoil, the rammer levers are driven forward by the springs, and the cartridge is catapulted into the firing chamber. The gun fires and the cycle repeats. When either the loader or trigger check levers are released the rammer will be held in the cocked position.

7B10. Recoil-counterrecoil system. The recoil-counterrecoil system for each gun consists of a recoil spring fitted over the aft end of the barrel assembly and a hydraulic recoil unit secured to the underside of the slide with the piston rod attached to the housing. The spring is compressed during recoil and returns the gun to battery. As the recoil piston moves back and forth in the hydraulic cylinder a fluid mixture of glycerine and water is forced through a series of orifices. Both recoil braking and counterrecoil buffing are accomplished in this single cylinder unit. A needle valve adjusts orifice size, and its setting determines the time required for counterrecoil. Thus the rate of firing can be controlled to some extent. The needle-valve adjustment has no effect on length of recoil. The normal recoil is from $7\frac{1}{2}$ inches to 8 inches, and for satisfactory operation it should be within this range.

7B11. Trigger mechanism. The trigger mechanism is that part of the firing attachments mounted within the slide. It is operated by the *firing plunger* (fig. 7B6) which passes through the trunnion. The firing plunger, in turn, is actuated by the firing-control mechanism described below.

Motion of the *trigger*, shown in figure 7B9, is controlled by the *firing selector lever* shown in figure 7B6. When this lever is at *stop fire* the trigger mechanism is locked so that the trigger cannot be moved to depress the trigger catch lever. On *auto* the trigger can be set for and maintained in the firing position. On *single* the trigger can be moved enough to depress the trigger catch lever and then it is automatically released so that the trigger catch lever can return to its non-firing position.

7B12. Firing-control mechanism. The firing-control mechanism includes those devices which are used to control the operation of the trigger mechanism. It consists of a pointer's *foot-firing pedal*, a *firing motor*, a *firing solenoid*, a *firing stop mechanism* and the associated mechanical linkages which connect these components to the trigger mechanism. The functions of these components are described in the three methods of firing discussed below:

1. *Firing from the pointer's station with firing motor running.* The foot pedal is depressed a short distance and this motion closes the circuit to the firing solenoid. The solenoid operates a clutch which connects the firing motor to the trigger mechanism through the connecting linkages. The firing motor provides the power necessary to operate the firing-control mechanism.

2. *Firing from the pointer's station without firing motor.* The foot pedal is depressed to the limit of its travel and this motion is transmitted through the mechanical linkage to the trigger mechanism.

3. *Firing from a remote point.* This is accomplished by means of the firing solenoid. When the solenoid is energized by closing a firing key, it operates the clutch and connects the firing motor to the trigger mechanism as in 1 above.

The firing stop mechanism functions to make the firing-control mechanism inoperative when the gun is pointed toward any part of the ship's structure. When clear of the structure, the mechanism is automatically reset by releasing the firing pedal or deenergizing the solenoid, and the gun can again be fired by one of the above methods.

7B13. Hand-operating mechanism. This mechanism, located on the same side of the slide as the trigger mechanism, provides the means for performing manually the operations required to prepare the gun for firing; i.e., operations which are otherwise accomplished automatically during recoil and counterrecoil. It consists of the *hand operating lever* and its internally connected levers, as shown in figure 7B9. One lever provides an interlock with the trigger mechanism to prevent firing when the hand operating lever is pulled to the rear. A second lever is used to cock the rammer. A hook on the hand operating lever engages the toe of the breech-mechanism outer crank and cocks the breech mechanism when the hand operating lever is pulled to the rear.

7B14. **Summary of operation.** Assume that the gun is to be put into operation starting with the gun unloaded and the breech closed. The firing mechanism and the rammer are cocked by means of the hand operating lever. When this lever is pulled aft the breechblock is lowered, the cylinder catches are released, and the rammer is cocked. With the lever to the rear, cartridges are placed in the loader and the bottom one is forced past the star wheels to the rammer tray. Since the loader is full, the feed-control check lever is depressed. The tray catch lever is also depressed since the rammer tray is in battery. However, the trigger mechanism is not actuated so the trigger catch lever is in position to hold the rammer shoe. The hand operating lever is then returned to its forward position. The firing selector lever is set for automatic and the operating sequence is as follows:

1. When the firing key is pressed the trigger catch lever is depressed and the rammer forces the cartridge forward.

2. At the forward end of rammer travel the rammer levers are spread apart and the cartridge is catapulted into the firing chamber.

3. Just before the cartridge is fully seated in the chamber its rim strikes the extractor lips and forces them forward. This releases the breechblock so that the operating spring can raise it to close the breech.

4. When the breechblock has closed the sear is actuated to release the firing pin and the gun fires.

5. After the gun fires, the barrel, housing, rammer tray, and rammer move in recoil. This motion compresses the barrel spring.

6. As the gun recoils, the roller on the outer crank rides along the operating cam on the side door. The outer crank turns the crankshaft and forces the breechblock down. Rotation of the crankshaft also winds the operating spring.

7. The firing mechanism is cocked before the breechblock drops.

8. In its downward movement the breechblock actuates the extractors which eject the empty case and hold the breechblock down. The extracted case passes back over the rammer tray and out the aft end of the slide to the case chutes. The chutes carry the empty cases to the front end of the mount where they are discharged to the deck.

9. As the rammer tray moves in recoil the rammer rocker arm raises the tray catch lever so that it will engage the rammer shoe during counterrecoil.

10. During the rearward movement of the rammer tray the feed pawls are forced upward by the guides along the sides of the tray.

11. At the end of recoil, the rammer shoe, which is still in the forward position in the tray, is just to the rear of the catch levers. When counterrecoil begins the rammer shoe is caught by the tray catch lever and held to the rear.

12. During counterrecoil the rammer tray pawls release the star-wheel catches, the feed pawls are forced downward, and the bottom cartridge is forced past the star wheels onto the rammer tray.

13. As the gun returns to battery, the cam on the underside of the rammer tray actuates the rammer rocker arm to depress the tray catch lever. The rammer is then free to catapult another cartridge into the firing chamber and a new operating cycle begins.

14. This cycle continues until: (1) the loader catch lever is raised due to insufficient ammunition in the feeding device, or (2) the trigger catch lever is raised due to releasing the firing key or foot-firing pedal. In either case the rammer shoe is held in the cocked position and thus the operation of the gun is stopped.

7B15. **Mount.** The 40 mm. twin and quad gun assemblies have conventional type open mounts. The training circle is secured to the stand and the carriage is supported on radial and

FIGURE 7B11. 40 mm. twin mount and operating crew.

thrust roller bearings. A platform on the aft end of the carriage is used by the loaders when the gun is in operation. The cooling-system tanks and circulating pumps are mounted on the aft end of this platform. The firing-control mechanism is mounted on the forward face of the carriage. The slide trunnions are supported in roller bearings.

Individual electric or electric-hydraulic power drives are used for the elevating and training gear. These drives are arranged for either director control or local control at the mount. As was mentioned in Article 7B1, there may be a single control lever (joystick) provided on some mounts for both training and elevating drives in addition to control by means of the pointer's and trainer's handwheels. The handwheels can also be geared directly to the elevating and training racks for manual drive.

The mount can be trained 360° either direction from the locked position. A training stop is provided to prevent training beyond these limits and a power drive cut-off switch operates, when near the limit, to shut off the driving motor. The elevation limits are 15° depression and 90° elevation. Elevation and train centering pins lock the gun and mount in place when the gun is secured.

Open peep and ring sights are provided for the pointer and trainer. If tracer control is employed, these are used initially in leading the target just prior to opening fire, after which the gun is positioned in accordance with the path of the tracers near the target.

7B16. Personnel. The basic gun crew for the twin or quad mount includes the following: (1) mount captain, (2) pointer, (3) trainer, (4) one loader for each gun, and (5) at least one ammunition passer for each gun; more will be needed if the ready boxes are some distance from the mount.

The mount captain is in charge of the gun and crew. In local control he designates targets and gives the orders for firing. The trainer is responsible for releasing the train centering-pin and for starting the circulating pumps in the cooling system in addition to his regular duty of training the mount. The pointer releases the elevation centering-pin, starts the firing motor, elevates the gun with his handwheels and operates the foot-firing pedal (joystick control in both elevation and train may be given to the pointer, if such control is desirable). The loaders set the firing-selector lever at the desired position (*stop fire, auto,* or *single*), operate the hand operating lever, and place the ammunition in the loader. The ammunition passers bring the clipped cartridges from the ready boxes or magazines and hand them to the loaders on the gun platform. The crew of a 40 mm. twin mount is shown in figure 7B11.

C. 20 MM. GUN AND MOUNTS

7C1. General. The 20 mm. gun assembly, shown in figure 7C1, is an AA machine gun mounted on a pedestal-type, free-swinging mount. The mount is so arranged that the trunnion height can be adjusted for the convenience of the gunner when the gun is used at different angles of fire. The elevation limits are 5° depression and 87° elevation. There are no limits in train.

The gun is designed for automatic firing only and, like other automatic guns, it uses some of the force developed by the explosion of the propellent charge to eject the empty case, cock the gun, reload, and fire the next round. It embodies, however, certain features of gun design which are not found in other automatic and semi-automatic guns. The most important of these are:

1. The gun barrel does not recoil.
2. The breechblock is never locked against the breech and is actually in motion at the moment the gun is fired.
3. There is no counterrecoil brake, the force of counterrecoil being checked by the explosion of the following round of ammunition.

The 20 mm. fires fixed ammunition which is fed into the gun from a magazine having a capacity of 60 rounds. The cyclic rate is about 450 rounds per minute. When the normal time required to change magazines is considered, the rate will vary between 250 and 320 rounds per minute depending upon the skill of the gun crew. The projectile used is a high-capacity type and is fitted with an impact-type nose fuze. Usually every other projectile or every third projectile in the magazine carries a tracer.

CHAPTER 7

The 20 mm. gun is being used extensively on all classes of naval vessels, on merchantmen, and at many shore stations. It has been designed for and is particularly effective against aircraft targets at ranges up to about 2,000 yards. For this purpose it has almost entirely replaced the caliber .50 machine gun, principally because the heavier 20 mm. projectile with its high-explosive charge causes much more damage than the non-explosive caliber .50 projectile. The gun may also be used against lightly-armored targets.

7C2. Mounts. There are several kinds of pedestal-type, free-swinging mounts in use. The major difference between these mounts is in the method of adjusting the trunnion height. The mechanical mount shown in figure 7C1 is the one which is most widely employed at the present time.

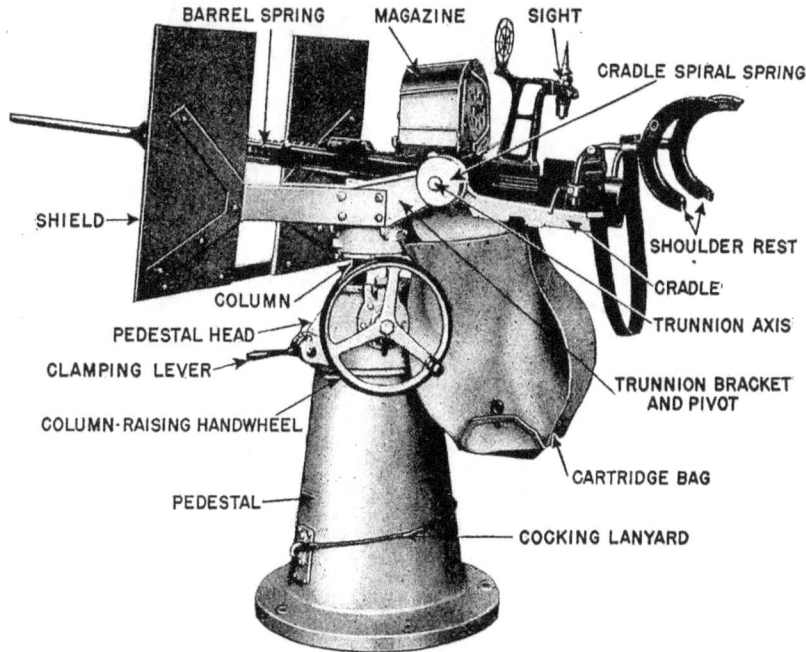

FIGURE 7C1. 20 mm. gun assembly; mechanical mount.

The fixed *pedestal* is bolted to the deck. The *pedestal head*, through which the *column* rises, is so mounted that it can be rotated around the top of the pedestal and locked in any desired position by means of the *clamping lever*. The *column-raising handwheel*, mounted on the pedestal head, is used to raise and lower the column. By this means the trunnions can be raised about 15 inches to better position the gun for the gunner when firing at elevated targets.

The *trunnion bracket and pivot* is free to rotate around the top of the column on ball bearings, and there are no limits to this training motion. This part supports the *shield*, the *cradle spiral spring*, and the *cradle*. The gun is carried in two grooved slides in the cradle, and is held in place by a bolt. The cradle spiral spring, mounted around the left trunnion, has one end attached to the trunnion and the other to the spring case. This spring functions to counterbalance the weight of the gun.

From the above description it is evident that the gunner, strapped to the aft end of the gun, can swing the gun to any position within elevation limits, and that the trunnion height can be adjusted for his convenience and comfort. As the gun is moved in train, the column-raising handwheel can also be moved into position for easier operation.

Two other mounts which are used to a limited extent are shown in figure 7C2. The *fixed*

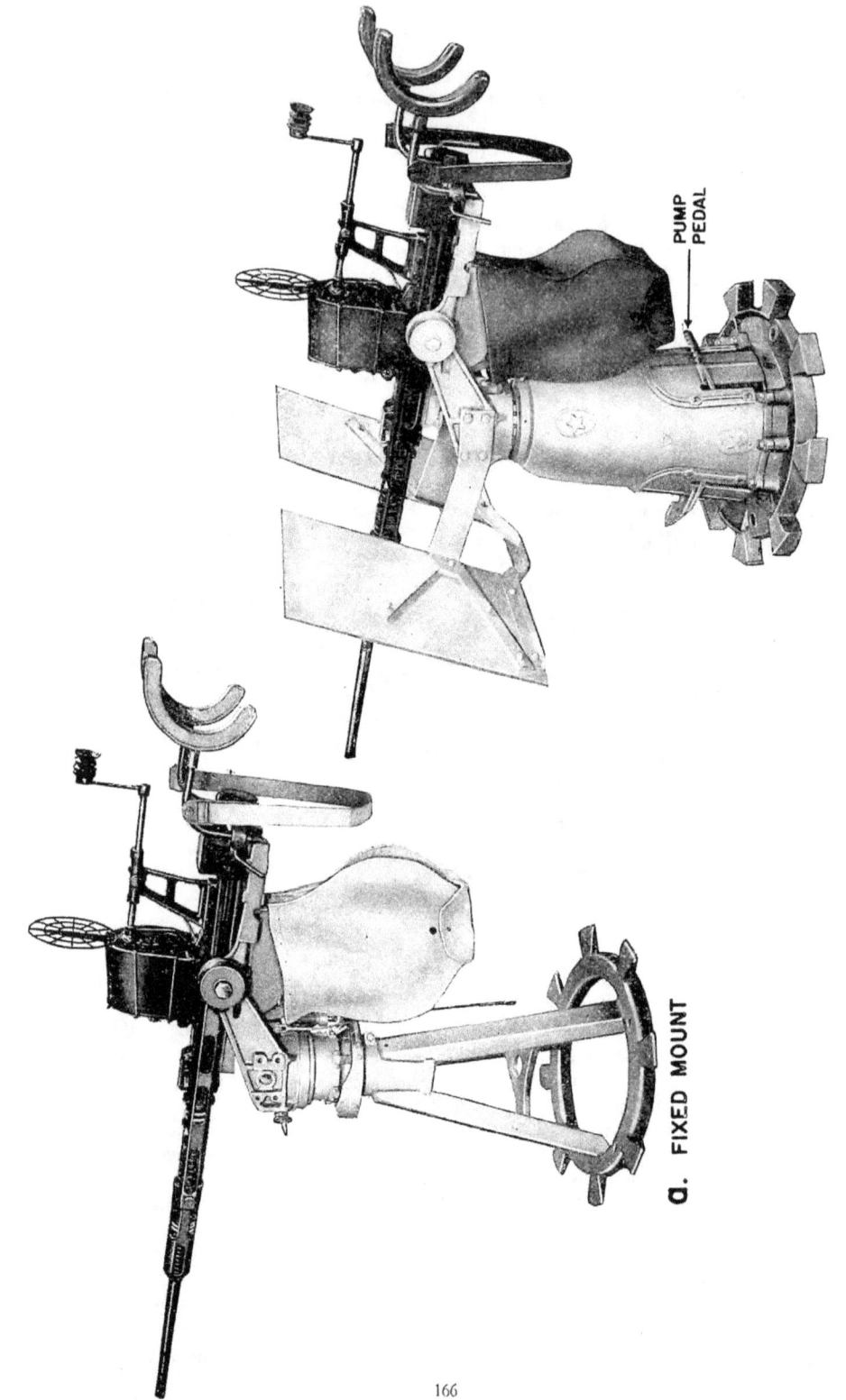

FIGURE 7C2. 20 mm. gun assemblies; fixed and hydraulic mounts.

CHAPTER 7

mount, so called because it does not have an adjustable trunnion height, has a higher pedestal than the mechanical mount. The gunner is provided with a circular stepped platform around the mount to enable him to utilize the full range of elevation and depression available. The elevation limits are the same as for the mechanical mount and there is no limit in train.

The *hydraulic mount* has a foot-operated hydraulic mechanism to raise and lower the column. The three equally spaced *pedals* around the pedestal are connected to the hydraulic pump and control valves within the pedestal. Any one of these pedals can be pumped to raise the column or fully depressed when the column is to be lowered. The column can be raised 24 inches, and the gun elevation limits are 15° depression and 90° elevation. There are no limits in train.

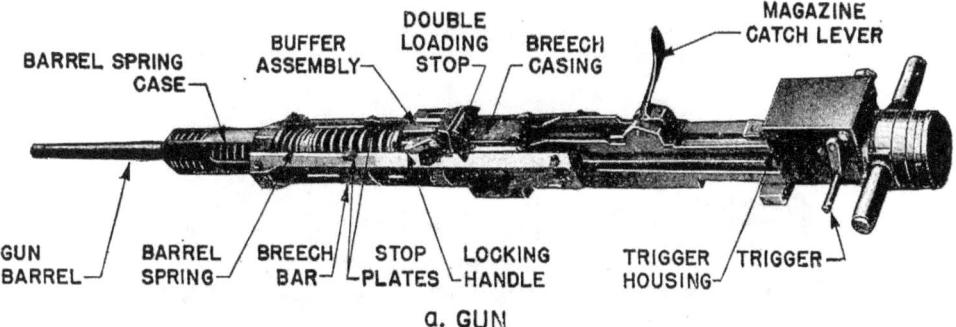

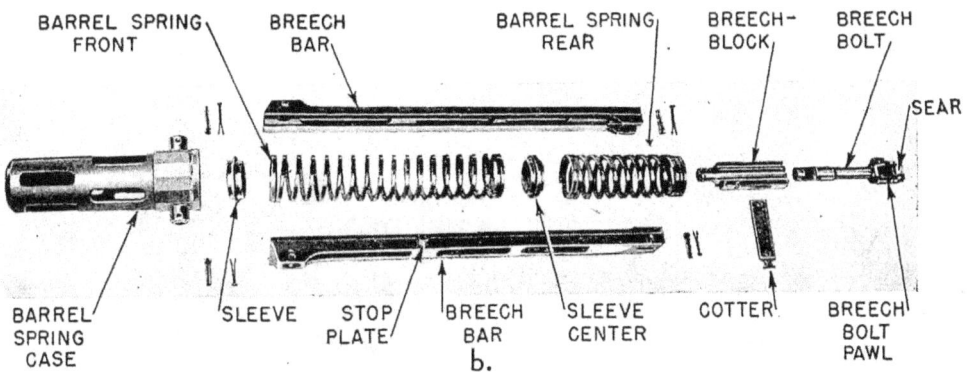

FIGURE 7C3. 20 mm. gun and recoiling parts.

7C3. The gun. The 20 mm. gun, shown in figure 7C3a, can be divided into four main groups of parts:
  1. The gun barrel and breech casing.
  2. The breechblock and the other recoiling parts attached to it.
  3. The recoil and counterrecoil system.
  4. The trigger mechanism and locking devices inter-connected with it.

7C4. Barrel and breech casing. The *barrel* is a forged steel piece a little more than 57 inches long. The bore has a diameter of 20 mm. (approximately 0.8 inches) and has uniform twist rifling. The forward end of the gun barrel flares slightly outward to form a flash guard. The after end is secured to the *breech casing* by means of a bayonet joint, which makes barrel replacement quick and easy. The barrel is air-cooled, but additional cooling is often obtained by spraying it with water.

The breech casing extends backward from the rear of the barrel to the after end of the gun. It carries the *trigger housing* on its aft end and serves to support the magazine when it is shipped on the gun. The breech casing also guides the breechblock and the other recoiling parts of the gun as

167 **RESTRICTED**

they move in recoil and counterrecoil. In this respect it corresponds to the slide on the 40 mm. and 3"/50 cal. guns. The breech casing is attached to the gun mount, and since the gun barrel is secured to the breech casing, neither moves in recoil.

7C5. Breechblock and other recoiling parts. Figure 7C3b shows the following principal recoiling parts: (1) the *breechblock*, (2) the *breech bolt*, (3) the *cotter*, (4) the two *breech bars*, and (5) the *barrel spring case*.

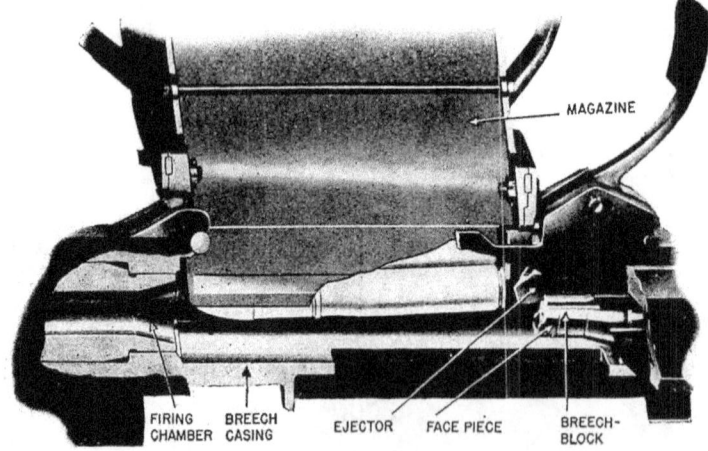

a. COCKED POSITION (BEGINNING OF COUNTERRECOIL)

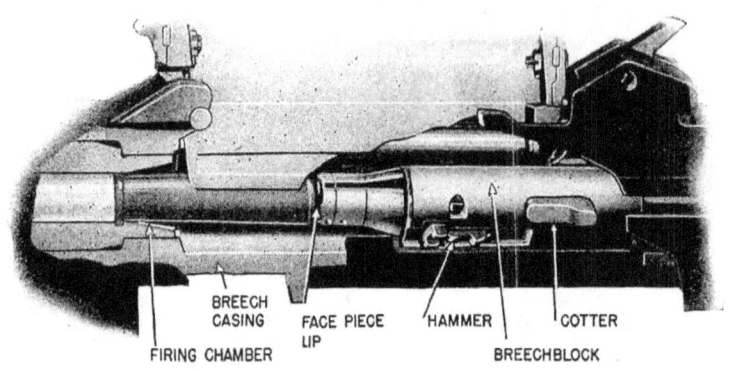

b. CARTRIDGE ENTERING FIRING CHAMBER

FIGURE 7C4. Operation of the 20 mm. gun; loading.

The breech bolt is fitted into the aft end of the breechblock. It is attached to the breechblock by the cotter, a heavy flat pin which passes crosswise through the block and secures the block not only to the breech bolt but to the two breech bars. These breech bars extend forward along the outside of the breech casing on either side of the gun and connect the breechblock to the barrel spring case.

7C6. Recoil and counterrecoil system. The recoil and counterrecoil system of the 20 mm. gun consists of two parts: (1) the *barrel spring*, and (2) the *recoil buffer*. The barrel spring surrounds the gun barrel and is divided into two sections separated by a *sleeve*. The after end of the spring seats against the recoil buffer assembly at the forward end of the breech casing, and the forward end is held by the barrel spring case, an elongated collar which fits around the gun barrel. The barrel spring case is drawn backward and the barrel spring is compressed when the breechblock

CHAPTER 7

recoils. When the force of recoil is spent, the barrel spring expands and forces the barrel spring case and the breechblock forward in counterrecoil.

The recoil buffer is located in the forward end of the breech casing. It consists of (1) the buffer proper, a metal collar fitted into the breech casing, and (2) the *buffer springs*, 12 small, very strong springs mounted in the breech casing and lying fore-and-aft around the rear of the gun barrel. The recoil buffer operates near the end of recoil when the barrel spring is almost fully compressed. At this point the barrel spring case is brought up against the buffer and forces it back into the breech casing, compressing the buffer springs to check the last inch of recoil. When recoil is fully checked, the buffer springs expand and start the movement of the recoiling parts forward in counterrecoil.

The 20 mm. gun has no counterrecoil buffer. The gun fires when the breechlock is almost at the

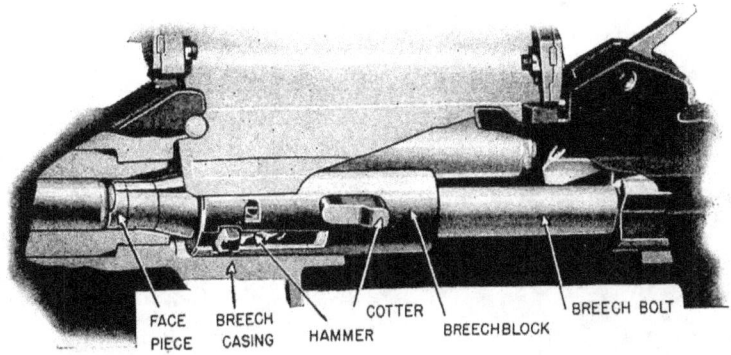

a. CARTRIDGE FIRED

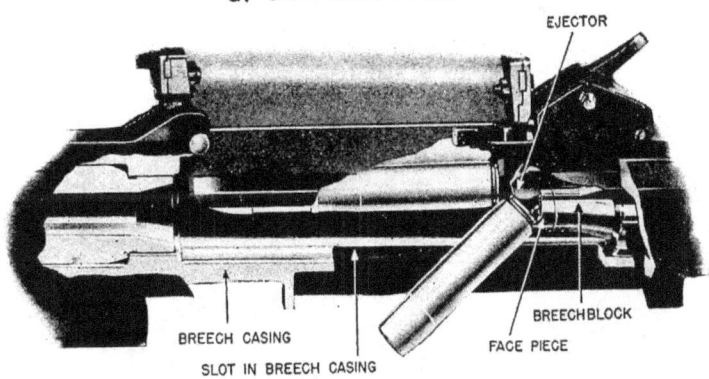

b. EMPTY CASE BEING EJECTED DURING RECOIL

FIGURE 7C5. Operation of the 20 mm. gun; ejection.

end of its counterrecoil travel. Part of the explosive force of the propellent charge is exerted rearward, checking counterrecoil and forcing the block to the rear in recoil.

7C7. **Operation of the breech mechanism.** The breech mechanism operates on the blowback principle. The explosion of the propelling charge forces the empty case out of the chamber, and hence, the sliding breechblock to the rear in recoil. The operation of the gun is illustrated in figures 7C4 and 7C5.

Figure 7C4a shows the gun cocked and the loaded *magazine* in position to commence firing. The breechblock is held to the rear against the action of the barrel spring and will be driven forward when the trigger is pulled. Figure 7C4b shows the breechblock on its way forward shortly after the trigger has been pulled. A cartridge is being forced out of the magazine and toward the *firing chamber*.

RESTRICTED

MACHINE GUNS

The downward pressure from the cartridge above forces the base of the first cartridge down into the *face piece lip* as the breechblock moves forward.

Figure 7C5a shows the position in which the gun is automatically fired just before the breechblock ends its forward motion. The force of the explosion on the case stops the forward motion and forces the case, breechblock and other recoiling parts to the rear. The breechblock slides back through the breech casing and carries the empty case to the rear. Near the end of recoil, as shown in figure 7C5b, the *ejector* strikes the base of the cartridge and tips it off the face piece. The case then flips out through an opening in the under side of the breech casing.

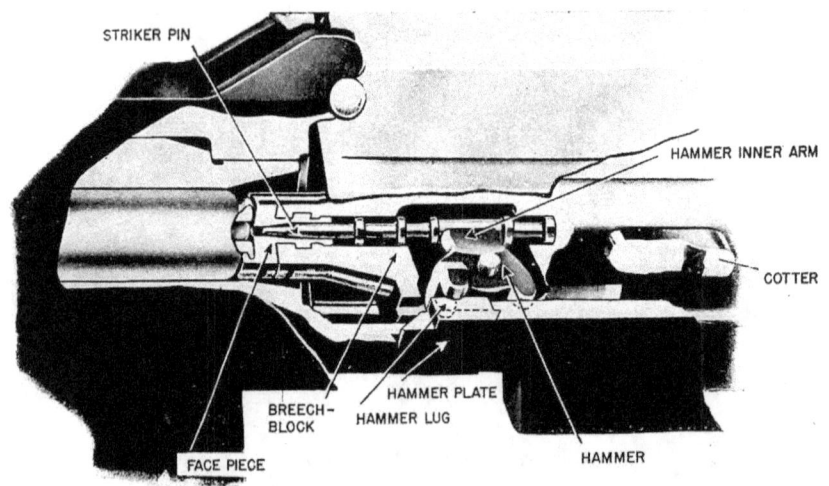

a. POSITION OF STRIKER AND ROUND BEFORE FIRING

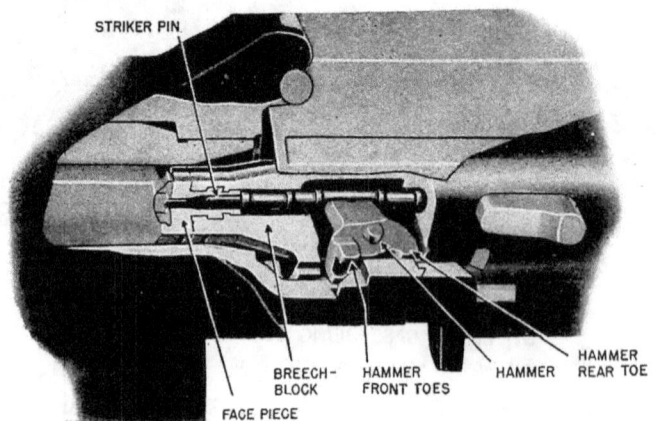

b. FIRING POSITION OF STRIKER AND ROUND

FIGURE 7C6. Operation of the firing mechanism; 20 mm. gun.

This sequence completes an operating cycle, and this cycle will be repeated for each round fired. It is seen that the breechblock moves back and forth through the breech casing continuously while the gun is firing and is never locked in the breech.

7C8. Firing mechanism. The firing mechanism for the 20 mm. gun consists of a *striker pin* and *hammer* housed within the breechblock and a *hammer plate* secured to the breech casing. Figure 7C6 shows these parts and illustrates the action of the firing mechanism. The hammer is pivoted in

CHAPTER 7

the left side of the breechblock and the *hammer inner arm* fits in a slot in the striker pin. The outboard edge of the hammer is shaped into two *front toes,* one on each side of the hammer, and a *rear toe* which is an extension of the center plate of the hammer. The hammer plate has a *hammer lug* on its inner surface and its forward edge is beveled to provide a cam surface for the front toes.

When the breechblock moves forward to carry a cartridge into the firing chamber, the striker pin is back. The hammer toes project from the side of the block and ride along the inside of the breech casing. As the breechblock continues its forward motion, the front toes pass over the hammer lug as shown in figure 7C6a. On the other hand, the rear toe strikes the hammer lug and the hammer is rotated about its pivot. This action forces the hammer inner arm forward, and the resulting thrust on the striker pin drives it against the primer in the cartridge case; this position is shown in figure 7C6b.

At the moment the gun fires, the rear toe of the hammer has been forced into the breechblock. The front toes project outward and through an opening in the breech casing just forward of the hammer plate. This location is the only point at which the gun can fire; elsewhere in the travel of the breechblock, the absence of an opening in the breech casing prevents the hammer from being rocked. As the breechblock recoils after firing, the front toes of the hammer ride up the cam surface on the forward edge of the hammer plate and pass on either side of the hammer lug. This action returns the hammer

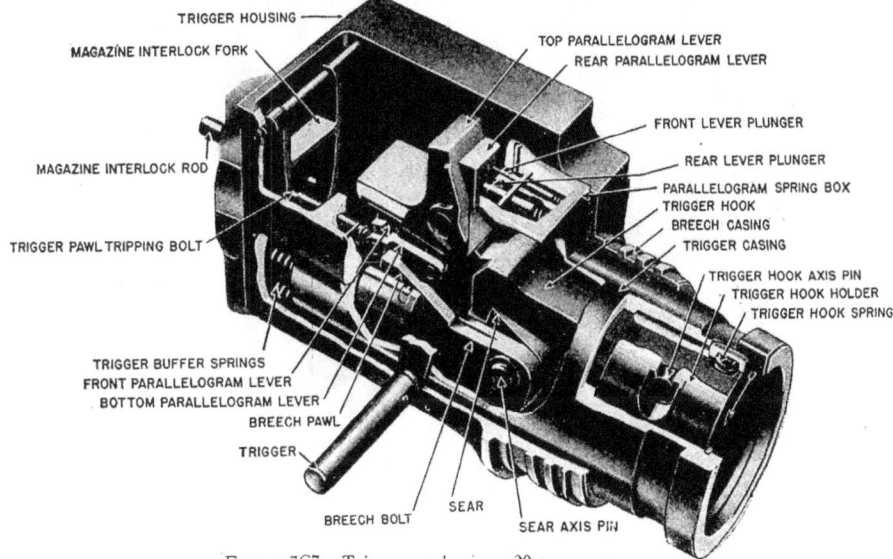

FIGURE 7C7. Trigger mechanism; 20 mm. gun.

to its original position, in which the striker pin is drawn back into the block and the rear toe again projects from the side of the block. It is to be noted that there is no firing spring in the 20 mm. firing mechanism. Force to drive the striker pin against the primer comes only from the forward motion of the breechblock in counterrecoil.

7C9. Trigger mechanism. The trigger mechanism and the magazine interlock device associated with it control the operation of the gun. The trigger mechanism, shown in figure 7C7, is mounted in the *trigger housing* on the aft end of the breech casing. The sole function of this arrangement of levers and springs is to actuate the *trigger hook* in response to trigger and magazine interlock action. When the trigger hook is forced downward it engages the *sear* on the breech bolt and thus holds the breechblock in the cocked position. When the trigger hook is raised, the sear is released and the breechblock is free to move forward to start the action of the gun. The trigger hook automatically intercepts the sear and holds the breechblock in the cocked position when the trigger is released or when the last round in the magazine has been fired.

MACHINE GUNS

The trigger mechanism has three principal operating groups: (1) the *trigger group*, (2) the *parallelogram*, and (3) the *trigger hook*.

The trigger group includes the following parts which are shown in figure 7C8: (1) *trigger intermediate lever*, (2) *trigger pawl holder*, (3) *trigger pawl*, (4) *trigger crank*, and (5) *trigger pawl spring*. The *trigger crank toe*, which projects inward under the top parallelogram lever serves as the connecting link between the trigger group and the parallelogram.

The parallelogram consists of the top, bottom, front, and rear parallelogram levers shown in figure 7C7. It is mounted on top of the *trigger casing*, which is a cylindrical tube within the trigger housing so positioned that the breech bolt moves into it at the end of recoil. The paral-

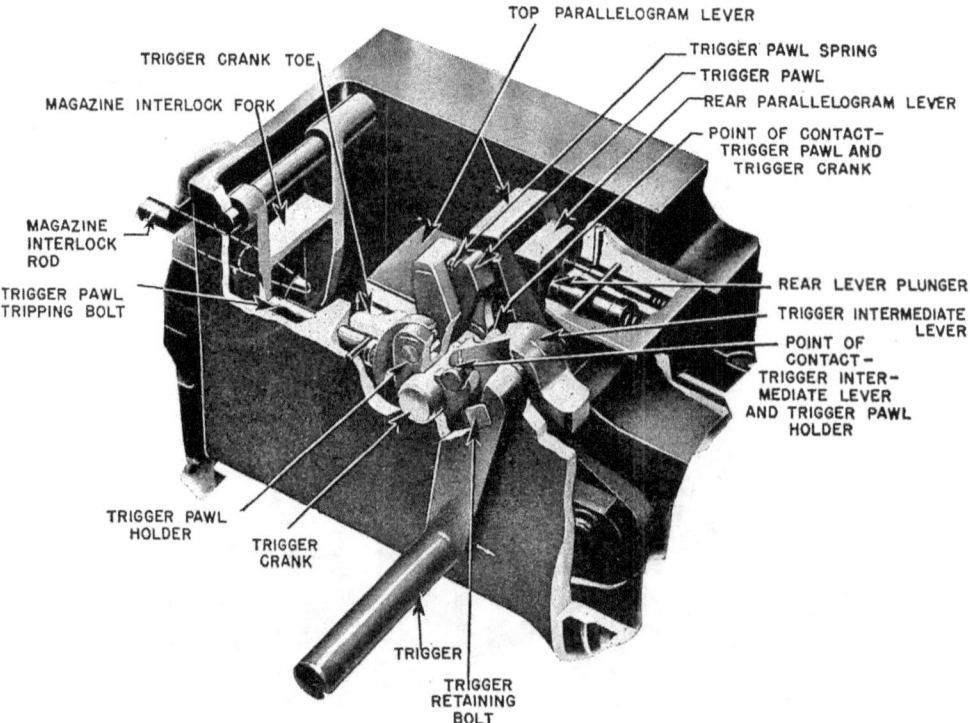

FIGURE 7C8. Trigger action on parallelogram; 20 mm. gun.

lelogram extends down inside the trigger casing so that its bottom levers lie just above the path traveled by the breech bolt. The parallelogram transmits the action of the trigger group to the trigger hook.

The trigger hook, shown in figure 7C7, is mounted inside the trigger casing just above the path traveled by the breech bolt. The trigger hook engages the sear whenever it is forced downward by the action of the parallelogram. The *trigger hook spring* forces the trigger hook upward to disengage the sear whenever the trigger hook is not held down by the parallelogram.

In brief, the action of the trigger mechanism is as follows: When the gun is cocked before firing, the breechblock is drawn to the rear until the breech-bolt pawls engage the bottom levers of the trigger parallelogram as shown in figure 7C9a. The lower parallelogram lever is drawn forward by the forward pull of the barrel spring on the breech bolt. This movement draws the lower part of the rear parallelogram lever forward and down against the trigger hook. The trigger hook is thus forced downward to engage the sear and hold the breechblock to the rear. This is shown in figure 7C9b.

CHAPTER 7

When the trigger is pulled, the parts in the trigger group move in the directions indicated by the arrows in figure 7C8. If the trigger pawl engages the trigger crank the trigger crank will lift the forward end of the parallelogram as shown in figure 7C10a. The breech-bolt pawls then clear the bottom parallelogram levers and the spring-loaded *rear lever plunger* forces the upper end of the rear parallelogram lever forward. The resulting rotation of the rear parallelogram lever about its pivot frees the trigger hook so that it rises to release the sear (see fig. 7C10b). The breech-

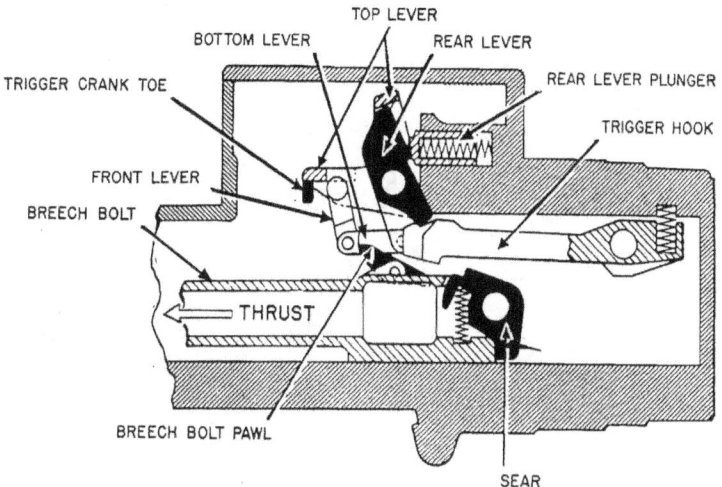

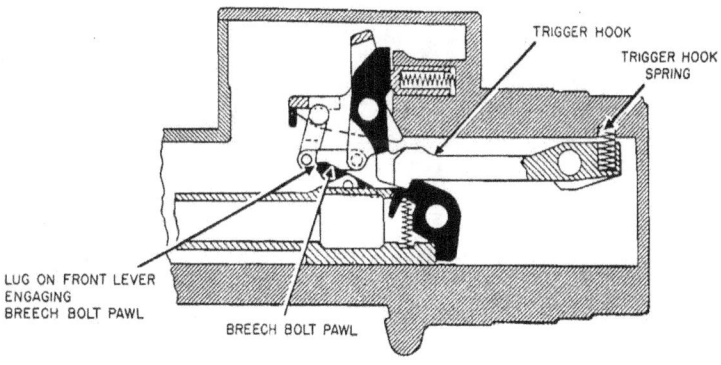

FIGURE 7C9. 20 mm. trigger mechanism; cocking operation.

block then moves forward to start the gun action. Figure 7C10b shows the position of parts just after the sear has been released.

When the trigger is released, or the trigger pawl and trigger crank are disengaged by the magazine interlock device, the upper parallelogram lever is forced back into the position shown in figure 7C9a by means of two spring-loaded *front lever plungers* (not shown). The parallelogram is then in position to again engage the breech bolt pawls as the breechblock starts to counterrecoil. Thus, through the action of the parallelogram, the trigger hook automatically engages the sear and holds the breechblock in the cocked position whenever the trigger is released or the magazine interlock device functions.

The shock of stopping the forward movement of the bolt might be sufficient to break parts of the trigger mechanism if the full force of counterrecoil had to be taken by the trigger mechanism alone. For this reason 15 small *trigger buffer springs*, similar to those in the recoil buffer, are mounted in the forward end of the trigger casing. Two of these are shown in figure 7C7. The forward ends of these springs seat against the breech casing. When the trigger hook catches the sear as it moves forward, the trigger casing, which is free to slide forward a short distance in the trigger housing, is drawn forward compressing the buffer springs. In this way the shock of stopping the bolt is cushioned.

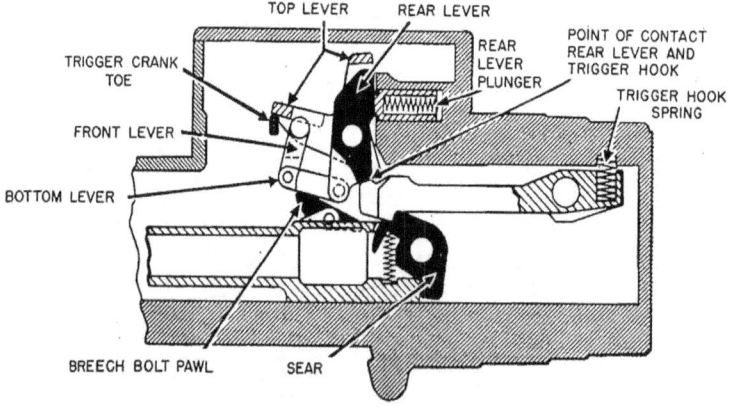

a. JUST BEFORE RELEASE

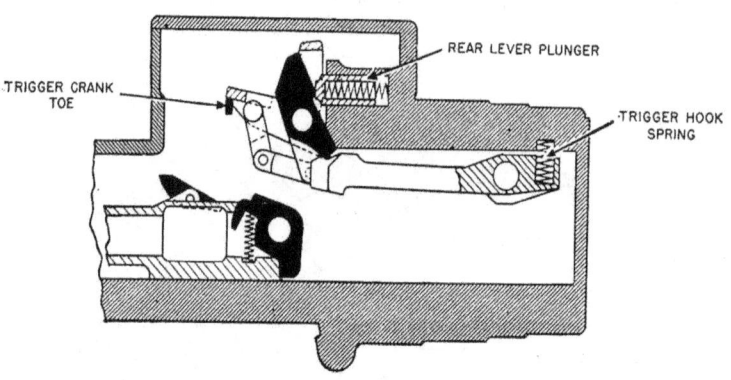

b. JUST AFTER RELEASE

FIGURE 7C10. 20 mm. trigger mechanism; releasing operation.

7C10. Magazine interlock gear. Since the force to check the forward motion of the breechblock in counterrecoil comes only from the explosion of the propelling charge in a cartridge, the breechblock must not be allowed to return to battery unless it carries a cartridge into the gun chamber. If this precaution were not taken, the block would ram against the end of the chamber with enough force to break or damage parts of the gun. It is necessary, therefore, to have some device which automatically disengages the trigger so that the breechblock will be caught in the cocked position, either when the magazine is empty or when it is unshipped during firing. This is the function of the magazine interlock gear shown in figures 7C11 and 7C12. This arrangement also makes it unnecessary to recock the gun by hand when changing magazines.

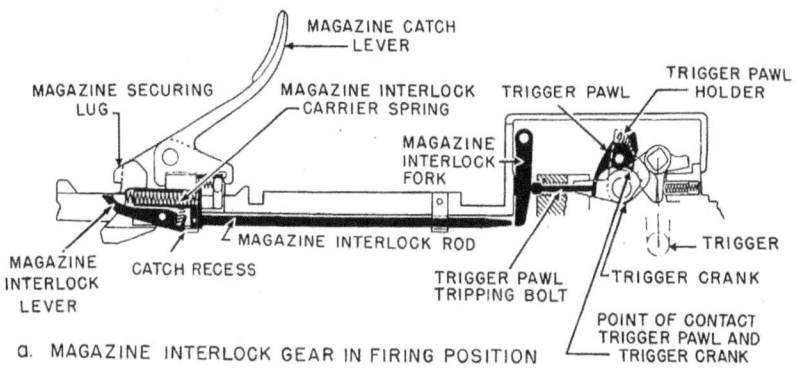

a. MAGAZINE INTERLOCK GEAR IN FIRING POSITION

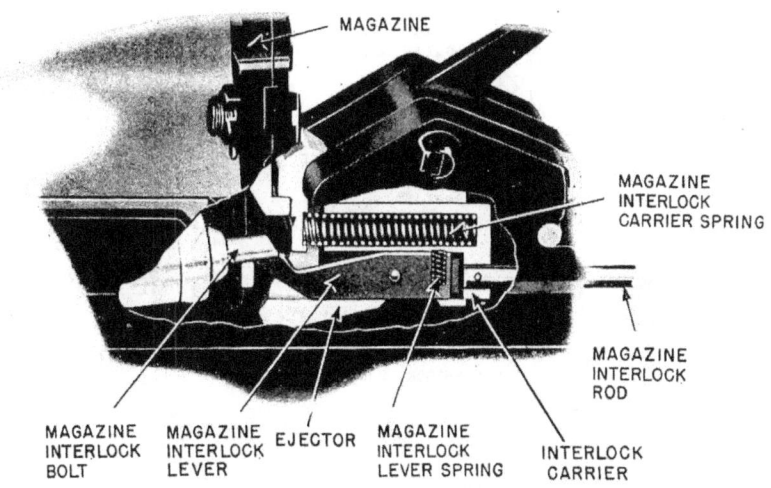

b. INTERLOCK LEVER—LAST ROUND FIRED

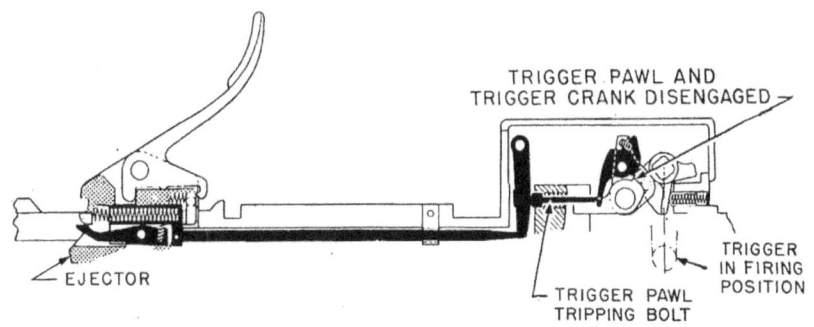

c. MAGAZINE EMPTY—LAST ROUND FIRED

FIGURE 7C11. Magazine interlock gear; 20 mm. gun.

MACHINE GUNS

The *magazine interlock* includes the following principal parts (fig. 7C11a): (1) *magazine interlock lever,* (2) *magazine interlock carrier spring,* (3) *magazine interlock rod,* (4) *magazine interlock fork,* and (5) *trigger pawl tripping bolt.* These parts function to disengage the trigger pawl and trigger crank when:

1. The *magazine interlock bolt,* a spring-loaded plunger in the magazine cartridge feeder shown in figure 7C11b strikes the magazine interlock lever after the last round has been fed into the gun.

2. The *magazine catch lever* is thrown forward to remove the magazine.

Briefly the action is as follows: When the magazine interlock lever is depressed by the magazine interlock bolt, the aft end clears the *catch recess* and the magazine interlock rod is forced aft by the interlock carrier spring. This position is shown in figure 7C11c. The motion of the rod, transmitted through the magazine interlock fork and trigger pawl tripping bolt, causes the trigger pawl to rotate about its pivot enough to disengage the trigger pawl and trigger crank. Thus the trigger automatically becomes inoperative.

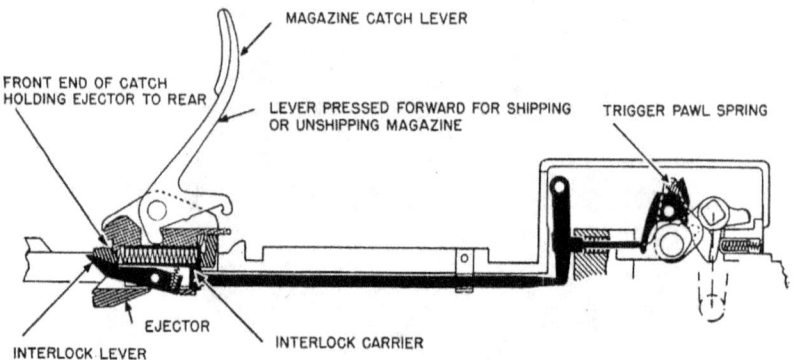

FIGURE 7C12. Magazine catch gear; 20 mm. gun.

Forcing the magazine catch lever forward to unship a magazine also moves the magazine interlock rod aft to disengage the trigger pawl. This is shown in figure 7C12. A magazine catch holds the magazine interlock rod in this position until another magazine is shipped, at which time the rod returns to its forward position and the *trigger pawl spring* forces the trigger pawl back into engagement with the trigger crank.

7C11. Double-loading stop. The *double-loading stop,* shown in figures 7C13 and 7C14, is a safety attachment which prevents the jam that would occur if a live cartridge were loaded when a torn cartridge case from the preceding round remained in the firing chamber. The double-loading stop blocks the forward motion of the breech bars, and hence the breechblock, when such a condition exists.

The device consists of: (1) the double-loading stop, mounted on top of the breech casing just above the firing chamber, with arms, or horns, which project out over the breech bars, (2) a system of plungers, levers, and springs to actuate the double-loading stop, and (3) *stop plates* on the breech bars. The operation of this device is briefly described below.

Figure 7C13a is a sectional view of the operating parts of the double-loading stop. The *lower plunger* projects a short distance into the firing chamber when the latter is empty. The following actions occur when a cartridge case is inserted in the chamber as in figure 7C13b: (1) both *lower* and *upper plungers* are forced upward, (2) the *stop lever* is turned counterclockwise about its axis (looking at the gun from the left side), (3) the forward arm of the stop lever is lowered and the *lever spring* is compressed, (4) the double-loading stop is turned clockwise by the *stop spring,* and (5) the double-loading stop arms are lowered enough so that they will engage the

176

CHAPTER 7

stop plates on the breech bars during counterrecoil if the case still remains in the chamber at that time.

Just as soon as the cartridge case is extracted, the strong lever spring forces the forward arm of the stop lever upward and the double loading stop arms are raised high enough to clear the stop plates. When the gun is operating normally, the empty case is ejected before the stop plates have

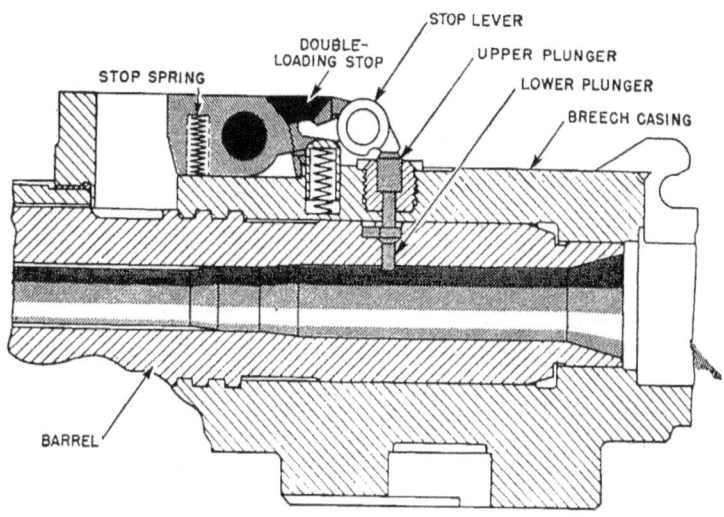

a. FIRING CHAMBER EMPTY

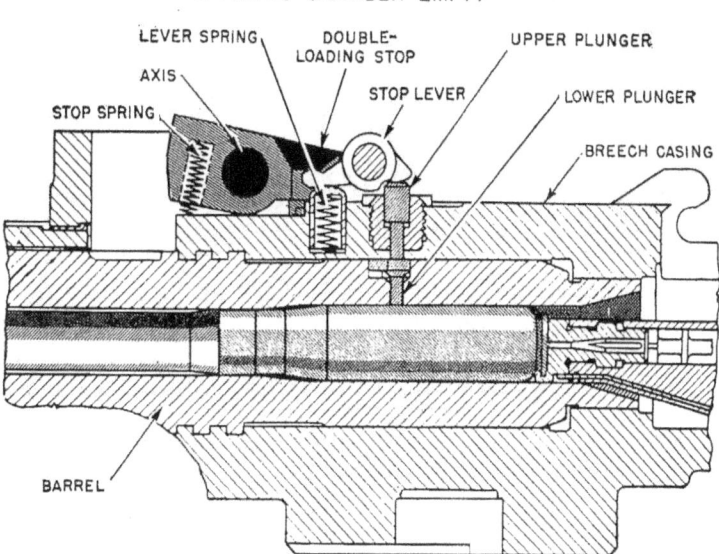

b. CARTRIDGE CASE IN FIRING CHAMBER

FIGURE 7C13. 20 mm. double-loading stop gear; double-loading stop.

recoiled far enough to reach the arms and they are raised enough to clear the plates. However, if part of the case remains in the chamber, the arms will be forced upward during recoil, against the action of the stop spring, by the beveled surface on the rear edge of the stop plates as shown in figure 7C14a. When the stop plates have passed by, the arms immediately move back into position to

RESTRICTED

intercept the plates as they move in counterrecoil. This is shown in figure 7C14b. Thus the breechblock is prevented from returning to battery.

The gun must be cocked and the torn cartridge case drawn out with a cartridge-extractor tool before firing can be resumed. If it is necessary to return the gun to action quickly, the entire barrel may be replaced.

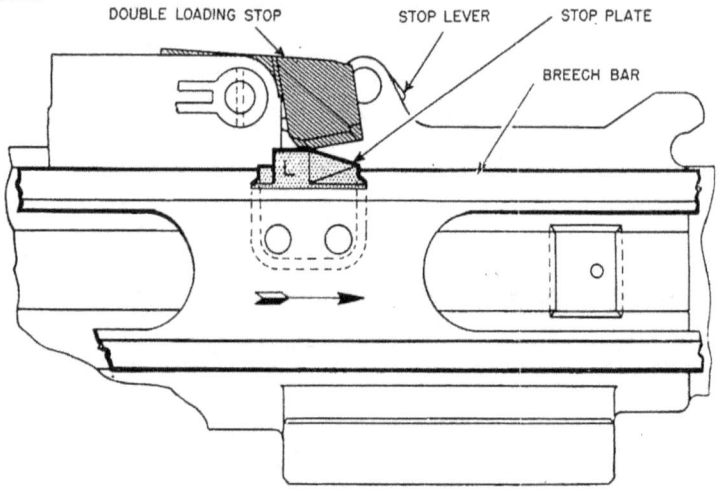

a. PART OF CARTRIDGE CASE LEFT IN CHAMBER DURING RECOIL

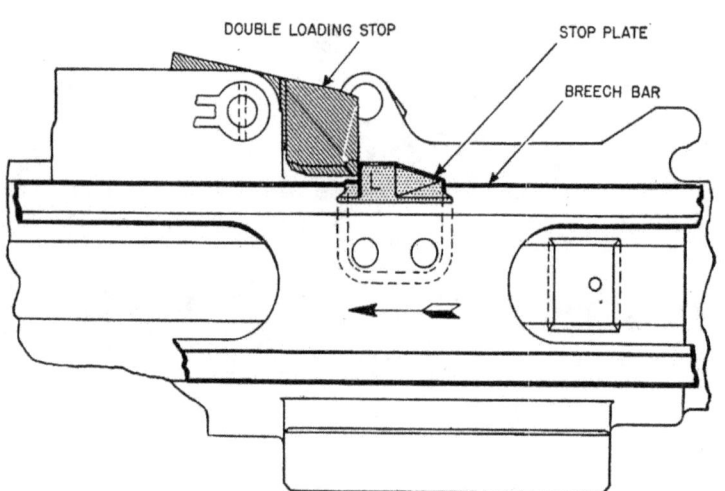

b. DOUBLE LOADING STOP ENGAGING STOP PLATE ON COUNTERRECOIL

FIGURE 7C14. 20 mm. double-loading stop gear; stop plates.

7C12. *Summary of operation.* The gun is secured uncocked, and it must be cocked by hand before firing. The breechblock is drawn back by one of the cocking methods described below, until the trigger hook engages the sear. A loaded magazine is shipped and from this point the cycle of operation is as follows:

CHAPTER 7

1. When the trigger is pulled, the parallelogram is raised and the trigger hook is disengaged from the sear.
2. The breechblock is driven forward by the compressed barrel springs.
3. As the breechblock moves forward, it picks up a round from the magazine and forces it into the firing chamber.
4. Just before the breechblock reaches the end of its forward motion, the hammer strikes the hammer lug and the resulting motion of the hammer drives the striker pin forward to fire the gun.
5. The explosion of the propellent charge stops the forward motion of the breechblock and forces it and the other recoiling parts to the rear in recoil, compressing the barrel spring. Near the end of recoil, the buffer springs also are compressed.
6. As the breechblock recoils, the front hammer toes ride up on the hammer plate and the striker pin is withdrawn into the breechblock.
7. The breechblock carries the empty case to the rear until the ejector strikes the upper edge of the case and ejects it from the gun.
8. At the end of recoil the compressed barrel and buffer springs force the breechblock forward again and the cycle is repeated.

Normally the automatic operation of the gun continues until: (1) the trigger is released, (2) the magazine interlock device is tripped when the last cartridge is fed out of the magazine, or (3) the magazine is unshipped. Any of these actions releases the trigger mechanism so that the sear is caught at the end of recoil and the breechblock is held in the cocked position. Since the gun stops firing in the cocked position, it is necessary to uncock it by hand before the gun is secured so as to relieve the compression on the barrel springs when the gun is not in use.

7C13. Ammunition. The 20 mm. gun uses fixed ammunition which is designed to give an I. V. of 2,725 f. s., and a maximum horizontal range of about 5,000 yards. The following projectiles are used: (1) high-explosive bursting charge with tracer, (2) high-explosive bursting charge without tracer, (3) blind loaded with tracer, (4) blind loaded and plugged, and (5) incendiary. The high-explosive bursting charge is either tetryl or a high explosive known as pentolite. The tracer charge is effective for about $3\frac{3}{4}$ seconds duration. It occupies about one-half of the projectile cavity, and hence projectiles with tracers carry approximately one-half as much bursting charge as those without tracers.

The projectiles are equipped with a simple impact-type nose fuze. On impact a closing disk is displaced into the fuze body and an air column within is instantaneously compressed and forced through an inner disk, thus heating the air column. The heated air ignites a small pellet of lead azide which in turn sets off pressed tetryl in a detonator, exploding the burster charge. The 20 mm. projectiles do not have the self-destructive arrangement found in 40 mm. ammunition.

7C14. Magazine. A sectional view of the magazine is shown in figure 7C15. The *spring* is hand wound from the outside and its tension is indicated on the *indicator block*. The *cartridge feeder*, driven by the spring, exerts a pressure on the innermost round and tends to force the cartridges around the spiral guides and out of the magazine mouthpiece. The lips of the magazine mouthpiece are partly closed so that a round can pass through only by being pushed out longitudinally at right angles to the direction of the pressure from within. A gap is cut in the forward end of the mouthpiece to allow the rounds to slide in or out. The magazine has a capacity of 60 rounds; a new-type magazine holds 100 rounds.

7C15. Gun sights. In the past, fire control of the 20 mm. gun has been primarily by means of tracers, ring sights having been used to get on the target. However, the present tendency is to rely upon the Mark 14 sight, supplemented by spotter's observations of tracers. At present, then, available sights for this weapon include the following:

1. The Mark 14 sight, a gyroscopic sight mechanism which automatically computes the lead angle necessary for different speeds and motions of the target. The gunner sights on the target through a small window through which luminous crosslines are projected. The cross-

RESTRICTED

MACHINE GUNS

lines are kept on the target, and, as the gun swings following the target's movement, the crosslines are offset to give the proper lead angle for the gun.

2. The older cartwheel sight, or peep and ring sight, consists of a rear peep and a forward circular grid, with concentric rings to indicate the amount of lead necessary for different target speeds. This sight enables the gunner to get on the target before opening fire; thereafter, tracer control normally is used. Several types of ring sights have been developed.

7C16. Gun crew. The gun crew for the 20 mm. mechanical mount consists of at least three men (four if the Mark 14 sight is used):

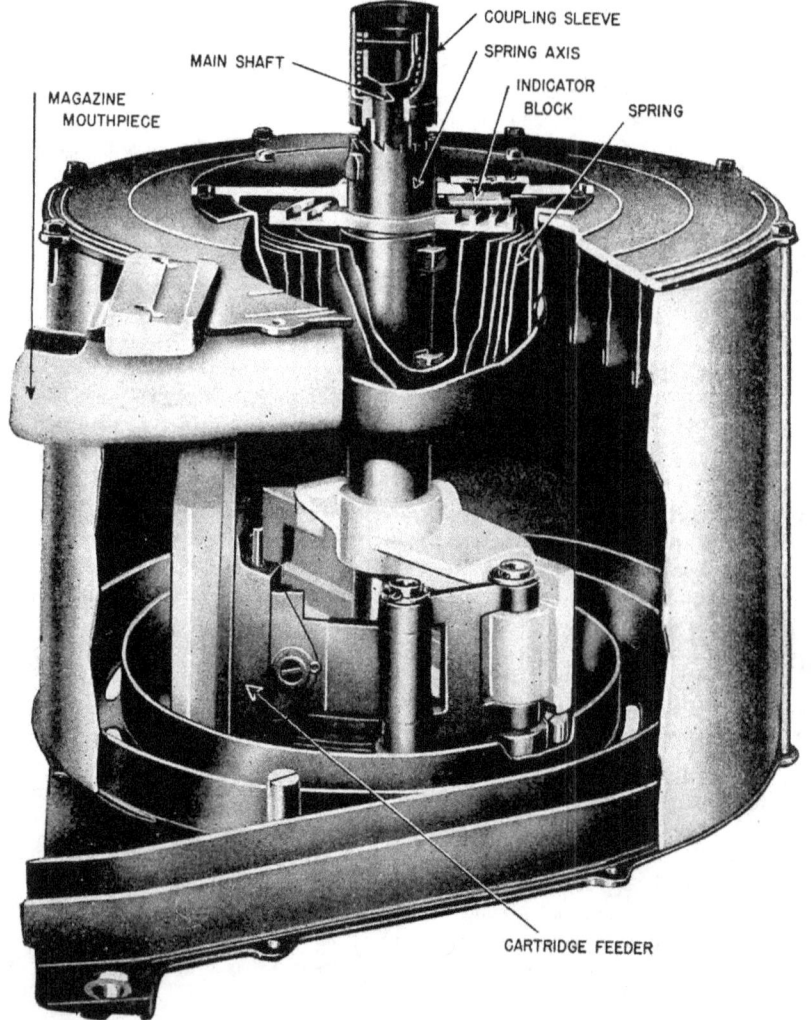

FIGURE 7C15. 20 mm. magazine.

1. The *gunner,* who is strapped to the shoulder rests on the after end of the gun elevates, trains and fires the gun.

2. The *trunnion operator,* who raises and lowers the column on which the gun is mounted.

3. The *loader,* who places loaded magazines on the gun and unships them when they are empty.

4. The *range setter,* who observes the effects of fire and changes range setting on the Mark 14 sight.

CHAPTER 7

The gun crew may also include *ammunition passers*, who hand loaded magazines to the loader and take the empty magazines after they are removed from the gun. The number of ammunition passers used depends upon the location of the ready boxes.

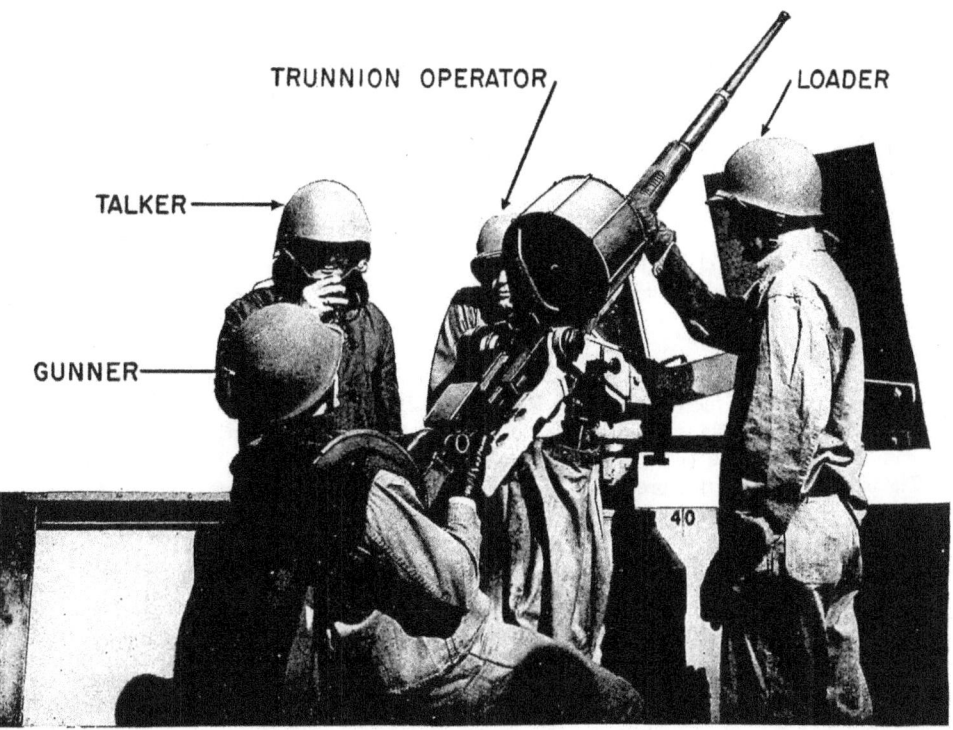

FIGURE 7C16. 20 mm. gun and operating crew.

The gun is entirely free-swinging, and thus can be elevated or trained merely by the movement of the gunner's body. The trunnion operator stands to the left of the gun, faces the gunner, and keeps his back to the shield on the mount. He moves his handwheel around the pedestal of the mount as the gunner swings the gun in train. He watches the gunner's knees and adjusts the gun height so as to keep the gunner's knees slightly bent during firing. The loader stands to the right of the gun and, operating the magazine catch lever, removes the spent magazine and replaces it with a filled one. A gun and operating crew is shown in figure 7C16.

7C17. **Cocking the gun.** Since the breechblock must be drawn back against more than 500 pounds pressure exerted by the barrel spring, cocking the gun is a relatively difficult task. The methods for cocking the gun described below are those in common use.

One method involves a *collar*, or *stirrup*, which slips over the barrel and comes up against the barrel-spring case. The collar has a *lanyard* attached to each side and these are knotted to provide hand holds. The gun is secured in either the vertical or horizontal position and it is cocked by the pull of two or three men on each lanyard. A variation of this method employs a single lanyard with a ring on one end which is hooked over the left end of the *cotter*. Four or five men can cock the gun by pulling on the lanyard.

Another method makes use of the *cocking lanyard* attached to the pedestal of the mount. The gun is first locked at maximum elevation, and the lanyard is hooked over the notched end of the cotter. The column is then cranked up by the column-raising handwheel. This forces the gun barrel and breech casing upward. Since the block is held back by the lanyard, it passes aft with

RESTRICTED

respect to the breech casing as the column is raised until the sear locks in the cocked position. The column is then cranked down again and the lanyard is unhooked from the cotter.

A third method requires a *roller* mounted in the aft end of the cradle. The mount-cocking lanyard is passed over the roller and is attached to the cotter. One or two men then raise the aft end of the gun far enough to cock the gun.

7C18. Jams and breakages. If the gun jams during firing, the cause may be faulty ammunition, insufficient grease on the ammunition, or breakage of some part of the gun mechanism. For satisfactory operation of guns using the blowback principle, it is necessary that the ammunition be completely covered with a light coat of suitable grease before being loaded into the magazine. If this is not done, the empty case may not be blown out of the chamber and a jam will result. Broken hammers and striker pins also cause jams. When a jam occurs, the following procedure is recommended:

    1. Cool the barrel by turning a hose on it. This is necessary since the nose fuze on the projectile can be exploded by heat, and the heat of the gun barrel may be sufficient to cause detonation.

    2. Cock the gun.

    3. Set the trigger on "Safe" so that an accidental blow on the trigger will not release the breechblock while the cartridge is being removed from the barrel.

    4. If the cartridge is not drawn out of the chamber when the breechblock is pulled back, it must be rammed out (using a special rammer issued with the gun) and thrown overboard.

The parts of the gun that break most frequently are the barrel springs, the breechblock face piece, the striker pin, and the hammer. Although ordinarily long-lived and dependable, the gun barrels sometimes develop cracks when they are overheated. The useful life of the barrels is increased considerably by changing them after every 300 rounds continuous fire to prevent overheating.

7C19. General data.

Caliber	20 mm. (.79 inch)
Length of barrel (calibers)	72 (approximately)
Limits on elevation (mechanical mount)	5° depression and 87° elevation
Arc of train	unlimited
Firing rate (assuming continuous ammunition supply)	450 rounds per minute
Firing rate (considering time to change magazines)	250-320 rounds per minute
Magazine capacity (rounds)	60 (100)
Muzzle velocity (f. s.)	2,725
Maximum range (36° elevation)	approximately 5,000 yards
Commence firing range (with Mark 14 sight)	
High elevation	1,000 yards
Horizontal	2,000 yards
Weights:	
Gun	141 pounds
Mount (mechanical)	1,578 pounds
Magazine—loaded	63 pounds
Magazine—unloaded	31 pounds
Round of ammunition	approximately 0.5 pound

D. CALIBER .50 MACHINE GUN

7D1. Introduction. The Browning machine gun, caliber .50, is a type of automatic weapon. It is produced with three kinds of barrels as shown in figure 7D1. The *tank and field model* is an air-cooled gun having a heavy barrel (designation, HB). This gun is normally fired in short bursts or in single shots. When used in this manner, firing may be continued for an appreciable length of time because the heavy barrel retards overheating. The *antiaircraft* water-cooled gun has

CHAPTER 7

a water jacket surrounding the barrel for the purpose of preventing overheating when firing for long periods. The barrel recoils during firing and the water jacket is stationary. Therefore, packing glands are provided near the breech and muzzle ends of the barrel to prevent water from escaping where the barrel slides in and out of the jacket. The *aircraft* gun is an air-cooled model which has a perforated barrel jacket to permit the circulation of air around the barrel.

Each of these models is alternate-feed, which means that by repositioning some of the parts, the ammunition may be fed to the gun from either the right or left side. Any of the guns may be mounted on a fixed or flexible (free swinging) mount. Manual cocking and loading of the gun is

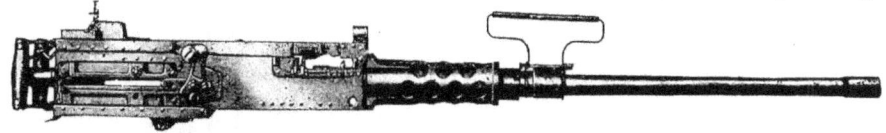

a. TANK AND FIELD –(AIR COOLED) HB WT. 81 LB.

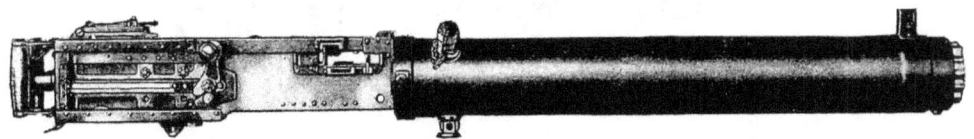

b. ANTIAIRCRAFT (WATER-COOLED) WT. 121.5LB.

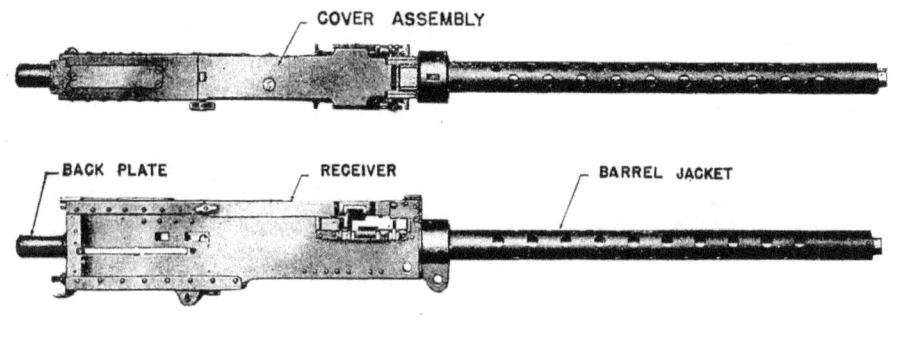

c. AIRCRAFT (AIR COOLED)   WT. 64 LB.

FIGURE 7D1. Caliber .50 Browning machine guns.

necessary before it is ready to function automatically. Then it may be fired by a mechanical or electrical accessory, or by a manual trigger and trigger bar. While firing, all mechanical action is automatically performed by the gun itself which handles 400 to 850 rounds per minute, depending on the gun model and the rate desired. More specifically, the tank and field gun has a rate of from 400 to 500 rounds per minute, the water-cooled gun from 600 to 750, and the aircraft gun from 750 to 850 rounds per minute. The I. V. for both the tank and antiaircraft gun is about 2,900 f.s., and approximately 2,700 f.s., for the aircraft gun. The maximum range of all models is about 7,200 yards.

7D2. Ammunition. The caliber .50 machine gun employs small-arms cartridge ammunition. Single cartridges or rounds are assembled into a series of nested links as shown in figure 7D2. This forms a flexible ammunition belt with a single empty link trailing on one end, and a double-filled link on the other. The double link end is fed into the gun.

Based upon use, the principal classifications of the ammunition used in the caliber .50 machine gun are as listed in the following table. Belts may be loaded with various combinations of these cartridges.

MACHINE GUNS

TYPE	PROJECTILE MARKING	USES
1. Ball	None (bare copper)	Against personnel and light-material targets.
2. Armor piercing	Nose painted black $\frac{7}{16}$ inch from tip	Against armored vehicles and planes, and other bullet-resisting targets.
3. Tracer	Nose painted red $\frac{7}{16}$ inch from tip	Observation of fire.
4. Incendiary	Nose painted blue $\frac{7}{16}$ inch from tip	Incendiary purposes.
5. Dummy	None (identified by holes in the body of the cartridge case)	Drill purposes.

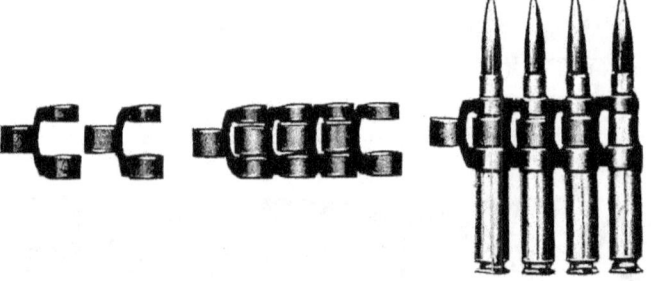

FIGURE 7D2. Ammunition belt links; caliber .50 Browning machine gun.

7D3. General description. The working parts, which automatically perform the numerous mechanical operations while the gun is firing, are the same in all three of the different models. The only major differences, already pointed out, are in the barrel and barrel-jacket assembly. There are minor differences in the firing attachments but all of the guns function identically. The following discussion of the caliber .50 gun is related to only one of the three types, namely: the air-cooled, caliber .50, M2, aircraft gun. In studying the description and functioning of this weapon it is suggested that figure 7D18 be kept available for reference.

The *receiver and barrel jacket group* forms the main exterior portion of the gun, and is non-recoiling during operation. The *back plate group* is assembled to the rear of the receiver, forming an end cover. The *driving spring* runs lengthwise in the receiver, with one end resting against the back plate and the other against a stop inside the *bolt group*. The bolt group is housed in the upper forward portion of the receiver, and slides backward and forward during operation. The *oil buffer body and oil buffer group* is contained in the lower-rear portion of the receiver. The bolt slides over the top of the oil buffer body during the back portion of the stroke. In the *barrel extension and barrel group*, the breech end of the barrel is screwed into the barrel extension to form a single unit. The barrel slides inside the barrel jacket. The normal position of the barrel extension is in the lower forward portion of the receiver. The bolt slides in grooves within the barrel extension. The *cover and belt feed group* is hinged at its forward end to the top front portion of the receiver.

7D4. General functioning. Although this gun is an automatic weapon, it is necessary to *cock* it manually to start the operating sequence. Assume that the gun is cocked, and that the first cartridge is in the chamber.

When a cartridge is fired, the pressure created by the gases reaches about 50,000 pounds p.s.i. Since the pressure pushes against the base of the ½-inch projectile, which up to this moment is still within the case, a driving force of about five tons propels the projectile from the barrel. This same force is exerted on the cartridge case and tends to drive it out of the chamber toward the rear. Such action is prevented by the bolt which is positively locked against the base of the cartridge case at the instant of firing.

When a round is fired the force of recoil drives the recoiling parts, which are the barrel, barrel extension and bolt to the rear a short distance. This motion unlocks the bolt from the bar-

CHAPTER 7

rel and barrel extension, and the bolt continues its rearward travel against the action of a spring. The spring serves to drive the bolt forward again; hence it is called a driving spring. During recoil the empty case is withdrawn from the chamber by a *T-slot* on the bolt, and the next cartridge is automatically extracted from the ammunition belt by the *extractor* on the bolt.

The *long* rearward motion of the bolt is stopped by the back plate. The short rearward motion of the barrel and barrel extension is checked by the oil buffer and its spring. The bolt is moved forward by the action of the driving spring, the empty case is ejected, and the next cartridge is moved into the barrel chamber. The buffer spring serves to drive the barrel and barrel extension forward. At the end of this forward motion, the bolt is locked to the barrel and barrel

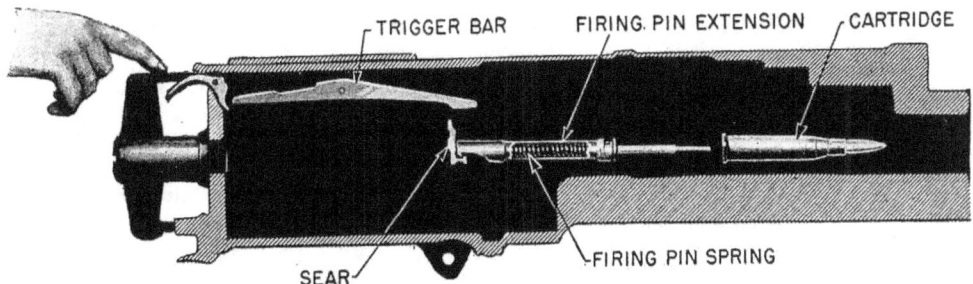

a. GUN READY TO FIRE

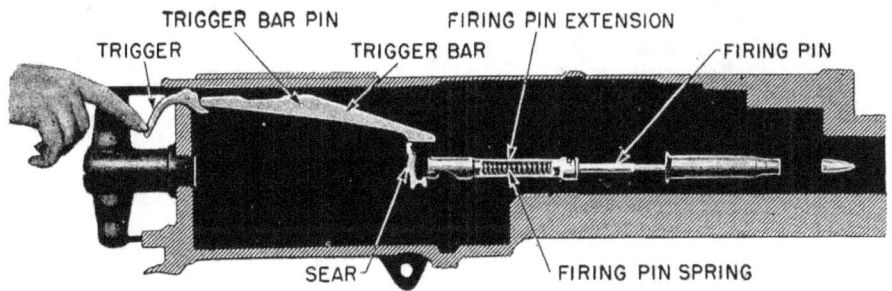

b. GUN FIRED

FIGURE 7D3. Firing action; caliber .50 Browning machine gun.

extension by the *breech lock*, thus again preventing the case from being driven to the rear. When the bolt and barrel are in battery, the firing pin can strike the primer, provided the trigger is pressed so that the *sear* releases the firing pin. This cycle continues as long as the trigger is pressed, and as long as ammunition is supplied to the gun.

7D5. Detailed functioning. In the description of the detailed functioning of the gun which follows it is assumed that, first, the ammunition belt has been properly started into the gun and the cover has been closed and latched, second, that the gun has been manually cocked and that a cartridge is in its proper position in the chamber and ready to be fired, and third, that a manual trigger and trigger bar are to be used to fire the gun.

Each time a cartridge is fired, the mechanical action within the gun involves many parts moving simultaneously or in their proper order. For convenience in studying the operation of these parts and their relationship to each other the action has been separated into seven phases. These are described in the following order:

1. Firing
2. Recoiling
3. Counterrecoiling
4. Cocking
5. Automatic firing
6. Feeding
7. Extracting and ejecting

7D6. Firing. When the gun has been loaded and the firing pin spring has been cocked or compressed manually, the firing mechanism is as shown in figure 7D3a. The gun is now ready to fire.

When the trigger is pressed it raises the back end of the trigger bar. The trigger bar pivots on the trigger-bar pin, causing the front end to press down on the top of the sear. The sear is forced down until the notch in the sear is disengaged from the shoulder of the firing-pin extension. The firing pin and firing-pin extension are driven forward by the firing-pin spring to fire the cartridge (see fig. 7D3b).

7D7. Recoiling. The complete cycle of the recoiling portion of the gun, which takes place as each cartridge is fired, consists of the recoil stroke when the barrel, barrel extension and bolt move rearward, and of the counterrecoil stroke when these same parts move forward. At the instant of firing, the recoiling parts are in the forward position in the gun, as shown in figure 7D4.

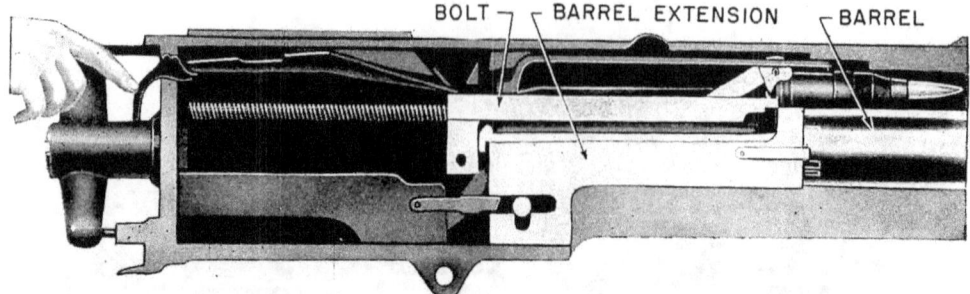

FIGURE 7D4. Forward position of recoiling parts; caliber .50 Browning machine gun.

At this time the bolt is held securely against the base of the cartridge by the breech lock which extends up from the barrel extension into a notch in the underside of the bolt (see fig. 7D5a).

After the cartridge explodes, and as the bullet travels out of the barrel, the force of recoil drives the recoiling portion rearward. During the first three-quarters inch of travel the breech lock is pushed back off the breech-lock cam step (see fig. 7D5b). This permits the breech lock to be forced down out of the notch in the bolt by the breech-lock depressors engaging the breech-lock pin. Thus the bolt is unlocked.

As the recoiling parts move toward the rear, the barrel extension rolls the accelerator rearward. The tip of the accelerator strikes the lower projection on the bolt and flips or accelerates the bolt to the rear (see fig. 7D5c). Note that the breech lock is completely disengaged from the bolt notch.

The barrel and barrel extension have a total rearward travel of $1\frac{1}{8}$ inches at which point they are completely stopped by the oil-buffer body assembly (see fig. 7D6).

During this recoil of $1\frac{1}{8}$ inches the oil-buffer spring is compressed in the oil-buffer body by the barrel extension shank. This spring is locked in the compressed position by the claws of the accelerator which are moved against the shoulders of the barrel extension shank (see fig. 7D7a).

The oil buffer assists the oil-buffer spring in bringing the barrel and barrel extension to rest during the recoil stroke, as shown in figure 7D7b. During the $1\frac{1}{8}$ inch of rearward travel the piston rod head is forced from the forward end of the oil-buffer tube to the rear. The oil at the rear

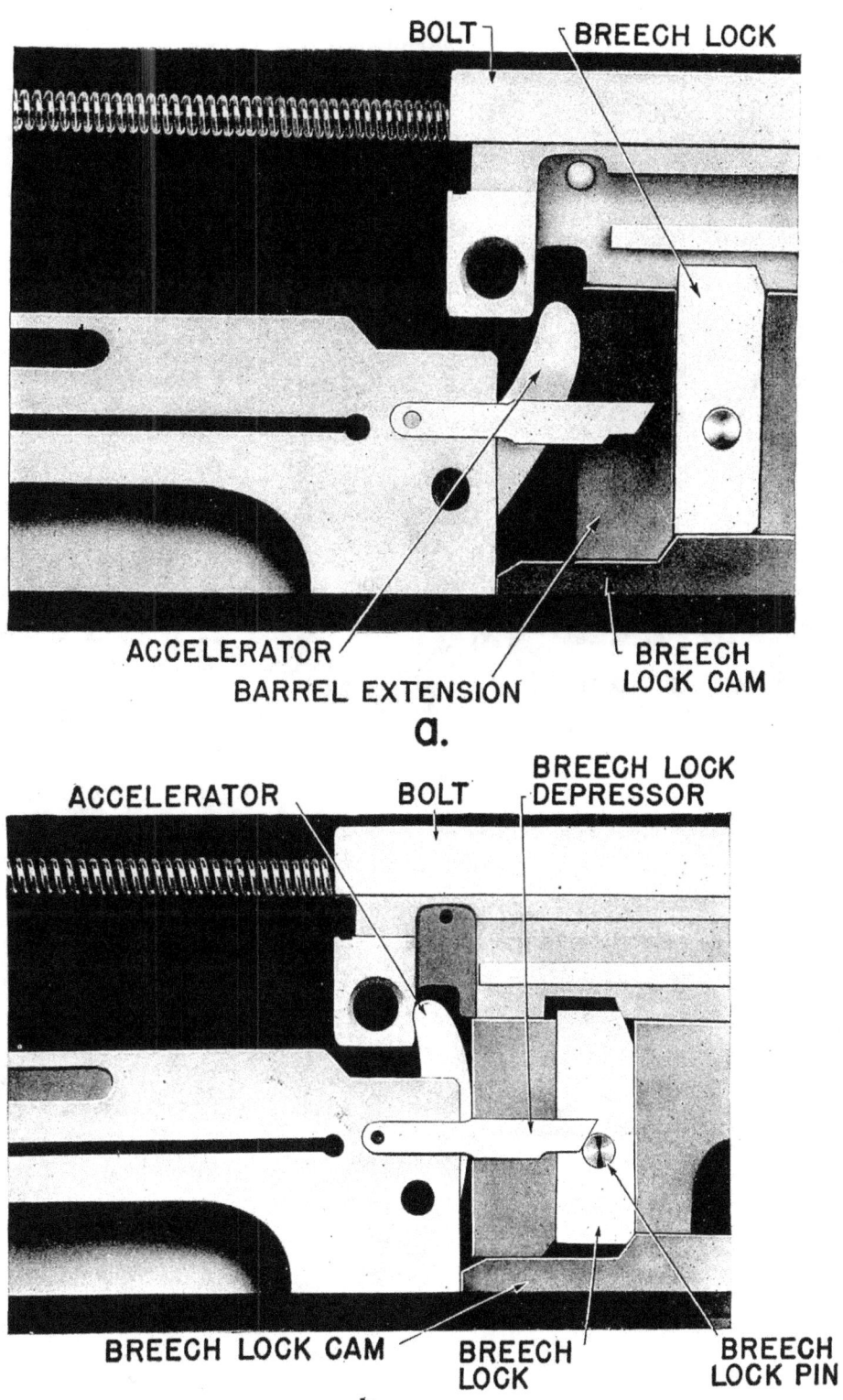

Figure 7D5. Unlocking the bolt by recoil; caliber .50 Browning machine gun.

MACHINE GUNS

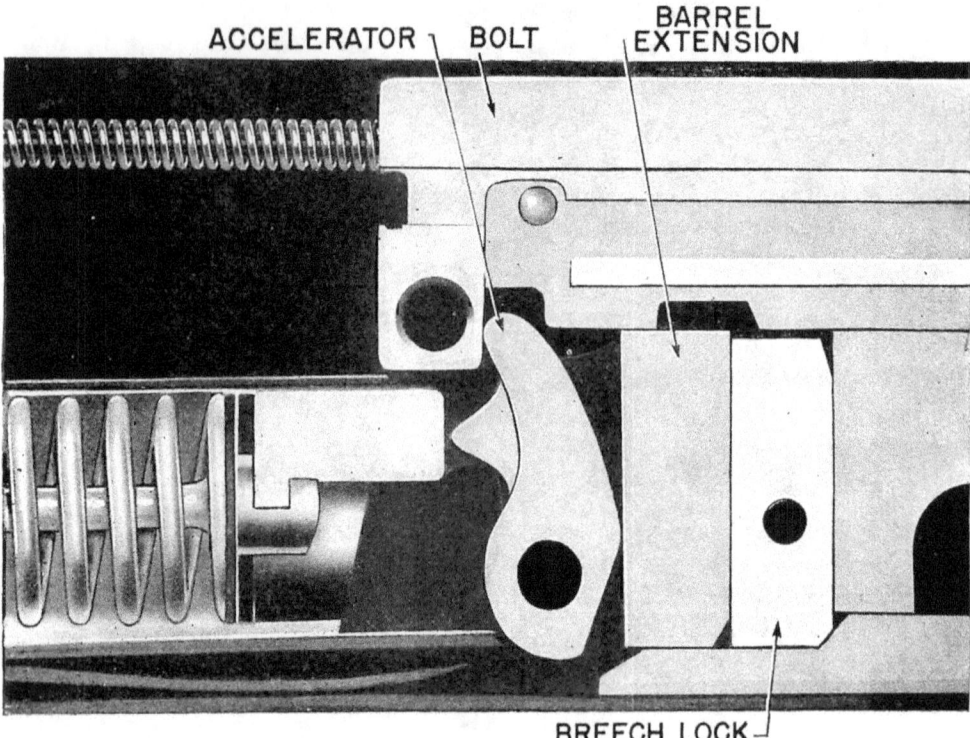

FIGURE 7D5c. Unlocking the bolt by recoil; caliber .50 machine gun.

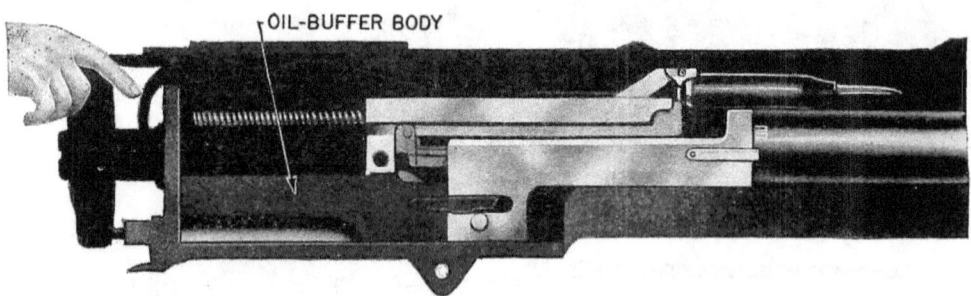

FIGURE 7D6. Barrel and barrel extension at end of recoil; caliber .50 Browning machine gun.

of the oil-buffer tube under pressure of the piston escapes to the front side of the piston. Its only path is through restricted notches between the edge of the piston-rod head and the oil-buffer tube.

The bolt travels rearward for a total of $7\frac{1}{8}$ inches. During this movement the driving spring is compressed. The rearward stroke of the bolt is finally stopped as the bolt strikes the buffer plate, as shown in figure 7D8. Thus, part of the recoil energy of the bolt is stored in the driving spring, and the remainder is absorbed by the buffer disks in the back plate.

CHAPTER 7

7D8. Counterrecoiling. Upon completion of the recoil stroke, the bolt is forced forward by the energy stored in the driving spring and the compressed buffer disks. When the bolt has moved forward about 5 inches the tip of the accelerator is struck by a projection on the bottom of the bolt (see fig. 7D9a). This rolls the accelerator forward.

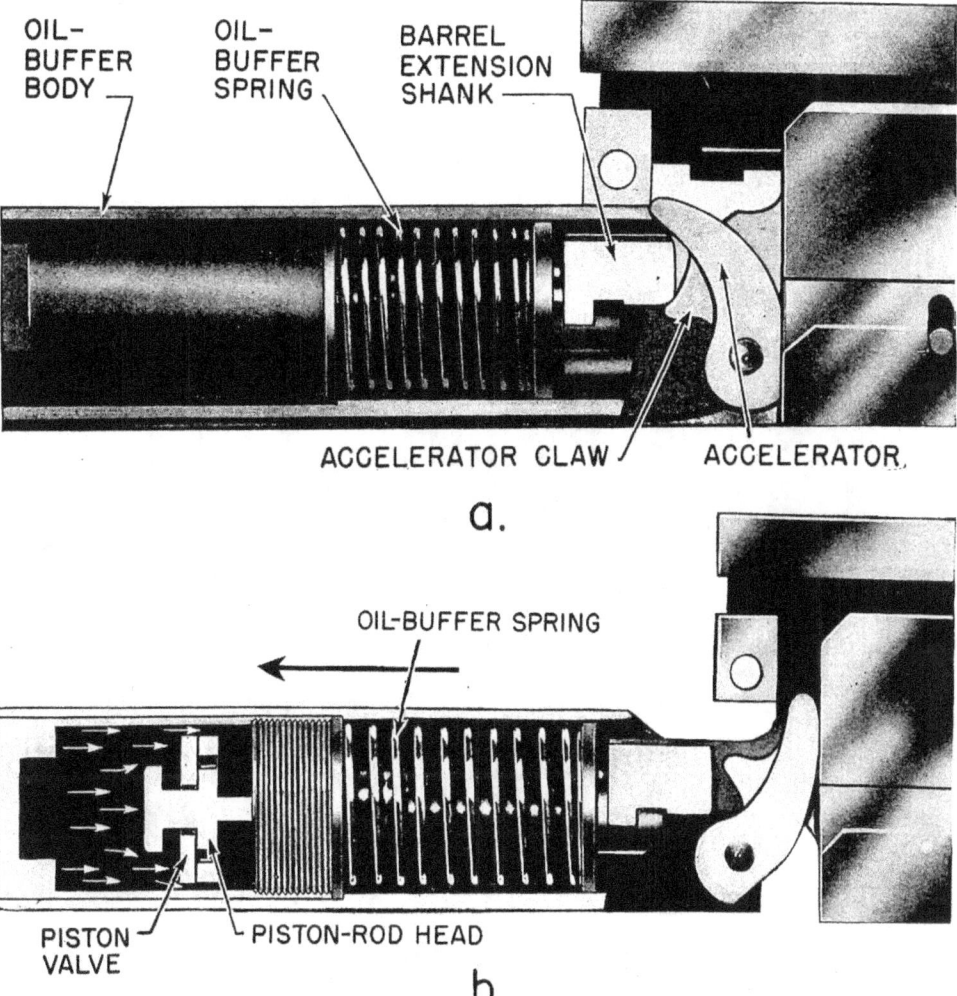

FIGURE 7D7. Action of oil buffer and spring in recoil; caliber .50 Browning machine gun.

As the accelerator rolls forward the accelerator claws are moved away from the shoulders of the barrel extension shank. This releases the oil-buffer spring. The energy stored in the spring shoves the barrel extension and barrel forward (see fig. 7D9b).

No restriction to motion is desired on the forward or counterrecoil stroke of the barrel and barrel extension; therefore, on the forward stroke additional openings for oil flow are provided in the piston-rod head of the oil-buffer assembly. The piston valve is forced away from the piston-rod head as the parts move forward, uncovering these additional openings. This provides an added path and permits oil to escape freely at the opening in the center of the piston valve as well as at the edge of the piston valve next to the tube wall (see fig. 7D10a).

RESTRICTED

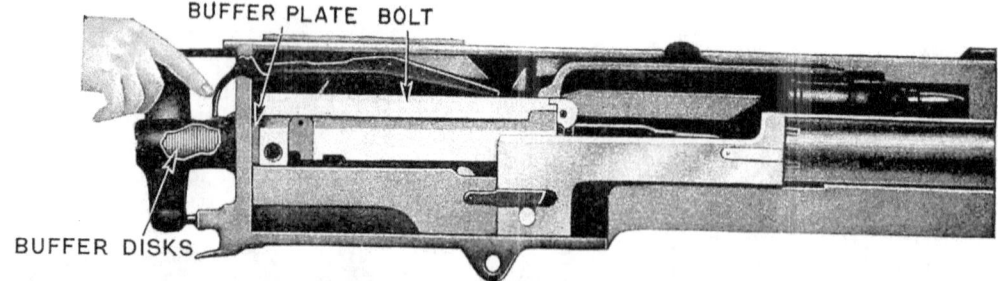

FIGURE 7D8. Full recoil position of bolt; caliber .50 Browning machine gun.

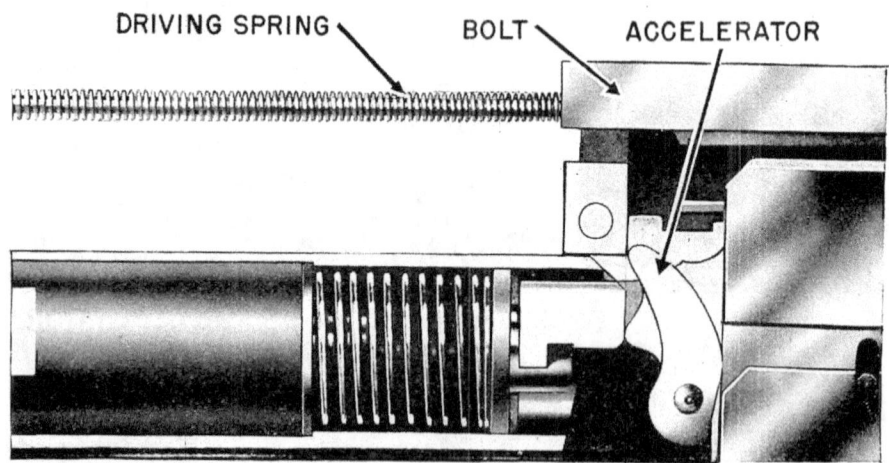

a. BOLT COUNTERRECOILING

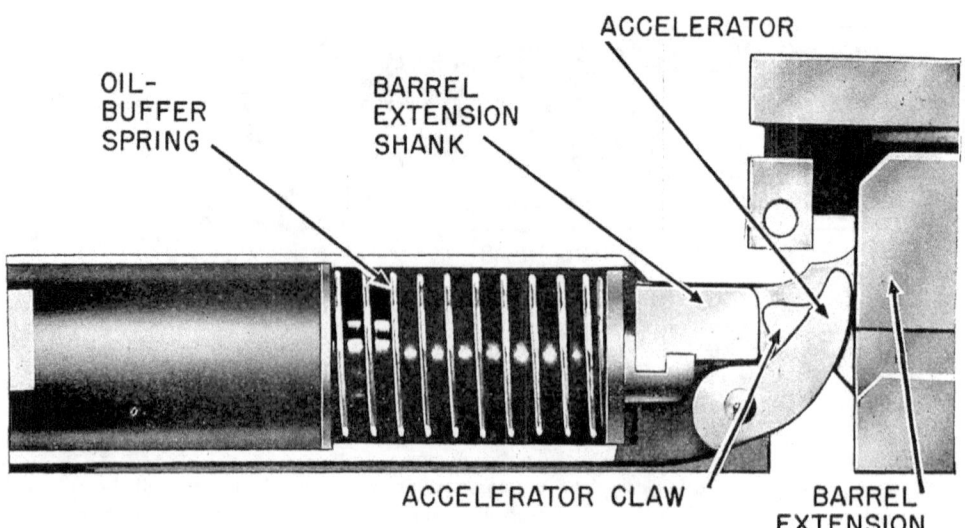

b. BARREL AND BARREL EXTENSION COUNTERRECOILING

FIGURE 7D9. Action of the accelerator in counterrecoil; caliber .50 Browning machine gun.

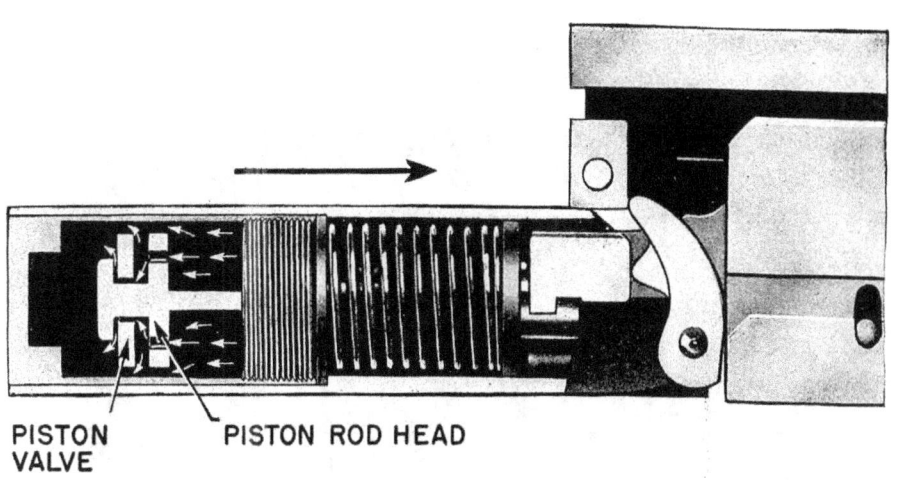

a. ACTION OF OIL BUFFER AND SPRING IN COUNTERRECOIL

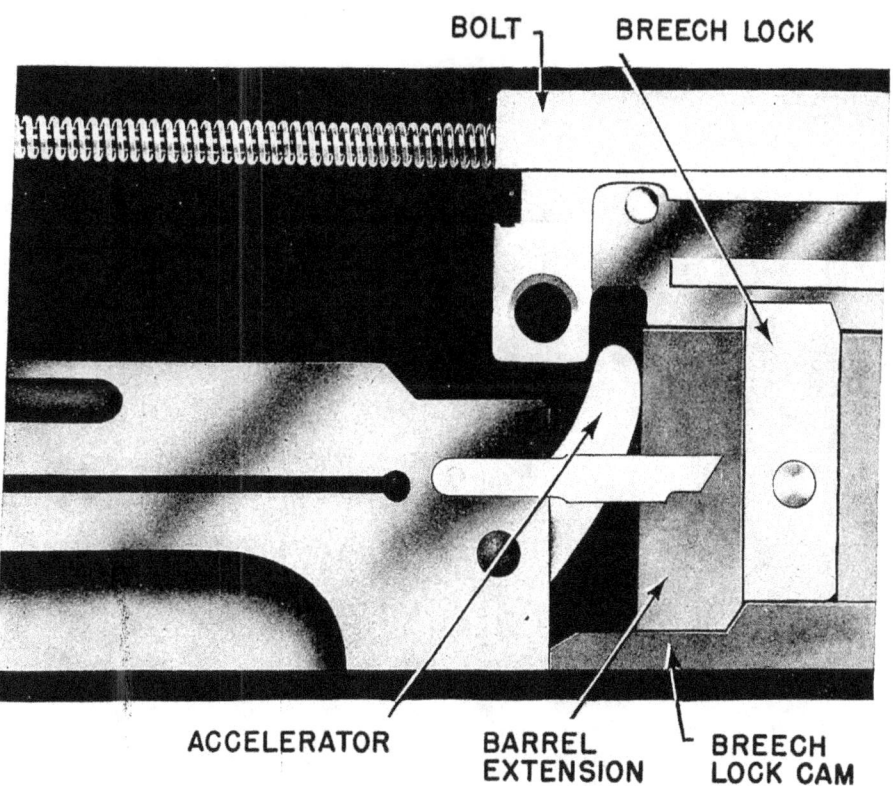

b. LOCKING OF BOLT AFTER COUNTERRECOIL

FIGURE 7D10. Buffer action and locking of bolt; caliber .50 Browning machine gun.

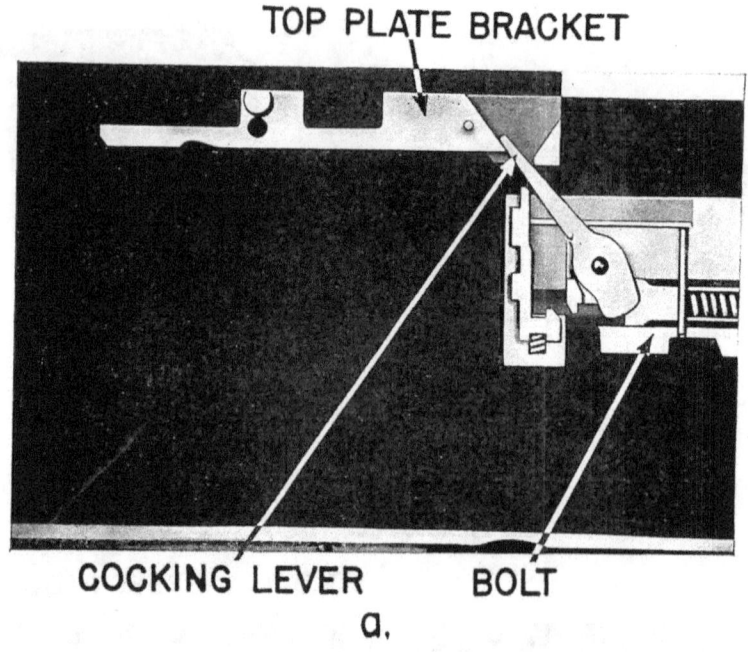

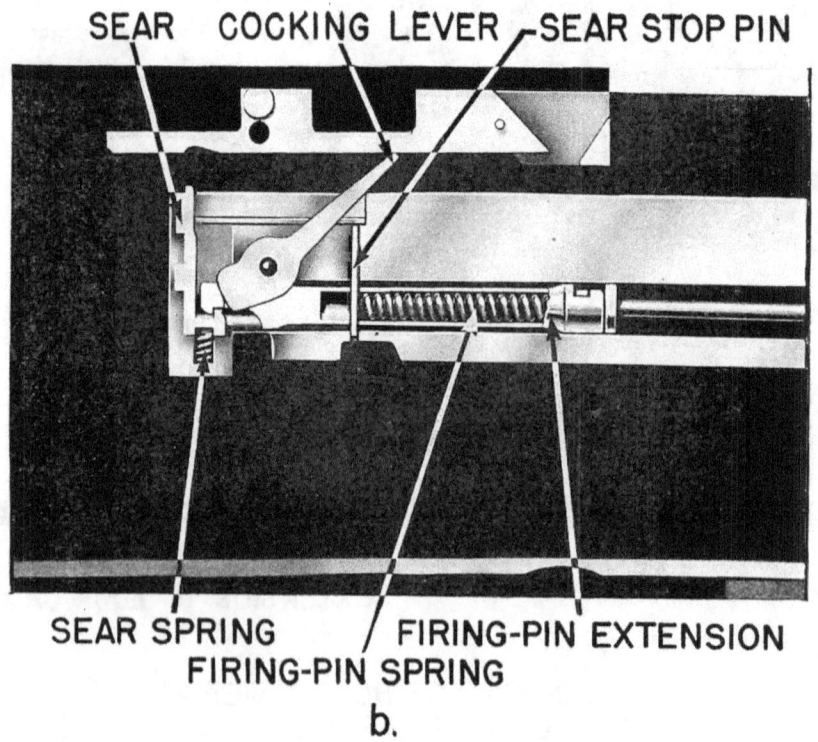

Figure 7D11. Cocking action; caliber .50 Browning machine gun.

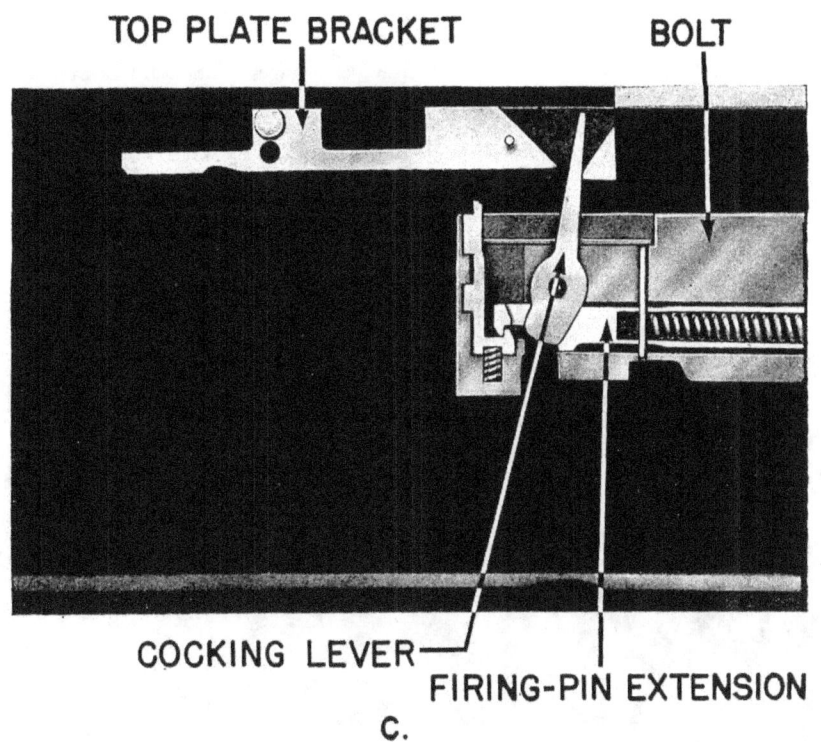

c.

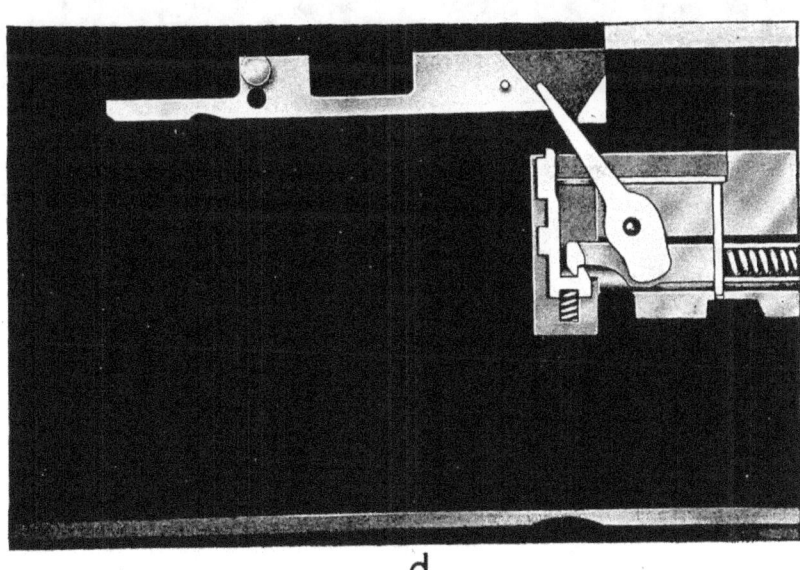

d.

Figure 7D11. Cocking action; caliber .50 Browning machine gun.

MACHINE GUNS

As the barrel extension moves forward the breech lock engages the breech-lock cam and is forced upward. The bolt has been continuing its forward motion since striking the accelerator, and has at this instant reached a position where the notch on the underside is directly above the breech lock. Thus the breech lock is permitted to engage the bolt (see fig. 7D10b). The bolt is thereby locked to the breech end of the barrel just before the recoiling parts reach the firing position.

7D9. Cocking. The act of cocking the gun is begun as the bolt starts to recoil immediately after firing. Thus the tip of the cocking lever which is in the V-slot in the top plate bracket, as shown in figure 7D11a, is forced forward.

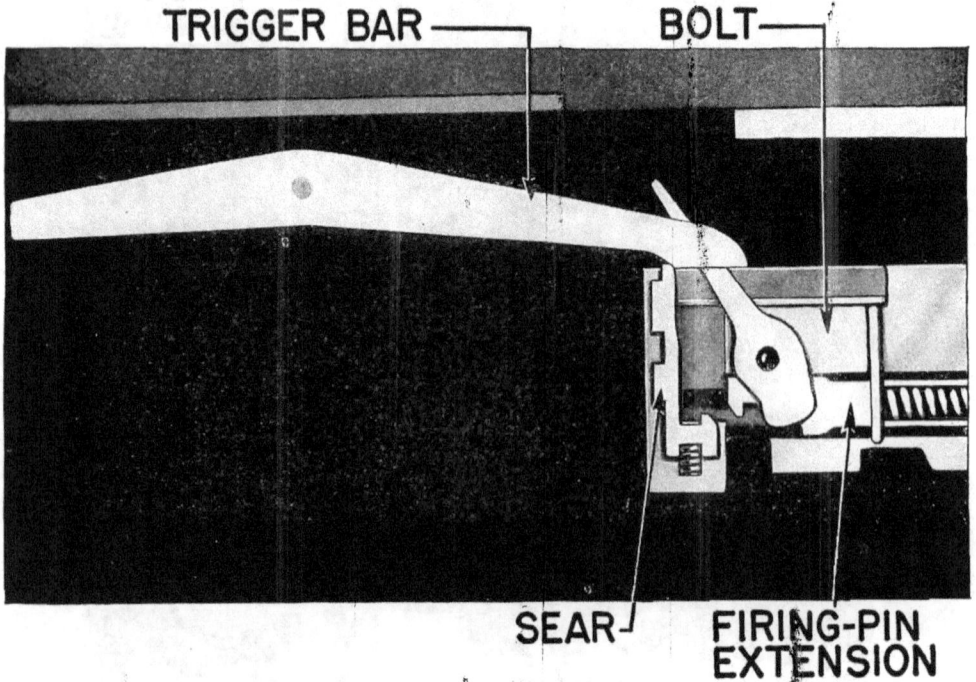

FIGURE 7D12. Trigger pressed for automatic firing; caliber .50 Browning machine gun.

The cocking lever is pivoted so that the lower end forces the firing pin extension rearward. The firing-pin spring is thus compressed against the sear stop pin. The shoulder at the back end of the firing-pin extension is hooked over the notch at the bottom of the sear under pressure of the sear spring (see fig. 7D11b).

During the forward motion of the bolt, the tip of the cocking lever enters the V-slot of the top plate bracket (see fig. 7D11c). This action swings the bottom of the cocking lever out of the path of the firing-pin extension, as illustrated in figure 7D11d, thus permitting the firing pin to snap forward to fire the cartridge.

When the recoiling portion is almost in the forward position, the gun is ready to fire. If no trigger action occurs at this instant, the recoiling portion assumes its final forward position, as shown in figure 7D11d, and the gun ceases to fire. The parts are now in such position that the gun is again ready to fire, as shown in figure 7D3a.

7D10. Automatic firing. For automatic firing the trigger is pressed and held down. The sear is depressed as its tip is carried against the cam surface of the trigger bar by the forward movement of the bolt near the end of the counterrecoil stroke (see fig. 7D12). The notch in the bottom of the sear releases the firing-pin extension and the firing pin, thus automatically firing the next cartridge at the completion of the forward stroke. The gun fires automatically as long as trigger action is maintained, and until the ammunition supply is exhausted.

7D11. **Feeding.** The belt-feed mechanism is actuated by the bolt. When the bolt is in the forward position the belt-feed slide is within the confines of the gun. Figure 7D13a shows the mechanism from above with the cover removed. A stud at the rear of the belt-feed lever is engaged in the diagonal groove or way in the top of the bolt.

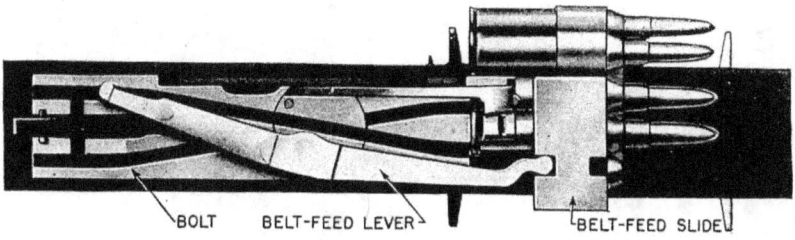

a. BOLT FORWARD

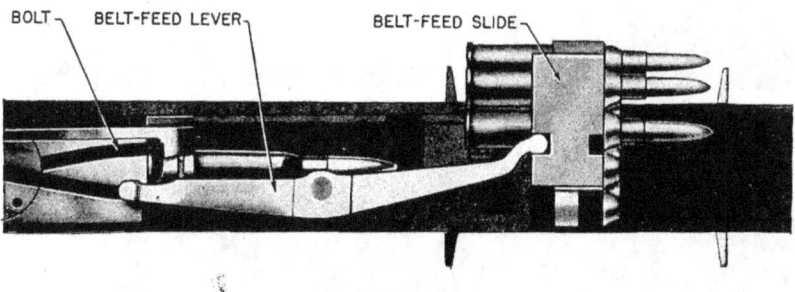

b. BOLT AFT

FIGURE 7D13. Belt feed mechanism—top view; caliber .50 Browning machine gun.

As the bolt moves rearward during recoil the belt-feed lever is pivoted. The forward end of the belt-feed lever moves the belt-feed slide out of the side of the gun and over the ammunition belt. (Note: Ammunition feed in figure 7D13b is from the left side of the gun. Feed from either side is possible with all caliber .50, M2 guns.)

The ammunition belt is pulled into the gun by the belt-feed pawl which is attached to the belt-feed slide. When the bolt is forward the belt-feed pawl has positioned a cartridge directly above the chamber. The belt-holding pawl is in a raised position to prevent the ammunition belt from falling out of the gun (see fig. 7D14a).

As the bolt recoils the belt-feed slide is moved out over the belt, and the belt-feed pawl pivots so as to ride over the next cartridge, as shown in figure 7D14b.

At the end of the recoil stroke the travel of the belt-feed slide is sufficient to permit the belt-feed pawl to snap down behind the next cartridge in order to pull the belt into the gun (see fig. 7D14c).

As the bolt moves forward on the counterrecoil stroke the belt is pulled into the gun by the belt-feed pawl. The belt-holding pawl is forced downward as a cartridge is pulled over it, as shown in figure 7D14d. When the forward stroke of the bolt is completed the belt-holding pawl snaps up behind the cartridge, as shown in figure 7D14a.

7D12. **Extracting and ejecting.** When recoil starts, a cartridge is drawn from the ammunition belt by the extractor. The empty case is withdrawn from the chamber by the T slot in the front face of the bolt (see fig. 7D15a).

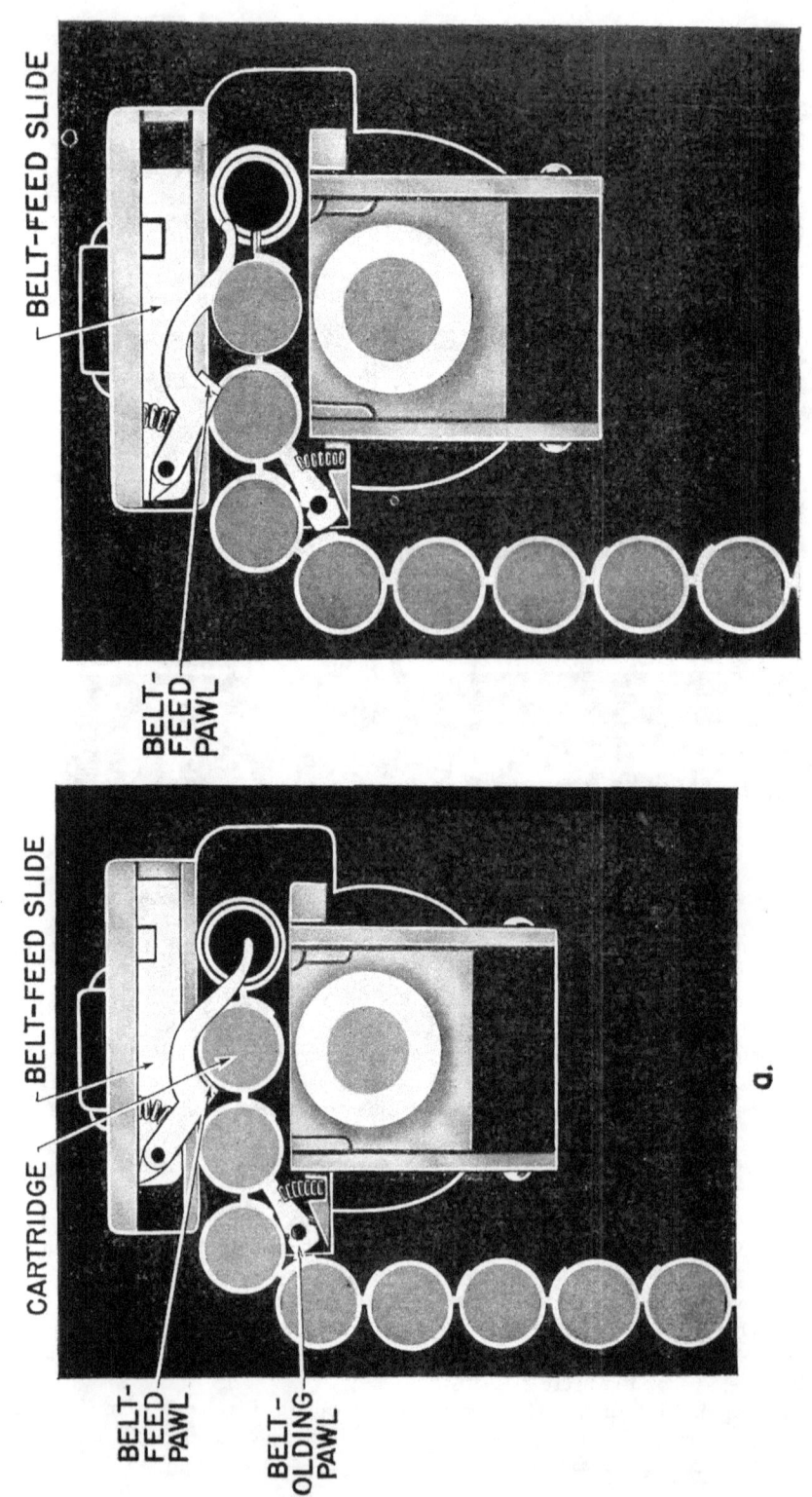

Figure 7D14. Feeding sequence; caliber .50 Browning machine gun.

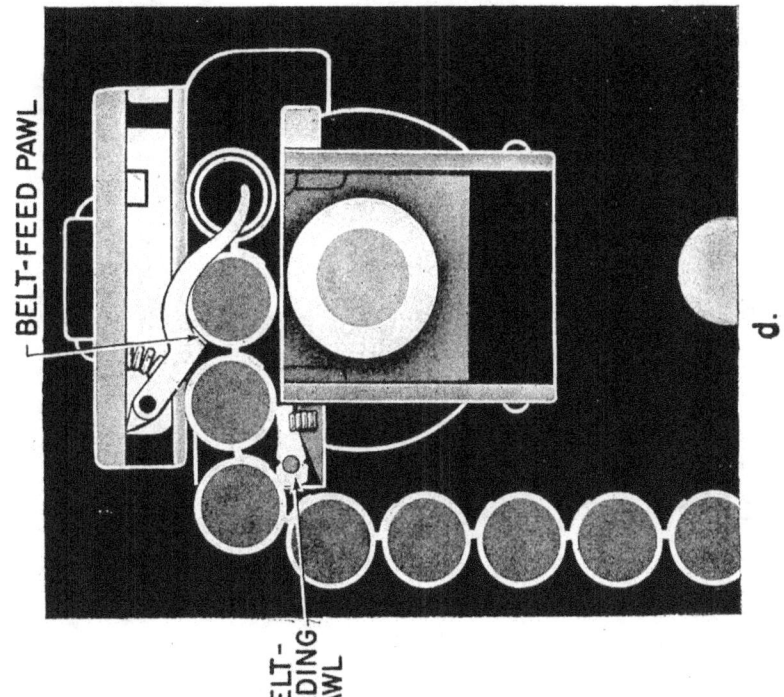
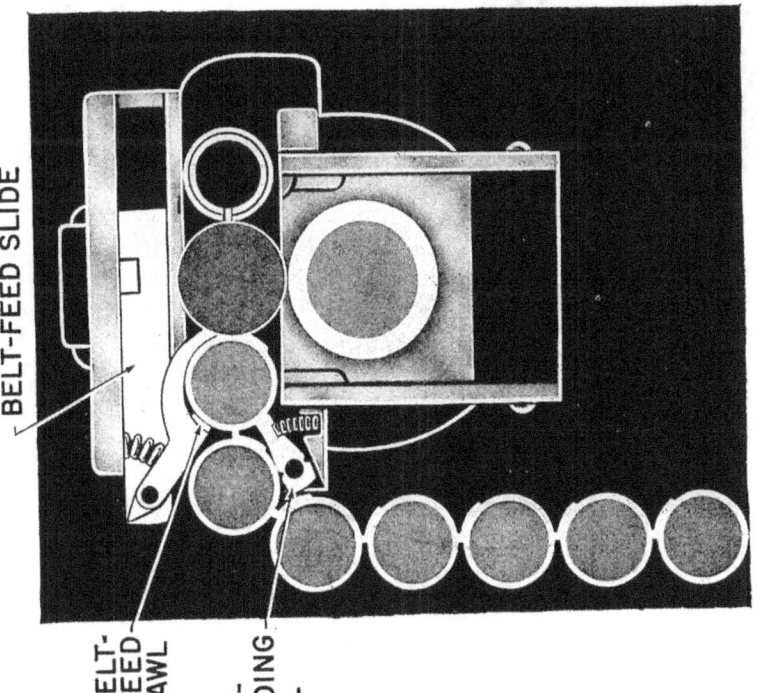

Figure 7D14. Feeding sequence; caliber .50 Browning machine gun.

MACHINE GUNS

The empty case, having been expanded by the force of explosion, fits the chamber very snugly, and the possibility of tearing the case exists if the withdrawal is too rapid. To prevent this, and to insure slow initial withdrawal, the top front edge of the breech lock and front side of the notch in the bolt are beveled, as shown in figure 7D15b. Thus, as the breech lock is disengaged, the bolt moves away from the barrel and barrel extension in a gradual manner.

As the bolt moves to the rear the cover extractor cam forces the extractor down, causing the cartridge to enter the T-slot, as shown in figure 7D16.

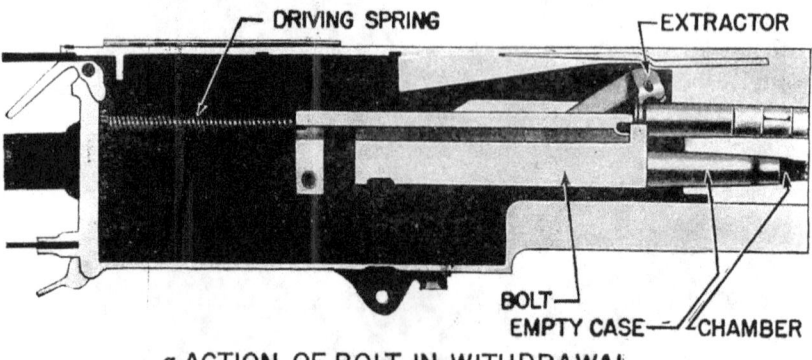

a. ACTION OF BOLT IN WITHDRAWAL.

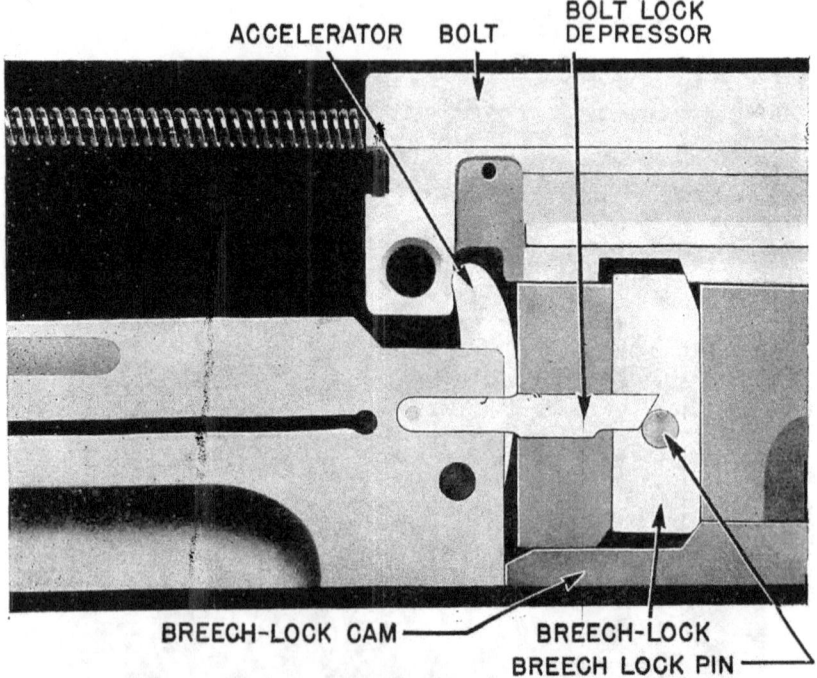

b. ACTION OF BREECH LOCK IN WITHDRAWAL.

FIGURE 7D15. Empty case withdrawal; caliber .50 Browning machine gun.

198

CHAPTER 7

As the extractor is forced down a lug on the side of the extractor rides against the top of the switch causing the switch to pivot downward at the rear, as can be seen in figure 7D17a. Near the end of the rearward movement of the bolt the lug on the extractor overrides the end of the switch, and the switch snaps up to its normal position.

On counterrecoil the extractor and cartridge are forced farther downward by the extractor lug riding on the under side of the switch. The cartridge pushes the empty case out of the T-slot.

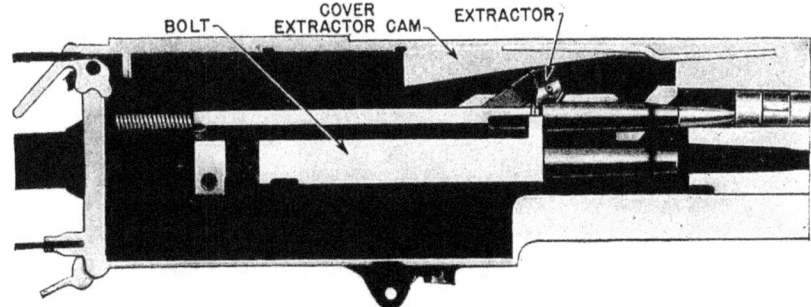

FIGURE 7D16. Entering new cartridge; caliber .50 Browning machine gun.

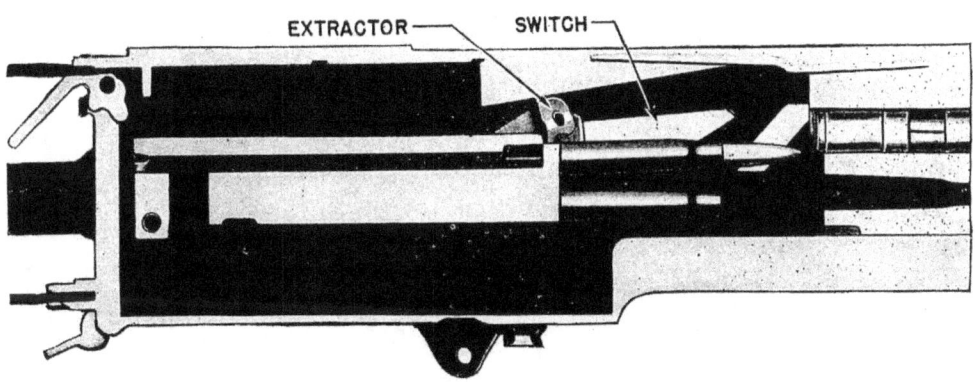

a. EJECTION

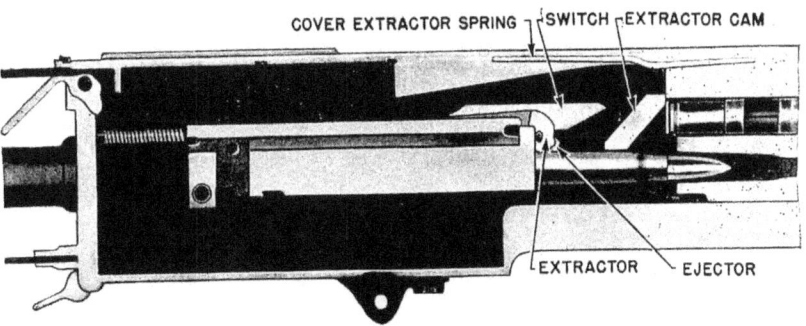

b. CHAMBERING

FIGURE 7D17. Action of extractor, switch, and cam; caliber .50 Browning machine gun.

199   **RESTRICTED**

The extractor stop pin in the bolt limits the downward travel of the extractor so that the cartridge, assisted by the curvature of the ejector, enters the chamber (see fig. 7D17b). The ejector also ejects the last empty case. When the cartridge is practically chambered the extractor rides up on the extractor cam, compresses the cover extractor spring, and snaps into the groove of the next cartridge in the belt.

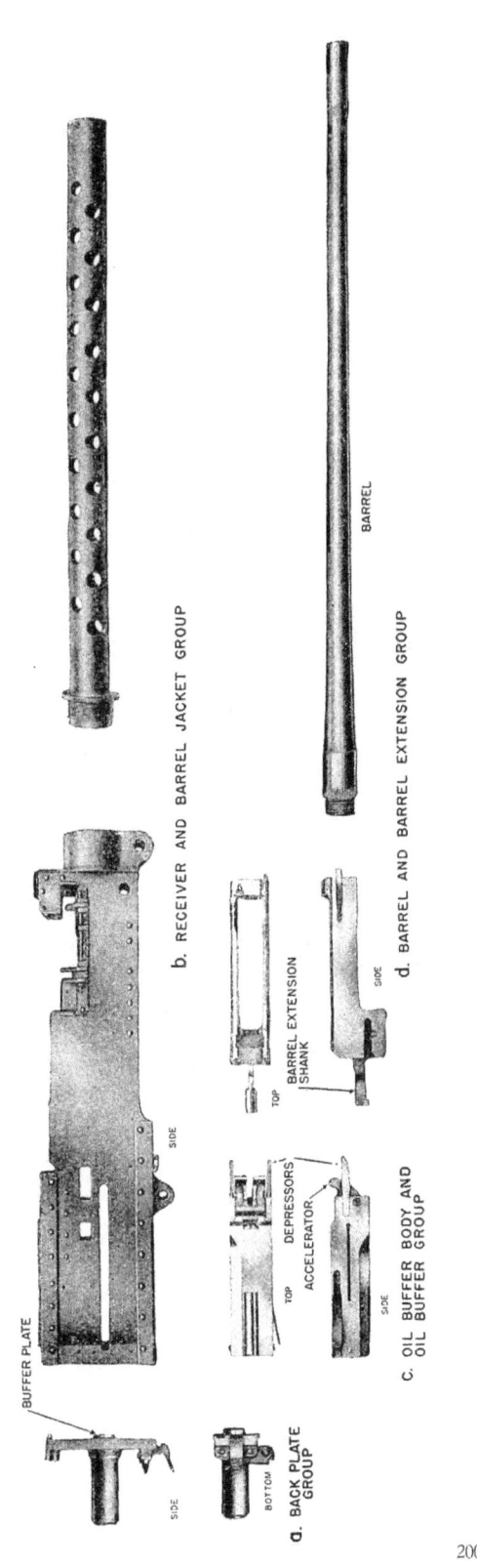

FIGURE 7D18. Group assemblies of a Browning caliber .50 machine gun.

CHAPTER 8

SMALL ARMS

A. INTRODUCTION

8A1. General. Small arms have been defined previously as guns of caliber .60 and smaller. However, the term small arms is generally considered to mean weapons that are carried in the hand or on the shoulder.

Compared with the Army, the Navy requires a relatively small number of small-arms weapons and ammunition. Hence, the Navy generally uses the same types that are used by the Army, and obtains them from the War Department. This policy makes for interchangeability of matériel and facilitates the solution of supply problems. Though the caliber .50 guns are small arms by definition, certain types thereof are especially designed for naval use and are not generally regarded as being small arms. They are discussed in Chapter 7.

8A2. Small-arms types. There are many variations of small arms in current use, and no attempt will be made to list or describe all of them. We may arbitrarily assign naval small arms to four general types: Examples are as follows:

1. Rifles:
    Caliber .30 Springfield.
    Caliber .30 Browning automatic rifle.
2. Hand weapons:
    Caliber .45 Colt automatic.
    Caliber .38 revolver.
3. Submachine guns:
    Caliber .45 Thompson ("Tommy gun").
4. Carbines:
    Caliber .30 carbine.

8A3. Operating principles. Only one example of each of the foregoing types is described in the succeeding sections. These examples and the principles upon which they operate are as follows:

1. Bolt operated—Springfield rifle.
2. Recoil operated—Colt caliber .45 pistol and Thompson submachine gun.
3. Gas operated—carbine.

Bolt-operated guns are those in which actuation of the bolt *by hand* performs all the operations of unlocking the bolt, ejecting an empty cartridge case, cocking the gun, loading a new cartridge from the magazine into the chamber, and closing and locking the bolt. Bolt-action guns utilize only manual power for these operations, whereas recoil- and gas-operated guns employ some of the energy of explosion for ejecting the empty cartridge case, cocking the gun, and loading another cartridge.

Recoil-operated guns are those in which the barrel and block (bolt) recoil together for a short distance until the projectile is clear of the muzzle and the pressure is off the gun. Then the recoil of the barrel is stopped, but the block, on being unlocked from the barrel, continues to recoil against spring pressure until it has moved a sufficient distance to the rear where the empty case is ejected. The force of recoil also cocks the gun and compresses a spring which then closes the block (loading a new cartridge), and returns the barrel and block to normal, or firing, position.

Gas-operated guns are those which utilize the force of a small amount of the explosion gases which escape through a small port near the muzzle for opening the block, ejecting the empty

SMALL ARMS

case, cocking the gun, and compressing a spring sufficiently to load a new cartridge and close the block.

8A4. Description. A thorough and detailed description of the many parts and the complex functioning of each of these weapons would be too extensive for inclusion here. Rather, the purpose is to acquaint the student with the general characteristics of each type, together with the method of loading and firing. Detailed descriptions of functioning, maintenance, and operation may be obtained from appropriate Ordnance Pamphlets and War Department publications.

8A5. Safety precautions, selected from Article 972, Navy Regulations.

"110. In firing small arms, machine guns, and submachine guns, whenever a blowback occurs the bore shall be examined for foul bore before firing another round."

"111. When a misfire occurs in small arms, machine guns, or submachine guns, another attempt may be made to fire the weapon provided it can be recocked without opening the bolt; the bolt should not be opened until at least ten seconds have elapsed after the last attempt to fire the weapon, except in those installations, such as distant control in aircraft, where the opening of the bolt of the weapon would not endanger personnel."

B. CALIBER .30 SPRINGFIELD RIFLE

8B1. Description. The caliber .30 Springfield rifle is a bolt-operated, magazine-fed rifle. Its magazine capacity is one clip of five rounds. The rifle is fitted with open and peep sights and a knife bayonet. It is fired in single shots, the bolt being operated by hand for each round. The maximum rate of fire depends upon the skill and the position of the operator, and varies from 10 to 15 shots per minute. The sights are graduated from 100 to 2,850 yards.

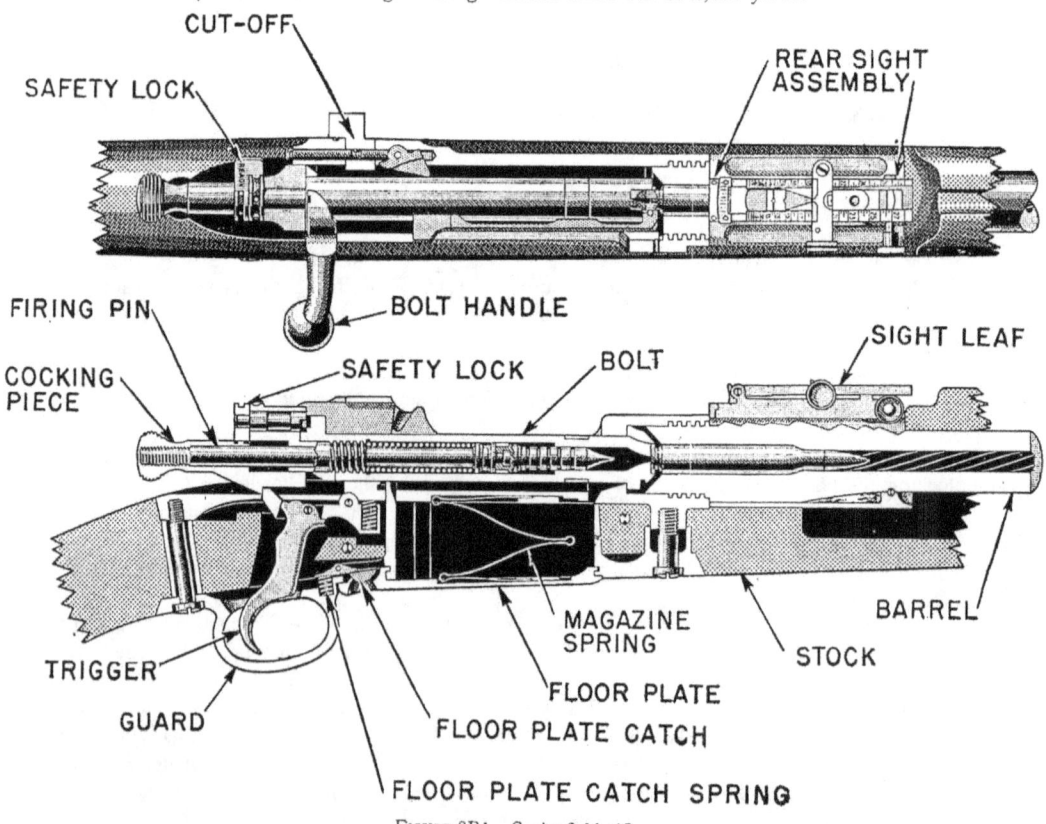

FIGURE 8B1. Springfield rifle.

CHAPTER 8

8B2. Operating the gun. To load the magazine, turn the *cut off* up showing "on", draw the *bolt handle* fully to the rear, and insert the cartridges into the *magazine* from a clip or singly from the hand. To load from a clip, place one end of a loaded clip into its seat in the *receiver*, and with the fingers of the right hand under the rifle against the *floor plate* and the base of the thumb on the top cartridge case, press the cartridges firmly down into the magazine with the thumb, until the top cartridge is caught by the right edge of the receiver. The empty clip is then removed. After loading the magazine, a round is placed in the chamber by closing the bolt.

If it is desired to carry the gun cocked, with a cartridge in the chamber, the *safety lock* should be turned to the right to "safe". To fire the gun, turn the safety lock to "ready" and squeeze the trigger. After firing, pull the bolt handle up and all the way to the rear. This ejects the fired cartridge case. Closing and locking the bolt enters a new cartridge and cocks the gun for firing.

When the rifle is used as a single-loader, cartridges are inserted directly into the chamber by hand, the cut-off being turned down. In this case the magazine remains inoperative, and a full magazine can be held in reserve. Closing the bolt loads and cocks the gun, which may then be fired by squeezing the trigger.

8B3. General data. The following are the essential features of the Springfield rifle.

Caliber	.30	Maximum powder pressure in chamber	51,000 pounds
Weight	9 pounds	Rifling	4 grooves
Weight of bayonet	1 pound	Twist, right hand	1 turn in 10 inches
I. V.	2,700 f.s.	Maximum range at 45° elevation	5,500 yards

C. CALIBER .45 AUTOMATIC PISTOL

8C1. Description. The caliber .45 automatic pistol is usually known as the Colt .45 or the Colt automatic. This pistol is recoil operated and magazine fed. The principal parts of the pistol are the *barrel,* the *slide,* and the *receiver.* The block or bolt is part of the slide. The recoil spring is located under the barrel. Front and rear open sights are used.

The magazine holds seven cartridges and is inserted as a separate unit in the bottom of the handle. Though the pistol is called automatic, it requires a separate squeeze of the trigger for each shot. When the last cartridge in the magazine has been fired the slide remains aft in the open position. The *magazine catch* is then pressed and the empty magazine falls out. The rate of fire is largely dependent upon the skill of the operator in firing and in inserting a new magazine.

8C2. Operating the pistol. A loaded magazine is placed in the handle and the slide drawn fully aft and released so as to fly forward. If the slide is locked aft by the *slide stop,* push the slide stop down and allow the slide to go forward. This action loads and cocks the pistol ready for firing.

If it is desired to carry the gun loaded and cocked, press the *safety lock* upward. To fire the gun, simply press the safety lock down, and then squeeze the trigger with the fore finger and the *grip safety* with the palm of the hand. The trigger cannot be pulled unless the grip safety is squeezed simultaneously. Firing the pistol ejects the empty case, loads a new cartridge, and cocks the gun ready for firing.

8C3. General data. The following are the essential features of the caliber .45 automatic:

Caliber	.45	Rifling	6 grooves
Weight	2½ pounds	Twist, left hand	1 turn in 16 inches
I. V.	800 f.s.	Maximum range at 45° elevation	1,955 yards
Maximum powder pressure in chamber	16,000 pounds		

SMALL ARMS

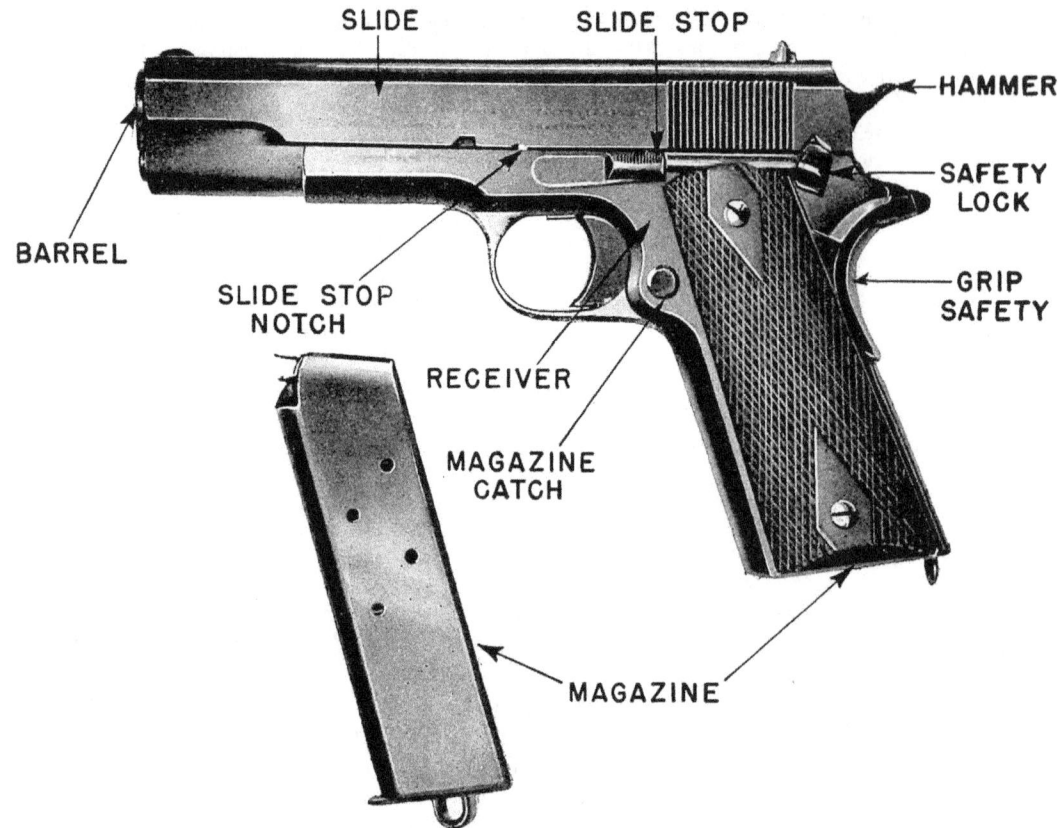

FIGURE 8C1. Automatic pistol, caliber .45.

D. THOMPSON SUBMACHINE GUN

8D1. Description. The Thompson submachine gun, caliber .45, is a shoulder firing, automatic weapon. It is commonly called a "Tommy gun". The gun is air cooled, magazine fed, and recoil operated. It is composed of four principal groups of parts; the *magazine, stock, frame,* and *receiver*. The magazine may be of two types; box, or drum. The box magazine holds 20 rounds, and the drum holds 50 rounds of caliber .45 cartridges. The magazine is secured to the forward end of the frame by a *magazine catch*. The gun may be fired in single shots or fully automatic.

The gun can be provided with a *Cutts compensator*. This is an attachment for the muzzle of a machine gun which decreases the tendency of the muzzle to rise during automatic firing, in addition to steadying the gun and dissipating some of the force of recoil. It consists essentially of a chamber which has orifices on its upper side that are directed toward the rear of the gun. Forcing of gases through these orifices tends to hold the muzzle of the gun down and to take up some of the recoil.

8D2. Operating the gun. The normal method of firing the gun is in single shots. However, the gun may be fired automatically by setting the *rocker pivot* (or fire-control lever) to point forward toward the muzzle.

To fire single shots, first cock the bolt by pulling the *actuator* (or bolt handle) all the way to the rear. Turn the *safety lever* to the "safe" position (pointing rearward) and then turn the fire-control lever so that it also points to the rear to the "single" position. The gun is now cocked for single-shot fire but locked by the safety lever. To fire, turn the safety lever to "fire" position;

CHAPTER 8

that is, so that it points forward toward the muzzle, and pull and release the trigger quickly for each shot.

To fire fully automatic, the fire-control lever is set to "full auto" position; that is, so that it points forward toward the muzzle. The gun will now fire as long as the trigger is held down.

The bolt remains in a retracted position after the last shot is fired from the magazine. It must be in its rearward or retracted position in order to insert another magazine, to set the safety lever, or to set the fire-control lever.

8D3. General data. The following are the characteristics of the Thompson submachine gun:

Caliber	.45	Weight of drum magazine loaded	6 pounds
Weight	9¾ pounds without magazine	Maximum range (effective)	300 yards
Magazine capacity	box, 20; drum, 50	Rate of fire, single shot	up to 100 shots per minute
I. V.	800 f.s.	Rate of fire, automatic	600 shots per minute

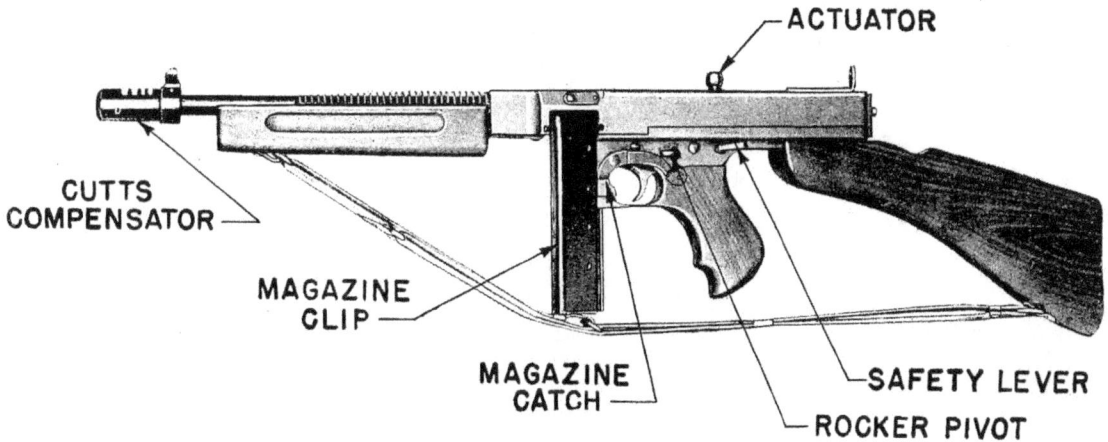

FIGURE 8D1. Thompson submachine gun, caliber .45.

E. U. S. CARBINE, CALIBER .30

8E1. Description. The U. S. carbine, caliber .30, is a gas-operated shoulder weapon. The gun is air cooled, magazine fed, and fires a single shot for each squeeze of the trigger. It is of light construction weighing about five pounds. The gun is fed from a box-type magazine having a capacity of 15 rounds. There are four principal assemblies of parts: The *magazine; handguard; stock;* and the assembly consisting of the *barrel, receiver, operating slide,* and the *trigger group*.

8E2. Operating the gun. To load the carbine, lock the gun by pushing the *safety lock* to the right. Then insert the magazine and snap it into position, making certain that the *magazine lock* has latched. With the fore finger of the right hand pull the *operating-slide handle* all the way to the rear and release it to close the bolt. This action loads and cocks the gun ready for firing. When ready to fire, push the safety lock to the left and fire the gun by squeezing the trigger, releasing the pressure on the trigger after each shot. The firing rate is dependent upon the skill of the operator and the speed at which magazines are changed. An empty magazine may be removed by pressing the magazine lock from right to left and withdrawing the magazine. When a new magazine is inserted, the gun is made ready to fire by repeating the sequence above.

SMALL ARMS

The carbine can also be operated as a single loader without having a magazine in the gun. The operating slide is pulled to the rear and the operating slide catch pressed down locking the slide in this position. A single cartridge is inserted in the chamber, and the operating-slide handle pulled back and released. This allows the bolt to go forward, and, after the safety lock is pushed to the left, the gun can be fired.

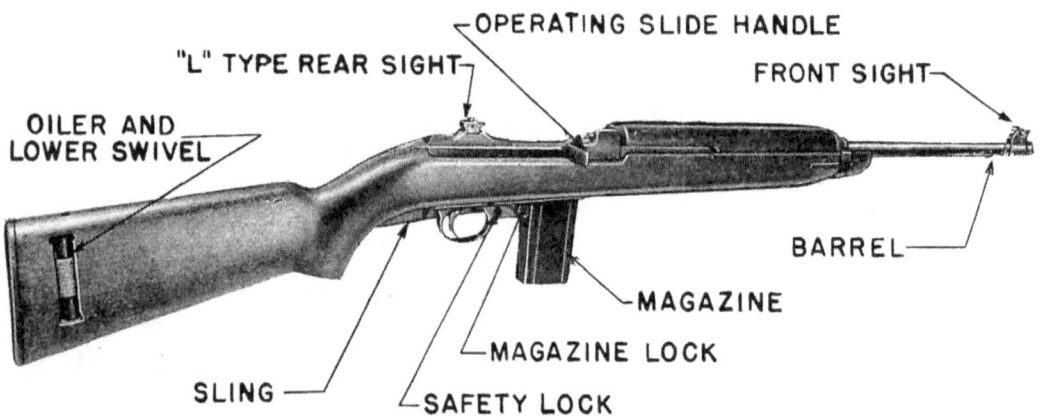

FIGURE 8E1. U. S. Carbine, caliber .30

8E3. **General data.** The following are the characteristics of the U. S. Carbine:

Caliber	.30	Maximum powder pressure in chamber	40,000 pounds
Weight	5 pounds	Rifling, right hand twist	4 grooves
I. V.	1,900 f.s.	Maximum range	2,000 yards
Magazine capacity	15		

CHAPTER 9

BOMB-TYPE AMMUNITION AND RELEASE GEAR

A. INTRODUCTION

9A1. General. Attacks on submerged submarines and the underwater portions of surface ships are usually made with high-explosive filled, bomb-type ammunition such as torpedoes, mines and depth charges. The destructive effects of such weapons are produced by the forces of explosions occurring outside the target, and do not depend upon previous penetration of the hull.

9A2. Underwater explosions. The force of explosion produced by the underwater detonation of a high explosive appears in two forms. The first is the nearly instantaneous crushing shock of the *percussive wave* which travels through the water *before* the gases of explosion have had time to expand appreciably. The second form is the physical disruption produced as the gases expand.

The explosive is so quickly transformed to gases that extremely high pressure exists until expansion occurs. This pressure is transmitted in all directions by the water. A submerged submarine in the vicinity is subjected on all sides to this pressure, which tends to crush the plates inward. It is this pressure which makes a depth charge effective against a submarine at moderate distances, even though a direct hit is not obtained.

The physical force exerted against the hull of a ship by an adjacent explosion is much greater underwater than in air. While air is compressible and "gives" readily under the influence of pressure, water is practically incompressible and resists the expansion of explosion gases. Consequently, the pressure developed in an explosion reaches a higher value underwater than in air, and causes more damage. The resistance offered by the water to expansion of the gases increases with the depth, hence it is desirable to explode a charge against a hull as far beneath the surface as possible. It has been demonstrated that an explosion is several times more effective immediately underneath a ship than against its side.

9A3. Explosives used. The main charge in torpedoes, mines, and depth charges is usually cast TNT, although torpex is coming into use in some torpedoes. The booster required to detonate the main charge is generally a small charge of tetryl (several ounces) or granular TNT (several pounds). Fulminate of mercury is employed in the detonators. In general, the boosters are arranged so that they will be approximately in the center of the main charge, where they will produce the most uniform detonation. In the booster container is a recess into which the detonator is inserted automatically by an arming device after the charge has been launched.

B. TORPEDOES

9B1. General. Torpedoes are self-propelling submarine weapons designed to carry a large charge of high explosive to a target. They are driven through the water by gas and steam, or by electricity. They are controlled to run on a fixed course, at a constant speed and at a set depth. All torpedoes are similar in appearance to the one shown in figure 9B1. The diameter is about 21 inches, the length about 21 feet. Some torpedoes, launched from aircraft or PT boats, are about 22½ inches in diameter and 13½ feet long.

Speeds of torpedoes vary between 27 and 47 knots, and ranges from 4,000 to 16,000 yards. The distance a torpedo can travel depends on the speed, because the energy required for propulsion increases rapidly as the speed increases. High speeds are preferable to low speeds, because they allow the target less time to maneuver. In addition to single-speed torpedoes, two-and three-speed torpedoes are in use, permitting a choice for different firing conditions.

BOMB-TYPE AMMUNITION AND RELEASE GEAR

Torpedoes are characteristic weapons of destroyer escorts, destroyers, some cruisers, PT boats, submarines and torpedo planes. The number carried varies from 1 in planes and 2 or 4 on PT boats to 8 to 16 on destroyers. Launching is accomplished by dropping from a rack on a plane, or ejection from tubes of water-borne ships. Expulsion from tubes is effected by use of compressed air in the case of submarines, or by torpedo impulse-charges of black powder on most other ships.

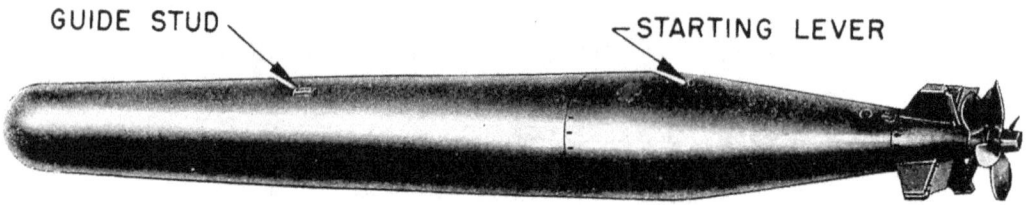

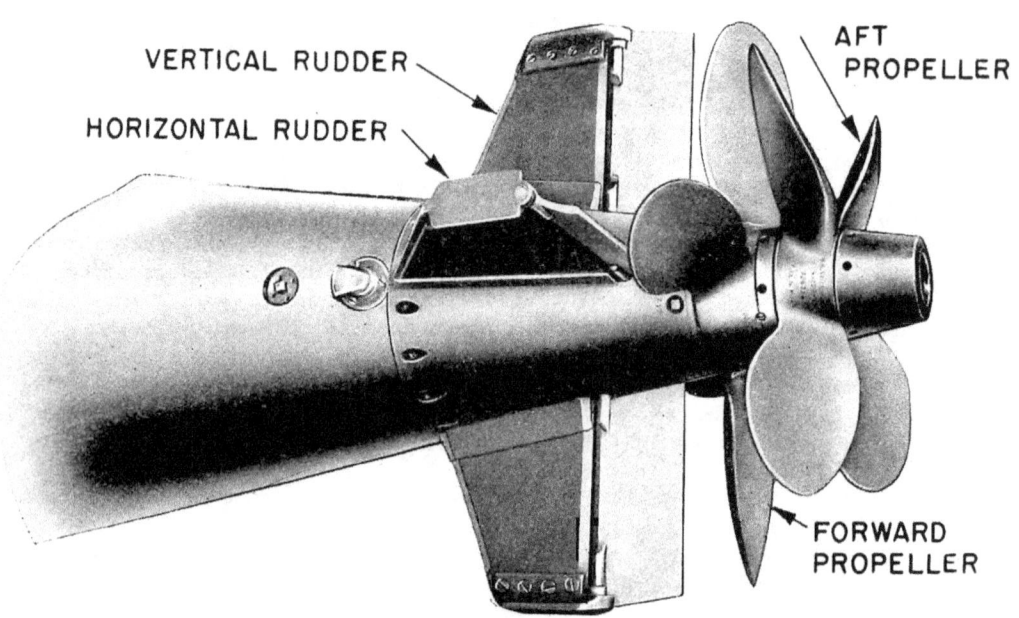

FIGURE 9B1. A torpedo.

9B2. Above-water torpedo tubes. Torpedo tubes are either fixed or trainable, depending on the installation. Submerged tubes used in submarines are fixed, and must be aimed by turning the ship. The following discussion considers only the trainable, above-water type typical of the tubes carried by destroyers.

This type of tube consists of 3, 4, or 5 barrels supported in a single mount. An example is shown in figure 9B2. Destroyer torpedo tubes are mounted on deck either outboard or on the centerline. The barrels are supported in a *saddle* which functions like the carriage of a gun mount except that no provision is made for elevating the barrels. The saddle rests on a roller bearing assembly supported by a *stand*. The stand, like that of a gun, carriers a training circle. The

CHAPTER 9

tube is normally trained by an electric-hydraulic drive, controlled by the operation of the *training hand-wheels*. Provision is made for manual training in case of power-drive failure.

Each barrel is an assembly of a *main barrel*, a *spoon*, and a *spoon extension*. The main barrel is cylindrical in shape, but the spoon and spoon extension are open on the underside. The spoon extension is hinged at the top so that it can be folded back along the spoon to save deck space. The breech end of the barrel is closed by a door to form a chamber for the powder gases which expel the torpedo from the tube. Mounted on top of the barrels are a seat for the trainer and gyro setter, part of the training gear, a tube sight, the firing mechanism, and other devices.

Within the barrel, at the bottom, are rollers which facilitate loading the torpedo. Running the length of the top of the barrel to the end of the spoon extension is a T-shaped guide slot. The *guide stud* on the torpedo (see fig. 9B1) rides in this slot as the torpedo is fired from the tube. The stud prevents the torpedo from dropping downward before the tail has cleared the main barrel, thereby preventing damage to the tail assembly and helping to keep the torpedo horizontal as it is launched.

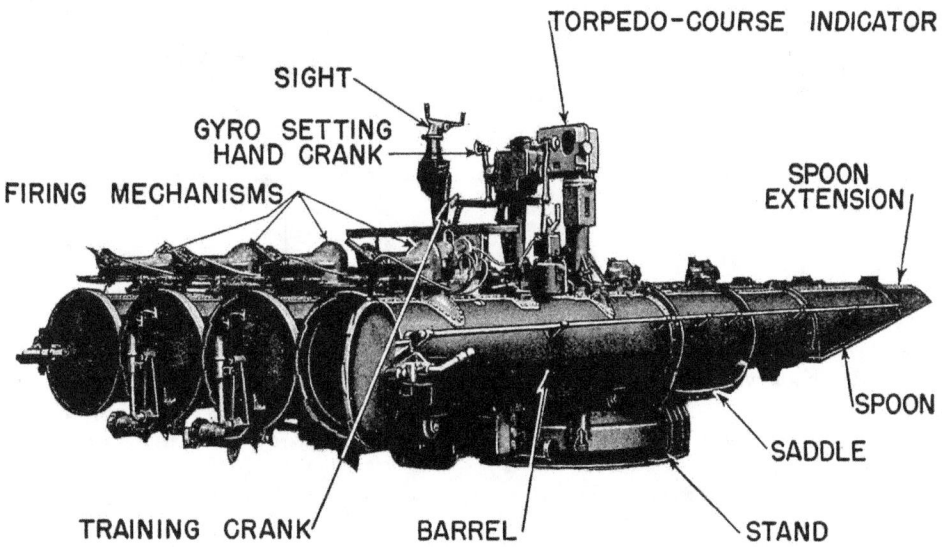

FIGURE 9B2. A quadruple tube.

Torpedoes are expelled from one barrel at a time by impulse charges fired in the *firing mechanisms* on top of the barrels. Normally, the charges are fired electrically from the bridge, but they can be fired by percussion at the tube if necessary. Each black-powder impulse charge is contained in a special 3-inch case about 13 inches long. Sphero-hexagonal black powder is used as the impulse charge for wing-mounted tubes. Sodium-nitrate black powder, which produces a greater ejection velocity, is used in the charges for centerline mounts.

The tubes are arranged so that speed, depth and gyro settings can be made while the torpedoes are in the barrels. These settings may be introduced up to a few moments prior to firing. Torpedoes are kept in the barrels, hence a torpedo stop is required to keep each of them in place. The stop is held against the guide stud on the torpedo by a link which parts when the torpedo is fired. A new link is installed whenever another torpedo is loaded into the barrel.

A tripping latch is mounted on each barrel. Its function is to trip the *starting lever* (see fig. 9B1) when the torpedo is fired. Tripping of the starting lever initiates operation of the torpedo mechanisms so that they will take control of the torpedo as soon as it enters the water.

RESTRICTED

Normally the tube is trained as directed by a visual torpedo-course indicator mounted on the tube in front of the trainer. The use of this indicator is explained in Chapter 20. If the indicator system fails, the trainer uses an open sight mounted on top of the hand-wheel bracket. This sight may be offset laterally to provide the lead angle required to compensate for movement of the target during the run of the torpedo.

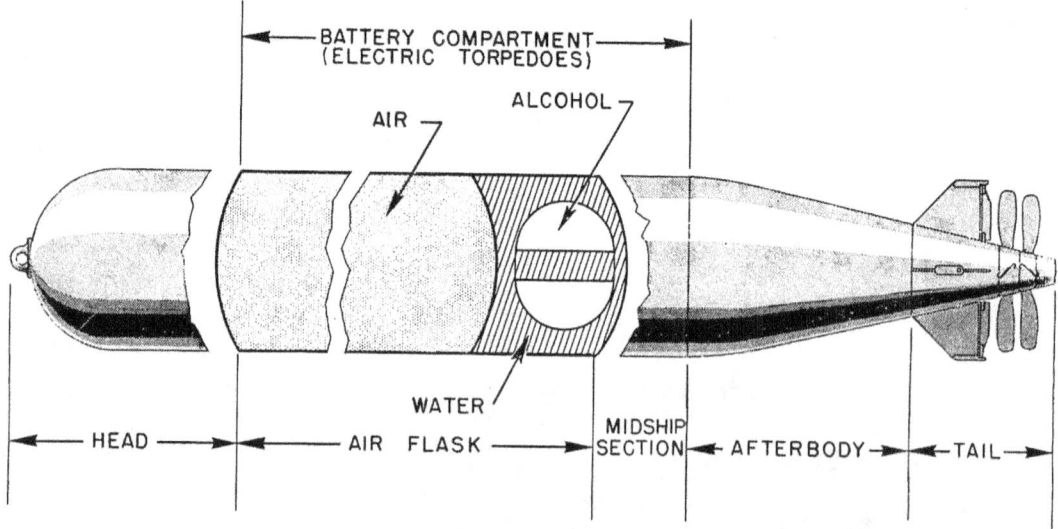

FIGURE 9B3. Divisions of a torpedo.

9B3. Major divisions. The major divisions of torpedoes are shown in figure 9B3. In a gas- and steam-driven torpedo, they are:

1. The *head*. It may be a *warhead* containing the explosive, or an *exercise head* used for practice shots.

2. The *air flask*. This section contains the potential energy required for propulsion and control.

3. The *midship section*. The valves and combustion flask needed to release the energy are in this part.

4. The *afterbody*. This is the "engine room" of a torpedo, housing nearly all components of the propelling, depth, and steering mechanisms.

5. The *tail*. It supports stabilizing vanes, propellers and rudders.

Electric torpedoes have many features in common with the gas- and steam-driven type. The principal difference is the replacement of the air flask and midship section with a battery compartment, and the use of an electric motor rather than a turbine.

9B4. General functional essentials. Any torpedo must have (1) an *explosive charge*, (2) an *energy supply* for operation, (3) a *propelling mechanism*, (4) a *depth mechanism*, and (5) a *steering mechanism*. The following articles are applicable in general to all gas- and steam-driven torpedoes. With respect to the explosive charge, and depth and steering mechanisms, they are also descriptive of electric torpedoes.

9B5. Explosive charge. The explosive charge, varying from 400 to 800 pounds, is TNT or torpex cast into the warhead as shown in figure 9B4. The head is a relatively thin-walled container strengthened by ribs and the solidity of the charge itself. To increase the underwater stability of the torpedo, the charge is cast on a bias, and lead ballast is secured to the bottom, thus lowering the center of gravity.

The *exploder mechanism* is housed in a cavity at the bottom of the warhead. It has the functions of (1) arming the exploder during the first part of the run, (2) preventing counter-mining due to shock of a nearby explosion in the water, and (3) detonating the main charge upon impact. The latest type is also equipped with electrical devices which actuate the exploder when the torpedo passes through the magnetic field of the target.

Practice firings required to test torpedoes and train personnel are conducted with *exercise heads.* They have the same dimensions as warheads, and are partially filled with water to give them the same weight. The water is automatically expelled by air from the air flask at the end of the run, so that the torpedo will rise to the surface. A torch pot or lamp is carried in the exercise head to produce smoke or light which will serve to guide the recovery crew.

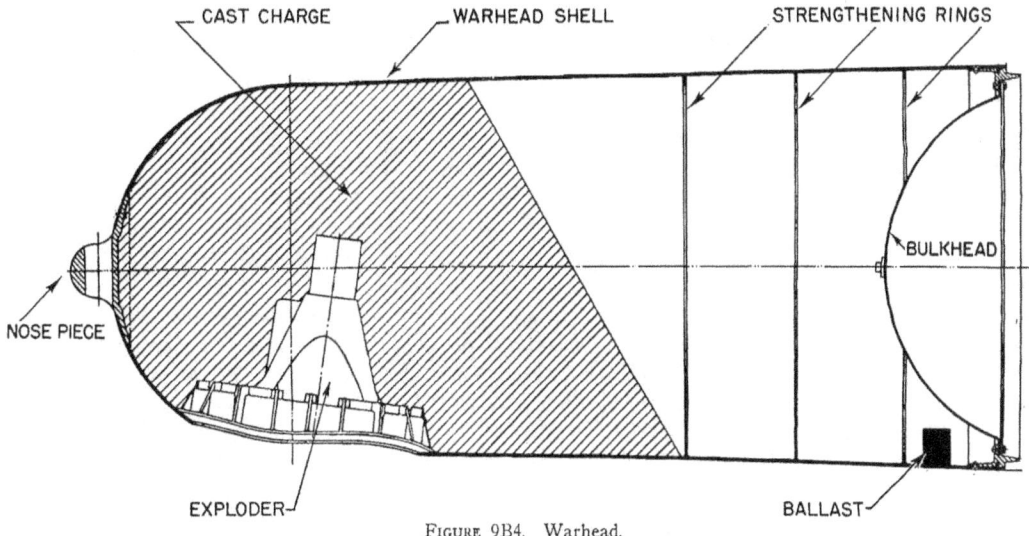

FIGURE 9B4. Warhead.

9B6. **Energy supply.** The energy needed to propel and control the torpedo is stored in the *air flask*. The larger forward part of the flask is charged with compressed air at 2,800 pounds p.s.i. pressure. Within the small compartment at the rear of the air flask is a tank containing alcohol. The space surrounding the tank is filled with water. The air, alcohol and water are led to valves in the midship section.

When the *starting lever* is tripped at launching, a starting valve opens, admitting air to various mechanisms of the torpedo. A small part of the air is used at high pressure in the starting and steering mechanisms. The rest of the air is throttled to a pressure of about 400 pounds p.s.i, and goes to the fuel and water tanks, the combustion flask, and the depth and steering mechanisms.

The air admitted to the fuel and water tanks forces the liquids through separate pipes to the combustion flask. At the same time, an ignition primer in the combustion flask is fired by means of a compressed-air actuated firing pin. The alcohol is mixed with air, and the mixture enters one end of the combustion flask, where it is ignited by the burning primer. The burning material is transformed to hot gases, into which water is sprayed. The resulting mixture of gas and steam is led to the propelling mechanism.

9B7. **Propelling mechanism.** The gas-steam mixture from the combustion flask is directed through nozzles against the blades of a *turbine* contained in the afterbody. The turbine has two wheels, revolving in opposite directions, connected by gears and shafts to the two *propellers*. The propellers, of opposite pitch, also revolve in opposite directions and drive the torpedo through the water.

9B8. **Depth mechanism.** The depth at which the torpedo runs is controlled by the *depth mechanism* located in the afterbody. This mechanism contains a spring-loaded diaphragm which is actuated by the pressure of the sea water. Movement of the diaphragm, caused by variations in the outside pressure, controls a valve which in turn causes the depth engine to move the rudders. The depth of run is set by adjusting the spring before the torpedo is launched.

9B9. Steering mechanism. The course a torpedo steers after launching is controlled by a *steering mechanism* contained in the afterbody. It consists of a gyroscope, linkage, and steering engine. At the moment of launching, a blast of high-pressure air spins the gyroscope, and from then on, low-pressure air keeps it spinning.

The rapidly-spinning gyroscope maintains its axis of spin parallel to its original position throughout the run of the torpedo. Variations in the angle between this axis of spin and the axis of the torpedo are transmitted by the linkage to the valve of the steering engine. The engine moves the vertical rudders to steer the torpedo so that the angle between the two axes remains constant at the value set before the torpedo was launched. The gyro can be set, before launching, to cause the torpedo to travel straight from the tube, or it can be set to cause the torpedo to turn through an angle before it takes up its fixed course.

9B10. Detailed functional operation. The operations of the chief functional units of a torpedo are described in the following articles. The explanations are functional, and will serve to acquaint the student with the principles involved. The details of specific torpedoes are too involved for presentation here, and may be obtained from appropriate Ordnance Pamphlets. The schematic diagram in figure 9B10 shows the general relationships of the many parts of a typical torpedo.

9B11. Exploder operation. A section through a typical exploder in the unarmed condition is shown in figure 9B5a. The *detonator,* contained within the *detonator holder,* is completely housed within the *safety chamber.* The *firing spring,* held between the *arming screw* and *firing-pin holder,* is not compressed. In addition, a pair of safety balls in the annular groove at the base of the firing-pin holder prevents motion of the pin. (The safety balls fit in the same groove as the firing balls, but actually are in a plane at right angles to the section shown, and are released when the mechanism is armed.)

As the torpedo enters the water after launching, the arming impeller (not shown) revolves and drives gears in mesh with the *arming gear.* Rotation of the arming gear revolves the safety chamber which is secured to the arming gear. The detonator, prevented from rotating by guides, is moved upward into the booster cavity by the rotation of the safety chamber. At the same time the arming screw, threaded within the arming gear, moves upward compressing the firing spring, and the exploder appears as shown in figure 9B5b.

The exploder is now ready to operate on impact. An impact, either direct or glancing, causes the *firing ring* to wedge itself between the *trigger plate* and the *fingers* as shown in figure 9B5c. The upward motion of the trigger plate raises the *trigger cap* until the annular groove within the cap is aligned with the *firing balls.* The firing-pin holder, pulled upward by the firing spring, pushes the firing balls outward and drives the firing pins into the primer caps, exploding the warhead.

An anti-countermining (ACM) device is part of exploder mechanisms now in use. It consists of a diaphragm exposed to sea water, and connected by a linkage to a lug which fits between the *top plate* and *trigger cap.* The detonating wave from a nearby explosion, pressing against the diaphragm, forces the lug between the trigger cap and top plate, thereby preventing the firing ring from wedging between the trigger plate and fingers. The ACM device automatically unlocks a short time after the detonating wave subsides.

The exploder mechanisms operated magnetically contain a mechanical element similar to that described above, and in addition, the necessary electrical devices. These electrical elements are actuated by passage of the torpedo through a magnetic field, and cause a solenoid to displace the *firing ring* mechanically.

9B12. Conversion of energy supply. The energy available in the compressed air and alcohol must be transformed into a usable form before it can be applied. The conversion takes place in the valves and *combustion flask* located in the midship section, represented in figure 9B6.

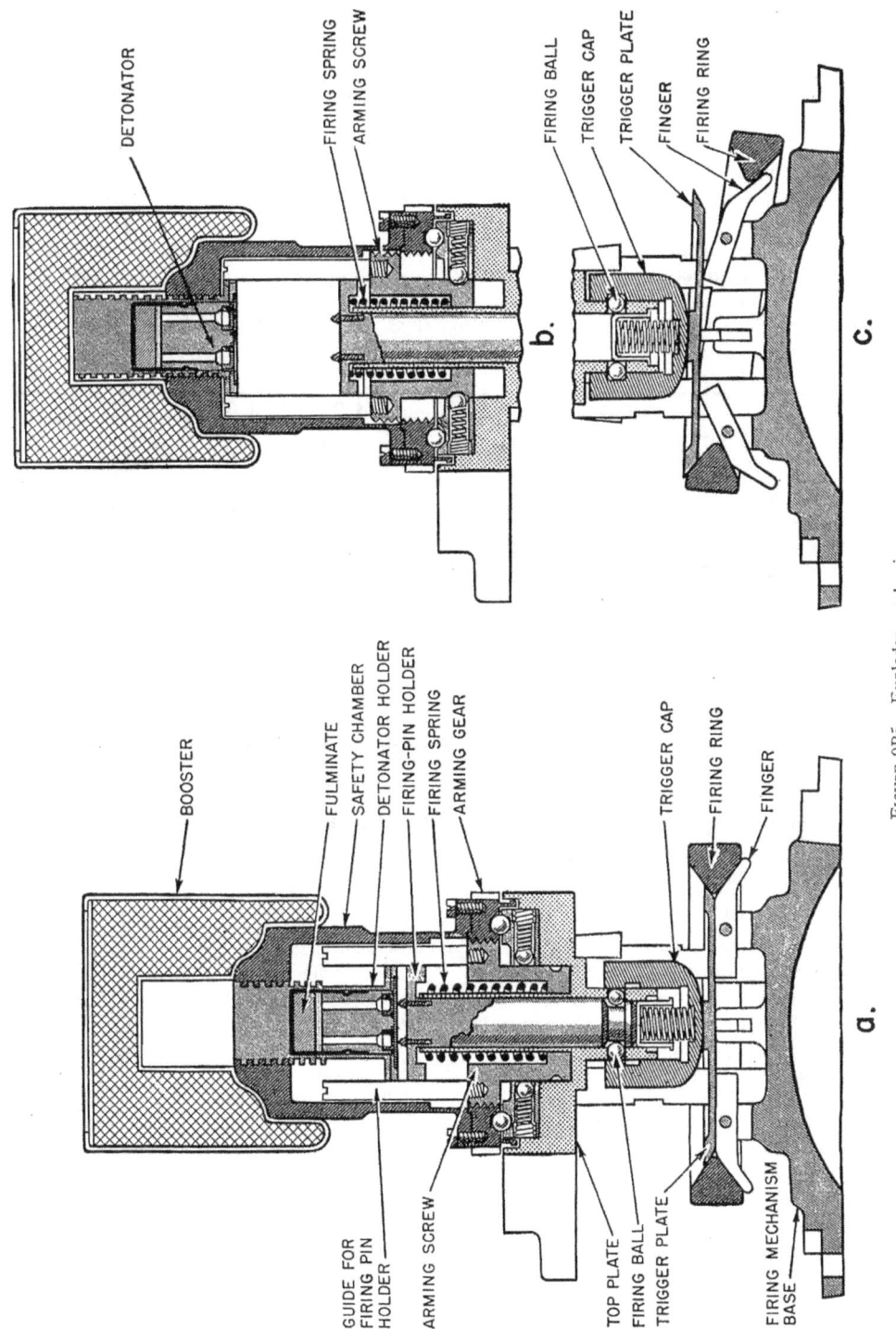

FIGURE 9B5. Exploder mechanism.

The *starting valve* has the function of initially releasing the air when the torpedo is launched. This valve is opened by air pressure when the starting lever (see fig. 9B1) is tripped. The lever operates a mechanism in the afterbody, and this mechanism controls the starting valve through the agency of an air line.

After the starting valve opens, a small quantity of high-pressure air goes to the gyro to bring it up to speed quickly. This small supply is automatically shut off after the gyro has been accelerated. The rest of the high-pressure air goes to the *reducing valve* which automatically throttles the air.

It is the function of the reducing valve to maintain a constant air pressure so that the turbine speed, and hence the torpedo speed, will be uniform during a run. The reducing valve receives the high-pressure air and throttles it to about 400 pounds p.s.i. pressure which is maintained as the flask pressure drops. Most of the low-pressure air goes directly to the combustion flask, but some of it is used to force alcohol and water into the combustion flask, and to operate the depth and steering mechanisms described in subsequent paragraphs.

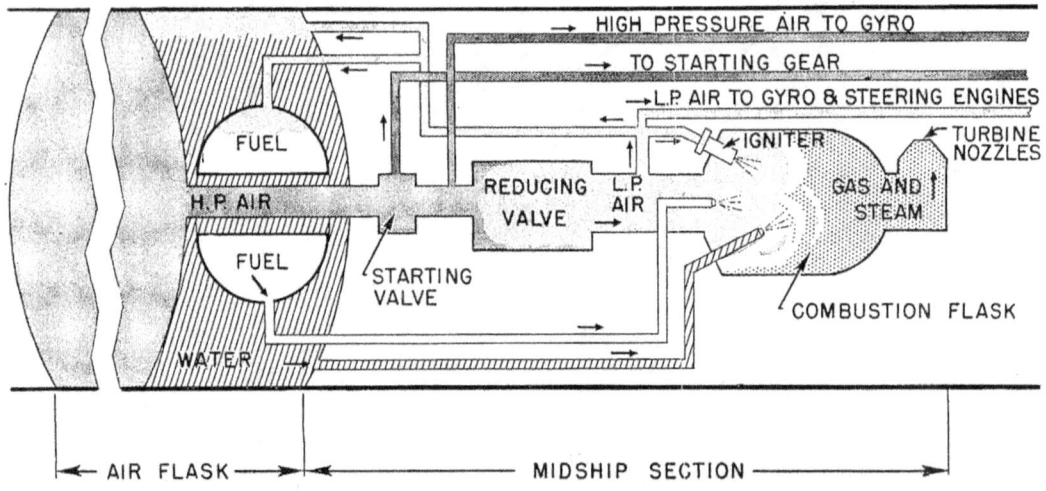

FIGURE 9B6. Energy conversion; schematic.

A torpedo will run "cold" on the energy of the compressed air alone. However, it can be made to run farther by supplying additional energy in the form of heat. The air and alcohol are intimately mixed as they enter the flask at one end. The mixture burns, forming various gases, which are too hot to be used in the turbines. A spray of water injected into the flask turns to steam, and cools the mixture to a safe temperature. The water is not a source of energy, but only a means of making the energy of the hot gases usable. The mixture of gas and steam goes to the turbines to drive the propellers.

The ignition of the air-alcohol mixture is accomplished by the *igniter*. When the starting valve opens, a charge of low-pressure air from the reducing valve operates a piston within the igniter. This piston drives firing pins into two caps, setting off two separate ignition charges which project their flame into the air-alcohol mixture for about 10 seconds. The use of double ignition elements greatly reduces the chance of misfires with resultant "cold shots".

9B13. Propelling mechanism. The propelling mechanism is illustrated in figure 9B7. The gas-steam mixture expands through nozzles, acquiring a high velocity. It then strikes the blades of the lower turbine wheel, giving up part of its energy. After passing the first wheel, the mixture strikes the blades of the upper wheel, giving up more energy. The spent mixture then passes through exhaust pipes (not shown), leaves the turbine, and forms the bubbles which appear in a torpedo's wake.

CHAPTER 9

As indicated in the diagram, the pairs of turbine wheels, gears, shafts and propellers revolve in opposite directions to eliminate gyroscopic effects. Any rapidly revolving part exhibits the gyroscopic property of resisting a change which tends to alter its axis of rotation. Single turbine wheels, gears and shafts in a torpedo would exhibit this property and develop resistance to steering and depth changes. Other reasons for using two propellers are to avoid sidewise creep caused by one propeller, and to eliminate the tendency of the torque reaction of a single propeller to roll the torpedo about its longitudinal axis.

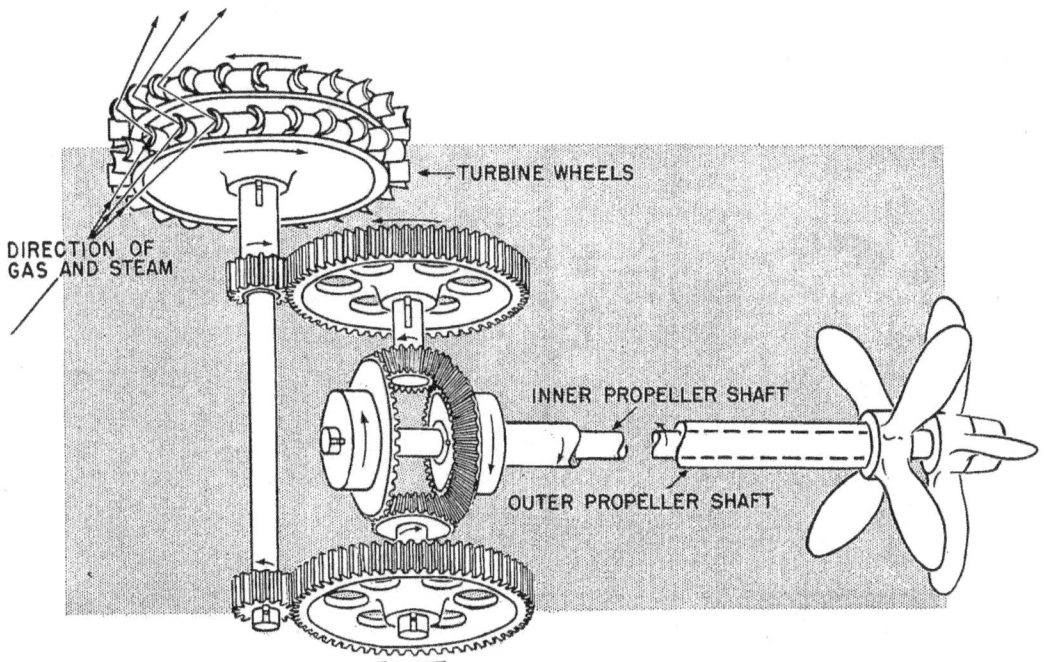

FIGURE 9B7. Propelling mechanism; schematic.

9B14. Depth mechanism—operation. There are two reasons why the horizontal rudders may need to be shifted. One is incorrect depth, the other is too much inclination of the torpedo to the horizontal. Assuming a torpedo is below depth, the rudders will be thrown up, and the torpedo will begin to ascend. If the rudders are left up, the torpedo will continue to turn upward, taking a sharp inclination to the horizontal. Upon arrival at the correct depth, the rate of climb will be so great that the torpedo will over-shoot, and appreciable down rudder will be required to bring the torpedo down again. Satisfactory runs require control of *depth*, and the *rate of depth change*.

Depth-control mechanisms contain a hydrostatic *diaphragm* to determine depth, and a *pendulum* to control rate of depth change, arranged as indicated in figure 9B8. These two devices must be relatively sensitive, and therefore do not have sufficient strength to move the rudders directly. They operate the valve of the *depth engine*, which in turn moves the horizontal rudders.

The diaphragm is subjected on one side to atmospheric pressure of the air trapped in the *chamber* when the mechanism was assembled. The other side is subjected to sea-water pressure through suitable passages. Attached to the diaphragm is a *spring* held in tension by the *depth screw* which may be adjusted outside the torpedo to make the required depth setting. The diaphragm is connected to the pendulum through a *lever* and a *link*.

The depth engine is a simple cylinder containing a piston, and a valve concentric with the piston. The arrangement is such that the piston, driven by low-pressure air, follows the movement of the

BOMB-TYPE AMMUNITION AND RELEASE GEAR

valve. The valve is attached to the pendulum and the piston is attached to the rudders. Thus, movement of the pendulum causes movement of the rudders.

Assume the torpedo is below its set depth, with its axis horizontal. The pressure of the water will force the diaphragm down, against the action of the spring and trapped air. This movement rotates the lever counterclockwise, moving the link, pendulum and valve to the left. The piston following the valve, applies up rudder.

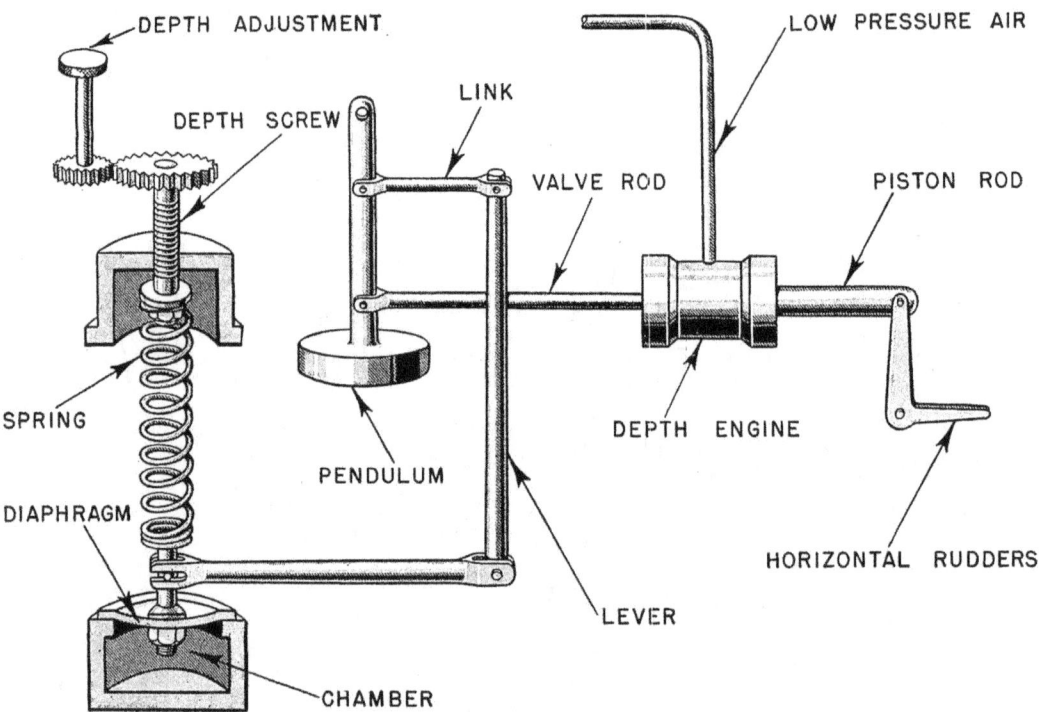

FIGURE 9B8. Depth control mechanism; schematic.

As the torpedo begins to incline upward, gravity tends to pull the pendulum down (to the right in the diagram) against the action of the diaphragm. This results in a reduction in the amount of up rudder. When the torpedo inclines enough to cause the pendulum action to just balance the diaphragm action, the rudders will be in neutral, but the torpedo will be climbing.

As the torpedo climbs, the diaphragm action decreases, and the pendulum moves farther to the right, applying down rudder. This causes the torpedo to begin levelling off, and it arrives at its set depth at a slow rate of climb.

The action is similar when the torpedo is running above its set depth, except that the control initially causes down rudder. The pendulum thus tends to eliminate overshooting, and the torpedo runs at a fairly uniform depth, established by the tension set in the spring before the run.

9B15. Steering mechanism; operation. The steering mechanism is illustrated in figure 9B9. The *gyroscope* determines the rudder action required and the *pallet mechanism* transmits the gyro indication to the control valve of the *steering engine*. The engine in turn moves the *vertical rudders* as required.

9B16. Gyroscope. The gyroscope, with its spin axis horizontal, is supported in bearings in the *inner gimbal ring*. The inner gimbal ring is maintained on a horizontal axis at right angles to the gyro axis, in bearings carried by the *outer gimbal ring*. The outer gimbal ring is supported on a vertical axis, in bearings contained in a housing (not shown). The gyro is thus

CHAPTER 9

free to rotate about three mutually-perpendicular axes, and is said to have three degrees of freedom.

The gyro is spun initially by high-pressure air operating a small turbine (not shown) geared to the gyro axis. The high-pressure air is shut off, and the turbine disconnected from the gyro before the torpedo leaves the tube. The spin is maintained by low-pressure air acting on small blades cut on the periphery of the gyro. Once the gyro is spinning, the inertia of the rapidly spinning wheel keeps the axis of spin parallel to itself.

The *cam plate* is secured to the outer gimbal ring. On its periphery are a *cam* and two grooves. Each groove extends from the cam toward the back of the plate. One groove is on the upper edge, the other on the lower edge. The cam is over the axis of spin, hence it indicates the direction of the axis.

The gyroscope, being very sensitive, cannot be used to operate the rudders directly, or even the control valve of the steering engine. The pallet mechanism receives the gyro indication from the cam, and operates the control valve as necessary.

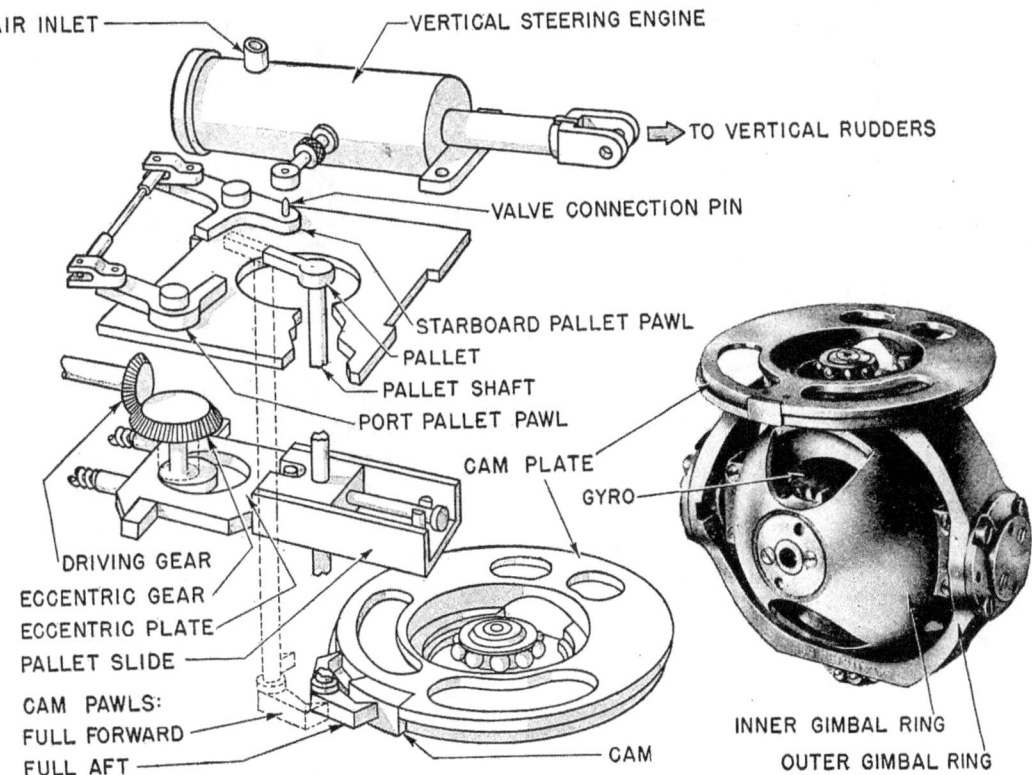

FIGURE 9B9. Steering mechanism; schematic.

9B17. Pallet mechanism. The pallet mechanism includes a reciprocating *eccentric plate*, a *pallet shaft*, a *cam pawl*, and a *pallet*. The cam pawl and pallet are secured to the pallet shaft, so that rotation of the cam pawl rotates the pallet. The eccentric plate is driven away from the cam plate by an *eccentric* which is rotated by gearing driven by the main propeller shaft, and toward the cam plate by the *springs*. The pallet shaft, supported vertically in the eccentric plate, moves back and forth with the plate.

One finger on the cam pawl is vertically aligned with the left groove on the cam plate; the other finger is similarly aligned with the right groove. The mechanism is adjusted so that the

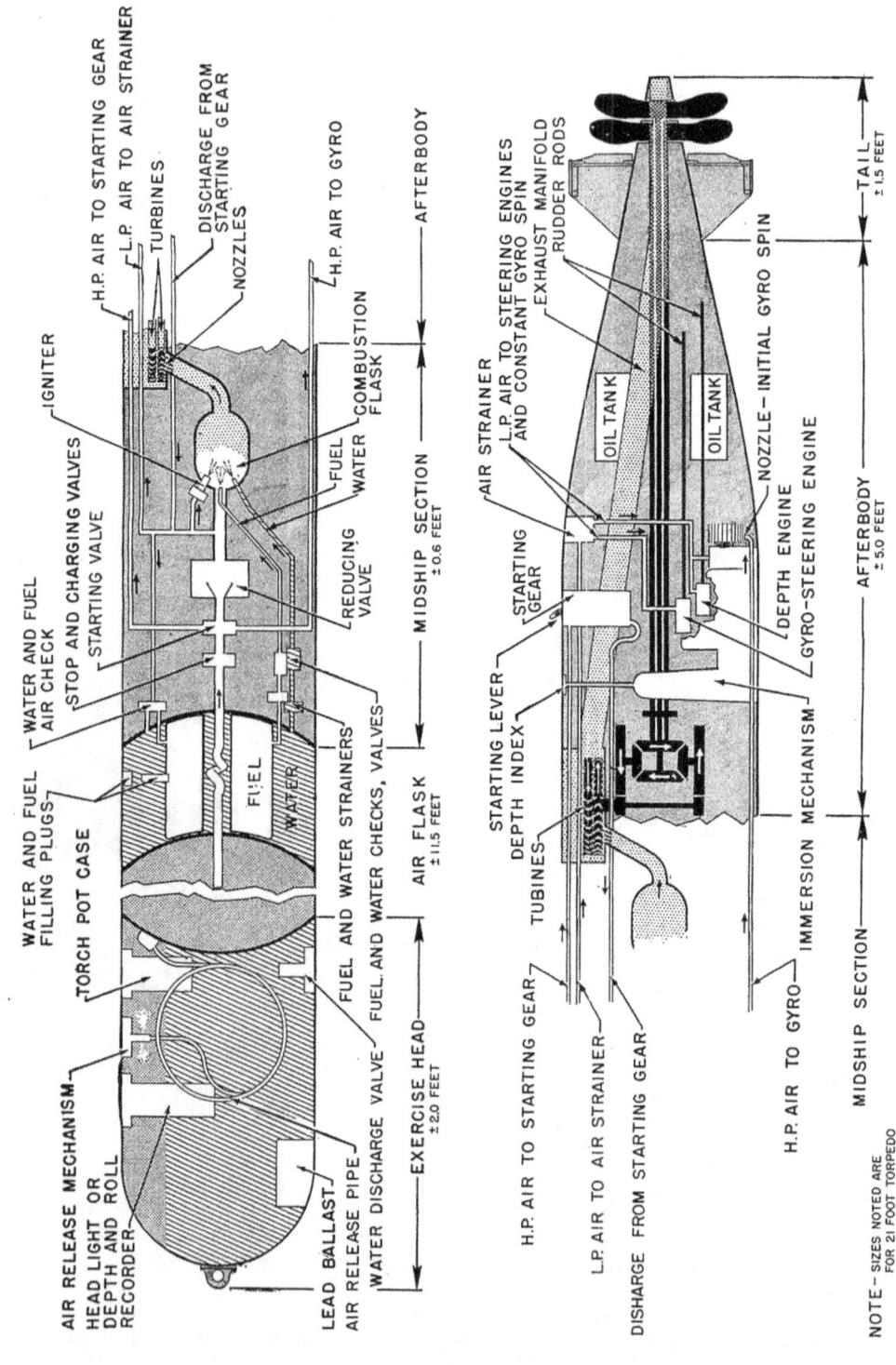

FIGURE 9B10. Torpedo with exercise head in place; schematic of single-speed torpedo.

CHAPTER 9

two fingers will straddle the cam when the cam pawl moves toward the cam plate if the torpedo is on its course. When the torpedo is off its course, the pallet mechanism attached to the torpedo, has rotated around the cam plate which is held in position by the gyroscope. As the cam pawl moves toward the cam plate, one of the fingers strikes the cam, and the other finger enters its corresponding groove. This action rotates the cam pawl and pallet shaft, thereby shifting the pallet to the right or to the left. The direction of the shift depends on whether the torpedo is heading to the right or to the left of the course established by the axis of the gyro.

9B18. Steering engine and valve. The pallet is vertically aligned with two *pallet pawls* which are attached to the *control valve*. When the eccentric plate moves away from the cam plate, one of the pallet pawls is struck by the pallet if it has been shifted by the cam pawl. This action rotates the pallet pawl which shifts the valve in the steering engine. The valve remains in this position until the pallet shifts in the opposite direction and strikes the other pallet pawl.

The steering engine, like the depth engine, consists of an air-operated piston within a cylinder. However, the piston in the steering engine is always either full forward or full aft, and the rudder is full right or full left. If the torpedo is heading to the right of its required course, the cam causes the pallet mechanism to shift the valve to produce left rudder. The torpedo then swings to the left until the mechanism applies full right rudder. Thus, the steering is a succession of slight shifts right and left of the required course, but the net result is equivalent to a straight course.

9B19. Initial turn. The gyro and pallet mechanism are arranged so that an initial offset can be made between the axis of the eccentric plate and the axis of the gyro. When the torpedo is launched, the rudders will remain hard over until the torpedo has turned from the direction of launching by the angle of initial offset. This offset is made, before launching, from outside the torpedo tube through an arrangement of gears. Thus, it is possible to launch a torpedo on a course other than the one it is to follow, and it will automatically take up the correct course as it starts travelling through the water.

9B20. Safety precautions, selected from Article 972, Navy Regulations.

"95. Torpedo air flasks shall never be charged above the pressure designated on the flask, even when subsequent cooling will reduce the pressure to that designated. When the working pressure is for any reason reduced, the new pressure designated shall be stamped on the flask near the charging valve."

"96. The artificial cooling of torpedo air flasks after charging by spraying with water or by flooding the torpedoes in the tubes is prohibited."

"97. Any cutting of torpedo air flasks, accumulators, piping, or other receptacles for compressed air is prohibited."

"98. Torpedo air flasks shall not be hoisted from one deck to another or struck below in a charged condition unless on an authorized torpedo or airplane elevator. Torpedo air flasks may be transferred from a tender to a seaplane or patrol boat in a charged condition but will not be so transferred to activities able to charge their own torpedoes except in an emergency. Torpedo air flasks shall never be shipped in a charged condition."

"99. In recovering a torpedo in the water the propeller lock shall be put on at the first opportunity and kept on until the torpedo is safely landed."

"100. Leaky or punctured torpedo torch pots may supply the flame to ignite combustible gases. Therefore,

(1) Torch pots of any sort shall not be stowed below decks.

(2) Torch pots on vessels with submerged tubes shall not be taken below until just prior to firing.

(3) Torch pots shall not be taken on board submarines except when it is contemplated to fire torpedoes. They shall be habitually stowed on the tender at the base.

(4) Torch pots shall, when practicable, be kept at least 20 feet from gasoline containers."

RESTRICTED

C. MINES

9C1. General. Naval mines are high-explosive filled, thin-walled cases designed to explode underwater upon being touched or closely approached by a ship. They are planted to await any enemy submarine or surface ship, and are used in offense and defense. Depending upon their position in the water, mines may be:

    1. *Drifting.* This type floats free, usually at a set depth beneath the surface.

    2. *Moored.* These are held a fixed distance beneath the surface by a line attached to an anchor.

    3. *Grounded.* Mines in this group are heavier than water and rest on the bottom. They must be planted in relatively shallow areas.

Mine shapes vary considerably, but most of them are spherical, cylindrical with more or less rounded ends, or bomb-shaped. Their diameters vary from about 16 inches to 46 inches. Cylindrical mines may exceed 10 feet in length. Many mines have fittings such as contact horns, mooring cables, and antennae, but some of the latest have no prominent external identifying features. The explosive charge may vary from 300 to over 1,000 pounds of TNT.

The main charge is cast in the *case* around a tube which provides a recess for the *booster* and the *detonator-extender mechanism*. The extender holds the detonator a safe distance away from the booster until after the mine has been laid. Extender mechanisms are operated hydrostatically and function to seat the detonator within the booster when the mine reaches a predetermined depth. This action is sometimes intentionally delayed for a few minutes to several hours by a soluble *washer* attached to the mechanism. The extender can not operate until this washer has been dissolved by the water.

9C2. Firing mechanisms. The *detonator* in all mines is fired electrically by a current from batteries within the mine case. The firing circuit is closed by a relay (an electromagnetically operated switch) within a firing mechanism. The several types of firing mechanisms are actuated by different means, and the means classifies a mine as one of the following types:

    1. *Contact.* Contact between a ship and horns on the mine case or an antenna trailing from the case produces a galvanic action which causes a current to flow in the firing mechanism, closing the relay.

    2. *Magnetic.* The motion or presence of a ship in the vicinity of the mine changes the magnetic field around the firing mechanism, causing the relay to close.

    3. *Acoustic.* The sound disturbance from the propellers and machinery of a ship is received by a microphone arranged to control the relay.

    4. *Magnetic-acoustic.* The combination of magnetic and sound disturbances actuates the relay. Generally both types of influence must occur simultaneously, but one or the other alone may actuate the firing mechanism of some mines.

Various safety devices are incorporated in the mine to make the firing mechanism operative only after the mine has been in the water a definite time. Often special features are provided to delay firing until a few seconds after the initial impulse, or to require a succession of impulses before firing will occur.

9C3. Laying and sweeping. Mines may be laid from torpedo or bomb racks of planes, from torpedo tubes or special release chambers of submarines, and from the decks of minelayers and some minesweepers. When laid from surface ships, the mines are supported on carriages which roll on tracks attached to the deck. The mine and the carriage, which also serves as the anchor for moored mines, are simply rolled off the stern of the ship. Mines are usually planted in fields which may be used to block channels, protect harbors, harass trade routes, and embarrass the enemy during Fleet actions.

The fact that a mine awaits its target and seldom destroys itself presents a problem not usually posed by other weapons. Projectiles, bombs, torpedoes and depth charges ordinarily destroy themselves or descend to depths at which they are harmless. Mine sweeping, the antidote for mines, is a complex subject in itself. Effective sweeping methods are developed almost as fast as new

CHAPTER 9

types of mines. In general, moored mines are swept by sweep wires extended between several minesweepers. Magnetic mines are exploded harmlessly by producing strong magnetic fields at a safe distance from the sweeping ship. Acoustic mines are combatted by intentionally creating underwater sounds of the appropriate quality.

The firing mechanisms of some mines are so sensitive that great care must be taken in dealing with a mine that has been cast up on the beach. The magnetic material in an object as small as a pocket knife, the sound of a voice, or the scraping of feet through pebbles may cause an explosion. Consequently, only personnel fully acquainted with safe methods of approaching mines should be allowed near them.

9C4. A moored, contact mine. A brief description of the Mark 6 mine is presented here to acquaint the student with one of the less complex types. This mine, illustrated in figure 9C1, is of the moored, contact-operated variety. The anchor is built to moor the mine in areas where the depth of water does not exceed about 3,000 feet. The mine itself may be held at varying distances beneath the surface, usually not exceeding 300 feet.

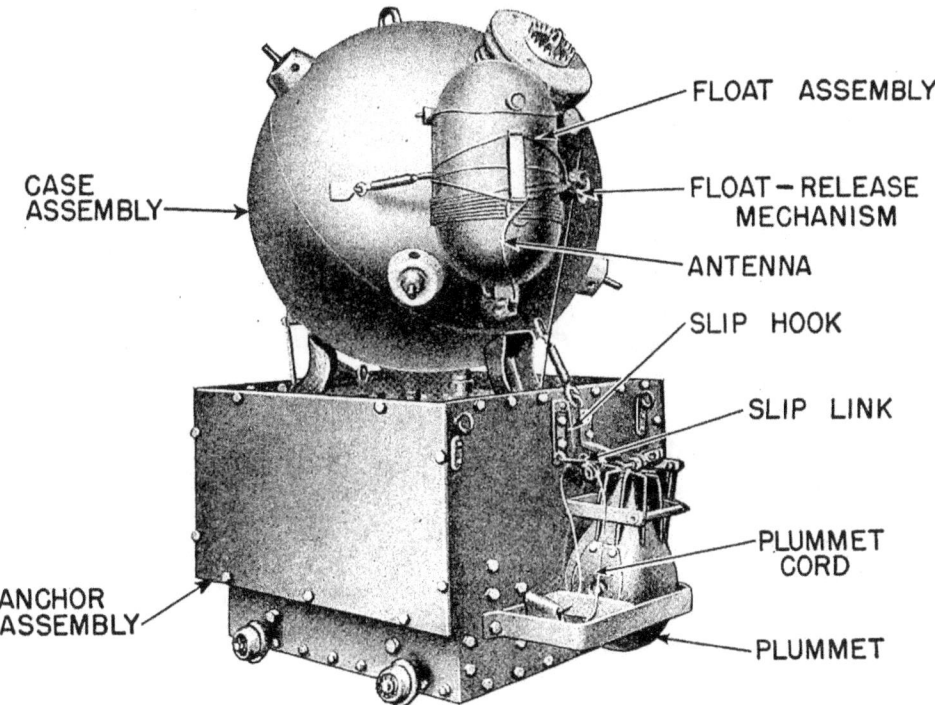

FIGURE 9C1. Mine assembly; Mark 6.

The mine assembly consists of an *anchor assembly*, a *case assembly* and a *float assembly*. These three components, secured together before launching, are automatically separated after the mine is in the water. The anchor assembly, mounted on wheels, provides a carriage for rolling the mine off the deck tracks. In the water it operates to automatically release the case, pull it under the surface to the prescribed depth, and then moor it in position. Within the anchor housing is a drum on which is wound the mooring cable. Attached to the drum is a mechanically operated locking device which locks the drum after the proper length of cable has been released.

The case, about 34 inches in diameter, contains a 300 pound charge of TNT which fills about one-third of the interior. The case also houses the booster, the detonator and its extender mechanism, a battery, four exterior contact horns, a firing mechanism, and the necessary wiring and securing

RESTRICTED

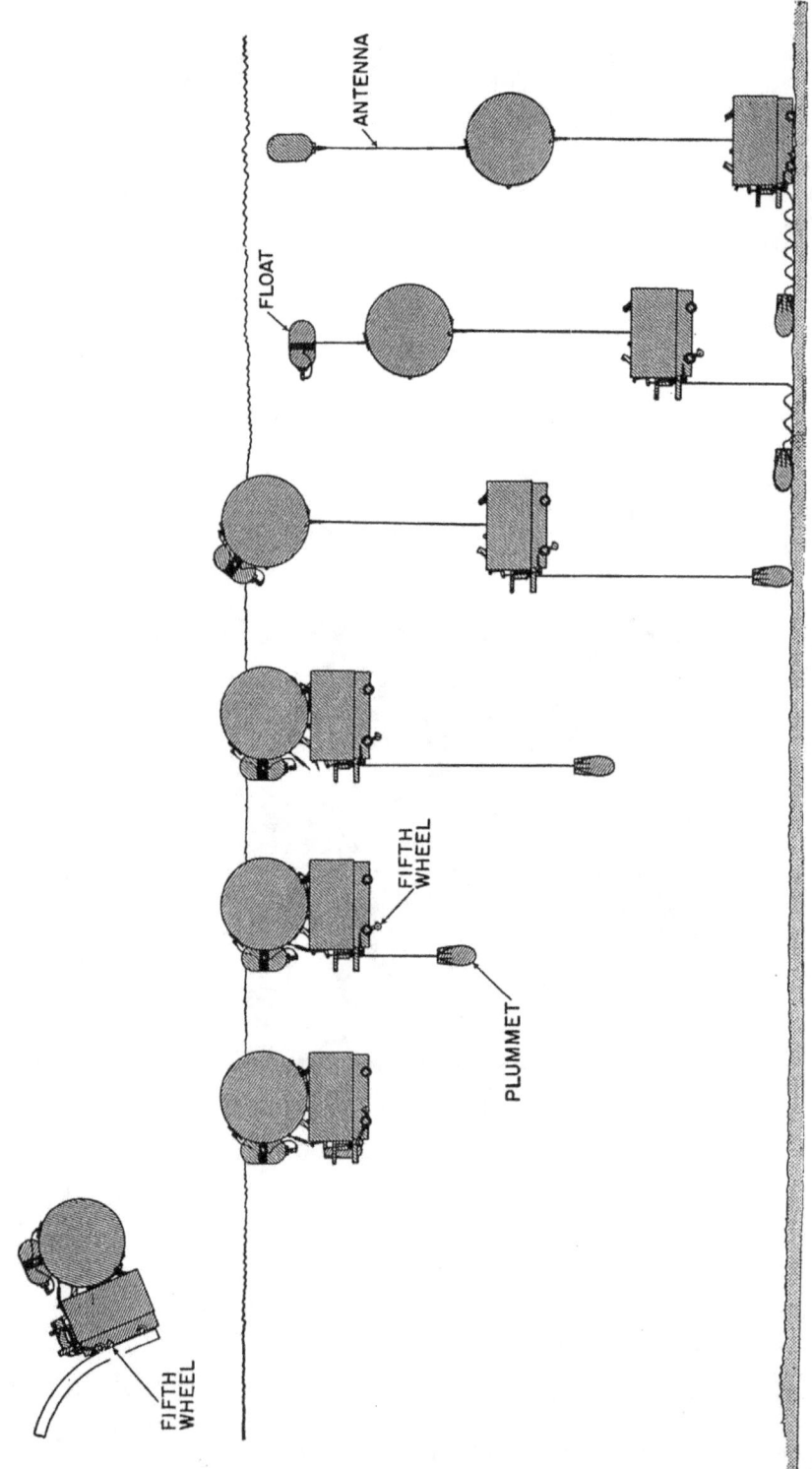

FIGURE 9C2. Operation of mine assembly during planting.

devices. The float is a small steel chamber wrapped with a copper antenna and secured to the case by a suitable release mechanism. The float streams the antenna from the mine to increase the chances of contact with a target.

9C5. **Operation of the mine assembly.** The sequence of operations which takes place as the mine is launched is shown schematically in figure 9C2. Prior to launching, the mine assembly is buoyant and returns to the surface immediately after dropping. When the assembly is launched, a *fifth wheel* which rested on one of the tracks unlocks the hook which supports the *plummet* on the anchor. The actual release of the plummet is delayed by a dashpot for about eight seconds after launching to allow the mine to right itself in the water.

The plummet then sinks, unwinding cord from a spool within itself. When the end of this cord is reached, the jerk pulls the *slip link* off the *slip hook* (fig. 9C1), which allows the case and anchor to separate. A flooding valve on the anchor opens, and allows the anchor to flood and sink. The buoyant case remains on the surface, pulling cable from the drum within the anchor.

The cable drum-locking device is held in a released position by the weight of the plummet until it strikes bottom. Then the locking device locks the drum, no more cable is payed out, and the case is pulled beneath the surface a distance equal to the length of the plummet cord. The float remains secured to the case until it is pulled under the surface. Hydrostatic pressure then operates the float-release mechanism and frees the float. The float streams upward until it reaches the end of the antenna. The antenna, secured to the case, pulls the float under the surface.

9C6. **Safety features.** The mine is prevented from exploding prematurely by several devices. A hydrostatic mechanism, which may be delayed a few minutes to several hours by a soluble washer, prevents the firing mechanism from operating until the case is submerged a distance of from 10 feet to 20 feet. The extender mechanism is also hydrostatically operated at a depth of about 25 feet, and may be delayed an hour or two by a soluble washer. When the safety devices have functioned the mine is armed and ready to fire. It will remain in this condition unless the case breaks its mooring and rises to the surface, where the hydrostatic mechanisms will normally disarm the mine.

9C7. **Explosion of the mine.** Contact of the steel hull of a ship or submarine with the copper antenna or any of the horns will cause a current to flow in the firing mechanism. This operates the relay, which connects the batteries to the detonator, and fires the mine. Slight bending of the horns by contact with a wooden ship will also do the job. Sometimes horns are added to the float to increase the area of effective contact, especially with wooden hulls which will not actuate the mine by contact with the antenna.

D. DEPTH CHARGES

9D1. **General.** Depth charges are thin-walled containers filled with a heavy charge of TNT, designed to explode at a predetermined depth in the water. They are dropped by surface ships in attacks on submarines. Airplanes drop depth *bombs*, which are functionally similar to depth *charges*, but are not considered here.

The two most common depth charges are cylindrical in shape and about 28 inches long. The diameter of one is about 18 inches, of the other about 25 inches. The smaller contains 300 pounds of TNT, the larger, 600 pounds. Another recently designed depth charge has a tear-drop shape to increase the sinking rate and improve the underwater trajectory. It contains 200 pounds of TNT, has the same overall dimensions as the 300-pound cylindrical charge, and results in more accurate attacks. Representative depth charges are shown in figure 9D1.

The depth at which explosion will occur can be controlled down to 600 feet. The charge is fired by the action of water pressure on the exploder mechanism, and does not require contact with a submarine. It is extremely difficult to obtain direct contact with depth charges because the exact position of the submarine usually is not known. Consequently, depth charges depend on the percussive wave of the explosion for their effectiveness, although contact explosion is more desirable.

BOMB-TYPE AMMUNITION AND RELEASE GEAR

The effective radius of the percussive wave depends upon the structural strength of the attacked vessel, and no definite values can be stated. Approximate information indicates that a 600-pound charge may cause moderate damage at 100 feet, but to be fatal it must explode within about 40 feet. The 300-pound charge may prove fatal within 30 feet. It is to be noted that doubling the weight of charge does not double the effective radius.

Destroyers, destroyer escorts, sub-chasers, PT boats, and other ships likely to engage submarines carry depth charges. Generally, several charges are released in rapid succession to form a pattern in depth, width, and length, and thus increase the probability of destroying the submarine. The pattern is obtained by dropping some charges from release gear on the stern, and firing others abeam from projectors. Appropriate depth settings are made on the depth charges before launching. The charges are never thrown far from the attacking vessel, hence care must be taken that the ship is making sufficient speed to be out of the way of the violent surface effects produced.

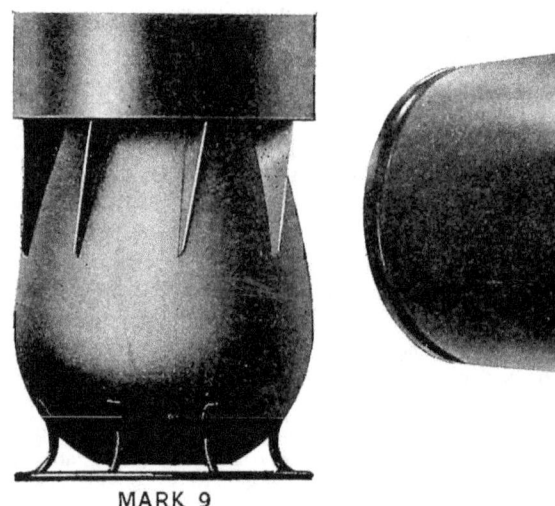

MARK 9

MARK 6

FIGURE 9D1. Depth charges.

9D2. Release gear. The tracks used to drop depth charges from the stern of a vessel are illustrated in figure 9D2. Essentially, they consist of inclined rails on which the charges rest, with suitable mechanical devices arranged to release one charge at a time. The track shown will hold eight of the 300 pound charges, but can be rearranged to hold six of the 600 pound charges by raising the upper rails. Modifications of some of the parts will make the track suitable for dropping tear-drop charges.

The charges roll aft by their own weight and rest against the *after detents*. When the *track control lever* is operated, the after detents depress, allowing the first charge to roll off the track into the water. At the same time, *forward detents* (not shown) rise and hold the second and following charges steady on the rails. When the track control lever is returned to its original position, the detents move in reverse directions, and the second charge rolls down to the dropping position against the after detents. The track control unit is connected hydraulically to a bridge control unit by means of which the charges may also be released.

9D3. Projectors. The currently used depth charge projector, commonly called a K-gun, is illustrated in figure 9D3. It is used to launch the 300 pound cylindrical depth charge or the 200 pound tear-drop charge. The old Y-gun which has two barrels and launches two charges simultaneously, has been superseded by the K-gun.

The K-gun consists of a smooth-bore barrel attached to an expansion chamber fitted with a breech mechanism. The breech plug is the interrupted-screw type, housing a firing mechanism which provides for local percussion firing by lanyard, or firing from the bridge by electricity. The depth

CHAPTER 9

charge is launched with the *arbor* attached, the action of the propellent gas being against the base of the arbor. The arbor used with a cylindrical depth charge remains attached to the charge after firing. The arbor used with a tear-drop charge detaches after firing to permit the charge to descend at a more nearly uniform rate.

The projector is permanently secured to the ship, and cannot be trained or elevated. Variations in range are obtained by altering the weight of the propellent charges, which are assembled in 3-inch cases similar to those used for torpedo impulse charges. Three standard weights of spherohexagonal black-powder charges are used to obtain ranges of 50, 75, or 120 yards.

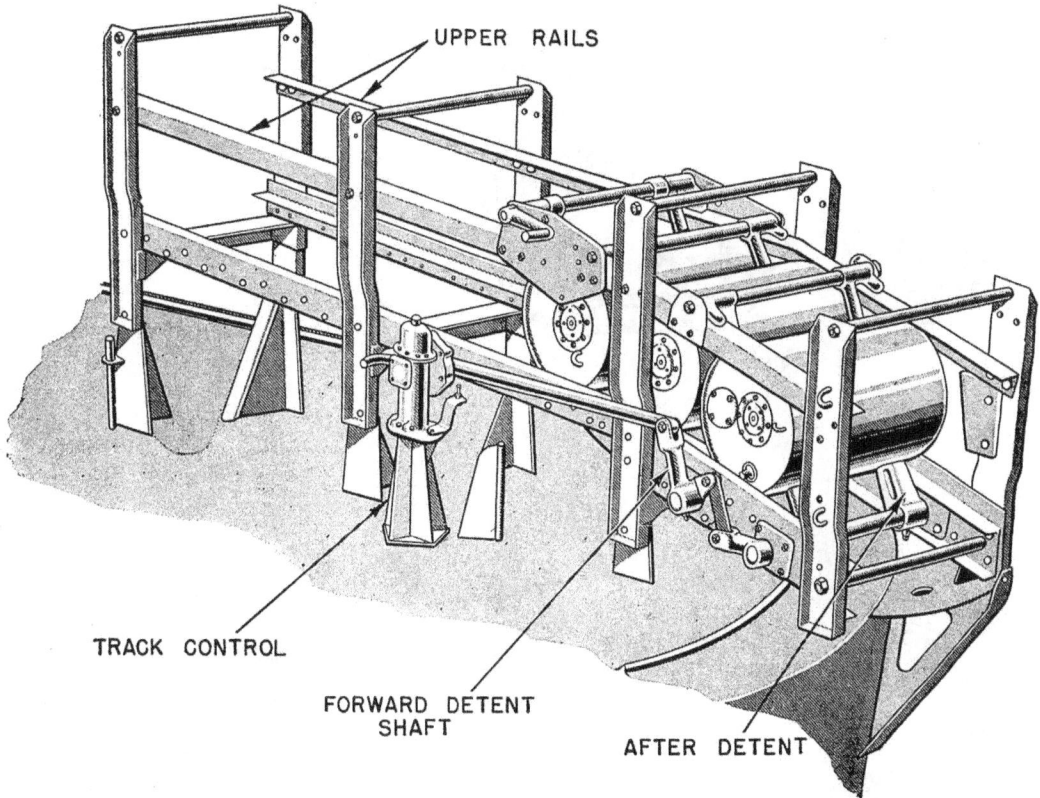

FIGURE 9D2. Depth-charge release gear.

9D4. Depth-charge operation. Depth charges, regardless of size and shape, are similar in functional operation. The general arrangement of a typical charge is shown in figure 9D4. The principal components are:

The main charge and case. The outer shell is a sheet-metal case, through which passes a *central tube*. The space between the tube and case is completely filled with cast TNT.

The booster and booster extender. The booster is attached to the booster extender, and is free to move longitudinally in the tube. A central recess in the inboard end of the booster receives the detonator when the booster is moved against the *centering flange*. The booster extender, operated by hydrostatic pressure, pushes the booster against the centering flange when the charge reaches a depth of between 12 feet and 15 feet.

The detonator and pistol. The detonator is secured to the inner end of the pistol, and is supported in the center of the tube by the centering flange. The pistol mechanism operates by hydrostatic pressure to release a firing pin which strikes the cap of the detonator. The depth at which the pistol

225 **RESTRICTED**

is to operate is set before the charge is launched by rotating the index pointer to the chosen depth on the dial. Calibrations are in steps of 50 feet from 50 feet to 600 feet with additional positions of *safe* and 30 feet (see dials, fig. 9D5).

Accidental detonation of the depth charge is prevented by setting the dials on "safe", thereby locking the pistol, and by the *safety fork* which prevents the booster from moving inward. The booster will not detonate unless it partially surrounds the detonator. The *inlet valve cover* protects the water inlets to the pistol against dirt, and unless removed, will prevent proper entrance of water after a charge is launched.

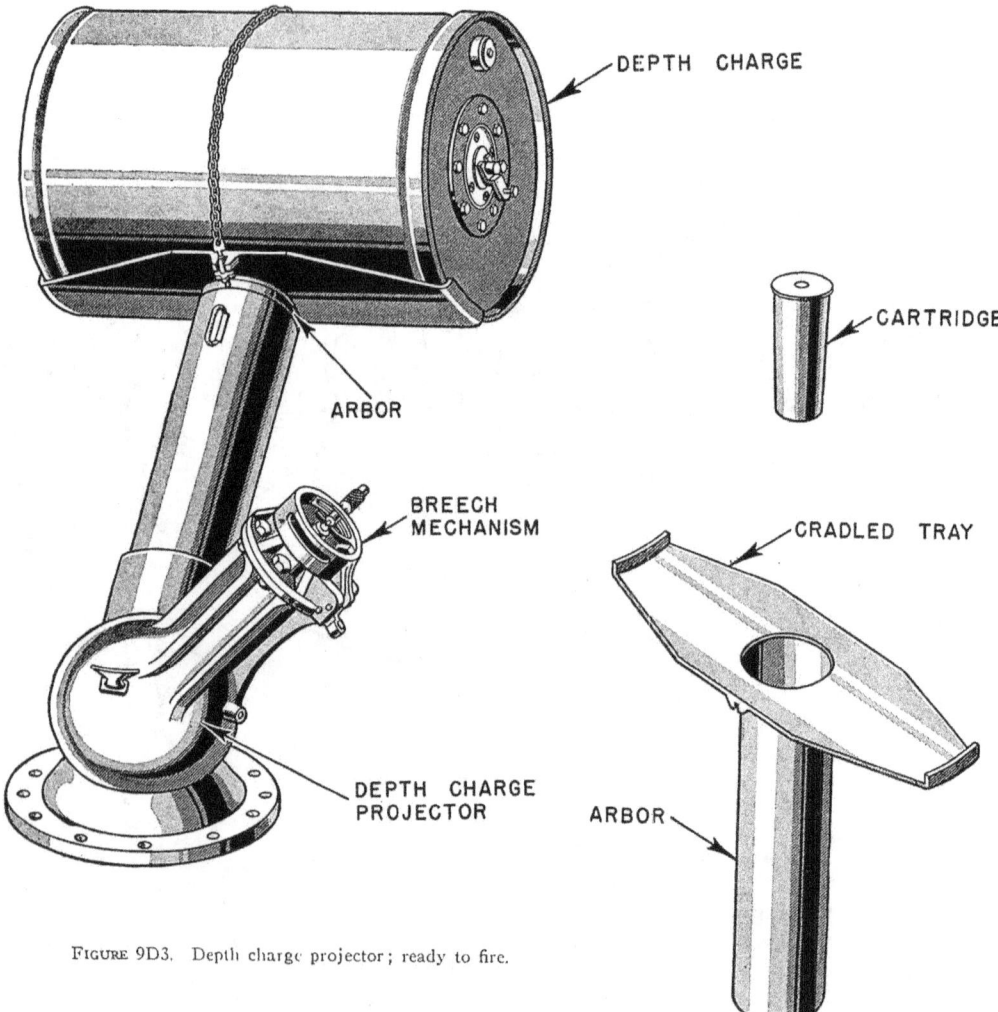

FIGURE 9D3. Depth charge projector; ready to fire.

Just prior to launching, the firing depth is set on the dials. If the charge is dropped from a rack, the safety fork and inlet valve cover are automatically wiped off by wiping plates on the rack. The fork and cover must be removed by hand if the charge is launched from a projector.

9D5. Pistol operation. The construction of a pistol is illustrated in figure 9D5. The *firing pin* (fig. 9D5a) is secured to the *firing plunger*. Around the plunger are three holes, each of which contains a *lock ball*. The *release plunger* is free to slide within the firing plunger, and is held by a

CHAPTER 9

spring in the position shown. The lock balls, held outward by the release plunger, rest against the end of the *guide tube bushing* and prevent the firing plunger from moving toward the detonator.

When water flows through the *inlet,* it fills the *bellows.* Water pressure then expands the bellows, pushing the *piston* and *piston stem* toward the release plunger. This action compresses the *depth spring* and the *firing spring.* As the movement continues, the piston stem contacts the release plunger and moves it toward the firing pin. When the annular recess on the release plunger has moved under the lock balls, they move inward. The firing plunger, driven by the firing spring, then drives the firing pin into the detonator.

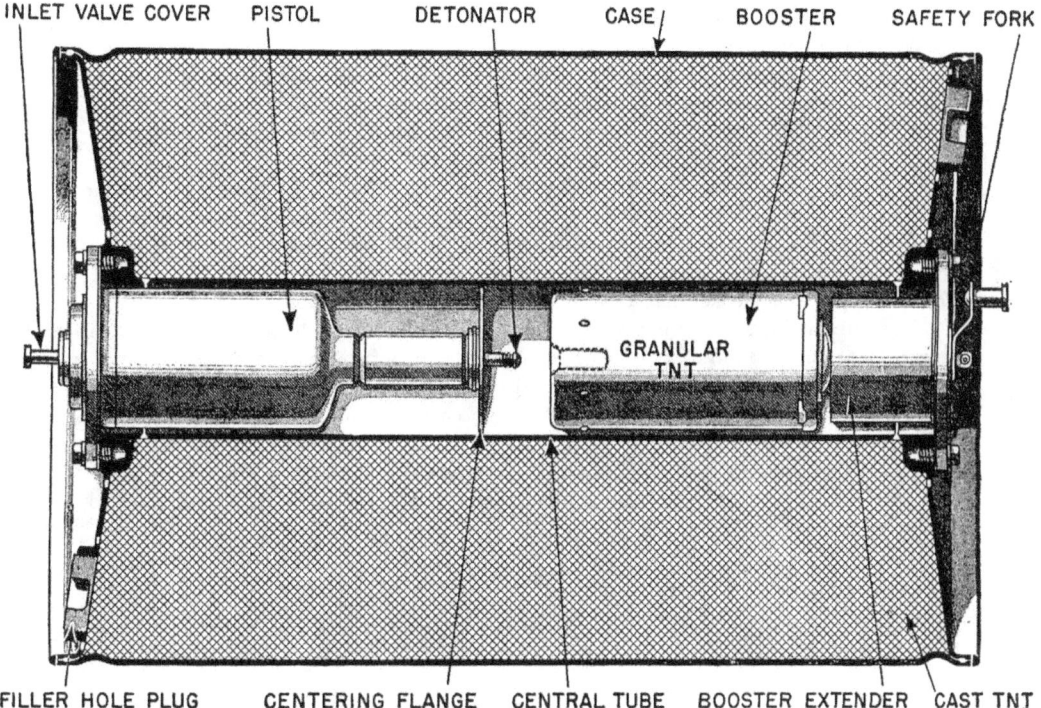

FIGURE 9D4. Depth charge assembled.

The depth spring is seated in the *adjusting bushing,* and is compressed between it and the *spring collar* as the bellows expands. The adjusting bushing, threaded within the *depth-setting sleeve* and keyed to the piston stem, is positioned by rotation of the sleeve when the *index plate* is turned. The position of the bushing controls the resistance offered by the depth spring to the expansion of the bellows, and thus determines the water pressure required to operate the pistol. When the pistol is set on "safe", the adjusting sleeve is so close to the spring collar that the piston stem is prevented from moving far enough to release the firing balls.

The 300-foot setting uses the maximum strength of the depth spring; hence it cannot keep the pistol from firing at greater depths. The inlet valve provides the necessary control for depths between 300 feet and 600 feet. The *valve ball* is held against its seat by the *valve spring.* The compression of this spring is set by rotating the *adjusting sleeve* (with the inlet valve-cover off). When the valve is set for depths between 0 feet and 300 feet, it admits water freely as soon as the charge is launched. When set for greater depths, the valve admits no water until the set depth is reached. Then the pressure of the water opens the valve, and the water pressure immediately expands the bellows, firing the detonator.

227 **RESTRICTED**

The depth settings are made with a special wrench (see fig. 9D5b). For depths up to 300 feet, the *deep-firing pointer* on the adjusting sleeve, is set at *0 to 300*, and the *index pointer* on the index plate is set at the required depth. The *dial plate* has opposite each depth mark a small recess which receives the *index pointer plunger*, which acts as a detent to maintain the setting. For depths greater than 300 feet, the index pointer is set at *100*, and the deep-firing pointer is set opposite the required depth.

9D6. Booster-extender operation. A booster-extender mechanism, ready for launching, is shown in figure 9D5c. The operating parts are enclosed within a watertight *bellows*, which fills with water and extends the booster to the detonator as the depth charge descends.

The booster and the *piston* are secured to the *spindle*, which is locked in position by the *safety fork*. When the safety fork is removed at launching, the bellows, assembled under slight compression, elongates enough to open the *inlet* around the spindle. The spindle moves inboard until the outboard shoulder of the annular groove contacts the *locking balls* held in the *locking slide*. The *spring* prevents further inboard movement of the locking slide until water begins to operate the bellows.

As the water pressure builds up, the locking sleeve is gradually drawn inboard by the spindle. When the charge reaches a depth between 12 feet and 15 feet, the locking balls move outward into the enlarged recess in the *spindle guide*, and release the spindle. The existing water pressure then quickly extends the bellows, arming the charge.

9D7. Safety precautions, selected from Article 972, Navy Regulations.

"101. Current instructions prescribe effective measures to prevent the accidental arming or launching of mines, depth charges, and aircraft bombs in storage or in handling. Mines, depth charges, and aircraft bombs shall at all times be handled and treated as if armed."

"102. Fuzes, firing mechanisms, or primer mechanisms normally kept in bombs, depth charges, or mines shall not, except as covered by special orders or current instructions of the Bureau of Ordnance, be removed, disassembled, repaired or in any way altered."

"106. Defective bombs, depth charges, and mines shall be turned in to an ammunition depot at the first opportunity."

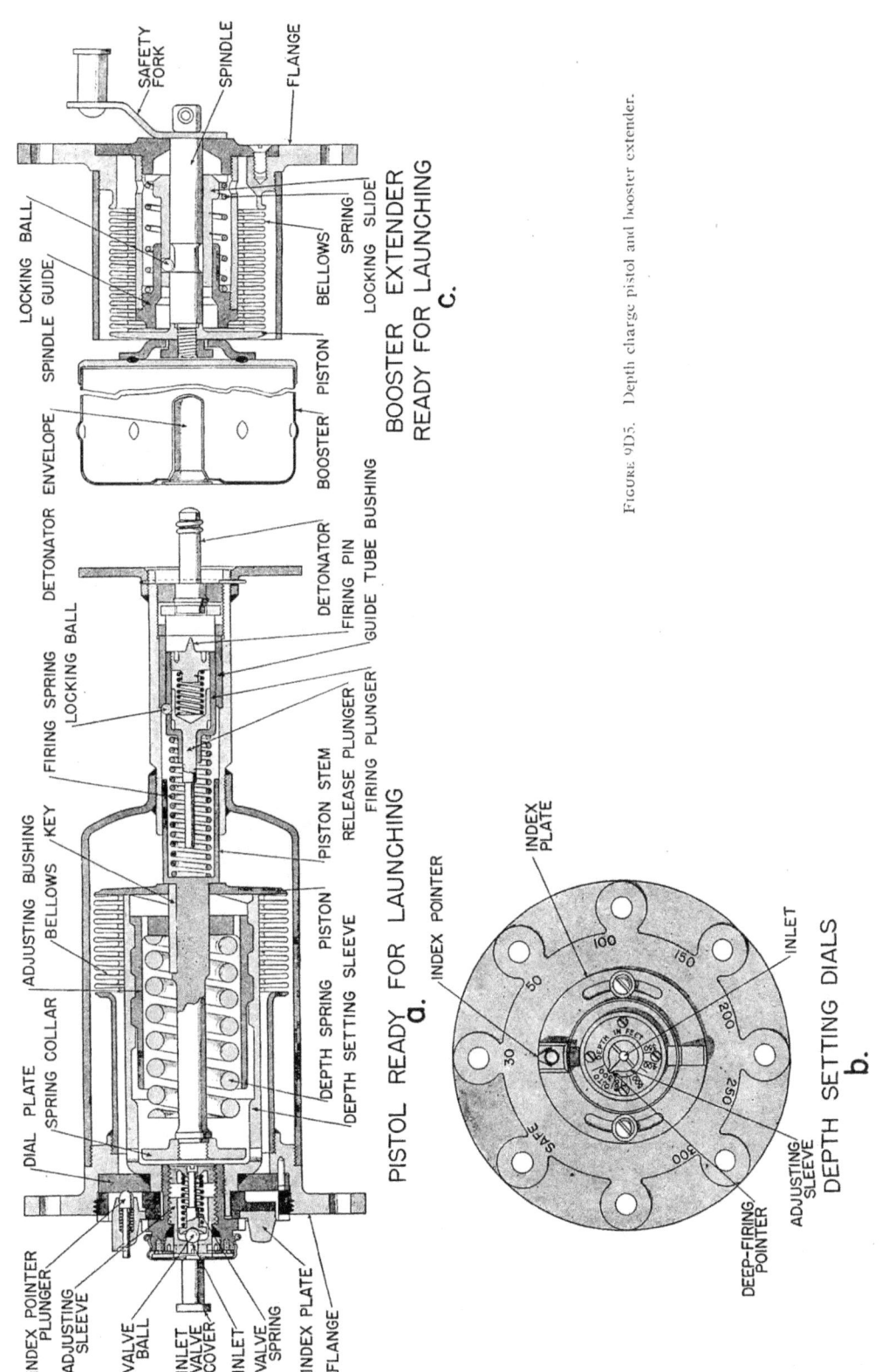

FIGURE 9D5. Depth charge pistol and booster extender.

CHAPTER 10

INTRODUCTION TO FIRE CONTROL

A. HISTORICAL

10A1. General. In a broad sense *fire control* refers to the entire process of utilizing a ship's armament. It involves the matériel, personnel, methods, communications, and organization necessary to inflict upon the enemy maximum destruction in minimum time. By custom, fire control refers only to the control of guns, while more specific terms such as *torpedo control, mine control,* and *depth-charge control* are used in designating the regulation of other weapons.

Fire control is a science in itself. Its phenomenal development has occupied the minds of some of the world's most famous scientists, manufacturers, designers, engineers, and naval officers. Progress has been the result of intimate cooperation between civilian and service personnel.

The historical development of this science is both interesting and educational; knowledge of the older parent techniques inevitably aids in understanding modern methods. It should be understood, however, that this text cannot include a complete history of fire control, nor can it even completely describe every method, device, or combination of devices included in the various fire-control systems of modern fighting ships.

10A2. Early fire control. Although cannon have been used for centuries, fire control in anything approaching its present form is a comparatively new development in gunnery. Prior to about the year 1800 guns were not even equipped with sights, and it may be said that no form of fire control existed. Naval gunnery at that time depended upon bringing the guns as close to the enemy vessel as possible, and literally firing muzzle to muzzle. Favorite ranges of the period were *pistol shot* (about 50 yards) and *half pistol shot* (about 20 yards).

Sighting consisted chiefly of setting the guns in azimuth, and of leveling them by eye. Sometimes allowance for the curvature of the trajectory was made by sighting by the line of metal; i.e., by bringing the top of the breech and the top of the muzzle in line with the point of aim, which caused the gun to be elevated by the amount of the taper of the gun from breech to muzzle.

Practically no attempt was made to regulate initial velocity. Powder charges were estimated, and the projectile load was one shot or several shots. At close quarters, guns sometimes were loaded up to the muzzle with grape. It appears that no notable progress in fire control had been made even at the end of the War Between the States, battle ranges having remained so short that there was no real need for special methods.

10A3. Development of gun sights. Gun sights as first introduced during the early years of the nineteenth century consisted merely of fixed front and rear sight points so adjusted that the line of sight (LOS) across their tips was parallel to the bore of the gun. Apparently the only advantage sought was a definite means of leveling the gun at the point of aim, or of adjusting elevation within small limits by regulating the coarseness or fullness of the sight taken across the points.

Developments in cannon and powder during and after the War Between the States resulted in such greatly increased ranges and accuracy that graduated gun sights became a necessity. Sight design was improved by fitting into the rear sight a movable leaf adjustable both vertically and laterally. The introduction of rifled guns, with consequent drift of the projectile, brought forth sights which automatically made an approximate allowance for drift by inclination of the rear sight bracket.

The next outstanding development in fire control was the *sight telescope*, which toward the end of the nineteenth century was adapted to guns in such a manner that compensation could be made for range, drift, and speeds of own ship and target ship. The effects of the motions of the firing vessel and the target were investigated and provision was made for offsetting the line of sight laterally to allow for these effects in deflection. Elevation scales were graduated in accordance with data determined at proving grounds, and powder charges were carefully regulated as to weight and index.

10A4. Development of fire control. At the close of the century, fire control was beginning to take on the aspects of a science. However the only fire-control instrument in use other than sights was the *stadimeter,* adopted in 1898 as a range-finding instrument. Previously ranges were determined by estimates, using the dip below the horizon, or the vertical angle subtended by the target, or the technique of firing ranging shots. Fire-control communication systems were non-existent, transmission of information being accomplished by word of mouth; a shouting task in battle.

At the beginning of the present century, interest in the improvement of fire-control methods led to the establishment of an efficient, competitive, target-practice system. Rules for practices were standardized, and complete analyses and reports of methods were required. Elementary and advanced practices under simulated battle conditions were scheduled for individuals and units of the Fleet.

The competitive spirit of the practices stimulated among both officers and men an interest in increasing the efficiency of matériel and personnel, and in devising new methods to improve the ship's score. New fire-control devices were introduced, including optical range finders, graphic and mechanical devices to assist in finding and keeping the range, visuals, and voice tubes and telephones for communications. Many developments represented the initiative of younger officers, who designed and built makeshift assemblies to demonstrate their practicability.

10A5. Development of director firing. The simplest and original method of firing guns is *pointer fire,* in which each gun is fired independent of all others at the will of the pointer. In this procedure, the gun sights are set for range and deflection, the pointer and trainer direct the line of sight at the target, and the gun is fired by the pointer.

One of the most important advances in fire control was the centralization of control at a common point. The early forms of centralized control provided for little more than the announcement of range and spots by the gunnery officer, all other functions being performed at the guns. Advances in both visual- and oral-communication systems came with attempts at further centralization, and soon the control station began to perform other functions.

It was nearly impossible to intelligently spot the fall of shot from several guns firing independently at the same target. The shots would not land at the same time, nor would they land as close to the same point as they should. The correcting spot to be applied to the group of guns could be little more than a guess. Early improvements were obtained by sounding a buzzer as a firing signal to the various pointers; later, a single master firing key was inserted in series with the keys at all of the guns. The master key was placed at the control point, from which all the guns could be fired simultaneously.

Shortly after 1910, *director firing systems* were introduced to eliminate certain undesirable errors which are likely to exist in any form of firing where the guns are laid independently with respect to their own lines of sight. In director systems, which today are the primary means of controlling guns, the elevating, training and firing of a group of guns is controlled from a single advantageous, removed position. The key instrument in these systems originally was the *director,* which for the present may be considered as a sort of master gun whose movements are duplicated by all the guns under its control. There are numerous other instruments in most fire-control systems, and in modern installations some of them perform many of the functions which were formerly associated with the director.

10A6. Increased ranges. As improved fire-control methods and equipment developed, the accuracy of gunfire, and consequently the effective ranges, increased. In the engagement between the Monitor and the Merrimac, the latter vessel is reported to have landed a projectile on her

CHAPTER 10

opponent from a distance of about a mile. Practically the entire engagement, however, was conducted at very short range, probably within 100 yards.

At Tsushima, in 1905, the range varied between about 4,000 and 6,000 yards. By 1910 the Navy was holding battle practices at 12,000 yards with reasonable accuracy. At the Dogger Bank engagement early in 1915, the opening range was between 18,000 and 20,000 yards. In 1916 long-range practices were conducted at 20,000 yards, and at the end of World War I, fire control was efficient enough to cope with the ranges of about 24,000 yards possible with the maximum elevation of 15° to which our turrets were then limited. Recent developments in ranging and computing instruments, and in gun construction, have raised the range limits to over 40,000 yards.

10A7. Improvements in spotting. In order to efficiently advance the maximum range to more than 20,000 yards, it was necessary to extend the fire-control system beyond the limits of the firing ship. Spotting, even from the loftiest shipboard position available, becomes exceedingly difficult and uncertain as the target approaches the sea horizon. With the target near the horizon, about the best the ship's spotter can do is to correct the deflection and then determine whether impacts are *short* or *over;* even the counting of shorts and overs is uncertain. The distance to the horizon for most spotting heights on capital ships is about 25,000 yards.

During World War I *kite balloons* were introduced in order to secure a higher spotting station. These balloons were towed by battleships at a height of from 1,000 feet to 3,000 feet, and carried a spotter who was in communication with the ship by telephone. This expedient was soon abandoned, however, for it was found that the balloon was difficult to manage and limited the maneuverability of the ship. Aircraft spotters took the place of balloon spotters, and today aircraft spotting is an important part of fire control.

Aircraft further extended the scope of fire control by providing a means for controlling fire at a target invisible from the firing ship. The feasibility of this procedure was first demonstrated in 1922-23 when fire was directed against a target below the horizon, and was again shown in 1925-26 when fire was successfully directed against a target which was obscured from the firing vessel by an island.

10A8. Range keeping. It is obvious that the problem of controlling the fire of a gun is considerably more complicated aboard a moving, pitching and rolling ship than it is on a fixed shore installation. Nearly always the target as well as the firing ship is moving. The distance between the two is continuously changing, and since this distance governs the elevation of the guns, it is, perhaps, the most important factor in fire control. There are numerous computations involved in keeping track of target distance, and in determining correct elevation and train. Mechanical *range keepers,* or *computers,* have been developed to handle the mathematics of the fire-control problem practically automatically.

10A9. Flexibility of control. Among the advantages provided by modern fire-control installations is the great increase in flexibility. Formerly it was accepted that tactics were limited by fire control which could not function efficiently while the ship was maneuvering. To some extent this will always be true. In order that maneuvers of the firing ship may not result in loss of the hitting range, provisions are now made to compensate instantly for changes in the ship's course and speed. It may be said, therefore, that tactics have become largely independent of fire control.

Structural features of a ship will always cause a ship's volume of fire on certain bearings to be less effective than on others. Such limitations, however, have been reduced by arranging batteries to permit maximum fire on all probable bearings on which they may be engaged.

B. MODERN ARMAMENT

10B1. Batteries. Before the advant of modern gunnery a ship's battery often consisted of various gun types. The difficulties of controlling and spotting the fire from several gun sizes quickly showed the desirability of having identical guns in one battery. Other advantages are a reduction in the variety of replacement parts and simplification of ammunition-supply problems.

INTRODUCTION TO FIRE CONTROL

The size of the guns in any particular battery depends upon the type of ship. The classifications of batteries are:

Main battery. The guns of largest caliber aboard ship.

Secondary battery. Guns smaller than those of the main battery and not specifically designated for use against aircraft. Only those ships mounting turrets have a secondary battery, and that battery may be composed of dual-purpose guns.

Antiaircraft battery. All guns installed primarily for use against aircraft, such as the AA machine-gun battery.

A typical battleship, such as that illustrated in figure 10B1, may have a main battery of nine 16"/50 cal. guns, a secondary battery of sixteen or twenty 5"/38 cal. guns in twin mounts, and an AA battery of about 15 quad mount 40 mm. guns and about eighty 20 mm. guns.

Heavy cruisers (CA's) generally have nine 8"/55 cal. guns as a main battery, twelve 5"/38 cal. guns in twin mounts as a secondary battery, and numerous 20 and 40 mm. guns as an AA battery. Light cruisers (CL's) may have a main battery of nine 6"/47 cal. guns, with secondary and AA batteries similar to those of the heavy cruisers.

The main battery of many destroyers, and of some light cruisers, is composed of 5"/38 cal. guns, while on destroyer escorts, subchasers and some patrol craft it usually consists of 3-inch guns. On these ships there is no secondary battery, but there is an AA battery of 20 and 40 mm. guns.

It is evident that it is not practicable to enumerate the guns included in each battery of the many ships in service. The classification by batteries is merely a matter of convenience for distinguishing the several groups aboard a particular vessel.

10B2. Fire-control systems. There are many variations among fire-control systems, but their underlying principles and much of their equipment have common features. The arrangement of a system depends upon various factors, including the type of ship, the size, number, and effective range of its guns, and whether the guns are for use against air or surface targets. In general, most of the following features are provided in the fire-control systems of any combatant ship:

1. The entire armament is capable of being brought into action simultaneously.

2. Director fire, where provided, is the primary method of firing all guns, but provision is made for taking up pointer fire without delay.

3. Where the number and arrangement of guns in a battery permits, provision is made to group the guns for divided fire, with the control of each group centralized.

4. Provision is made for controlling the armament at night.

5. All guns of a battery are of the same caliber. (Major subdivisions of AA batteries may have different sizes of machine guns.)

6. Provision is made for taking observations for the control of gunfire, including range finding, spotting, and the proper application of these observations for hitting the target.

7. There is a system of keeping range so that changes due to own ship movements are supplied automatically insofar as practicable; this permits the ship to change course and speed without loss of accuracy in fire.

8. Where practicable, a ready means is provided for continuation of range keeping in the event that the primary range-keeping instruments fail.

9. Transmission of fire-control information is rapid and continuous.

10B3. Battleship systems. A typical modern battleship is shown in figure 10B1, in which a few of the principal director emplacements are indicated. The two primary directors of the main battery are the highest on the ship, the forward one being about 120 feet above the waterline, the after one about 75 feet. Within the fire-control tower, each turret and each of two plotting rooms, are the equivalent of six more directors, any one of which can control all or part of the battery.

Five range finders serve the main battery. One is in each of the elevated directors and in the after part of each turret. The ends protruding from the turret and director enclosures are protected by hoods.

Below the waterline are forward and after plotting rooms, each of which contains range keepers,

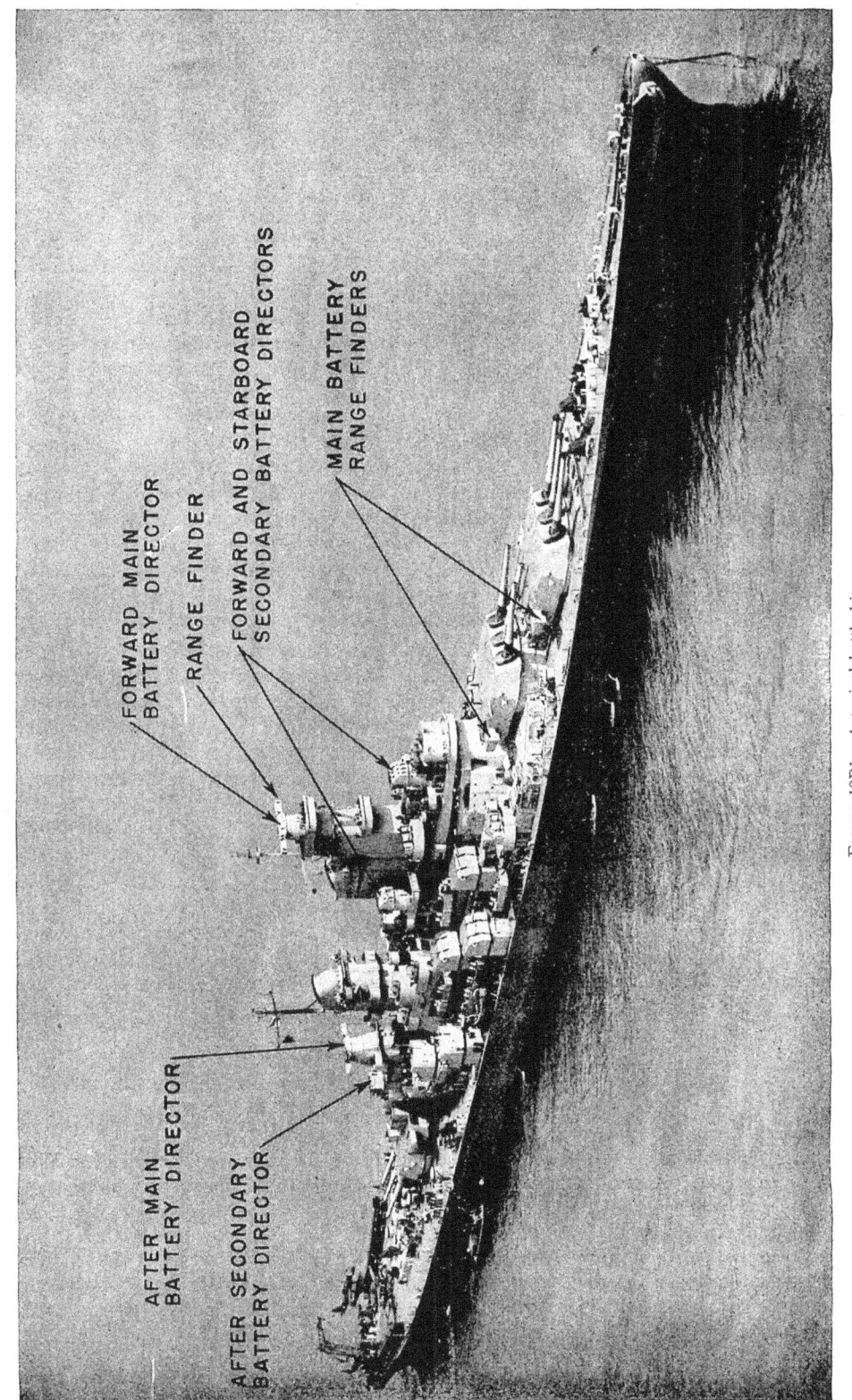

FIGURE 10B1. A typical battleship.

graphic range-keeping devices, switchboards and numerous other instruments. Information is received in these rooms from the range finders and directors, and combined with other information obtained from various sources. Computed quantities are transmitted to the guns for their control. Either plotting room may control any part of the battery. Auxiliary computing equipment is provided in the fire-control tower and each of the three turrets so that any of those four stations may control all or part of the battery in the event of casualty to both plotting rooms.

The secondary battery has four directors, three of which show in the illustration. The fourth is hidden from view by the forward stack. Each of these directors has its own range finder, and may control any part of the secondary battery. There are two secondary battery plotting rooms similar to those of the main battery, except that they contain no graphic plotting devices. Each is equipped with two computers, two stable elements, the necessary indicating instruments, and switching equipment. It is possible to engage four secondary battery targets simultaneously with director-controlled fire, with independent illumination control. In some installations, switching arrangements are provided to permit the secondary battery directors to supply information to the main-battery system.

The trend in control systems for AA machine-gun batteries is to provide a single director with its own computer, or a simple computing sight for each gun. No plotting rooms are provided, for the duration of an AA engagement is so short that simplified methods of control are necessary. In addition, the AA battery must be capable of engaging many targets at once, and the best means of doing so is with individual director control of each gun. However, modern installations in larger combatant ships include cross connections for control of the 40 mm. batteries by Mark 37 directors, and for control of 5"/38 cal. guns by adjacent 40 mm. gun directors.

10B4. Cruiser and destroyer systems. Cruisers having both main and secondary batteries have fire-control systems similar to those of battleships. Their main-battery systems are nearly identical in flexibility, varying only in relatively minor details. The secondary batteries usually have no more than two directors, and somewhat reduced flexibility, chiefly because they have fewer guns.

Main batteries of destroyers such as the one shown in figure 10B2 ordinarily have one director and one plotting room, although some have two directors which permit divided director fire. The control of AA batteries on destroyers and cruisers is very similar to that on battleships. The principal difference is the reduced number of targets which may be engaged simultaneously because of the smaller number of guns and directors provided.

10B5. The fire-control problem. Destruction of the enemy by gunfire requires solution of a *fire-control problem*. This problem, which is the principal subject of the subsequent chapters, begins with the designation of the target and the type of fire, and ends when the target is destroyed. Regardless of the manner in which fire is opened, the fire-control problem is solved in these five steps:

1. Positioning the target.
2. Determining sight settings.
3. Applying sight settings and firing.
4. Correcting sight settings.
5. Maintaining correct sight settings.

Actually, all five steps are carried out continuously and simultaneously, for the entire fire-control party operates as a team, with each member doing his assigned task. In analyzing what happens, however, it is simpler to study the separate operations. The student should avoid losing sight of the relationship of any given step to the whole problem.

Positioning the target. Target position is established by range, bearing, and elevation. These quantities may be obtained by estimates, repeated measurements, or computations based on previous measurements. Target position is the foundation upon which the remainder of the problem is based.

Determining sight settings. After the target is positioned, the values of sight angle and sight deflection must be computed for that position. The correct determination of these two angles

FIGURE 10B2. A typical destroyer.

INTRODUCTION TO FIRE CONTROL

is the most important part of the problem, for unless the angles are properly established no hits will be obtained. It is possible to find the equivalent of the angles by trial and error, as is sometimes done by tracer fire. However, various computing devices are usually employed to aid in positioning nearly all naval guns.

Applying sight settings and firing. The fundamental method of applying sight settings is to set the sights at the gun and let the pointer fire when the line of sight (LOS) is on the target. The use of a director provides another method, for the sight settings, while computed, are not employed directly at the guns. They are included in the signals transmitted to the guns, so that in effect they are indirectly applied. In director fire, the guns are fired by a master key at a remote point.

Correcting sight settings. In spite of all the computations, sight angle and sight deflection may contain small errors which are eliminated by applying corrections based on observations of the fall of shot, bursts, or tracers. The computations serve to provide initial values of the angles, but as the student will subsequently learn, there are certain approximations which enter into them. Any errors remaining are simply removed by spotting.

Maintaining sight settings. After the hitting angles have been established by computation and spotting, it is necessary to maintain them in order to continue effective fire. This part of the problem involves the continuous determination of (1) relative motion between own ship and target, and (2) the computed part of sight settings.

10B6. Presentation of the fire-control problem. The problem as outlined is fundamentally the same for surface or air targets. There are a number of factors which influence AA fire more than surface fire, and for this reason the problem is presented first as it applies to surface fire. Actually, the treatment of the AA problem is the general situation, for it can be applied to a surface problem, while the reverse is not possible. It will be seen that a system designed to control AA fire can also control surface fire.

Most of the following chapters are devoted to principles, divorced as much as possible from specific installations. An understanding of these principles will enable the young officer to undertake further studies of the construction and use of the equipment with which he must work aboard ship. Descriptions of some of the most common and currently used equipment are provided in certain chapters to acquaint the student with modern methods. It is again emphasized that the discussions provided cannot be all-inclusive, and other publications must be studied to fully qualify an officer.

Methods of use are subject to change; consequently they cannot be set forth as doctrine in this publication. Those procedures which are described in the text are illustrative. In actual practice numerous variations therefrom may be encountered, and it should be remembered that there is more than one way to accomplish results.

No student should feel that it is necessary to know all about every system in existence, for all systems have common features which are fundamental. However, a gunnery officer will never cease to study the subject of fire control. He must know the systems on his ship thoroughly, and the more he learns about other ship's systems, the more broadened his experience will become.

10B7. Symbols and definitions. There are many definitions and symbols which are used in fire-control work to provide a common language and a descriptive means of presenting the subject. Gunnery definitions as standardized by the Division of Fleet Training of the Office of the Chief of Naval Operations are provided in Appendix B. For ordinary use they are adequate, but for descriptions of specific fire-control systems they are not sufficiently rigid. The Bureau of Ordnance has established standard fire-control and torpedo-control symbols and definitions of a more rigid nature, and they are reproduced partially in Appendix A. Throughout this text, symbols and definitions are provided as the subject matter requires. In some cases the standard definitions have been somewhat modified to make them easier to understand.

CHAPTER 11

RANGE MEASUREMENT

A. RANGE FINDERS

11A1. Introduction. In general terms, range is the distance to the target, and its measurement is the first step in the solution of the fire-control problem. Eye estimates of target distances were the earliest form of range measurement, and are still used by crews manning guns on many small ships and by spotters on all classes of ships. The ability to estimate distance by eye with some degree of accuracy is a valuable asset for any naval officer, especially one in a gunnery department.

Advances in range finding have kept pace with improvements in fire-control methods, and have led to the development of various measuring devices. Among the instruments used to measure range for fire-control purposes is the *range finder*, which is essentially a system of optical units assembled in a long, cylindrical tube. The tube is supported by a mount which permits training the range finder on the target. Early forms of mounts were little more than simple tripods, but modern range finders are generally supported on mounts integral with turrets or director housings, leaving only the protruding ends exposed.

11A2. The range triangle. Range finders determine range by solving a right triangle in which one side and two angles are known, range being one of the unknown sides. Referring to figure 11A1, the range from a range finder PS to a target T is R. The distance between P and S is the range-finder *base length* (B). The lines of sight from P and S form the *parallactic angle* (θ) at the target. The line SI is parallel to PT, hence $\angle IST = \theta$. The instrument, built so that $\angle TPS$ is always 90°, measures θ at S, which is equivalent to measuring θ at T. The value of R corresponding to any value of θ is obtained from the trigonometric relation $R = B \cot \theta$. The range-finder scale, calibrated according to this equation, reads range directly for any measured value of θ.

An understanding of the practical range-triangle solution requires an introduction to a few basic optical principles and their applications in range finders. The range triangle is formed by light rays, which may be considered to travel through a uniform medium in straight

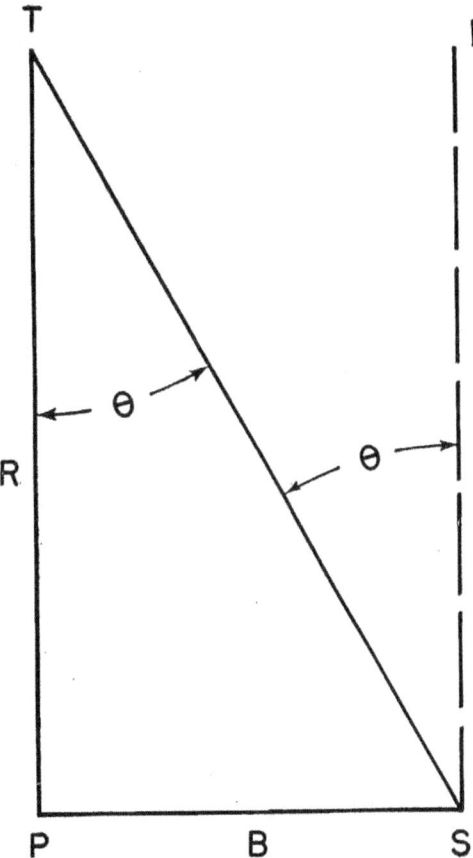

FIGURE 11A1. *The range triangle.*

lines. These rays may be deviated (i.e., have their direction of travel changed) by *refraction* or *reflection* as explained in succeeding articles.

11A3. Refraction. The optical density of a medium is a measure of the resistance it offers to the passage of light; a denser medium is one in which the light travels more slowly. A ray which strikes the surface of a *denser* medium obliquely is bent *toward* the perpendicular to that surface. Conversely, a ray striking the surface of a *less dense* medium obliquely is bent *away* from the perpendicular to that surface.

The application of this principle is illustrated by the effect on a light ray as it passes through a wedge-shaped piece of optical glass, or a *prism*. Since glass has greater optical density than air, a ray is bent as shown in figure 11A2a. As it strikes the face AB, the ray is inclined to the perpendicular by the angle a. As it enters the glass, the ray is bent toward the perpendicular to a smaller angle b. The ray meets face AC at angle c, and as it leaves the glass and re-enters the air, is bent away from the second perpendicular to angle d. Note that the resulting deviation is toward the prism base as the ray enters and again as it leaves.

The amount of refraction as light passes from one medium to another depends (in part) on the angle at which the light strikes the face. In the prism of figure 11A2a, this angle can be varied for any given light ray by rotating the prism about an axis such as that indicated. This application of a prism is employed in range finders to measure the parallactic angle. However, rotation of a single prism deflects the rays out of the triangulation plane (the plane containing the range triangle).

11A4. Compensator wedges. The optical unit which measures the parallactic angle is a pair of *compensator wedges* similar to those shown in figure 11A2b. The two wedges are identical, being nothing more than simple prisms made circular for convenience in instrument construction, and are so mounted that they rotate in opposite directions.

When these wedges are turned so that the thin side of one is opposite the thick side of the other, a ray passing through the pair emerges parallel to itself as shown in the illustration. Although the ray is deviated by the first prism, the second gives it an equal but opposite deviation which keeps the ray parallel to itself (although it is displaced vertically by a very small and negligible amount).

When the wedges are rotated in opposite directions as in figure 11A2c, their combined action produces deviation in the triangulation plane, but their effects at right angles to the triangulation plane still cancel each other. The ray is thus deviated in the triangulation plane and kept parallel to it.

The amount of wedge rotation determines the value of the deviation, which varies between zero and the maximum that occurs when the thick faces are opposite each other. Thus, the angular rotation of the wedges is a direct measure of the light-ray deviation.

11A5. Reflection. A light ray striking a polished surface is reflected at an angle equal to that at which it strikes the surface. This is generally expressed by the statement that the angle of incidence equals the angle of reflection, as illustrated in figure 11A3a. The reflection principle is applied in range finders to reflect light rays from the target through various optical elements of the instrument. The devices used to accomplish this are *reflecting prisms,* of which there are many types. The reflecting surfaces of such prisms may or may not be silvered, depending upon the design, for under certain conditions a glass surface will totally reflect the light falling upon it. Although refraction exists in reflecting prisms, their function is that of reflection.

11A6. The single-reflecting prism. Figure 11A3b shows a 90° prism, the inclined face AB being silvered. A light ray PL entering the prism at right angles to the face AC is reflected from the face AB and emerges at an angle of 90° from the entering ray. Following the light rays entering the prism from the arrow on the left-hand side of the drawing indicates how the image of an object can be reflected by a prism.

The angle between the entering and emerging rays is dependent upon the angle at which the entering ray strikes the face of the prism. Figure 11A3c shows a light ray falling obliquely upon the prism face AC. The ray is refracted toward the perpendicular DE, is then reflected by face AB, and is finally refracted away from the perpendicular FG. The angle (a) between the emerging ray

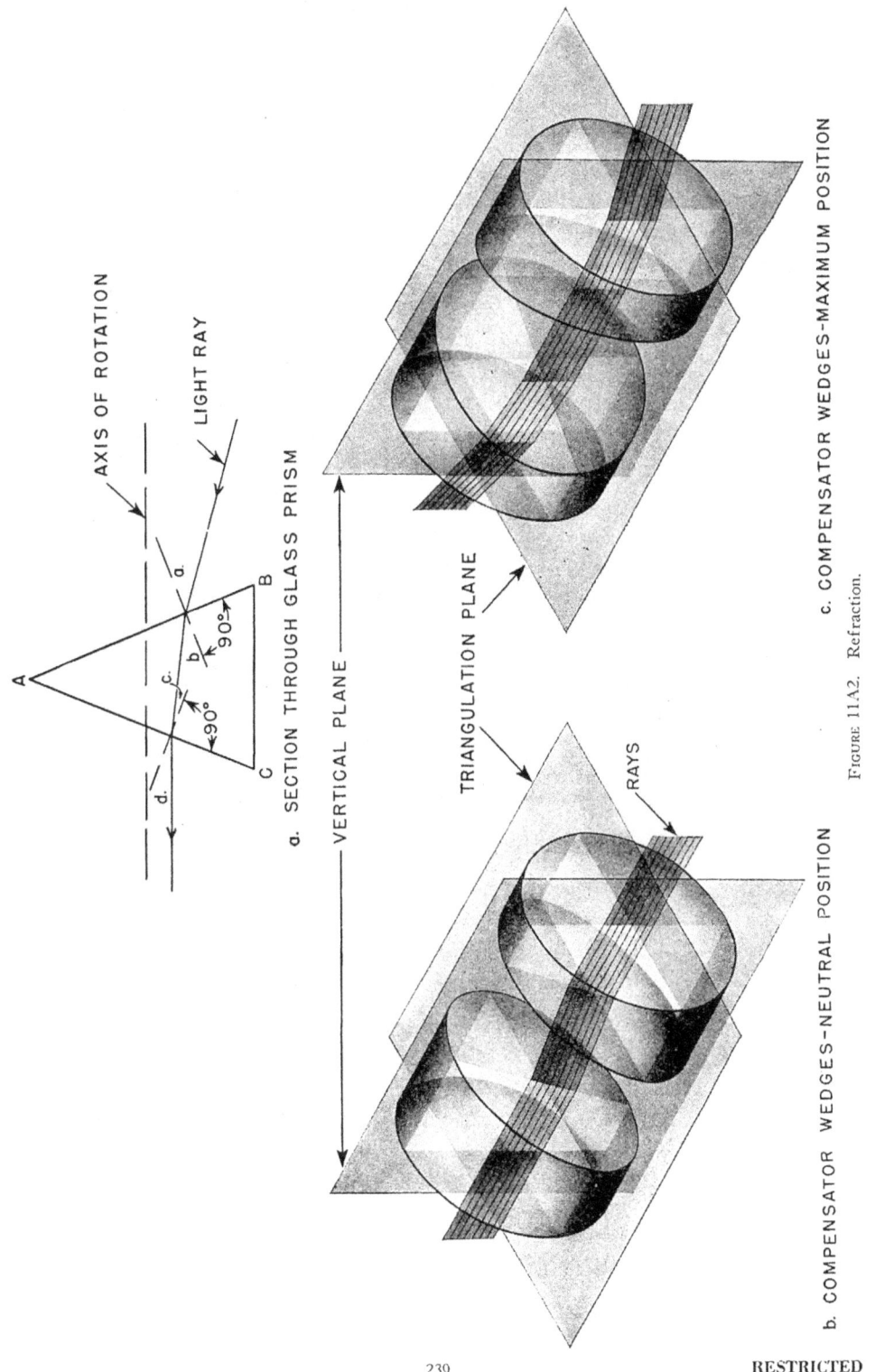

Figure 11A2. Refraction.

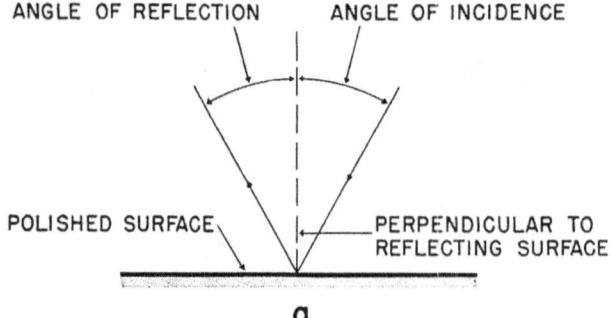

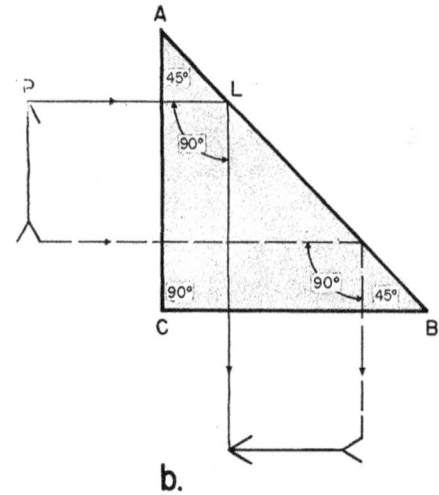

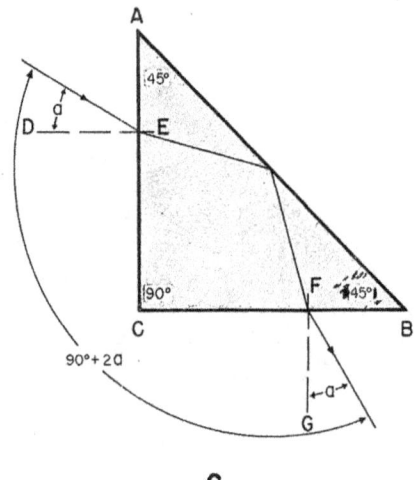

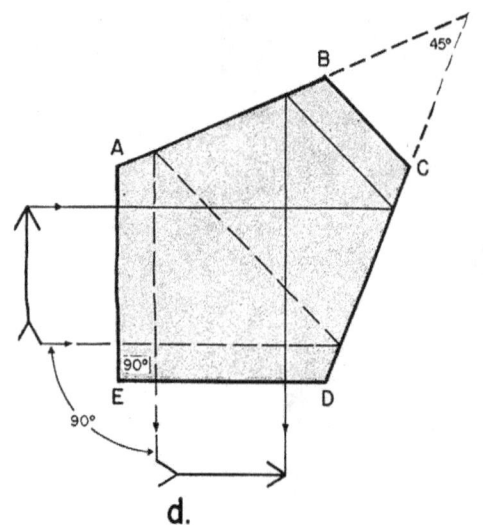

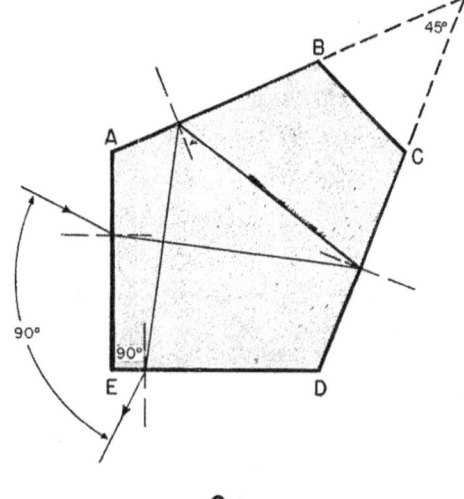

FIGURE 11A3. Reflection.

CHAPTER 11

and FG is the same as the angle between the entering ray and DE. However, these two angles lie on opposite sides of their perpendiculars, and the total angle through which the ray is deviated is $90° + 2a$.

Obviously, this type of prism cannot be used for range-finder end reflectors because it is essential that the angle between the entering and emerging rays be constant, regardless of the angle at which the rays strike the entrance face. The single-reflecting prism is useful, however, in deflecting the light (target image) from within the range-finder tube to the operator's eye.

11A7. The double-reflecting prism. The double-reflection principle is expressed by the statement that if a light ray is reflected by each of two plane surfaces, its deviation is equal to double the angle between the two reflecting surfaces. If the angle between the two surfaces is 45° as in the pentaprism ABCDE shown in figure 11A3d, the ray deviation is 90°. In this case, the surfaces AB and CD are silvered to provide reflection.

Figure 11A3e shows the great advantage of the pentaprism. The entering and emerging rays always make the same angle with each other, in this case 90°, regardless of the angle (within certain limits) at which the entering ray strikes the face AE. This principle is applied in range-finder end reflectors to eliminate the shortcoming of the single-reflecting prism. In large range finders, two separate mirrors are mounted with their faces at 45° to each other in order to accomplish the same results.

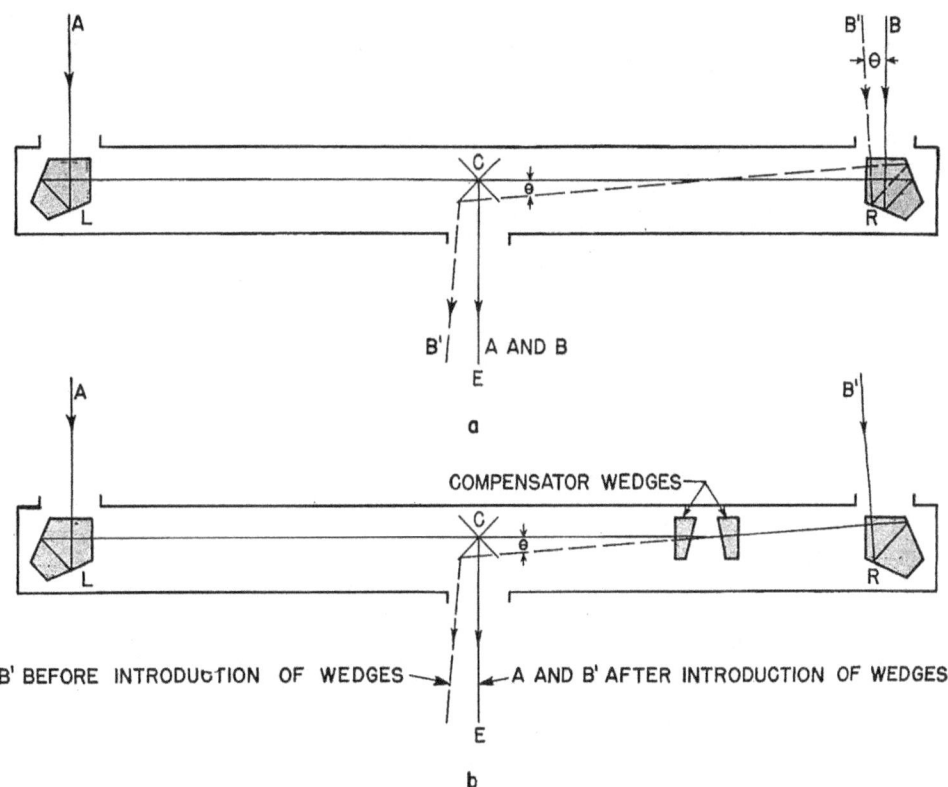

FIGURE 11A4. The range-finder principle.

11A8. The range-finder principle. The fundamental principle of all naval range finders is shown in figure 11A4. This figure represents a simplified instrument containing two pentaprisms (L and R) so placed that light rays striking their forward faces are reflected 90° toward the center of the tube. Between these prisms are two reflecting surfaces set at 90° to each other and arranged to reflect the rays from the prisms to an eye at E. An understanding of range-finder operation is aided by considering the action when the target is (1) at infinity, and (2) at a finite range.

Light rays from a point at infinity are parallel. If the range finder of figure 11A4a were directed at an object at infinity, rays A and B from that object would be parallel, and when reflected 90° by the pentaprisms as shown, would meet, or coincide at the point C. If the two reflecting surfaces were viewed from position E, the reflected images would merge and appear as one.

In the same figure, A and B' represent rays from a target at a finite distance. These rays, reflected by L and R, fail to coincide by the amount of the parallactic angle (θ). When the reflected rays are viewed from position E, they appear displaced as shown, and present a double image of the object.

However, if compensator wedges are placed between R and C as in figure 11A4b, the ray B' can be refracted so that it meets or coincides with A on the surface at C. Viewed from E the two rays would again appear as one. Obviously, the amount of rotation of the compensator wedges depends upon the value of the parallactic angle, and therefore, upon the target distance. A scale attached to the compensators can be calibrated to read the range in yards corresponding to the wedge position when the rays A and B' are brought into coincidence.

11A9. Range-finder types. The naval service employs two types of range finders; the *coincidence* and the *stereoscopic*. The latter is more widely used, since it is better suited to ranging on aircraft. Both types are similar in general appearance and have optical systems which are practically alike except for the elements near the center of the instrument. The names of the two types are derived from the principles involved in determining that the light rays entering the right end reflector have been deviated through the parallactic angle.

In the simple range finder shown in figure 11A4 the compensator wedges enable the operator to deviate the ray from the right pentaprism until it merges at the center of the tube with the ray from the left pentaprism. When thus deflected through the parallactic angle, the rays are superimposed and form a single image of the target. The essential difference between the two range-finder types is in the devices used to enable the operator to recognize this superposition and thus decide when the wedges are set correctly.

11A10. The coincidence range finder. An elementary coincidence range finder is shown in figure 11A5. Rays from the target enter the left pentaprism, are reflected through a 90° angle, pass through the left objective, and enter the coincidence prism. Rays entering the right pentaprism are also reflected 90°, pass through the compensator wedges and the right objective, and then enter the coincidence prism.

The coincidence prism shown, while not typical of currently used prisms, serves to illustrate the principle involved. The purpose of this unit is to discard the rays forming the *top* half of the image from one end of the instrument and those forming the *lower* half from the other end, and deliver the remaining halves so that they can be fitted together to form a single image.

The prism is an assembly of two separate parts held together by transparent cement along the surface AC. The section BC is silvered and reflects rays striking it from either side, while rays are transmitted through section AB without interference. The edge of the silvered surface at B is called the *halving line*.

The lower half of the rays entering from the right strikes the silvered surface and is reflected upward through the prism top. The upper half of the rays entering from the left misses the silvered surface and also passes through the top. These two groups of rays are discarded. Rays forming the remaining image halves are turned into the eyepiece group by a single-reflecting prism.

The roof shape on the longer part of the coincidence prism reverses the image entering the right half so that the right side of the upper image is over the right side of the lower image. This step is necessary since the pentaprisms turn the target images in opposite directions in reflecting the rays into the range-finder tube.

The objectives simply focus the light rays to form target images at the halving line. The eye-

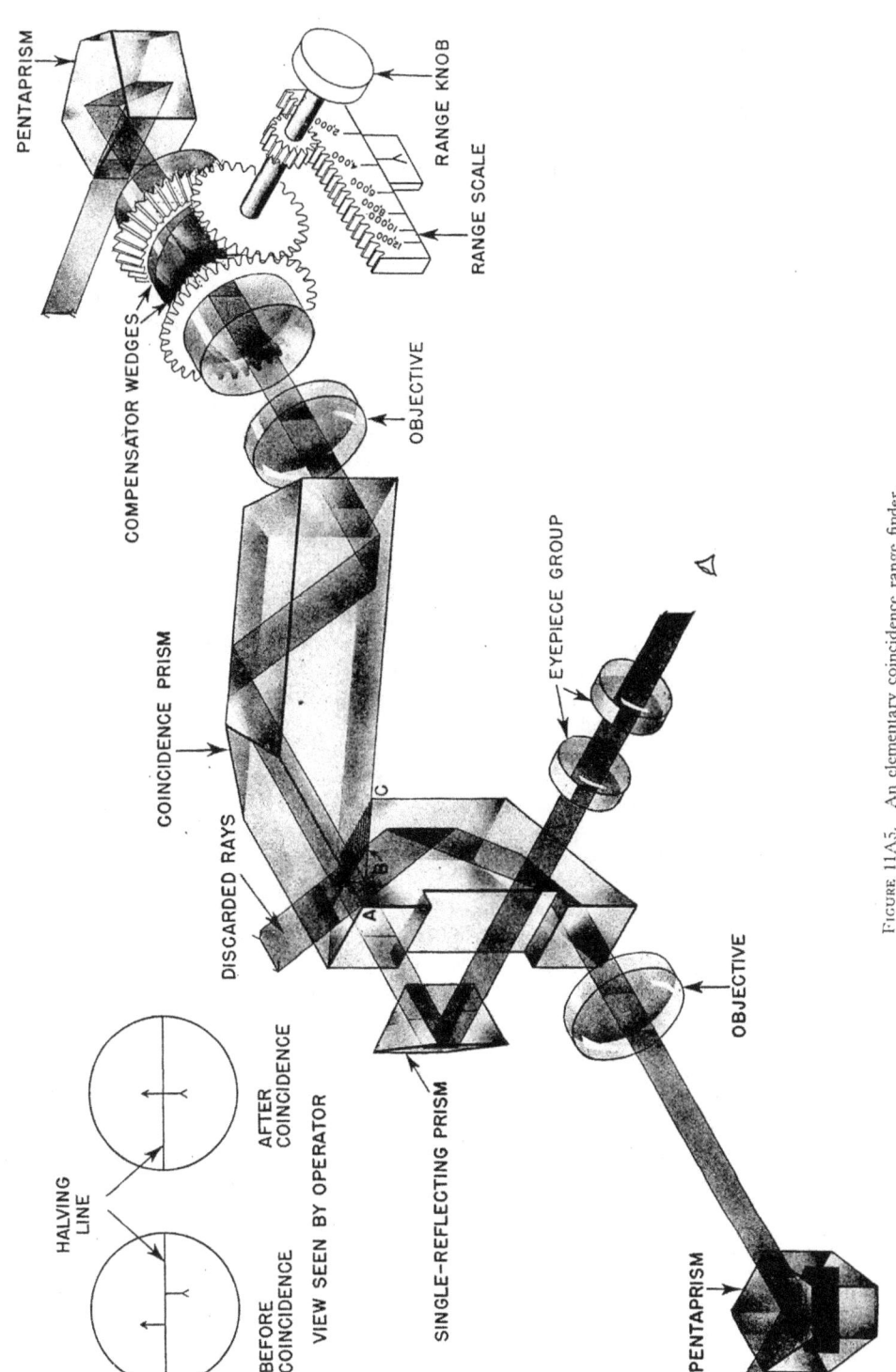

FIGURE 11A.5. An elementary coincidence range finder.

piece group magnifies these images, making the target appear closer to the operator than it actually is. Provision is made for focusing the eyepiece group to obtain a clear, well-defined target image. In order to simplify the diagram, the effects of the various lenses upon the light rays are not shown.

The operator, using one eye as in looking through a telescope, sees a target which appears cut in two by the halving line, as indicated in the circles in figure 11A5. Before the compensator wedges are adjusted, the two halves appear separated. When the wedges are set properly, the two halves join together to form a single image.

In use, the range finder is trained on the target, preferably directed at a clearly defined vertical line which is bisected by the halving line. The range knob, geared to the compensator wedges, is turned to bring the image halves into coincidence and the range is read directly from a scale driven by the range knob.

11A11. Depth perception. Stereoscopic range finders employ stereoscopic vision, or depth perception as the means of determining when the compensator wedges are set correctly. The principle involved is well illustrated by the old-fashioned stereoscope and the modern novelty devices used to view two separate pictures which blend into one that appears to have depth.

Figure 11A6a represents an observer viewing an object at O. Although each eye receives an independent image of the object, the brain fuses the two so that the observer is conscious of only one image. Since the left eye looks in the direction LO, it judges the object to be somewhere along this line. Similarly the right eye judges the object to be somewhere on the line RO. By fusing the two images the brain locates the object at the intersection of LO and RO.

In figure 11A6b transparent screens or reticles have been placed between the eyes and the object at O. On each screen is a reference mark which lies on the line of sight passing through the reticle to the object at O. The right eye judges the mark to be on line RO_R and the left eye judges it to be on LO_L. Hence, in fusing the reticle images the brain locates the reference mark at the intersection RO_R and LO_L extended, and the mark appears to be at the same distance as the object at O.

If the object is moved to O' as in figure 11A6c, the brain still locates the reference mark at O and it appears closer to the observer than the object at O'. The object at O' can be made to *appear* to be at O by inserting a pair of compensator wedges between O' and the right reticle, as in figure 11A6d, and then adjusting the wedges so that the ray from O' passes through the reference mark on O_R.

This principle is used in the stereoscopic range finder by placing two reticles between the pentaprisms and the observer's eyes. Actually, the left and right target rays need not pass through the reference marks to make the target appear at the same distance as the mark. As long as the rays from the reference marks make the same parallactic angle as the rays from the target, the mark and the target appear at the same distance.

11A12. The stereoscopic range finder. The basic elements of a stereoscopic range finder, shown in figure 11A7, are similar to those of the coincidence type except for the central group. Note that the coincidence prism is replaced by two reticles, and that there are two single-reflecting prisms to provide images for both eyes.

The reticles are permanently set in the instrument. The operator, seeing one diamond-shaped reference mark with each eye, is conscious of a single mark which always appears at a fixed distance. Each eye also sees an image of the target, which is judged to be at a distance which depends upon the angle between the reflected target rays as they enter the eyes. The compensator wedges are rotated to deviate the right target ray until the target appears to be at the same distance as the reference mark. Under this condition the position of the compensator wedges is a measure of the parallactic angle subtended by the target, and a suitably graduated scale geared to the wedges indicates range in yards.

The objectives focus the rays to form target images on the planes containing the reticle reference marks. While these images cannot actually be seen without suitable eyepiece groups, they

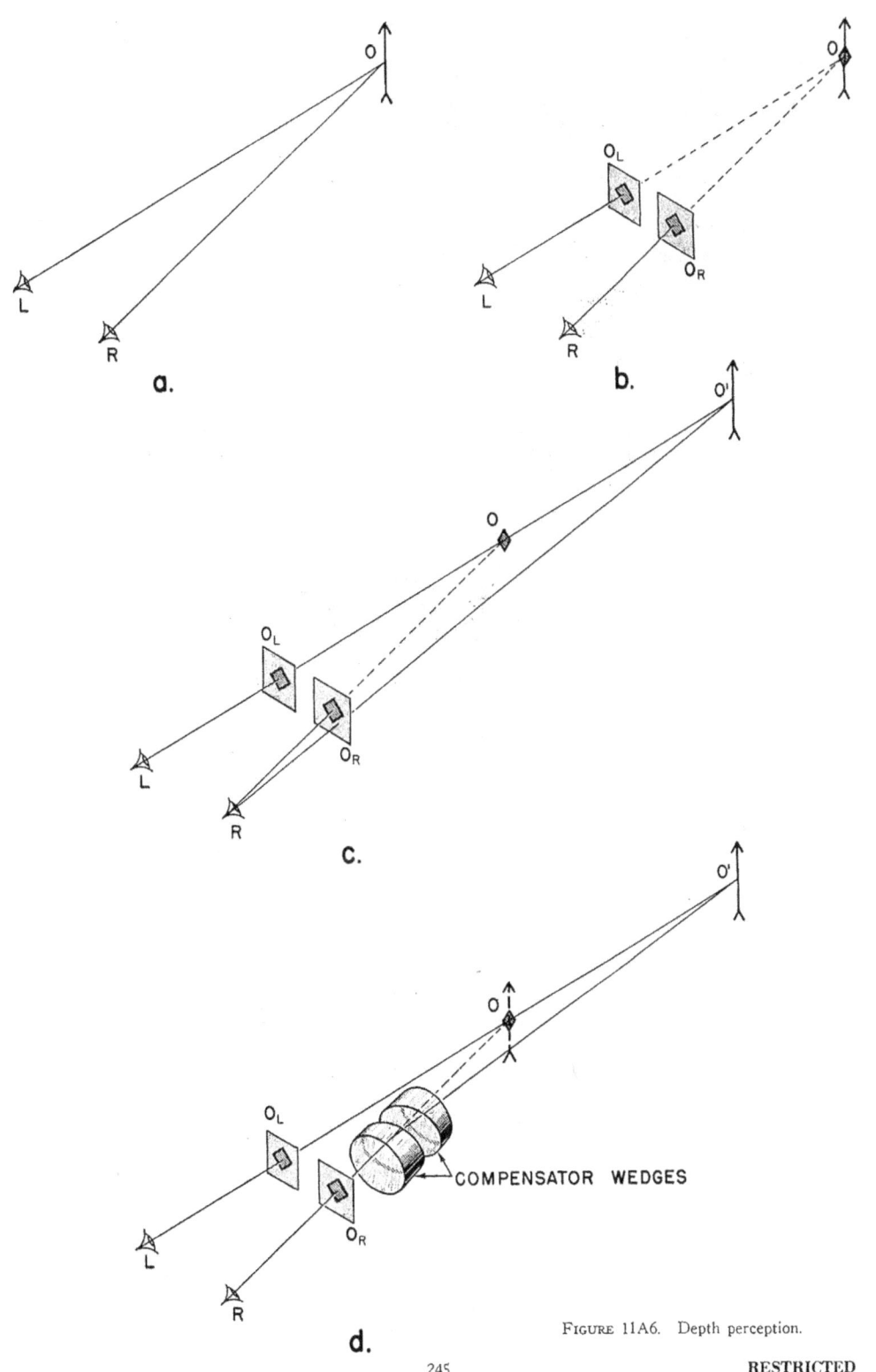

FIGURE 11A6. Depth perception.

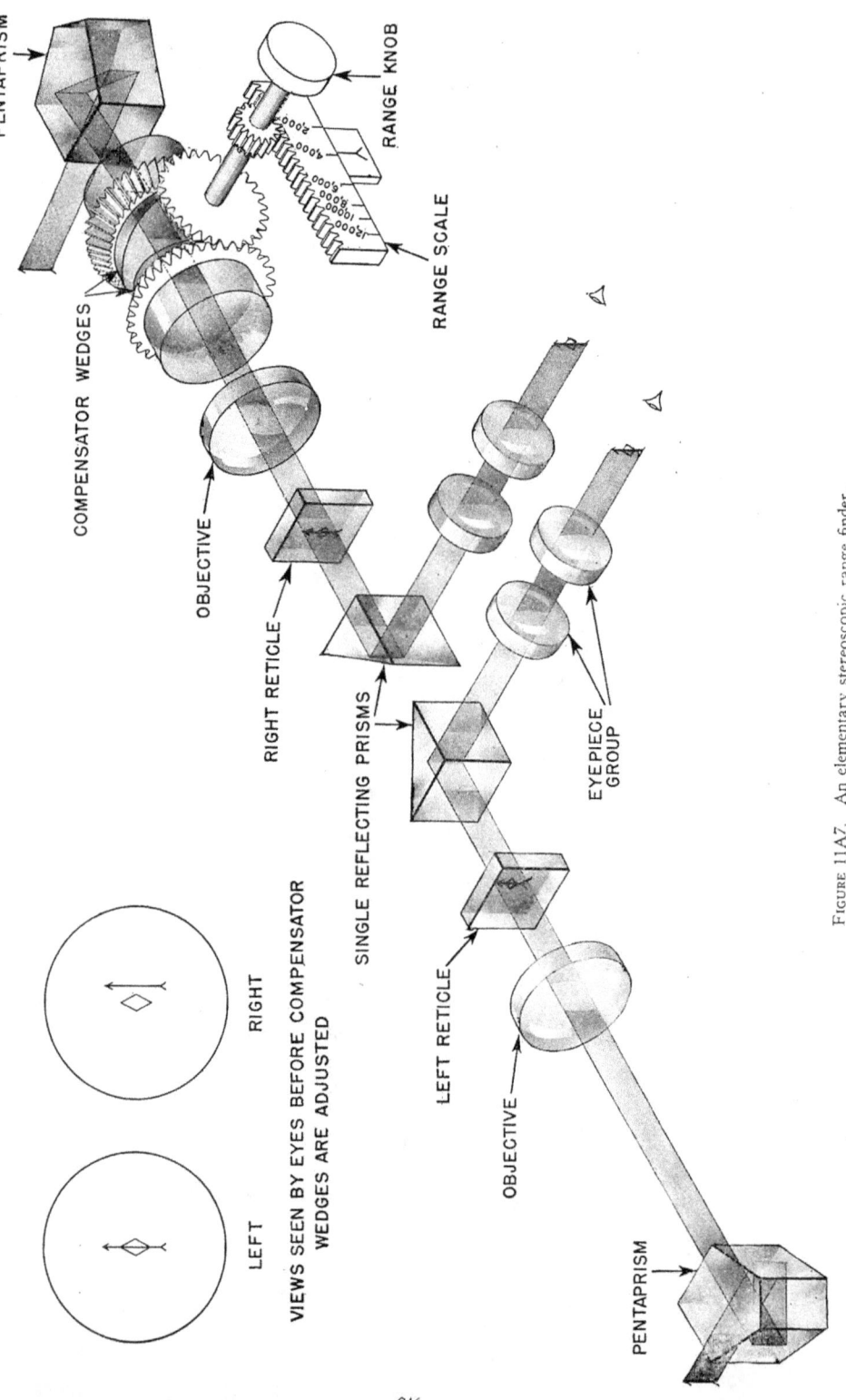

FIGURE 11A7. An elementary stereoscopic range finder.

are shown in the diagram to indicate their positions with respect to the reference marks. As in the coincidence range finder, the eyepiece groups simply magnify these images and the reference marks.

The views in the upper-left corner are shown as they might appear with the compensators not yet adjusted. Assuming the range finder is so trained that the left reference mark lies on the left target image, the right reference mark lies to one side of its corresponding target image. If the operator looks at the reference marks, he will be conscious of the fact that the target appears farther away than the mark, and will make the necessary wedge setting.

A typical stereoscopic range finder (Mark 42) is shown in figure 11A8. The details of its construction are too involved for discussion in this section, but may be found in the appropriate Ordnance Pamphlet. This type of range finder may be used to judge *differences* in distances, hence it is useful in spotting as explained later. This advantage, combined with the fact that ranges can be taken on irregularly shaped, fast-moving targets make the stereoscopic range finder superior to the coincidence type, especially for aerial targets.

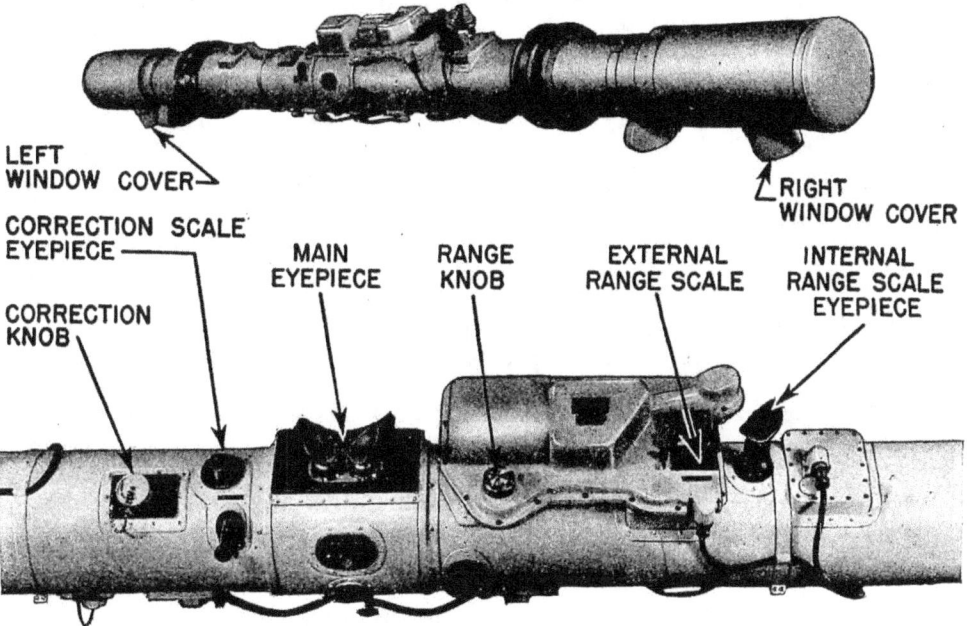

FIGURE 11A8. Stereoscopic range finder, Mark 42.

11A13. Limitations of the eyes. The resolving power of the eye is its ability to distinguish fine lines, one from another. It has been determined experimentally that with so called "normal" vision and maximum contrast between the lines and background, two lines can be distinguished when they subtend at least 12 seconds of arc at the eye. When closer than this the two lines blend into one and the eye no longer recognizes the separation.

The eyes, since they are separated by the *interpupillary distance,* act as a range finder in which distance is judged in terms of the parallactic angle. An observed difference between the parallactic angles for two objects results in the conclusion that one is farther away than the other. On the basis of about 2.5 inches between the eyes and 12 seconds of arc for the limiting parallactic angle, the maximum distance at which the unaided eye can distinguish differences in distance to an object is about 1,200 yards. The limiting distance can be extended by artificially increasing the parallactic angle by (1) extending the base length, and (2) magnifying the target image.

Base length. Obviously the distance to an object, for which the parallactic angle is 12 seconds of arc, depends upon the distance between the two points of view. By increasing the spacing between viewpoints, or the *base length*, the distance at which changes of range can be detected is correspondingly increased. Range finders vary in length from a few feet to over 40 feet for battleship main batteries, depending on the average ranges for which they are to be used.

Magnification. Magnification in the optical system is equivalent to bringing the object closer to the eyes and thus increasing the parallactic angle. The higher the magnification, the greater the distance at which the parallactic angle *appears* to be within the limiting 12 seconds of arc. However, magnification reduces the field of vision and is usually limited to 30 power, commonly used values being 12 and 24 power.

Magnification is accomplished by building the left and right halves of the instrument like two telescopes. Each half has an objective lens which brings the image of the target to a focus. This focused image is magnified in turn by an eyepiece group so that the operator sees an enlarged image which appears closer than the actual target.

11A14. Unit of error. The errors resulting from inability of the eye to detect small differences in distance are measured in terms of the *unit of error (e)*. The equation below gives the theoretical unit of error which may be expected under given circumstances, assuming that a well-trained range-finder operator can detect differences in the parallactic angle no smaller than 12 seconds of arc:

$$e = \frac{58.2 \, R^2}{BM}, \text{ where}$$

R = range in thousands of yards.
B = base length in yards.
M = magnifying power of the optical system.

It is evident from this equation that increases in the base length or in magnification cause corresponding decreases in the unit of error for a given range, and that as the range increases, the unit of error increases. As an example, consider the base length as one yard, magnifying power unity, and range 1,000 yards. The unit of error is 58.2 yards. This means that at 1,000 yards a change of 58.2 yards is the minimum that can be detected. If the base length is increased to 5 yards and the magnification to 24, the unit of error at 1,000 yards is less than one-half yard. In other words, the accuracy has been increased so that it is possible to detect changes of one-half yard at this range.

11A15. Reticles and reference marks. The unit of error has a practical application in making range spots with a stereoscopic range finder. Some of the later instruments such as the one shown in Figure 11A8, have several sets of reference marks engraved on their *reticles*. As an example, one type of reticle is shown in figure 11A9. The diamonds are used for measuring range to the target, while the straight lines are used in estimating spots.

The diamonds all appear to be the same distance from the eyes, hence any one of them can be used as a reference mark when adjusting for range. The straight lines also remain stationary, but the different rows appear to be at different distances from the observer. The rows above the diamonds on the reticles appear in space to be farther from the observer than the diamonds, while the lower rows appear closer. The marks in the first rows above and below the diamonds are so spaced that their apparent distance from the diamonds is 50 units of error. The spacing of the second rows from the diamonds corresponds to an apparent range difference of 100 units of error.

In ranging, the operator sets the compensator wedges to make the diamonds appear at the same distance as the target. Instead of being forced to change the compensator adjustment to obtain the range to a burst or a splash, the operator can use the straight line reference marks to estimate the spot in *units of error*. This estimate can then be converted to its equivalent in yards for the existing range.

All the reference marks are so spaced that the center-to-center distance between any two adjacent marks in the *same row* corresponds to 5 mils in deflection. Hence deflection spots can be made conveniently without training the range finder off the target.

CHAPTER 11

11A16. Auxiliary optical systems. Most range finders contain several auxiliary optical systems or units which are used in ranging or when adjustments are made on the main optical system. Among these auxiliaries are the following:

Main scale reading system. This system is used in reading internal range scales, which are usually connected to the compensator wedges. The optical system may be arranged so that the range-finder operator sees the scale along one side of the field, or a separate eyepiece may be provided to permit the assistant range-finder operator to read the ranges.

Internal adjuster system. This system provides an internal artificial target which is used in adjusting the main optical system. The optics are so arranged that light rays from an internal target introduced into the main optical system appear to come from infinity or some selected standard range. The range scale is set at this known distance and the instrument adjusted for this range, usually by turning a knob which rotates a *correction wedge* in the main optical system.

Adjuster scale reading system. The correction wedge used in making the internal adjustment is similar to a single compensator wedge. The setting of the correction wedge is indicated by a scale attached to it. Some range finders have a separate optical system to facilitate reading this scale.

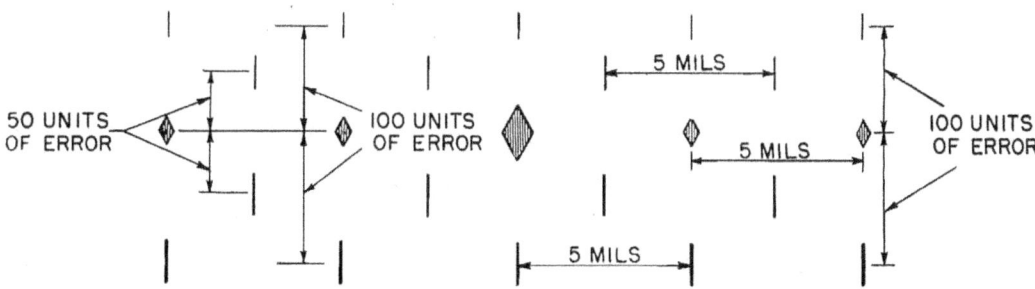

FIGURE 11A9. Reticle marks.

Height and halving adjustment. Light rays reflected by the two end reflectors toward the center of the range finder may be out of vertical alignment due to distortion of parts in the instrument. This error in height is corrected by rotating a knob which actuates a prism in the left half of the optical system. This is the *height adjustment* in stereoscopic range finders, and results in raising or lowering the left target image until both left and right images appear at the same height in the field of view. In the coincidence range finder, the same operation is called the *halving adjustment*, and adjusts the amount of the left target image discarded by the coincidence prism so that a complete image is seen when coincidence is made.

Trainer and finder telescope. Some range finders are equipped with a low-power telescope which is used by an assistant operator to locate the target and position the range finder. Such a telescope has crosswires which are used in centering the target in the range-finder field.

Ray filters. These are provided to enable the operator to regulate the intensity of the light entering the eyepiece group.

Astigmatizers. Since the coincidence range finder requires a vertical line to effect accurate coincidence, it usually is equipped with astigmatizers. They are used at night when no well defined vertical lines are visible on the target and ranging is confined to lights. The astigmatizers (not shown in the illustrations), located between the coincidence prism and each objective, consist of cylindrical lenses which may be swung into the optical axis by means of a knob on the shell of the instrument. They transform the images from the objectives into narrow, vertical bands of light which are brought into coincidence in the normal manner.

11A17. Errors and calibration. It should be obvious that range finders are precision instruments, for the parallactic angles which they measure are very small. Like all mechanical measuring devices, range finders are subject to certain errors which must be considered when mea-

RESTRICTED

suring distances. The errors are of two general types: (1) inherent instrumental errors, and (2) errors of observation, or personnel errors.

Instrumental errors are detected by frequently checking and calibrating the apparatus. Calibration consists essentially of comparing ranges measured by the instrument against more-accurately measured or known ranges. When checking on a target at measurable range, readings can be compared with those of a more accurate range finder or with known target distances.

The data obtained are plotted on curves which are used for reference purposes during subsequent range taking. One form of curve (A curve) shows ranges vs. the change in range reading caused by an adjuster-scale setting change of one unit. Another curve (B curve) shows ranges vs. the *change* required in the adjuster-scale reading to make the range finder read true ranges.

Personnel errors can be reduced to a minimum only by using great care in operating the range finders. Ranging requires a considerable amount of skill and operating technique which can be acquired only by constant practice. All individuals are not equally adept with range finders, consequently operators should be selected carefully. Details of calibration techniques, personnel selection, care, and operation are all provided in appropriate Ordnance Pamphlets carried aboard ship.

CHAPTER 12

GUN SIGHT PRINCIPLES

A. TRAJECTORY ANALYSIS

12A1. General. Before examining the problem of positioning a gun in space so that a projectile fired from it will hit the target, it is necessary to understand certain characteristics of the *trajectory*, which is the path followed by the projectile. Theoretically, the curvature of this path, projected in a vertical plane, is caused by the force of gravity acting to modify the velocity of the projectile after it leaves the gun. The path projected on a horizontal plane is also a curve, due to gyroscopic properties imparted to the projectile by rotation received from the rifling in the gun. The effect of this gyroscopic force is called *drift*, and it is always to the right because of the fact that naval guns are rifled with right-hand twist. Actually, there are other factors (atmosphere, projectile characteristics, and initial velocity) which modify the curvature of a trajectory, as explained later.

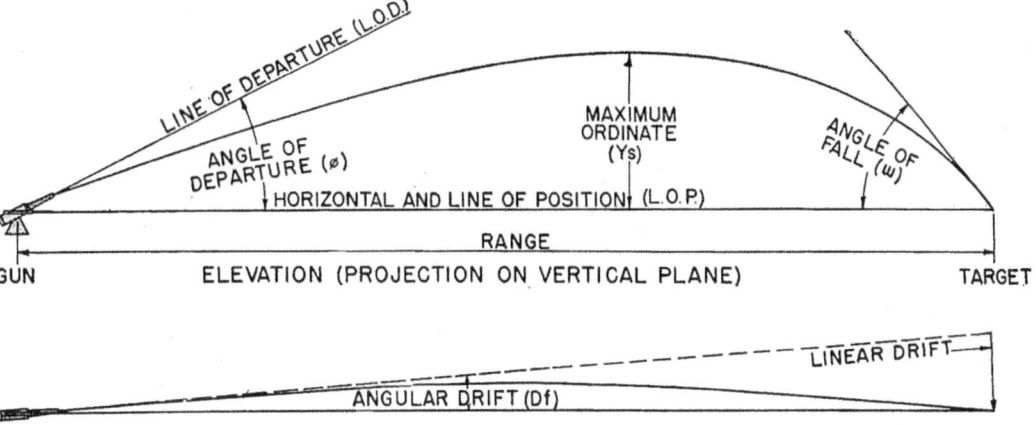

FIGURE 12A1. A simple trajectory.

12A2. Elements of a trajectory. Since the trajectory is curved, it is obvious that the bore of the gun cannot be pointed directly at the target (except at point blank ranges), but that it must be directed at some other position in space as shown in figure 12A1. The projectile must describe a definite trajectory in order to hit the target. To obtain this trajectory, the gun must be laid accurately with respect to some reference system. The most accurate reference for the vertical angle is the horizontal plane, but it is difficult to establish on a rolling and pitching ship.

The most convenient reference is a line joining the positions of the gun and target, and this is called the *line of position*. Since it is normally the line sighted from the gun to the target, it is also called the *line of sight*. The two terms are used interchangeably in these discussions.

Considering that the correct vertical and horizontal angular offsets have been computed and that the gun has been elevated by the vertical angle above the line of position and offset laterally from the vertical plane containing the line of position, the bore of the gun is so directed in space that a projectile fired from it will hit the target.

251 RESTRICTED

GUN SIGHT PRINCIPLES

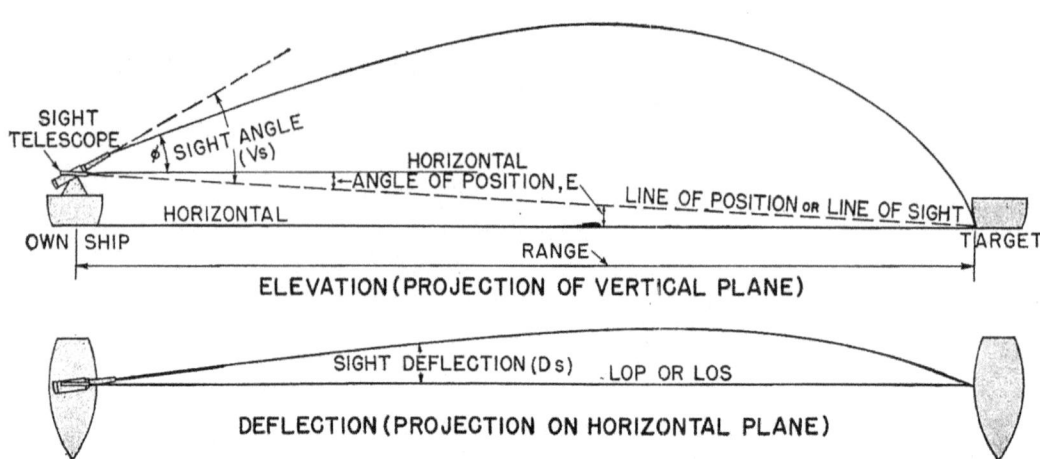

FIGURE 12A2. Positioning a gun with respect to a line of position.

12A3. Symbols and definitions. The symbols and definitions used in analyzing trajectory characteristics are included in the following list. Many of these terms are freely employed in subsequent chapters, and summary of all definitions and symbols is provided in Appendixes A and B.

SYMBOL   DEFINITION

Horizontal plane. The imaginary plane tangent to the earth's surface at a point instantaneously occupied by own ship. Any plane parallel to this one is also horizontal. Angles or distances measured in a horizontal plane are called *horizontal* angles or distances and are usually identified by the letter (h).

Vertical plane. A plane perpendicular to the horizontal. Angles and distances measured in a vertical plane are called *vertical* angles or distances and are usually identified by the letter (v).

Slant plane. The plane (1) containing the trunnion axis, and (2) parallel to the LOS. Angles or distances measured in or parallel to the slant plane are sometimes called *slant* (plane) angles or distances, and are usually identified by the letter (s).

Trajectory. The path described by a projectile in flight.

R *Range.* The distance from the origin of the trajectory to the target.

Rh *Horizontal range.* The projection of the range in a horizontal plane.

LOP *Line of position.* The line joining the gun and target.

LOS *Line of sight.* Usually the same as LOP, and so considered in this chapter.

E *Angle of position* or *target elevation.* The vertical angle between the LOP and the horizontal plane.

LOD *Line of departure.* The extension of the bore axis; it is tangent to the trajectory at the instant the projectile emerges.

$\emptyset$, Eg *Angle of departure* ($\emptyset$) or *vertical gun elevation* (Eg). The vertical angle between the horizontal plane and the LOD.

Vs *Sight angle.* The angle between the LOD and the slant plane, measured in the plane containing the LOD and perpendicular to the slant plane. This is a vertical angle when the trunnion axis is horizontal.

ω *Angle of fall.* The vertical angle between the horizontal and the tangent to the trajectory at the point of fall.

Maximum ordinate. The vertical distance to the highest point of the trajectory.

252

CHAPTER 12

SYMBOL	DEFINITION
H	*Altitude (height).* The vertical distance of the target above the horizontal (AA fire).
Tf	*Time of flight.* The total time which elapses during the flight of the projectile from the gun to the point of fall (or burst).
Df	*Drift.* The horizontal angle between the LOS and the LOD to compensate for drift.
Ds	*Sight deflection.* The slant plane angle between the LOS and the LOD.

12A4. Range table trajectory characteristics. A range table is a tabulation of the trajectory characteristics for a particular gun when fired under standard conditions. The table also includes data used in compensating for variations from standard conditions. The student is referred to the range table in Appendix D at the back of the book. This table is for a 5"/38 cal. gun firing a 54 pound projectile at an I. V. of 2,600 f.s. Entering the table for a range of 15,000 yards, for example, it is seen that the characteristics of the trajectory at this range and under standard conditions (discussed later) are as follows:

Col. 1: Range 15,000 yards.
Col. 2: Angle of departure 23 degrees 30 minutes.
Col. 3: Angle of fall 39 degrees 53 minutes (note that the angle of fall is materially greater than the angle of departure).
Col. 4: Time of flight 40.34 seconds.
Col. 5: Striking velocity 950 f.s. (much less than I. V.).
Col. 6: Drift 248 yards (always to the right).
Col. 8: Maximum ordinate 7,165 feet.

It must be clearly understood that these characteristics will be obtained only if the gun is fired under standard conditions. In order to correct for variations from standard, columns 10 to 18 are used in the manner described in Chapter 13.

12A5. Use of a local line of sight. If a gun is mounted on a platform ashore it can be laid to the desired elevation by the use of a gunner's quadrant, which measures angles of elevation with respect to the true horizontal. At sea such a method is impracticable because of ship's motion. It is more convenient to offset the gun, vertically and laterally, from a line of sight directed at the target as shown in figure 12A3. It is apparent from the figure that the gun elevation Eg remains constant as the ship rolls, as long as the LOS is held on the target.

The vertical angle through which the gun is elevated with respect to the *horizontal* is defined as the angle of departure ($\emptyset$). The term $\emptyset$ represents an exterior ballistics quantity and, in practical fire control, is known as vertical gun elevation (Eg). The vertical angle between the axis of the bore and the line of sight is defined as the sight angle (Vs). If the LOS is horizontal, $\emptyset = Eg = Vs$. The lateral angular offset of the bore axis from the LOS due to the drift of the projectile is called drift (Df). The *total* lateral angle, due to all factors causing deflection, is called sight deflection (Ds).

12A6. The principle of trajectory rigidity. When a gun is elevated with respect to its LOS it ordinarily is not elevated with respect to the horizontal because the gun and target usually do not lie in the same horizontal plane. Assuming that the water's surface is the horizontal, the gun is placed above this surface a distance depending on its vertical location aboard ship. The LOS from the gun is normally directed at the waterline when the target is a ship. In this case, the LOS dips below the horizontal by the amount of the position angle E.

In surface fire this angle is small and may be neglected because of the principle of trajectory rigidity. This principle states that for small angles of position, a gun which obtains a certain inclined range along a line of position when elevated the required angle above that line, will have the same range along the horizontal when elevated the same angle above the horizontal.

In surface fire it is unusual to encounter an angle of position greater than about 35 minutes, and a value as great as this occurs only at extremely short range, in which case the angle of eleva-

253   **RESTRICTED**

tion is also small. For example, with a 5"/38 cal. gun mounted 50 feet above the waterline and firing at the waterline of a target at a slant range along the LOS of 1,600 yards, the angle of position E is about −35 minutes and Eg listed in the range table is 43.6 minutes. For these values of E and Eg, the ratio of the inclined range R to the horizontal range Rh is about 1.0001, showing that the difference in ranges is negligible. Therefore, even though the angles of elevation listed in

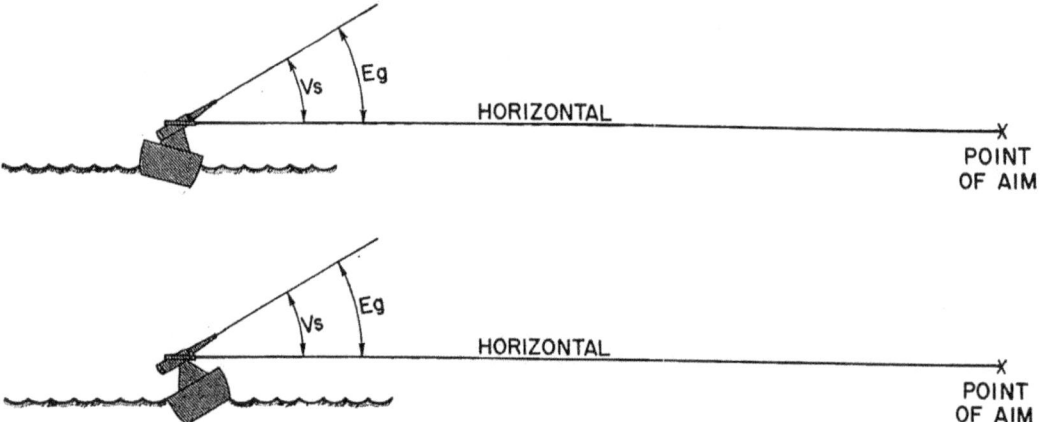

FIGURE 12A3. Positioning a gun on a moving ship.

the range tables are values of Eg (vertical angles measured with respect to the horizontal), these angles when set on the gun sights as Vs (measured with respect to the line of position) result in the required ranges along the LOS with insignificant error. This principle is applied in surface fire; hence, in the following chapters relating to surface fire the LOS, and thus the slant plane, are assumed to be horizontal.

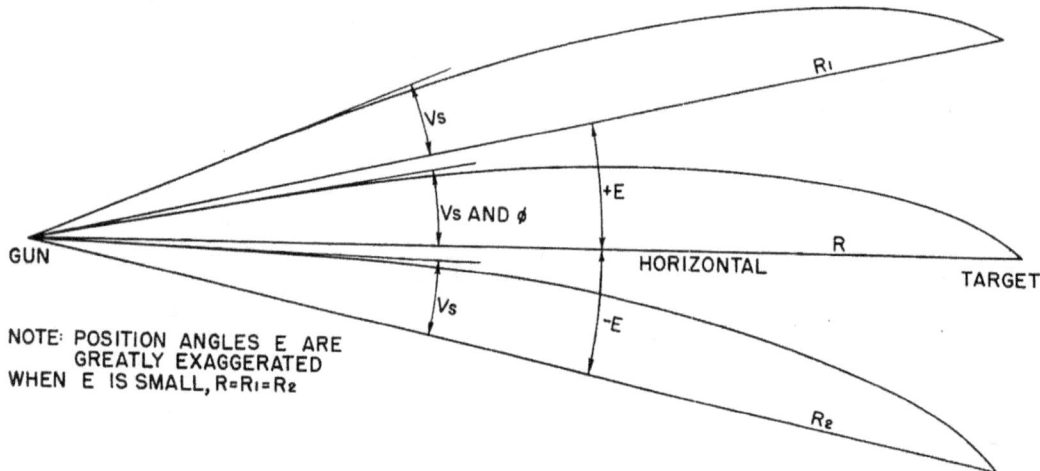

FIGURE 12A4. Rigidity of the trajectory for small position angles.

It is to be noted that the principle of trajectory rigidity does not hold in AA fire because normally the angles of position are large, approaching 90°. Figure 12A5 shows a gun positioned with respect to both the line of position and the horizontal, for firing on an elevated target.

12A7. Exterior ballistics. It is the duty of the *fire-control party* to determine the values of sight angle and sight deflection required to hit the target. It is the function of the *gun sights* to provide a means for establishing the line of sight, and a function of the *sight scales* to provide an

CHAPTER 12

accurate means for setting and measuring the angular offsets which are required. The determination of Vs and Ds is based upon the subject of exterior ballistics, which deals with projectile motion after the projectile leaves a gun. A study of the entire subject includes not only the theoretical aspects of this motion and the forces affecting it, but also the practical problem of laying a gun so that a projectile fired from it will hit the target. In a study of exterior ballistics, it is necessary to examine the characteristics of trajectories in general as well as particular characteristics obtained under specific firing conditions.

A thorough study of exterior ballistics, even from the practical aspect, involves four phases, any one of which is an extensive subject in itself. These phases are:
1. Examination of forces affecting trajectories.
2. Preparation of a ballistic table.
3. Preparation of range tables.
4. Use of a range table.

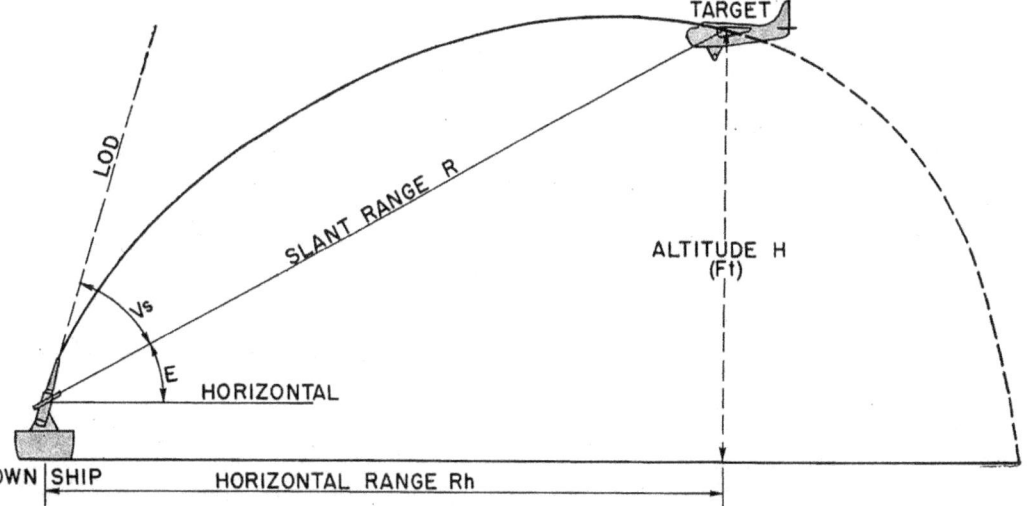

FIGURE 12A5. Trajectory to intercept an elevated target.

12A8. Other forces acting upon the projectile. So far, the only forces which have been considered as acting upon the projectile after it leaves the gun are gravity and the precessional force which causes the projectile to curve to the right. If the projectile traveled in a vacuum the only force acting upon it would be that of gravity. Under such circumstances, the result would be vertical deceleration in the ascending branch until the summit was reached, followed by an exactly equal acceleration in the descending branch to the point of fall. This would result in:
1. A trajectory in the form of a parabola with a vertical axis.
2. A maximum ordinate located half way between the gun and the point of fall.
3. An angle of fall equal to the angle of departure.
4. A striking velocity equal to the initial velocity.
5. A maximum range obtained with an angle of departure equal to 45°.

However, the firing of a projectile in *air* changes the above characteristics of the trajectory because of the following factors:

1. *The density of the atmosphere.* The atmosphere offers resistance to the projectile which materially alters the characteristics of the trajectory. Since atmospheric density differs from hour to hour with changes in temperature and barometric pressure, and also according to altitude, the resistance varies not only from time to time, but in the different altitude zones through which the projectile travels.

GUN SIGHT PRINCIPLES

2. *Characteristics of the projectile.* The characteristics of a projectile which influence the amount of resistance to which it is subjected in its passage through air of a given density are:

    a. Weight.
    b. Cross-sectional area, which is proportional to the square of the projectile's diameter.
    c. Shape. A projectile which has a streamlined front end encounters less resistance than one which has a short, blunt nose.

3. *The initial velocity.* With air density and projectile design constant, initial velocity affects the characteristics of the trajectory, because the amount of resistance offered by air varies with velocity.

12A9. Preparation of a ballistic table. A study of the several factors which affect the flight of a projectile would show that they bear a complex relationship to one another. Such a study cannot be undertaken here, but a brief description of the vast amount of work which must precede the preparation of range tables will prove useful to the student.

There is no single, simple equation whose solution will give the characteristics of a trajectory at any point, principally because of the complications introduced by the air medium in which a projectile travels. To facilitate the solution of trajectories, certain arbitrary values are assigned to some of the factors for the purpose of developing a *ballistic coefficient* (C). This coefficient is a measure of comparison between the retardation of a given projectile in air of a particular surface density to the retardation of a standard projectile in air of standard surface density.

Based on the variables $\emptyset$, I. V., and C, formulas are developed and used to compute solutions for a great many trajectories for all values of $\emptyset$, I. V. and C that are likely to occur in practice. The resulting trajectory solutions are tabulated in a *ballistic table* with values of $\emptyset$, I. V. and C serving as entering arguments. As a matter of interest, such a table consists of many pages of closely-spaced data, which obviously cannot be included here.

There are certain other factors which affect a trajectory, such as the forces introduced by rotation of the projectile, rotation of the earth, wind, and gun motion. These factors are not included in a ballistic table as affecting standard trajectories, and are dealt with by separate computations to determine their effect upon trajectories obtained under specific firing conditions.

Although the study of a ballistic table is important if the student hopes to gain a thorough understanding of the ballistics problem, even an elementary consideration is not attempted in this manual. For more complete treatment, the student is referred to books which deal primarily with this subject.

12A10. Preparation of range tables. It has been stated that a ballistic table constitutes a basis for the solution of trajectories corresponding to any combination of $\emptyset$, I. V. and C that is likely to occur, and is applicable to many velocities and many types of projectiles. In order that the data for a *given* gun may be available in convenient form, a *range table* is prepared for each gun. Such a table is a tabulation of elements for various trajectories corresponding to the particular initial velocity and projectile of a given gun, as contrasted with a ballistic table, which may be considered as a "master range table" applicable to many velocities and projectiles.

In preparing a range table, the particular initial velocity and type of projectile are determined for the specific gun. The ballistic table is entered with corresponding values of I. V. and C, and with various angles of departure ($\emptyset$) up to the maximum attainable by the gun. For each angle of departure, the values of range, angle of fall, time of flight, striking velocity, and maximum ordinate are determined. Note again that these data apply only to a particular gun, projectile, and initial velocity. Further discussion of range-table elements is found in Section 12B.

12A11. Use of a range table. Finally, there is the practical employment of the range table aboard ship for the determination of sight angle (V_s) and sight deflection (D_s) to be used in setting the sights of the gun in elevation and deflection in order to hit the target under the conditions existing at the time of firing. Use of the range table is explained in Chapter 13.

CHAPTER 12

B. RANGE TABLES

12B1. Standard conditions for range tables. In the computations upon which the ballistic table and range tables are based, certain arbitrary values are assigned to the several variables that must be considered. These arbitrary values are generally spoken of as *range table standard conditions*. In order to use the range table under conditions other than standard, it is necessary to provide in the range table corrections for variations from these conditions.

Most of the standard conditions assumed for range-table values are that:

1. The projectile leaves the gun with the designed velocity.
2. The projectile is of the designed weight.
3. The atmosphere is of an accepted, arbitrarily-chosen standard density.
4. There is no wind.
5. The gun is motionless.
6. The target is motionless.*
7. The earth does not rotate.
8. The gun and target are in the same horizontal plane.
9. The gun is elevated in the vertical plane; that is, the axis of the gun trunnions is horizontal at the time of firing.
10. The gun is warm from previous firings.

It is obvious that inasmuch as the range table is based upon these conditions, variation from any one of them will cause the projectile from an otherwise correctly-aimed gun to miss the target. In addition to indicating characteristics of trajectories under standard conditions, it is also one of the functions of the range table to provide necessary corrections for variations from the first seven of these standard conditions. Corrections for variations from any of the latter three require separate consideration or computation.

12B2. Method of preparing a range table. As previously stated, the entering arguments of a ballistic table are angle of departure ($\emptyset$), initial velocity, and the ballistic coefficient (C). It is now desired to construct a range table for a given gun, I. V., and projectile. Since the I. V. is known and $\emptyset$ must vary from zero to the maximum elevation attainable for the gun, it is only necessary to determine the value of C in order to complete the arguments for entry into the ballistic table. The ballistic coefficient depends on values of atmospheric density, projectile weight and diameter, and a coefficient of form (i). The coefficient of form is defined as the ratio of retardation of the given projectile to that of a projectile of arbitrarily chosen standard characteristics, and its value varies with the angle of departure.

All of these factors which make up the ballistic coefficient (C), are fixed with the exception of the coefficient of form, whose value for each angle of departure must be determined by *experimental ranging* at the Naval Proving Ground. A brief description of the method of ranging and computation follows.

12B3. Experimental ranging. Let it be assumed that the maximum elevation of the gun, as mounted on shipboard, is to be 45°. A number of shots (from four to seven) are fired at each angle of departure in a series such as the following: 5°, 10°, 15°, 20°, 25°, 30°, 45°, and again at 15°. The velocity at each angle is measured, or, if impossible to measure at high angles, is computed by utilizing the results for low angles.

Throughout the firing, meteorological observations are made, both at the surface and aloft, to determine wind and atmospheric density. On the basis of these observations computations are made for *ballistic density* and *ballistic wind* for the predicted height of maximum ordinate. Since the projectile travels through various altitude zones it is obvious that values of wind and air density do not remain the same throughout the trajectory. Therefore, fictitious values of wind and air density are computed which represent a constant value for each over the entire trajectory. This is

---

* Although motion of the target does not affect the trajectory itself, it enters into the problem of determining the trajectory that will reach the target.

done by means of weighting factors which are applicable in each zone. The fictitious values obtained are called ballistic density and ballistic wind; they are used for correcting the trajectories derived from observed data to those which would exist in a standard atmosphere.

Uncorrected observed ranges are recorded and averaged for each angle of departure. These ranges are then corrected for errors due to (1) the height of the gun above tide level, (2) the curvature of the earth, and (3) the effect of the component of wind acting parallel to the line of fire. The first two corrections must be made if the range table is to conform to the requirement that the point of fall be in the horizontal plane tangent to the earth at the gun.

The observed lateral deviation is corrected for the effect of the component of the wind acting at right angles to the line of fire: the resultant deviation thus corrected is the *observed drift*.

12B4. Computations. These corrected, observed values are *not* suitable for inclusion in a range table due to probable small variations in initial velocity and projectile weight, and to non-standard, atmospheric density. Therefore, such values are used in entering the ballistic table to obtain a *final* coefficient of form (i) for each of the various angles of departure. Since it is the only variable in the expression for the value of the ballistic coefficient (C), and since its final value has been determined, the value of C may be computed for the *designed* weight and diameter of the projectile.

The final step in the computation of that portion of the range table dealing with trajectory characteristics for standard conditions is the determination of the values to be tabulated in columns 1, 2, 3, 4, 5, 6, and 8. This is performed by entering the ballistics table with the designed I. V., with angles of departure in increments of 5°, and with the values of C corresponding to the value of (i) determined for each angle of departure. Then from the table are taken corresponding values of range, time of flight, angle of fall, striking velocity, and maximum ordinate.

The computation of drift for standard conditions is a separate calculation involving the use of the observed drift in a special formula to solve for drift at each of the 5° increments of angle of departure.

12B5. Compilation of range-table data. It is necessary to compile the data obtained in convenient range-table form with the argument for entry being 100 yard increments of range rather than 5° increments of angle of departure. A large scale graph is constructed, much larger than the example shown in figure 12B1. The 5° increments of angle of departure, from zero to maximum (over 43° in the example), are used as abcissas. The other values (range, angle of fall, time of flight, striking velocity, drift, and maximum ordinate) are used as ordinates. A smooth curve is then faired in for each element, making the data available for intermediate values of angle of departure.

A range table may now be compiled by selecting points on the range curve in increments of 100 yards and picking off the corresponding values of all the other elements, including angle of departure. The values obtained from the curve are tabulated against the increments of range. The entering argument of range determines a point on the range curve. A vertical line through this point intersects the other curves, the points of intersection read against the proper scales determining the corresponding values of these elements.

12B6. Range-table corrections. It must be clearly understood that the characteristics of trajectories listed in columns 1 to 6 and 8 of a range table only hold under the standard conditions listed in Article 12B1. For a final determination of Vs and Ds necessary for properly setting the sights of a gun under the *actual* conditions at the instant of firing, it is necessary to compute corrections for any variations from standard. These computations are called *ballistics*.

The corrections provided in columns 10 to 18 of the range table afford a means of compensating variations in I. V., weight of projectile, air density, and wind, and for movement of gun and target (items 1-6, standard conditions). The correction for rotation of the earth, (item 7, standard conditions listed in Article 12B1), is included as a separate tabulation at the back of most range tables. The methods of computing ballistics required for accurate determination of V_s and D_s are described in subsequent chapters.

CHAPTER 12

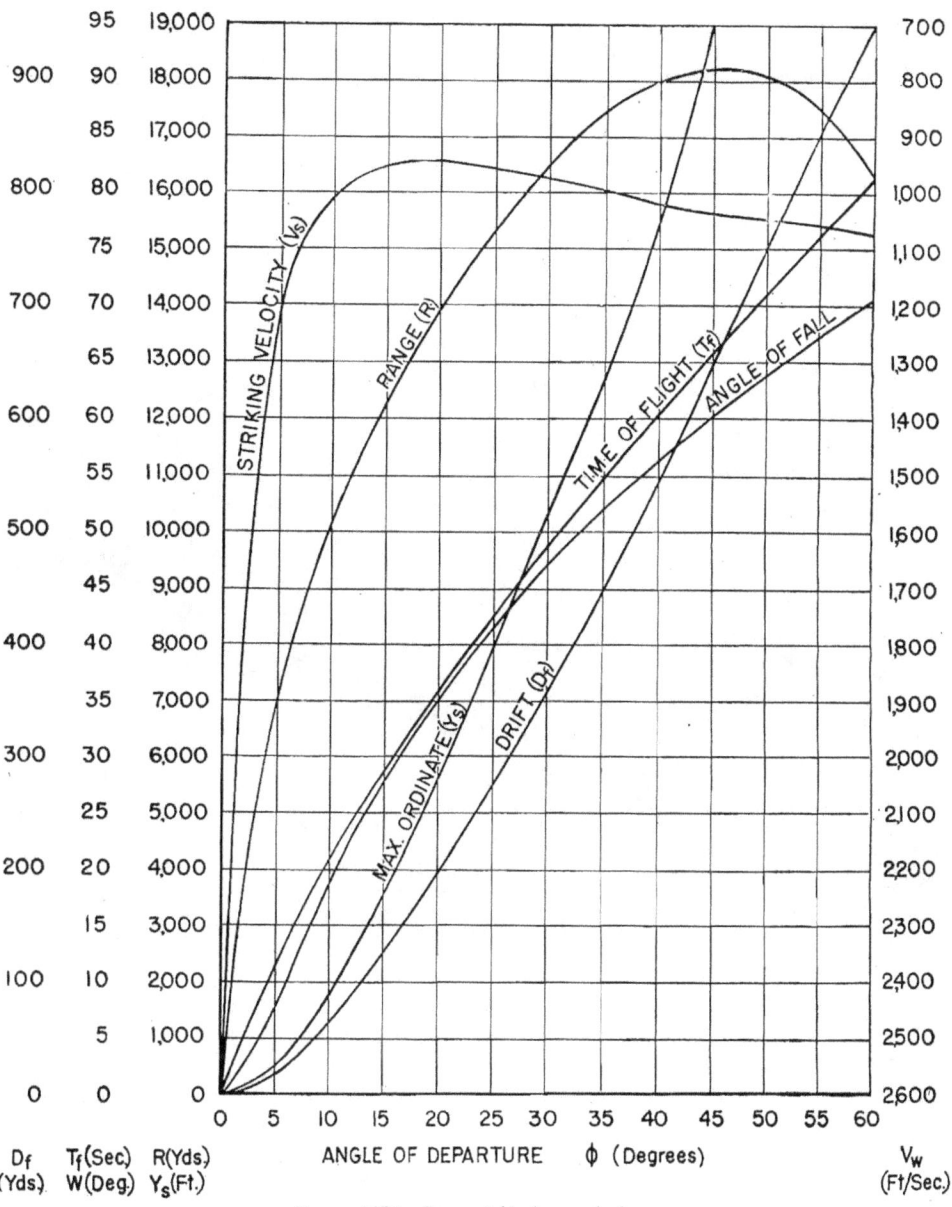

FIGURE 12B1. Range table in graph form.

C. GUN SIGHTS AND SIGHT SETTING

12C1. General. The necessity for laying a gun so that a projectile fired from it will describe a trajectory which terminates at the target is obvious. The most convenient manner of positioning the gun is with respect to a line of position which is the sighting axis of the telescopes at the gun. When the required sight angle (Vs) and sight deflection (Ds) are set between the axis of the gun bore and the sighting axis, the gun may be laid in the proper direction by pointing the sighting axis

GUN SIGHT PRINCIPLES

of the gun sight at the target. If the sighting axis is kept on the target and correct values of Vs and Ds are continuously set on the sights, the proper positioning of the gun in both train and elevation will be maintained regardless of ship motion. The elevation of the gun with respect to the deck may vary considerably due to the roll and pitch of the ship, but its elevation above the LOS depends only upon the setting of the gun sight.

12C2. Principle of an elementary gun sight. For the various types of sight mechanisms described in Chapter 4, Section D, there are several means of determining the sighting axis. In all the various mechanisms, the principles involved in offsetting the bore axis from the sighting axis by the required Vs and Ds are similar. These principles may be illustrated by a description of a simple open sight as shown in figure 12C1.

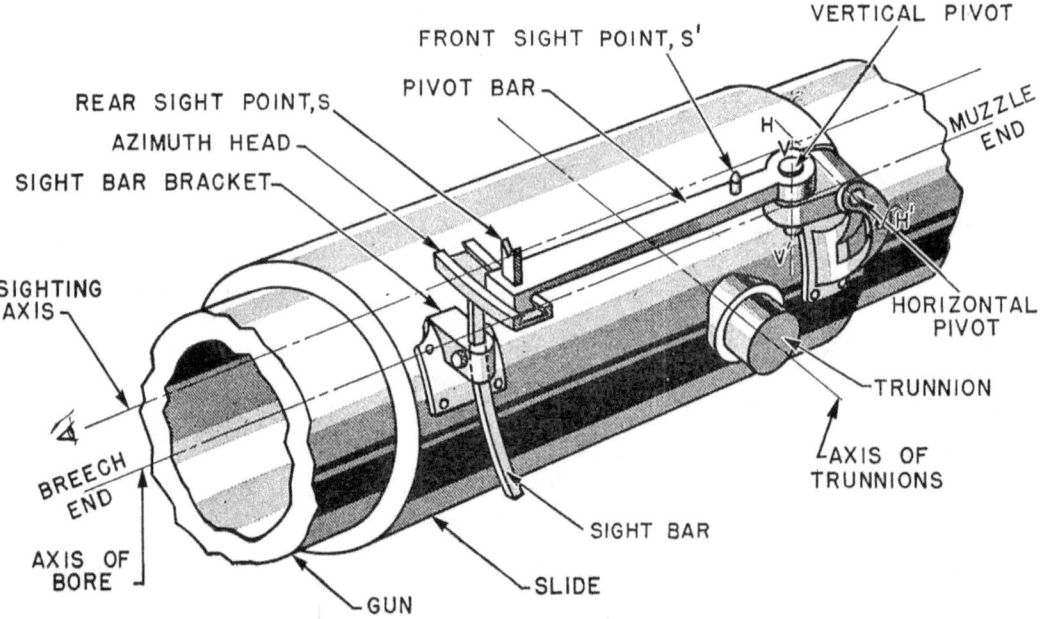

FIGURE 12C1. An elementary gun sight.

The *front* and *rear sight points* determine the sighting axis (SS'), at the end of which the eye is placed so as to sight the target. In order that the sighting axis may be set at various angles with the axis of the bore in the vertical and horizontal planes, the two sight points are placed on the *pivot bar*, which is pivoted to swing about the *vertical sight pivot* (VV'), and about the *horizontal sight pivot* (HH'). These pivots are mounted in a pivot block which is attached to the gun slide. The after end of the pivot bar slides in a curved groove in the *azimuth head*, the center of groove curvature being in the vertical sight pivot (VV'). The azimuth head is at the top of, and is an integral part of the *sight bar*, which slides vertically in the *sight bar bracket* attached to the gun slide. The sight bar is shaped to the arc of a circle whose center is in the horizontal sight pivot (HH').

If sight deflection (Ds) is set so that the rear end of the pivot bar is moved horizontally in the azimuth head, the sighting axis is offset laterally from the bore axis by the angle Ds. Similarly, if sight angle Vs is set so that the sight bar is raised in the sight-bar bracket, the sighting axis is offset vertically from the bore axis by the angle Vs. Now, when the gun is elevated and trained so that the LOS is directed at the target (assumed to be a surface target), the bore axis will be offset horizontally and vertically from the LOS by these angles.

12C3. Trunnion tilt. The *trunnion axis* is the axis about which the gun elevates. This axis intersects the bore at right angles. The *train axis* is perpendicular to the trunnion axis and is the

CHAPTER 12

axis about which the gun or mount rotates in azimuth. The gun is so installed that when the ship is on an even keel and has even trim, the trunnion axis is horizontal and the train axis is vertical.

Thus, when the sights are set, Vs is in a vertical plane and Ds is in a horizontal plane (assuming the LOS to be horizontal). However, these angles will lie in the true horizontal and vertical only when the axis of trunnions is not inclined or tilted. The inclination of the trunnion axis to the horizontal, caused by the rolling and pitching of the ship, is called *trunnion tilt*. Its effect is to move the planes in which Vs and Ds are set out of the true vertical and horizontal planes, respectively.

One of the standard conditions for range tables is that the gun is elevated in a vertical plane; that is, the trunnion axis is horizontal at the time of firing. If the trunnion axis is tilted at the instant of firing, the computed values of Vs and Ds will not produce correct horizontal and vertical offsets unless they are corrected for trunnion tilt. The magnitude and compensation of the error due to this cause are discussed in subsequent chapters.

12C4. Sight scales. A general description of sight scales is given in Section D of Chapter 4. It is only necessary here to show the relationship between setting sights in yards and in angular measure.

The range scale is so calibrated that the reading of range in yards on the range strip corresponds to the angular elevation of the gun in minutes under standard conditions. In other words, the range and minute graduations on the scale are just like the first two columns of the range table for the particular gun. Therefore, if a 5"/38 cal. gun is fired at 15,000 yards, corresponding to an angle of departure of 23 degrees 30 minutes (or 1,410 minutes), it will send the projectile this distance only if the conditions under which the shot is fired are the standard ones upon which the range table (and range scale) is based. Under the actual conditions of a particular problem the range setting (Vs), for the actual range of 15,000 yards might be 15,500 yards, corresponding to 25 degrees, 23 minutes, or 1,523 minutes.

The reading set on the deflection scale is the angular sight deflection (Ds) which includes not only the correction for drift as found in the range table for the existing range, but also corrections for variations from standard conditions. To set Ds on the scale, the sum of the corrections in yards must be converted into units of angular measure called *mils* by dividing the total correction in yards by 1/1,000 of the existing range. To move the point of fall to the right, the scale reading is raised (right, raise), and to move the point of fall to the left, the scale reading is lowered (left, lower).

D. GUN-SIGHT ALIGNMENT

12D1. General. The theoretically-perfect sight installation would be (1) a single telescope whose sighting axis intersected the axis of the bore at the trunnion axis, (2) whose sighting axis moved so that exact values of Ds and Vs were set without inaccuracies due to lost motion, and (3) whose sighting axis had only vertical and horizontal movement with respect to the bore axis.

Obviously, however, a sight telescope cannot be mounted inside the bore. Moreover, two telescopes are necessary; one for the pointer, the other for the trainer. Therefore, there are three sighting axes: the actual ones from the pointer's and trainer's telescopes, and the theoretical sighting axis intersecting the gun's axis at the trunnions. The theoretical axis cannot be directed precisely at the point of aim unless the pointer's and trainer's telescopes are converged on that point for each range. Since the error of convergence depends on the distance of the telescopes from the theoretical axis, the amount of error cannot be large because the distance between axes is only a matter of a few feet at most. However, the error can be large if the sighting axes depart excessively from parallelism, or from convergence at the desired range. The object of *bore sighting* is to establish convergence of these axes at a specified distance, such as the mean battle range for a particular battery, with the range scale set at zero and the deflection scale set at the midpoint, or no deflection.

The second desired condition is that the readings of Vs and Ds as set on the scales be actually introduced between the sighting axis and the bore of the gun. Any differences between the readings of the scales and the actual offsets are the result of *lost motion* in the gearing of the sight-setting mechanism. The effect of lost motion is to introduce an error in Vs and Ds, and such error must be compensated or its cause (lost motion) be removed.

The third desired condition is that (1) when Vs is set, no horizontal component is introduced, and (2) when Ds is set, no vertical component is introduced. Naval sight construction, typified in principle by the elementary sight in figure 12C1, fulfills this condition only when the trunnion axis and sighting axis are horizontal. The angles Vs and Ds are set in planes perpendicular to the horizontal and vertical sight pivots, which in turn are respectively parallel and perpendicular to the trunnion axis. Consequently, Vs and Ds will be truly vertical and horizontal only when the trunnion axis is horizontal.

By definition, the slant plane contains the trunnion axis and is at the same time parallel to the LOS. It can be seen in figure 12C1 that the vertical pivot is perpendicular to the LOS, the trunnion axis, and therefore, to the slant plane, regardless of the gun elevation or line of sight elevation. Thus, Ds is actually set in the slant plane, and is horizontal only when the LOS and the trunnion axis are both horizontal. In surface fire the LOS is so nearly horizontal that the setting of Ds may be assumed to occur in the horizontal. Sight installations are checked for *parallelism* to see that a Vs setting does not alter Ds and that a Ds setting does not change Vs when the slant plane is horizontal.

However, in AA fire the LOS and the slant plane usually are inclined to the horizontal, and the computation of Ds must be governed accordingly. In addition, a setting of Ds occurring when the LOS is elevated alters the effective value of Vs that is actually set on the scales. These considerations are treated more completely in Chapter 15.

12D2. Boresight apparatus. The equipment used to boresight a gun consists of the following apparatus: *a breech bar, a boresight telescope,* and *a muzzle disk.* The equipment and the method of mounting on a gun are shown in figure 12D1.

The breech bar is a precision-machined bar which can be attached to the face of the breech by two screws. There is a hole through the mid-section to receive the outer tube of the boresight telescope.

The boresight *telescope tube* is mounted within an *outside adjusting tube* which screws into the breech bar and is locked by means of the *locking ring.* Within the adjusting tube, the telescope is mounted in a *spherical bearing* which permits the telescope to be adjusted in both the horizontal and vertical planes by means of four adjusting screws so that the axis of the telescope can be made to coincide with the axis of the bore. The telescope has three rings near the eye piece: (1) the *focusing ring* for focusing the eyepiece to the individual eye, (2) the *parallax-adjusting ring* which permits the optics to be adjusted so that objects can be viewed at any distance without parallax (apparent displacement) should the eye be shifted from the optical axis of the eyepiece, and (3) the *rotating ring* which permits rotation of the telescope about its axis within the outer adjusting tube.

The muzzle disk is a circular casting designed to fit snugly in the muzzle of the gun. Through the center of the disk is a small hole and around it are four larger holes, arranged as shown in figure 12D1. Etched rings around the periphery of the disk provide means for fitting the disk in the muzzle so that it lies in a plane perpendicular to the axis of the bore. A notch is engraved on both disk and gun as index marks; one row of holes is aligned vertically, and the other horizontally with respect to the gun. The purpose of the disk is merely to assist in the alignment of the boresight telescope axis with the bore axis. After this alignment is effected, the muzzle disk is removed and has no further part in the boresighting of the gun.

12D3. Boresighting preparation. The general steps necessary in getting the gun ready for boresighting are as follows:

    1. Remove all evident lost motion from the sight mechanisms.

    2. Lash back the breech plug so that motion of the ship will not swing the plug against the boresight apparatus. Make sure that the breechblock is securely held down on gun having this type of mechanism.

    3. Ship the breech bar, boresight telescope, and muzzle disk.

CHAPTER 12

4. Remove all parallax from the boresight telescope and from the pointer's and trainer's telescopes.

5. Focus the boresight and center the crosshairs on the small center hole in the muzzle disk, using the four outer holes to align the crosshairs vertically and horizontally.

6. Remove the muzzle disk.

7. Set the range scale at zero and deflection scale at midpoint (100 or 500).

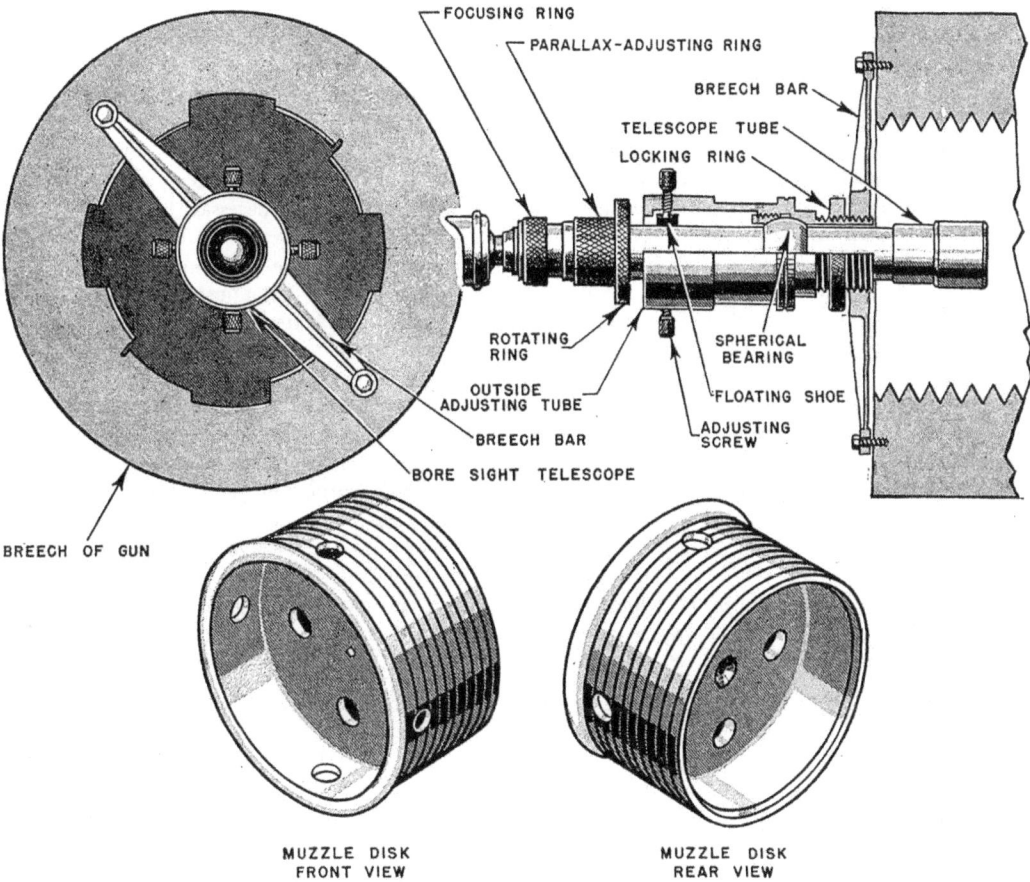

FIGURE 12D1. Boresight apparatus.

12D4. Boresighting the gun. The gun is now ready to be boresighted. The boresighting operation is accomplished simply by bringing the gun to bear on a target at the selected range so that the vertical crosshair of the boresight telescope may be aligned with a vertical mark on the target for a check in *train*. When the boresight is on, "mark" is called. If the pointer's and trainer's telescope are not on, they are adjusted in their brackets until proper correction has been made. Similarly, the horizontal crosshair of the boresight telescope is aligned with a horizontal mark on the target for a check in *elevation*. When the boresight is on, "mark" is called. If the other telescopes are not on, they are again adjusted in their brackets.

Upon completing the boresighting and tightening all parts, a check is made for looseness of parts and lost motion. The check for looseness is made by shaking all the adjustable parts and then rechecking the crosshairs on the target. If the sights are off, the loose parts must be tightened and the telescopes must again be converged on the target.

RESTRICTED

GUN SIGHT PRINCIPLES

The sights are now calibrated so that, when range and deflection are set and the gun trained and elevated, the gun is offset by the correct angles from the theoretical line of sight established by the boresight telescope, provided that there is no lost motion in the sight-setting mechanism, and that parallelism exists.

12D5. Lost motion. A final check on the calibration of the sight mechanism is made for lost motion. The sights are set to maximum range on the range scale and then returned to zero. A check is made again in elevation with the horizontal crosshairs of the telescopes on the target. If all crosshairs are not on, there is lost motion in the range-setting mechanism, and the lost motion can be removed by taking up excessive play or clearance between moving parts.

Similarly, maximum right deflection is set and then removed. A check is made with the vertical crosshairs and any necessary adjustments made. The same process is repeated, setting to maximum left deflection, returning the scale to midpoint, and checking for lost motion.

12D6. Parallelism. The check for parallelism is usually made while in a navy yard and the usual procedure is to use the batten method. Navy yards have available the necessary battens, which are merely flat boards forming a framework about 20 feet high. Such a framework is placed on deck so that the plane of the batten is parallel to a vertical plane through the gun trunnions.

The gun is secured in train, accurately leveled, the boresight installed and adjusted, and the sight scales set at zero range and no deflection. Points are spotted (on the battens) where the lines from the boresight and the gunsight telescopes appear to intersect the battens (three separate points). Next, the guns are elevated and a point is spotted near the top of the battens where the line of sight from the boresight telescope intersects. The upper and lower points for the gun determine the *elevation line* of the gun; a fine piano wire is run between the pair of points to represent this line. From the lower points established by the pointer's and trainer's telescopes, *sight lines* are marked off by stringing piano wire on either side of and parallel to the elevation line of the gun. With the gun elevated near the top of the battens, the vertical crosshairs of the pointer's and trainer's telescopes should, when range is set on the sights, move along their respective sight lines on the battens. If they do not, the sights and gun do not elevate in parallel planes.

A similar test for parallelism of the motion of the sight and gun in azimuth is made by establishing a *training line* for the gun on a horizontal batten and checking to see that the sight lines follow this line when deflection is set. It is necessary that the training line be located on the same level as the sights to be checked in order to keep the slant plane horizontal. Otherwise when deflection is set the LOS will drop and fail to follow the horizontal line even though the sights are in correct adjustment.

CHAPTER 13

THE SURFACE PROBLEM: ANALYTICAL DETERMINATION OF SIGHT SETTINGS

A. RANGE KEEPING

13A1. General. The *sight-setting problem* is the quantitative determination of sight angle (Vs), and sight deflection (Ds). It consists of two parts: (1) *the range-keeping problem,* and (2) *the ballistics problem.* The sight-setting problem is the principal part of the *fire-control problem,* for while other factors are involved, no hits will be obtained unless the angles Vs and Ds have been correctly determined.

The best available estimate of the target distance at the present instant is *present range,* and its continuous determination is the range-keeping problem. Calculation of the corrections which must be included in Vs and Ds to compensate for deviations from standard conditions and for other factors, is the ballistics problem, and its solution is based on present range. Note that the data in the range table are tabulated with respect to range.

The purpose of this chapter is to acquaint the student with the theoretical determination of sight settings for the surface fire-control problem. An analysis of the sight-setting problem is essential to an understanding of the other parts of the fire-control problem and the functions and uses of fire-control equipment.

13A2. Fundamental definitions. The following terms and symbols are defined for use in this and subsequent sections. The quantities are illustrated in figure 13A1.

SYMBOL	DEFINITION
R	*Present range.* The best available estimate of target distance at the present instant.
	True. An adjective used in describing angles. It means the angle is measured from north, clockwise from 0° to 360°.
	Relative. An adjective used in describing angles. It means the angle is measured from the bow of own ship, clockwise from 0° to 360°.
A	*Target angle.* The horizontal angle between the vertical plane containing the fore-and-aft axis of target and the vertical plane containing the LOS, measured clockwise from the bow of target.
B	*True target bearing.* The horizontal angle between a north and south vertical plane and the vertical plane containing the LOS, measured clockwise from north.
Br	*Relative target bearing.* The horizontal angle between the vertical plane containing the fore-and-aft axis of own ship and the vertical plane containing the LOS, measured clockwise from the bow of own ship.
Co	*Own-ship course.* The horizontal angle between a north and south vertical plane and the vertical plane containing the fore-and-aft axis of own ship, measured clockwise from north.
Ct	*Target course.* The horizontal angle between a north and south vertical plane and the vertical plane containing the fore-and-aft axis of target, measured clockwise from north.
Kts.	(Abbreviation for knots) *Knot.* The unit of speed equivalent to one nautical mile per hour. The nautical mile is 6,080.27 feet (about 2,027 yards).
S	*Target speed.* The speed of target with respect to the earth.
So	*Own-ship speed.* The speed of own ship with respect to the earth.

13A3. Relative target motion. In almost all cases the distance to the target is *continually* changing. There are two reasons why this distance, or range, must be kept. First, the instant of measurement comes before the instant of firing, and during the interval, the range varies. Second,

THE SURFACE PROBLEM: ANALYTICAL SOLUTION

range measurement may be interrupted. In either event, present range is required for the correct determination of Vs and Ds.

Relative target motion is the apparent motion of the target with respect to own ship, and is due to motions of both own ship and target. The target position with respect to own ship is established by the distance to the target along the line of sight, or *present range* (R), and by the target bearing from own ship, or *relative target bearing* (Br).

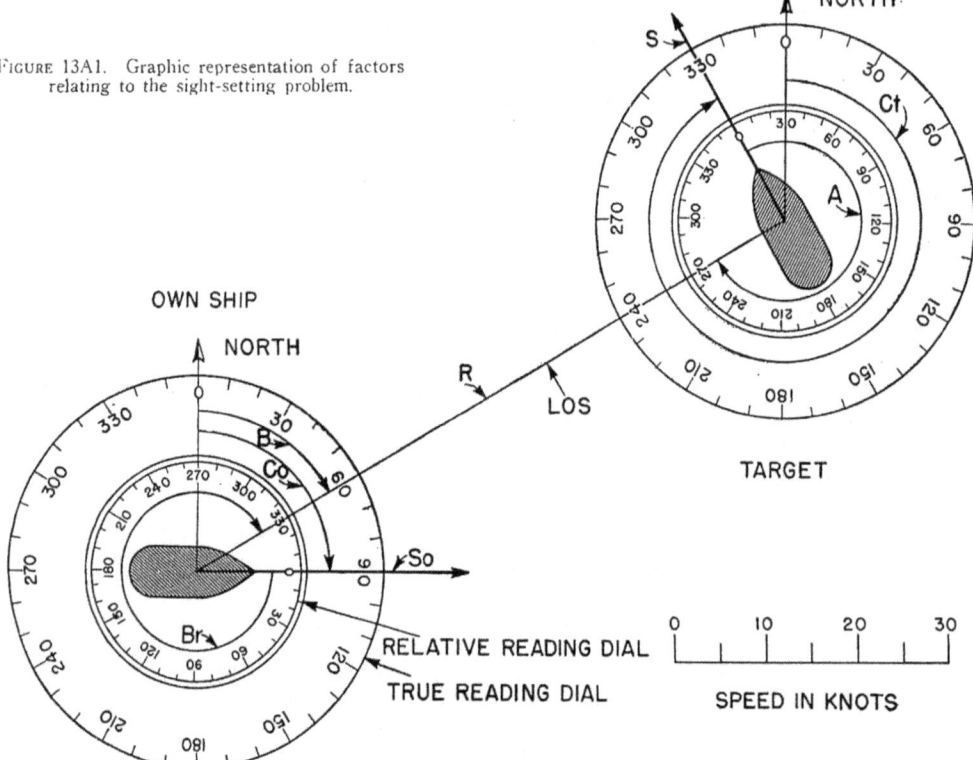

FIGURE 13A1. Graphic representation of factors relating to the sight-setting problem.

The problem of range keeping is to continuously position the target by keeping track of the values of R and Br, which are always changing except in very special and rare circumstances. While R is the quantity of principal importance in the determination of sight settings, the continuous determinations of R and Br are so intimately related that they must both be examined together.

Relative target motion is illustrated in figure 13A2. The situation represented is based on the assumption that each ship maintains its course and speed. At position 1, the relative target bearing is Br; present range is R. During time interval $\triangle T$, the ships move to position 2, where the new present range is R_1 and the new relative target bearing is Br_1. Both ships have individually moved considerable distances, but the change in relative positions has been comparatively small. The range has decreased by $\triangle R$; the bearing has been changed by $\triangle Br$. Over a period of time, these small changes accumulate and result in a large change in relative position.

13A4. Evaluating relative target motion. The velocities of the two ships are represented by arrows called vectors. A velocity has *speed* and *direction*, hence:

1. The direction of each vector is along the course of the ship it represents (Ct and Co).
2. The length of each vector is proportional to the speed of the ship it represents (S and So).

Any suitable scale may be used.

CHAPTER 13

Components. Each velocity has components *in* and *across* the LOS; that is, the target motion shown in figure 13A2 is partly toward own ship and partly to the right as seen from own ship, and the own-ship motion is partly away from and partly to the right of the target. The components and their symbols are:

SYMBOL	DEFINITION
Yo	*Line component of own-ship velocity.* The horizontal component of own-ship velocity in the vertical plane containing the LOS.
Yt	*Line component of target velocity.* The horizontal component of target velocity in the vertical plane containing the LOS.
Xo	*Cross component of own-ship velocity.* The horizontal component of own-ship velocity perpendicular to the vertical plane containing the LOS.
Xt	*Cross component of target velocity.* The horizontal component of target velocity perpendicular to the vertical plane containing the LOS.

Rates. These components are combined to obtain the rates at which range and bearing are changing. Thus:

SYMBOL	DEFINITION
dR	*Range rate.* The time rate of change of range. It is the algebraic sum of Yo and Yt.
RdBr	*Linear relative-bearing rate.* The horizontal component of relative target motion perpendicular to the vertical plane containing the LOS. It is the algebraic sum of Xo and Xt.
dBr	*Relative angular-bearing rate.* The time of change of relative target bearing. It is the angular rate corresponding to the linear rate.

These rates have only *instantaneous* values because they are continuously changing. Inspection of figure 13A2 shows that the values of X and Y components at position 2 are different from the corresponding values at position 1. Since the components vary, the rates which are based upon them necessarily are not constant. Knowing dR and dBr, changes in R and Br during a time interval can be computed.

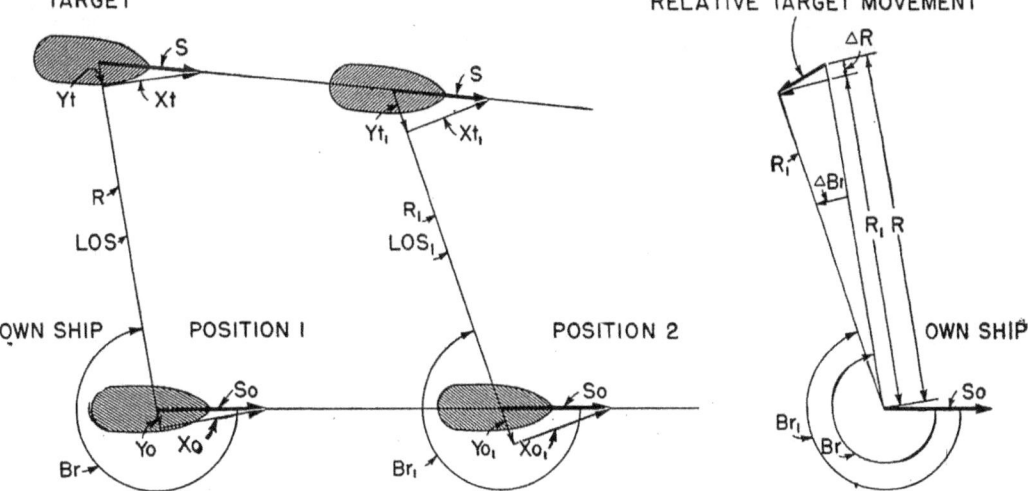

FIGURE 13A2. Relative target motion.

13A5. Plotting the problem. The analysis of rates is most conveniently made with the aid of a vector diagram in which the motions of own ship and the target are represented by vectors plotted with respect to the line of sight. The type of diagram required for the range-keeping

THE SURFACE PROBLEM: ANALYTICAL SOLUTION

problem is shown in figure 13A3. As a matter of practice, the LOS is drawn up and down on the paper, and all other parts of the diagram are placed with respect to the LOS. Own ship is always placed at the lower end; target at the top.

The diagram for a problem may be plotted by carrying out the following steps:

1. Draw and label the LOS and place a dot at each end to indicate own ship (below) and target (above).

2. Through own ship dot draw a line which makes the angle Br with the LOS. To establish Br, measure counterclockwise from the LOS, but label it with a clockwise arrow in accordance with the definition of Br.

3. Using any suitable scale, make the length of this line proportional to So. Thus, if the scale is 5 knots to ¼ inch, the length for So = 20 knots is 1 inch. At the end of this line place an arrowhead and label the arrow So. The arrow then becomes the vector representing own-ship motion, and a ship outline may be added to the diagram. (If the drawing is made to scale, computations may be checked by measurement.)

4. Establish the direction of true north by drawing an arrow which makes the *clockwise* angle Co with the own-ship vector, measured from the north arrow. This angle also may be conveniently established by measuring counterclockwise, but it must be labelled with a clockwise arrow. (This step merely provides a convenient means of determining Ct when A is known.)

5. Establish the direction of target motion by one of the following methods:

   a. Draw a line through the target dot making the *clockwise* angle A from that line to the LOS, following the procedure used for Br.

   b. Establish a true north arrow through target and draw a line making the *clockwise* angle Ct with it, measured from north.

6. Employing the same scale used for So, complete the target vector by making its length proportional to S, and label it S. Add the target outline if desired.

13A6. Determining X and Y components. The usual procedure in determining the components in and across the LOS is:

1. Drop perpendiculars to the LOS from the ends of the vectors (extending the LOS if necessary). Place arrowheads on the components pointing in the directions corresponding to the directions of the vectors*.

2. Determine the acute angles between the vectors and the LOS. These angles are θo and θt shown on the diagram. They are obtained by inspection, taking the difference between 180° or 360° and A or Br as conditions may require. If either A or Br is the acute angle, use its value for θ.

3. Compute the values of the components from these equations:

$$Yo = So \cos \theta o \qquad Xo = So \sin \theta o$$
$$Yt = S \cos \theta t \qquad Xt = S \sin \theta t$$

13A7. Range rate. Range rate (dR) is the algebraic sum of Yo and Yt. A line component is called *decreasing* when it tends to decrease the range; *increasing* when it tends to increase the range. Range rate is similarly called decreasing or increasing, depending upon its net effect on the range. Thus in figure 13A3a, own ship is opening the range, Yo is increasing, the target is closing the range, and Yt is decreasing. Since Yt is greater than Yo, dR is decreasing. It should be noted that *decreasing* and *increasing* as used here do not refer to the change which is taking place in the instantaneous value of the component or rate. The terms simply refer to the effect that the component or rate has on range.

The rate obtained by combining Yo and Yt in knots should be converted to yards per second for reasons which will be apparent later. The transition is made by using this relationship (see fig. 13A3):   dR (yards per second) = .563(Yo + Yt) (knots)

---

* The diagram up to this point is sufficient to permit the determination of the components from these equations:
   Yo=So cos Br   Xo=So sin Br
   Yt=S cos A   Xt=S sin A

CHAPTER 13

13A8. Bearing rate. The linear bearing rate RdBr is the algebraic sum of Xo and Xt, and must be established before the angular rate can be determined. Each X component is decreasing when it causes a decrease in Br. and increasing when it causes an increase in Br. If the algebraic sum of Xo and Xt is decreasing, RdBr is decreasing and Br is decreasing. If the sum is increasing, Br is increasing.

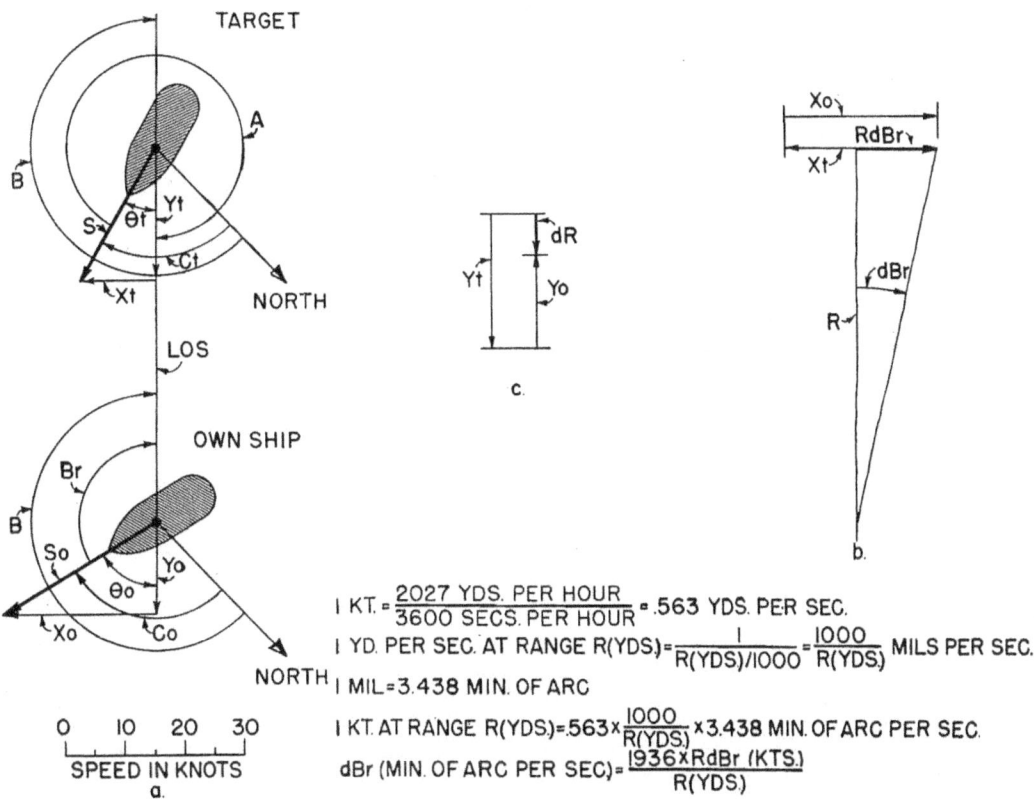

FIGURE 13A3. Range-keeping diagrams.

The effect of each component is most readily determined by considering the motion of one ship at a time, and assuming the other ship to be motionless. Thus, in figure 13A3a, assuming target motionless for the moment, own ship is moving to the left, Xo causes the LOS to shift to the right. This increases Br, hence Xo is increasing. Again, assuming own ship to be motionless, target moves left, the LOS moves left, Br decreases, and Xt is decreasing. The effect of the larger of Xo and Xt becomes the effect of RdBr.

The angular rate should be expressed in minutes of arc per second. The transition from linear to angular measure is accomplished by dividing the linear rate by the range, using the proper units. This fact explains the reason for using R in the symbol RdBr, for dBr = RdBr ÷ R as indicated in figure 13A3. The relationship o use is:

$$\text{dBr (minutes of arc per second)} = 1936 \times \text{RdBr (knots)} \div \text{R(yards)}$$

13A9. Use of range and bearing rates. Range can be *kept* by multiplying the rates by elapsed time to obtain increments of range ($\triangle R$) and bearing ($\triangle Br$). When these increments are added algebraically to the initial values of R and Br, new values are obtained. New rates, and corresponding new increments can be computed on the basis of the new R and Br to *keep* the range and bearing during the next interval of time. By making the process continuous, present range

THE SURFACE PROBLEM: ANALYTICAL SOLUTION

can be generated with no further measurements after the initial determination. Thus, present range will be continuously available even though there may be interruptions in ranging.

The precise determination of the changes by this process requires integration or careful plotting and measurement. Neither of these methods is essential to the elementary analysis of the problem, and an approximate method is employed as explained in the next article. The rates for a given instant are computed and then multiplied by the time interval to obtain the changes. Since the rates are instantaneous, an error is introduced in proportion to the length of the time interval used. The shorter the time interval, the smaller will be the error.

13A10. Example of range keeping. The following example illustrates some of the basic principles involved in the problem of "keeping" range, and will serve as a guide in working range-keeping exercises:

Given: $R = 15,000$ yards; $A = 45°$; $Br = 300°$; $So = 30$ knots; $S = 20$ knots.

Required: 1. dR in yards per second and dBr in minutes of arc per second.
2. R and Br 30 seconds later and 60 seconds later, assuming dR and dBr remain constant for 60 seconds.
3. R and Br 60 seconds later, assuming dR and dBr remain constant for 30 seconds.
4. The magnitude of the errors introduced in the values of R and Br computed for 60 seconds later by assuming dR and dBr remain constant for 60 seconds instead of 30 seconds.

Solution to (1): Plot the vector diagram as shown in figure 13A4, indicating the effect of each component by (D) for decreasing and (I) for increasing. Then compute the components as follows:

$Yo = 30 \cos 60° = 30 \times .500 = 15.00$ knots (D)
$Yt = 20 \cos 45° = 20 \times .707 = 14.14$ knots (D)

$dR = \quad 29.14$ knots (D)
$dR = .563(29.14) = \quad 16.4$ yards per second (D)
$Xo = 30 \sin 60° = 30 \times .866 = 25.98$ knots (D)
$Xt = 20 \sin 45° = 20 \times .707 = 14.14$ knots (I)

$RdBr = \quad 11.84$ knots (D)
$dBr = 1936(11.84) \div 15,000 = 1.5'$ per second (D)

Solution to (2): compute the changes in R and Br during the time intervals, and apply them to the present values:

$\triangle R_{30} = 30(16.4) = 492$ yards (D)
$R_{30} = 15,000 - 492 = 14,508$ yards
$\triangle Br_{30} = 30(1.5) = 45'$ (D)
$Br_{30} = 300° - 45' = 299°15'$
$R_{60} = 14,508 - 492 = 14,016$ yards
$Br_{60} = 299°15' - 45' = 298°30'$

Solution to (3): Using R_{30} and Br_{30} from the solution to (2), determine the rates existing at the end of the first 30 seconds. Use these new rates to compute the changes during the second 30 second period. Apply these changes to R_{30} and Br_{30}.

$Yo = 30 \cos 60°45' = 30 \times .489 = 14.67$ knots (D)
$Yt = 20 \cos 44°15' = 20 \times .716 = 14.32$ knots (D)

$dR_{30} = \quad 28.99$ knots (D)
$dR_{30} = .563(28.99) = \quad 16.3$ yards per second (D)
$\triangle R_{60} = 30(16.3) = \quad 489$ yards (D)
$R_{60} = 14,508 - 489 = \quad 14,019$ yards
$Xo_{30} = 30 \sin 60°45' = 30 \times .873 = 26.19$ knots (D)
$Xt_{30} = 20 \sin 44°15' = 20 \times .698 = 13.96$ knots (I)

CHAPTER 13

$RdBr_{30} =$ 12.23 knots (D)
$dBr_{30} = 1936(12.23) \div 14{,}508 =$ 1.6' per second (D)
$\triangle Br_{60} = 30(1.6) =$ 48' (D)
$Br_{60} = 299°15' - 48' =$ 298°27'

Solution to (4): Obtain the differences between the values computed for R_{60} and Br_{60} in solutions to (2) and (3). The differences are the errors.

Error in $R_{60} = 14{,}019 - 14{,}016 = 3$ yards
Error in $Br_{60} = 298°30' - 298°27' = 3'$

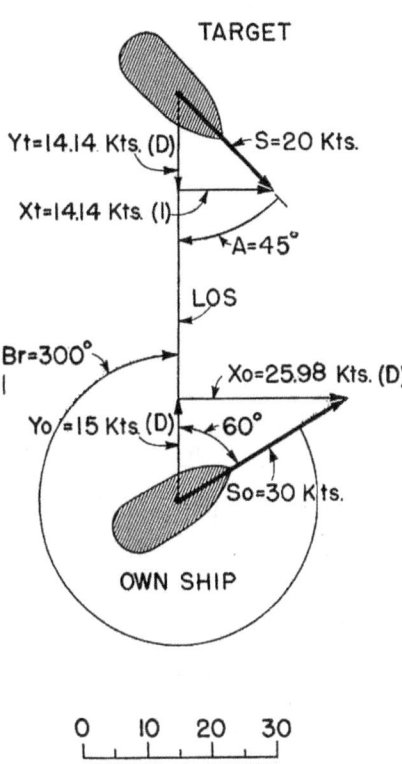

FIGURE 13A4. Range-keeping example.

13A11. Examination of errors. While it is not sound practice to establish rules on the basis of results from a single example, it appears that the above errors are small enough to be neglected for purposes of *predicting* range for short intervals such as the time of projectile flight in the surface fire-control problem. Obviously, higher rates due to faster target speeds will cause greater errors if rates are assumed to be constant, and in the antiaircraft fire-control problem, these errors cannot be neglected.

It should be noted that the errors are cumulative, and that with each additional time interval the total error will increase. For example, if $\triangle R$ and $\triangle Br$ are determined for a 5 minute interval using rates constant for one minute intervals, the error will be much greater than if the rates are revised every 30 seconds.

The errors can be reduced by shortening the time intervals to 10 seconds or even to 1 second. They can be eliminated only by reducing the interval to an infinitesimal value. Such a procedure becomes a form of integration, which cannot be examined here. However, mechanical range keepers perform this process of integration, which is essential in *keeping* range. The manner in which they perform this function is explained in the next chapter.

B. THE GUN BALLISTIC

13B1. The ballistics problem. The determination of sight angle (Vs) and sight deflection (Ds), based on present range, is the *ballistics problem*. This involves the evaluation of the elements which, when combined, give the values of Vs and Ds that will result in a hit.

There is an angle of departure for *standard conditions* corresponding to each possible target distance, or present range, within the limits of the maximum range of a gun. Standard conditions are never obtained, and compensating corrections must be applied to the standard angle of departure to obtain the value of Vs which will provide the trajectory required to hit a target at a given range. Similarly, corrections must be computed to obtain the correct value of Ds.

13B2. Ballistics definitions. The general meaning of *ballistics* has been previously defined as the science of projectile motion. In the Navy the term has been given this specific meaning for use in fire-control work:

RESTRICTED

THE SURFACE PROBLEM: ANALYTICAL SOLUTION

Ballistics are computed corrections to be applied to the measured or estimated values of range and to the midpoint of the deflection scale to compensate for known or predicted errors and for variations from selected standard conditions.

Each of the several individual corrections is a ballistic, hence the word *ballistic* alone needs modification before it has specific meaning. There are several types of ballistics which are grouped together to form the *gun ballistic, control ballistic,* and *arbitrary ballistic.*

Total ballistic is the correction determined prior to firing to be applied to the present range and in deflection to the midpoint of the deflection scale to obtain the sight-bar range, deflection, and, in antiaircraft fire, the fuze setting. It includes the gun ballistic, arbitrary ballistic, and control ballistic.

All ballistics are, for convenience, initially computed as linear quantities expressed in yards. It must be remembered that they are for inclusion in the angles Vs and Ds, and must eventually be converted into angles. Thus, Vs is an angle representing the sum of standard angle of departure *and* the angular equivalent of the *total ballistic in range.* The deflection angle Ds is composed entirely of corrections, and is therefore the angular equivalent of the *total ballistic in deflection.* These relations are illustrated in figure 13B1.

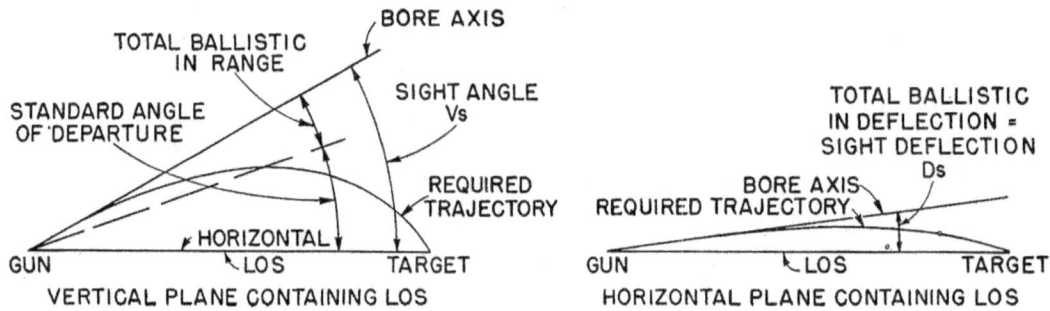

FIGURE 13B1. Angular equivalents of total ballistics.

13B3. The gun ballistic. The gun ballistic takes into account the calculable corrections which must be included in Vs and Ds, under the normal conditions of laying a gun with respect to a line of sight which coincides with the line of position. Incalculable corrections are considered in the arbitrary ballistic, while abnormal conditions of laying the gun are considered in the control ballistic. Thus:

Gun ballistic consist of the corrections to compensate for the following variations from standard conditions: (1) initial velocity errors due to temperature of powder, gun erosion, and variation of projectile weight; (2) errors due to atmospheric conditions of barometric pressure, temperature, and wind; (3) errors produced by relative movements of gun and target; (4) drift.

Those elements of the gun ballistic usually considered may be divided into two groups for convenience. Group A includes elements causing errors which are *independent* of target bearing. Group B includes elements causing errors which are *dependent* on target bearing. The errors are examined individually and then combined to determine the compensating corrections. It is helpful in studying each error, to consider that all other conditions are standard.

Errors in range are *shorts* or *overs,* depending on whether they tend to cause the projectile to fall short of or beyond the target. Errors in deflection are *rights* or *lefts,* depending on whether they tend to cause the projectile to fall to the right or left of the target.

13B4. Use of range tables. Range tables provide the data required for the computation of the gun ballistic (and part of the control ballistic which is considered later). The 5"/38 cal. gun

CHAPTER 13

is the basis for the examples in this section. The range table for this gun is provided in Appendix D at the back of the book and should be referred to when the examples are studied.

The range to use in entering the range table is the best available estimate of the target distance at the instant of firing. It is assumed for the present that the ballistics computations can be made, the sight settings determined and applied, and the guns fired instantaneously. Therefore, present range is used in the examples. The problem used as an example is a continuation of that presented in the preceding section, in which R is 15,000 yards.

13B5. Group A errors. The elements causing errors not dependent on target bearing are :*

1. Variations in I. V. due to powder temperatures.
2. Variations in I. V. due to erosion.
3. Variations in air density.
4. Drift.

The first three of these elements cause errors in range, the fourth is an error in deflection.

13B6. Variations in I. V. due to powder temperature. The standard powder temperature upon which the range tables are based is 90° F. If the powder used has a lower temperature, as it normally does, the projectile will leave the gun at an I. V. lower than standard. This will cause a *short*, for the projectile will fall short of the range indicated in the range table. Similarly, a powder temperature higher than 90° F. will cause an *over*. The temperature to use is the average magazine temperature over a period of several days, for the powder changes temperature slowly.

It has been determined experimentally that for guns larger than 3″, each 1° F. difference between actual and standard powder temperature produces a change of approximately 2 f.s. I. V. Column 10 of the range table supplies the data needed to convert this loss in I. V. to an error in yards. The values in this column are obtained from the ballistic table and tabulated for a 10 f.s. variation in I. V. The data in this column are sufficiently accurate for determination of the error in point of fall for reasonable variations from standard I. V., but cannot be used for such large variations as those between service and reduced charges.

For the 15,000 yard range of the selected example, the range table indicates an increase of 50 yards for each 10 f.s. increase in I. V. A decrease in I. V. results in the same amount of error, but in the opposite direction. Thus, if the powder temperature is 73°, the range error is $(90 - 73) \times 2 \times 50 \div 10 = 170$ yards *short*.

13B7. Variations in I. V. due to erosion. Erosion always causes a loss in I. V., which results in a *short*. The amount of this error is also obtained from column 10, after the loss in I. V. has been determined from either a *rounds-fired curve* or *bore-enlargement curve*, both of which are prepared by the Naval Proving Ground. These curves, while not standard parts of range tables, have been included in Appendix D for purposes of illustration.

Rounds-fired curve. A record of rounds fired is kept for each gun through its useful life. A service charge is considered as one round, while a reduced charge is counted as less than one round. A target charge, for example, causes but a fraction of the erosion produced by a service round. The equivalent value of reduced charges depends upon the gun and charge, and may be in the neighborhood of one-fourth of a service charge. The values to use are established by the Naval Proving Ground.

Using the *equivalent* rounds fired, the rounds-fired curve is employed to determine the approximate velocity loss. For example, if the equivalent rounds fired for the 5″/38 cal. gun is 450, the velocity loss from the curve is 60 f.s. The error in range due to this loss is $60 \times 50 \div 10 = 300$ yards *short*.

Bore-enlargement curve. A more accurate indication of the loss in I. V. is provided by the amount of bore enlargement because a given number of rounds fired does not always produce the

---

*Variations in projectile weight would also belong in this group. A correction for this element is usually unnecessary; hence this variation is discussed in Article 13D4 with other less-frequently used corrections.

RESTRICTED

same erosion in similar guns. Bore enlargement is measured with a star gauge when facilities are available, and the velocity loss at that time is established from the bore-enlargement curve. Loss from subsequent firings is determined from the rounds-fired curve, until the bore can again be star gauged. As an illustration, if the bore enlargement is .074 inches, the velocity loss is 60 f.s., and the error in range due to this cause is $60 \times 50 \div 10 = 300$ yards *short*.

13B8. Variations in air density. The data in the range table are based on an arbitrarily chosen ballistic density corresponding to the surface density existing when the temperature is 59° F., the barometric pressure is 29.53 inches of mercury, and the air is 78 per cent saturated with moisture. Variations in pressure, temperature, or moisture content cause changes in air density. Consideration is given to all three factors when suitable aerological observations are available, but the effects of moisture variations need not be discussed here.

Any deviation from the assumed standard density will cause an error in range. A density higher than standard will cause a *short*, while a density lower than standard will cause an *over*. The standard density is assigned a numerical value of 1.000, and actual densities are expressed in terms of this standard. Thus, a density of 1.043 is 4.3 per cent higher than standard, and will cause a *short*.

Column 12 of the range table, obtained from computations based on the ballistic table, lists the increase in range for a —10 per cent variation in air density. The data is applicable to both plus and minus variations, but is listed for the minus variation to avoid minus signs in the column. Thus, the error at 15,000 yards due to a density of 1.043 is $4.3 \times 792 \div (-10) = -340$ yards, or 340 yards *short*.

Ballistic density. The density which should be used in computing the error from column 12 is the *ballistic density*. Since temperatures and pressures are different at various altitudes, the density in each altitude zone through which the projectile travels is different from that at the surface. The error in range in each zone due to variation of zone density from standard density depends upon the time the projectile remains in that zone. This time, in turn, is determined by the maximum ordinate of the trajectory. The ballistic density takes into account the various zone densities and the time the projectile remains in each zone. It is the fictitious density of constant magnitude which would have the same effect on the projectile as the sum of the effects of the actual densities at the various altitudes.

The aerological party assigned to certain ships or Fleet units makes daily aloft observations for the data needed to determine the ballistic density. When aloft observations are not available, the ballistic density is established from the surface density on the basis of the means of many previously-established densities. It is present practice to incorporate these data in a *monogram*, or alignment chart which provides a convenient approximation of the range error due to variations in air density.

Use of the nomogram. The nomogram provided in the range table (Appendix D) is an example of the type of chart used. Instructions for its use are also included. Assuming the surface temperature is 47° F. and the surface pressure is 30.20 inches for the range of 15,000 yards, the chart gives —340 yards error in range. Alignment of 47° and 30.20 inches provides a point on the support, which is the equivalent of determining the surface density. Alignment of that point with 15,000 on scale (R) gives a multiplier of —.43 on scale M, which is the equivalent of determining the ballistic density corresponding to the surface density. Alignment of the multiplier with 15,000 on scale (R_1) indicates the error in range is —340 yards. The last step is the equivalent of multiplying the error from column 12 by the ratio of the per cent variation from standard to the 10 per cent variation upon which the column is based, or $4.3 \times 732 \div (-10) = -340$ yards.

13B9. Drift. The lateral deviation due to drift is tabulated in column 6 of the range table. Drift is always to the right, and is simply picked from the table for the appropriate range. Thus at 15,000 yards the error due to drift is 248.0 yards *right*.

CHAPTER 13

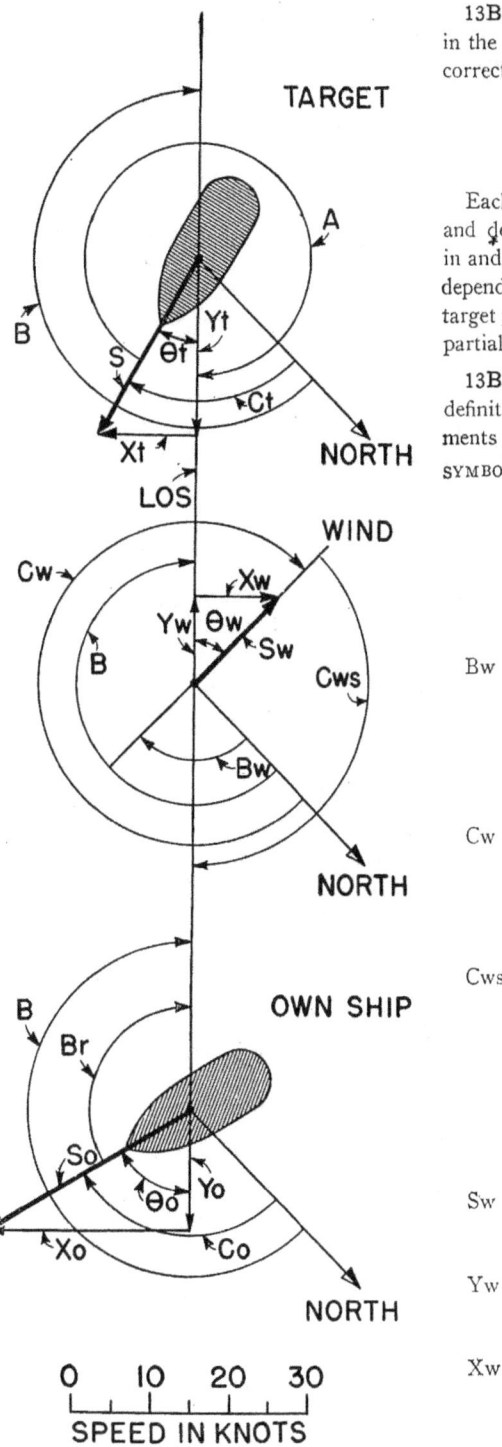

Figure 13B2. Line-of-fire diagram.

13B10. Group B errors. The remaining factors in the gun ballistic usually considered in computing corrections are:
1. Wind.
2. Motion of target.
3. Motion of own ship.

Each of these factors causes an error in range and deflection, for each generally has components in and across the LOS. The size of the components depends to some extent upon the bearing of the target; hence the errors due to such influences are partially dependent upon target bearing.

13B11. Definitions. The following additional definitions are needed before the analysis of elements in group B can be made:

SYMBOL	DEFINITION
	True wind. The actual wind blowing with respect to the earth. This is the usual method of describing wind.
	Wind course. The direction toward which the wind is blowing.
Bw	*True direction true wind.* The horizontal angle between north and the direction *from* which the true wind is blowing, measured clockwise from north. Thus, a south wind blows *from* the south, or 180°.
Cw	*True course true wind.* The horizontal angle between north and the direction *toward* which the true wind is blowing, measured clockwise from north.
Cws	*Wind angle.* The horizontal angle between the direction *toward* which the wind is blowing (Cw) and the vertical plane containing the LOS, measured clockwise from the direction *toward* which the wind is blowing to that part of the LOS nearest own ship (See fig. 13B3).
Sw	*True wind speed.* The horizontal speed of wind with respect to the earth. (Speed of true wind.)
Yw	*Range wind.* The horizontal component of true wind in the vertical plane containing the LOS.
Xw	*Cross wind.* The horizontal component of true wind perpendicular to the vertical plane containing the LOS.

275   **RESTRICTED**

THE SURFACE PROBLEM: ANALYTICAL SOLUTION

13B12. Line-of-fire diagram. An examination of the effects of the motions of wind, target and own ship on the gun ballistic requires a *line-of-fire diagram*. It is a vector diagram, the same as that used in the range-keeping problem, except that the vector representing wind must be added to it. Using figure 13A3 as a basis, the completed line-of-fire diagram appears like that in figure 13B2.

Knowing true direction true wind, the wind vector may be added in these steps:

1. Place a dot labeled *wind* on the LOS near its center, and establish a true north arrow through the dot.
2. Draw *through* the dot a line making the angle Bw with the north arrow, measured *clockwise* from north. Label the angle with a clockwise arrow.
3. Along this line, 180° from Bw, draw the wind vector, preferably to scale, and label it Sw. Indicate the angle from north to the vector with a clockwise arrow, labeling it Cw. This vector now represents the *course* of the wind, and thus gives an indication of the action of the wind on the projectile.
4. Determine Cws by subtracting Cw from B+180° (or B+540°), and label it with a clockwise arrow from the wind vector to the end of the LOS nearest own ship.
5. Draw the line and cross components Xw and Yw, with arrowheads, just as the own-ship and target components were drawn*.
6. Determine the acute angle θw between the vector and the LOS by inspection, taking the difference between 180° or 360° and Cws as necessary. If Cws is the acute angle, use its value for θw.
7. Compute the value of Xw and Yw from these equations:

$$Yw = Sw \cos \theta w \qquad Xw = Sw \sin \theta w.$$

13B13. Example. The example used in the discussion of the group B elements is a continuation of that used for the range-keeping problem (see Article 13A10). Data required for the additional element of wind is included in the summary below. The appropriate line-of-fire diagram is shown in figure 13B3, which is figure 13A4 with the wind added.

Given: R = 15,000 yards   Co = 325°
A = 45°   So = 30 knots
Br = 300°   S = 20 knots
Bw = 320°   Sw = 15 knots

Required:
The errors in point of fall due to wind, motion of target and motion of own ship (gun).

13B14. Wind. Range wind blowing in the direction of projectile travel (toward target) increases the range attained, or causes an *over*. When blowing against the direction of projectile travel (toward own ship), it causes a *short*. Similarly, a cross wind blowing from left to right causes a *right*, and one blowing from right to left causes a *left*. After the components have been added to the line of fire diagram, it is advisable to indicate the *errors* they will cause. Thus, in figure 13B3, wind will cause a *short* and a *left*.

The determination of the effects of wind is not difficult, but it is too involved a process for discussion in this text. It is sufficient to state that evaluation of the effects is made by considering (1) the motion of the projectile with respect to the moving air, and (2) the motion of the air itself. The effects of these separate motions can be computed from data already provided in other columns of the range table, without recourse to the ballistic table. The results are combined, then tabulated for 10 knot components for range wind in column 13 and for cross wind in column 16.

Thus, the *errors* due to wind in the example are:

1. In range: Yw = 15 cos 55° = 15 × .574 = 8.6 knots.
               8.6 × 188(Col. 13) ÷ 10 = 162 yards *short*.
2. In deflection: Xw = 15 sin 55° = 15 × .819 = 12.3 knots.
               12.3 × 120(Col. 16) ÷ 10 = 148 yards *left*.

---
*The components may be determined from these equations:
   Yw = Sw cos Cws   Xw = Sw sin Cws

CHAPTER 13

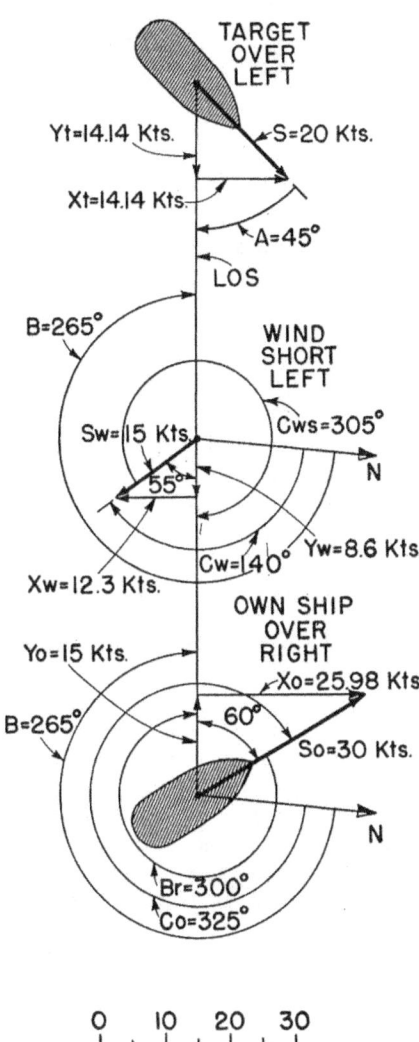

FIGURE 13B3. Line-of-fire diagram for example.

Ballistic wind. The data in columns 13 and 16 are based on horizontal wind considered to be of uniform speed and direction throughout the trajectory. *Ballistic wind* should be used to compute the errors due to wind in order to account for variations in wind speed and direction in different altitude zones. Ballistic wind is a fictitious wind of constant magnitude which would have the same effect on the projectile as the sum of the effects of the actual winds throughout the trajectory. Obviously, different maximum ordinates associated with appreciably different ranges require separate determinations of ballistic wind. While vertical wind components exist, their effect is not considered because of the difficulty of measuring and correcting for them.

The process of establishing ballistic wind is somewhat similar to that of determining ballistic density. It involves aloft observations, weighting of the zone winds, and resolution of the weighted winds into a single equivalent wind. However, while aloft densities have a reasonably definite relation to surface density, aloft winds have no such relation to surface wind. Ballistic wind, therefore, can not be approximated from measurement of surface wind alone. It is quite possible that ballistic wind will have at least one component whose direction is opposite to the corresponding component of surface wind.

Ballistic wind is established by the aerological unit if one is available, or it may be approximated by tracking a balloon with an AA director. It may even be based on estimates of aloft winds. If no ballistic wind is available, surface wind is used.

13B15. Motion of target. Motion of the target in no way affects the trajectory or flight of the projectile. However, during the time of flight, the target moves away from the position it occupied at the instant of firing. This motion, unless compensated in the gun ballistic, will cause the actual point of fall to be in error with respect to the desired point of fall.

If target motion in the LOS (Yt) is toward own ship, it will cause an *over*, or if away from own ship, it will cause a *short*. If target motion across the LOS (Xt) is toward the right of the LOS, it will cause a *left*. Similarly, if Xt is to the left, it will cause a *right*. The simplest way to determine the direction of these errors is to examine the line-of-fire diagram and indicate on it the direction of the *errors*, as in figure 13B3.

The errors for 10 knot components are identical in magnitude for both line and cross components, and are tabulated in columns 15 and 18 of the range table. The data in these columns are obtained by multiplying 10 knots by the time of flight (Tf) for each range. Thus, at 15,000 yards, Tf = 40.34 seconds, and the movement of the target during Tf is $40.34 \times 10 \times .563 = 227$ yards, as shown in the table.

THE SURFACE PROBLEM: ANALYTICAL SOLUTION

The *errors* due to target motion in the example are:
1. In range:
$$Yt = 20 \cos 45° = 20 \times .707 = 14.14 \text{ knots.}$$
$$14.14 \times 227(\text{Col. 15}) \div 10 = 320 \text{ yards } over.$$
2. In deflection:
$$Xt = 20 \sin 45° = 20 \times .707 = 14.14 \text{ knots.}$$
$$14.14 \times 227(\text{Col. 18}) \div 10 = 320 \text{ yards } left.$$

13B16. Motion of own ship. The motion of own ship imparts component horizontal velocities to the projectile in proportion to the components of own-ship motion in and across the LOS. The component in the LOS changes the initial velocity produced by the powder charge, but it cannot be directly added to I. V., which is measured along the elevated line of departure.

If own-ship motion in the LOS (Yo) is toward the target, it will cause an *over,* or if away from the target, it will cause a *short.* The sidewise motion imparted to the projectile by own-ship motion across the LOS will cause a *left* if Xo is toward the left, a *right* if Xo is toward the right. The effects of Yo and Xo also should be indicated on the diagram, as in figure 13B3.

The errors for 10-knot components are tabulated in column 14 for motion in the LOS and column 17 for motion across the LOS. The values are computed by subtracting the error due to a 10-knot wind from the error due to a 10-knot target motion. Thus, at 15,000 yards:

$$\text{Column 14} = \text{Column 15} - \text{Column 13} = 227 - 188 = 39 \text{ yards.}$$
$$\text{Column 17} = \text{Column 18} - \text{Column 16} = 227 - 120 = 107 \text{ yards.}$$

The discrepancies of a yard or so are due to the "rounding off" process used in the preparation of these data.

If the gun were fired in a vacuum, the additional horizontal velocity would remain constant throughout the time of flight, and the effect of own-ship motion would be equal to the effect of target motion.

However, the gun is fired in air, which for purposes of this discussion is assumed to be motionless. The motion of own ship creates a wind apparent to a person on the ship, which also influences the projectile flight. The apparent motion of this wind is equal in speed but opposite in direction to ship motion. Its effect is to partially nullify the extra velocity imparted by own ship, for it acts on the projectile just as though it were an actual wind. Thus, the effect of a component of own-ship motion is *numerically* equivalent to the effect of an equal component of target motion less the effect of an equal component of wind. This may be expressed algebraically as (omitting the divisor 10):

$$Y_o(\text{Col. 14}) = Y_o(\text{Col. 15}) - Y_o(\text{Col. 13})$$
$$X_o(\text{Col. 17}) = X_o(\text{Col. 18}) - X_o(\text{Col. 16})$$

Note that target motion has no influence on this error, and that motion of own ship *after* the projectile leaves the gun cannot alter the trajectory.

The *errors* due to own ship motion in the example are:
1. In range:
$$Y_o = 30 \cos 60° = 30 \times .500 = 15.0 \text{ knots.}$$
$$15.0 \times 41(\text{Col. 14}) \div 10 = 62 \text{ yards } over.$$
2. In deflection:
$$X_o = 30 \sin 60° = 30 \times .866 = 26.0 \text{ knots.}$$
$$26.0 \times 107(\text{Col. 17}) \div 10 = 278 \text{ yards } right.$$

13B17. Summary of gun ballistic. The seven factors discussed in the preceding articles are those which will always be considered in the gun ballistic in any practical case. It is possible that the errors due to some of the elements may be zero, but they must be examined to establish that fact. There are several other factors discussed later, which, if considered, are logically part of the gun ballistic. It is assumed for the present that these other factors are negligible, and that the arbitrary and control ballistics are zero. Under these conditions, all the elements affecting Vs and Ds have been considered, and the sight settings may be determined. The solution of the

CHAPTER 13

problem should be made in some condensed form for practical purposes, in a manner such as shown in the accompanying tabulation. Sufficient accuracy is obtained by computing components to the nearest tenth of a knot; errors and corrections to the nearest yard.

SAMPLE COMPUTATION OF GUN BALLISTICS

Given:

5″/38 cal. 2,600 f.s. gun	$A = 45°$
$R = 15,000$ yards	$Br = 300$
48 hour avg. mag. temp. 73° F.	$Bw = 320°$ (ballistic wind)
Surface barometer 30.20 inches	$Co = 325°$
Surface air temperature 47° F.	$So = 30$ knots
Rounds fired 450	$S = 20$ knots
	$Sw = 15$ knots (ballistic wind)

Required:
    (1) Gun ballistics in range and deflection.
    (2) Sight-bar range and scale.

CAUSE	COMPUTATION		
		\multicolumn{2}{c}{ERROR}	
\multicolumn{2}{c}{RANGE}	OVER	SHORT
Powder temperature $\quad (90 - 73) \times 2 \times 5.0 =$		X	170
Erosion (I. V. loss 60 f.s. from R-F curve) $60 \times 5.0 =$		X	300
Air density $\quad$ From nomogram		X	340
Target in LOS $\quad Yt = 20 \cos 45° = 14.1; 14.1 \times 22.7 =$		320	X
Wind in LOS $\quad Yw = 15 \cos 55° = 8.6; 8.6 \times 18.8 =$		X	162
Own ship in LOS $\quad Yo = 30 \cos 60° = 15.0; 15.0 \times 4.1 =$		62	X
	sum	382	972
		\multicolumn{2}{c}{CORRECTION}
		DOWN	UP
*Computed gun ballistic in range		X	590
Sight-bar range $15,000 + 600 = \boxed{15,600}$			use 600
		\multicolumn{2}{c}{ERROR}
\multicolumn{2}{c}{DEFLECTION}	RIGHT	LEFT
Drift $\hfill 248 =$		248	X
Target across LOS $\quad Xt = 20 \sin 45° = 14.1; 14.1 \times 22.7 =$		X	320
Wind across LOS $\quad Xw = 15 \sin 55° = 12.3; 12.3 \times 12.0 =$		X	148
Own ship across LOS: $Xo = 30 \sin 60° = 26.0; 26.0 \times 10.7 =$		278	X
	sum	526	468
		\multicolumn{2}{c}{CORRECTION}
		LEFT	RIGHT
*Computed gun ballistic in deflection (yards)		58	X
Deflection $= 58 \div 15,000$ (mils)		4	
Scale $= 500 - 4 \quad = \boxed{496}$			

* Since arbitrary and control ballistics are zero, these are equivalent to the total ballistics.

THE SURFACE PROBLEM: ANALYTICAL SOLUTION

Note that each element is considered as an error, and that the errors are summarized before the corrections are determined. The headings of the right-hand columns are labeled to provide an automatic transition from errors to corrections. Obviously, the gun must be pointed higher to compensate for a net short, and the *correction* to apply to present range is *up*. Also, the gun must be pointed farther to the left to compensate for a net right error, and the *correction* to apply to the midpoint of the deflection scale is *left*.

Sight angle. The computation of the gun ballistic in range is in yards, a linear measure. It is the usual practice to set the vertical angle in terms of range, hence the sight angle may be expressed as a sight-bar range*. By definition:

Sight-bar range is the range shown on the sight scale at the instant of firing. It is the algebraic sum of present range, the total ballistic, and spots.

Since the assumption was made that all factors not considered in the example are zero, sight-bar range for the example is the sum of the gun ballistic and present range. The act of setting sight-bar range on the sights automatically converts the linear range into the corresponding Vs (see Article 12C4). Because range scales are generally graduated in 50 yard increments, it is customary to compute sight-bar range to the nearest 50 yards.

Sight deflection. The angular value of deflection depends on the target distance, and not upon sight-bar range which is but the expression of an angle. Hence, deflection in mils is obtained by dividing deflection in yards by one-thousandth of the present range (expressed in yards as in the example). Deflection is always expressed to the nearest mil. Since the midpoint of the deflection scale on the 5"/38 cal. gun is 500, the scale setting is $500 - 4 = 496$ (left, lower).

13B18. Steps in computing gun ballistic. The steps involved in the computation of the gun ballistic are listed below as a guide. It should be emphasized that the steps are only a guide, and that variations are admissible so long as they do not introduce errors in the results.

1. Plot the situation, preferably drawing the vectors approximately to scale. Draw the components and indicate on the diagram the direction of the error due to each component.

2. Determine the acute angles between the vectors and the LOS by any convenient method, and compute the value of each component.

3. Opposite each element in the tabulation put a cross in the error space which does not apply in order to avoid entering an error in the wrong column.

4. Obtain the values needed from the range table, dividing all except those for air density and drift by 10. It is convenient to simply move the decimal point one place to the left when recording the range-table values.

5. Compute the individual errors, summarize them, and determine the compensating corrections.

6. Determine sight-bar range to the nearest 50 yard scale to the nearest mil (assuming no other corrections are required in the total ballistic).

C. CONTROL, ARBITRARY, TOTAL, AND INITIAL BALLISTICS

13C1. The control ballistic. The normal conditions for gun control are that the line of sight is directed along the line of position and that the sight scales are properly graduated. Departure from these conditions necessitates corrections which are included in the control ballistic, defined as follows:

The *control ballistic* consists of corrections to compensate point-of-aim error and sight-scale range-table error.

13C2. Range-table sight-scale corrections. A gun's range scale may not be graduated in agreement with the range table applying to the projectile and initial velocity used. For example, target charges may be fired with service charge range scales on the sights. The student should

---

*In many recent installations, when Vs is transmitted electrically to the guns, it is measured in minutes of arc, and the minute scale is used in making the setting.

CHAPTER 13

realize that the sight-bar range computed from the correct range table, when set upon a range scale not graduated to the same table, will introduce an incorrect value of the angle Vs.

If it is necessary to set sight-bar range on an incorrectly graduated scale, the procedure is to compute sight-bar range and determine its angular equivalent from column 2, using the range table which applies to the actual conditions of firing. Using the angle thus determined, enter the range table for which the range scale is graduated and read the corresponding range. The difference between this range and the computed sight-bar range is the *range-table sight-scale correction*. If the range shown in the incorrect table is greater, the correction is up.

13C3. Point-of-aim correction. There are circumstances when it is desirable to direct the line of sight at a point on the target removed from the desired point of impact. In such cases, the LOS and line of position do not coincide. Generally, impact is desired at the center of the waterline. If this point is not clearly visible because of peculiar conditions of visibility or dip below the horizon, a suitable point of aim must be selected. The correction may be required in either range or deflection, or both.

Correction in deflection is readily determined in yards and included in the total ballistic in deflection before it is converted to mils. The vertical correction must be expressed as its horizontal equivalent in yards of range. Column 19 of the range table provides the information required.

Assume for the sake of example that the target is a destroyer and that the point of aim is to be the foretop 75 feet above the waterline, with impact desired at the waterline. Column 19 indicates that a 100 yard change in sight-bar range will shift the point of impact 250 feet vertically. The correction in range required to obtain a hit at the waterline is:

$$(75 \div 250) \times 100 = 30 \text{ yards } down.$$

13C4. The arbitrary ballistic. There are certain incalculable errors for which corrections cannot be made by direct means. Some of these errors may have been noted consistently in past gunnery practices, where accurate data as to ranges, target course and speed, and other elements were available for an exhaustive post-firing analysis. After all errors which can be accounted for (such as errors in range-finder ranges and in estimates of target course and speed) have been eliminated, there may remain a certain error in the point of fall for which no definite cause can be assigned. This error, which may be in both range and deflection, is recorded and compensated in the next firing by an arbitrary ballistic. Thus:

Arbitrary ballistic is a correction for the indeterminate errors in the total ballistic used on previous firings, obtained by analysis of those firings.

Cold-gun correction. A cold-gun correction must be applied to the first salvo due to the fact that the range table is compiled on the basis of a warm gun. This correction is part of the arbitrary ballistic, but is treated separately so that it may be removed after the first salvo. It is generally an *up* correction in range.

13C5. The total ballistic. After the factors of gun, control, and arbitrary ballistics have been determined, they are combined to obtain the total ballistic. For a post-firing analysis after target practice, ballistics are computed on a prepared form known as Gunnery Sheet 10. One side of this sheet is provided with suitable spaces for entering the gun and ship for which the sheet is prepared, the sources of some of the data used, and various other information needed to identify the sheet. The other side is a ballistics computation form which is arranged to facilitate solution of the problem.

Figure 13C1 is a reproduction of this computation form filled in for the example used in this chapter. No discussion of the details of the use of this sheet are necessary here. Its principal use is for post-firing analyses, but it may also be used in computing the initial ballistic described later.

13C6. Practical solutions for Vs and Ds. The preceding articles have outlined the nature of most of the factors entering the determination of sight settings for the surface fire-control problem. It is obvious that any practical solution must be rapid if a high rate of fire is to be obtained, and certainly the analytical method is not adequate. The best that can be done analytically is to compute the total ballistic for the opening salvo and then spot to obtain the hitting gun range.

BALLISTIC COMPUTATION FOR	~~Determination of initial spot~~ Post-firing analysis								IDENTIFICATION _SAMPLE_ (Ship, Station, Run, Date, etc.)							
Sine.	0°	5°	10°	15°	20°	25°	30°	35°	40°	45°	50°	55°	60°	65°	70°	75°	80°	85°	90°
Cos...	90°	85°	80°	75°	70°	65°	60°	55°	50°	45°	40°	35°	30°	25°	20°	15°	10°	5°	0°
	.000	.087	.174	.259	.342	.423	.500	.574	.643	.707	.766	.819	.866	.906	.940	.966	.985	.996	1.000

					Visibility (in yards)		—
			_5_ Inch _38_ Cal. _2600_ F. S.		Weather conditions		
h	Target Bearing True or Line of Fire	265°	A	Range ① G. I. art. 635 _15000_	Index of powder		—
	OVER LEFT		B	Column 10 {For 10 f.s. ~~For 60 f.s.~~} _50_	a	Aver. mag. temp.	73°F
			C	Column 12 —	b	Veloc. change (±)	-34
			D	Column 13 _188_	c_1	Equiv. service rounds	450
			E	Column 14 ~~Per 10 knots~~ For 10 knots _41_	c_2 ②	Ave. bore enlargement	—
			F	Column 15 _227_	d	Corresp. I. V. loss	60 f.s.
	TARGET 45°		G	Column 2 —	d_1	Uncompensated I. V. loss	—
			H	Column 6 _248_	e_1	Barometer	30.20"
			I	Column 16 _120_	e_2	Thermometer	47°F
i	True course	—	J	Column 17 ~~For 10 knots~~ For 10 knots _107_	f	Ballistic density	—
j	Speed	20K	K	Column 18 _227_	g	Multiplier for col. 12	—
	Target angle	45°		RANGE			
k	Acute angle *	45°		Cause	Gun ballistic computation	Over	Short
l	Cosine	.707	1	Velocity error (temp. powder)	[b _34_ ×B _50_] 10 +or -10	×	170
m	Sine	.707	2	Velocity error (erosion) ②	[c _60_ ×B _50_]	×	300
	SHORT LEFT		3	Air density	[g ....×C _NOMOGRAM_]	×	340
			4	Target's motion	[j _20_ ×F _227_ ×l _707_]	320	×
			5	True wind ⑥ Apparent wind	[s _15_ ×D _188_ ×q _574_] +or -10	×	162
			6	Own ships' motion	[t _30_ ×E _41_ ×v _500_]	62	×
	55° TRUE ~~APPARENT~~ WIND ⑥		7		sum	382	972
						Down	Up
		Ballistic	8	⑤ Computed gun ballistic		×	590
	~~Surface~~			Control ballistic computation		Down	Up
n	True course	140°	9	Sight scale or range converter R. T. correction		×	×
o	Velocity	15K	10	Point of aim correction (point of aim................)		30	×
n_1	Rel. course	—	11	Vertical parallax correction (if not applied at director)		×	×
o_1	Rel. velocity	—	12		sum	30	×
p	Acute angle *	55°	13	Computed control ballistic		30	×
q	Cosine	.574	14	Arbitrary ballistic (from analyses of previous practices)		×	×
r	Sine	.819	15			×	×
	OVER RIGHT		16	③ TOTAL BALLISTIC (8)+(13)+(14)+(15)		×	560
			17	④ Correction for first salvo only		×	40
	60°		18	COMPUTED TOTAL BALLISTIC, FIRST SALVO		×	600
				DEFLECTION			
				Cause	Gun ballistic computation	Right	Left
	OWN SHIP		19	Drift	(H)	248	×
			20	Target's motion	[j _20_ ×K _227_ ×m _707_]	×	320
s	True course	325°	21	True wind ⑥ Apparent wind	[s _15_ ×I _120_ ×r _819_] +or -10	×	148
t	Speed	30K	22	Own ship's motion	[t _30_ ×J _107_ ×w _866_]	278	×
u	Acute angle *	60°				Left	Right
v	Cosine	.500	23		sum	526	468
w	Sine	.866	24	⑤ Computed gun ballistic (Yds.)		58	×
				Control ballistic computation		Left	Right
	MAKE 3 SKETCHES (One in each space provided)		25	Point of aim correction		×	×
	Indicate direction of motion to line of fire with arrow through dot in proper square.		26	Horizontal parallax correction (if not applied automatically)		×	×
			27	Computed control ballistic (Yds.)		×	×
	Indicate true North with arrow through dot in circle right of wind square.		28	Arbitrary ballistic (from analyses of previous practices)		×	×
			29			×	×
	* k, p, and u are acute angles between direction of motion and line of fire.		30	TOTAL BALLISTIC (24)+(27)+(28)+(29) (Yds.)		58	×
			31	TOTAL BALLISTIC [(30) _58_ ×1000]÷A _15000_ (Mils)		4	×

FIGURE 13C1. Gunnery Sheet 10.

CHAPTER 13

The early methods of computing ballistics included the use of various charts and diagrams which were employed to facilitate determination of several of the elements. While these graphical aids are still in existence, they are not considered here. The problem of keeping range was another major consideration which led to the development of graphical, and then mechanical range-keeping devices.

The total ballistic is a varying quantity, as examination of its several elements indicates. In a given total ballistic, errors due to certain factors such as I. V. loss, air density, and drift change slowly, being dependent upon range alone. Errors due to true wind are also relatively slow-changing, but are dependent upon range and target bearing. Errors due to the motions of own ship and target are subject to very rapid changes, for they not only depend upon range and target bearing, but also upon course and speed. Any sudden change in the motion of either ship introduces an abrupt change in the total ballistic. It is logical to suppose that a mechanical device which handles some of the ballistic elements will at least compute the corrections required for motion of the ships, and probably the corrections for wind. Automatic computation of corrections for the slowly changing factors, while not so necessary, is also desirable.

Modern range keepers are the result of developments over a period of many years. They solve both the range-keeping problem, and, in varying degrees of completeness, the ballistics problem. They perform so many functions that the term *range keeper* no longer is descriptive, and the newer designs of such machines are called *computers*. An introduction to these mechanisms is provided in Chapter 14.

13C7. Initial ballistic. Since many ballistic corrections are computed automatically, only those which are not so handled must be obtained by other means. Thus:

> *Initial ballistic* is that part of the total ballistic which is not determined automatically. If none of the total ballistic is computed automatically, then the initial ballistic is equal to the total ballistic.

It is essential that the elements handled by the range keeper be known before the initial ballistic is computed. For example, assume that a range keeper is used to solve the problem of Article 13B17, and that it makes automatic corrections for these elements:

IN RANGE	IN DEFLECTION
Motion of own ship.	Motion of own ship.
Motion of target.	Motion of target.
	Wind.
	Drift.

All of the usual deflection elements are automatically corrected; hence the initial ballistic in deflection is zero. Erosion is often handled automatically at the guns, and it is assumed that it is in this case. The remaining elements are considered as follows:

CAUSE	COMPUTATION	ERROR	
		OVER	SHORT
Powder temperature	(90 — 73) x 2 x 5.0	X	170
Air density (from nomogram)		X	340
Wind in LOS	8.6 × 18.8	X	162
	sum		672
		CORRECTION	
		DOWN	UP
Computed initial ballistic in range			672

283

RESTRICTED

THE SURFACE PROBLEM: ANALYTICAL SOLUTION

The computed initial ballistic is applied to the range keeper, which automatically adds it into the total ballistic.

13C8. Sources for measured quantities. The entire solution of the problem depends upon measurement of the various quantities which enter into it. A summary of the sources cannot be complete, for the methods of measuring are many and varied, depending on the size and age of the ship, and the type of equipment with which it is fitted. Briefly, however, some of the sources are:

Present range. Spotter's estimate, or range finder and other mechanical means of measurement.

Relative target bearing. Director and target designator.

Own-ship course. Gyrocompass repeater.

Wind. Aerological party, measurements from own ship, or spotter's estimate.

Air density. Aerological party aboard, or by signal from other sources, and from thermometer and barometer on the bridge.

Magazine temperature. Average temperatures of the magazines containing the powder that is to be fired.

Target angle. Spotter's estimate, or by tracking. (Target angle is easier to estimate than target course. Either quantity is the equivalent of the other as far as information is concerned).

Target speed. Spotter's estimate, or by tracking.

Generally, all the information required is known with a fair degree of accuracy except target angle and target speed. The latter two quantities are the principle considerations in the rate-control problem, considered in the next chapter. Graphical and mechanical methods are available to aid in determining them.

D. OTHER CONSIDERATIONS

13D1. Apparent wind. It has been noted that motion of own ship creates a wind apparent to a person on that ship, even if the air actually is still with respect to the earth. If a true wind is blowing, the wind apparent aboard the ship will be neither the true wind nor the wind due to own ship motion, but the resultant of the two, or apparent wind. Definitions pertaining to apparent wind are:

SYMBOL	DEFINITION
	Apparent wind. The wind apparent to the observing station. It is the resultant of true wind and reversed motion of the observing station, indicated by apparent direction and apparent speed.
Bwa	*True direction apparent wind.* The horizontal angle between north and the direction *from* which the apparent wind is blowing, measured clockwise from north.
Cwa	*True course apparent wind.* The horizontal angle between north and the direction *toward* which the apparent wind is blowing, measured clockwise from north.
Swr	*Apparent wind speed.* The horizontal speed of apparent wind with respect to own ship.
Ywr	*Apparent range wind.* The horizontal component of apparent wind in the vertical plane containing the LOS. $Ywr = Yw - Yo$.
Xwr	*Apparent cross wind.* The horizontal component of apparent wind perpendicular to the vertical plane containing the LOS. $Xwr = Xw - Xo$.

The relation between true wind and apparent wind is illustrated in figure 13D1. The components of wind due to motion of own ship are equal in magnitude to the components of own-ship motion, but their direction is reversed. The values of Swr and Cwa can be computed readily from simple trigonometric relations. In practice, the wind actually measured aboard ship is apparent wind. True wind must be computed. The obvious advantage of using true wind is that its value does not

CHAPTER 13

change when own ship changes course or speed. Apparent wind is subject to every change in own-ship motion.

13D2. Apparent wind in the gun ballistic. Apparent wind may be used in computing the gun ballistic. In fact, some range keepers make their computations on the basis of apparent wind. It has been shown that the effect of a line component of own-ship motion is: Yo(Col. 14) = Yo(Col. 15) — Yo(Col. 13). The expression Yo(Col. 13) is the apparent wind due to motion of

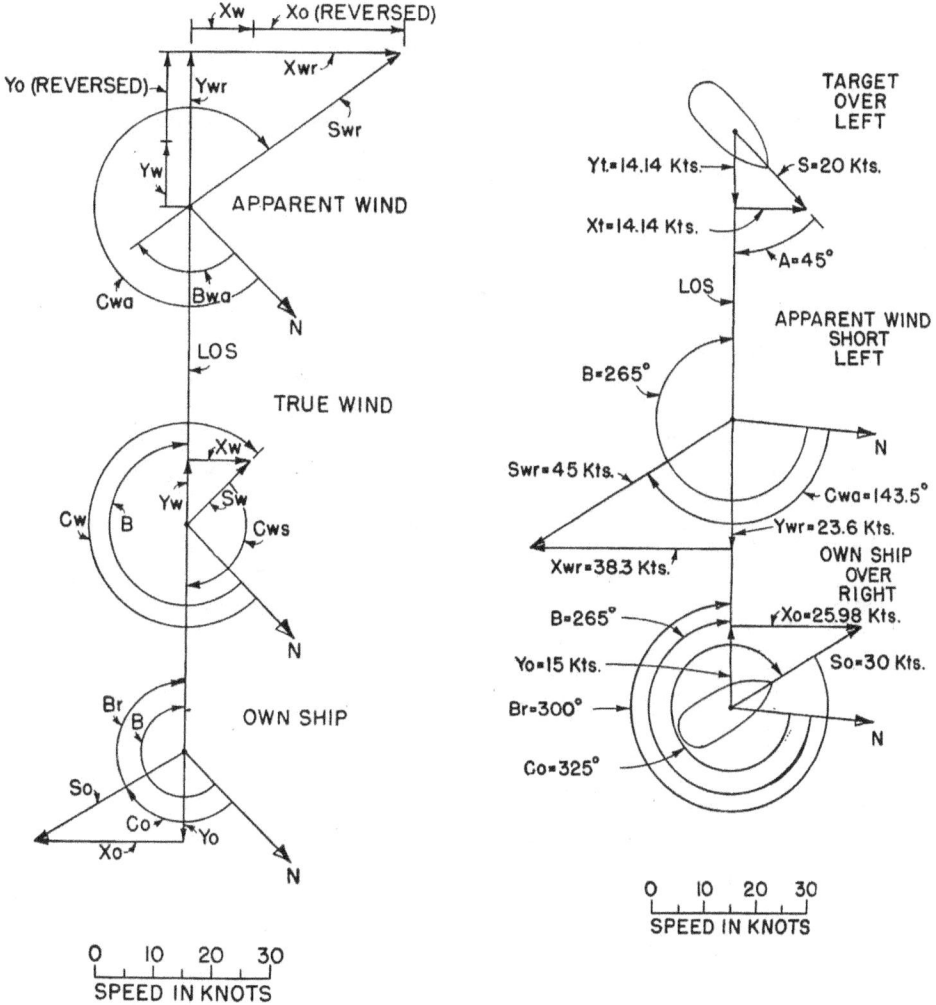

FIGURE 13D1. Relation between true and apparent wind.

FIGURE 13D2. Line-of-fire diagram using apparent wind.

own ship. It makes no difference in the final gun ballistic whether this expression is evaluated when considering motion of own ship or when considering wind, but it must not be included twice. If apparent wind is used in computing the error due to wind, then the error due to motion of own ship must be computed from column 15. Since the effect of target motion is also computed from column 15, the components of own ship and target may be combined, and a single correction determined for both elements. Considering only the three elements of motion, and dropping the divisor 10 for convenience, the following relations are seen to be true for range:

285   RESTRICTED

THE SURFACE PROBLEM: ANALYTICAL SOLUTION

Error due to own-ship motion = Yo(Col. 15) — Yo(Col. 13)
Error due to *true* wind   = Yw(Col. 13)
Error due to target motion   = Yt(Col. 15)

Combined error = Yo(Col. 15) + Yt(Col. 15) + Yw(Col. 13) — Yo(Col. 13)
= (Yo + Yt) (Col. 15) + (Yw — Yo) (Col. 13)
Error due to relative target motion = (Yo + Yt)(Col. 15) = dR(Col. 15)
Error due to *apparent* wind = (Yw — Yo)(Col. 13) = Ywr(Col. 13)

Similar reasoning may be applied to the deflection effects. The net error tabulated in Article 13B17 for the three elements of motion for range is $320 + 62 - 162 = 220$ yards over, and for deflection is $320 + 148 - 278 = 190$ yards left. The same results are obtained when apparent wind is used. Figure 13D2 is the same as figure 13B3 except that apparent wind is shown in the former. Apparent wind under the conditions of the problem is Swr = 45 knots. Cwa = 143.5°. The components are Ywr = Yw — Yo = 23.6 knots, and Xwr = Xw — Xo = 38.3 knots. The errors are:

1. In range:
   Motion of target and own ship: $29.1 \times 22.7 = 661$ yards *over*
   Apparent wind: $23.6 \times 18.8 = 444$ yards *short*
   Net error*:   217 yards *over*
2. In deflection:
   Motion of target and own ship: $11.9 \times 22.7 = 270$ yards *right*
   Apparent wind: $38.3 \times 12.0 = 460$ yards *left*
   Net error   190 yards *left*

13D3. Dead time. It was assumed earlier that the ballistics computations could be made and applied, and the guns fired instantaneously. Such is not the case, and the student should be acquainted with the error caused by the delay, which in fire-control language is *dead time*, defined thus:

Dead time (Tg). The sum of the time lags that accumulate from the beginning of the solution of a particular fire-control problem to the instant the guns are fired on the basis of that solution.

Time lag enters the problem the instant measurements of target position are begun. It takes time to enter the measurements into the problem, whether it is solved automatically or by other means. After the solution is obtained, it takes more time to transmit it to the guns. If fuzed projectiles are used, they cannot be loaded until after the fuzes have been set in agreement with the solution. Even in the modern automatic fire-control systems, when time-fuzed projectiles are used, a correction must be made for dead time.

During dead time, relative target motion causes a change in range and relative target bearing. The change in bearing is of no consequence at this point, for the LOS follows the target, and the guns are laid with respect to the LOS. Change in range is important, however, for present range at the instant of firing will be different from present range at the instant the target is positioned and the solution of the problem is begun.

In the example used in Article 13B17, the range rate Yo + Yt = dR at the present instant is $.563 \times 29.1 = 16.4$ yards per second decreasing. If Tg is 10 seconds, the range will decrease 164 yards. Unless a correction is made in computing sight-bar range, the motion of target and own ship will cause an *over* of 164 yards. In practice, any dead time which exists is not taken care of in the computation of the gun ballistic, but by graphical or mechanical means, as is shown in subsequent chapters.

13D4. Other elements. There are several other elements which either affect the trajectory itself, or influence the selection of the trajectory that is to be used. Among these factors are:

---

* The disparity of 3 yards is accounted for by the slight discrepancy in the range table data as noted in Article 13B16.

CHAPTER 13

Variations in weight of projectile. Any variation of projectile weight from the standard upon which the range-table data are based will result in a range change. Column 11 in the range table provides the information needed to make corrections for this element. Note the change in sign between the 7,500 and 7,700 yard ranges in the 5-inch table (Appendix D).

Since projectiles are required to be of designed weight within close tolerances, corrections for variations in weight are not generally required aboard ship. However, this computation might be necessary if projectiles of a slightly different designed weight were supplied to the ship to meet special conditions. It is obviously impracticable to consider making a correction for each projectile fired, but variations within allowed limits are one cause for the dispersion of the shots in a salvo.

Curvature of earth and height of gun. One of the assumed conditions for range tables is that the gun and target are in the same horizontal plane. This is seldom the case, for both the height of the gun above the waterline and the curvature of the earth generally result in an inclined line of position. As long as the gun is laid with respect to this line of position, the theory of trajectory rigidity indicates that no corrections need be applied for these elements. However, if the gun is laid with respect to a horizontal datum plane, as it is in some of the modern fire-control systems, an error in range may be expected due to these causes.

Rotation of the earth. The motion of the earth in space causes errors in both range and deflection. Compensating corrections may be computed from formulae, but they are rather involved. Tabulated corrections are provided in some range tables (not in the range table provided with this text) for the effects at various ranges at each 10° of latitude and 15° of azimuth of the line of fire for range errors, and each 30° of azimuth for deflection errors. The values for intermediate conditions are obtained by interpolation.

In general, the errors in range increase as the range increases, and decrease as the latitude increases. The errors in deflection increase as the range increases, and also as the latitude increases.

Cross wind effect on range. The apparent wind blowing at right angles to the line of fire has an effect on the range of projectiles. If the apparent cross wind blows from the left to the right of the LOS, the effect is a decrease in range, or a *short*. When blowing from right to left, the wind causes an *over*. When a correction is made for this factor, it is usually included in the arbitrary ballistic.

13D5. Trunnion tilt. The value of Vs as determined in the computation of sight-bar range must be set in a vertical plane if the gun is to be properly laid. Sight mechanisms are so constructed that Vs is set in a plane parallel to that plane which contains the bore axis and is at the same time perpendicular to the trunnion axis. If the trunnion axis is tilted out of the horizontal, the plane containing Vs is not vertical, and the component of Vs remaining in the vertical is smaller than Vs.

Similarly, Ds as determined in the computation of deflection for surface fire must be set in a horizontal plane. Actually, it is set in a plane parallel to both the trunnion axis and the sighting axis. This plane is called the *slant plane*, but in the surface problem the angle of position is very small, and the slant plane may be considered horizontal if the trunnion axis is horizontal. If the trunnion axis is tilted, the slant plane is no longer horizontal and the component of Ds remaining in the horizontal plane is smaller than Ds.

When the trunnions are tilted, Vs has a component in the horizontal, and Ds has a component in the vertical. For example, in an exaggerated case, assuming Ds is 50 mils and Vs is 20°, if the trunnions tilt 90°, Vs is wholly contained in the horizontal plane and becomes a deflection of 20°, while Ds is in the vertical plane and becomes a 50 mil angle of elevation. In any case of trunnion tilt, the component of Vs in the horizontal is introduced into the deflection, and the component of Ds in the vertical is introduced into the angle of elevation. Generally, the error in elevation is small, but the error in deflection may be quite large, as shown in figure 13D3.

The angle between the LOD and the LOS is the set value of Vs. It is assumed that Ds is zero. The plane GHH' is horizontal and contains the LOS. The plane HSH' is perpendicular to the LOS. Before the trunnions are tilted, the bore axis intersects HSH' at S. Planes GTS and GPS' are vertical. Obviously the LOS will move away from the target T as the trunnions tilt, but after the

287   **RESTRICTED**

THE SURFACE PROBLEM: ANALYTICAL SOLUTION

pointer and trainer bring the LOS back on, it occupies essentially the same position it occupied before the tilt occurred.

After the tilt (Z) occurs, the *vertical* angle remaining between the bore axis and the LOS is Vsv, which is less than Vs by the amount of Vsz. The *horizontal* angle is introduced between the LOS and the bore axis is Vsh. It is apparent from the diagram that Vsh is considerably greater than Vsz, and that this will always be the case within the limits of the amount of tilt that is likely to occur during firing.

Firing without regard to trunnion tilt may result in considerable errors, especially in deflection. The deflection error in mils for surface fire is approximately one-third of the product of Vs in degrees and Z in degrees. Thus, if the gun is fired at an elevation of 20° and the trunnion tilt is

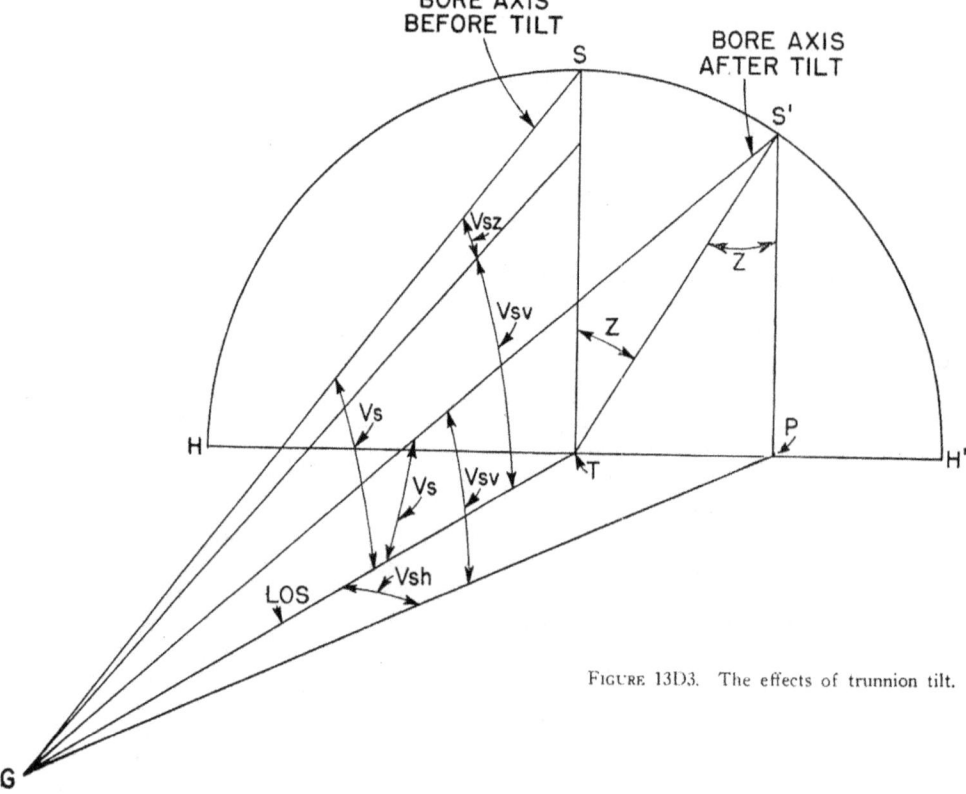

FIGURE 13D3. The effects of trunnion tilt.

6°, the error in deflection is 20 × 6 ÷ 3 = 40 mils. At 15,000 yards this is equivalent to a deflection error of 600 yards, which certainly must not be neglected. A more accurate determination of the *errors* may be computed from these equations:

$$\tan Vsh = \frac{TP}{TG} = \frac{TS' \sin Z}{TS' \cot Vs} = \tan Vs \sin Z$$

$$\sin Vsv = \frac{PS'}{GS'} = \frac{TS' \cos Z}{TS' \csc Vs} = \sin Vs \cos Z$$

It should be noted that these equations do not give the corrections which must be set on the sights when the trunnions are tilted. In order to obtain Vsh in the horizontal, the deflection set in the slant plane must be greater than Vsh. Furthermore, when the gun is trained to bring the LOS back on

CHAPTER 13

the target after the correction is set on the sights, the gun will be trained "uphill" and will partake of additional elevation which will more than offset the loss of elevation due to the original error.

It should be quite evident that corrections for trunnion tilt cannot be effectively treated by mathematical means, for the amount of tilt is continuously changing. Corrections can be made by various mechanical devices, known as *trunnion-tilt correctors,* which are incorporated in fire-control systems. If correctors are not available, the long range guns should be fired only when the trunnions are approximately horizontal, with a consequent decrease in the rate of fire.

13D6. Indirect fire. Thus far it has been assumed that the point of aim is on the target although not necessarily at the desired point of impact. Relative motion between the point of aim and the point of impact is small in magnitude in such a case. There are occasions when it is necessary to employ indirect fire, such as when it is desired to lay a barrage on shore with respect to some prominent object. In such cases, the point-of-aim corrections may assume large and varying proportions. A thorough discussion of indirect fire cannot be included in this text, but the nature of the problem involved can be seen from figure 13D4.

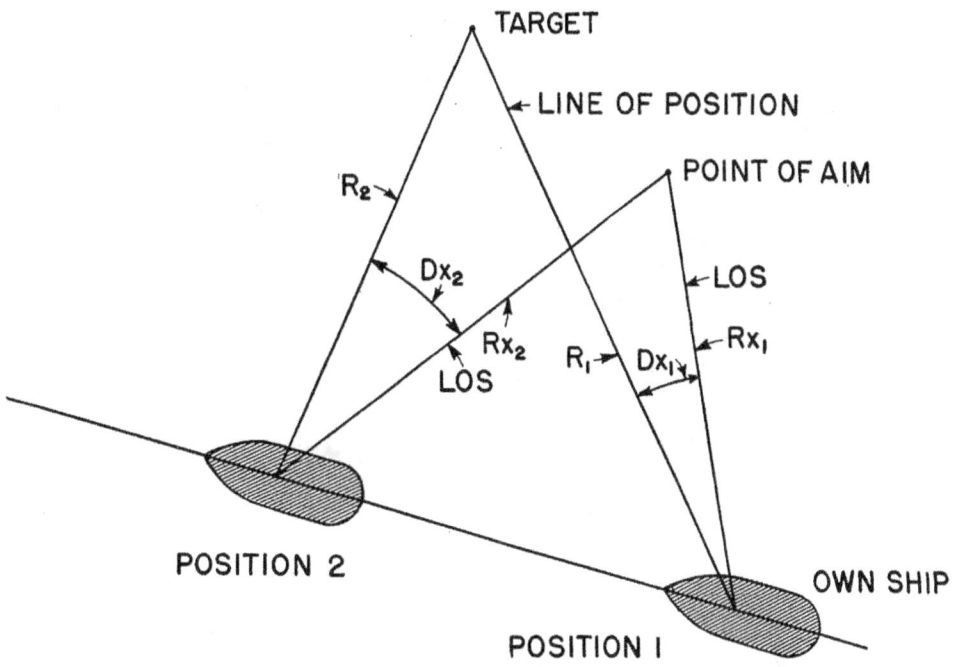

FIGURE 13D4. Indirect fire.

It is assumed that the target is not visible to own ship, but a prominent point of aim is available. At position 1, the present range to the target is known to be R_1, and the distance to the point of aim is Rx_1. The sight-bar range and deflection must be computed on the basis of R_1, and an appreciable lateral angle Dx_1 must be included in deflection to allow for the difference between bearings of the line of sight and the line of position.

It is apparent that as the ship continues on its course, both R and Rx will change, and not in direct proportion to each other. Also, Dx will change according to a definite relation established by R and Rx. The problem, of course, is to keep track of R, Rx, and Dx so that Vs and Ds will be continuously correct. If the target happens to be a moving ship, the problem becomes more complex.

CHAPTER 14

THE SURFACE PROBLEM: PRACTICAL DETERMINATION OF SIGHT SETTINGS

A. GRAPHIC TRACKING

14A1. Purpose. The tracking board is a graphic device used to determine target course (Ct) and target speed (S). The importance of these two quantities should be evident from the analytical study of the range-keeping and ballistics problems. It should also be plain that target course is just as useful as target angle, for either determines the direction of target motion with respect to the line of sight.

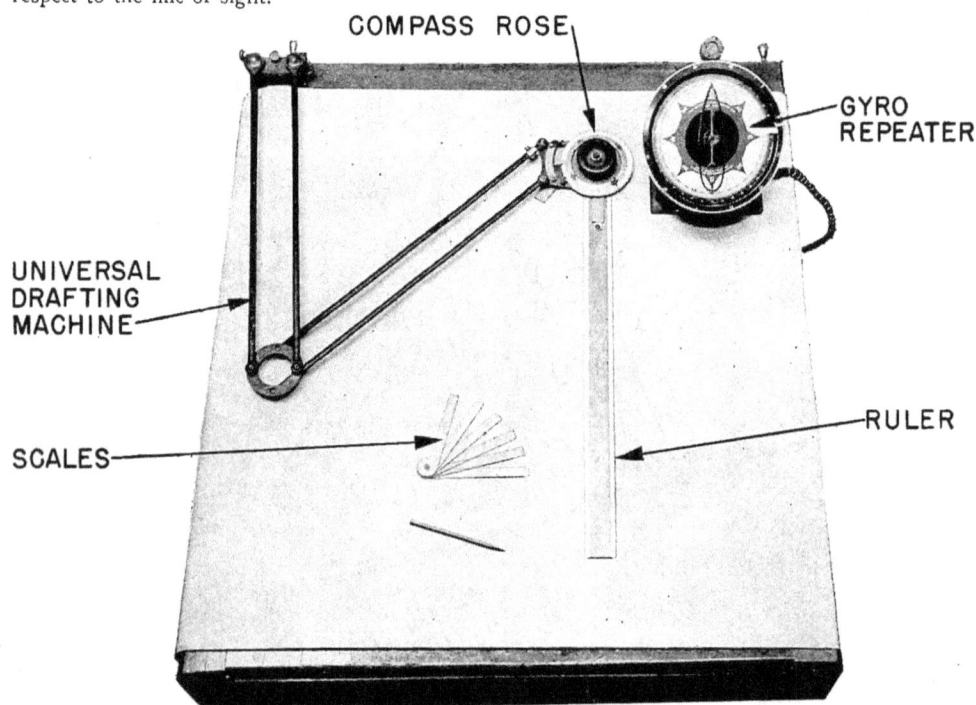

Figure 14A1. Tracking board.

14A2. Description. A typical tracking board is illustrated in figure 14A1. The table is about four feet square and is mounted on a stand. Plain paper is supplied from a roll at one side and accumulated after use on a roll at the other side. Attached to the top of the table is a universal drafting machine, which is nothing more than a parallel-motion mechanism carrying a compass rose and a ruler. The ruler can be clamped at any angle with respect to the compass rose, and both remain parallel to themselves as they are moved about the table. The rule is graduated in tenths of an inch for convenience in plotting.

14A3. Tracking. The drafting machine is used to plot to scale the motion of own ship and the position of the target relative to own ship at successive intervals. The successive target posi-

CHAPTER 14

tions lie along the *track* of the target, and establish its course and speed. It is customary to use 1 minute intervals, and to represent 1,000 yards by 1 inch on the ruler.

The plot shown in figure 14A2 illustrates the principles of tracking. The points are usually numbered as shown, but the rest of the labeling of the tracks is provided only for illustrative purposes. Short ranges are used in the example for convenience.

The track of own ship is laid out to true course with respect to the compass rose. Each numbered point represents the end of a time interval, and is therefore spaced from the preceding point by a distance proportional to own-ship speed. The constant $100 \div 3$ may be used to convert knots to yards per minute.*

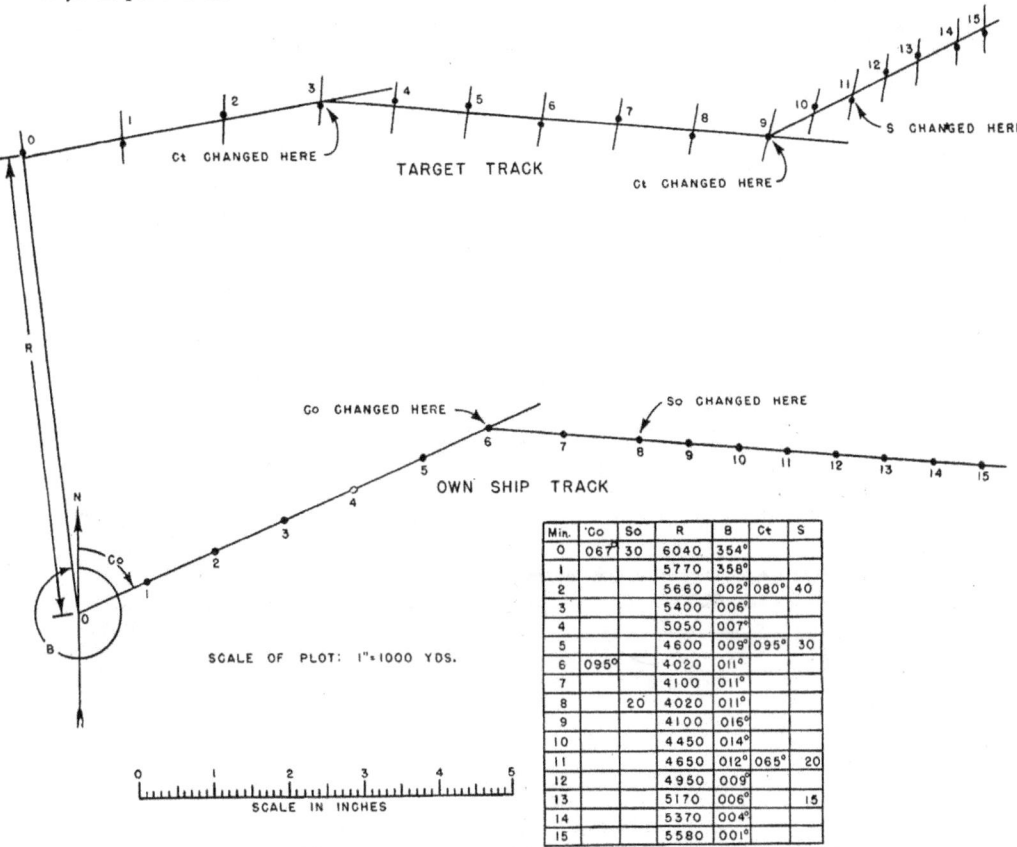

Min.	Co	So	R	B	Ct	S
0	067°	30	6040	354°		
1			5770	358°		
2			5660	002°	080°	40
3			5400	006°		
4			5050	007°		
5			4600	009°	095°	30
6	095°		4020	011°		
7			4100	011°		
8		20	4020	011°		
9			4100	016°		
10			4450	014°		
11			4650	012°	065°	20
12			4950	009°		
13			5170	006°		15
14			5370	004°		
15			5580	001°		

FIGURE 14A2. Tracking a target.

As each point is plotted along own-ship track, a corresponding point is plotted for the target. The ruler is set on true target bearing, and the present range is measured to scale from own-ship position. Theoretically, two successive plots of target position establish its track. In practice, the errors in measurement of range and bearing and the errors in plotting result in irregularities which make it necessary to plot several points and then fair a line through them to establish a track.

Changes in course of either ship appear as changes in direction of the tracks. Changes in speed appear as changes in distances between successive points. Ships do not change speed or course instantaneously, and plotting at one-minute intervals obviously admits errors during changes.

The data used in the plot is ordinarily tabulated for use in subsequent analyses and for convenience in operating the board. Ranges are obtained from a plotting board or from range finders.

---

* 1 knot = $2027 \div 60$, or approximately $100 \div 3$ yards per minute.

THE SURFACE PROBLEM: PRACTICAL SOLUTION

If from the range finders, the ranges must be averaged at the tracking board to obtain a mean value. Target bearing and own-ship course are derived from a ship's bearing indicator, and own-ship speed from a speed indicator.

14A4. Measuring Ct and S. As soon as the target track has been faired in, the ruler is placed along it and Ct is read from the compass rose. The distance between successive target positions is measured, and S is determined on the basis of the scale employed. It is convenient to measure the distance for three succeeding intervals when using 1 inch = 1,000 yards, for then the reading of the ruler in inches multiplied by 10 is the value of S in knots.* Also, the irregularities between any two points will be less likely to affect the results.

Changes in Ct and S are visible in the track of target just as they are in the track of own ship. It is apparent that changes may not be clearly evident until an appreciable time after they occur because of the time it takes to establish sufficient points to provide a fair line.

B. GRAPHIC RANGE KEEPING

14B1. Purpose. The plotting board is a graphic range-keeping device whose principal use is keeping present range, and in predicting range for short intervals. It also provides a convenient means of (1) applying the initial ballistic and spots to present range to obtain sight-bar range, and (2) keeping a record of scale. This device was first introduced in 1911, when it was an important part of the main battery fire-control system on capital ships. Its basic functions are performed by range keepers in modern systems, and by improved plotting devices, but it is still provided as a stand-by in main-battery plotting rooms on some ships.

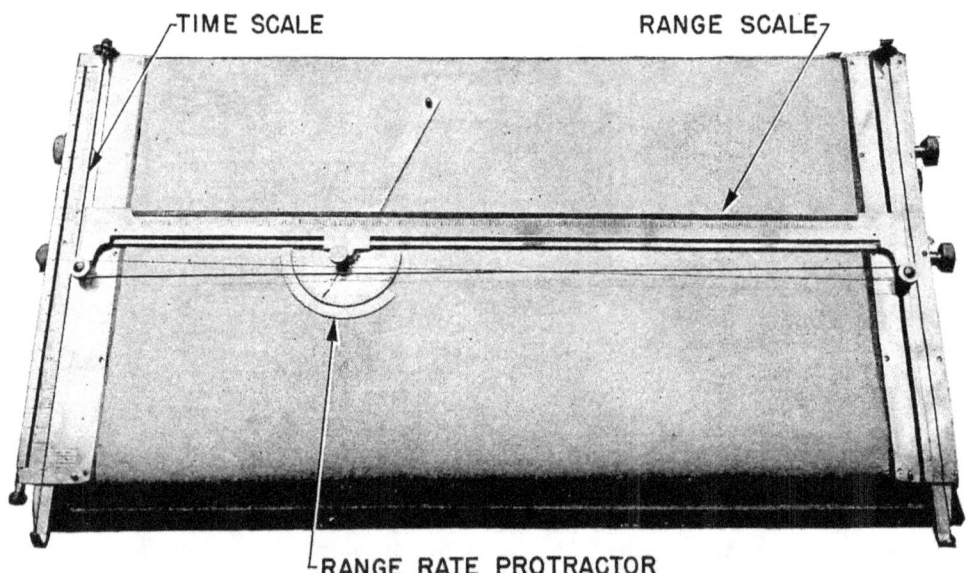

FIGURE 14B1. Plotting board, Mark 2.

14B2. Description. The Mark 2 plotting board is illustrated in figure 14B1. It is about 2 feet wide and 4 feet long, and is mounted on a stand at a convenient working height. The paper is plain, and although carried on rolls beneath the board, is not moved during plotting operations. The ruler running lengthwise on the board is graduated to read ranges. It is moved down the board by hand and is so mounted that it remains parallel to itself. The protractor carried on the

---

* Yards per minute may be converted to knots by multiplying by 3 ÷ 100. The 3 is introduced automatically if three intervals are used.

CHAPTER 14

ruler is free to slide sideways so that it may be moved out of the way when not in use. Parallel to the left edge of the board is a time scale graduated in 10 second increments from 0 to 20 minutes. The board is thus equivalent to a large sheet of graph paper on which the abscissas represent range and the ordinates represent time.

14B3. Personnel. Four individuals are needed to operate the board. At the top is the range talker who plots ranges received by phone from the range finders. At the left is the spot talker who moves the ruler down the board according to time indicated by a stop watch attached to the board, and who also indicates at the left edge of the paper the time and value of spots received, and the time of firing each salvo. The range and deflection talker, at the right of the board, indicates the time and value of each sight setting announced by the plotter. The plotter, at the front of the board, plots range lines, determines sight settings and exercises general control over the board.

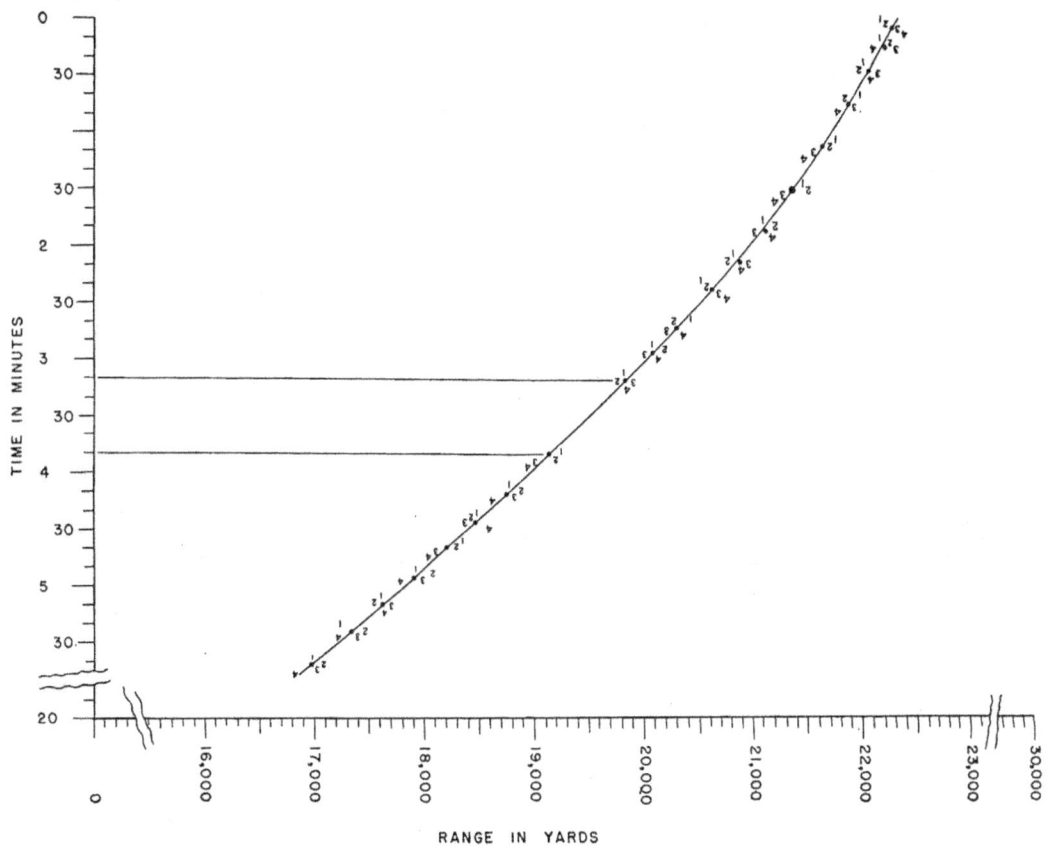

FIGURE 14B2. Part of a mean range line.

14B4. Mean range line. Each range received from the range finders is plotted against the ruler, and given the number of the range finder from which it is received, as shown in figure 14B2. Generally, there are several range finders ranging on the target, and their readings will come to the board in a regular sequence. Since all the range-finder data is subject to inherent errors, the plotted ranges for any particular instant are not likely to coincide. The average of the several ranges is estimated by eye and denoted by a dot or small circle, which then represents *mean range-finder range* for that particular instant.

The *mean range line* is faired through the successive mean ranges, and thus becomes a graphic

293 **RESTRICTED**

THE SURFACE PROBLEM: PRACTICAL SOLUTION

representation of the best available estimate of the target distance. Actually, the range indicated by the line lags behind present range by the time it takes to transmit the ranges to the board and subsequently plot the line. It must be remembered that the operation of the board is progressive, and nothing appears on the paper below the present position of the ruler as it moves down the board.

14B5. Interruptions in ranging. The mean range line is a smooth curve, its direction depending upon whether the range is opening or closing. Normally, the line is faired in so that it is continuously against the edge of the ruler. The fairing process can be continued through temporary interruptions in range without appreciable error, as it was from 3 minutes 10 seconds to 3 minutes 50 seconds in the illustration. The extension of the curve through interruptions is nothing more than a graphic method of multiplying range rate dR by time increments.

14B6. Range rate. Range rate can be measured directly from the curve by placing the arm of the protractor tangent to the curve and reading dR in knots on the scale opposite the pointer (see fig. 14B1). The range rate thus determined can be used in computing the initial ballistic in range if apparent wind is used, and it can also be employed as a partial check on the set-up of the range keeper. In modern installations there is little use for this feature of the plotting board, as an automatic plotting board, or graphic plotter, is used for the graphic determination of range rate.

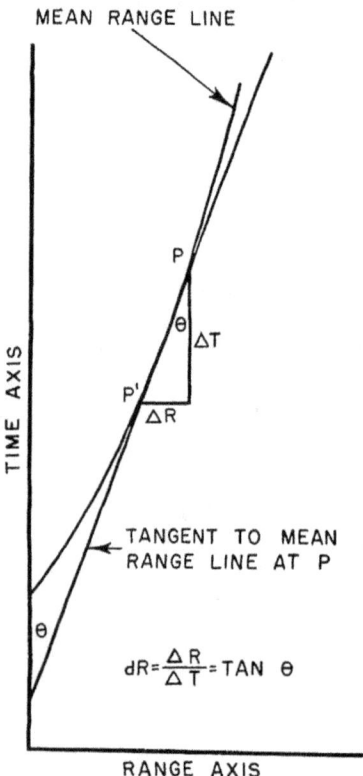

FIGURE 14B3. Range rate; the slope of mean range line.

The slope of the mean range line at any point with respect to the time axis is a direct indication of the range rate at that point. The slope is determined by the angle formed between the tangent and the time axis, as indicated in figure 14B3. The curvature of the mean range line is usually so gradual that a tangent at any point practically coincides with the curve for a distance of an inch or more. Stated another way, the curve is essentially straight for short distances, and the range rate is the same at P, P', or any point between P and P'. Recalling the process of determining a change in range ($\triangle R$) during any short time interval ($\triangle T$), $\triangle R = dR \times \triangle T$.

In the diagram, $\triangle R$ is the change in range and $\triangle T$ is the time interval between points P and P'. The angle between the tangent at P and the time axis is θ. Assuming that $\triangle T$ is small, P' on the curve is also on the tangent, and $\tan \theta = \triangle R \div \triangle T = dR$. Thus, it is only necessary to measure θ, and dR can be determined from $\tan \theta$. The protractor measures θ, but the scale is so graduated that it reads the value of dR in knots corresponding to the value of θ which the protractor measures. The range rate is negative, or decreasing, when the pointer is left of center, positive, or increasing, when the pointer is right of center.

14B7. Gun range line. It is convenient to represent sight-bar range on the plotting board by means of a gun range line, so named to distinguish it from the mean range line. The opening sight-bar range, as determined in the ballistics problem, is the sum of present range plus the total ballistic in range. Hence, the initial ballistic, equivalent in this case to the total ballistic, is added graphically to the present range represented by the mean range line in order to start the gun range line, as shown in figure 14B4. The plotter extends the gun range line parallel to the mean range line, making adjustments for spots as firing progresses. Any point on the gun range line then represents the sight-bar range required for firing at the corresponding present range shown on the mean range line. By the time the sight-bar range indicated at any point can be applied to the guns, it is incorrect because of the range change during dead time, and suitable allowance must be made.

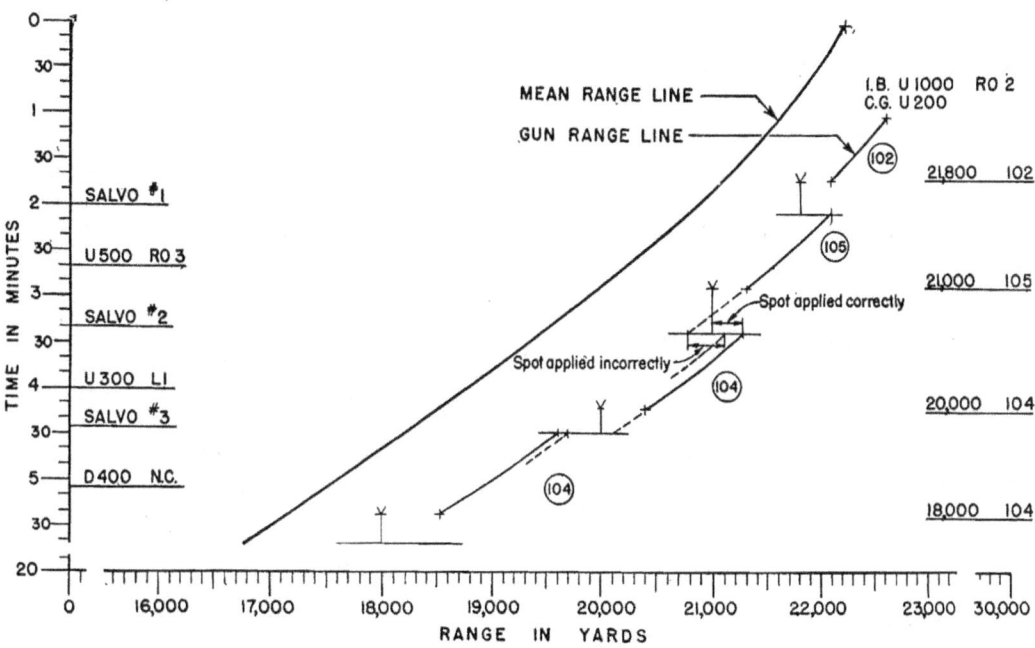

FIGURE 14B4. A completed plot.

14B8. Dead time. This method of determining sight-bar ranges introduces an appreciable dead time into the problem. It was previously mentioned that the mean range line slightly lags the actual present range. That lag is part of the dead time. Additional delay occurs between the instant the plotter decides to send out a range and the instant he states it to the range and deflection talker. More time is lost in transmitting the settings to the guns or director, and in setting the sights. Still more time is lost before the guns can actually be fired after the sights are set. The sum of all these delays is sufficient to require compensation. Obviously the total dead time depends on the efficiency of the entire fire-control party, and must be established for each ship by analyses of actual performances at drills. For the sake of illustration, it is arbitrarily assumed to be 20 seconds.

14B9. Advance range. The sight-bar range predicted for the instant of firing by graphical or mechanical devices is called *advance range* in the surface fire-control problem. It is present range "advanced" to compensate movements of own ship and target, and for other ballistics, and spots, and is in principle no different from the sight-bar range determined analytically.

A point on the gun-range line for its corresponding time instant represents present range plus spots and all the ballistics except that for dead time. The correction for dead time is made by advancing the gun range line by eye estimate through a distance corresponding to dead time, and then reading the corresponding advance range. A fairly realistic representation of the actual conditions may be obtained by covering figure 14B4 with a sheet of paper, and aligning its upper edge with the time scale as though it were the range ruler.

THE SURFACE PROBLEM: PRACTICAL SOLUTION

Thus, if at 1 minute 40 seconds the plotter decides to announce sight settings, he sees only the lines above this time line. The plotter estimates by eye that 20 seconds later the gun-range line will show 21,800 yards, and he announces this range. As a record of his estimate, he places a check or other suitable mark opposite the range announced. He also announces the scale setting, recording it nearby. The gun range line may be discontinued at this point, for it must be corrected for the spot before the next range is sent out.

14B10. Application of spots. At the instant the salvo is fired, as at 2 minutes 00 seconds, the plotter draws a line along the ruler across the general vicinity of the gun range line. He then drops a line from the check perpendicular to this salvo line. The point thus established on the salvo line indicates the range which was presumably set on the sights at the instant of firing. When the spot is received at the end of the time of flight, the ruler will be at some lower position such as 2 minutes 40 seconds, and there will be no gun range line beyond the point where range was announced. The spot is graphically applied on the salvo line to determine the range which should have been set on the sights, and the point thus determined lies on the new gun range line. Then the new gun range line is continued to the ruler, and a new advance range is sent out. (If a cold gun correction is included in the initial ballistic, it is removed when the first spot is applied, as was done at 2 minutes in figure 14B4.

It should be noted that the spots are applied to the range which was announced. This procedure automatically eliminates any errors which might be introduced by applying the spots to the gun range line if it were continued to the salvo line and the salvo were fired either before or after the estimated dead time expired. The example at 3 minutes 20 seconds shows what would occur if the guns fired 10 seconds late, and the spot were applied to the extended gun range line. The new gun range line would be the dotted one, and would give a range 200 yards less than required as shown by actual firing. The situation at 4 minutes 25 seconds shows what would happen if the guns fired 5 seconds early. In this case, the gun range line would indicate about 75 yards too much range. The solid lines show the correct application of the spot.

14B11. Time and data records. Running records of the data received and transmitted are kept at the left and right edges of the board. The spot talker draws a line along the ruler at the left edge of the paper each time a spot is received or a salvo is fired. He records the value of each spot and the number of each salvo. The range and deflection talker draws a line and records the announced advance range and deflection at the right side of the paper. The lines provide a time record, and with the written data give a running check on the plotting and sight setting, and serve as a basis for post-firing analysis.

14B12. Limitations. Compared to a range keeper, the plotting board appreciably retards the firing rate of the battery because of the dead time involved. Furthermore, the mean range showing on the plotting board is dependent on the range finders, although it can be approximated during short interruptions in ranging. If ranging ceases after action begins, the plotting board cannot be used to keep present range. Even when the range finders are supplying ranges, the present range line is subjected to inaccuracies, especially at the critical times when own ship changes course.

The plotting board does not compute ballistics in either range or deflection. It merely provides a convenient means of keeping a record of sight-bar range and deflection. The ballistics in both range and deflection are continuously changing quantities, and theoretically should be recomputed for each salvo. The spotting automatically takes care of this part of the problem, but obviously the spots must in general be larger than they would be if the ballistics were continuously computed for the existing present range as is done by modern range keepers.

C. BASIC RANGE-KEEPING MECHANISMS

14C1. Introduction. Working a range-keeping problem mathematically is sufficient demonstration that for practical purposes, range cannot be kept analytically. The graphical solution is partially adequate until the range-measuring instruments fail, and then it cannot be used success-

CHAPTER 14

fully. One of the earliest devices introduced as an aid in solving this problem was a *range projector* which computed range rate. It was used in conjunction with a *range clock* whose hands indicated range and were driven by a motor at a speed adjusted in accordance with the range rate obtained from the range projector.

The next logical development was the consolidation of these two devices into a single mechanism to provide generated values of present range. Further additions provided for computation of some of the ballistic corrections. This section provides an elementary discussion of the devices required to keep range. Ballistics-computing units are discussed subsequently.

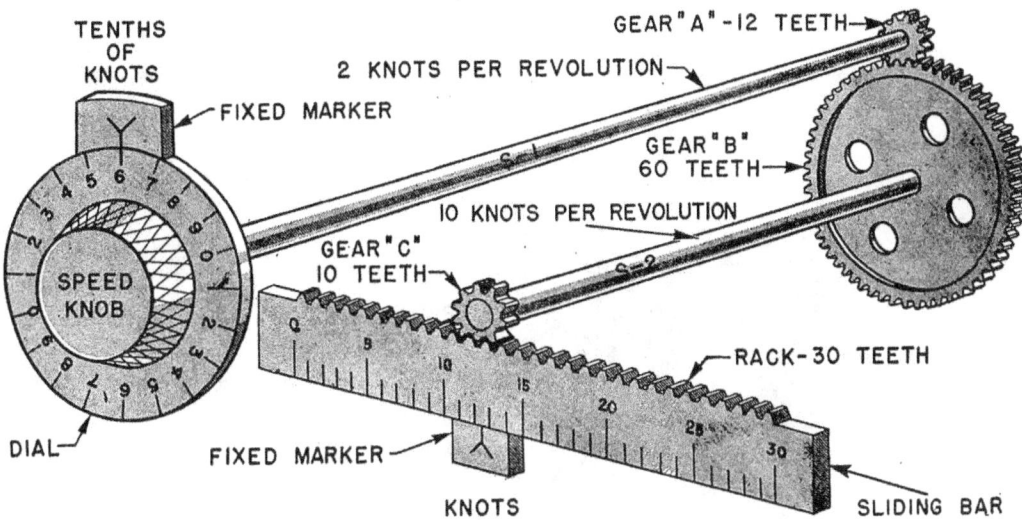

FIGURE 14C1. Mechanical representation of quantities.

14C2. Mechanical representation of quantities. In any mechanical computing instrument, linear motions of bars and rotary motion of shafts, gears and dials are used to represent quantities. In effect, these motions represent actual quantities to some suitable scale, just as the vectors in a line-of-fire diagram represent ship motions to scale (assuming that they are drawn to scale).

Suppose, for example, that it is desired to have the sliding bar in figure 14C1 indicate speed in knots introduced by rotation of the knob. On the upper edge of the bar is a rack having 30 teeth, each tooth being equivalent to one knot as shown on the speed scale. The number of teeth on gear A is twelve, on B sixty, and on C ten. It is evident that one revolution of shaft S-2 represents 10 knots, for one revolution of S-2 will move the bar 10 teeth. One revolution of S-1 represents 2 knots, for S-1 must revolve five times to turn S-2 once.

Since the knob is attached to S-1, it must be turned 180° to enter one knot, or 18° for each tenth of a knot. The scale divisions on the knob are spaced 18° apart, hence each represents one-tenth of a knot, and the speed setting can be entered with ease to the nearest tenth.

14C3. Multiplication by a constant. The input through the knob, hence the rotation of S-1, is a variable quantity, depending on the speed setting of the knob. The ratio between gears A and B is a constant. The output of S-2 is the product of the variable and the *gear constant* of .2, for (rotation of S-2) $= .2 \times$ (rotation of S-1).

Another example of the use of a gear constant is the conversion of knots to yards per second. Suppose that in figure 14C1 one revolution of S-1 were to represent one knot, and one revolution of S-2 were to represent one yard per second. The gear constant would have to be .563 and the output of S-2 would be $.563 \times$ (input to S-1). The ratio between the gears would be difficult to establish exactly, but a close approximation could be obtained with 22 teeth on gear A and 39 teeth on gear B. The revolutions of S-2 would then be $22 \div 39$ or .564 times the revo-

RESTRICTED

lutions of S-1, and the output of S-2 would be very nearly the yards per second corresponding to the knots input to S-1.

Gear constants are frequently used in range keepers and other fire-control instruments as will be seen in subsequent sections. The fact that there are many gears in an instrument makes it possible to introduce the constants in a number of different places. Any single constant may be the result of several different constants obtained in a series or train of gears. The student should not be concerned with the actual gear ratios used, but should appreciate the ease with which a variable can be multiplied by a constant.

14C4. Differential gears. The differential gear is a device used for obtaining the sum or difference of two quantities. As shown in figure 14C2, it consists of a left side, a right side, and a spider. The spider is a frame secured to the central shaft and holding two pinions which are geared together. Also, the left pinion is geared to the left side, and the right pinion is geared to the right side. The gears constituting the left and right sides are free to revolve about the central shaft.

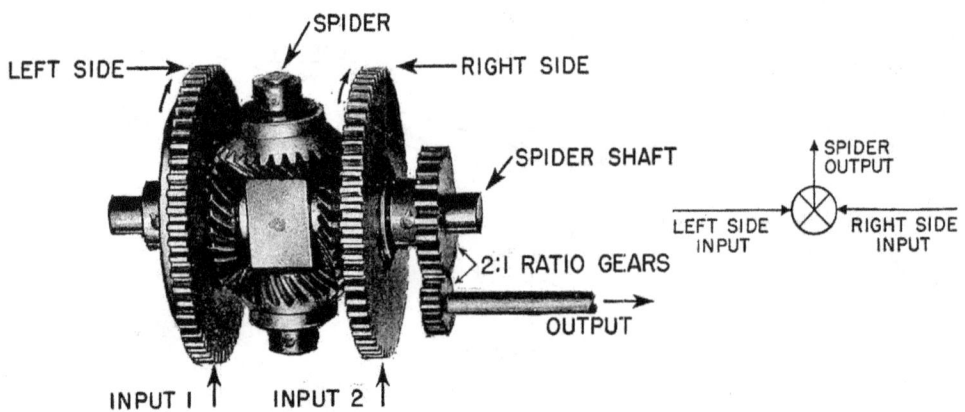

FIGURE 14C2. The differential gear.

Addition. Assume that two inputs are to be supplied to the sides as shown in the figure. Let input 1 rotate the left side through one revolution in the direction indicated by the arrow, while the right side is held. The gearing arrangement will cause the spider and the spider shaft to rotate one-half revolution in the same direction. Now let input 2 rotate the right side through one revolution in the direction indicated by the arrow, while the left side remains stationary. This will cause the spider and the spider shaft to move another half revolution, and this element now will have made a complete revolution in the indicated direction. The spider shaft can be connected to 2:1 ratio gears as shown, so that the output shaft makes two revolutions, which is the sum of the movements of the two inputs.

Subtraction. The differential will subtract as readily as it adds. Thus, if the two sides rotate in opposite directions, the spider will represent the difference between the two inputs. In other words, the spider will continuously give the algebraic sum of the movements of the two side gears.

Variations. As actually used, the spider is not necessarily the output element. Any two of the three elements of the differential can be geared to the inputs, and the third element will furnish the output. The arrangement of inputs and output is simply a matter of convenience in gearing. There are various constructions used in different types, but the principle of operation is always the same.

Schematic representation. The conventional symbol used to represent a differential on schematic drawings is a crossed circle as shown in figure 14C2. The spider is represented by the cross. The directions of the arrows indicate which are the inputs and output. In the illustration, the spider

CHAPTER 14

supplies the output. Each differential in an instrument is assigned a number prefixed by D, such as D-1, D-2 etc., and these designations are shown on schematic drawings.

14C5. Component solvers. The component solver is a device for resolving a vector into components with respect to a reference line. The most common application in range keepers is the resolution of own ship, target and wind motions into line and cross components with respect to the line of sight to the target.

Mechanical vector representation. The basic principle underlying a component solver is illustrated in figure 14C3. The dial carrying the compass rose and own-ship outline is arranged to rotate about its center. The vertical pin carried by the dial is disposed so that it may be moved radially in the radial slot. The two reference lines are perpendicular to each other at the center of the dial, and one of the reference lines corresponds to the line of sight.

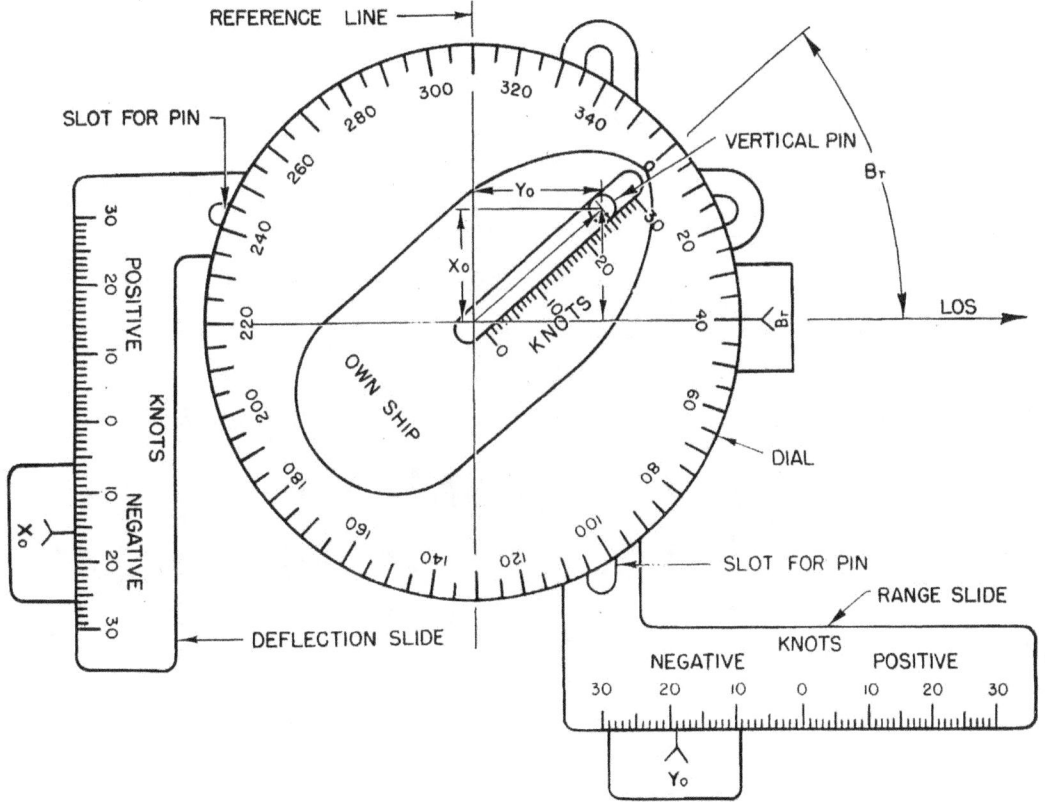

FIGURE 14C3. Principle of the component solver.

The center of the vertical pin corresponds to the arrowhead of a vector representing the motion of own ship. The distance of the pin from the center of the dial represents the speed of own ship to scale. The angular position of the radial slot with respect to the LOS corresponds to relative target bearing. If Br and So are set into the dial and pin respectively, the displacement of the pin from the reference lines will be a measure of Xo and Yo to the same scale as that used for setting S.

For *any* values of Br and So, as long as they are correctly set, the displacements of the pin will be direct measures of Xo and Yo. In other words, the component solver can continuously solve the equations $Xo = So \sin Br$, and $Yo = So \cos Br$.

Measuring the components. To make the values of Xo and Yo available for use, the position of the pin must be measured or transformed into mechanical displacements. The device in figure 14C3 schematically illustrates a simple way to do this. The range slide is mounted so that it moves parallel to the LOS, while the deflection slide moves across the LOS. The vertical pin extends below the dial and fits in the slots of the slides. The calibrations on the slides are to the same scale as the speed calibrations on the dial; hence, the components may be read directly against the fixed indexes.

To make this mechanism useful in a range keeper, the scales on the slides are changed to gear teeth. A similar device is provided to resolve target motion into components. The movements of the two range slides are combined by a differential to obtain range rate, and the movements of the two deflection slides are combined in another differential to obtain bearing rate. This elementary arrangement is used in the Mark 2 range keeper described in the next section.

14C6. The basic component solver. Component solvers used in modern range keepers are somewhat more refined than the type just described, but the principle of operation is unchanged. Figure 14C4 shows the type now in general use for resolution of ship, target, and wind motions.

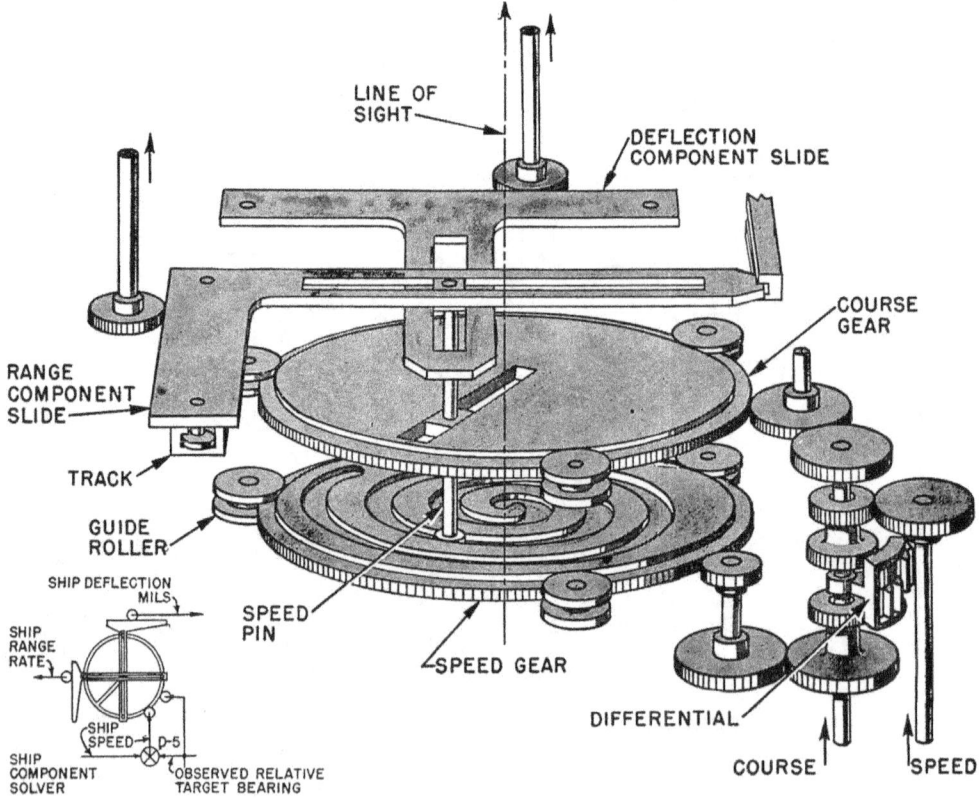

FIGURE 14C4. The component solver.

The dials indicating the speed and angle settings are remotely located, and connected by suitable shafts and gears to the setting mechanisms. The position of the course gear determines the direction of the vector with respect to the reference line. The radial position, or speed setting of the pin is controlled by the spiral-shaped groove cut in the speed gear. When the speed gear is rotated *relative* to the course gear, the spiral slot causes the pin to move in or out, depending on the direction of rotation. When the speed and course gear are rotated together, there is no radial movement of the pin.

Differential. It is apparent that when the angle is set, the speed gear must be rotated the same amount as the course gear to prevent changing the speed setting. Also, when the speed is changed,

CHAPTER 14

the speed gear must rotate while the course gear stands still. Hence, a differential is provided to prevent a change in the radial setting whenever the course gear is rotated.

Course and speed are the two inputs to the differential, the first rotating the central shaft and the second revolving the upper side gear. The central shaft carries the spider and extends through the differential so that it can drive the course gear. If the speed remains constant while the course changes, the upper side gear will be held stationary while the central shaft and spider rotate. The stationary upper gear will cause revolution of the spider pinions, which will in turn rotate the lower side gear and drive the speed gear. Thus a change in course will rotate the course and speed gears in unison, thereby preserving the radial setting of the pin.

A change in speed will rotate only the speed gear. In this case the central shaft is stationary so there is no revolution of the course gear. The change in speed rotates the upper side gear, causing rotation of the spider pinions but not of the spider frame. Rotation of the pinions rotates the lower side gear and drives the speed gear so as to make the appropriate change in setting of the speed pin.

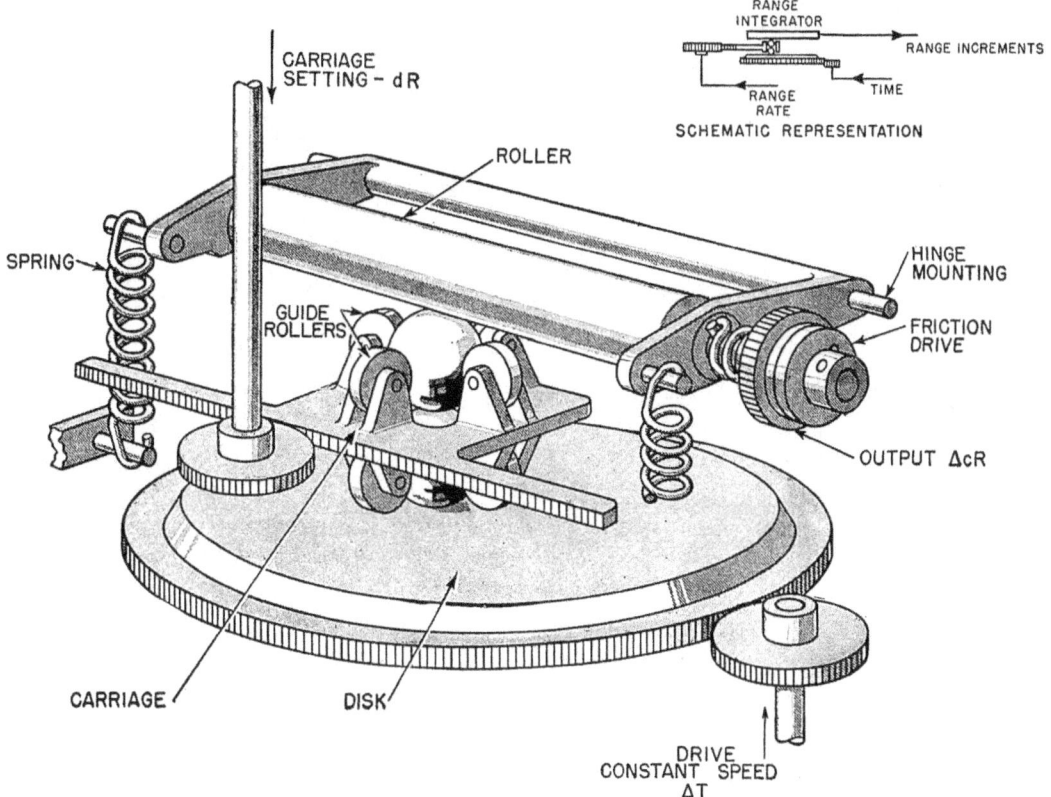

FIGURE 14C5. The disk-type integrator.

Schematic representation. The standard schematic representation of component solvers is shown in the lower left corner of figure 14C4. The symbol used in some of the drawings in this text is a plain rectangle, which is normally understood to include the speed and course gears, the differential and the component slides.

14C7. Integrators. There are a number of different mechanical integrators used for various applications in range keepers and computers. Basically, they are all multiplying devices. The integrators employed to keep range are of the disk type, which is commonly used to multiply a rate by time.

The disk-type integrator, as shown in figure 14C5, is simply a flat circular disk which revolves at

301

RESTRICTED

THE SURFACE PROBLEM: PRACTICAL SOLUTION

a constant speed, and by friction drives a pair of balls which in turn drive a roller. Rotation of the disk rotates the lower ball, which rotates the upper ball, which in turn rotates the roller. The balls are supported in a carriage so that the point of contact between the lower ball and the disk can be shifted by moving the carriage along a straight line to any radius on the disk, on either side of the center. Spring tension placed on the roller puts sufficient contact pressure between the roller, balls and disk to prevent slipping. Two balls are used to prevent sliding when the carriage is moved.

Roller speed. The speed of roller rotation depends upon the speed at which the balls rotate. The speed of the balls, in turn, depends upon the number of times the circumference of the lower ball is contained in the circumference of the circle of the disk upon which the ball is riding. If the carriage is set so that the lower ball is at the center of the disk, the roller will stand still. If the carriage is moved to the left, the roller will rotate in one direction, while if the carriage is moved to the right, the roller will rotate in the opposite direction. The farther the carriage is from the center, the greater the circumference of the contact circle, and the faster the roller will turn.

Range increments. The constant speed of the disk represents time, and the position of the carriage represents range rate. The output of the roller is generated range increments ($\triangle cR$), representing the product of time increments ($\triangle T$) and range rate (dR). The value of dR is obtained from the differential which combines Yo and Yt, and time is supplied by a constant-speed motor.

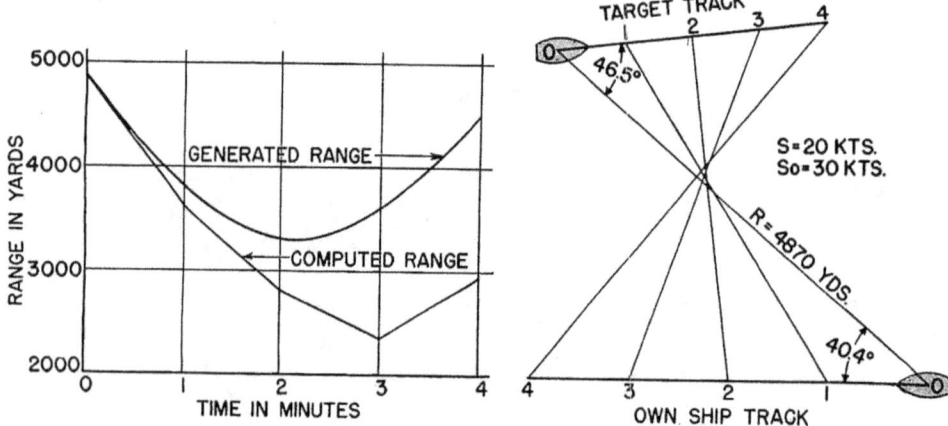

FIGURE 14C6. Output of the integrator.

How to keep range by multiplying range rate by time increments was shown in Chapter 13. The integrator does the same thing, but without the errors that are introduced by assuming dR constant during finite time intervals. The component solvers supply continuously correct values of dR, and the integrator carriage is continuously positioned accordingly. The roller turns at a speed proportional to the instantaneous value of dR. This of course presupposes that the angles (A and Br) and speeds (S and So) are always correctly set in the component solvers.

The integrator roller revolves continuously, and its total revolutions from the time of starting are a direct measure of the total change in range during that time. This total change may be added to the initial range in a differential, and the output of the differential will be *generated present range*.

Schematic representation. Disk-type integrators are normally shown on schematic diagrams in the manner indicated in the upper-right corner of figure 14C5. In this text they are sometimes represented by a suitably labeled rectangle in order to eliminate minor details from the drawings.

Accuracy. An illustration of the usefulness of the integrator is provided in figure 14C6. A short 4-minute problem is indicated by the tracks of own-ship and target motion. At zero time, indicated by 0 in the diagram, A = 46.5°, S = 20 knots, Br = 40.4°, So = 30 knots, and R = 4,870 yards. Assuming the rates to remain constant for one-minute intervals, and calculating new values of range,

CHAPTER 14

target angle, and relative bearing each minute, the range at the end of 4 minutes is computed to be 2,995 yards, as indicated by the *computed range* line on the graph. By actual measurement of the ranges from the scale drawing, the ranges are found to follow the *generated-range line* which is a smooth curve, and the range at the end of 4 minutes is 4,510 yards. The integrator generates range according to the smooth curve, and consequently its results do not suffer the large error introduced in the analytical process where 1-minute intervals are used.

D. AN ELEMENTARY RANGE KEEPER

14D1. General. An introduction to mechanical range keeping is provided by the Mark 2, or Baby Ford range keeper, shown in figure 14D1. This instrument was designed to perform for

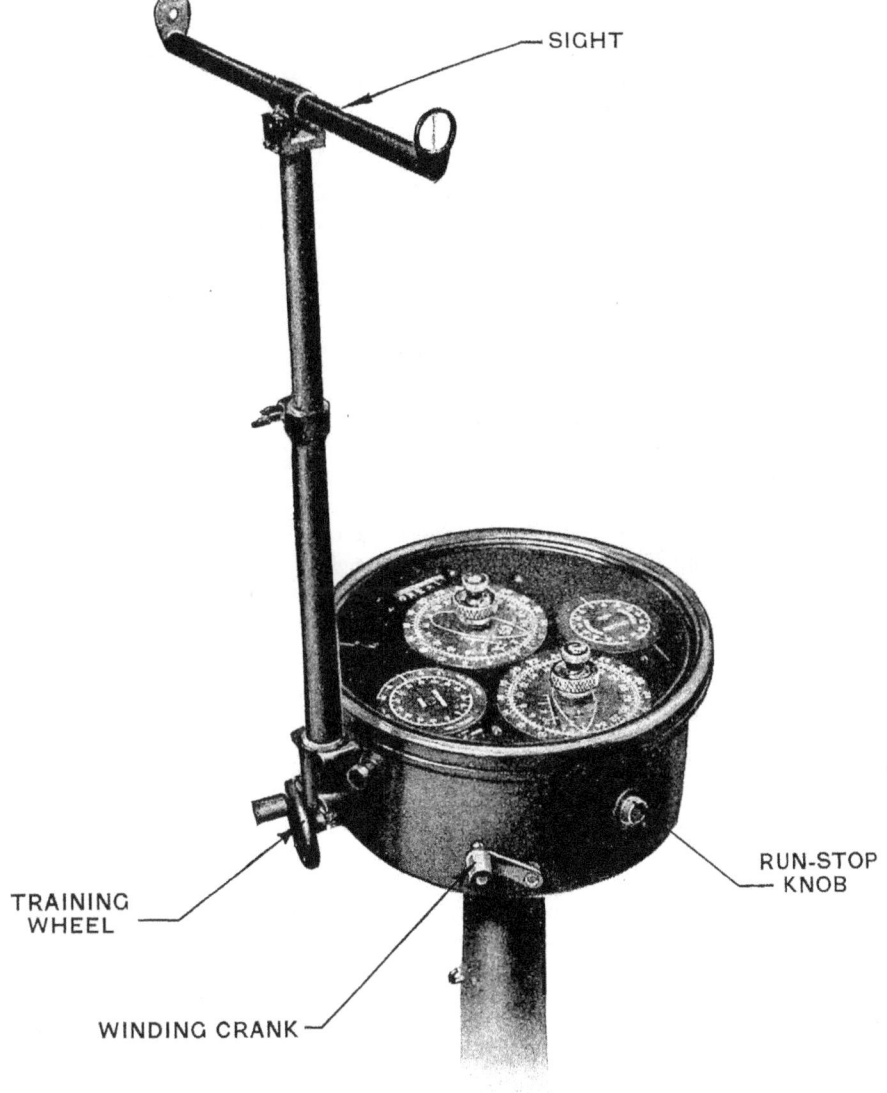

FIGURE 14D1. The Baby Ford range keeper.

THE SURFACE PROBLEM: PRACTICAL SOLUTION

minor- and intermediate-caliber batteries the same functions as the plotting board performs for the main battery. Every range keeper or computer generates present range. Although obsolete, the Baby Ford is the only one which is not complicated by computing devices that also calculate ballistic quantities. It does just these three things:

1. Computes range rate in knots.
2. Computes bearing rate in knots.
3. Generates present range.

This range keeper can be employed to maintain approximations of advance range and deflection. However, the manner in which this is done is not included in this section. The principles of the mechanical determination of advance range and deflection are considered in a subsequent section.

14D2. General description. The instrument is mounted on its own pedestal so that it may be trained on the target by means of the *training wheel* and *sight*. The mechanism is driven by a clockwork motor which is wound by the *winding crank* and controlled by the *run-stop knob*.

The face of the instrument, shown in figure 14D2, carries the various dials and indicators. In the upper-right sector is a *timing pointer* which revolves once every 20 seconds when the motor speed is correct. Speed is checked against a stop watch, and may be adjusted if necessary. The *winding pointer* next to the timing pointer indicates when the spring should be rewound.

The generation of present range is simply the continuous process of adding increments of range to the initial value of range. The *range counter* indicates the sum of initial range entered through the *range crank* and the cumulative total of the range increments computed by the mechanism.

Range rate dR is indicated in knots on the *range rate dial* opposite its *fixed index*. The dial is graduated from 0 to 70 knots for both increasing and decreasing ranges.

Linear relative bearing rate (RdBr) can be read from the *bearing rate dial*. At the time the instrument was first produced, some sight-deflection scales were calibrated in knots, and the reading of the dial was used directly on the sights. Consequently, the graduations of the dial are arranged so that a reading of 50 knots represents a zero bearing rate, and the dial is labeled *deflection*. The bearing rate is determined by taking the difference between 50 and the actual dial reading opposite the *fixed index*.

The ship dial groups are arranged to present a graphic picture of the vector diagram for the problem set on the instrument. The *target dial* (T) and the *own-ship dial* (OS) are each inscribed with a ship outline, a compass rose and a speed scale. Surrounding each ship dial is a *compass ring*. Between the two ship dial groups on the line joining their centers is a *fixed pointer* which represents the line of sight.

14D3. Inputs. Relative target bearing (Br) is introduced automatically when the range keeper is trained on the target. The own-ship dial is so assembled in the mechanism that the ship outline fore-and-aft axis always remains parallel to the actual own-ship fore-and-aft axis. The sighting axis is parallel to the line joining the centers of the ship dials, hence the fixed pointer reads Br on the own ship dial.

Own-ship course (Co) is introduced through the *own-ship course knob* over the own-ship dial. When the knob is depressed, it engages the own-ship compass ring which is in turn geared to the target dial group. The setting is made by turning the knob until the own-ship compass ring reads Co opposite the bow of the own-ship outline. Then true target bearing (B) may be read from the own-ship compass ring opposite the fixed pointer. This setting also rotates the target dial for reasons explained in the next article.

Target angle A is set after Co is set, by means of the *target angle knob* over the target dial. When the knob is depressed, it engages the target dial. The knob is turned until the target dial reads A against the fixed pointer. If Ct were known, it could be used instead of A by setting the bow of the target outline opposite Ct on the target compass ring. However, it is customary to set A, for this value is more conveniently estimated by the spotter.

Target speed S is set by depressing and turning the *target speed knob* until the pointer on the

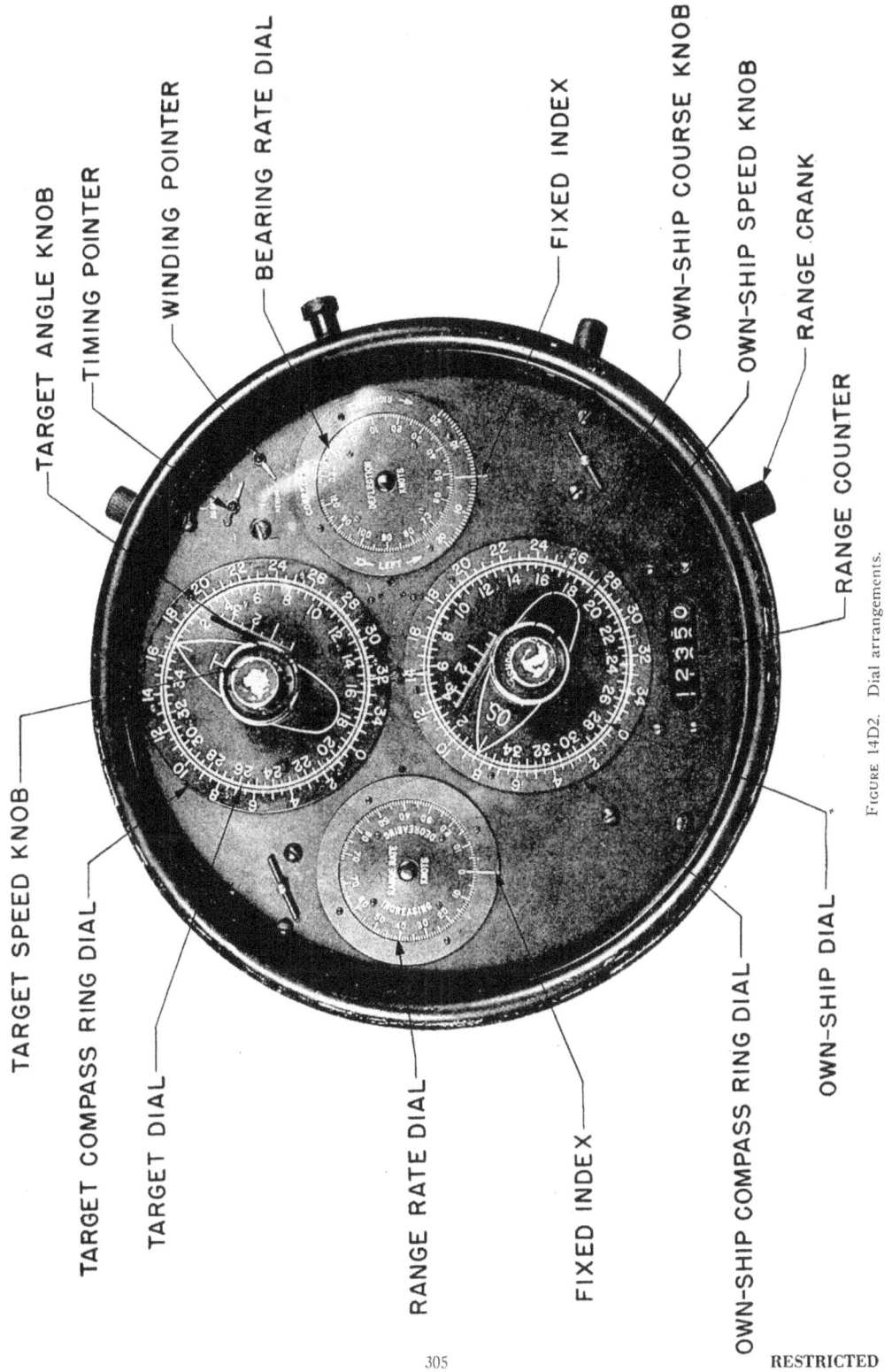

Figure 14D2. Dial arrangements.

speed scale indicates the value of S. Similarly, own-ship speed So is introduced through the *own-ship speed knob*.

14D4. Maintenance of set-up. The set-up of the angular quantities is subject to change due to three factors. Mere continuation of ship movements along their initial courses causes equal changes in A, B, and Br. A change in Ct causes a change in A, but no instantaneous change in B or Br. A change in Co causes a change in Br, but no instantaneous change in A or B.

The own-ship and target dial groups are geared together in such a manner that changes due to continued movement of the actual ships are automatically entered into the set-up. Continuous training of the range keeper on the target keeps A, B, and Br in agreement with actual conditions, assuming that the set-up was correct in the first place. For example, a 5° increase in Br causes a 5° increase in both A and B. In the range keeper, the own-ship dial with its compass ring, and the target dial with its compass ring will all rotate counterclockwise 5° with respect to the fixed pointer, and thus will read the correct values of A, B, and Br.

If the target changes course, the change must be entered through the target-angle knob. Only the target dial will rotate, for the change in Ct causes an equal but opposite change in A, and the target compass ring must stand still to meet this condition.

If own ship changes course, there is an equal but opposite modification in Br. The instrument receives the change in Br automatically, and both dial groups rotate with respect to the fixed pointer by the amount of the change. As explained above, this movement will change the set-up of A and B, while the change in Co causes no alteration in the actual values of A or B. The error thus introduced in the set-up of A and B is eliminated by setting Co on the own-ship compass ring. This setting rotates the two compass rings and the target dial, but not the own-ship dial, and the net result is that the compass rings and target dial stand still with respect to the fixed pointer.

14D5. Summary of set-up. It is to be noted that dR and RdBr do not depend upon Co and Ct, but upon A and Br. The use of Co is merely for convenience in maintaining the correct values of A and Br set in the range keeper. If Co were not used, it would be necessary to alter the setting of A each time own ship changed course, and the operator would be confronted with the problem of having to remember the value of A at the instant own ship began such a change.

The introduction of the angles and speeds into the range keeper is the equivalent of drawing the vector diagram *to scale*. Under each ship dial is a vertical pin which corresponds to the arrowhead of the vector in the diagram. The distance of the pin from the center of the dial is equivalent to the length of the vector, and is determined by the speed setting. The direction of the pin from the center represents the direction of the vector in the diagram, and is established by the setting of A or Br. The positioning of these pins is thus the introduction of two vectors into the mechanism. The only remaining element needed for the generation of present range is the initial range.

14D6. Symbols and definitions. Some additional symbols and terms are useful in discussions of range keepers, for they are commonly used in Bureau of Ordnance publications. Terms needed in the remainder of this section are:

SYMBOL	DEFINITION
cR	*Generated present range.* The value of present range as computed by the range keeper.
$\triangle$cR	*Increments of generated present range.* The changes in range during increments of time as computed by the range keeper.
$\triangle$T	*Increments of time.* The increments of time during which $\triangle$cR occurs.
jR	*Initial range.* The range initially set on the range keeper when it is started.

The letter c before a quantity is used to distinguish between the mechanically-computed value of the quantity and the observed value of the same quantity. Thus, at any instant the observed target distance is R, and the target distance computed by the range keeper is cR. The letter j is used before a quantity to indicate an initial, or corrective setting of that quantity, to distinguish that

CHAPTER 14

setting from the continuously changing value of the quantity. Thus, R is always changing, but jR is one particular value of R which existed at some previous instant.

14D7. The range-keeping equation. Expressed in words, the range at any instant is equal to the range at some previous instant plus the sum of all the increments of change in range during the time elapsed from the previous instant. This relation may be expressed as *this simple equation*:

(Generated present range) = (Initial range) + (Increments of generated present range), or
$$cR = jR + \triangle cR$$

From the analysis of the range-keeping problem in Chapter 13, Section A, the student will recall that :*

$$\triangle cR = \triangle T \times dR = \triangle T(Y_o + Y_t) = \triangle T(S_o \cos Br + S \cos A)$$

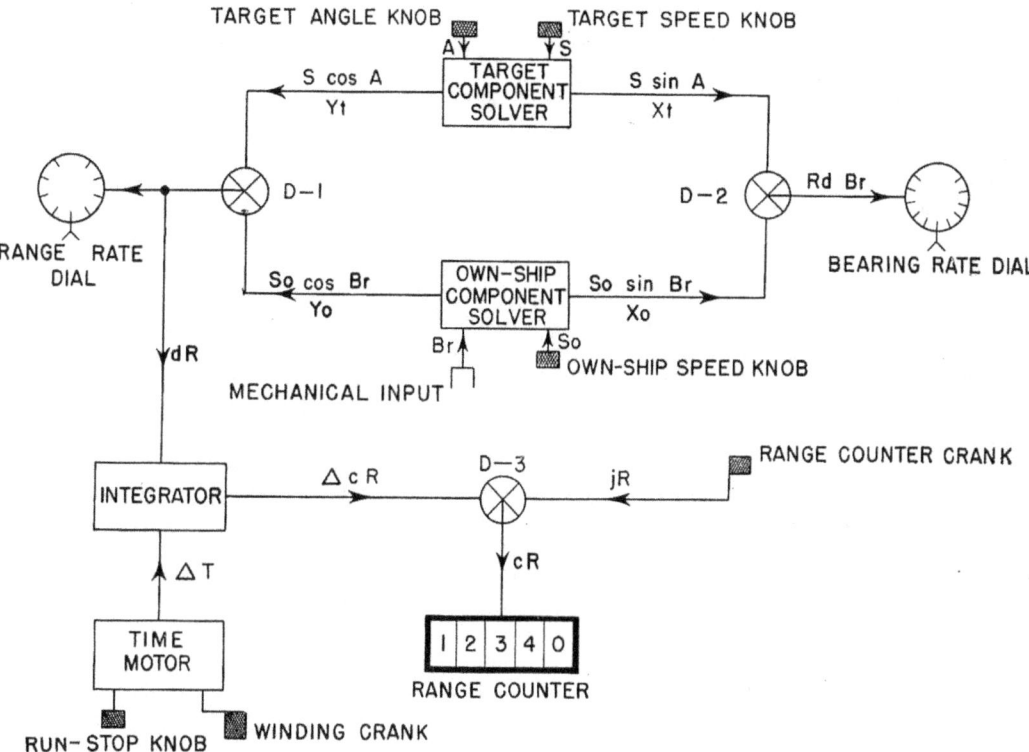

FIGURE 14D3. Simplified schematic diagram of range keeper, Mark 2.

Substituting this value of $\triangle cR$ in the first equation:
$$cR = jR + \triangle T(S_o \cos Br + S \cos A)$$

This equation may be called the range-keeping equation, for it tells the complete story of the process of keeping range. There are three basic mechanisms which are suitably combined to solve this equation continuously. Once the individual mechanisms are understood, a glance at the equation will show how the mechanisms must be arranged to solve it. Furthermore, the equation shows the inputs that are required.

14D8. Mechanisms. The principal computing units in the Baby Ford, and in the present range section of any range keeper built for surface problems are (1) component solvers, (2) differentials, and (3) integrators. The details of these units may vary among different range keepers, but

---
*For purposes of simplification, the signs used in these equations and others in subsequent sections of the text are not always strictly correct. In general, plus signs are used, whereas in certain cases minus signs are in order. A student should refer to a diagram of the situation to determine what signs actually apply when evaluating any of the quantities involved.

the basic principles do not. A simplified schematic drawing of the arrangements of the mechanisms in the Baby Ford is shown in figure 14D3.

Component solvers. The component solver is a device for resolving a vector into components with respect to a reference line. In the Baby Ford there are two component slides under each dial group. One of the slides (range slide) under each dial is free to move parallel to the line joining the dial centers, or in other words, it can move only in the LOS. The other slide (cross slide) under each dial can move only across the LOS. The pin under each dial, representing the vector as explained in Article 14D5, fits in crossheads in these slides. The position of each slide then represents a component of the vector. The positions of the range slides therefore correspond to So cos Br and S cos A, two of the quantities in the range-keeping equation. The positions of the cross slides represent So sin Br and S sin A, but these quantities are not needed to keep range, since A and Br are kept in continual agreement with observed conditions by training the range keeper.

Differentials. A differential is an arrangement of gears which can perform algebraic addition. It can add positive and negative quantities, or in other words, it can determine the sum or difference of two quantities. The next step in the solution of the equation is to add So cos Br and S cos A; hence these quantities go to differential D-1, which adds them. The output of D-1 is dR, which goes to the range rate dial, and to the integrator. Similarly, So sin Br and S sin A go to D-2, which adds them. The output of D-2 is RdBr, which goes to the bearing rate dial for indication, but is not used elsewhere in the mechanism.

Integrator. An integrator is a device which continuously multiplies a rate by time increments to obtain increments of change due to the rate. The disk type is used in generating increments of range. It consists essentially of a flat circular disk revolving at a constant speed, an output roller, and a pair of steel balls which transmit the rotation of the disk by friction drive to the output roller. The position of the balls with respect to the center of the disk determines the rotational speed of the output roller. The balls are positioned by dR, the constant speed of the disk represents time, and the output of the roller represents $\triangle T \times dR$, or $\triangle T (So \cos Br + S \cos A)$ in the equation.

Time motor. The time motor in the Baby Ford is a clockwork mechanism, which runs at a constant speed. A governor on the motor holds the speed constant, and a regulator on the governor provides for adjustment of the speed when necessary.

Generated present range. The integrator output cR goes to D-3. The other input to D-3 is jR which must be introduced by hand. The output of D-3 is cR, the solution of the equation. It is made available for use in the range counter which is a group of cylindrical drums that indicate cR to the nearest 50 yards.

14D9. Elementary rate control. The values of A, Br, S, So, and jR are the inputs upon which this range keeper bases its computations. Of these, Br, So and jR are usually known quite accurately. The spotter's estimate is generally the source for A and S, and obviously the accuracy of these observed quantities is limited. While a tracking board eventually produces values of A and S, it is too slow. The range keeper provides a quicker means of checking A and S.

Error in generated range. It is apparent that cR will not long remain in agreement with R if A and S are incorrectly set on the range keeper. The operator compares the range generated by the range keeper with the measured range. When disagreement exists, he adjusts the settings of A and S to find values which will cause cR to remain in agreement with R. In doing this he is controlling the rate dR. There is only one combination of A and S that will *maintain* the correct value of dR, and that combination is the one which agrees exactly with the actual motion of the target.

Error in rates. Suppose that a situation exists as shown in figure 14D4. At position 1, own ship is on course Co at speed So, and its vector is correctly set in the range keeper. The target is *actually* on course Ct at speed S, but due to an error in estimating the target motion, its vector is set in the instrument as though it were on course Ct' at speed S'. The respective target angles are A_1 and A_1' and the line component of the vector S' is equal to the line component of S. Under these conditions, dR computed by the range keeper agrees with the *actual* value of dR. It is apparent that there are many combinations of A_1', Ct' and S' which will have the same line component.

After an interval of time has elapsed, the two ships will have advanced along their actual courses

CHAPTER 14

distances proportional to their actual speeds, and will be at position 2. The actual target vector is still S; the target vector set in the instrument is still S'. The target angles between S and the LOS and S' and the LOS will have changed by the same amount that Br has changed, and the line components will be Yt_2 and Yt_2'. Evidently the computed value of dR will no longer agree with the actual value of dR, cR will be out of agreement with R, and it will be necessary to further adjust the values of target angle and target speed set on the instrument.

At position 1, computed dR was correct. But computed RdBr was in error by the difference between Xt_1 and Xt_1'. The only set-up of the range keeper which would produce the correct values of RdBr and dR at the same time is that in which A and S are used for the settings. Had both range and bearing rates been correctly set on the instrument at position 1, both of these rates would be correct at position 2, and cR would be in agreement with R. Thus, if the correct target track is to be established, the operator must control both rates.

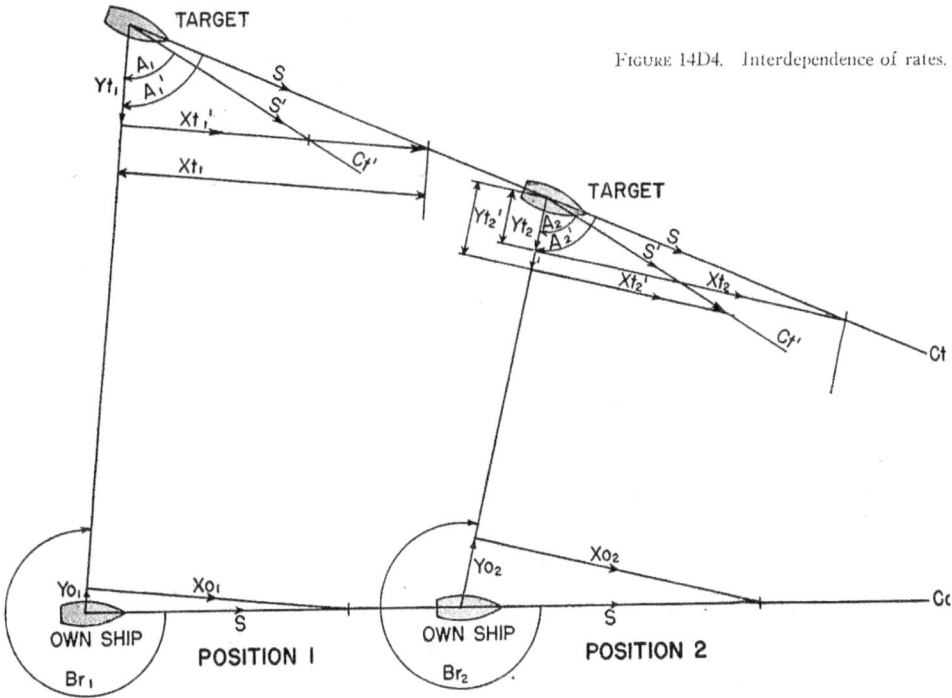

FIGURE 14D4. Interdependence of rates.

Rate control on the Baby Ford. The Baby Ford gives no indication that computed RdBr is right or wrong. The best the operator can do is to use his judgment in changing A and S, following a few simple rules. The target angle set on the instrument indicates whether target speed or target angle should be changed to correct dR. For example, if A is 0° or 180°, a change in S will have the greatest effect on dR. If A is 90° or 270°, a change in A will have the greatest effect. For intercardinal values of A, both A and S should be changed.

Use of generated bearing. When there is no indication of the correctness of computed RdBr, the operator will have difficulty in establishing the correct settings of A and S. If target bearing is generated in a manner similar to the generation of range, then a disagreement between generated bearing (cBr) and Br will indicate that computed RdBr is wrong. When RdBr is wrong and dR is correct as at position 1 in the diagram, the operator can adjust A and S to correct RdBr without changing dR by maintaining the dial reading of dR at the value indicated before the adjustment is made. Similarly, dR can be corrected without changing RdBr if conditions require such an adjustment.

RESTRICTED

THE SURFACE PROBLEM: PRACTICAL SOLUTION

Even this process contains an element of trial and error, but the indications of both cBr and cR reduce to a minimum the doubt as to what corrections are needed. In the installations where the Baby Ford was used, the errors resulting from a slightly incorrect set-up were not serious. However, the Mark 1 range keeper for the main battery and all modern main-battery and dual-purpose range keepers and computers generate bearing for use in rate control among other things.

E. BASIC BALLISTICS COMPUTING MECHANISMS

14E1. Introduction. It has been mentioned previously that the total ballistics are not fixed quantities any more than is present range. In fact, the ballistics may change quite suddenly in situations where the corrections for ship motions are large parts of the totals. The mechanical computation of ballistics is an integral part of all modern range keepers and computers.

Since the term *range keeper* does not fully convey the meaning of what modern instruments accomplish, the more descriptive term *computer* is applied to the latest designs. There are still various marks and modifications of range keepers being manufactured, in addition to the computers, but from the standpoint of purposes accomplished, the difference in names is not significant.

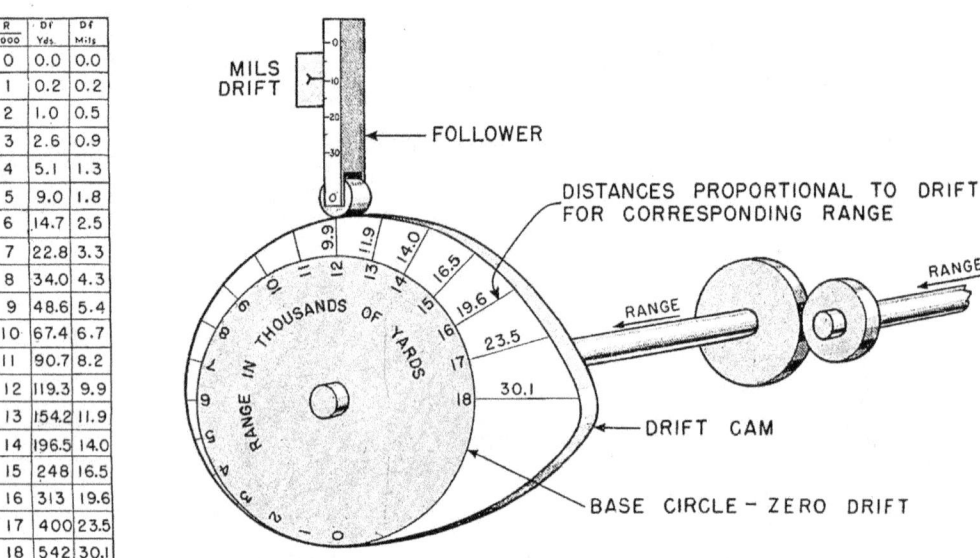

FIGURE 14E1. Plate-type cam.

Those elements of the ballistics problem which are computed mechanically differ among the various marks and modifications. The accuracy of the automatic corrections also varies, for approximations are frequently made to simplify the mechanisms. Among the devices used to compute ballistics are gears, cams, and multipliers.

14E2. Gears. It has been explained that gears can be used to multiply a variable quantity by a constant. Gear constants play an important part in computing ballistics, either in obtaining part of a correction, or in obtaining an approximate value of a whole correction. The employment of gear constants is best illustrated by the example of a ballistics-computing range keeper, which is presented in the next section.

14E3. Cams. A *cam* is a device, rotating or sliding, which imparts a peculiar motion to a part called the *follower*. The surface of the cam can be cut to any one of an infinite number of shapes, so that any desired motion of the follower may be developed within practical limits. As used in range keepers, the cam is generally a computing device which delivers a function of a variable. For example, drift and time of flight are *functions* of range; sines and cosines are *functions* of angles.

CHAPTER 14

It may be helpful to the student to think of a cam as the mechanical equivalent of the data in a column of the range table.

Plate cams. One of the simplest types is the plate cam, such as is shown in figure 14E1. The cam is rotated by range, a *variable*, and the motion imparted to the follower is drift, a *function* of the variable. The scale on the follower indicates drift against the fixed index. The surface of the cam is cut to the shape determined by the lengths of the radial lines. The data used is simply that shown in the range table.

It is apparent that the follower of this type of cam must be held against the cam surface by some positive acting force, such as that provided by a spring. In computing mechanisms, a slightly different type of cam is often used to supply positive motion to the follower.

Positive motion cams. The cam shown in figure 14E2 provides positive motion to its follower. Secured to the underside of the follower is a roller which is constrained to ride in the cam groove cut in the circular plate. The cam contour is determined in essentially the same manner as it is for a plate cam, making due allowance for the circular movement of the follower. The output of the cam is delivered from the gear sector of the follower to a gear on a shaft; hence the rotation of that shaft is a measure of the *function* of the variable which rotates the cam.

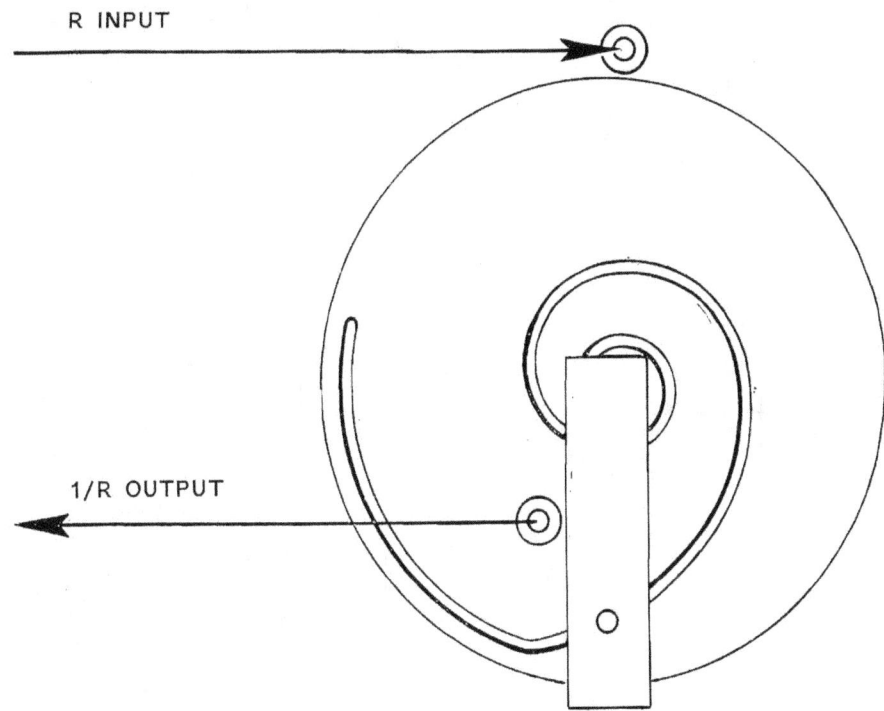

FIGURE 14E2. Positive-motion cam.

This type of cam is used to compute time of flight, drift, trigonometric functions, reciprocals such as $1 \div R$, and many other quantities. The construction is not necessarily always the same as that illustrated. Sometimes the follower is arranged to move in a straight line, and a rack attached to it drives the output receiving gear.

14E4. Reducing cam displacement. If a cam is cut to supply the full value of a function of a variable, it must necessarily be fairly large to provide accurate outputs. The size of the cam can be reduced if it is made to supply only part of the function. Figure 14E3 shows a drift cam which takes care of only part of the total drift correction.

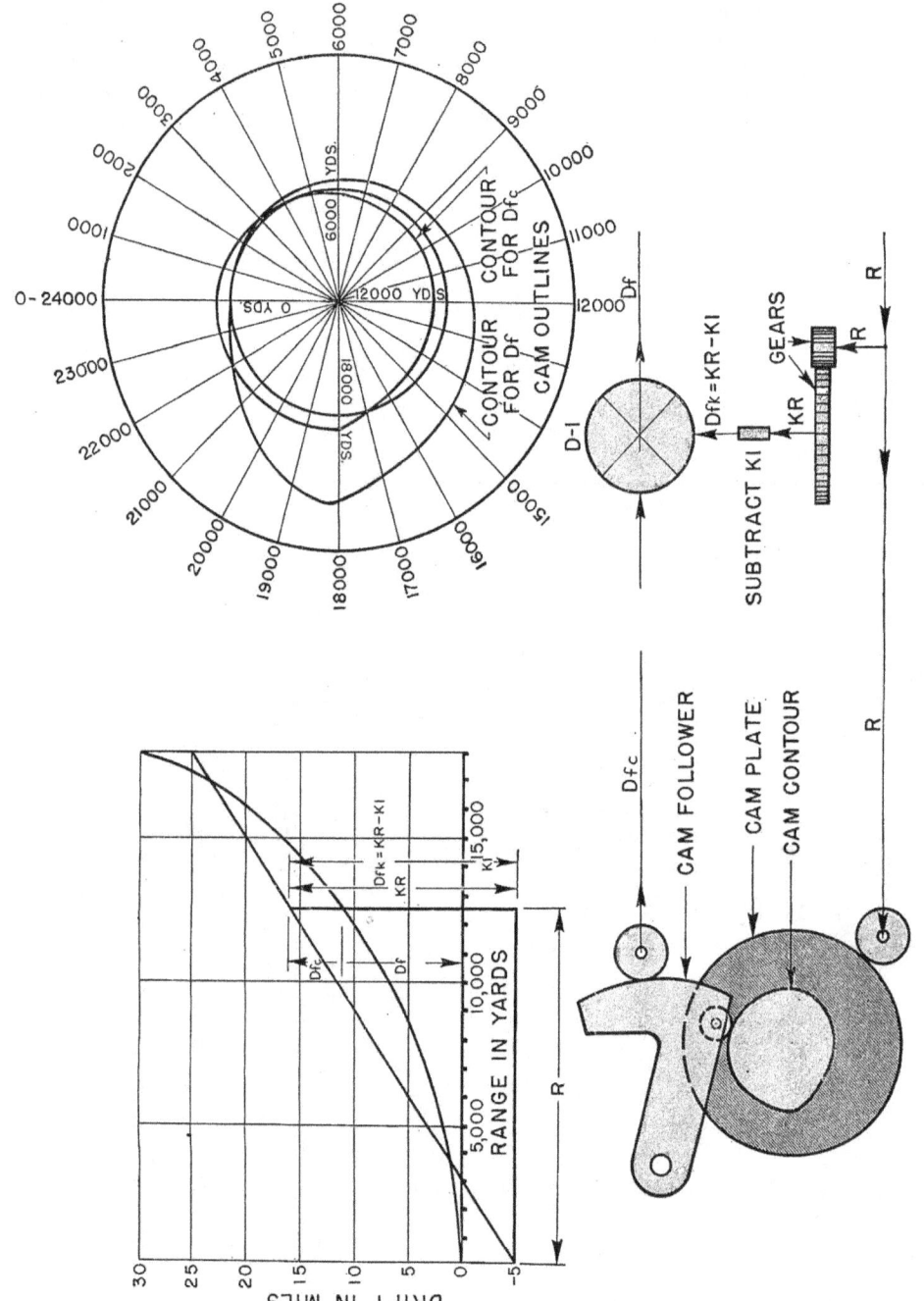

Figure 14E3. Partial-correction cam.

CHAPTER 14

The drift curve is shown in the graph. The straight line is drawn so that its maximum displacement from the curve is as small as possible. The drift at any range is equal to the drift (Dfk) indicated by the straight line, plus the difference (Dfc) between the drift indicated by the curve and the drift indicated by the straight line. (The difference is positive when the straight line is below the curve; negative when the straight line is above.) The cam computes the difference Dfc, and since it is considerably less than the total value of drift Df, the size of the cam can be reduced with no loss in accuracy.

The value of Dfk is computed by multiplying range by a gear constant, and subtracting a second constant from the product. It is evident that any ordinate Dfk of the straight line is equal to $KR - K1$, for the ratio between the sides of the triangle is fixed, and remains the same for any value of R. Thus,

$$Df = Dfk + Dfc = KR - K1 + Dfc.$$

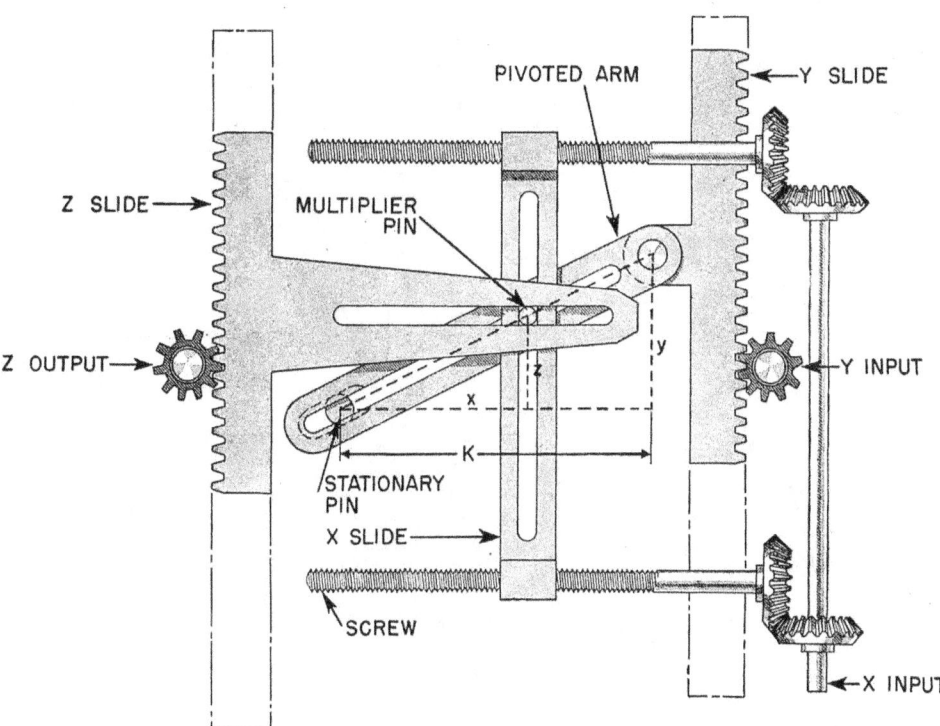

FIGURE 14E4. Screw-type multiplier.

The mechanism required to solve this equation is also shown in figure 14E3. Range R rotates the cam, and the follower delivers Dfc to one side of differential D-1. Range R likewise goes through gears where it is multiplied by K. When the mechanism is assembled, an angular or rotational offset is introduced to remove K1. Actually, both of the constants may be taken care of in the gearing of the shaft to the differential, but for illustrative purposes a separate set of gears is shown, and the rotational offset is indicated by a small rectangle in the shaft. Actual schematic drawings do not show these symbols representing the introduction of constants, for there is no separate computing mechanism involved.

14E5. Multipliers. Most of the ballistic corrections involve the multiplication of two variables. For example, the correction for motion of the target in the LOS during time of flight is the product of the target line component and time of flight. Both of these quantities are variables, the first depending upon the target speed and target angle, the second upon range. While gears can be

RESTRICTED

THE SURFACE PROBLEM: PRACTICAL SOLUTION

used to multiply a variable and a constant, another type of computing device is needed to multiply one variable by another variable.

Screw-type multipliers. The various types of multipliers operate on the principle of similar triangles as illustrated by the screw-type multiplier in figure 14E4. The two inputs are X and Y; the output is Z. The stationary pin is a fixed distance K from the movement line of the pin attached to the Y slide. The multiplier pin is positioned sidewise by movement of the X slide, the distance of the X slide slot from the stationary pin being X. The up-and-down position of the multiplier pin is established by the location of the pivoted arm, which is set by the input Y.

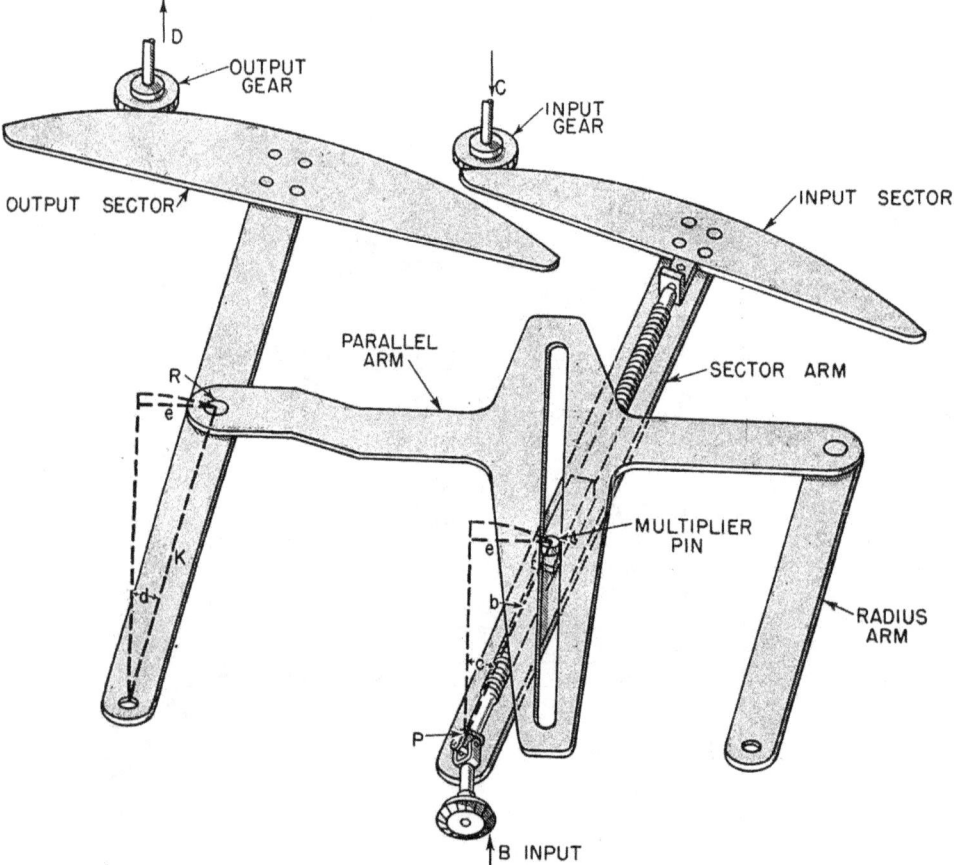

FIGURE 14E5. Sector-type multiplier.

The position of the multiplier pin established by X and Y determines the value of the distance Z, which is communicated to the output gear by the Z slide. The length of Z is a direct measure of the product of variables X and Y. In the similar triangles:

$$\frac{Z}{X} = \frac{Y}{K}, \text{ and}$$

$$KZ = XY$$

The distance K is fixed; hence K is a constant which may be removed by the selection of proper gear ratios in the introduction of the inputs or in the transmission of the output.

The multiplier as illustrated can handle both positive and negative Y inputs, but X inputs of only one sign. Provision for both positive and negative X inputs can be made by placing the stationary pin at the center of the pivoted arm.

Sector-type multipliers. Another multiplier is the sector-type illustrated in figure 14E5. This multiplier consists essentially of an input sector and an output sector connected by a parallel arm. The right sector, which receives the inputs, carries a long screw. Input B rotates the screw and thus locates the multiplier pin block which is threaded on the screw. The pin slides in a vertical slot in the parallel arm. This arm is attached to the radius arm and to the output sector so that it always remains parallel to its initial position. Input C sets the angular position of the input sector. Output D is the product of the two inputs, B and C.

The distance between P and the multiplier pin is b, the equivalent of the B input. The angle c is the equivalent of the C input, and the angle d is the equivalent of the output, or the product of b and c. The lateral movement of the parallel arm, and of the pin R on the output sector is e. The distance from R to the pivot of the output sector is a constant K. From the two right triangles established by the inputs:

$$e = b \sin c = K \sin d$$

$$\sin d = \frac{b \sin c}{K}$$

When c and d are sufficiently small, they are proportional to their sine functions, and the equation becomes:

$$d = \frac{bc}{K1}, \text{ and } D = \frac{BC}{K1}.$$

The difference between K and K1 accounts for the constant which expresses the proportion between an angle and its sine function for small values of the angle. Since K1 is a constant, it may be removed by selection of the gear ratios through which the inputs and outputs are transmitted. The sector-type multiplier is not as accurate as the screw-type because of the approximation used in the equations above. However, the errors introduced are within allowable limits where the sector type is used.

Division with multipliers. The sector-type multiplier can be employed for direct division if one of the inputs is delivered to the output sector, and the output is taken from the input sector. Other types of multipliers may be used for division if the reciprocal of the divisor is employed as a multiplier. The input, or divisor, is first put through a reciprocal cam, which delivers the reciprocal to the multiplier. Thus, if the division $P = M \div R$ is required, R is inverted by a cam to obtain $1 \div R$, which is then supplied to the multiplier as an input. The multiplier then performs the multiplication $P = M \times (1 \div R)$.

F. AN ELEMENTARY COMPUTER

14F1. Functions. This discussion of the Mark 7, mod. 5 range keeper, shown in figures 14F1 and 14F2, is presented to illustrate the mechanical determination of sight angle and sight deflection. An understanding of its operation will provide a foundation for future study of all modern range keepers and computers. This range keeper was designed for use with the 5"/51 cal. gun, and for mounting adjacent to the director. It performs the following functions for either service or target charge initial velocities:

1. *Generates present range* which appears on the *present range counter.* This value is used as the basis for the ballistic computations. It provides the operator a means of checking the set-up of the instrument by comparing generated present range with measured present range.

2. *Computes advance range* which appears on the *advance range counter.* This value is normally used as the basis for automatic computation of sight angle in minutes, but it may be read in yards from the counter and transmitted verbally to the guns and director for use as sight-bar range.

THE SURFACE PROBLEM: PRACTICAL SOLUTION

3. *Computes sight deflection* which appears on the *mils deflection counter* as a scale reading. This value may be transmitted verbally to the guns and director for use as the scale setting.

4. *Electrically transmits sight angle and sight deflection* to the guns, where it appears on dials which indicate the sight settings to the sight setter. Sight angle, computed from the value of advance range in yards, is transmitted in minutes. Sight deflection is transmitted in mils as a scale setting.

5. *Sounds a time-of-flight signal* at the end of the time of flight to identify salvos for the spotter.

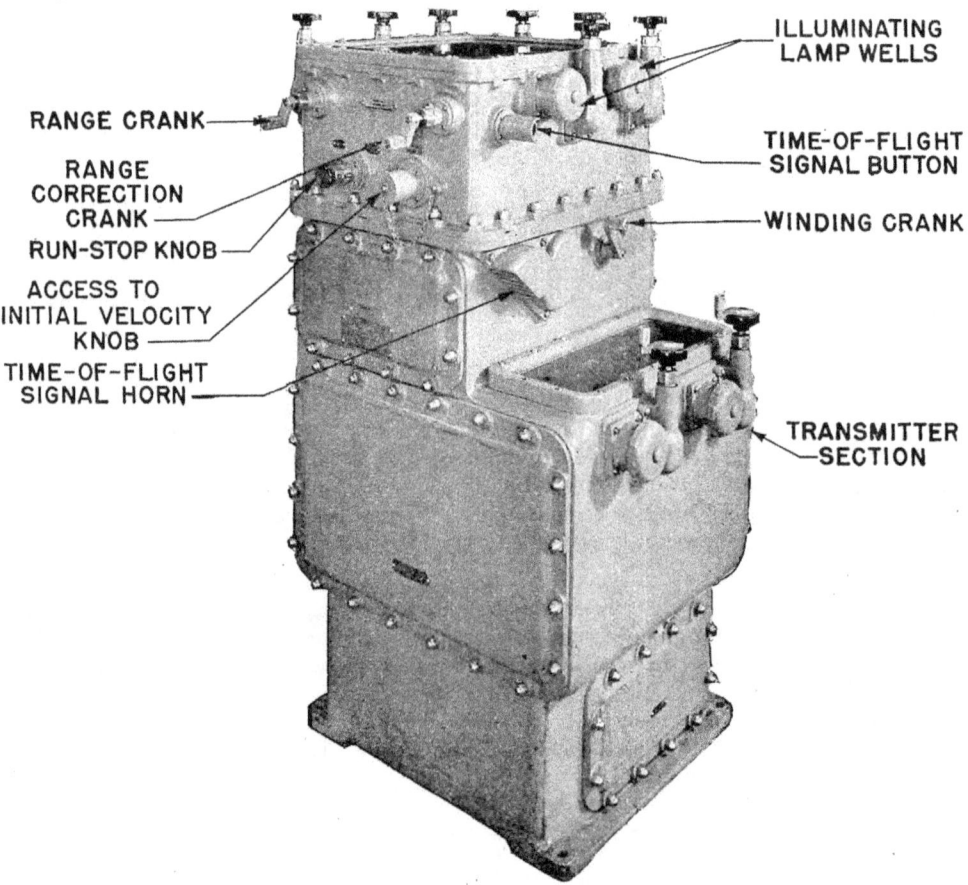

FIGURE 14F1. Range keeper, Mark 7, mod. 5; front right-side view.

14F2. Automatic ballistic corrections. The ballistic corrections computed by this instrument are close approximations of the values which would be obtained by computations based on the range table. The factors automatically handled are:

1. Motion of own ship in the LOS.
2. Motion of target in the LOS.
3. Motion of own ship across the LOS.
4. Motion of target across the LOS.
5. Cross wind.
6. Drift.

14F3. Initial ballistic. The elements not corrected automatically must be introduced by

CHAPTER 14

hand as the *initial ballistic*. The initial ballistic includes the calculated corrections for air density, range wind, change in I. V. due to powder temperature, and any arbitrary (including cold-gun correction) or control ballistic. Erosion is not normally considered in the initial ballistic, as it usually is automatically compensated at the guns. The continuous computation of advance range and deflection, and the rapid transmission of sight angle and sight deflection, make dead time practically zero; hence it is not considered. All the corrections normally included in the deflection gun ballistic are automatically handled, and no initial ballistic is required in deflection other than for arbitrary or control ballistics.

14F4. Inputs. The inputs required by the instrument are essentially the same quantities as those used in solving the ballistics problem analytically, plus the initial ballistic to compensate the elements not automatically handled. These inputs are:

1. *Relative target bearing* (Br). Supplied automatically and continuously from the director through an input shaft not shown in the figure. The value is indicated on the *own-ship dial* opposite the *fixed pointer* which corresponds to the LOS.

2. *Ship course* (Co). Normally received electrically from the ship's gyrocompass by a course receiver within the range keeper. If the receiver is energized, the *receiver indicator* reads *on*. If the receiver fails, own-ship course must be introduced through the *true-bearing knob*, setting Co on the *own-ship compass ring* opposite the bow of the own-ship outline.

3. *Target angle* (A). Set by hand through the *target-angle knob,* and indicated on the *target dial* opposite the fixed pointer. Target course (Ct), if known, may be set instead of A by placing the bow of the target outline opposite Ct on the *target compass ring.*

4. *True wind direction* (Bw). Set by hand through the *wind-angle knob,* and indicated by the *wind pointer* on the target compass ring. The pointer indicates the direction from which the wind is blowing.

5. *Speeds of ship, target, and wind* (So, S, and Sw). Set through their respective speed knobs and indicated on the corresponding speed dials.

6. *Initial range* (jR). Initially set through the *range crank,* and indicated on the present-range counter. Subsequent differences between measured present range (R) and generated present range (cR) are corrected through the range crank.

7. *Initial ballistic.* Introduced through the *range-correction crank,* and indicated against the *fixed range correction dial.* When the crank is turned, both the *total pointer* and the *individual pointer* move together and indicate the input. After the setting is made, the crank is pushed inward and the individual pointer snaps back to zero, ready to indicate the amount of the next individual correction.

8. *Range spots.* Introduced through the range-correction crank in the same manner as the initial ballistic. The amount of the spot is read on the dial opposite the individual pointer. The sum of the initial ballistic and all spots applied is shown on the dial opposite the total pointer.

9. *Deflection spots.* Introduced through the *deflection-correction knob,* and indicated cumulatively on the *deflection-correction dial.*

10. *Initial velocity.* Selected by the *initial-velocity knob* and indicated on the *initial-velocity dial.* The choice is either service (3,150 f.s.) or target (2,300 f.s.) velocity, which corresponds in the analytical solution to selecting either range table O. P. 185 or O. P. 183, upon which the design of the range keeper is based. There is no adjustment for loss of velocity from either of these rated values.

14F5. Operation. When action is imminent or when drill begins, the course receiver and illuminating lamps for the dials will be energized at the fire-control switchboard. The range-keeper operator checks own-ship course (Co) indicated on the range keeper with the reading of the ship compass. As soon as the director trains on the target, Br is introduced automatically.

With the *run-stop knob* in the *run* position, the operator starts the instrument by winding the clockwork driving motor through the *winding crank* until the *winding indicator* shows *full.* (The

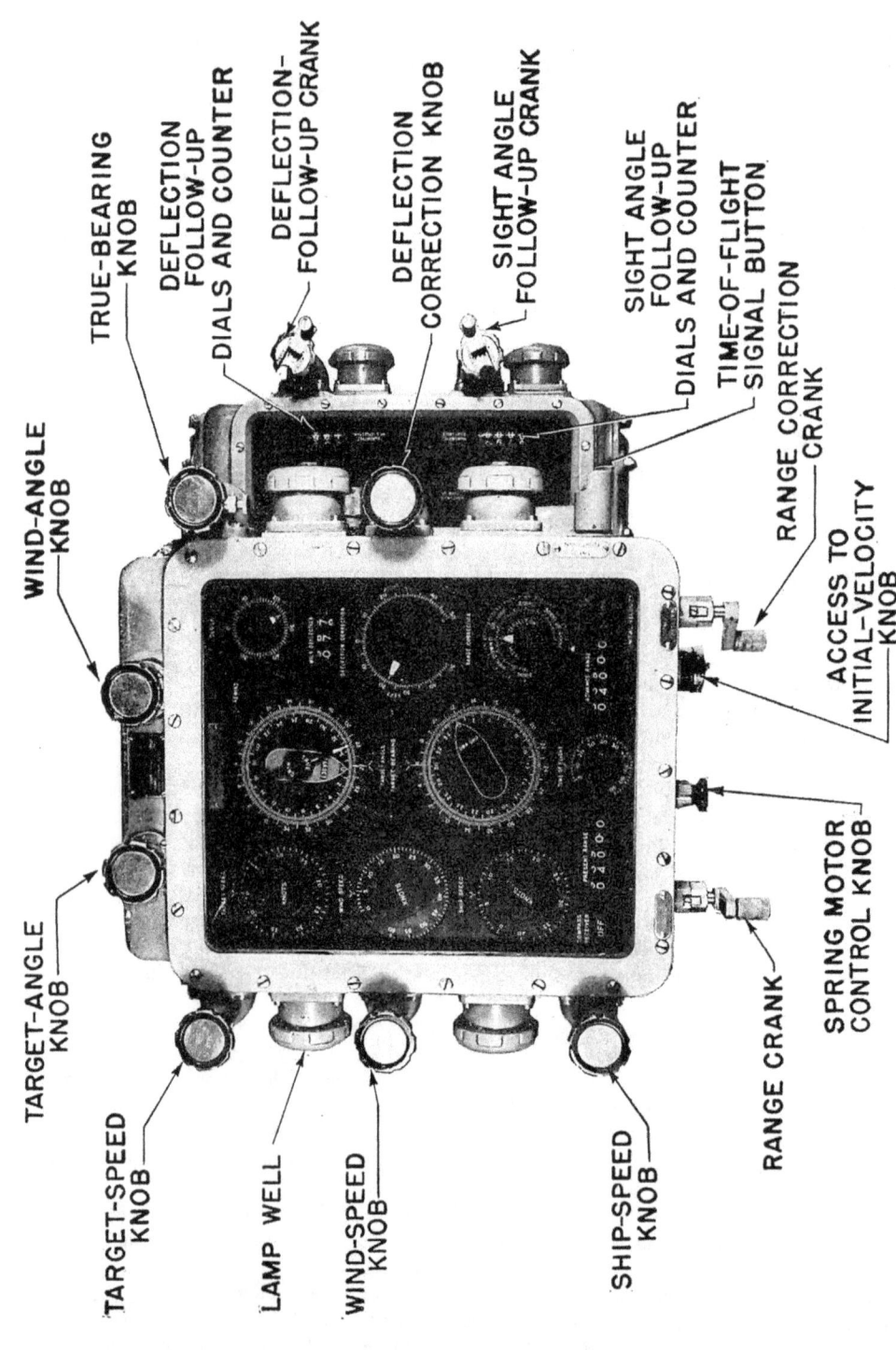

FIGURE 14F2. Range keeper, Mark 7, mod. 5; top view.

CHAPTER 14

motor should be allowed to run down when securing, so that the subsequent winding required for the next operation will free the grease on the driving spring.) The speed of the motor is checked for accuracy by comparing the *clock* on the instrument with a stop watch.

The range-keeper operator enters the hand inputs described above, using information from various sources. Ship speed may usually be read from an indicator near the range keeper. Wind direction and speed are usually available from indicators, but may be estimated by the spotter. The spotter supplies estimated target angle and speed, and estimated initial range, which is used unless a range-finder range is available. The initial ballistic is calculated by one of the control party.

The sight angle and sight deflection transmitter section, integral with the range keeper, is manned by an assistant operator. He operates the electrical transmitters by rotating the two *follow-up cranks* to keep index marks on concentric *follow-up dials* matched against fixed indexes. The rotation of the respective cranks is a direct measure of sight angle in minutes and sight deflection in mils, and causes the transmitters to send corresponding electrical signals to indicators at the guns. The indicators show the settings required on the sights, and the sight setters at the guns set the sights accordingly. The transmitted values may be read from the *transmitted angle counters* near the follow-up cranks. If the electrical transmission fails, these values may be communicated by telephone or other means.

Each time a salvo is fired, the operator pushes the *time-of-flight signal button*. Shortly before the splash occurs, the signal horn sounds. Time of flight computed by the instrument is indicated by the *time-of-flight dial*.

14F6. Rate control. The rate-control process of adjusting target angle and target speed to make generated present range agree with actual present range is carried out continuously by the operator, both before and after commencing fire. When generated present range showing on the present range counter is out of agreement with measured present range, the difference is supplied to the instrument through the range crank, and target angle and target speed are adjusted by trial to make the generated present range remain in agreement with measured present range.

Disagreement between generated and measured values will be evidenced by discrepancies between range-finder ranges and present range readings, or by unusually large spots. A large spot in range is generally applied through the range crank, since it is probably due to the use of an incorrect present range. Smaller spots, due to accidental errors in the total ballistic, are entered through the range-correction crank.

14F7. Functional operation. The arrangement of the mechanism is shown schematically in figure 14F5, which should be kept open for reference while the operation is being studied. Although the apparatus is designed to make computations for both service and target-charge initial velocities, for purposes of simplification the provisions for handling target-charge velocity are not shown on the diagram, nor are they included in this discussion. Further simplification is provided by grouping the principal units in sections according to their functions, rather than as they are actually located in the instrument.

The types of mechanical computing units in the instrument are:

1. The *differential*, which receives two quantities, adds them algebraically, and delivers the sum.

2. The *component solver*, which resolves a vector into its components with respect to a reference line (LOS). The inputs required are the length of the vector (speed) and the angle between the vector and the reference line (target angle, wind angle, or relative bearing).

3. The *integrator*, which continuously multiplies a rate (range rate) by time and delivers increments of change (change of range) due to the rate.

4. The *multiplier*, which receives two quantities, multiplies them and delivers their product.

5. The *cam*, which receives a quantity and delivers a function of that quantity.

14F8. Component section. The component section provides means for setting up the line-of-fire diagram and computes components of wind, own-ship motion, and target motion with respect to the line of sight. The outlines of the ships on the dials and the wind pointer represent the direc-

RESTRICTED

THE SURFACE PROBLEM: PRACTICAL SOLUTION

tion of the vectors, while the speed dials indicate their lengths. The angular quantities are illustrated in figure 14F3.

Introduction of angles. The own-ship dial, rotated by Br received mechanically from the director, indicates Br opposite the fixed pointer. One input to differential D-1 is Br, the other is Co received from the course receiver. The output of D-1 is B, the solution of the equation $B = Br + Co$. The compass rings of both ship-dial groups are driven by B, and serve as compass roses for setting true

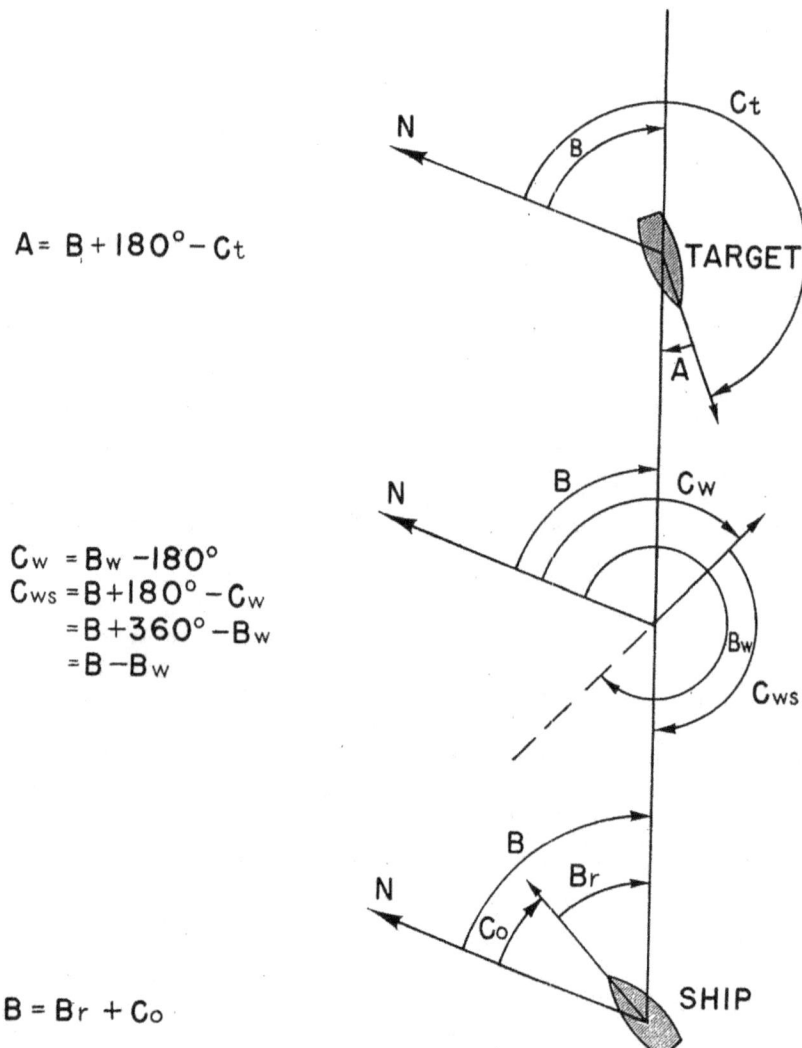

$A = B + 180° - Ct$

$Cw = Bw - 180°$
$Cws = B + 180° - Cw$
$ = B + 360° - Bw$
$ = B - Bw$

$B = Br + Co$

FIGURE 14F3. Determination of A, B, and Cws.

courses. The fixed pointer indicates B on the own-ship compass ring. If the course receiver fails, B must be supplied manually through the true-bearing knob by setting Co on the compass ring opposite the bow of the outline of own ship.

Differential D-2 receives B from D-1, Bw from the wind-angle knob, and delivers wind angle Cws to the wind pointer as the solution to $Cws = B - Bw$. The wind pointer is offset 180° so that

CHAPTER 14

it points to Bw on the target compass ring. Since the input through the wind-angle knob is Bw, the knob need be reset only when the true direction of true wind changes, which it is not likely to do during an engagement.

During manufacture, a rotational offset equivalent to 180° is added to the value of B supplied to D-3. The other input to D-3 is target course Ct from the target-angle knob. The output of D-3 is A, the solution to $A = B + 180° - Ct$. The target dial is driven by A, which is indicated opposite the fixed pointer. The bow of the target outline points to Ct on the target compass ring. The operator ordinarily uses A to establish the setting, but may use Ct if it is known. In any event, the input through the knob is Ct, and does not need changing unless the target changes course.

It is apparent that as Br changes while the ships move along their courses, the set values of A and Cws will be automatically altered by the same amount. Assuming A and Cws are precisely set in the first place, the setting of A will be continuously correct unless Ct changes, and the setting of Cws will be continuously correct unless Bw changes. The computed value of B remains constant during any variation in Co, since any alteration in Co causes an equal but opposite change in Br.

Component solvers. The values of A, Cws, and Br used to position the dials also enter the target, wind, and own-ship component solvers respectively. Speeds S, Sw, and So are introduced to the component solvers from the speed knobs, and are indicated on the speed dials. The components computed are Xo, Xw, and Xt, which go to the deflection section, and Yo and Yt, which go to the present range section. Range wind Yw is not computed.

14F9. Present range section. The present range section solves this equation:*
$$cR = jR + \triangle T(Yo + Yt)$$
Line components Yo and Yt are combined in D-4 to produce range rate (dR). The integrator multiplies dR by time increments ($\triangle T$) received from the time section, producing increments of generated range ($\triangle cR$). Initial range (jR), entered through the range crank, is combined with $\triangle cR$ in D-5 to produce generated range (cR). The value of cR is indicated on the present range counter, which provides means for comparing cR with measured range. The quantity cR goes to the advance range and deflection sections.

14F10. Advance range section. The advance range section continuously computes an approximate value of advance range. Advance range R2 is cR plus the correction, or range prediction (Ro) for own-ship motion; the correction, or range prediction (Rt) for target motion; and the corrections Rj to advance range (initial ballistic and spots).

The solid lines in figure 14F4a are the range-table values of corrections for *one*-knot line components Yo and Yt. The broken lines are the corrections computed by the range keeper. It is apparent that the prediction for a one-knot component Yo is equal to a constant K0 multiplied by generated present range. In other words, the prediction for a one-knot Yo at any point on the straight line Ro is equal to $K0 \times cR$. If the range doubles, so does Ro. The prediction for *any* value of Yo is simply $Ro = K0cR \times Yo$. Similarly, the prediction for target motion is $Rt = K1cR \times Yt$. At any range, the value of cR is of course the same in both equations, and the only variables remaining are Yo and Yt.

The mechanism is simplified by combining the corrections Ro and Rt to obtain the single range prediction Rto due to target and own-ship motion. Thus:
$$Rto = Ro + Rt = K0cR \times Yo + K1cR \times Yt = cR(K0Yo + K1Yt)$$
$$= K1cR(\frac{K0}{K1}Yo + Yt)$$
Since both K0 and K1 are constants, their quotient is another constant K, which is less than 1, and
$$Rto = K1cR(KYo + Yt) = K1cR \times dRs$$
The expression (KYo+Yt) is nothing more than a rate which is called range prediction rate (dRs). It simply indicates the fact that the effect of own-ship motion is less than the effect

---
* See footnote to Article 14D7.

RESTRICTED

THE SURFACE PROBLEM: PRACTICAL SOLUTION

of target motion because of the apparent wind created by own ship. Range rate dR is already available in the present range section; hence it is used to obtain dRs according to this relationship:

$$dRs = (KYo + Yt) + KYt - KYt = K(Yo + Yt) + (1-K)Yt$$
$$= KdR + (1 - K)Yt$$

It is a simple matter to multiply a quantity by a constant, using a proper gear ratio. In the range keeper, dR from the present range section is multiplied by a gear constant K to obtain KdR which goes to D-6 in the advance range section. Similarly, Yt is multiplied by gear constant (1—K), and (1—K)Yt goes to D-6. Thus, D-6 delivers dRs.

The input cR to the advance range section is multiplied by the constant K1. Both K1cR and dRs go to the range multiplier, which delivers their product K1cr × dRs, shown above to be equal to Rto.

Differential D-7 receives and adds cR and Rto, and delivers the sum to D-8. Corrections to advance range Rj (initial ballistic and spots), entered through the range-correction crank, register on the dial and enter the other side of D-8. The output of D-8 is advance range R2 which goes to the advance range counter, and the time and transmitter sections. Thus, the equation for the advance range section is:

$$R2 = K1cR [KdR+(1-K)Yt]+cR+Rj$$

14F11. Deflection section. The deflection section continuously computes an approximate value of sight deflection Ds, including the corrections for wind, own-ship and target motions, and drift.

Mils deflection Dt due to target motion is practically proportional to present range, plus a constant as shown in figure 14F4b. The curve is obtained by plotting the *mils* corresponding to the yards deflection for a *one*-knot Xt as obtained from the range table. The straight line approximation is:

$$Dt = K2cR \times Xt+K3Xt = (K2cR+K3)Xt$$

The quantity cR is multiplied by gear constant K2. The constant K3 is added to K2cR by a rotational offset provided during manufacture. One input to the target multiplier is (K2cR+K3), the other is Xt. The output is Dt, which goes to D-8.

Mils deflection Dw due to cross wind is nearly proportional to present range as shown in figure 14F4c. The straight-line approximation used in the range keeper is Dw = K4cR × Xw. One input to the wind multiplier is K4cR, obtained by multiplying cR by gear constant K4. The other input to the multiplier is Xw. The multiplier output Dw is combined in D-9 with Dt to provide mils deflection Dtw due to wind and target motion, which goes to D-10.

Mils deflection due to own ship motion Do is nearly independent of range, being equal to a constant K5 multiplied by Xo, as shown in figure 14F4d. Hence Xo from the own-ship component solver is simply multiplied by gear constant K5, and added to Dtw in D-10 to obtain mils deflection Dtwo due to wind, own-ship, and target motions.

Drift (Df) is a function of range, f(R2), but not directly proportional to it, as shown in figure 14F4e. Drift is computed by a simple cam cut to provide an output which agrees with the actual drift curve. Drift depends more nearly on advance range than on present range; hence the cam is rotated by R2 from the advance range section. The cam output Df goes to D-11 where it is combined with Dtwo to obtain Dtwof.

Corrections to deflection (Dj) (initial ballistic and spots) are entered through the deflection-correction crank, indicated on the dial, and combined in D-12 with Dtwof to provide sight deflection (Ds). The mils deflection counter, reading 100 when Ds is zero, receives Ds and indicates scale setting instead of actual deflection. The equation for the computations of the section is:

$$Ds = Dt+Dw+Do+Df+Dj =$$
$$(K2cR+K3)Xt+K4cR+Xw+K5Xo+f(R2)+Dj$$

14F12. Time section. The time section contains the driving motor and its controlling mechanisms, and the time-of-flight signal. The motor drives the integrator in the present-range section, and a toothed wheel in the time-of-flight signal. When the signal button is pressed, the

CURVES BASED ON DATA FROM O.P. 185

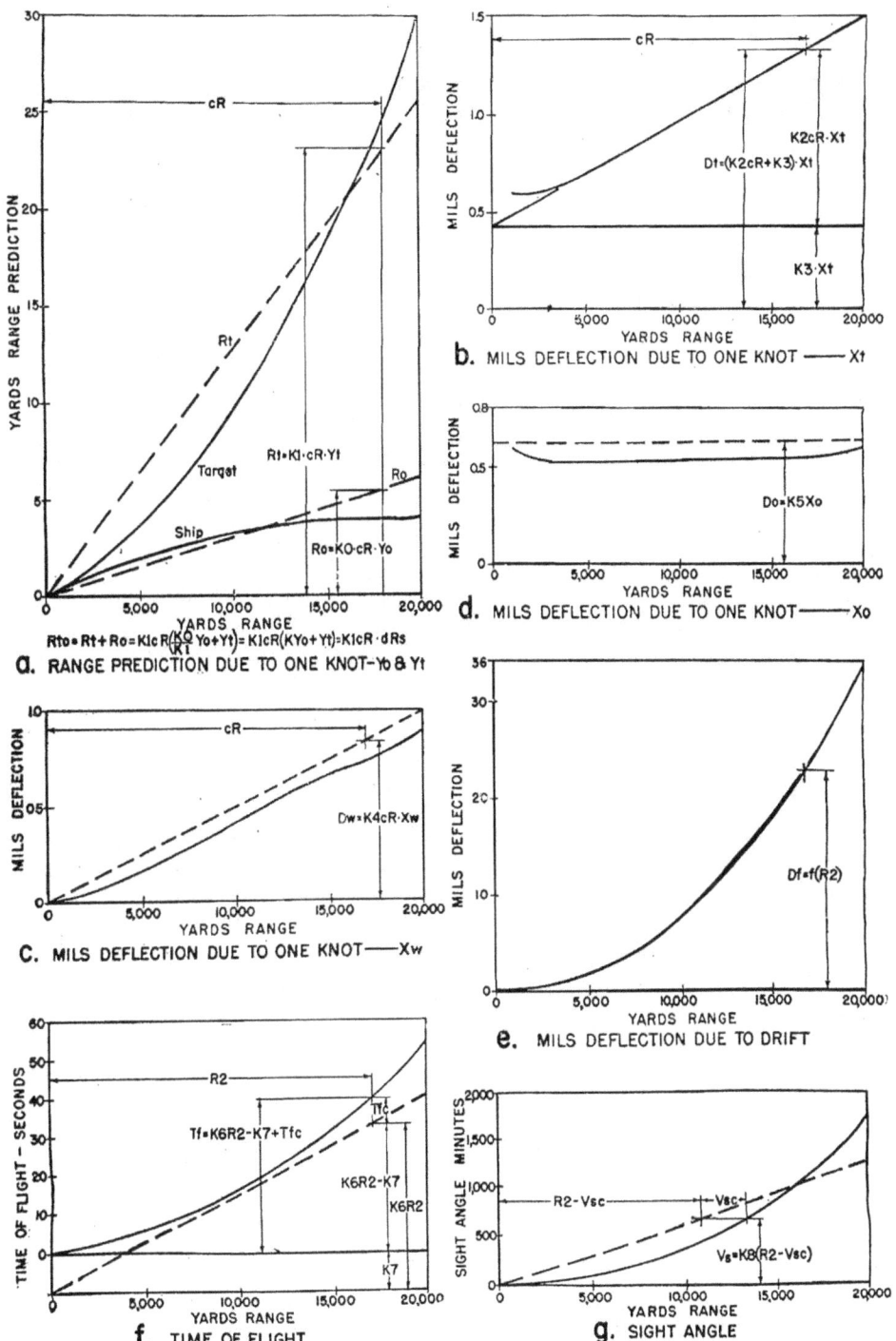

FIGURE 14F4. Ballistics computations of range keeper, Mark 7, mod. 5.

signal mechanism draws the revolving toothed wheel against a diaphragm which makes a horn-like noise shortly before time of flight (Tf) has elapsed.

The timing mechanism requires an accurate value of Tf, determined in part by a cam. The curve of figure 14F4f shows that Tf is a function of range. (Advance range is used to obtain a more-nearly correct result.) To reduce the size of the cam, a straight line approximation is made, and the cam is cut to supply only the difference Tfc between the actual curve and the straight line. The straight line part of Tf is K6R2—K7. It is supplied to D-13 by multiplying R2 by gear constant K6, and subtracting K7 by a rotational offset during manufacture. The cam is rotated by R2, and provides Tfc to D-13. The output of D-13 is Tf, which appears on a dial used for checking purposes, and controls the time-of-flight signal.

14F13. Transmitter section. The electrical transmitters, similar to small motors in appearance, are rotated by hand follow-up cranks. The position of each transmitter determines the value of the signal it sends to the guns, and must be carefully set if the signal is to be accurate. Zero reading follow-up dials are provided to show the operator when he has set the transmitters correctly. Basically, zero-reading dials indicate the *difference* between a *required* value and an *actual* value. Thus they read zero when the actual value is equal to the required value.

The deflection zero-reader dial is driven by differential D-14. One input to D-14 is Ds computed in the deflection section. The other input is *transmitted* Ds supplied by rotation of the follow-up knob. When transmitted Ds equals computed Ds, the index on the dial is opposite the fixed pointer. The follow-up knob drives Ds into the transmitter and the counter, as well as into D-14. The angle being transmitted can be read from the counter.

The sight-angle part of the transmitter is similar to the sight-deflection part, except that provision must be made to convert the linear quantity R2 into the angular quantity (Vs). Sight angle is a function of range, as shown in figure 14F4g. The value of Vs at any point on the curve is equal to the value of Vs at a point on the straight line having the same ordinate Vs. That point on the straight line has an abscissa equal to R2 less the difference between the abscissas of the two points, or Vsc. The value of Vs at any point on the straight line is equal to a constant times the abscissa of the point. Thus, $Vs = K8 (R2 - Vsc)$.

Sight angle Vs, set in by rotation of the follow-up crank, rotates the sight-angle cam which is cut to supply Vsc. The cam output Vsc goes to D-16. The constant K8 is removed from Vs by a gear ratio to obtain R2—Vsc which forms the other input to D-16. The output of D-16 is the value of R2 that corresponds to transmitted Vs, or in other words it is transmitted R2. The two inputs to D-15 are computed R2 from the advance range section and transmitted R2. Thus when transmitted Vs corresponds to computed R2, the dial pointer will be opposite the fixed pointer.

14F14. Range-keeper errors. In order to simplify the mechanism of fire-control instruments, the Bureau of Ordnance permits the introduction of class B errors. These are defined as errors resulting from the intentional substitution of approximate methods in the design. Most of the class B errors in the Mark 7 mod. 5 range keeper are indicated by the difference between the straight lines and curves in figure 14F4. Curves showing class B errors are available for all range keepers aboard ship.

It is evident that there may be conditions under which some of the errors will compensate each other. The errors are negligible in many cases, but at certain ranges they are appreciable, and must be compensated by corrections included in the initial ballistic.

There are also class A errors which result from imperfections in manufacture and *adjustment*. Consequently, range keepers are given regular routine tests to check their accuracy. The required tests are prescribed in Ordnance Pamphlets or Ordnance Data pertaining to the particular instrument.

14F15. General remarks. In the early stages of modern gunnery, graphical means were used in range keeping. The tracking board was employed to determine target course and speed, and plotting boards assisted in establishing range rate and advance range. The results obtained from these boards were used as arguments on curves, charts, or tables from which were obtained

CHAPTER 14

certain predictions or corrections. These corrections, suitably converted, were transmitted to sight setters at guns as range and scale.

It soon became apparent that the mechanisms available were inadequate for rapid, accurate, and continuous supply of required information, and in the course of time range keepers were developed. While the Mark 2 was the simplest range keeper, it was not the first. The Mark 1 was the first, and had many of the features found today on modern range keepers and computers.

Modern computers automatically handle nearly all the usual elements of the gun ballistic except the correction for air density. In addition, they make corrections for trunnion tilt, and compute fuze settings and gun-train and gun-elevation orders whose significance is made more apparent in subsequent chapters.

Main-battery range keepers are equipped with graphic plotters on which range-finder ranges are plotted automatically. Certain outputs of the range keeper are also plotted to provide convenient and accurate assistance to the operator in establishing correct target angle and speed. The dual-purpose computers have automatic rate-control features which make it possible to find the rates due to a fast-moving air target by simply following it with a director line of sight.

THE SURFACE PROBLEM: PRACTICAL SOLUTION

RANGE KEEPER, MARK 7, MOD. 5
SYMBOLS AND MEANINGS

A	Target angle ($B+180° - Ct$).	jR	Initial or corrective settings of generated present range.
B	True target bearing ($Br+Co$).		
Br	Relative target bearing.	R2	Advance range ($cR+Rto+Rj$).
Bw	True direction true wind.	Rj	Range spots.
Co	Own-ship course.	Ro	Range prediction due to own-ship motion ($K0cR \times Yo$).
Ct	Target course.		
Cws	Wind angle ($B - Bw$).	Rt	Range prediction due to target motion ($K1cR \times Yt$).
Df	Mils deflection due to drift $f(R2)$.	Rto	Range prediction due to target and own-ship motion ($K1cR \times dRs$).
Dj	Deflection spots.		
Do	Mils deflection due to own-ship motion ($K5Xo$).	S	Target speed.
Ds	Sight deflection ($Dt+Dw+Do+Df+Dj$).	So	Own-ship speed.
		Sw	True wind speed.
Dt	Mils deflection due to target motion ($K2cR+K3)Xt$.	T	Time generated by motor.
		ΔT	Increments of time.
Dw	Mils deflection due to cross wind ($K4cR \times Xw$).	Tf	Time of flight ($K6R2-K7+Tfc$).
Dtw	$Dt+Dw$.	Tfc	That part of Tf generated by cam.
Dtwo	$Dt+Dw+Do$.		
Dtwof	$Dt+Dw+Do+Df$.	Vs	Sight angle $K8(R2 - Vsc)$.
K, K1, etc.	Constants.	Vsc	That part of Vs generated by cam.
cR	Generated present range ($jR+\Delta cR$).		
		Xo	So sin Br.
ΔcR	Increments of generated range ($\Delta T \times dR$).	Xt	S sin A.
		Xw	Sw sin Cws.
dR	Range rate ($Yo+Yt$).	Yo	So cos Br.
dRs	Range prediction rate ($KYo+Yt$).	Yt	S cos A.

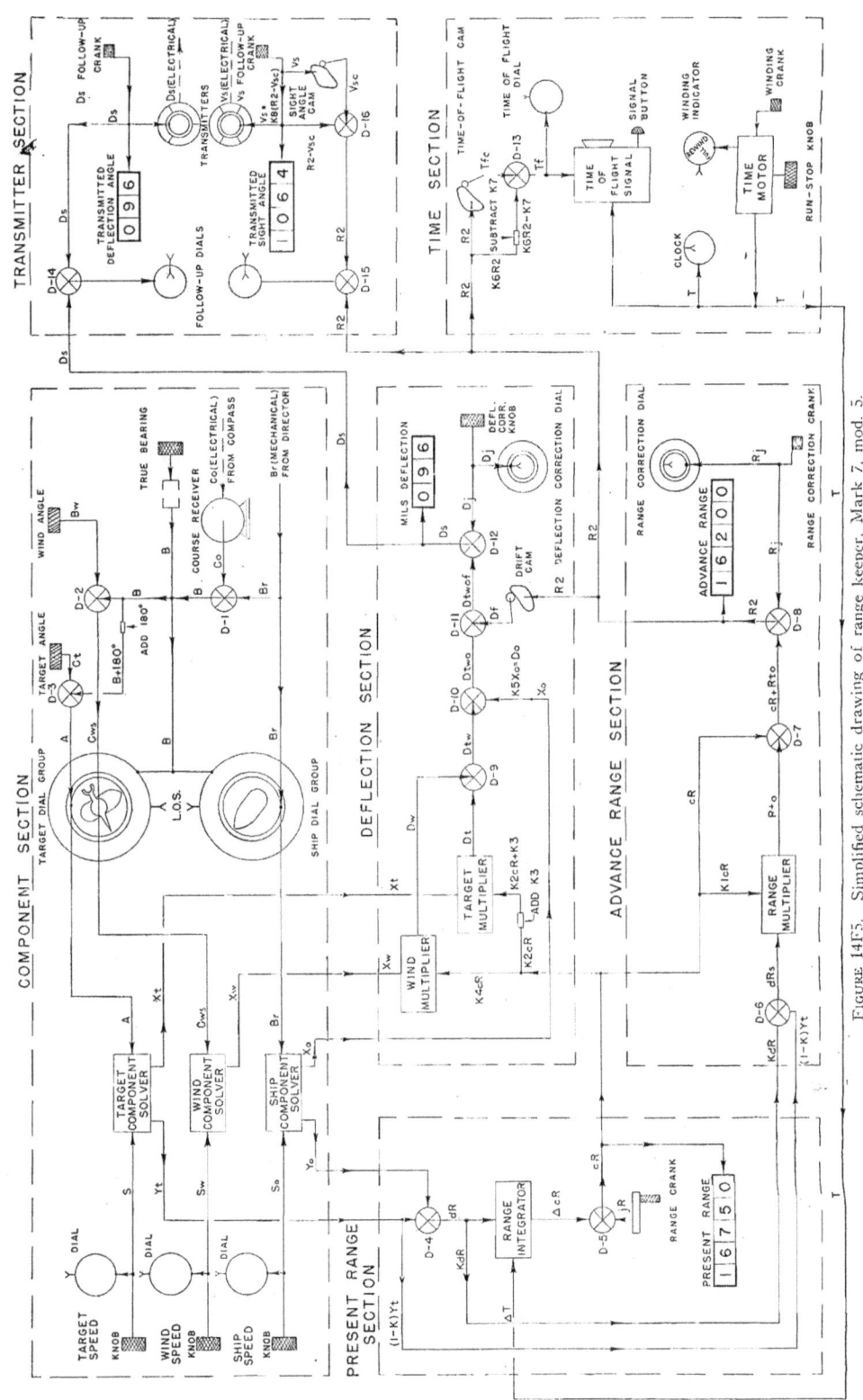

FIGURE 14F5. Simplified schematic drawing of range keeper, Mark 7, mod. 5.

CHAPTER 15
THE AA PROBLEM: ANALYSIS OF SIGHT AND FUZE SETTINGS

A. POSITIONING COMPUTATIONS

15A1. Introductory. The calculations of sight angle and sight deflection in the AA problem are similar in many respects to the corresponding calculations in the surface problem. The AA problem has the same distinct parts, namely: (1) continuous target positioning, and (2) continuous ballistics computing. The target's freedom of movement in three dimensions and the use of time-fuzed projectiles introduces additional complications. Since sight and fuze settings are based on the instantaneous relative target position, the positioning part of the AA problem is presented first.

15A2. Symbols and definitions. The quantities used in continuous target positioning are summarized in the following list:

SYMBOL	DEFINITION
A	*Target angle.* The horizontal angle between the vertical plane containing the direction of target motion and the vertical plane containing the LOS, measured clockwise from the direction of motion.
Br	*Relative target bearing.* The horizontal angle between the vertical plane containing the own ship fore-and-aft axis and the vertical plane containing the LOS, measured clockwise from own-ship bow.
cBr	*Generated relative target bearing.* The computed bearing as opposed to the actual value (Br). $cBr = jBr + \triangle cBr$.
$\triangle cBr$	*Generated bearing increments.* $\triangle cBr = \triangle T \times dBr$.
dBr	*Relative angular bearing rate.* The time rate of change of relative target bearing. $dBr = (RdBs \sec E) \div R$.
jBr	*Initial relative target bearing.*
$RdBs$	*Linear deflection rate.* The horizontal component of relative target motion perpendicular to the vertical plane containing the LOS. $RdBs = Xo + Xt$.
E	*Target elevation.* The vertical angle between the LOS and the horizontal.
cE	*Generated target elevation.* The computed elevation as opposed to the actual value (E). $cE = jE + \triangle cE$.
$\triangle cE$	*Generated elevation increments.* $\triangle cE = \triangle T \times dE$.
dE	*Angular elevation rate.* The time rate of change of target elevation. $dE = RdE \div R$.
RdE	*Linear elevation rate.* The component of relative target motion perpendicular to the LOS and in the vertical plane containing the LOS. $RdE = dH \cos E - dRh \sin E$.
jE	*Initial target elevation.*
H	*Height.* The vertical distance of the target above the horizontal plane through own ship. $H = R \sin E$.
dH	*Rate of climb.* The time rate of change of height, or the vertical component of target velocity.
R	*Present range.* The distance of the target from own ship measured along the LOS.
Rh	*Horizontal range.* The projection of present range on a horizontal plane. $Rh = R \cos E$.

CHAPTER 15

SYMBOL	DEFINITION
cR	*Generated present range.* The computed range as opposed to the actual value (R). cR = jR + $\triangle$cR.
$\triangle$cR	*Generated range increments.* $\triangle$cR = $\triangle$T × dR.
dR	*Range rate.* The time rate of change of present range. dR = dH sin E + dRh cos E.
dRh	*Horizontal range rate.* The time rate of change of horizontal range (Rh).
jR	*Initial range.*
S	*Air target speed relative to ground.* The speed of the target with respect to the earth.
So	*Own-ship speed.* The speed of own ship with respect to the earth.
Sh	*Air target horizontal ground speed.* (target horizontal speed). The horizontal component of target velocity.
Sr	*Relative target speed.* The speed of the target relative to own ship.
$\triangle$T	*Time increments.*
Xo	*Cross component of own-ship velocity.* The horizontal component of own-ship velocity perpendicular to the vertical plane containing the LOS. Xo = So sin Br.
Xt	*Cross component of target velocity.* The horizontal component of target velocity perpendicular to the vertical plane containing the LOS. Xt = Sh sin A.
Yo	*Line component of own-ship velocity.* The horizontal component of own-ship velocity in the vertical plane containing the LOS. Yo = So cos Br.
Yt	*Line component of target velocity.* The horizontal component of target velocity in the vertical plane containing the LOS. Yt = Sh cos A.

15A3. Relative target position. Present target position with respect to own ship is established by the three coordinates: present range (R), relative target bearing (Br), and target elevation (E). These quantities are illustrated in figure 15A1a, which should be available while this section is studied. Both Br and E can be determined by measuring the position of a LOS to the target, while R can be measured by any standard range-finding method. When there is motion of own ship and the target, as in the usual case, successive instantaneous positions may be established by simply continuing to direct the LOS at the target and by measuring present range.

It is not practicable to obtain the basic coordinates by direct measurement at each instant, for any interruptions or errors in individual measurements would be reflected in the computed values of sight angle (Vs), sight deflection (Ds), and fuze setting (F). It is only logical that the relative target position be "kept" by a process similar to that employed in surface fire control. Modern computers generate relative target position, and this position is used as the basis of the solution for Vs, Ds, and F.

While height (H) could be used as a substitute for either R or E in establishing target position, it is not so employed. However, it serves a useful purpose in analyzing ballistics computations as is shown later. It is evident from the illustration that H = R sin E.

15A4. Target-motion designation. Since an airplane may be climbing or diving, target motion is not necessarily horizontal, and three quantities are needed to define it, as shown in figure 15A1a. The direction of motion in a horizontal plane is described by target angle (A), which is measured in a horizontal plane just as in the surface problem. Target speed is defined by vertical speed, or rate of climb (dH), and air target horizontal ground speed (Sh)—usually called target horizontal speed. These two speeds are simply components whose resultant is actual target speed relative to the earth. The use of Sh and dH simplifies the positioning-problem solution, for they indirectly define the inclination of target motion to the horizontal. It should be noted that dH is a vertical

RESTRICTED

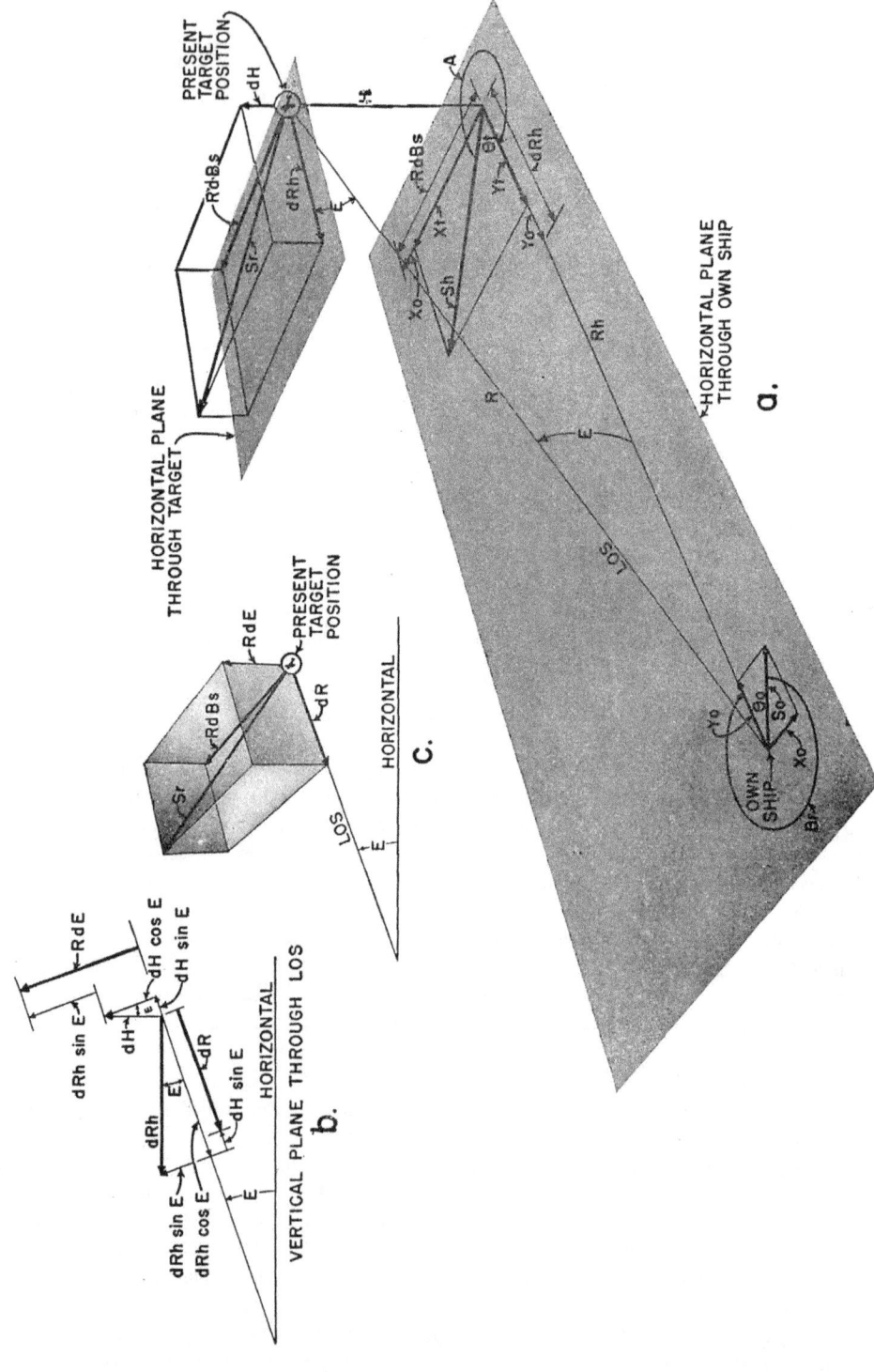

FIGURE 15.A1. Relative target motion resolution.

CHAPTER 15

speed, being equivalent to the rate of change of height (H). It also represents the total vertical relative motion, since own ship moves only in the horizontal.

Initial values of A, dH and Sh are estimated, and of course are subject to error. For the present, it is assumed that these quantities are known. Mechanical computers are equipped with rate-control mechanisms which quickly determine their correct values according to basic principles that are discussed in the section on rate control.

15A5. Relative target motion. For AA fire, relative target motion is used to keep track of relative target position in a manner similar to that employed for surface fire. In other words, the present values of R, Br, and E are generated. This requires the determination of range rate (dR), relative angular bearing rate (dBr), and angular elevation rate (dE). These rates are multiplied by time increments ($\triangle$T) to obtain generated-range increments ($\triangle$cR), generated-bearing increments ($\triangle$cBr), and generated-elevation increments ($\triangle$cE). The generated increments are added to the initial values of their corresponding coordinates (jR, jBr, and jE) to obtain the generated values cR, cBr, and cE. Thus:

$$cR = jR + \triangle T \times dR = jR + \triangle cR.$$
$$cBr = jBr + \triangle T \times dBr = jBr + \triangle cBr.$$
$$cE = jE + \triangle T \times dE = jE + \triangle cE.$$

The initial values, of course, are obtained by measurement. The generated increments are computed by integrators to which the change rates must be supplied. The three rates dR, dBr, and dE completely describe relative target motion and are determined by following these two broad steps:

1. Resolve own-ship and target motions into horizontal and vertical components of relative target motion.
2. Resolve the horizontal and vertical components into the three rates with respect to the LOS.

15A6. Horizontal and vertical rates. Since own-ship motion and part of target motion are defined by horizontal speeds and directions, it is a simple process to determine relative *horizontal* target motion. The process is essentially the same as that used in the surface problem, and is illustrated in figure 15A1a.

The line components (Yo and Yt) and cross components (Xo and Xt) can be computed by using the acute angles θo and θt. Their effects upon range and bearing may be considered as increasing or decreasing as was done in the surface problem. The components are combined to produce horizontal range rate (dRh) and linear deflection rate (RdBs) as shown in the illustration.

Mathematically, the components may be computed from these equations:

$$Yo = So \cos Br. \qquad Xo = So \sin Br.$$
$$Yt = Sh \cos A. \qquad Xt = Sh \sin A.$$

Since the cosine of an angle is positive for angles from 0° to 90° and 270° to 360°, and negative for angles from 90° to 270°, Yo and Yt are positive when they are decreasing the range as in the figure. Range rate is conventionally considered to be negative when it decreases the range, consequently the expression for horizontal range rate is:

$$dRh = -(Yo + Yt) = -(So \cos Br + Sh \cos A).$$

The sine of an angle is positive for angles from 0° to 180° and negative for those from 180° to 360°. When Br lies between 0° and 180°, Xo is positive. Further, a positive value of Xo causes the LOS to shift to the right, thereby increasing Br. When A lies between 0° and 180°, Xt is positive and causes Br to increase. In the illustration, both Xo and Xt have negative values, and so does RdBs. Thus:

$$RdBs = Xo + Xt = So \sin Br + Sh \sin A.$$

The vertical relative motion is rate of climb (dH) as previously mentioned, and is positive when the target is climbing, negative when diving. Hence total vertical relative target motion is defined by dRh, RdBs, and dH. When these three rates are combined about the target position, their resultant is the vector Sr, whose direction represents the path and whose length represents the speed of the target with respect to own ship.

15A7. Rates in and about the LOS. The horizontal and vertical rates are simply employed in determining the rates at which the basic coordinates R, Br, and E are changing. It is helpful to

THE AA PROBLEM: ANALYTICAL SOLUTION

the student to recognize the existence of the vector Sr, although it is not used directly in maintaining target position. This single vector can be resolved into any number of components, but those needed in generating target position are (1) range rate, measured in or along the LOS, (2) bearing rate, measured perpendicular to the LOS in a horizontal plane, and (3) elevation rate, measured perpendicular to the LOS in a vertical plane. The relations of the linear rates to Sr are shown in figure 15A1c.

Range rate. Range rate along the LOS is direct range rate (dR). Since the LOS is elevated, dR is not the same as dRh, but has a component due to dRh and another due to dH, as shown in figure 15A1b. Both dRh and dH lie in the vertical plane through the LOS and are inclined to the LOS by E and (90° — E), respectively. It can be seen from the diagram that:

$$dR = dH \sin E + dRh \cos E.$$

Elevation rate. The linear elevation rate (RdE) also has components due to dRh and dH. It may be expressed as:

$$RdE = dH \cos E - dRh \sin E.$$

The quantity dRh sin E is subtractive because the effect of a negative or decreasing value of dRh is to increase RdE. In other words, the LOS must be elevated as an airborne target, flying horizontally, draws closer to own ship. In the illustration dRh has a negative value, dH a positive value. Both dRh and dH have components tending to increase E, and RdE is positive since it tends to increase E. The angular elevation rate is obtained by dividing the linear rate by the slant, or present range, (R). Thus:

$$dE = \frac{RdE}{R}$$

Bearing rate. The linear relative bearing rate (RdBs) is measured perpendicular to the vertical plane through the LOS, and has the same value whether measured in the slant plane through the target or in the horizontal plane through own ship. However, the corresponding angular rate has a different value in the slant plane than in the horizontal. Since relative target bearing (Br) is measured in the horizontal, relative angular bearing rate (dBr) must be computed for the horizontal by dividing RdBs by horizontal range (Rh). It is evident that $Rh = R \cos E$, hence:

$$dBr = \frac{RdBs}{R \cos E} = \frac{RdBs \sec E}{R}$$

15A8. Generated target position. When the equivalent values of the rates are substituted in the equations of Article 15A5, the generated target position is described by these three equations:

$$cR = jR + \Delta T \,[dH \sin E - (So \cos Br + Sh \cos A) \cos E].$$

$$cBr = jBr + \frac{\Delta T}{cR} \sec E \,(So \sin Br + Sh \sin A).*$$

$$cE = jE + \frac{\Delta T}{cR} \,[dH \cos E + (So \cos Br + Sh \cos A) \sin E].$$

These three equations are the ones solved by modern computers in generating target position. It is evident from the equations that a series of component solvers, integrators, and differentials is required. The quantities 1/cR and sec E are computed by cams. Since the mechanical integrations are continuous, the generated values of R, Br, and E will agree with actual values as long as correct inputs are supplied to the computer.

B. RATE CONTROL

15B1. Error source. The primary data used in generating target position are the quantities included in the three equations of the preceding article. It is evident that these elements are of two classes:

---

* This equation does not take into account variations in cBr due to changes in own-ship course.

CHAPTER 15

1. *Estimated:* Target angle (A), target horizontal speed (Sh), and rate of climb (dH).
2. *Measured:* Present range (R), relative target bearing (Br), target elevation (E), own-ship speed (So), and time (T).

Since the generated target position is the foundation upon which all the subsequent calculations are based, it is highly important that the generated quantities cR, cBr, and cE be accurate. When correct values of all eight elements listed above are used, the generated coordinates cR, cBr, and cE will continuously agree with the measured values of R, Br, and E. When there is disagreement, one or more of the primary elements is in error. Since the elements in the second group are measured, their accuracy is relatively high, assuming proper measuring devices are used. The likeliest source of error is in one or more of the three estimated quantities.

15B2. Symbols and definitions. In addition to the quantities used in the preceding section, the following are used in the analysis of rate control:

SYMBOL	DEFINITION
jBc	*Linear deflection-rate correction.* The rate-control correction primarily affecting linear deflection rate (RdBs).
jEc	*Linear elevation-rate correction.* The rate-control correction primarily affecting linear elevation rate (RdE).
jHc	*Rate-of-climb correction.* The rate-control correction to rate of climb (dH) jHc = jdR sin E + jEc cos E.
jdR	*Direct range-rate correction.* The rate-control correction primarily affecting direct range rate (dR).
jdRh	*Horizontal range-rate correction.* The rate-control correction primarily affecting horizontal range rate (dRh). jdRh = jdR cos E − jEc sin E.

15B3. Rate-control corrections. The problem of determining which of the three elements of target motion (A, Sh, and dH) should be adjusted, and how much, is complicated by the fact that a change in any one of them affects all three of the rates dR, RdBs, and RdE. The dominating element depends upon the values of target angle and target elevation, as indicated in the table below:

Rate considered	Target Elevation	Target angle near	Dominating target motion element
dR	0°–45°	0° or 180°	Sh
	45°–90°	0° or 180°	dH
	0°–45°	90° or 270°	A
	45°–90°	90° or 270°	dH
RdE	0°–45°	0° or 180°	dH
	45°–90°	0° or 180°	Sh
	0°–45°	90° or 270°	dH
	45°–90°	90° or 270°	A
RdBs	0°–45°	0° or 180°	A
	45°–90°	0° or 180°	A
	0°–45°	90° or 270°	Sh
	45°–90°	90° or 270°	Sh

In the surface problem, the rules used to indicate whether target angle or target speed is the element most likely in error are simple (see Article 14D9). In addition, the time available in surface fire allows a computer or range-keeper operator an opportunity to adjust the settings of target angle and target speed to obtain the correct set-up. An air attack is of such high speed and short duration that an operator does not have time to consider and apply rules such as those outlined in the table. Hence, computers for AA fire are equipped with rate-control mechanisms which

RESTRICTED

THE AA PROBLEM: ANALYTICAL SOLUTION

establish correct values of A, Sh, and dH rapidly and accurately.

An error in any one of the generated coordinates usually requires an adjustment to all three target-motion elements. What the operator needs is a mechanism which permits him to correct the rate of change of the particular coordinate by means of a single knob. In other words, he needs three knobs to correct the three rates: one for dR, another for RdBs and a third for RdE. Turning one of these knobs should theoretically correct A, Sh, or dH by the amount needed to bring the computed rate into agreement with the actual rate. The inputs to these knobs are the corrections to the corresponding rates, and are known as direct range-rate correction (jdR), linear deflection-rate correction (jBc), and linear elevation-rate correction (jEc).

15B4. Detecting rate errors. It is a simple matter to arrange a set of dials that will indicate disagreement between the actual and computed values of range, bearing, and elevation. For the sake of example, consider target elevation, and assume a dial group like that in figure 15B1. The outer ring is driven by measured or actual target elevation. That is, it is mechanically (or electrically) connected to a telescope which establishes the LOS to the target. The inner dial is driven by generated target elevation (cE).

As the LOS follows the target, changes in target elevation will obviously cause the ring to rotate. The speed of rotation will be proportional to the *actual* elevation rate. Changes in generated target elevation drive the inner dial at a speed proportional to the *computed* elevation rate. Equal rotational speeds of the ring and dial are apparent when the dial marks remain stationary with respect to the ring marks. This condition signifies that the *computed* rate equals the *actual* rate. When the dials turn at different speeds, the computed rate (RdE) is evidently wrong and must be corrected. The same process can be applied to compare computed and actual range and bearing rates.

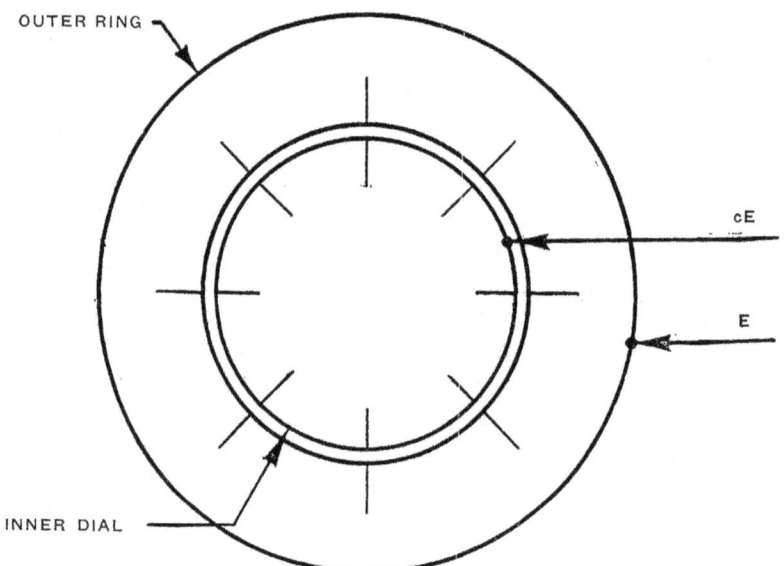

FIGURE 15B1. Comparison dials.

15B5. Target-motion corrections. The difference between the rotational speeds of the comparison dials indicates the existence of an error in the computed rates. Suppose that in a typical situation, the three computed rates dR, RdBs, and RdE are in error by the amounts jdR, jBc, and jEc, as indicated by the heavy black arrows in figure 15B2. The primary elements A, Sh, and dH must be adjusted so that the computed rates are (dR + jdR), (RdBs + jBc), and (RdE + jEc). The quantities jdR, jBc, and jEc are simply equivalent to the required rate corrections. The

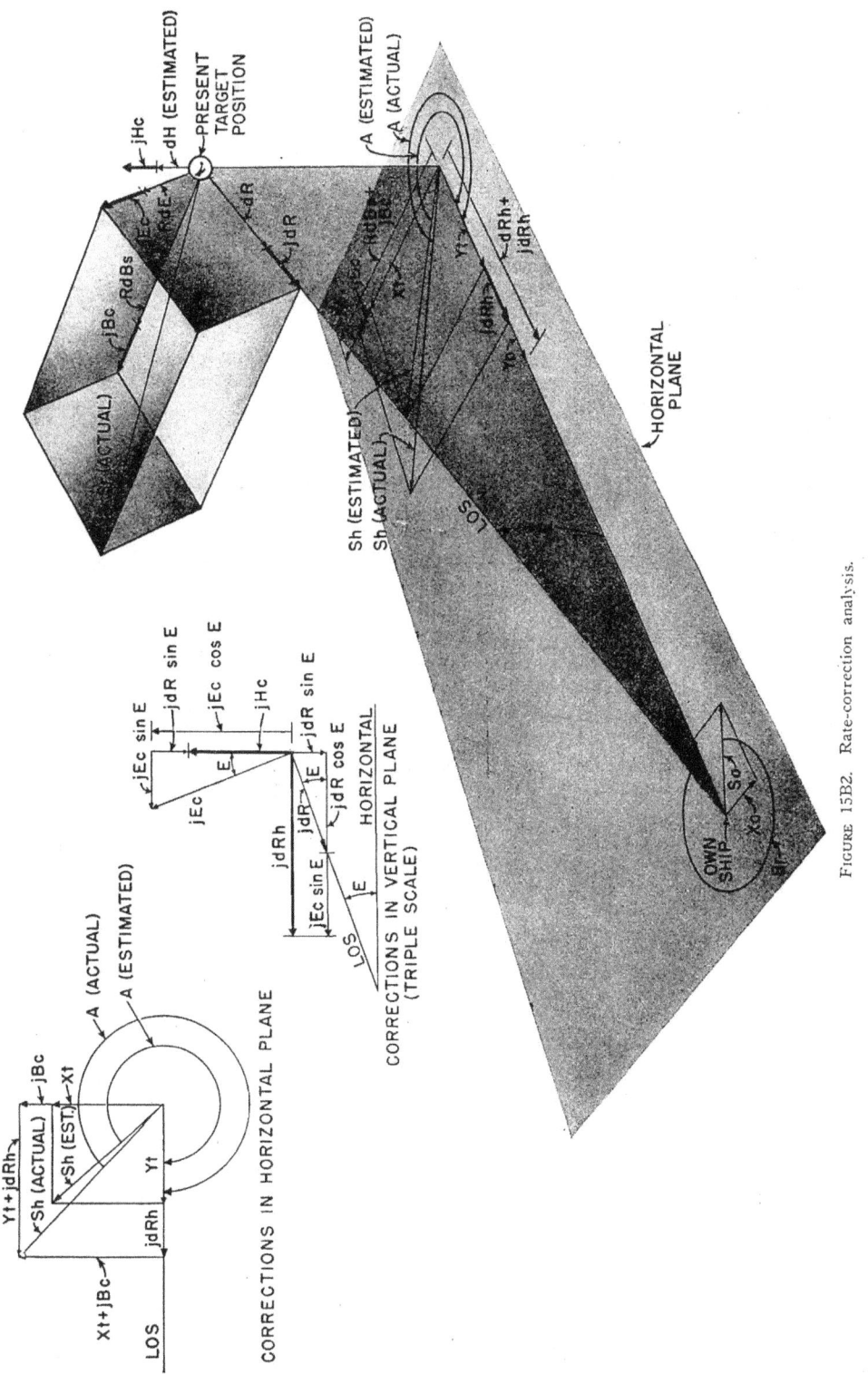

FIGURE 15B2. Rate-correction analysis.

next step in the problem is to resolve these corrections into components which will indicate the changes required in the primary elements of target motion.

It can be seen from the diagram that both jdR and jEc have vertical and horizontal components whose values are jEc sin E, jEc cos E, jdR sin E, and jdR cos E. The summation of the vertical components is evidently the net required correction to dH, or rate-of-climb correction jHc, for jBc has no vertical component. Since values of jdR and jEc which tend to increase the values of dR and RdE both tend to increase dH, the equation for jHc is:

$$jHc = jdR \sin E + jEc \cos E.$$

The resultant of the horizontal components is the net required correction to dRh, or horizontal range rate correction (jdRh). An upward value of jEc produces a decreasing effect on horizontal range, hence:

$$jdRh = jdR \cos E - jEc \sin E.$$

This is the horizontal correction required in the vertical plane through the LOS. The horizontal correction required across the same plane is jBc. Since it is assumed that only the target-motion elements are in error, jdRh and jBc are the changes required in Yt and Xt, respectively. In other words, the actual horizontal components of target motion are (Yt + jdRh) and (Xt + jBc), where Yt and Xt are computed values based on the original estimates of A and Sh. It is possible to find the actual values of A and Sh by solving these two equations simultaneously:

$$Sh \cos A = (Yt + jdRh).$$
$$Sh \sin A = (Xt + jBc).$$

15B6. Practical rate correction. A computer operator, seeing a discrepancy on the comparison dials, turns the appropriate rate-control knob until the dials turn at the same speed. The knob rotation is a measure of the rate correction.

Component integrators, similar in principle but not in construction to component solvers, break jdR and jEc into horizontal and vertical components. Differentials combine the components to obtain jdRh and jHc. Theoretically, it is only necessary to solve the last two equations of the preceding article to obtain the correct values of A and Sh. This can be done in a vector solver, which is similar to a component solver but capable of working backwards. When the quantities (Yt + jdRh) and (Xt + jBc) are fed to its slides, the outputs are A and Sh.

Actually, a computer must use target course (Ct) as a basic quantity instead of target angle (A). The rate-control mechanism must correct Ct, and in order to do so, it must break jdRh and jBc into north-south and east-west components. This requires two additional component integrators, but does not alter the fundamental principle of reducing rate corrections to changes in the target-motion set-up. The vector solver then receives north-south and east-west components and delivers Ct and Sh.

The correct value of A is readily obtained from Ct. Then the new values of A and Sh are used automatically to set the target component solver. The rate-of-climb correction (jHc) is added automatically to the initial set-up of dH to produce the corrected value of dH.

15B7. Rate-correction interdependence. It is to be noted that a correction to any one of the rates will affect the other two. An adjustment to RdBs requires a change in the settings of A and Sh, which in turn affects the computed values of dR and RdE. Corrections to either dR or RdE require changes in A, Sh, and dH, hence adjustment of one affects the other, and also influences RdBs.

In other words, in a hypothetical case where two rates might be correct and the third in error, a correction to the erroneous rate would inadvertently produce a change in the other two rates. In general, any case of rate control will require adjustment of all three rates to produce correct values of A, Sh, and dH. By making suitable adjustments to all three rates simultaneously, correct values of target motion are quickly established.

C. BALLISTICS COMPUTATIONS

15C1. General. The purpose of generating present target position as outlined in the preceding sections is to provide a foundation upon which to base the calculations of sight angle, sight

CHAPTER 15

deflection, and fuze setting. It is the purpose of this section to examine these quantities to indicate how they are developed from present range (R), relative target bearing (Br), and target elevation (E). Although an analytical solution consumes considerable time, the student should bear in mind that an effective solution must be continuous and instantaneous; such a solution is provided by mechanical computers.

In order to simplify the presentation, certain factors in the problem are described in a manner which does not exactly agree with the process carried out by a computer. In particular, the automatic solution involves so many empirical expressions and mathematical tricks that certain refinements and allowable approximations in the solution are ignored here. However, many of the steps outlined in the following articles are similar to those represented in mechanical instruments.

15C2. Symbols and definitions. The quantities used in this section and not defined in Section A are summarized below. Most of them are illustrated in figure 15C7, which should be referred to as this section is studied. No direct references are made to this drawing, but it will help the reader to see how each component fits into the over-all picture.

SYMBOL	DEFINITION
	Slant plane. The plane (1) containing the trunnion axis, and (2) parallel to the LOS.
	Line of fire. The line from the gun to the target at the instant of burst; i.e., the line from the point where the gun will be at the end of time of flight to the point where the target will be at the end of time of flight.
2	*Plane of fire.* The vertical plane containing the line of fire. The numeral 2 following a symbol signifies the value of the indicated quantity that will exist at the end of time of flight.
B	*True target bearing.* The horizontal angle between a north-south vertical plane and the vertical plane containing the LOS, measured clockwise from north. $B = Co + Br$.
B2	*Predicted true target bearing.* $B2 = Co + Br2$.
Br2	*Predicted relative target bearing.* $Br2 = Br + Dth$.
Bw	*True direction true wind.* The horizontal angle between north and the direction *from* which the true wind is blowing, measured clockwise from north.
Co	*Own-ship course.* The horizontal angle between a north-south vertical plane and the vertical plane containing the fore-and-aft axis of own ship measured clockwise from north.
Cw	*True course true wind.* The horizontal angle between north and the direction *toward* which the true wind is blowing, measured clockwise from north. $Cw = Bw \pm 180°$.
Cwg	*Predicted wind angle.* The horizontal angle between the direction toward which the true wind is blowing and the plane of fire, measured clockwise from the direction toward which the wind is blowing to that part of the plane of fire between the wind and own ship vectors. $Cwg = Co + Br2 - Bw$.
Dfs	*Drift.* The lateral deflection angle to correct for projectile drift, measured in the slant plane.
Dj	*Deflection spot.*
Ds	*Sight deflection.* The slant plane angle between the bore axis and the line of sight. $Ds = Dt + Dfs + Dw + Dj$.
Dt	*Deflection prediction.* The lateral deflection angle to correct for relative target movement. $Dt = \tan^{-1}(Tf \times RdBs/R2)$.
Dth	*Horizontal deflection prediction.* The predicted change in relative target bearing during time of flight. $Dth = \sin^{-1}(Tf \times RdBs/R2 \cos E2)$.

337 **RESTRICTED**

THE AA PROBLEM: ANALYTICAL SOLUTION

SYMBOL	DEFINITION
Dw	*Wind deflection correction.* The lateral deflection angle to correct for apparent cross wind.
E2	*Predicted target elevation.* $E2 = E + Vt$.
Eg	*Vertical gun elevation.* The vertical angle between the bore axis and the horizontal plane. $Eg = E + Vt + Vf + Vw + Vm + Vu + Vj$.
F	*Fuze setting.*
H2	*Predicted height.* $H2 = R \sin E + Tf \times dH$.
R2	*Advance (or predicted) range.* $R2 = R + Rt$.
R3	*Fuze range.* The value of present range that will exist at the end of dead time plus time of flight. $R3 = R2 + Tg \times dR + Rj$.
Rj	*Range spot.*
Rt	*Range prediction.* The predicted change in range during time of flight. $$Rt = Tf \times dR + \frac{(Tf \times RdE)^2 + (Tf \times RdBs)^2}{2(R + Tf \times dR)}$$
Sw	*True wind speed.* The horizontal speed of the wind with respect to the earth.
Tf	*Time of flight.* The total time elapsing during the projectile flight from the gun to the point of fall or burst. $Tf = Tfk + Tfm + Tfu + Tfw$.
Tfk	*Standard time of flight.* Uncorrected time of flight for the advance target position.
Tfm	*I. V. time-of-flight correction.* The correction to time of flight for initial velocity variations.
Tfu	*Air density time-of-flight correction.* The correction to time of flight for air-density variations.
Tfw	*Range wind time-of-flight correction.* The correction to time of flight for the effect of range wind.
Tg	*Dead time.* The time elapsing between the fuze-setting and firing instants.
Vf	*Superelevation.* The correction included in vertical gun elevation to allow for trajectory curvature in a vertical plane.
Vj	*Elevation spot.*
Vm	*I. V. elevation correction.* The correction included in vertical gun elevation to allow for initial-velocity variations.
Vs	*Sight angle.* The angle between the bore axis and the slant plane, measured in the plane through the bore axis perpendicular to the slant plane. This is a vertical angle only when the trunnion axis is horizontal. $Vs = Eg - E - Vx = Vt + Vf + Vw + Vm + Vu + Vj - Vx$.
Vt	*Elevation prediction.* The correction included in vertical gun elevation to allow for relative target movement. $Vt = \sin^{-1}(Tf \times RdE/R2)$.
Vu	*Air-density elevation correction.* The correction included in vertical gun elevation to allow for air-density variations.
Vw	*Wind elevation correction.* The correction included in vertical gun elevation to allow for apparent range wind.
Vx	*Complementary error correction.* The correction to allow for the vertical offset between the bore axis and the LOS produced by sight-deflection setting. $Vx = \sin^{-1}(\sin E \sec Ds) - E$.
Xog	*Cross component of own-ship velocity.* The horizontal component of own-ship velocity perpendicular to the plane of fire. $Xog = So \sin Br2$.

CHAPTER 15

SYMBOL	DEFINITION
Xwg	*Cross wind.* The horizontal component of true wind perpendicular to the plane of fire. Xwg = Sw sin Cwg.
Xwgr	*Apparent cross wind.* The horizontal component of apparent wind perpendicular to the plane of fire. Xwgr = Xog + Xwg.
Yog	*Line component of own-ship velocity.* The horizontal component of own-ship velocity in the plane of fire. Yog = So cos Br2.
Ywg	*Range wind.* The horizontal component of true wind in the plane of fire. Ywg = Sw cos Cwg.
Ywgr	*Apparent range wind.* The horizontal component of apparent wind in the plane of fire. Ywgr = Yog + Ywg.

15C3. Statement of the problem. A gun fired under a given set of conditions will produce just one trajectory that is in no way affected by target motion or by the point on the trajectory at which the hit, or burst, occurs. The student will recall that surface range-table values pertain principally to the point of fall. In other words, the data are for the point at which the trajectory returns to the horizontal plane from which the projectile was fired. It is apparent that surface range tables cannot be used in computing ballistics for AA fire, for the target may be at any point in a given trajectory.

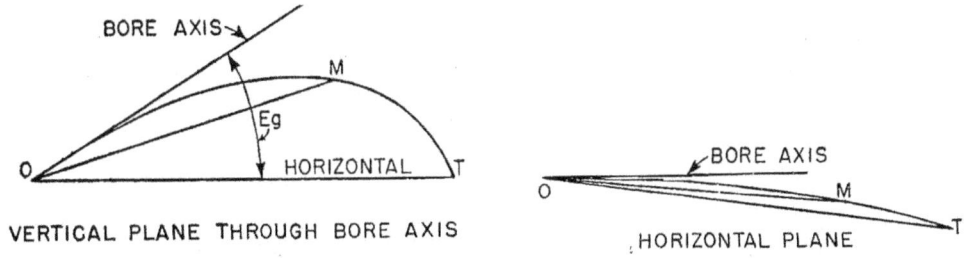

FIGURE 15C1. Trajectory for an elevated target.

The trajectory OMT in figure 15C1 is the one that would be used against a surface target at T or an air target at M or any other point on the entire trajectory. No matter where the target is on OMT, the angle of departure, or vertical gun elevation (Eg) is the same. However, sight angle (Vs) and sight deflection (Ds) will have different values for each point, since OM is the LOS for a target at M, while OT is the LOS for a target at T. Evidently Vs and Ds depend upon both present range (R) and target elevation (E). Since corrections for wind, and own-ship and target motions depend upon relative target bearing (Br), sight settings in any practical case also depend upon Br. The problem is to determine the values of Vs and Ds which will place the bore axis in the space position that will produce the trajectory required to hit the target.

While in theory the correct values of Vs and Ds will produce hits, a perfect evaluation of these two angles is impossible. Some errors inherent in the solution methods and others due to inaccurate estimates of such factors as air density, wind, and target motion are bound to exist. In addition, natural gunfire dispersion causes variations between shots fired under seemingly identical conditions. Consequently, projectiles are fuzed to burst near the plane to increase the hitting probability. When the fuze bursts the projectile close to the target, some of the fragments may hit the plane regardless of small discrepancies in sight settings. In addition, the nearby blast effect may either damage the plane or at least make it difficult to maintain a smooth, uninterrupted bombing run. So the determination of fuze setting (F) is also a part of the problem when time-fuzed projectiles are used.

In AA fire, ballistics are required for the same elements that are considered in surface fire. However, factors such as earth rotation, projectile-weight variation, and arbitrary ballistics are

THE AA PROBLEM: ANALYTICAL SOLUTION

not generally considered, and are therefore not discussed in this chapter. Trunnion-tilt corrections assume greater proportions in AA fire because of higher angles of departure, but due to the rapidity with which they change, they are not included in sight settings. Rather, they are separately applied to the guns in an indirect manner as explained in Chapter 16. It is therefore assumed that the trunnion axis is horizontal for the discussions in the following articles, which consider the effects of:

Gravity	Wind
Drift	I. V. variations
Target motion	Air-density variations
Own-ship motion	Complementary error

The bore axis must be offset laterally from the LOS by an angle measured in the slant plane which includes corrections for drift, own-ship and target motions, wind, and a deflection spot. Since the slant plane is normally elevated in AA fire, the lateral offset introduces a vertical movement of the LOS called the complementary error. The bore axis must be offset from the LOS by a vertical angle that includes corrections for gravity (causing trajectory curvature), own-ship and target motions, wind, I. V. and air-density variations, and an elevation spot. The vertical angle set on the sights as Vs must be equal to the sum of these corrections, minus the complementary error introduced by the deflection setting.

15C4. Ballistic data. The trajectory graph included in Appendix D shows the trajectories that would be obtained with the 5"/38 cal. gun when fired at various angles of departure under the standard conditions listed in Article 12B1. Any variation from these conditions results in modifications of the trajectories. While it would be possible to make a similar graph for any other particular set of conditions, it is obviously impossible to consider making such a diagram for all conceivable variations that might occur. Instead, the graph as shown serves as the basis for many of the ballistics which account for variations. Among the items directly available from this graph for any point on a given trajectory are:

1. *Time of flight.* Read from time marks on each trajectory.

2. *Fuze setting* (for powder-train fuzes). Read from curves approximately centered about origin. This is different from time of flight because the powder burning rate varies with air density in the different altitude zones.

3. *Height.* Read from the vertical scale at the left edge of the drawing.

4. *Horizontal range.* Read from the range scale at the bottom.

5. *Slant range.* Obtained by using dividers and the horizontal range scale.

6. *Target elevation* (position angle). Measured by laying straight edge through origin and point on trajectory, and reading on the scale along the top and right side.

7. *Superelevation.* Explained in Article 15C11.

Some corrections for variations from standard conditions are provided on graphs, generally about 18 inches by 24 inches in size, issued by the Bureau of Ordnance. To acquaint the student with the type of information available, excerpts from some of the graphs for the 5"/38 cal. gun are reproduced in miniature in figure 15C6. This set of curves is referred to at appropriate points in this section.

15C5. Relative target motion. During time of flight (Tf) the target moves from the position it occupied at the instant of firing. It is obvious that the gun should be fired not at the present position, but at the advance position the target will occupy at the end of Tf. However, the LOS must be directed at the present position, hence the gun must lead the LOS (and the target) by correction in elevation and deflection.

Own ship is normally in motion, and while such motion after the projectile leaves the gun cannot alter the trajectory, it imparts horizontal velocity components to the projectile at the instant of firing. These components must be considered in computing sight settings, for they result in a trajectory different from that which would be obtained with the same angle of departure under standard

CHAPTER 15

conditions. If the projectile were fired in a vacuum, the velocity components due to own-ship motion would remain constant during Tf, for there would then be no air to cause retardation.

The effect of the air under actual conditions may be considered equivalent to that of a horizontal wind of magnitude equal to own-ship motion but of opposite direction. Such a wind is the apparent wind due to own-ship motion, and its effect may be taken into account by combining it with true wind to obtain the total apparent wind, which is then used in making the elevation and deflection corrections for wind. This is done in AA fire control, hence own-ship motion is here considered as equivalent target motion, and the combined effects of own-ship and target motions, or relative target motion, may be used in computing sight settings.

15C6. Relative target movement. The time of flight movement of the target relative to own ship is Tf $\times$ Sr, as illustrated in figure 15C2 (use as reference through Article 15C10). The velocity Sr is simply the resultant of actual own-ship velocity and actual target velocity, as obtained in section A. As long as own ship and target hold their courses and speeds, the vector Sr is fixed in space. In other words, its value is not instantaneous like the values of the rates dR, RdBs, and RdE. These three rates have instantaneous values only because they are measured with respect to the LOS, which is continuously changing position with respect to own ship and target.

The vector Sr can be resolved into many different sets of components that will completely describe it, and the components of any chosen set will remain fixed in value so long as Sr remains constant. The values of the rates dR, RdBs, and RdE for the present target position form one set of these components, and are already available from the positioning part of the problem. When the present values of the rates are each multiplied by Tf, their products (Tf $\times$ dR, Tf $\times$ RdBs, and Tf $\times$ RdE) are components of relative target movement during Tf. When these components are added vectorially as in the illustration, they terminate at the end of the vector Tf $\times$ Sr and define the advance target position with respect to the present target position.

15C7. Advance target position. The target position relative to own ship that will exist at the end of the time of flight is the advance (or predicted) target position. Like the present target position, it is defined by three coordinates: (1) advance (or predicted) range (R2), (2) predicted relative target bearing (Br2), and (3) predicted target elevation (E2). These quantities are predictions of the "present" values of R, Br, and E that will exist at the moment a projectile, fired at the present instant, hits or bursts; i.e., at the end of the time of flight. Note that the numeral 2 is used to indicate predicted values pertaining to the advance position. Thus, the predicted value of true target bearing is B2.

In this chapter, the line extending from the gun to the advance target position is called the line of fire. The vertical plane containing the line of fire is called the plane of fire. The actual flight of the projectile is from the own-ship position to the advance target position, or figuratively, along the line of fire. While in the surface problem ballistics were based on the line of sight, the change in target position during time of flight in the AA problem is of sufficient magnitude to warrant computing ballistics on the basis of the line of fire.

The advance position is obtained by the application of various *plane* geometry and *plane* trigonometry relations. The student is cautioned to avoid letting the three-dimensional aspects of the problem obscure the simplicity of the mathematics involved. Time of flight appears in many of the expressions in the following articles. Its value depends on a number of factors as explained subsequently in Article 15C17. For the present, it should be assumed that Tf is known.

15C8. Range prediction. Triangles OBC and OCD in figure 15C2 are right triangles and are reproduced in their true shapes in figure 15C2b. It is evident that OB = R + Tf $\times$ dR. (Tf $\times$ dR has a negative value in the illustration.) The correct value of R2 can be computed from OB by using the plane geometry theorem which states that in a right triangle, the square of the hypotenuse is equal to the sum of the squares of the two sides. Thus:

$$(OC)^2 = (BC)^2 + (OB)^2, \text{ and}$$
$$(R2)^2 = (CD)^2 + (OC)^2 = (CD)^2 + (BC)^2 + (OB)^2.$$

RESTRICTED

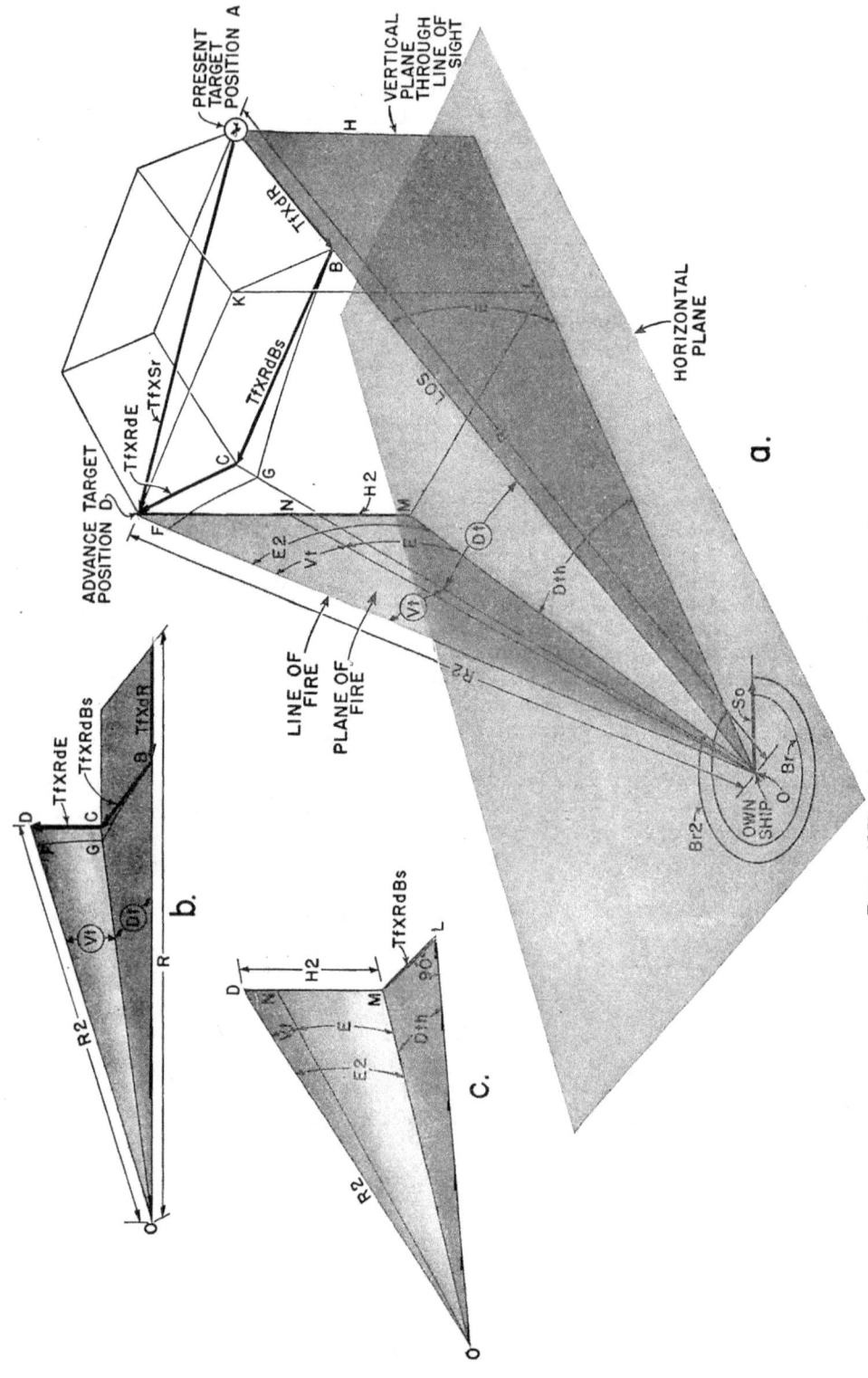

FIGURE 15C2. Advance target position and predictions.

CHAPTER 15

Substituting the equivalent values of the lines and taking the square root of both sides of the equation,
$$R2 = \sqrt{(Tf \times RdE)^2 + (Tf \times RdBs)^2 + (R + Tf \times dR)^2}$$

A more convenient method of finding R2 is to compute the range prediction (Rt), which when added to R, produces R2. The arcs BG and GF are drawn with a radius equal to OB, or $R + Tf \times dR$, using O as a center. It is evident from figure 15C2b that:
$$OB = OG = OF = R + Tf \times dR, \text{ and}$$
$$R2 = OD = OF + FD = R + Tf \times dR + FD.$$

Thus, the range prediction (Rt) is $Tf \times dR + FD$. Since $Tf \times dR$ is available from preceding computations, FD is the only remaining unknown, and can be obtained as follows:
$$(OD)^2 = (OC)^2 + (CD)^2, \text{ and}$$
$$(OB)^2 = (OC)^2 - (BC)^2.$$

Subtracting the second equation from the first,
$$(OD)^2 - (OB)^2 = (CD)^2 + (BC)^2.$$

Factoring the left-hand term,
$$(OD - OB)(OD + OB) = (CD)^2 + (BC)^2, \text{ and}$$
$$OD - OB = \frac{(CD)^2 + (BC)^2}{OD + OB}$$

Substituting equivalent values of the lines,
$$FD = \frac{(TF \times RdE)^2 + (Tf \times RdBs)^2}{R2 + R + Tf \times dR}$$

Little error will be introduced if $R + Tf \times dR$ is substituted for R2 in the denominator of the right-hand term. Thus,
$$Rt = Tf \times dR + FD = Tf \times dR + \frac{(Tf \times RdE)^2 + (Tf \times RdBs)^2}{2(R + Tf \times dR)}.$$

15C9. Elevation prediction. The change in height during time of flight is evidently $Tf \times dH$, since dH is the total vertical-motion component. Thus, the height of the advance target position, or H2, in figures 15C2a and c is $H + Tf \times dH$. Since $H = R \sin E$, $H2 = R \sin E + Tf \times dH$. Predicted target elevation (E2) is evidently the angle subtended at the gun by the predicted height (H2). Thus:
$$\sin E2 = \frac{H2}{R2} = \frac{R \sin E + Tf \times dH}{R2} \text{ and}$$
$$E2 = \sin^{-1} \frac{R \sin E + Tf \times dH}{R2} *$$

The elevation prediction (Vt) can be readily obtained from the relation $Vt = E2 - E$. The vertical gun elevation (Eg) must, of course, include the prediction Vt to allow for the vertical change in target position during the time of flight.

The value of Vt may be found directly by an approximation. The angle subtended at the gun by the line $Tf \times RdE$ is the circled angle Vt, and is nearly equal to Vt. The discrepancy is due to the fact that the plane OCD containing the gun position and $Tf \times RdE$ is not vertical, and the angle between OC and the horizontal is less than E. Ignoring these considerations:
$$\sin Vt = \frac{Tf \times RdE}{R2}, \text{ or}$$
$$Vt = \sin^{-1} \frac{Tf \times RdE}{R2}, \text{ and}$$
$$E2 = E + Vt = E + \sin^{-1} \frac{Tf \times RdE}{R2}$$

---
* The expression $\sin^{-1}$ is read "the angle whose sine is".

THE AA PROBLEM: ANALYTICAL SOLUTION

15C10. Deflection prediction. The change in relative target bearing during Tf is the horizontal angle Dth, subtended by the horizontal distance between the advance target position and the vertical plane through the LOS. In figure 15C2a, DKLM is a vertical plane through DK, intersecting the horizontal plane in line LM. Since DK is perpendicular to the vertical plane through the LOS, LM is perpendicular to OL, as shown in figure 15C2c. It is clear from the drawings that OM = R2 cos E2, and LM = Tf × RdBs, hence:

$$\sin \text{Dth} = \frac{\text{LM}}{\text{OM}} = \frac{\text{Tf} \times \text{RdBs}}{\text{R2} \cos \text{E2}}, \text{ and}$$

$$\text{Dth} = \sin^{-1} \frac{\text{Tf} \times \text{RdBs}}{\text{R2} \cos \text{E2}}$$

The sign of Dth is the same as that of RdBs, which has a negative value in the illustration, since it tends to decrease Br. Hence, the predicted relative bearing is:

$$\text{Br2} = \text{Br} + \text{Dth}.$$

The slant-plane lateral offset or deflection prediction (Dt) required to compensate relative movement during time of flight is approximately equal to the circled angle Dt in the figures. The error is due to the fact that Tf × RdBs does not lie exactly in the slant plane. The approximate value of Dt may be computed as follows:

$$\tan \text{Dt} = \frac{\text{BC}}{\text{OB}} = \frac{\text{Tf} \times \text{RdBs}}{\text{R} + \text{Tf} \times \text{dR}},$$

or, since R2 is nearly equal to R + Tf × dR,

$$\text{Dt} = \tan^{-1} \frac{\text{Tf} \times \text{RdBs}}{\text{R2}}$$

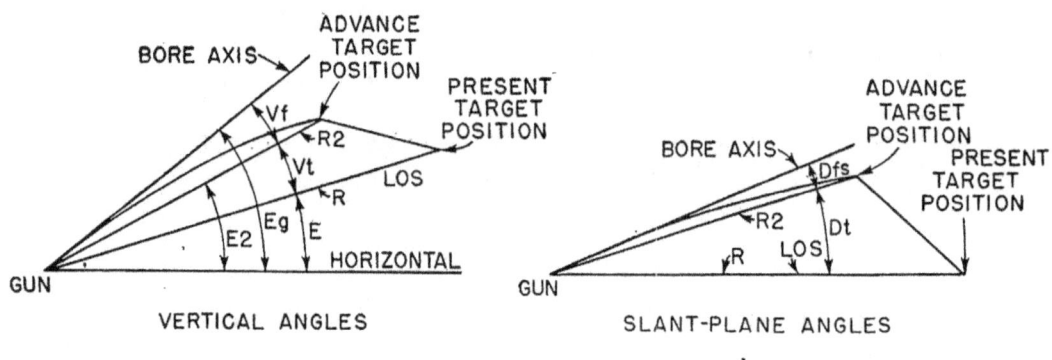

a.　　　　　　　　　　　　b.

FIGURE 15C3. Trajectory curvature compensation.

15C11. Trajectory curvature. Obviously the bore axis cannot be pointed directly at the advance target position because of the trajectory curvature. The curvature in a vertical plane is compensated by a correction called superelevation (Vf), as illustrated in figure 15C3a. Superelevation is sometimes referred to as the correction for the gravity effect on projectile flight. The value of Vf may be obtained from the trajectory graph (in Appendix D). The advance target position is plotted on the graph, using advance range (R2) and predicted target elevation (E2) as obtained in the preceding articles. The difference between E2 and the angle of departure (Eg) for the trajectory that passes through the plotted point is Vf.

Lateral deviation due to projectile rotation is compensated by drift (Dfs). It may be considered as the correction for trajectory curvature in the slant plane. Since sight deflection is set in the

CHAPTER 15

slant plane, Dfs must be computed for that plane and applied as in figure 15C3b. Drift varies from zero at the muzzle to the value at the point of fall, hence it varies from point to point along a given trajectory. The correction Dfs must vary similarly, and evidently depends upon the advance target position. Values of Dfs may be obtained from curves like those in figure 15C6a. The entering arguments are predicted target elevation (E2) and time of flight (Tf).

15C12. Wind. As in the surface problem, only the horizontal wind is considered, and its value is based upon the ballistic wind data obtained by aerological observations. However, it is considered with respect to the line of fire rather than with respect to the LOS. Since own-ship motion has been considered as equivalent target motion in obtaining relative motion, apparent wind must be used in making the wind corrections.

The components in and across the plane of fire are Yog and Xog for own ship, and Ywg and Xwg for true wind, as shown in figure 15C4a. The algebraic sums of the corresponding components are the components of apparent wind Ywgr and Xwgr with respect to the plane of fire. They may be computed from these equations:

$Yog = So \cos Br2.$ $Xog = So \sin Br2.$
$Ywg = Sw \cos Cwg.$ $Xwg = Sw \sin Cwg.$
$Ywgr = Yog + Ywg.$ $Xwgr = Xog + Xwg.$

The signs of the functions of Br2 and Cwg are such that the plus signs in the expressions for Ywgr and Xwgr yield the required values. If acute angles are used, care must be taken to reverse own-ship components before adding them to those of true wind. The value of Br2 is available from Article 15C10. Predicted wind angle (Cwg), as shown in figure 15C4b, is developed as follows:

$Cw = Bw - 180°,$ and
$Cwg + Cw = B2 + 180°.$ Therefore,
$Cwg = B2 - Bw + 360°.$ Substituting the equivalent of Cw,
$B2 = Co + Br2 - 360°.$ Hence,
$Cwg = Co + Br2 - Bw.$

The deflection component, or apparent cross wind (Xwgr), causes lateral deflection of the projectile which is compensated by wind deflection correction (Dw). The value of Dw may be obtained from curves like those in figure 15C6b. Since the cross-wind effect is given in yards vs. time of flight, it is necessary to establish the value of Tf before these curves can be used. The deflection is given in yards for a 10-knot cross wind, hence the correction from the curves should be multiplied by Xwgr ÷ 10, and then converted to angular measure.

The apparent range wind (Ywgr) has components WrR and WrE in and across the line of fire as shown in figure 15C4c. One effect of WrR is to alter the time of flight. When WrR opposes projectile motion, it takes longer for the projectile to travel a given distance. It is thus evident that Tf must include a correction for range wind.

Another effect of WrR is to alter the range attained by the projectile when fired on a given line of departure. The component WrE affects the elevation of the trajectory. To counteract the range and elevation changes, vertical gun elevation must include wind elevation correction (Vw), whose value may be found from curves such as those shown in figure 15C6c. These curves, which should be entered with R2 and E2, give the correction for a 10-knot rear range wind, hence the value obtained should be multiplied by Ywgr ÷ 10.

15C13. Initial-velocity loss. Any I. V. variation from standard due to erosion or powder temperature alters a trajectory. For a given angle of departure, a decrease in I. V. lowers the trajectory, while an increase raises it. An I. V. elevation correction (Vm) is required to counteract this effect. It may be obtained from curves like those in figure 15C6e, which gives corrections for a 100 f.s. I. V. loss. Predicted range (R2) and predicted target elevation (E2) should be used as the entering arguments. It should be noted also that a decrease in I.V. lengthens the time of flight (Tf), and vice-versa.

15C14. Air-density variations. A density higher than standard increases the time of flight and results in a lowering of the trajectory, and vice versa. The required air-density elevation correction (Vu) may be obtained from curves similar to those of figure 15C6f, entering them with R2 and E2. The correction from the curves is for a 10 per cent variation, hence it must be modified to agree with the actual per cent variation.

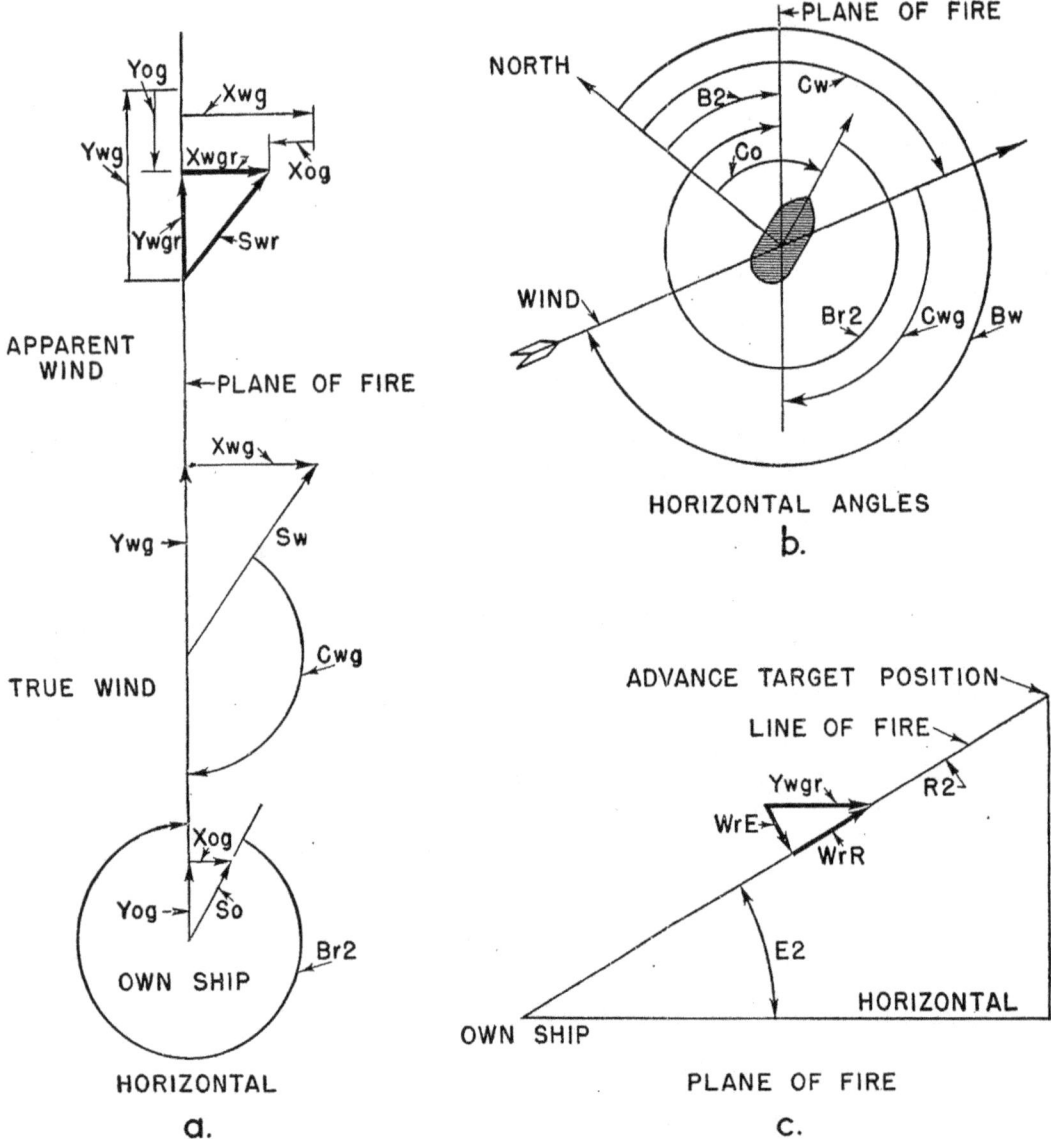

FIGURE 15C4. Wind components.

15C15. Gun-position summary. When allowance has been made for the various factors in the preceding articles, the theoretical bore-axis space position has been established. Any errors not accounted for by computation or due to inaccuracies are compensated by spots. An elevation spot (Vj) and a deflection spot (Dj) are sufficient in theory to adjust the trajectory so that the pro-

CHAPTER 15

jectile will hit the target. It is thus apparent that sight deflection (Ds) includes the following elements:
1. Deflection prediction (Dt).
2. Drift (Dfs).
3. Wind deflection correction (Dw).
4. Deflection spot (Dj).

Hence,
$$Ds = Dt + Dfs + Dw + Dj.$$

Vertical gun elevation is the sum of the following elements:
1. Target elevation (E).
2. Elevation prediction (Vt).
3. Superelevation (Vf).
4. Wind elevation correction (Vw).
5. I. V. elevation correction (Vm).
6. Air-density elevation correction (Vu).
7. Elevation spot (Vj).

It is evident that:
$$Eg - E = Vt + Vf + Vw + Vm + Vu + Vj.$$

15C16. Complementary error. If there were no deflection, sight angle (Vs) would be equal to Eg — E. However, as previously indicated, the setting of deflection introduces a vertical offset, or complementary error, between the LOS and the bore axis. Consequently, the actual value of Vs must be less than Eg — E by the complementary error correction (Vx).

The nature of the error is indicated by figure 15C5. By definition, the slant plane contains the trunnion axis and is parallel to the LOS. For practical purposes, the LOS may be assumed to lie in the slant plane, since its actual displacement is only a few feet at most. The student should review Section 12C to aid in visualizing the fact that Vs is set in the vertical plane through the bore axis. The vertical sight pivot is perpendicular to the slant plane, hence the LOS moves in the slant plane when Ds is set.

Assume for the moment that a target is at F in the drawing, and Ds has not been set. Then the LOS would extend from O to F. Next, asssume that Ds is set, thereby moving the LOS to B. Since the LOS is in the slant plane it obviously drops. It would become horizontal and coincide with the trunnion axis if 90° deflection were set. When the gun is trained to bring the LOS back to the target assumed to be at F, the LOS will actually extend from O to H, since the training occurs in the horizontal plane. In order to place the LOS back on F, it would be necessary to elevate the gun. This additional elevation would be entirely due to the deflection setting, and would result in too much vertical gun elevation.

The complementary error can be computed and compensated when determining Vs. In the diagram, OB (equal to R) is the LOS, the target is at B, and the vertical plane through the bore axis intersects the slant plane in OF, the horizontal in OG. By definition $\angle$ AOB is equal to target elevation (E). It is obvious that $\angle$ GOF is greater than E, the difference being Vx. The plane ABCD is vertical and parallel to the trunnion axis. It intersects the vertical plane through the LOS in the vertical line AB, and the vertical plane through the bore axis in the vertical line CD. It intersects the slant and horizontal planes in lines BC and AD, which are parallel to each other. It is therefore evident that triangles AOB, COD, and BOC are right triangles. From them the following relations are obtained:

$$CD = AB = R \sin E.$$
$$OC = R \cos Ds.$$
$$\sin (E + Vx) = \frac{CD}{OC} = \frac{R \sin E}{R \cos Ds} = \sin E \sec Ds.$$
$$E + Vx = \sin^{-1}(\sin E \sec Ds).$$
$$Vx = \sin^{-1}(\sin E \sec Ds) - E.$$

THE AA PROBLEM: ANALYTICAL SOLUTION

It is thus possible to express sight angle Vs as:
$$Vs = Eg - E - Vx = Vt + Vf + Vw + Vm + Vu + Vj - Vx.$$

15C17. Time of flight. It has been stated that range wind, I. V. variations, and air-density variations affect time of flight, hence Tf should include corrections for each of these factors. In other words, time of flight should be:
$$Tf = Tfk + Tfw + Tfm + Tfu, \text{ where}$$
Tfk = standard time of flight.
Tfw = range wind time-of-flight correction.
Tfm = I. V. time-of-flight correction.
Tfu = air-density time-of-flight correction.

Standard time of flight is that obtained from the trajectory graph when it is entered with R2 and E2. Curves for Tfw and Tfm are shown in figures 15C6d and 15C6g, and are likewise entered with R2 and E2. While similar curves can be made for Tfu, they are not shown.

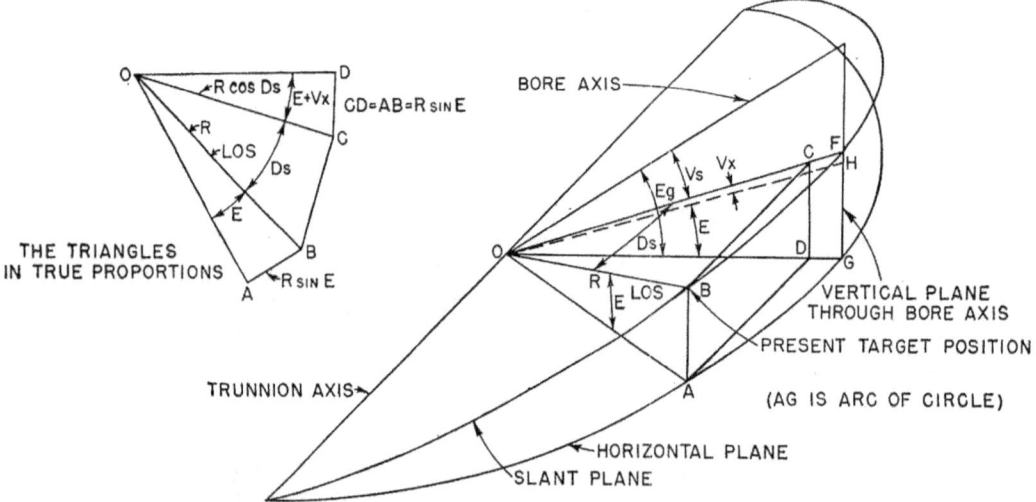

FIGURE 15C5. Complementary error.

The components of Tf are based on values of R2 and E2, which themselves depend upon Tf as shown in the equations of Articles 15C8 and 15C9. Consequently, from the analytical standpoint, it is necessary to assume a value of Tf such as that obtained by entering the trajectory graph with R and E, and then make a preliminary calculation of R2 and E2. Using these values of R2 and E2 to select Tfk and the time-of-flight corrections from the several graphs, a more-nearly correct value of Tf is obtained. Then R2 and E2 are recomputed and used to obtain a still more accurate value of Tf. If the process is repeated two or three times, the ultimate result will be a value of Tf that is sufficiently close for practical purposes. Mechanical computers perform the equivalent of this process practically instantaneously.

15C18. Fuze settings. It should be obvious that the setting of a time fuze must be based upon the time its projectile will be in flight. Air-density variations do not affect mechanical fuzes, hence their settings are the same as time of flight. However, air density affects the burning rate of powder and powder-train fuze settings must generally be different from time of flight. Since powder-train fuzes are no longer in general use, the ensuing discussion pertains to mechanical fuzes. Suitable powder-train fuze-setting curves may be obtained from the Bureau of Ordnance if necessary.

The calculations of Vs and Ds are based on the instantaneous values of R, Br, and E. It is, of course, impossible to compute Vs and Ds instantaneously by the analytical method, but mechanical

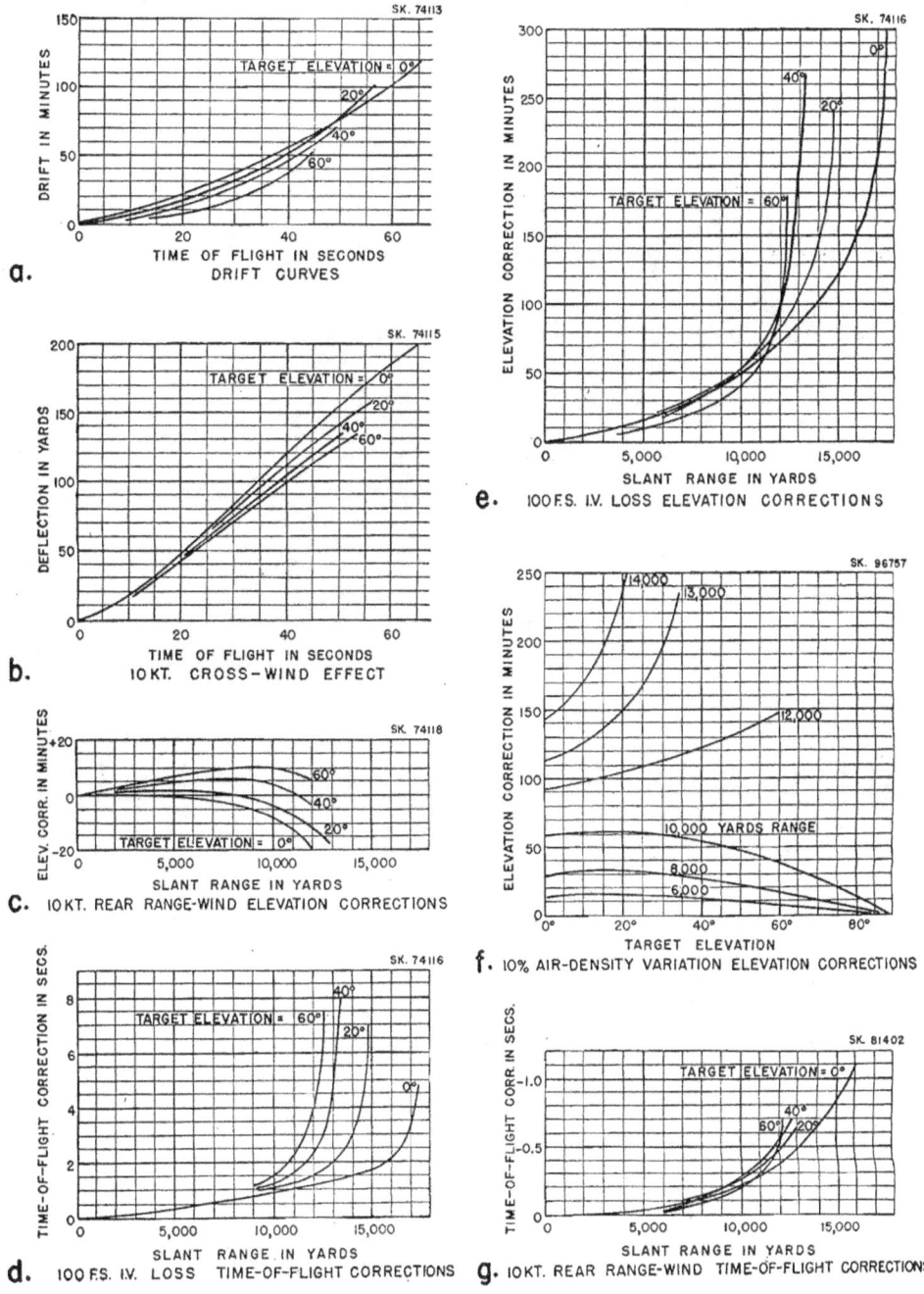

Figure 15C6. Ballistic data, 5″/38 cal. gun.

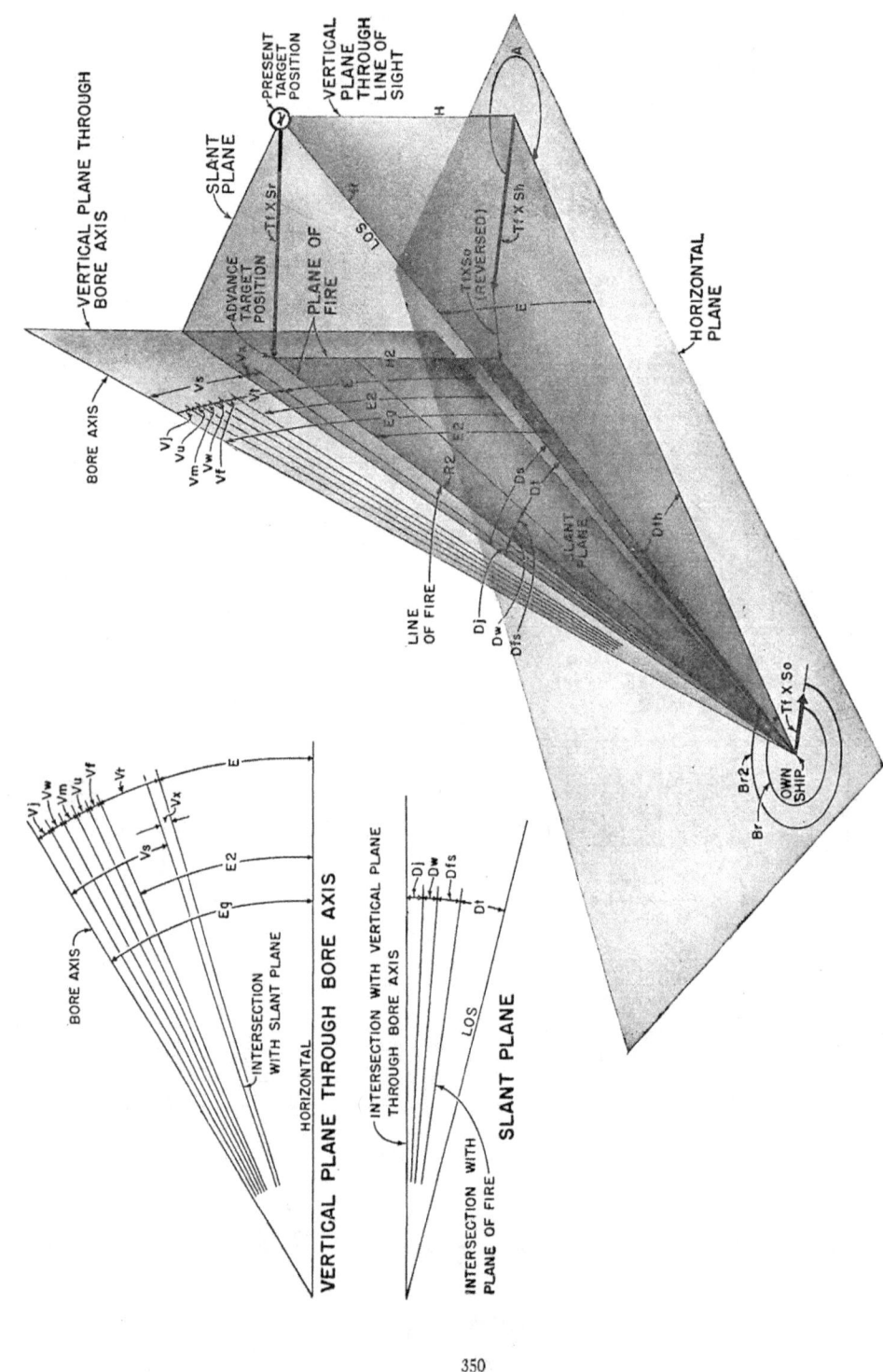

FIGURE 15C7. AA sight settings.

computers provide instantaneous values of these settings which can be continuously maintained on the gun sights by matching pointers. This means that when the LOS is held on the target, the gun may be fired at any time and the projectile will theoretically hit the target. The fuze setting on a projectile must be equal to the computed Tf that pertains to the instant at which that projectile is fired.

However, time fuzes must be set, or cut, before the projectiles are loaded. The time elapsed between cutting the fuze and firing the projectile is dead time (Tg). (Since computed values of Vs and Ds are continuously set upon the sights, there is no other element in dead time.) The value of Tg is determined by analysis of drills and firings, and is usually about 4 seconds. In order that the fuze may be set Tg seconds before the firing, it is necessary to predict the value of Tf that will exist when the projectile is fired.

The elements Tfk, Tfw, Tfm, and Tfu comprising Tf are all selected on the basis of the advance coordinates R2 and E2. Evidently, the time of flight that will exist Tg seconds later can be obtained by selecting its components on the basis of the advance range and predicted target elevation that will exist at the end of dead time. These coordinates would be fuze range (R3) and fuze target elevation (E3). The change during dead time in predicted target elevation (E2) is not considered in practice, as it is small enough to be neglected. However, the change in R2 is taken into account. Hence fuze setting (F) is equal to the time of flight selected from the graphs on the basis of R3 and E2. By the time the projectile is ultimately loaded and fired, the actual advance range has changed to a new value of R2, which is equal to the previous value of R3 used in computing F. The actual fuze setting is then the same as Tf for the instant of firing.

While the change in range during dead time could be computed exactly, it is adequate for this presentation to assume that the change is Tg x dR. If a time-fuzed projectile bursts behind or ahead of the target, a correction must be applied to the fuze setting (F). This is accomplished by a range spot (Rj), which for simplification, is here considered as affecting fuze range (R3) only. Thus,

$$R3 = R2 + Tg \times dR + Rj.$$

It should be understood that theoretically elevation spots (Vj) and deflection spots (Dj) can be made which will take care of the errors not accounted for in other corrections. But an air target is small, and the many variables entering into the positioning of a gun and natural dispersion would result in a low percentage of direct hits. Consequently, the fuze setting must be adjusted to burst the projectile close to the target. The change produced in R3 by the spot Rj obviously changes the values obtained from the graphs when selecting the components of F. The actual application of spots in a computer is more complex than indicated in this discussion, for they are applied to the advance coordinates R2 and E2 and therefore affect all of the ballistics computations.

D. PRACTICAL CONSIDERATIONS

15D1. Types of fire. Two broad classifications of AA fire are the *barrage method* and the *direct method*. In the former, projectiles are fired to burst ahead of target in such a position that the plane will pass through the bursts. This method is relatively simple, as it permits prearranged fuze and sight settings for various zones. The required settings for a particular zone may be quickly selected without the complications involved in computing all the possible corrections to Vs and Ds. The barrage must be shifted periodically to keep it continuously ahead of the target. While this method is wasteful of ammunition, it is an effective expedient when the direct method cannot be employed.

As its name implies, the direct method involves firing each projectile to hit or burst close to the target. This method is achieved with director firing systems having suitable computers, or with machine guns firing tracer ammunition. Tracer observation makes it possible to bring the stream of fire on the target by what might be called a trial-and-error process.

15D2. Graphic solution. The AA problem may be partially solved graphically when the attack is horizontal, or nearly so, and starts at sufficient range to permit the necessary tracking.

THE AA PROBLEM: ANALYTICAL SOLUTION

Figure 15D1 illustrates the problem of tracking and predicting the advance position of a plane. In this diagram, movements are relative to own ship. When the target is at P1 the range is R1 and target elevation, or position angle, is E1. The time elapsed between the actual measurement of R1 and E1 and the subsequent plotting is *lag*, during which the plane moves to P2. Sight and fuze settings based on P1 are transmitted to the gun, but during the dead time elapsing after the data are transmitted and before the gun fires, the plane moves to P3. Thus, when the gun fires, the LOS extends from O to P3, the range is R and target elevation is E.

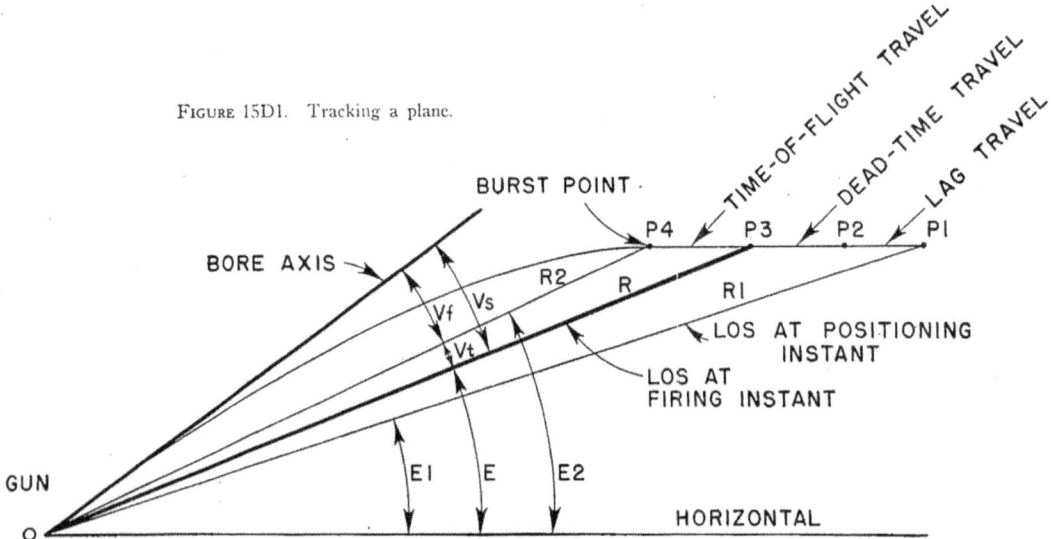

FIGURE 15D1. Tracking a plane.

During the time of flight, the plane moves to P4, which is the point at which the burst should occur. The point P4, where range is R2 and target elevation is E2, corresponds to the advance or predicted target position as discussed in the preceding section. Assuming all conditions standard except own-ship and target motion, it is evident that sight angle (Vs) must include superelevation (Vf) and elevation prediction (Vt), and that these elements, as well as fuze setting should be based on R2 and E2. This graphic method of obtaining R2 and E2 and subsequently determining the settings is too cumbersome for actual use, but in a modified form it provides the basis for barrage fire.

15D3. Fuze-setting zones. The graphic solution is simplified by dividing the space about the gun (or own ship) into a number of zones which are based on fuze settings. It is then possible to maintain ready ammunition with fuzes preset at certain selected values corresponding to the zones in which they are intended to function. Sight settings for various plane speeds and altitudes can be computed and tabulated for each zone, making the settings available in a simple form for speedy reference.

It is evident that for a single fuze setting, such as 10 seconds, there is a locus of burst points like that shown in figure 15D2. This locus, for mechanical fuzes, corresponds to a time-of-flight line on a trajectory graph. In other words, the time of flight is the same (10 seconds) to any point on the line. Assuming that it is desired to hit a plane traveling at 300 knots on line P1 to P5 when it reaches P5, the gun must be fired when the target is at P4. The distance between P4 and P5 is 10 seconds × 300 knots, or 1,689 yards. To hit a plane on the burst-point locus at any other altitude, the gun must be fired when the plane is 1,689 yards to the right of that locus. It is clear that a line can be drawn that is at every point displaced to the right of the burst-point locus by 1,689 yards. Such a line is a *zone locus*, in this case the zone 10, 300 knot locus. The zone 10 locus for a plane traveling at some other speed, such as 200 knots, can be obtained in a similar manner. In the practical approach, zone loci are provided for speeds varying in steps of 20 knots.

CHAPTER 15

It is to be noted that a zone locus pertains to a particular speed, hence the distance to a zone locus depends on the relative speed during the attack. Since the zone locus is based on horizontal travel, it can be used only for attacks that are horizontal or nearly so.

15D4. Advance time. Instead of opening fire just as the target arrives at a zone locus, firing begins when the plane is to the right at a point such as P3 in figure 15D2. The time required for the plane to travel from P3 to P4 is *advance time*. When the plotted target position is P1, the orders are given for opening fire for a particular zone. The sight settings actually made are those which are correct for firing when the plane is at P4. Due to lag, the actual target position is P2 when the plot shows it as P1. During the dead time before firing begins, the plane moves to P3, whereupon the first projectile leaves the gun. This projectile bursts on the burst-point locus at a point below P5, while the plane is somewhere between P3 and P5. During the advance time, the plane travels from P3 to P4. Theoretically, the projectile fired when the plane is at P4 bursts and meets the plane at P5. Successive bursts move upward on the burst-point locus, providing a barrage which partially compensates estimation errors, fuze dispersion and other unaccountable factors. This method assures that the gun will actually be firing with the desired settings when the target arrives at the zone locus.

15D5. Sight settings. Sight-setting tables are provided for use in the graphic solution. With fuze zones established, a series of sight settings can be tabulated for various target speeds at regular intervals of altitude. For the 5"/38 cal. gun, the tables include a section for each 1,000 feet of altitude up to 20,000 feet. The section for 9,000 feet is reproduced below. Intermediate zones are skipped in the higher settings because the plane travel during the greater time of flight would cause the zone loci to overlap if they were provided for every second. In addition, fuze dispersion and other factors make it undesirable to attempt zoning for every second.

	9,000 FOOT ALTITUDE												
	Zone	2	3	4	5	6	7	8	10	12	14	17	21
TARGET SPEED	180				460	445	445	445	470	505	555	635	760
	200				500	480	475	475	495	530	575	655	780
	220				540	515	505	500	520	550	595	675	795
	240				580	545	535	530	540	575	615	690	810
	260				615	580	560	555	565	595	630	710	825
	280				655	610	590	580	585	615	650	725	840
	300				690	640	615	605	605	635	665	740	855

SECTION FROM SIGHT SETTING TABLES 5"/38 CAL. GUN.

15D6. Tracking sheet. A tracking sheet is a graph about 2 feet by 3 feet used for tracking air targets making attacks horizontally or nearly so. One for the 5"/38 cal. gun has the following printed on it:

1. Radial position lines spaced 1° apart, radiating from the origin in the lower, left-hand corner, by means of which position angles, or target elevations of points are plotted.

2. Range circles in a series of arcs concentric about the origin. They are spaced 100 yards apart and are used to establish the plotted position in range on the position lines.

3. Altitude lines extending horizontally across the sheet, spaced at 1,000 foot intervals and used to determine target altitude.

4. Target travel scale in the margin, showing distance traveled at various speeds during different time intervals.

5. Zone loci in groups of colored lines for each zone. All loci of the same color are for the same plane speed, each speed being represented by a different color.

RESTRICTED

THE AA PROBLEM: ANALYTICAL SOLUTION

When the sheet is used, position angle and range are measured and plotted at 10-second intervals. As soon as several points are plotted, the distance between them can be compared to the target-travel scale by means of dividers, and the plane's relative speed determined. The altitude is read from the horizontal altitude lines. Knowing the speed, the dividers can be set for the target travel during lag, dead time, and advance time. When the plotted position is to the right of the properly colored zone locus by the distance thus set on the dividers, the order is given to commence firing, including the necessary zone description and sight-setting data. Information which must be sent to the gun includes speed, altitude, and zone, to permit setting Vs and either setting fuzes prior to loading or selecting projectiles with fuzes previously set for the zone. Similarly, scale must be provided either by estimate or from a suitable deflection diagram.

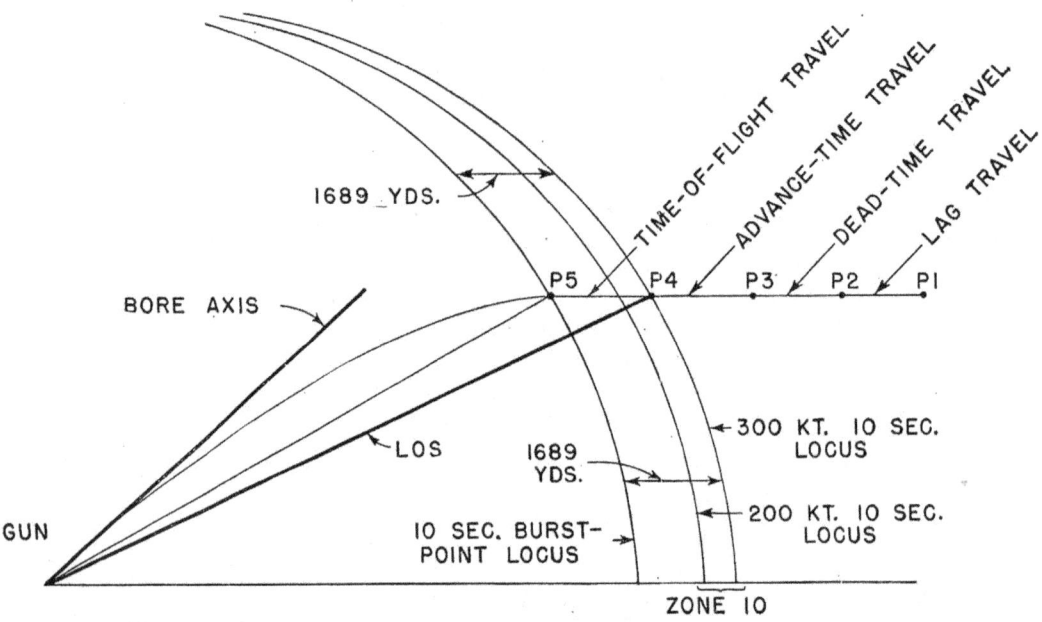

FIGURE 15D2. Zone loci.

The solution with the tracking sheet and sight-setting tables may be further refined if time permits. Sight deflection can be obtained from a deflection diagram that provides corrections for drift and relative target motion. Corrections to both Vs and Ds for wind, I. V. loss, and other factors may be determined from graphs. However, the rapidity with which an air attack is carried out ordinarily prohibits such elaborate preparations. Furthermore, inherent crudities in the graphic method seldom justify minor refinements.

Various shortcuts may be used to eliminate graphic tracking. Some AA firing is accomplished by using only two or three fuze settings, such as 2, 4 and 8 seconds, the sight setter observing the bursts and adjusting the sights as necessary after making a preliminary setting obtained from tables. Experience will indicate how many projectiles may be fired with each setting before it is necessary to shift to the next zone. In any approach to the AA problem, regardless of the refinements employed, it is probable that spotting will be required to obtain hits. However, the higher the accuracy of the initial computations, the easier will be the spotting problem.

15D7. Star-shell illumination. While star shells are used only for the illumination of surface targets, star-shell control has features in common with AA fire, since elevation, deflection and fuze range are involved. The desired burst point of a star shell is such that the star, in drifting downward and laterally with the wind, will cross a vertical plane through own ship and target at the mid-point of its burning period and at about 1,000 to 2,000 yards beyond the target, thereby silhouetting it. The burst should occur at an altitude that will permit the star to burn out shortly before hitting the water. This altitude is about 1,000 feet for three-inch and 1,500 feet for five-inch star shells.

The problem is similar to firing at a plane with zero speed at the desired burst altitude, and at a horizontal range equal to target distance plus 1,000 or 2,000 yards. Illumination-control data are available in the form of graphs and tables of various sorts. They provide sight-bar range, time of flight, fuze setting for powder-train fuzes where used, and corrections for variations like I. V. and air density.

15D8. Summary. The difficulties of actually establishing target angle, speed and rate of climb, and of compensating all the known factors quickly and accurately should be sufficient indication that mechanical instruments are required to provide a thorough and continuous solution. While mechanical computers solve the problem in a manner similar in its broader aspects to the analytical solution presented in this chapter, the specific methods of correcting for various elements may differ considerably. The practical construction of a computer involves the use of a number of empirical equations and mechanical and electrical expedients which tend to confuse the fundamental problem being solved.

Elaborate computers are incorporated in director firing systems for dual-purpose batteries of 5-inch guns and larger. For smaller guns, simpler computing devices such as gyroscopic sights are in use. These instruments are described in subsequent chapters to acquaint the student with the apparatus that is provided as the primary means of controlling gunfire.

CHAPTER 16

FUNDAMENTALS OF DIRECTOR CONTROL

A. APPLYING SIGHT SETTINGS AND FIRING

16A1. Local line of sight. No matter what method is used in laying a gun, its bore must be directed at a definite position in space at the instant of firing. It has been shown that this position may be established by vertical and lateral angles between the axis of the bore and a line of sight between the gun and the point of aim. In other words, a local line of sight originating at the gun is a reference with respect to which the gun may be laid. When each gun of a battery is properly offset by Vs and Ds from its own LOS, and each LOS is directed at the target, all of the guns are correctly positioned in space for firing. Obviously the use of a local LOS requires that the sights be correctly set at each gun, that each gun be laid by its own pointer and trainer, and that the pointer (or trainer) have a firing key by means of which the firing circuit may be closed when the LOS is on the point of aim.

16A2. Pointer key fire. The oldest and simplest method of firing is *pointer key*. In this process, the guns are laid by their own crews with respect to their own lines of sight. When the LOS of a given gun is on the target, the pointer closes his firing key, thereby firing that gun. There are occasions when the trainer actually does the firing, but this does not materially alter the method. Pointer fire is used today in cases where no director is provided, and it is the final stand-by method for all guns.

It is an established fact that successive shots fired from the same gun, or several shots fired at the same instant from identical guns all laid to the same position in space, will not fall in the same place. The spacing between the shots is *dispersion*, discussed more fully in Chapter 21.

Spotting is a difficult procedure when the guns of a battery are firing at will, as in pointer fire. The shots are not fired at the same instant, hence they do not land at the same time. The natural dispersion in gunfire, combined with additional dispersion inherent in pointer fire, makes the estimate of a suitable correction uncertain at best.

A better estimate of the spot necessary to correct Vs and Ds is obtained when several shots are fired at once to form a *salvo*. The geometric center, or *mean point of impact* of all the shots is an indication of the average point of fall. The spot to be applied is that which will place the mean point of impact on the target. Some of the earliest improvements in fire control were the result of efforts to reduce the dispersion in salvos in order to facilitate spotting and to increase the probability of obtaining hits.

16A3. Buzzer fire. The first advance from pointer fire was the introduction of *buzzer fire* in an effort to produce salvos. It is identical with pointer fire except that the pointer does not fire at will, but fires only during an interval when a buzzer is sounding. In this method, buzzers at each gun are sounded simultaneously by closing a key at the control station, hence those guns which are ready fire at approximately the same time to produce a salvo. The signal must be prolonged for several seconds to provide an opportunity for the various guns to fire, and this gives rise to an appreciable dispersion in time, for one gun may fire at the beginning of the interval, and another at the end.

Ship roll and pitch impart motion to the projectiles, in addition to the motion due to movement of the ship along its course. The velocity of roll or pitch is not uniform, for it must obviously be zero at the instant the motion reverses, and maximum at some intermediate point. Unless all the guns fire at the same instant, the velocities imparted by the rolling and pitching will vary among the projectiles of a salvo, and will contribute to the amount of dispersion.

16A4. Master key fire. A simple expedient known as *master key fire* was introduced to fire all guns of a battery at the same instant. Originally this method was the same as pointer fire, except

CHAPTER 16

that a master firing key at the control station controlled the guns. The master key was placed in series with the firing key at each gun. Under these conditions, the pointers close their keys when their guns are on the target, but their guns do not fire until the master key is also closed. The result is that all guns which are ready and on the target fire at the same instant, and dispersion due to firing at different times is eliminated.

In that form of master key fire in which the guns are laid with respect to their own, or local lines of sight, there are a number of difficulties, including the following:

1. The independent motions of the guns as they are held or brought on the target by their own crews at the instant of firing impart slightly different motions to the projectiles, contributing to dispersion.

2. The pointer and trainer may select the wrong target when there are several close together, or they may select a different point of aim when firing at the same target.

3. Interruptions in fire may be caused by haze, spray, or smoke interfering with visibility at the level of the guns.

4. Sighting may not be possible at maximum ranges because of the dip of the target below the horizon.

5. It is impracticable to hold the line of sight on the target continuously in the case of heavy guns, especially under unfavorable conditions of roll and pitch.

16A5. Remote line of sight. If all the guns of a battery were geared together in such a manner that the elevation and train of any one of them would automatically elevate and train each of the others, only one line of sight would be required for the entire battery. One gun could then be used as a master gun. Its sights would be set as in pointer fire, and its pointer and trainer would lay the master gun and all the other guns, and then fire the entire battery by means of a master firing key. Such an arrangement would eliminate the independent motion of separate guns, and prevent one gun from firing at a target different from that engaged by the other guns. If the master gun were then raised to a higher position on the ship, still being geared to the other guns, it would have a more distant horizon and would be subject to less interference from spray and smoke.

16A6. Director fire. Obviously, it is not practicable to gear all the guns together or to put a heavy gun in an elevated position. But the principle of the master gun can be applied by using electrical transmissions to the various guns, and by employing a comparatively light-weight instrument to establish the remote line of sight. Basically, this is just what is done in modern fire-control systems.

A director, mounted in an elevated position, transmits electrical signals to the guns through various modifying instruments. The signals are received and indicated at the guns, and followed by the pointers and trainers or by automatic power systems. The result is that the guns are laid with respect to a single, remote LOS, and as long as the signals are obeyed, the guns are not independent in motion. Using a director system it is possible to reduce the amount of pointing and training by laying the guns at predetermined angles and firing them as ship roll and pitch bring the guns into the required position.

16A7. Reference system. In pointer fire, the LOS is attached to the gun, and the angles between it and the bore axis are fixed by the sight mechanism. Neglecting trunnion tilt, the gun is continuously directed correctly in space as the ship rolls and pitches by simply holding the LOS on the target. In director fire, the director LOS and the gun bore axis are not connected, and the angles between them are maintained by using an intermediate reference system consisting of a reference plane and the fore-and-aft axis of the ship.

Fore-and-aft axis. The reference for measuring angles of train is the *fore-and-aft axis* of the ship. In director systems, the train angles necessarily are measured in the reference plane, clockwise from the bow.

Reference plane. The reference for measuring elevation angles with respect to the ship is the *reference plane*, sometimes called the *deck plane*, or *deck* for short. It is an imaginary plane, arbitrarily chosen for convenience, and never coincides with the actual deck, since a curved deck

RESTRICTED

FUNDAMENTALS OF DIRECTOR CONTROL

cannot be contained in one plane. It may be a plane which happens to lie in the horizontal when the ship is in drydock, the plane containing the roller path of one of the directors, or any other convenient plane approximately parallel to the deck and the roller paths of the various units of the system. No matter how it is originally selected, it remains fixed with respect to the ship, rolling and pitching with it. Gun and director roller paths are so nearly in the reference plane that train is considered to take place in that plane.

Perpendicular and vertical. In order to clearly indicate that an angle is measured in a plane perpendicular to the reference plane, the word *perpendicular* is sometimes used in describing the angle. Since guns, directors and other fire-control instruments train in the reference plane, they elevate through perpendicular angles, or in planes perpendicular to the reference plane. When *vertical* is used in this text, it means perpendicular to the true horizontal.

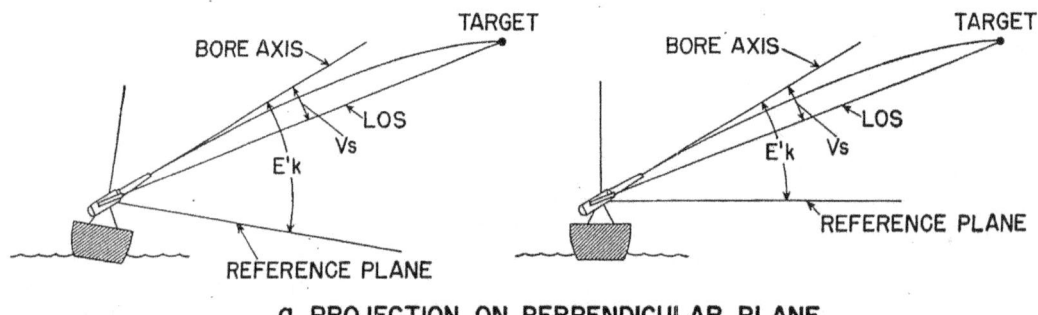

a. PROJECTION ON PERPENDICULAR PLANE

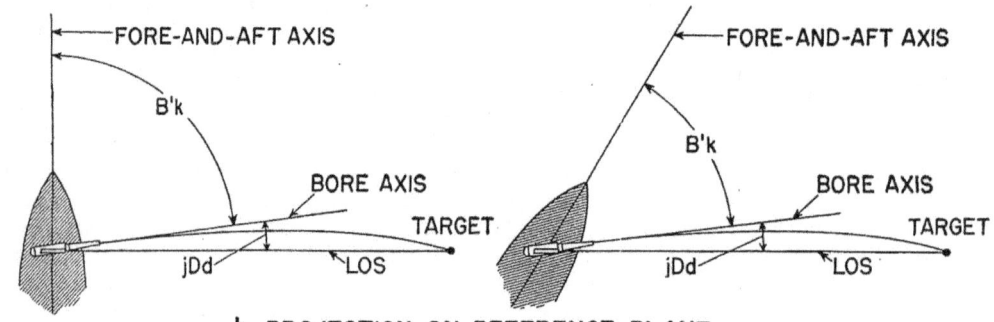

b. PROJECTION ON REFERENCE PLANE

FIGURE 16A1. Positioning a gun.

As as example, suppose that a gun is elevated 10° above its roller path when the reference plane is horizontal. The perpendicular and vertical angles between the bore axis and the reference plane are each 10°. When the reference plane tilts out of the horizontal, the perpendicular angle remains at 10°, but the vertical angle is changed.

16A8. Gun position. The position of the bore axis in space is established by two angles measured with respect to the reference system. One is actual *gun elevation* ($E'k$), which is the perpendicular angle between the bore axis and the reference plane; the other is *actual gun train* ($B'k$), the angle in the deck plane between the fore-and-aft axis and the perpendicular plane containing the bore axis. Note that angles measured in the reference or perpendicular planes are usually distinguished by a prime mark (') after the basic symbol.

For any given values of Vs and Ds, and any single position of the ship, there is just one combination of $E'k$ and $B'k$ that will place the gun in the correct position in space. The relation between Vs and $E'k$ is shown in figure 16A1a, which represents the projection on a plane containing the LOS and perpendicular to the reference plane.

CHAPTER 16

Sight deflection (Ds) is measured in the slant plane. The equivalent of Ds in the reference plane is *partial deck deflection* (jDd). The difference between these two quantities is not important at this point, but it is discussed later. The relation between B'k and jDd is shown in figure 16A1b, which is a projection on the reference plane.

As the ship rolls, pitches, or changes course, actual gun elevation and train must change to keep the bore axis fixed in space. It is one of the functions of director systems to continuously compute these angles. In any actual director system they are the resultant of a number of individual elements. The quantities transmitted to the gun to indicate the correct values of E'k and B'k are *gun-elevation order* (E'g) and *gun-train order* (B'gr). These orders from the director are not exactly equal to the angles to which the gun must be laid because of certain corrections which are made at the guns. These corrections are discussed in section 16D.

B. SYNCHROS AND SERVOS

16B1. Introduction. Frequent reference to the transmission of quantities by means of electrical signals has been made in preceding sections, and is made in subsequent chapters. It might be said that electrical transmitting devices are the nerves of a fire-control system. Without them, director fire would be far less efficient than it is today.

The chiefs units of modern transmission systems are *synchros* and *servos*. Both are somewhat similar in appearance to small electric motors such as those used in household fans. Synchros are used to transmit indications, and have the ability to position dials, but they are too weak to drive mechanisms. Servos are used to feed quantities into units of computers or other instruments in accordance with the indications of a synchro. Servos may be considered as amplifiers, for their effect is to increase the strength of a synchro indication to a working value.

The two main parts of either type of unit are the *rotor* and the *stator*. As the names imply, the rotor is the revolving part, the stator the stationary portion. The rotor is supported within the stator on a shaft carried in ball bearings, and is free to revolve.

16B2. Synchros. A synchro transmission requires at least two synchros connected to each other electrically, as indicated in figure 16B1. Each stator has three windings which are connected to the corresponding windings of the other stator by three wires. Each rotor has one winding connected through collector rings to terminals on the stator. The rotors are connected to each other and to a source of alternating current by two wires. The transmitting unit is the *generator* or *transmitter;* the receiving unit is the *motor* or *receiver*. Both are identical in principle, but usually there are mechanical differences in external construction.

The generator rotor is held in position by some mechanical means such as the knob shown in figure 16B1. The motor rotor assumes a position with respect to its stator identical to that of the generator rotor with respect to its stator. Suitable dials attached to the shafts indicate the rotor positions. Thus, the generator setting of 30° is reproduced by the motor, which indicates an identical reading. If the generator is turned to some other angle such as 50°, the motor will follow it and also read 50°.

It is possible to arrange the motor mechanically so that it can indicate the difference between an electrical input or signal and a mechanical movement in response to that signal. Suppose, for example, that it is desired to transmit relative bearing to a range keeper so that it may be introduced into the computing mechanism continuously. An arrangement for doing this is shown schematically in figure 16B2. The synchro stator is mounted so that it is rotated by the knob, which is used to introduce bearing to the computer. The motor is connected electrically to a synchro generator in the manner previously described. It is assumed in this case that the generator is mounted on a director in such a manner that it transmits measured relative bearing.

To illustrate the function of this type of synchro mounting, assume that the director and hence the generator is transmitting zero bearing, and that the knob input to the computer is also zero. Under these conditions, the single index mark on the *zero-reader dial* is opposite the fixed index. Next, assume that the director trains to 30° and hence that the generator transmits a bearing of 30°,

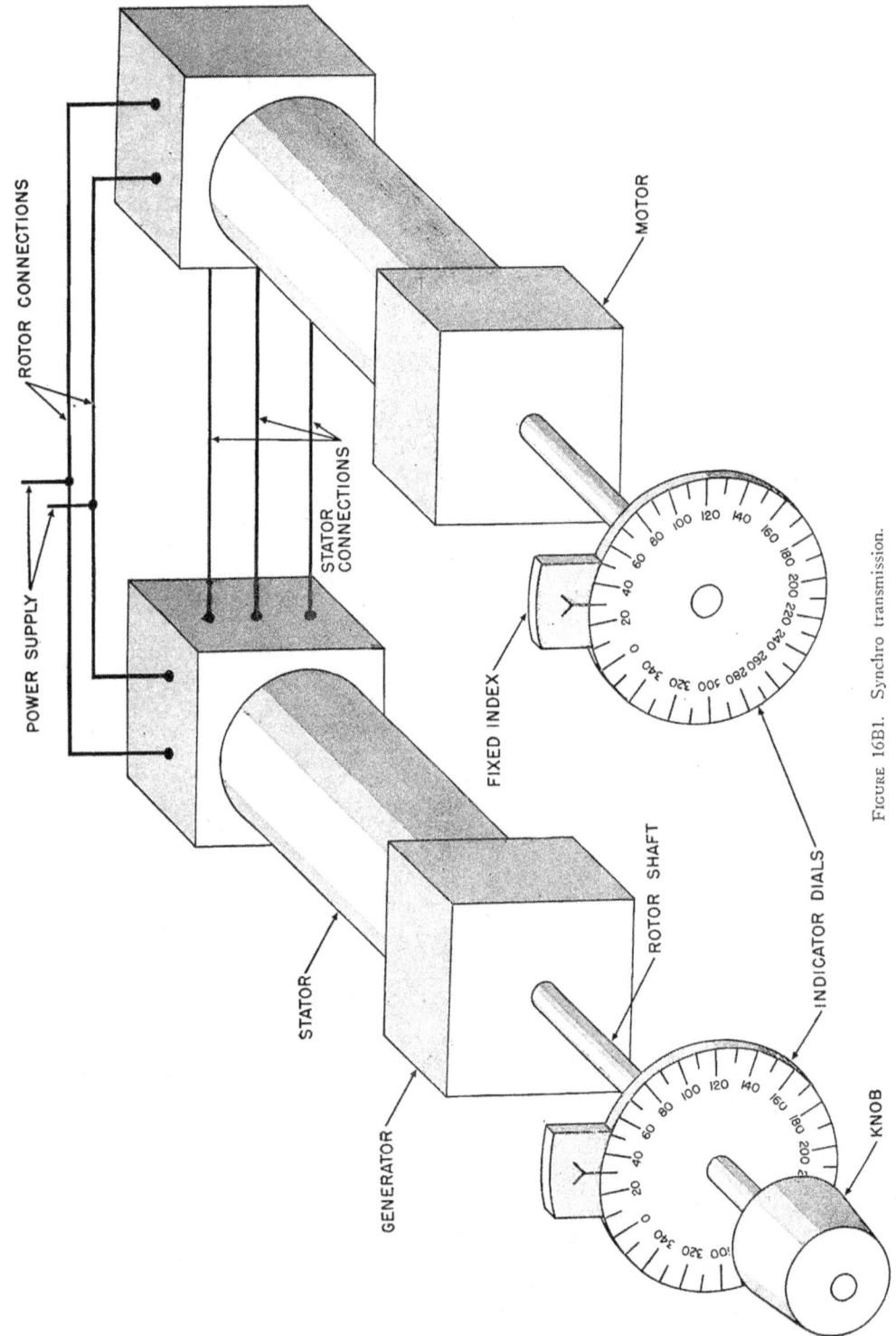

FIGURE 16B1. Synchro transmission.

CHAPTER 16

causing the rotor to turn 30° clockwise (CW) in its stator. The dial pointer has rotated 30° clockwise, and the word *increase* appears opposite the fixed index, indicating that the bearing input to the computer must be increased.

The knob is then turned CW to supply a 30° input. Rotation of the knob drives the stator counterclockwise (CCW) through the gears. The gearing is so arranged that the stator turns 30° when the input to the computer is 30°.

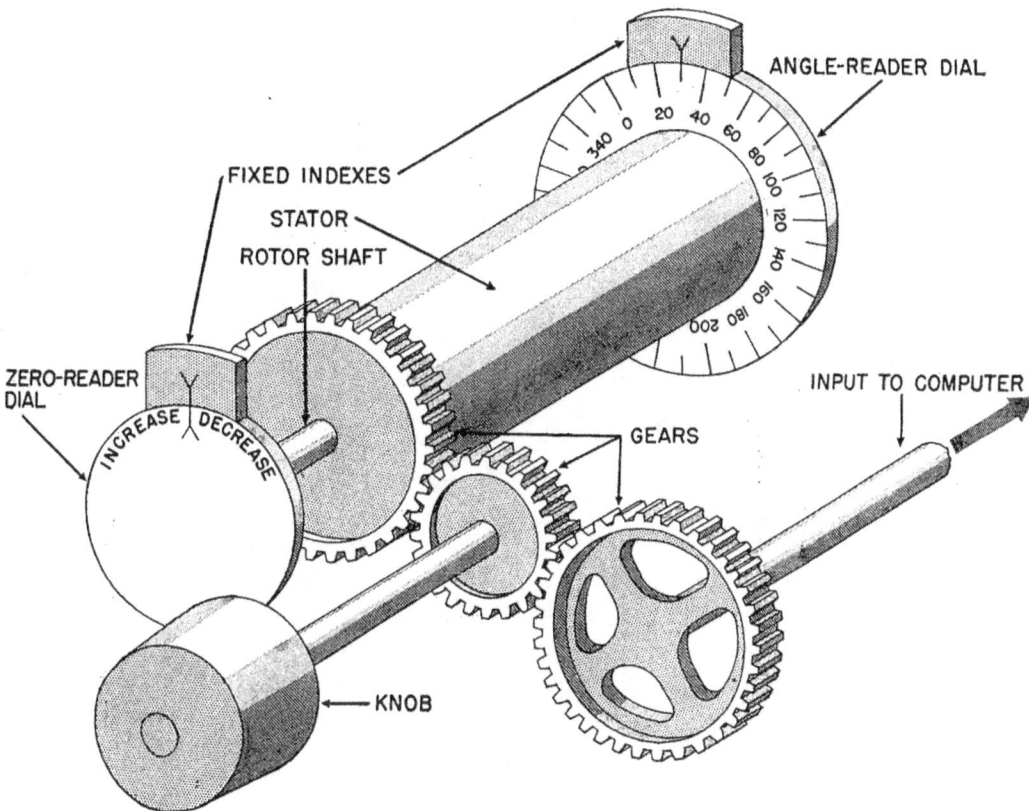

FIGURE 16B2. Zero-reader mechanism.

The CCW rotation of the stator turns the rotor CCW with it, since the rotor position is fixed with respect to the stator by the electrical signal. This brings the zero-reader dial index back to the fixed index, and makes the angle-reader dial indicate 30°. The zero-reader dial shows the *difference* between the order and the response. When the two are equal, their difference is of course zero, the indexes are matched, and the dial indicates "zero," whence comes the term *zero reader*. The process of matching the indexes is called *matching zero readers*.

16B3. Servos. One of the inherent characteristics of synchros is that they cannot exert a torque in their matched positions, hence they cannot be used directly to supply mechanical movement. In the example shown in figure 16B2, the synchro motor rotor, if connected to the knob, could not drive the shafting within the computer. There are many occasions where it is desirable to make the response to the indications automatic, and several methods are employed. Mechanisms which provide automatic response are called *follow-ups*. An electrical follow-up employs a servo motor to provide the additional power required. The application of a servo to the preceding example is illustrated schematically in figure 16B3.

RESTRICTED

FUNDAMENTALS OF DIRECTOR CONTROL

The servo motor stator has two separate windings, only one of which is energized at any given time. When one is energized, the servo rotor revolves CW; when the other winding is energized, the rotation is CCW. The servo is energized through the follow-up control which consists of two sets of sensitive contacts mounted on a fixed block. A light contact arm on the synchro rotor shaft operates one or the other set of contacts, depending on which way the rotor tends to turn. The CCW contacts are connected to the servo winding which produces CCW rotation (indicated by the arrows on the end of the servo), and the CW contacts are connected to the other winding.

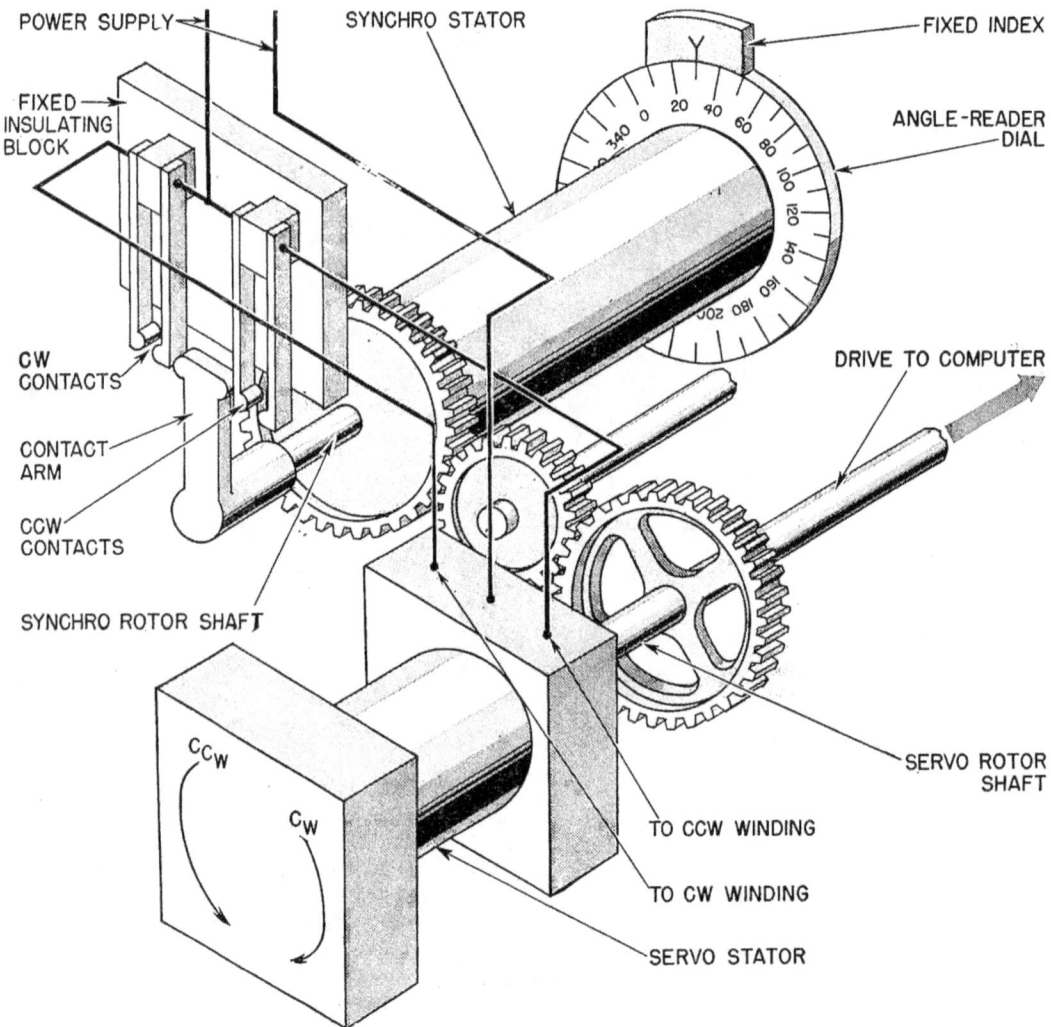

FIGURE 16B3. An electrical follow-up.

Again assume the remote synchro generator is sending a zero signal, and zero bearing is set into the computer. The contact arm is then in the neutral position, both sets of contacts are open, and the servo rotor is motionless. When the director trains to 30°, the generator sends a 30° signal to the synchro motor which rotates the contact arm CW far enough to close the CCW contacts. This energizes the servo CCW winding; the servo then turns the drive shaft CCW, and the synchro stator rotates CCW until the rotor places the contact arm in the neutral position. The servo then stops, and the input to the computer is 30°. If the signal then changes to 20°, the contact arm closes the CW contacts, and the servo removes 10° from the previous input to make the remaining input 20°.

In actuality, a follow-up mechanism has a number of refinements which are purposely omitted in this presentation. Usually dials are provided to indicate the settings made by the servos. In some cases the dials are zero readers, in others they are angle readers.

When accurate values of quantities are required, synchros are used in pairs geared to each other so that one transmits coarse readings, the other fine readings. For example, one might be used to transmit 360° per revolution, the other 10° per revolution. It is evident that the graduations on the fine dial can be more-widely spaced and therefore more easily read.

It is possible to connect several synchro motors to a single generator so that the transmitted quantity can be sent to several stations at the same time, and the values received will be synchronized at all stations.

C. AN ELEMENTARY DIRECTOR SYSTEM

16C1. General. Any director system at first glance appears to be a maze of electrical and mechanical units whose complexity is likely to obscure the basic fundamentals involved. As an aid to further study, a fictitious system is described here to acquaint the student with the principles which are employed in nearly all the actual installations. An understanding of this fictitious system will permit easy analysis of the more complex systems. In reality, fire-control systems are composed of neat and orderly mechanisms which can be studied without difficulty when they are approached in a logical manner, and when the fundamentals are previously understood.

The simplest arrangement to use as a basis for the fictitious system is a single director set up to control a single gun in train and elevation. For the present, it is assumed that the director is close to the gun; the problems introduced by mounting the director aloft and by adding more guns are considered later. The only connections between the two units are electrical. The essential construction features of the director and the equipment required at the gun are shown schematically in figure 16C1. Mechanisms such as the gun training and elevating gear and gun sights are omitted to avoid overcomplicating the diagram.

16C2. Director. The function of this director is to measure and transmit gun elevation order $E'g$ and gun train order $B'gr$. These two quantities are angles, as previously explained, and indicate required positions in elevation and train for the gun. For the present assume that $B'k = B'gr$, and $E'k = E'g$.

In operation, the simple director shown is similar to a gun. The *pilot*, corresponding to a gun barrel, is free to elevate in bearings supported by the *carriage*. The pointer controls the elevation of the pilot with the *elevating handwheel* through a suitable mechanism which is attached to the carriage. The carriage is supported on a roller path and arranged to turn in train. The trainer governs the train of the carriage, and hence of the pilot with the *training handwheel,* which is suitably geared to the *training rack*. The training rack, secured to the pedestal, is stationary.

Attached to the pilot is a sighting mechanism having telescopes for the pointer and trainer. It is similar to the sight mechanism of a gun, and is used to set Vs and Ds between the *pilot axis* and the sighting axis established by the telescopes. After Vs and Ds are set, the pointer and trainer bring the director LOS on the target, and the pilot axis is then positioned in space exactly as a gun would be if Vs and Ds had been set on its sights, and its line of sight were directed at the target.

The *elevation generator* is simply a synchro unit attached to the carriage. The generator rotor is so geared to the elevating mechanism that the angular position of the rotor is a measure of the pilot-axis elevation above the director roller path. In this simple system, it is assumed that this roller path lies in the reference plane; hence the position of the generator rotor is a measure of $E'g$. The generator transmits to the gun a continuous electrical signal representing $E'g$.

The *train generator* is similar to the elevation generator. Its rotor is geared to the training rack so that it measures the train of the carriage with respect to the fore-and-aft axis. Since the pilot must move in train with the carriage, this generator transmits an electrical signal representing $B'gr$, **the train of the pilot axis from the fore-and-aft axis.**

FUNDAMENTALS OF DIRECTOR CONTROL

16C3. Receivers. At the gun, receivers are needed to indicate the train and elevation orders. Actually the simplest devices would be synchro receivers carrying dials which would register $E'g$ and $B'gr$. If suitable scales were attached to the gun, the pointer and trainer could move the gun until the scales showed actual gun train and elevation equal to the orders. A more-convenient arrangement is provided by receivers like those shown in figure 16C1. The function of these instruments is to receive the orders to, and the response of the gun, and to indicate when the gun is properly laid by showing the difference between order and response.

Elevation receiver. The elevation receiver contains a synchro motor whose stator is mounted in bearings so that it can rotate. The stator is geared to the elevating rack of the gun; hence the angular position of the stator is a measure of the *actual* elevation of the gun above its roller path. It is assumed for the present that the gun roller path is in the reference plane.

The synchro motor is connected electrically to the elevation generator at the director. With respect to the stator, the motor rotor assumes an angular position identical to the angular position of the rotor in the elevation generator; therefore, its position is a measure of $E'g$. Attached to the rotor is a zero-reading dial which indicates the gun movement required to match $E'g$.

As an example, assume that the pilot axis and the gun-bore axis are parallel to the reference plane. Then $E'g$ is 0°, and the dial index is opposite the fixed pointer on the face of the receiver. If the director pointer elevates the gun pilot 10°, the elevation generator will transmit an order of 10°, causing the elevation motor rotor and the dial to rotate 10°. The gun pointer will then elevate the gun, which will rotate the elevation motor stator in a direction opposite to that of the original rotor rotation. The stator will rotate the rotor with it, since the position of the rotor *with respect to the stator* is established by the electrical signal from the generator, and the generator is stationary at this point. When the gun has elevated 10°, the stator has turned 10°, rotating the rotor 10°, and thereby bringing the dial back to the matched position.

The mechanical input to the elevation motor is the actual gun elevation, or *response;* the electrical input is the order. When the response ($E'k$) equals the order ($E'g$), the receiver dial is matched, indicating that no further movement of the gun is required. It is evident that as the director pilot is moved to keep the director LOS on the target while the ship rolls and pitches, the gun can be made to duplicate the motion of the pilot in elevation by simply matching the zero reader on the elevation receiver.

Train receiver. In this elementary system, the train receiver is identical with the elevation receiver, except that the mechanical input is actual gun train obtained from the training rack, and the electrical input is $B'gr$. The gun is matched in train with the director pilot by matching the zero reader on the train receiver.

16C4. Multiplicity of control arrangements. The chief reason for the introduction of director systems was the need for a means of controlling all the guns of a battery from a remote position. On battleships, cruisers, and even large destroyers, it is desirable to have facilities for dividing the battery into more than one group, each under director control. Since the director and gun of the fictitious system are connected only by wires, it is a relatively simple matter to connect additional guns and directors into the system by providing suitable switching arrangements.

A schematic arrangement of the wiring for two directors and two guns is shown in figure 16C2. The directors and guns are connected to switches at the switchboard, which is in the *plotting room* for protection and convenience. The connecting wires are shown as a single line for simplicity, whereas three wires are needed to connect the stators of synchro motors and generators. It should also be remembered that the rotors of all the synchros must be energized from a power source which is common to the entire system. The rotor connections, and the firing, illuminating and other circuits required by a complete system are not considered here.

On the schematic switchboard four busses are shown. They are nothing more than common conductors to which the switches for each gun and director are connected. The busses are supplied with orders from the directors, and the guns receive the orders from the busses to which they are connected. The switches in the diagram are set so that director 1 is controlling gun 1 and director

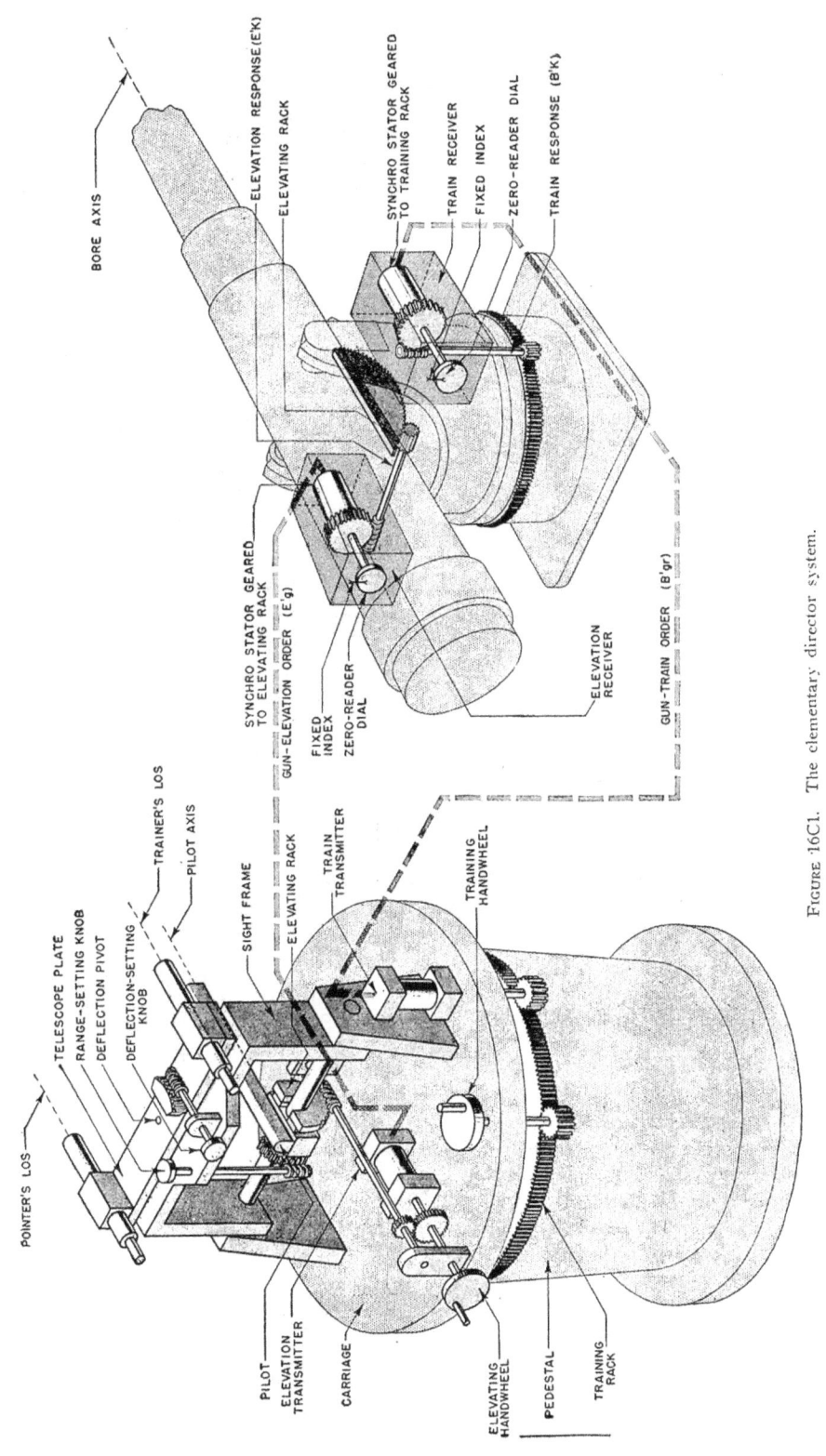

FIGURE 16C1. The elementary director system.

FUNDAMENTALS OF DIRECTOR CONTROL

2 is controlling gun 2. It can be seen that several other set-ups are possible, including:
1. Director 1 controlling both guns in train and elevation.
2. Director 2 controlling both guns in train and elevation.
3. Director 1 controlling gun 2; Director 2 controlling gun 1.
4. Director 1 controlling both guns in elevation, director 2 controlling both guns in train.

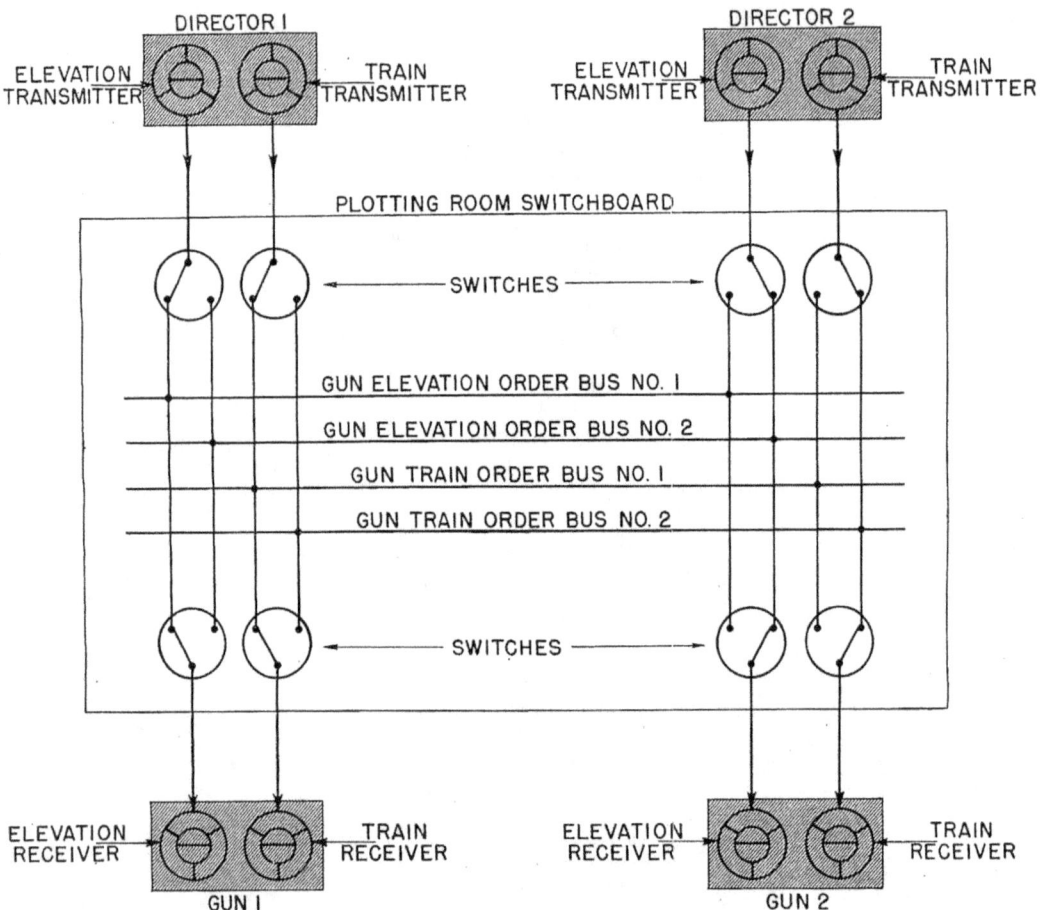

FIGURE 16C2. Schematic wiring diagram of director system.

More guns and more directors can easily be added to the system. Additional guns can be connected through the switchboard by simply adding the proper switches. If more directors are added, more busses are needed to permit all the directors to operate at the same time. Any modern director system provides switching arrangements which are in principle similar to those of this fictitious system. The handling of other quantities such as fuze and sight settings increases the number of wires and switches, but does not alter the principles involved.

16C5. Summary. Laying the guns with this system simply requires the setting of Vs and Ds on the director sights, pointing and training the director to place its LOS on the target, and pointing and training the guns to match the zero readers on the receivers. The pointers and trainers at the guns must keep their dials matched continuously to avoid losing opportunities to fire. The director pointer fires the guns with a master firing key when the director LOS is on the target. Firing keys at the guns are used to hold the firing circuits open when the dials are not matched, or when the guns are not ready to fire for other reasons. If more than one director is provided, fire may be divided, or one of the extra directors may take control in the event of casualty to the controlling director.

The essentials of this or any system are the development and application of gun elevation and train orders. The example given merely illustrates one way of obtaining these quantities. It is a simple way, for the director merely *measures* the orders and the bore axes are kept parallel to the pilot axis. A system like this is inadequate in practice because it admits certain errors. Mechanisms currently in use eliminate those errors, make corrections for some factors which cannot be handled readily by other means, and in general provide a more-nearly automatic solution of the entire fire-control problem.

D. DIRECTOR CONTROL WITH MEASURED GUN ORDERS

16D1. General. The use of a single line of sight removed from all the guns introduces certain elements which must be considered in establishing and applying gun-train and elevation orders. There are other factors having a definite effect upon the accuracy of gunfire, such as erosion and trunnion tilt, which are not caused by director systems but which may be compensated for by them. For the convenience of the student, the various elements that must be considered are divided into two groups. The first includes factors which require consideration when the gun orders are measured, while the second group, discussed in Section 16E, embraces the additional elements which must be taken into account when the orders are *computed*. The first group includes:

1. Horizontal parallax.
2. Vertical parallax.
3. Roller-path inclination.
4. Horizontal displacement.
5. Erosion.

16D2. Reference point. Guns and directors cannot occupy a common position, hence when they are all trained on a target independently, neither their lines of sight nor their lines of departure are parallel. In addition, the units are not all equi-distant from the target. To provide a common point to which angles and distances may be referred, a *reference point* is established for each battery having a director system.

It usually lies in the training axis of one of the directors, but may be located elsewhere. In elevation it may be at the mean trunnion height of the guns (but not necessarily so). Once the reference point is established, it must remain fixed, for the mechanical instruments are built to make some of their computations with respect to its position.

The distances from the reference point to the various units of a battery are measured in planes parallel to, or perpendicular to the reference plane. The distance between the reference point and a gun or director measured parallel to the reference plane is the *horizontal base* for that unit. Measured perpendicular to the reference plane, the distance is the *vertical base*. The base lines are horizontal or vertical only when the reference plane is horizontal, but the terms are used to identify the distances.

16D3. Horizontal parallax. By definition, *horizontal parallax* (Ph) is the horizontal angle formed at the target by a line of sight from each of two stations, such as two guns, two directors, or a gun and a director. In pointer fire, when each gun is laid with respect to its own LOS, the bore axes are automatically converged, since each LOS is placed on the same point of aim. In director fire, where all the guns receive the same gun-train order B'gr, the convergence in the reference plane must be obtained by applying at each gun an individual correction equal in magnitude to Ph for that gun, as shown in figure 16D1a. Unless the correction is applied, the shots will be separated just as much as the bore axes. This spread could be as much as 130 yards, as in the case of the forward and after turrets of some battleships.

FUNDAMENTALS OF DIRECTOR CONTROL

It is evident from figure 16D1a that the value of Ph depends on the horizontal base length. Range and bearing also affect the size of the correction, as shown in figure 16D1b. When the target is on the line of no convergence, Ph is zero, and when the target bears about 90° (or 270°) from that line, Ph is maximum. The greater the range, the smaller is Ph.

The range which must be used in determining Ph is the projection of the slant range on the reference plane. In the surface problem, this projection may be considered equal to the slant range for practical purposes. In the AA problem, the projected range may be considerably less than the slant range, hence the elevation of the target above the deck has an effect on horizontal parallax.

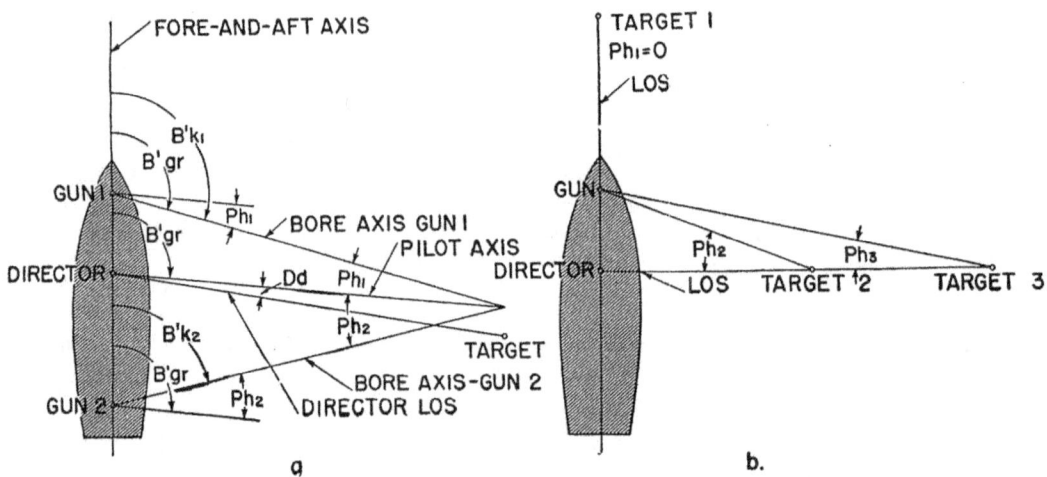

FIGURE 16D1. Horizontal parallax.

16D4. Compensating for horizontal parallax. Since Ph is dependent on the horizontal base, which is generally different for each gun, the correction to B'gr required to produce the correct value of gun train B'k must be applied separately at each gun. A schematic arrangement of a train receiver arranged to apply the horizontal parallax correction in the fictitious system is shown in figure 16D2. To simplify the mechanism, it is assumed that the receiver is intended only for use in surface fire.

The parallax mechanism is set for the base length of its gun when installed, but it must be provided continuously with range and bearing. Bearing is introduced automatically within the receiver, while range must be supplied by hand or other means. The computed value of Ph is combined in differential D-1 with B'k to obtain the mechanical input to the stator. The electrical input to the motor is B'gr, received from the director. When the gun is trained to the correct value of B'k, the input to the stator is also B'gr, and the dial attached to the rotor reads zero.

Any director not at the reference point has a parallax mechanism which automatically corrects the gun-train order so that the transmitted value of B'gr is the same as that which would be transmitted by a director at the reference point.

The plotting room computer in some systems calculates Ph for a "standard" horizontal base and transmits it to the guns and directors. The parallax mechanism at each of them converts Ph for the standard base length to Ph for the actual base length. This arrangement provides a convenient way of including the effects of elevation which must not be neglected in AA fire control. Elevation, range and bearing are continuously available in the AA computers, hence these three factors are automatically introduced into the computation of Ph. Horizontal base length, the only other factor, is accounted for at each gun or director.

16D5. Vertical parallax. The guns of a battery must be converged with their director to compensate for the *vertical parallax* (Pe), which is similar to horizontal parallax, except that it is measured in a plane perpendicular to the reference plane. In general, a gun must be elevated slightly more than the pilot axis of a director such as that of the fictitious system, by the equivalent of Pe for that gun, as shown in figure 16D3a.

The value of Pe depends upon the vertical base length, as the diagram indicates. Range and elevation affect the amount of convergence, as shown in figure 16D3b. When the target is directly overhead, Pe is a minimum; it is maximum when target elevation is nearly zero. As in the case of horizontal parallax, Pe decreases as the range increases.

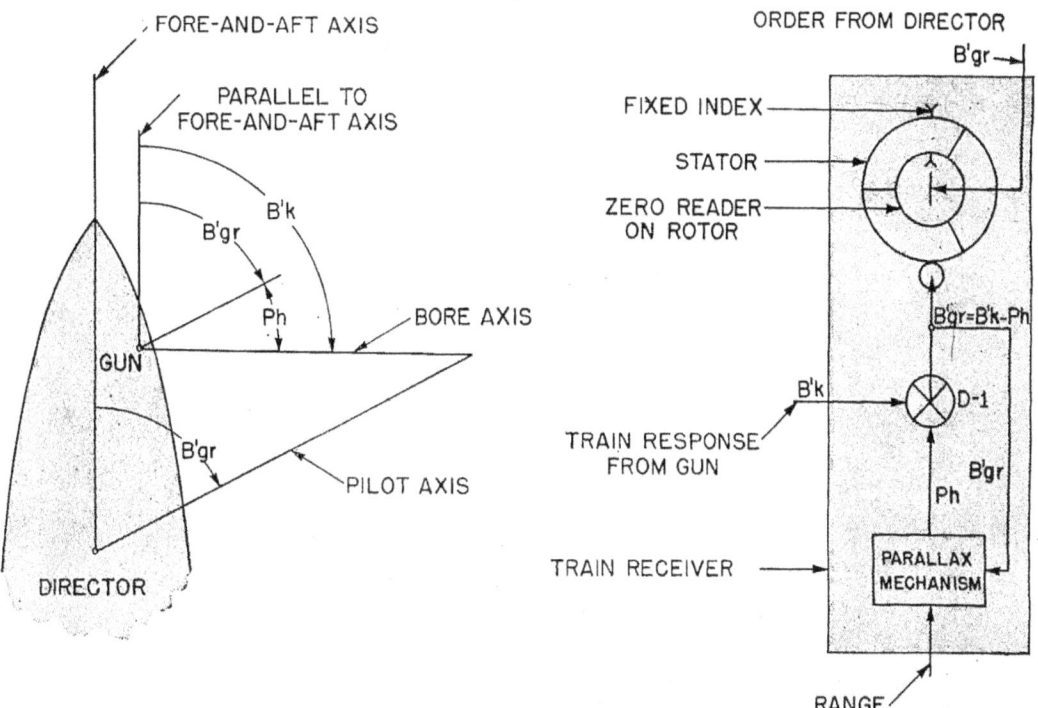

FIGURE 16D2. Compensating for horizontal parallax in surface fire.

Examination of figure 16D3c shows that vertical parallax is actually composed of two parts. One part, Pev, is due to the vertical base, while the other part, Peh, is due to the horizontal base. The horizontal displacement (Khv), which affects Peh, is the component of the horizontal base in the vertical plane through the LOS. In parts a and b of the figure, this component is zero. In part c, Khv is some value between zero and the full horizontal base length Kh. It is apparent that Khv, and hence Peh, depend upon target bearing. It is also apparent the Pe can never be zero, for when Pev is zero, Peh is maximum and vice-versa.

16D6. Compensating for vertical parallax. Since Pe depends upon both the horizontal and vertical base lengths, complete compensation can be made only at the guns, although corrections for a standard base length can be computed centrally. A mechanical device capable of making the complete correction would be quite elaborate, hence in practice approximate corrections are made. The vertical base length is nearly the same for all guns; therefore, the correction for Pev can be computed at a central point and included in the gun elevation order E'g.

Assuming that the fictitious system is for surface fire only, Pev can be applied simply by adding Pev to the value of Vs set upon the director sights. This puts the pilot axis in such position that when the guns are parallel to it, they are elevated the additional amount required to compensate for Pev as shown in figure 16D4. Since the system is used for surface fire only, no correction is needed for Peh, for it is essentially zero at low position angles. This method of correcting for

FUNDAMENTALS OF DIRECTOR CONTROL

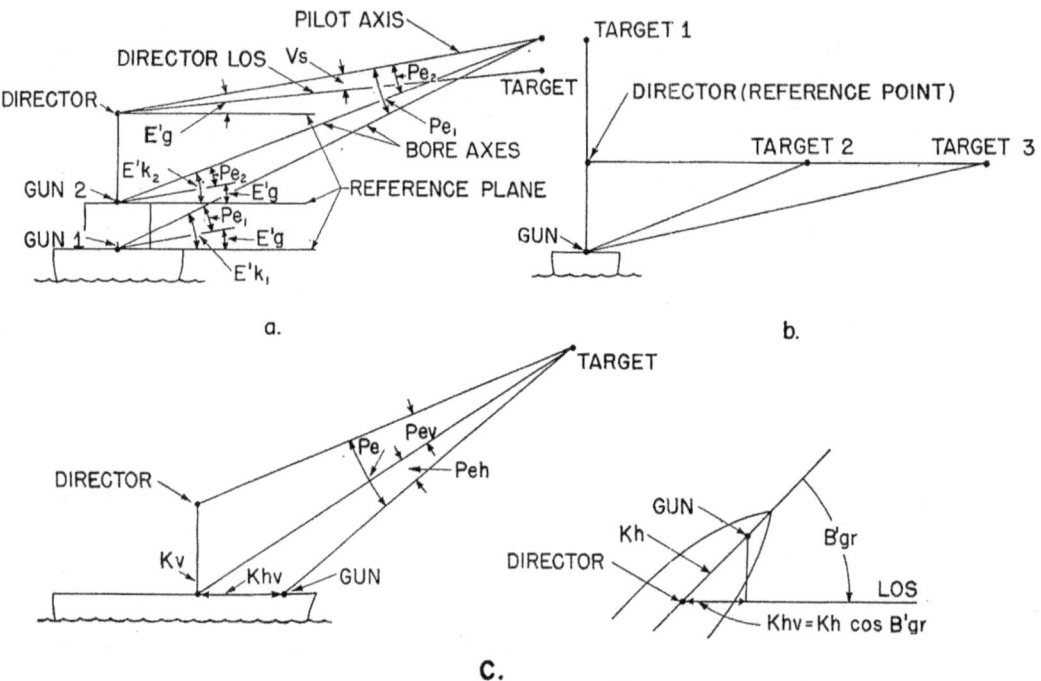

FIGURE 16D3. Vertical parallax.

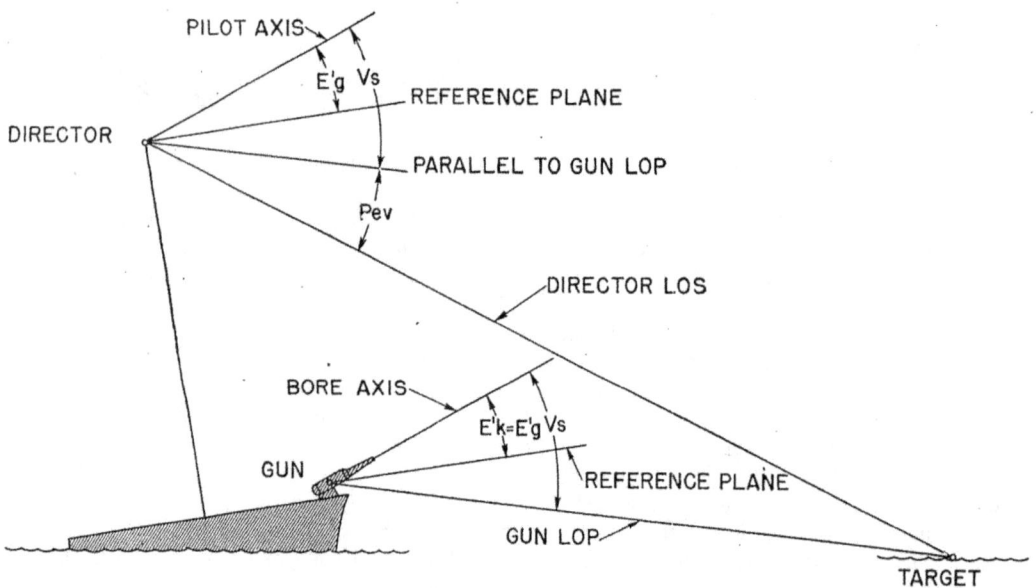

FIGURE 16D4. Compensating for vertical parallax in surface fire.

CHAPTER 16

vertical parallax was used in some older systems. In fact, Gunnery Sheet 10 (fig. 13C1) provides a space for adding the correction into the computation of sight-bar range.

Some computers for AA batteries are built to compute both Pev and Peh, although Peh is neglected in many systems. Generally, the horizontal base length for the secondary, dual-purpose batteries is relatively short, and the error due to Peh may be ignored in the interest of simplifying the mechanisms, but it is considered on some of the carrier installations.

16D7. Roller-path inclination. In this discussion it has heretofore been assumed that the roller paths of the several system units are parallel to the reference plane. However, it is a physical impossibility to install guns and directors so that this condition is fulfilled. By careful leveling of the roller paths when the ship is in drydock, the inclinations or tilts of the roller path planes to the reference plane can be reduced to small values. It is exceptional to have inclinations as high as 1°; the maximum for turrets is about 20 minutes. Every effort is made to keep the inclinations to a minimum, but even if the paths are perfectly leveled in drydock, subsequent settling of the ship's structure is likely to change the situation.

A mechanical instrument which measures angles in elevation makes such measurements with respect to the roller path of the gun or director upon which the measuring device is mounted. Suitable corrections must be applied to refer such measurements to the reference plane.

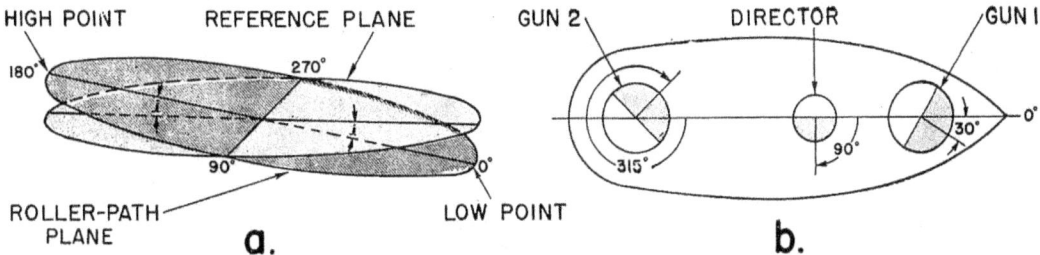

FIGURE 16D5. Roller-path inclination.

Suppose that a gun is laid with its bore axis parallel to the *roller path* of figure 16D5a. The roller-path plane is inclined to the reference plane by an angle (i), measured at the point of maximum inclination. When the gun is trained to 0°, the actual gun elevation is (—i). When trained to 180°, the actual gun elevation is (i). Actual gun elevation will be zero only when the gun is trained to 90° or 270°, the bearings of the intersection of the deck and roller-path planes and at which the roller-path inclination is zero. Obviously, the actual gun elevation, whether it be 0° or any other value, varies as the gun is trained, unless the elevation of the gun above its roller path is continuously compensated for the roller-path inclination.

The bearings used in the preceding illustration refer to train from the high point. The relative bearing of the high point with respect to the ship may be any value from 0° to 360°. It is just as unlikely that the high points of several roller paths will have the same relative bearings as it is that all the paths will be parallel. In figure 16D5b, the shaded portions of the several roller paths represent the sides which are tilted above the reference plane. The correction is maximum for gun 1 when it is trained to 30°, for gun 2 when it is trained to 315° and for the director when it is trained to 90°. The correction is also maximum (but opposite in sign) for each of these units when they are trained 180° from the high points of their roller paths. The correction varies from maximum to zero where the unit is trained from the high point to 90° on either side of the high point. So it is obvious that the correction depends on the train of the gun or director, and the inclination at the high point.

A gun elevates in a plane perpendicular to its roller path. When the gun is trained in any direction except along the line through the high or low points, the trunnions are inclined to the reference plane if the roller path is tilted. The gun then elevates in a plane which is no longer perpendicular to

FUNDAMENTALS OF DIRECTOR CONTROL

the reference plane, hence the measured gun angle is not a true criterion of the actual elevation above the reference plane. Similarly, the train of the gun no longer takes place in the reference plane, and the measurements of gun train are slightly in error.

In addition to causing minor errors in measurement, roller-path inclination produces a different value of trunnion tilt for each gun. Since it is impracticable to correct trunnion tilt for individual guns, this error can be eliminated only by keeping the roller-paths as nearly parallel to the reference plane as possible. The measurement errors and trunnion-tilt errors due to roller path inclination are neglected.

16D8. Compensating for roller-path inclination. The correction for roller-path tilt is made by mechanical devices called *inclination compensators* or *tilt correctors* at the guns and directors. One usually is associated with the sighting mechanism of each director for the purpose of correcting the director elevation measurements above its roller path to measurements above the reference plane. In some one-director installations, no compensator is provided (at the director) since the director roller path is used as the reference plane. In multi-director systems, compensators are usually installed at each director.

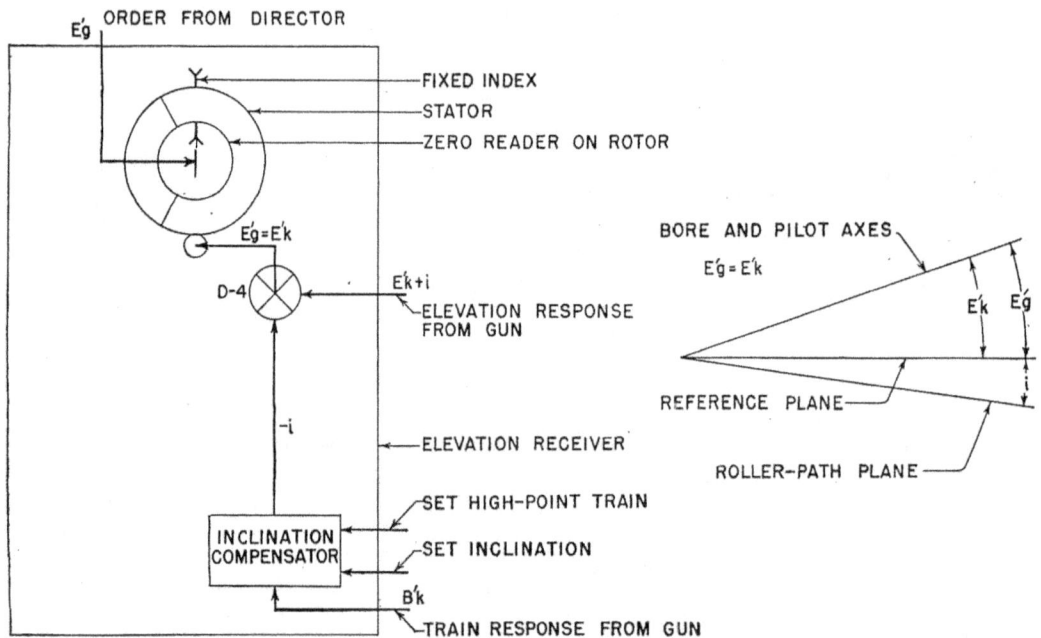

FIGURE 16D6. Compensating for roller-path inclination.

A schematic arrangement of an elevation receiver equipped with an inclination compensator for use in the fictitious system is shown in figure 16D6. Actual gun train $B'k$ is supplied mechanically from the same source that provides it to the train receiver. The inclination compensator is arranged so that the inclination of the high point and the train of the high point may be set into it, and changed when necessary. When the zero reader is matched, $E'k$ must be equal to $E'g$. The actual elevation measured by the receiver is $E'k + i$, which is the gun elevation above the roller path. The inclination compensator computes $(-i)$, which is combined with $E'k + i$ in D-4. The output of D-4 is $E'k$, which drives the stator. The electrical input is $E'g$; hence, when the gun is elevated the correct amount above the reference plane, the dial reads zero.

16D9. Horizontal displacement. The vertical and horizontal bases which produce the parallaxes also cause differences between the ranges to the target from the directors and guns. Obviously, these range variations exist in any form of fire, and theoretically should be taken into ac-

CHAPTER 16

count by computing Vs and Ds for each gun independently. However, the actual range differences for dual-purpose batteries are usually sufficiently small to be ignored in the interest of simplifying the fire-control installation. Corrections are sometimes made for the displacements between the turrets of battleships and cruisers, for their longer base lengths make their range differences assume greater importance.

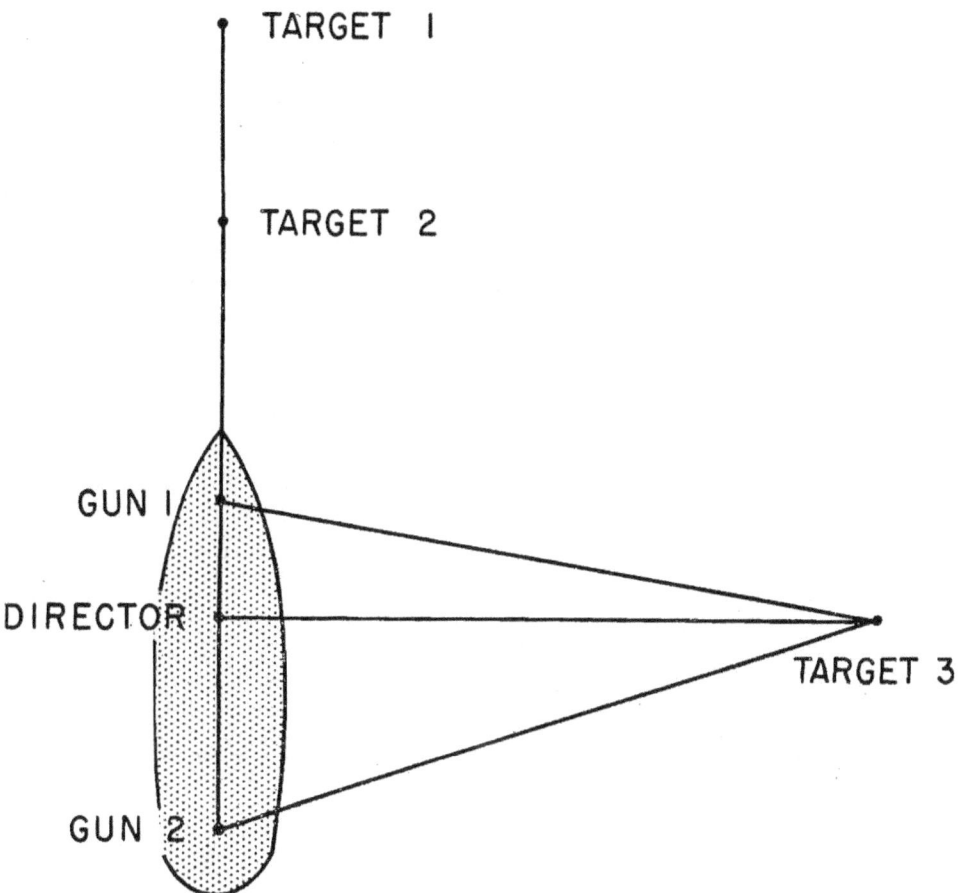

FIGURE 16D7. Horizontal displacement.

Since this factor is considered only in surface fire installations, the range differences between guns are due only to the horizontal base lengths, and the error is frequently referred to as *horizontal displacement*. Figure 16D7 illustrates the error existing in surface fire. Targets 1 and 2 are farther from gun 2 than from the reference point, while they are closer to gun 1 than to the reference point. The range differences for this bearing are equal to the corresponding base lengths, being greater for gun 2 than for gun 1. For practical purposes, target 3 is equi-distant from the reference point and both guns, and there is no horizontal displacement error. It is apparent, then, that the error varies with the bearing of the target and the horizontal base length.

The actual range to the target has no effect on the size of the range *differences*, but must be considered in determining the elevation *correction* to be applied to each gun. Compensation is made by elevating or depressing each gun by the angle necessary to produce the required change in range. This compensating change is smaller at short ranges than at long ranges. For example, to increase the range of the 5"/38 cal. gun by 100 yards when the range is 3,000 yards, requires

3.7 minutes extra elevation, while at 17,000 yards it requires 35 minutes as indicated in column 2b of the range table in Appendix D.

16D10. **Compensating for horizontal displacement.** To illustrate the incorporation of this correction in the elevation of a gun firing at a surface target a horizontal displacement corrector for the fictitious system is shown in figure 16D8. The horizontal displacement correction $E'r$ is not included in $E'g$ received from the director, hence the gun response $E'k + i$ must be decreased by $E'r$ before it goes to the stator. In other words, $E'k$ is no longer equal to $E'g$, for $E'k$ is the actual gun elevation for a single gun whose elevation must be slightly different from that of each other gun. The displacement corrector must be supplied continuously with range and bearing as indicated, but the base-length factor is permanently set at installation.

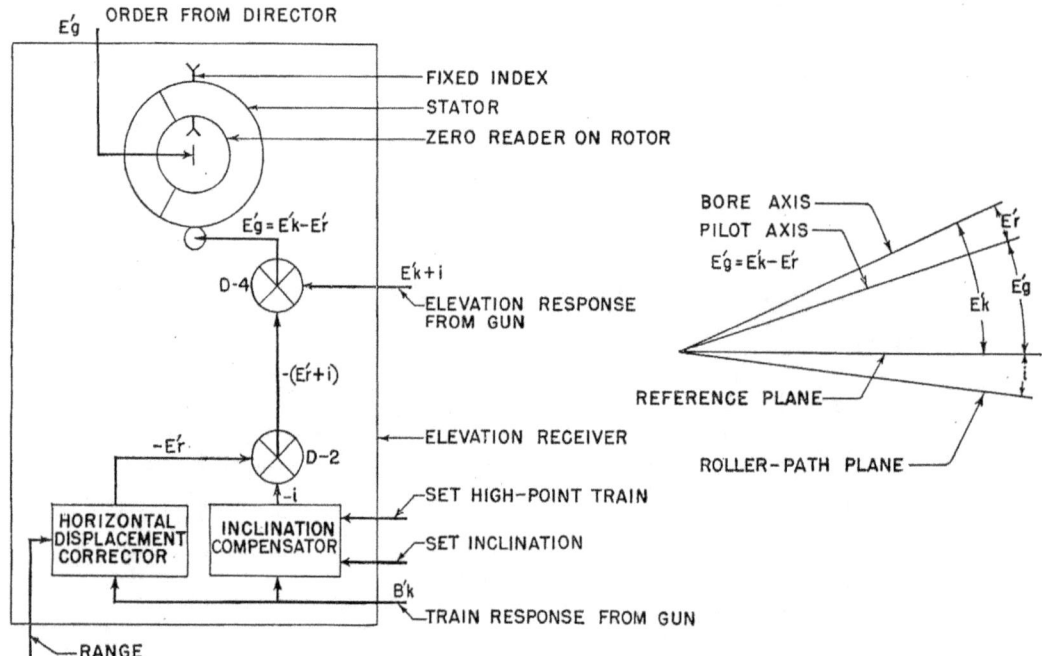

FIGURE 16D8. Compensating for horizontal displacement.

It is possible to adjust the roller-path inclination compensators so that they can supply a combined correction for roller-path tilt and horizontal displacement. When the latter correction is made in this manner, it must necessarily be based on a selected range. The correction is accurate only for that range, while for the other ranges it is an approximation.

16D11. **Erosion.** Compensation for the loss in I.V. due to erosion is automatically made in director systems. In some installations, the plotting-room computer makes a single correction which is applied to all guns alike, the assumption being that the erosion is the same in all units of the battery. For this reason it is desirable to equalize the rounds fired among the guns of a battery. Sometimes guns which are fired more frequently than others are interchanged with the latter to help keep the I.V. losses equal.

However, it is impracticable to shift heavy turret guns for this purpose, hence the elevation equipment at each turret contains a mechanism to compensate for erosion. In some systems this mechanism computes the entire erosion correction, while in others it simply gives the difference between the correction made by the plotting room computer and that which should be applied for the particular turret.

CHAPTER 16

The loss of range due to erosion when firing at a surface target depends upon the range and the loss in I.V., the two factors which are used in computing the correction from a range table. The I.V. loss is a function of rounds fired. Hence, one type of erosion corrector might be included

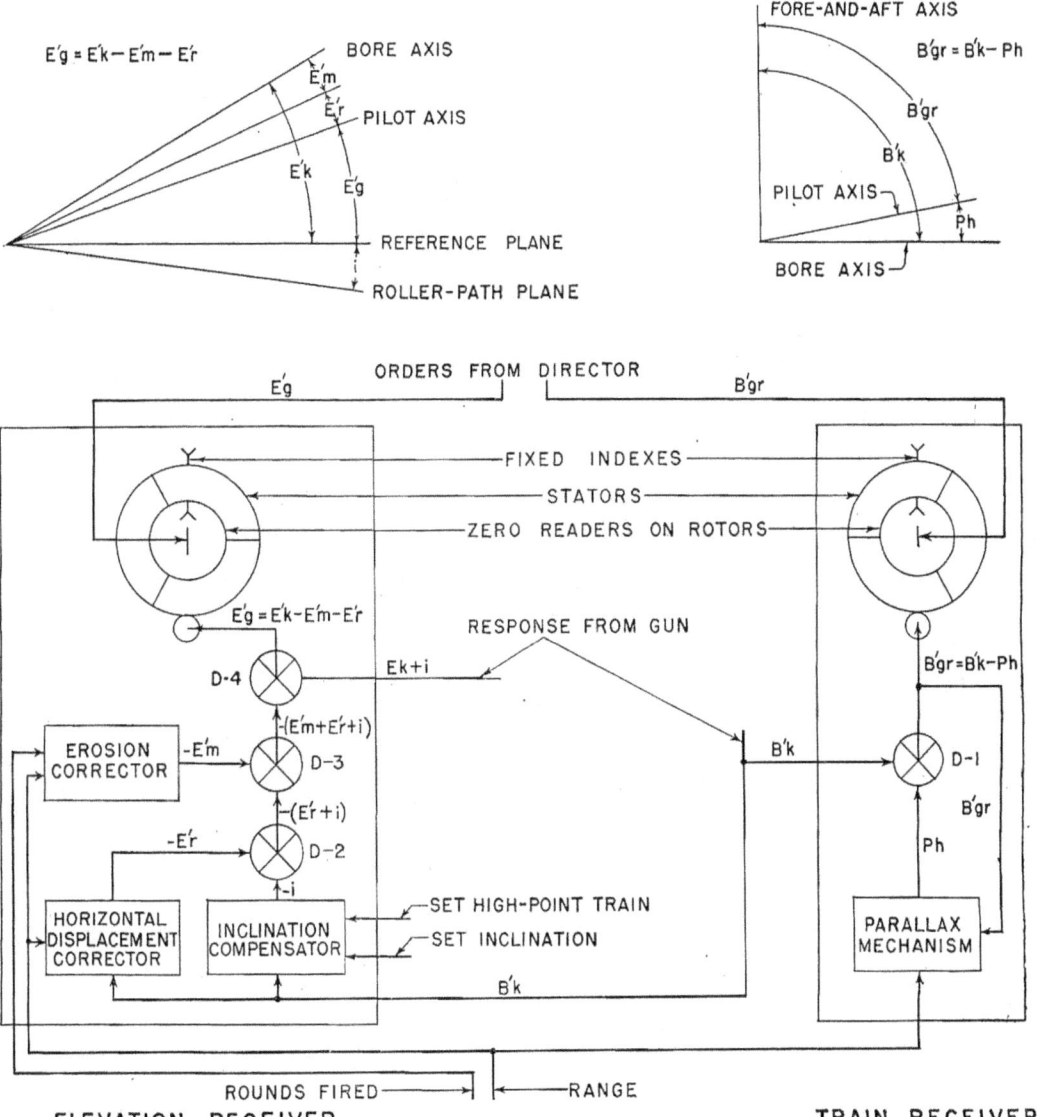

FIGURE 16D9. Completed gun-order receivers.

in an elevation receiver in the manner indicated in figure 16D9, which is a composite of the fictitious system equipped to handle the elements discussed to this point. The erosion correction E'_m is another factor which makes E'_k different from E'_g.

16D12. Summary. As explained in previous chapters, a gun's bore axis must be directed at a definite position in space at the instant of firing. That position is established by two coordinates, which are:

375   **RESTRICTED**

FUNDAMENTALS OF DIRECTOR CONTROL

1. Vs and Ds when the gun is laid with respect to its own LOS.
2. E′g and B′gr when the gun is laid with respect to the reference plane and the fore-and-aft axis.

Regardless of the coordinate system used, the bore axis must have a definite *vertical* elevation and a definite *horizontal* train.

The reference plane continuously rolls and pitches with the ship, and is not likely to be horizontal at any time, since roll and pitch are more or less independent of each other. Inclination of the reference plane across the line of sight causes trunnion tilt, which introduces errors in the space position of the bore axis unless Vs and Ds, or E′g and B′gr are properly adjusted. It is impractical to use values of Vs and Ds continuously corrected for the effects of trunnion tilt, and for this reason these angles do not include trunnion-tilt corrections. In the fictitious director system of the preceding sections, Vs and Ds were set on the director sights, and then the director measured E′g and B′gr. The gun orders obtained in this manner therefore could not include trunnion-tilt corrections, and the bore axis would be properly pointed in space only when the trunnion axes were horizontal.

Directors similar to the fictitious one have been used in the past for the control of surface fire. Some were provided with devices which automatically and continuously introduced trunnion-tilt corrections into the gun orders. Directors so equipped transmitted continuously-correct elevation and train orders for any position of the reference plane, hence the guns could be fired at practically any time during a cycle of ship roll or pitch. In these systems, as in the fictitious example, train and elevation orders were simply measurements of the gun-pilot position with respect to the reference system.

E. DIRECTOR CONTROL WITH COMPUTED GUN ORDERS

16E1. Symbols and definitions. The symbols, names and definitions of quantities used in this section are summarized below. Some previously-defined quantities are repeated for the convenience of the student:

SYMBOL	DEFINITION
B′gr	*Gun-train order.* The *ordered* angle between the fore-and-aft axis and the perpendicular plane through the bore axis. $B'gr = B'r + jDd + Dz = B'r + Dd$.
Br	*Relative target bearing.* Target bearing measured in the horizontal.
B′r	*Director train.* The angle between the fore-and-aft axis and the vertical plane through the LOS, measured in the deck plane clockwise from the bow.
jB′r	*Deck tilt correction.* The difference between director train and relative target bearing. $Br = B'r + jB'r$.
Dd	*Deck deflection.* Gun-train order minus director train. $Dd = B'gr - B'r = jDd + Dz$.
jDd	*Partial deck deflection.* The deck plane equivalent of Ds.
Ds	*Sight deflection.* Measured in slant plane. Does not include Dz.
Dz	*Trunnion-tilt correction in train.* The correction in train to compensate for trunnion tilt.
E	*Target elevation.* The vertical angle between the LOS and the horizontal plane.
Eb	*Director elevation.* The elevation of the director LOS above the deck, measured in the vertical plane through the LOS. $Eb = E + L$.
E′g	*Gun-elevation order.* The *ordered* perpendicular angle between the bore axis and the deck. $E'g = Eb + Vs - Vz$.
L	*Level angle.* The angle between the horizontal plane and the deck plane measured in the vertical plane containing the LOS.
Lj	*Selected-level angle.* Any arbitrary value of L.

CHAPTER 16

SYMBOL	DEFINITION
Vs	*Sight angle.* Does not include Vz.
Vz	*Trunnion-tilt correction in elevation.* The correction in elevation to compensate for trunnion tilt.
Zd	*Cross-level angle.* The angle between the horizontal plane and the deck plane measured in a plane which is at right angles both to the vertical plane containing the LOS and to the deck.
Zdj	*Selected cross-level angle.* Any arbitrary value of Zd.

16E2. Plotting-room systems. Modern fire-control systems (and old main-battery systems on cruisers and battleships) employ directors which measure, instead of gun orders, the LOS position with respect to the reference system. Since Vs and Ds are not set on the sights of this type of director, no sight-setting mechanism is required. In effect, the pilot is replaced by telescopes. The generators then measure and transmit director train (B'r) and director elevation (Eb) which define the LOS position.

The values of B'r and Eb are transmitted to a plotting room, which might be more accurately called a computing room. The plotting-room equipment, receiving information from several sources in addition to the director, *computes* gun orders and transmits them to the guns. Such a system provides an effective means of developing train and elevation orders that (1) are corrected automatically for the changes in Vs and Ds as the problem progresses, and (2) include the rapidly changing corrections for trunnion tilt.

Computation of gun orders by a modern system requires consideration of a number of factors in addition to the parallaxes and other elements mentioned in section 16D. In the following articles, it is assumed that the gun and director are at the reference point, so that the effects of parallaxes and range differences are eliminated. The additional factors are:

1. Partial deck deflection.
2. Trunnion-tilt corrections.
3. Deck-tilt correction.

16E3. Reference-plane inclination. In order to establish the correct bore-axis position in space with respect to the reference plane when gun orders are computed instead of measured, it is necessary to consider the reference-plane position with respect to the horizontal. The establishment of this position is also necessary in making trunnion-tilt corrections, since the trunnion axes are essentially parallel to the reference plane. The reference-plane inclination to the horizontal plane is established by measuring two angles between those planes. One is *level angle* (L), measured in a plane *through* the LOS; the other is *cross-level angle* (Zd), measured in a plane across the LOS. These angles may be measured in different planes in various fire-control installations. Throughout this section they are considered as illustrated in figure 16E1a, and as described in the following paragraphs:

Level angle. The level angle (L) between the reference and horizontal planes is measured in a vertical plane through the LOS. It is normally determined by a stable element which contains a gyroscope whose spin axis remains vertical as the ship rolls and pitches. The plane of gyro spin is thus horizontal and provides a reference from which L is measured.

A telescope mounted on a director may be used to determine L. The telescope-elevating axis must be maintained in a horizontal plane so that the angle measured is in a vertical plane. When directed at a surface target or the horizon, the sighting axis of such a telescope is inclined to the reference plane by an angle approximately equal to L. The difference between L measured by this method and the correct value is simply the dip of the sighting axis below the horizontal. The pointer's telescope of a surface-fire director provides a measure of L, but an AA director requires a separate leveler's telescope if L is to be measured in this manner, since the pointer's LOS is often elevated well above the horizontal.

RESTRICTED

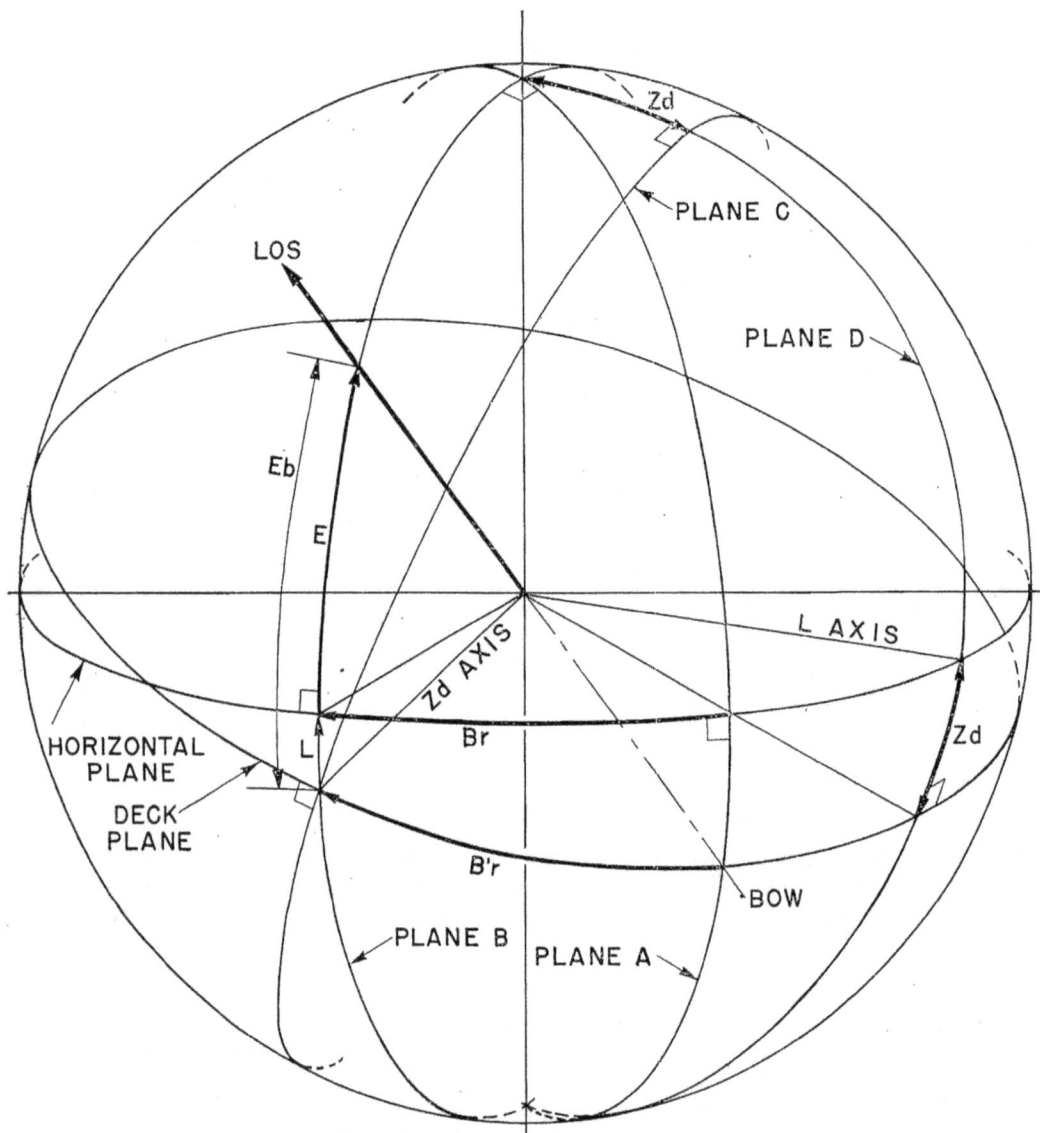

A. VERTICAL PLANE THROUGH FORE-AND-AFT AXIS
B. VERTICAL PLANE THROUGH LINE OF SIGHT
C. PLANE PERPENDICULAR TO DECK THROUGH INTERSECTION OF DECK PLANE AND PLANE B
D. PLANE PERPENDICULAR TO PLANE B AND DECK PLANE

$$Br = B'r + jB'r$$

a.

FIGURE 16E1a. Level, cross-level, and deck-tilt correction.

CHAPTER 16

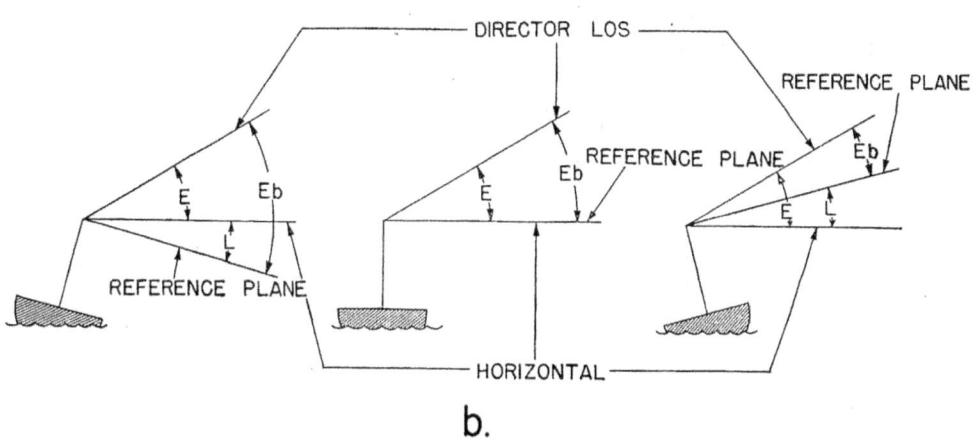

b.

FIGURE 16E1b. Level, cross-level, and deck-tilt correction, cont'd.

Cross-level angle. The cross-level angle (Zd) between the reference and horizontal planes is measured in a plane perpendicular to both the deck and the vertical plane through the LOS. In other words, Zd is measured about an axis formed by the intersection of the deck with the vertical plane through the LOS. It is determined by the same stable element that measures L. In some cases, directors are equipped with a cross-leveling telescope which provides an auxiliary means of measuring cross-level angle. Such a telescope is mounted so that LOS is turned 90° from the pointer's LOS, and when directed at the horizon, provides an approximate measure of Zd.

16E4. Stabilized director sights. The elevating axis of a director telescope is perpendicular to the plane in which the LOS elevates. If the telescope elevating axis is kept horizontal, the angle between that axis and the reference plane, measured in a plane perpendicular to the deck, is equal to Zd. It is apparent that the telescope-elevating axis can be maintained horizontal, or stabilized in cross-level, by rotating it through Zd about an axis parallel to the deck. This is accomplished in some modern directors by an electric motor which rotates the telescopes according to the value of Zd measured by the stable element. With a telescope thus stabilized in cross-level, it is capable of measuring director elevation (Eb) in a vertical plane, for elevation of the LOS then takes place about a horizontal axis.

It is clear from figure 16E1a that Eb is composed of target elevation (E) and level angle (L), hence $Eb = E + L$. Ship roll and pitch do not affect E, but cause rapid changes in L as indicated in figure 16E1b. When the sights are controlled by the director pointer, he has to turn his handwheels to keep the LOS on the target to compensate for ship roll and pitch. In doing this he is supplying the value of L to the telescope.

The telescopes are stabilized in level in some directors by an electric motor which supplies the value of L measured by the stable element. The same motor also supplies the movement required to compensate for changes in E which are generated by the computer as the problem progresses. Actually, the quantity controlling the motor is $E + L$, or Eb, which is developed in the computer. Thus the telescopes are driven by a computed value of Eb, and the LOS stays on the target in elevation with no help from the director pointer. Similarly, the LOS is held on the target in train by driving the director through changes in director train B'r, developed in the computer.

If the LOS drifts off the target in train, it indicates the computer is generating an incorrect value of B'r; if it drifts off in elevation, it indicates the generation of an incorrect value of E. Thus, stabilized sights provide a method of checking part of the computer set-up, and are part of an automatic rate-control system such as that used in AA director installations.

16E5. Partial deck deflection. Sight deflection, when set on the sights of a gun, provides the proper lateral offset between the bore axis and the LOS. Since sight mechanisms make the

FUNDAMENTALS OF DIRECTOR CONTROL

deflection setting in the slant plane, Ds is computed for that plane. Gun train order B'gr is computed in the deck plane, and the component of B'gr required to compensate for deflection is not equal to Ds. This component is called *partial deck deflection* (jDd), and is illustrated in figure 16E2.

In this diagram, the reference plane is assumed horizontal so that trunnion tilt need not be considered. The gun is at the center of the sphere and the angles are projected on its surface. All the arcs are in planes which pass through the center of the sphere, hence they are parts of great circles.

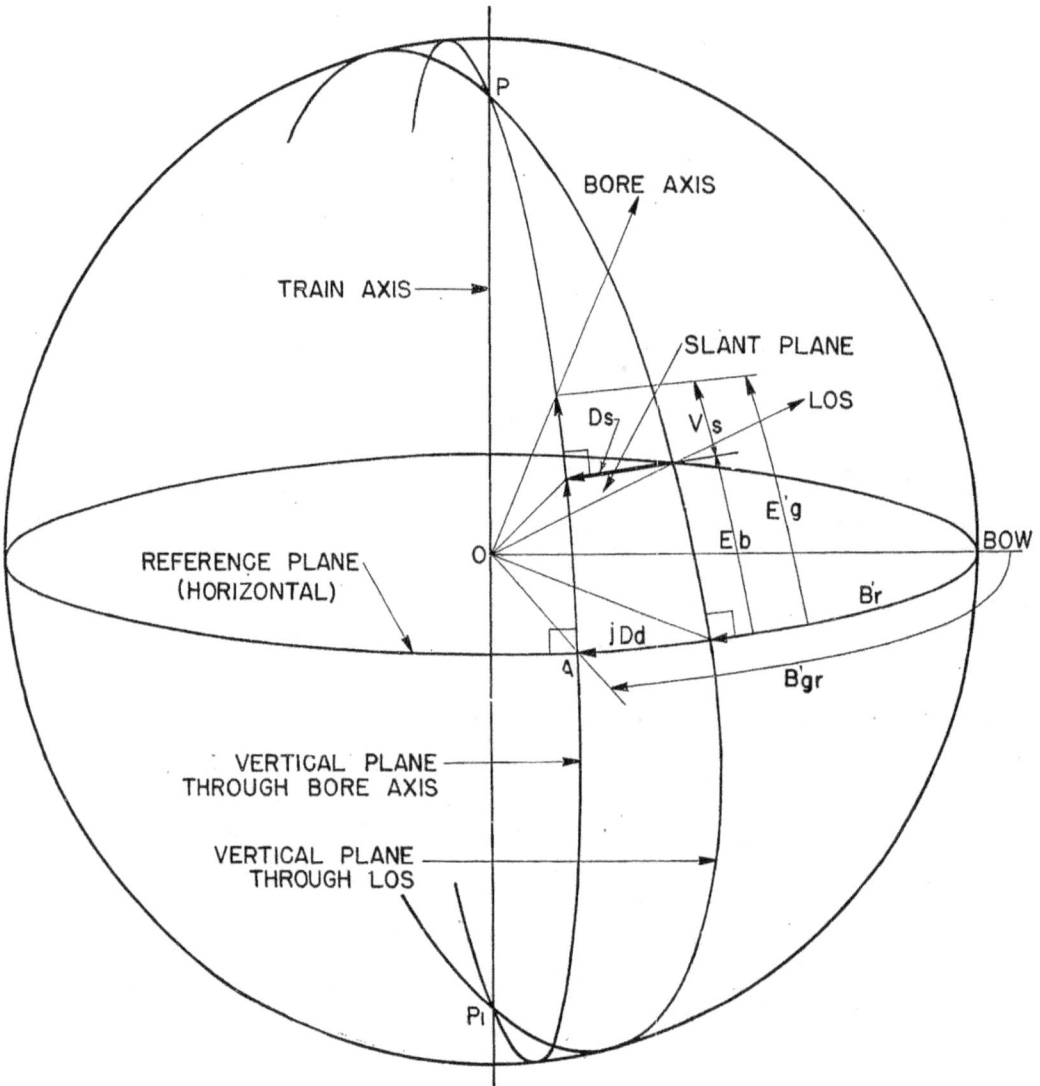

FIGURE 16E2. Gun orders; reference plane horizontal.

The planes through the bore axis and the LOS perpendicular to the deck intersect in line PP', which coincides with the training axis of the gun. Since the gun trains in the reference plane, it elevates in the plane of arc PAP'. It must be trained to B'gr to give it deflection Ds from the target. The diagram shows that jDd is always larger than Ds unless the LOS lies in the reference plane.

380

CHAPTER 16

It can be seen that as Eb increases (B′r and Ds remaining constant), jDd must become larger to keep the gun offset from the LOS by Ds. Since a variation in L causes a change in Eb, it also results in a variation in jDd. As long as B′gr is measured directly as in the fictitious system, jDd need not be calculated, but when B′gr is computed, the value of jDd corresponding to Ds must be provided continuously.

16E6. Trunnion-tilt corrections. The nature of the errors introduced by trunnion tilt is discussed in Chapter 13. The difficulties of computing Vs and Ds continuously corrected for trunnion tilt, of maintaining angles so corrected on the gun sights, and of subsequently laying the gun with respect to its own LOS, preclude the possibility of making the correction independently at each gun. In a director system, the trunnion-tilt corrections are incorporated in the gun orders, but not in Vs and Ds.

Inclination of the trunnion axis to the horizontal is approximately equal in value to cross-level angle (Zd). The plane of trunnion-tilt measurement should be the vertical plane through the trunnion axis. That plane is inclined to the Zd measurement plane by angles which vary with L and Ds, since Zd is measured in a plane perpendicular to both the deck and the plane through the LOS. Corrections for trunnion tilt are computed on the basis of Zd, hence they are approximations. Other factors entering the computations are Eb, Vs, and Ds. Figure 16E3 shows the elevation and train orders for a gun directed at a point G, (1) when there is no cross-level, and (2) when there is cross-level. It is assumed that there is no change in L, hence Eb is the same in both instances. The point G must in each case lie in a plane perpendicular to the deck, since the gun elevates in such a plane.

Trunnion-tilt correction in train. It is clear from the diagram that the gun must be trained an additional amount to place its elevating plane on G after Zd appears. The angle in the deck plane between the vertical plane through the LOS and the perpendicular plane through the gun is *deck deflection* (Dd). It differs from partial deck deflection (jDd) by the *trunnion-tilt correction in train* (Dz). In other words, Dd = jDd + Dz, and B′gr = B′r + jDd + Dz. The direction of the correction Dz is always toward the high side of the reference plane, making it necessary to train the gun "up hill."

Trunnion-tilt correction in elevation. The diagram also shows that after Zd enters the problem the gun elevation above the deck must be decreased to keep the bore axis on G. This decrease in gun elevation is the *trunnion-tilt correction in elevation* (Vz). It is evident that when the gun is trained "up hill" in applying Dz, the bore becomes elevated by too great an angle, and must be depressed by Vz. Trunnion tilt (before correction) actually causes a reduction in gun elevation above the horizontal, but the additional gun elevation resulting from the correction made in train more than compensates for this reduction, hence Vz is always negative, and E′g = Eb + Vs − Vz.

16E7. Computing gun orders. In the equations for the gun orders, Eb and B′r are measured, Vs, Vz, jDd, and Dz are computed. Again using a fictitious medium for illustrative purposes, the development of the orders is shown schematically in figure 16E4. No attempt is made to show all the inputs required by the various units. The purpose of the diagram is to indicate in a general way the method of developing gun orders corrected automatically and continuously for trunnion tilt and for changes in Vs and Ds.

The sight angle and deflection unit includes the mechanisms which continuously position the target and compute Vs and Ds based on the instantaneous target position. Outputs Vs and Ds are transmitted to the guns for sight-setting purposes, and do not include Vz and Dz. Thus, if necessary to shift to pointer fire, the sights are already set, but the settings are correct only when there is no trunnion tilt.

The partial deflection computer converts Ds to jDd, and the trunnion-tilt corrector supplies continuous values of Vz and Dz. The stable element supplies L and Zd for use in computing sight settings, partial deck deflection, and trunnion-tilt corrections. The computer simply solves continuously the equations B′gr = B′r + jDd + Dz and E′g = Eb + Vs − Vz. Since the solutions are continuous, the guns will remain directed at their correct space position as long as they follow the orders B′gr and E′g.

RESTRICTED

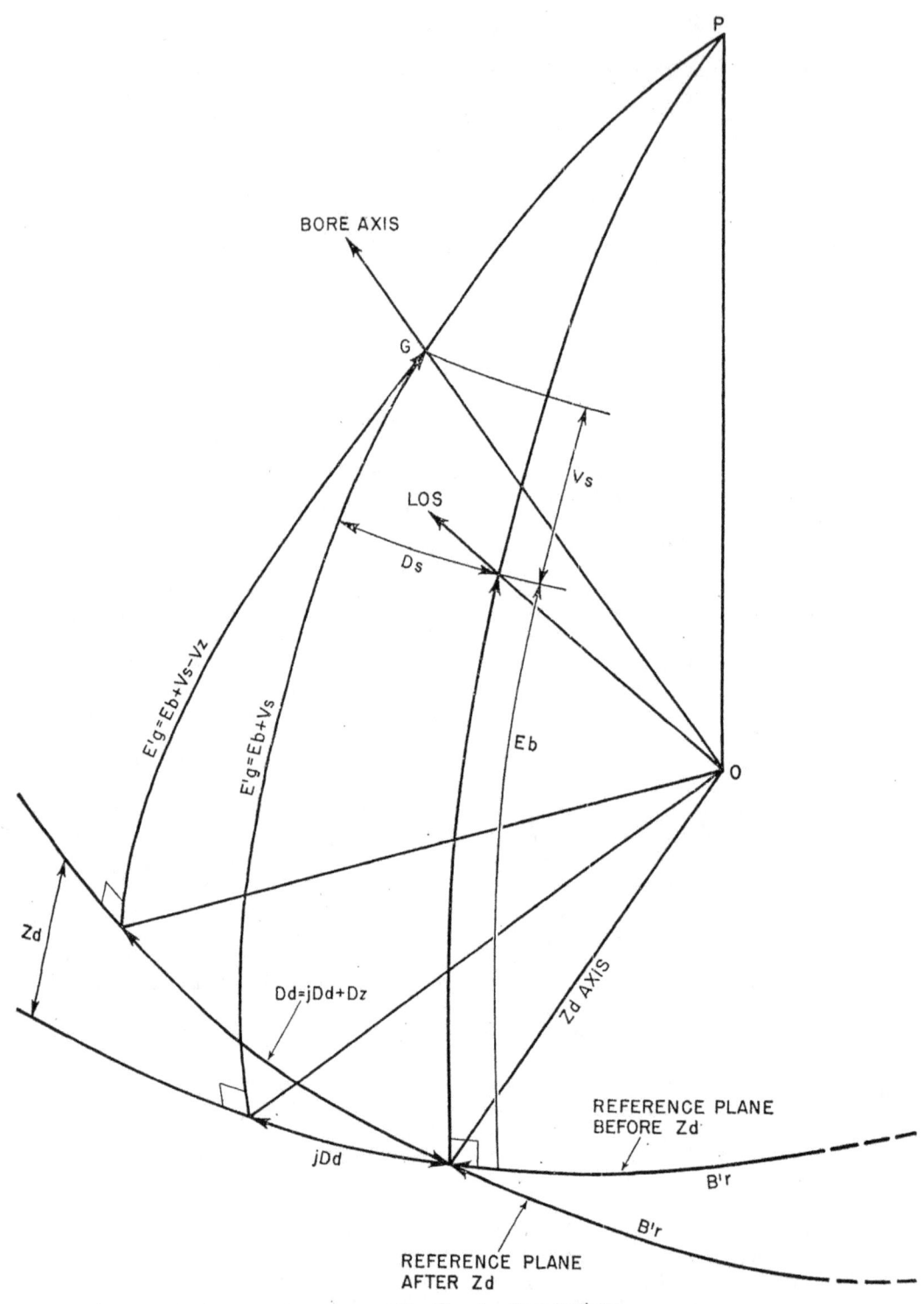

FIGURE 16E3. Trunnion-tilt corrections.

CHAPTER 16

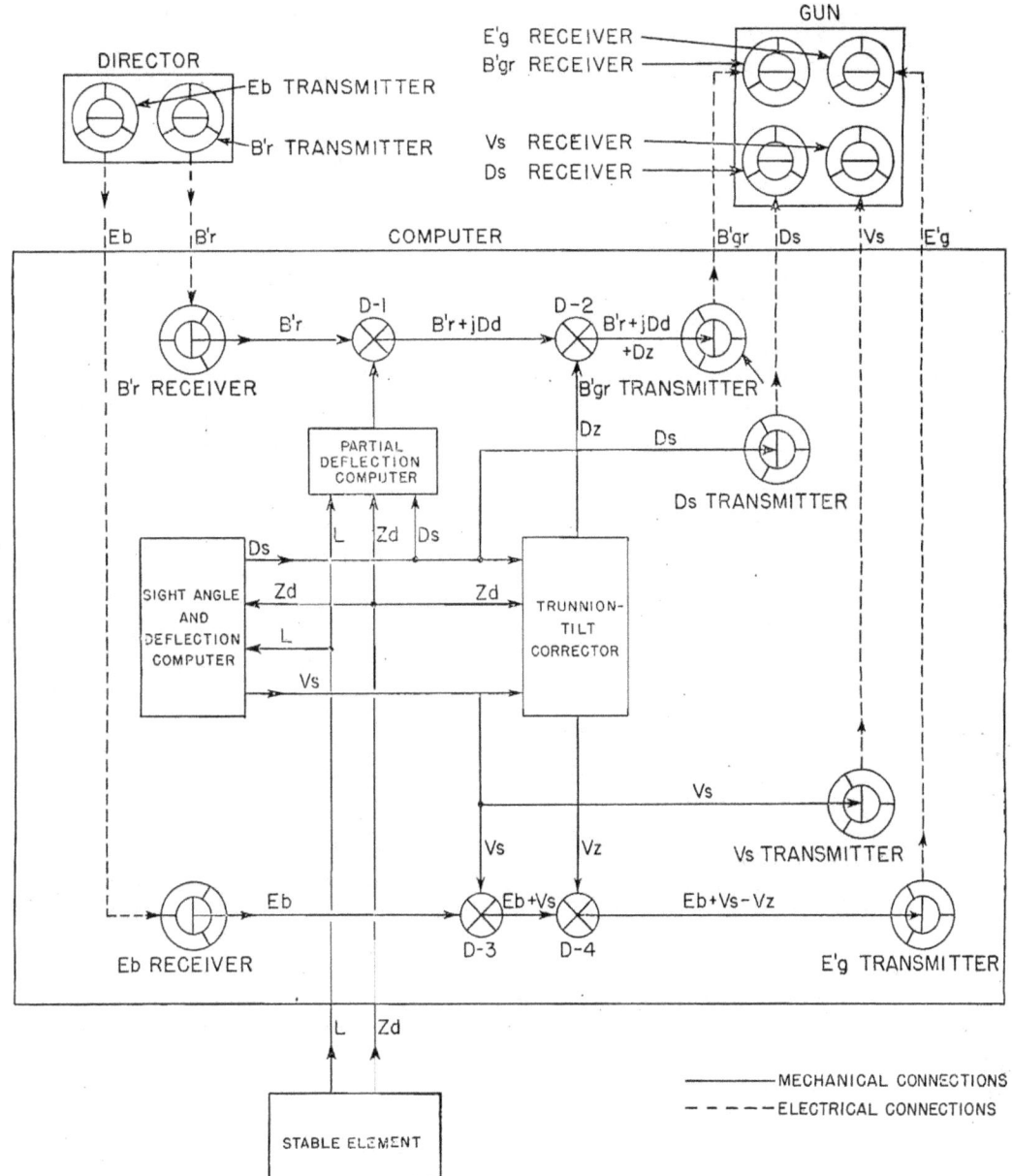

FIGURE 16E4. Schematic of gun-order computer.

16E8. Deck-tilt correction. The student will recall that the computations of range, bearing and elevation rates, sight angle, and sight deflection depend upon relative target bearing (Br) measured in the *horizontal*. A director necessarily measures bearing in the reference plane, and consequently does not produce true values of Br. The difference between director train (B'r) and Br is *deck-tilt correction* (jB′r) as indicated in figure 16E1a and b. It is the correction which must be applied to director train to obtain relative bearing in the horizontal plane.*

The value of jB′r is calculated by a deck-tilt computing unit within the plotting room computer, from a formula in which L and Zd are both factors. This correction is applied to B′r to obtain Br, which is supplied to other parts of the computer for use in calculating various quantities. Deck-tilt correction does not enter the gun-train order directly. It is simply a correction which must be applied to measured bearing (director train, B′r) to provide the computer with Br for use in other calculations.

16E9. Indirect fire. Modern systems in which target elevation, relative bearing, and range are generated in the computer, and level and cross-level angles are measured by a stable element, make possible a continuous computation of gun orders without inputs from the director. If the target is not visible, the computer can generate its position, and base the gun orders on that position. The complications in this method are the establishment of the initial values of E, Br, and R, and setting up the computer so that it calculates the correct rates. An AA target which is visible for a period long enough to permit tracking and setting up the computer, may subsequently be obscured by a cloud, but accurate fire can still be directed against it so long as it does not change course, speed, or rate of climb.

16E10. Methods of fire. The two main classes of firing methods are master key and pointer key. The latter has been described previously. Master key fire includes a number of different set-ups, in most of which a group of guns is controlled by some type of director system and firing is accomplished at the controlling director or at a stable element. The principal methods are:

Continuous aim. This method consists of keeping the director sights on the target at all times, and continuously laying the guns according to the computed gun orders. Under these conditions the bore axes remain pointed at the correct position in space. The master key may then be closed at any time. In AA fire, the master key is held closed to provide *continuous fire*, and each gun is fired when ready. Obviously, continuous fire can be accomplished only when continuous aim is effected.

Selected-level fire. This method consists of firing the guns at a pre-determined value of level angle. In this case, since Eb = E + L, gun elevation order becomes

$$E'g = E + L + Vs - Vz = E + Lj + Vs - Vz,$$

where Lj is a *selected-level angle*. The selected value is fixed, and the guns are fired as the reference plane arrives in the position where L = Lj. In the equation above, then, E and Vs are relatively slow-changing factors, Lj is constant, and Vz is the only element which varies because of roll and pitch. If cross-level angle Zd is relatively small, the changes in Vz will be minor and little gun motion in elevation will be required with respect to the deck.

Selected-level fire is useful when the combination of ship motion and gun angle is such that the guns would strike their elevation stops if continuous aim were employed. Obviously, the reduction in the amount of gun movement with respect to the deck makes it easier for the pointer and trainer to follow the gun orders.

Using this firing method, the effect on train is seen by examining the equation B′gr = B′r + jDd + Dz. Since the method is used when L is large and Zd is small, the Dz term has minor significance. The jDd term, representing Ds in the deck plane, changes slowly when selected level is used, since its value depends on Eb, which is in this case E + Lj. Except for changes in own-ship course, B′r changes slowly. Consequently, the training of the guns is not excessive in selected-level fire.

Selected-cross-level fire. This method consists of firing the guns at a predetermined value of cross-level angle. It is similar to selected-level fire, but is used when ship motion produces large values of Zd and relatively small values of L. A *selected-cross-level angle* (Zdj) reduces the variation in Dz and Vz in the gun-order equations, thereby reducing gun movement to a minimum when Zd is the predominant motion of the reference plane.

---

*Similarly, if the elevation of the target above the horizontal is measured in a plane perpendicular to the deck, it must be corrected to obtain target elevation (E) measured in a vertical plane.

CHAPTER 16

Selected-level and selected-cross-level fire are not used at the same time because it is highly improbable that selected values Lj and Zdj will occur simultaneously, and the guns would rarely if ever fire. In both methods, the values may be selected at the stable element and sometimes at the director. Firing is done by automatic contacts which are closed as the reference plane arrives at the selected position. Antiaircraft fire is so rapid that these selected methods are not used against other than surface targets.

16E11. Firing-delay compensation. When a gun is laid by continuous aim, the projectile travels along the same line of departure during the entire time it is in the bore of the gun. On the other hand, when a gun is fired while its bore axis is moving in space, as in selected level or cross-level fire, it changes position during the firing-delay period, which is the time elapsed between the closing of the firing circuit and the instant the projectile is ejected. It is a small time interval, but sufficient to permit the bore axis to be out of the required line of departure by the time the projectile leaves the gun if firing is not initiated before the bore axis arrives at its correct position.

Automatic firing-delay compensators are provided in stable elements and other instruments used to control the guns in the selected-angle methods. These devices simply fire the guns slightly before they arrive at their required positions, thereby permitting the projectile to leave the gun in the correct line of departure. Adjustments are provided to vary the length of time lead as required.

16E12. Battery alignment. It should be apparent from preceding discussions that the units of a director control system must be installed and adjusted so that all the measured and computed quantities are developed properly. For example, the mechanism of a director must be so adjusted that it actually measures bearing from the fore-and-aft axis and elevation from the reference plane. At the guns, the instruments must be so installed that they properly receive gun orders and permit applying the correct gun angle and gun train to the guns. The transmission systems must be properly adjusted, and in general, the entire system must be oriented or aligned before it can function in a satisfactory fashion.

The officer afloat is primarily concerned with *checking* the alignment to ascertain that it is correct. While minor adjustments can be effected aboard ship, original installations and subsequent major corrections are made in navy yards. The details of checking and correcting cannot be included in this text; they may be found in appropriate Ordnance publications. A few general comments on the method of checking a system are provided here to indicate the nature of the problem involved.

16E13. Director check. Obviously, if the system is put in operation, a check should show that the bore axes are all directed at the point in space designated by the director. When this condition is fulfilled, it is an indication that the system is properly aligned. Hence, system alignment is usually verified by a *director check*, which may be made by directing the bore axis of each gun and each director line of sight at a designated point of aim, preferably one at an infinite distance, such as a star. When the units are so directed, the bore and director sighting axes are parallel, and the dials on the various indicators should be matched, or indicating a zero difference between gun orders and gun response. If the dials do not register zero, their readings are an indication of the amount of existing misalignment.

Frequent director checks are made both afloat and in drydock by checking parallelism between units at various angles of train and elevation. Assuming that the guns have previously been bore-sighted, they may be laid by means of their sights instead of boresight telescopes.

16E14. Conditions for parallelism. Naturally, parallelism can exist only when there are no corrections introduced that will (1) cause the bore axes to be offset from the lines of sight, or (2) converge the lines of sight themselves. These conditions mean that all corrections are zero; that is, no sight angle and deflection are set, no parallax compensation and convergences are made, and erosion and trunnion-tilt corrections are not introduced. The only exception is that the compensation for roller-path inclination must still be made, because parallelism of the bore axes with the director lines of sight must be effected with reference to a common plane. If the guns and directors do not train in planes parallel to the common plane, parallelism between bore axes and the

RESTRICTED

director LOS cannot exist without roller-path tilt compensation. Conversely, if a check shows parallelism, the compensation for roller-path inclination is correct.

In brief, a parallelism check with all corrections zero should show that:

1. The director LOS and the bore axes are in vertical planes which remain parallel to one another when guns and directors are trained. A check for this condition is called a *director check in train*.

2. The director LOS and the bore axes are in parallel slant planes which remain parallel as the guns and directors are elevated through equal angles. A check for this condition is called a *director check in elevation*.

3. A satisfactory check in elevation will also show that correct compensation is made for roller-path inclination.

A test of system parallelism afloat admits inaccuracies due to ship motion, and also necessitates compensation for convergences introduced by the the choice of suitable reference points at which the lines of sight and bore axes can be directed. If the system is checked by using a target at infinity (star), the director LOS and the bore axes should be parallel both in elevation and train. However, it is generally impractical to make a director check by such means. It is more convenient to direct the system at the horizon for a check in elevation and at any suitable distant target for one in train; both reference points require that convergences be introduced to compensate for the finite distances.

16E15. System variations. Throughout the discussion of mechanisms involved in director control, reference has been made to methods used in making the measurements and computations. It is impossible to summarize the features of every system in a text of this nature; as a matter of fact, there are numerous variations among similar systems currently in use. Improved techniques and mechanisms are being developed constantly, which give rise to differences even among installations on ships of the same class. Angles may be measured in different planes and by different devices. Computations of many quantities involve such complicated mathematics that approximations are made in the interest of equipment simplification, and the approximations are not always the same for any given quantity.

None the less, it is recognized that familiarity with actual installations is desirable. To this end, descriptions of selected systems appear in Chapters 17 and 18.

CHAPTER 17

DUAL-PURPOSE BATTERY FIRE-CONTROL SYSTEM

A. GENERAL DESCRIPTION

17A1. Functions. The director firing system described in this chapter is used to control the fire of 5"/38 cal. dual-purpose guns and is frequently referred to as the Mark 37 director system. While this is not the only system employed to control these guns, it is the most-widely used at present. Its primary functions are to provide:
1. Continuous, automatic gun positioning.
2. Automatic fuze setting.
3. Continuous sight-angle and sight-deflection indication at the guns.
4. Continuous, selected-level and selected cross-level fire.
5. Star-shell fire control.
6. Searchlight control.

17A2. Presentation. The purpose of this section is to provide a general description of the entire system, while the next four sections present more of the details. In general, the first part of each subsequent section describes the appearance and location of the various mechanisms used directly by the operating personnel, and outlines the duties they must perform to make the equipment function. The remainder of each section explains the functional operation of the mechanisms with a view of showing the student the applications of mechanical and electrical devices to the solution of the fire-control problem.

It is reiterated that the duties of personnel as set forth must not be considered as standard doctrine, for many variations in the use of the system are possible. Furthermore, the information provided is not intended to qualify personnel fully in the proper care of the equipment. Bureau of Ordnance publications must be consulted for specific instructions regarding adjustments, repairs, tests, and other similar subjects.

The quantities whose names and symbols appear in this chapter are defined in Chapters 15 and 16 unless otherwise noted. In addition, the following terms are freely employed to avoid needless word repetition:

Sight plane. The vertical plane containing the line of sight or any plane parallel to that plane.

Deck plane. The battery reference plane or any plane parallel to it.

17A3. Installations. One complete director-firing installation capable of controlling the guns (and searchlights) comprises a modification of each of these three major units: (1) *Mark 37 director*, (2) *Mark 6 stable element*, and (3) *Mark 1 computer*. Sections B, C, and D of this chapter each explain one of these major units, while Section E describes the receiving equipment required at the guns to utilize the control information supplied by the director installation. In general, destroyers carry a single installation (one of each major unit), cruisers and aircraft carriers a double installation (two of each major unit), and battleships a quadruple installation (four of each major unit).

The directors are mounted in elevated positions to provide the maximum view consistent with ship-construction requirements. On destroyers and auxiliaries, the director is usually on top of the pilot house, as shown in figure 17A1. The two directors on carriers are at each end of the island, which is to one side of the ship centerline. On cruisers, the two are mounted forward and aft on the centerline. Battleships carry four secondary-battery directors arranged in a diamond-shaped pattern around the superstructure, as indicated in figure 10B1.

The stable elements and computers are housed below decks in protected plotting rooms. The interchange of data between a stable element and a computer is effected by mechanical shafting, hence the two stand adjacent to each other and always operate as a pair. The computer-stable

DUAL-PURPOSE BATTERY FIRE-CONTROL SYSTEM

element groups are connected electrically to the guns and directors by a synchro-transmission system which clears through one or more fire-control switchboards located in the plotting rooms. Destroyers, of course, have a single plotting room. Generally, the carriers and cruisers have a single plotting room containing two computer-stable element groups. Battleships may have one plotting room with four computer-stable element groups, or two separate plotting rooms, each containing two such groups.

FIGURE 17A1. Mark 37 director system; destroyer installation.

Usually each director in a multiple installation controls a designated group of guns in the battery. However, the switching arrangements on ships having double or quadruple installations make it possible for any one of the directors to feed information to any one of the computer-stable element groups. In turn, a selected director-computer-stable element group may control all or any portion of the battery. It is thus possible for multi-director installations to control fire simultaneously against more than one target, and to have standby arrangements available when all directors are not engaged.

On ships where the 5"/38 cal. guns form the secondary battery, the Mark 37 director can supply director train and director elevation to the main-battery computers. Furthermore, the secondary-battery director-computer-stable element groups can control main-battery fire against air targets when selected fuze settings are used.

17A4. Parallax and roller-path corrections. Since each Mark 37 director on a given ship may control any or all of the guns in its battery, all train and elevation orders and measurements are based on a common reference system. Differences in range due to the horizontal displacements between system units are not considered, but the following factors are accounted for:

Roller-path inclination. A correction for roller-path inclination is applied by a tilt corrector or compensator at each gun and searchlight, and at each director in multiple installations. When there is only one director, its roller path is the reference plane and it has no tilt corrector.

CHAPTER 17

Horizontal parallax. All train orders and measurements are based on a single horizontal reference point which for single installations is at the director, and for multiple installations is generally midway between the forward and aft directors on their centerline. Parallax correction for a 100 yard horizontal base is developed in the computer and transmitted as a separate quantity to all guns and searchlights, and in multiple installations to the directors not at or close to the reference point. The train-receiving instrument at each mount receiving the quantity has a parallax mechanism which, by means of a simple gear ratio, changes the 100 yard correction to that corresponding to the actual mount displacement. The mechanism then applies the proper correction to the mount response or measurement as required. In single installations such as on destroyers, the reference point is on the director training axis, hence the director needs no parallax correction and has no parallax mechanism.

Vertical parallax. The vertical base is the distance between the director telescopes and the mean gun-trunnion height, individual trunnion-height differences being small enough to neglect. In multiple installations, telescope-height variations between directors are also neglected. Parallax due to the vertical base is therefore approximated by a single calculation made in the computer and included in the gun-elevation order, and no separate mechanisms are required at the mounts for this factor. However, in cases where the horizontal base is considerable, as on certain aircraft carriers, vertical parallax due to a 100 yard *horizontal* base is developed in the computer and transmitted to the guns as a separate quantity. A parallax unit in the elevation-receiving instrument at the guns changes this quantity to agree with the actual base, and applies the necessary correction to the gun response.

17A5. Flow of information. A schematic diagram of the principal inter-connections in a single installation is shown in figure 17A2. It should be understood that all the electrical circuits between system units clear through the fire-control switchboard.

The director, which carries its own range finder in addition to telescopes, measures director elevation (Eb), director train ($B'r$), and present range (R), and transmits their values electrically to the computer. The control officer in the director estimates target angle (A), target horizontal speed (Sh), and rate of climb (dH), sending their values by telephone to the computer operators. Hand-operated transmitters in the director provide electrical transmission of elevation, deflection, and range spots to the computer. In some cases, a target-angle transmitter is provided so that target angle may be transmitted electrically.

The stable element is a gyroscopically controlled instrument that supplies the system with measurements of level angle (L) and cross-level angle (Zd). Both of these angles are transmitted mechanically to the computer for use in a number of calculations. A modified value of level, $L + Zd/30$, is also transmitted to the computer, where it is further modified to form the quantity $\triangle cE + L + Zd/30$. This quantity is then transmitted electrically to the director where it is used automatically to keep the LOS on the target in elevation. Cross-level angle (Zd) is transmitted electrically from the stable element to the director to stabilize the optical units in cross-level.

The computer is built to solve the antiaircraft problem, and of course, can handle the surface problem, since the latter is nothing more than a special case of the AA problem in which target elevation (E) is zero. Inputs to the computer are electrical from the director, ship's gyrocompass, and pitometer log (speed indicator); mechanical from the stable element; and manual from the computer operators. The quantities computed and electrically transmitted to the guns are: (1) gun-elevation order ($E'g$), (2) gun-train order ($B'gr$), (3) sight angle (Vs), (4) sight deflection (Ds), (5) fuze-setting order (F), (6) train parallax for a 100 yard horizontal base, and in some cases, (7) elevation parallax for a 100 yard horizontal base. These values are used for gun positioning, sight setting, and fuze setting. Indicator-regulators at the guns provide gun positioning and fuze setting either by fully automatic means or by matching pointers.

The computer also generates changes in range, elevation, and train for the director. These changes are used in conjunction with level and cross-level angles to hold the director telescopes and range finder on the target continuously. Not only is the range finder held on the target, but its

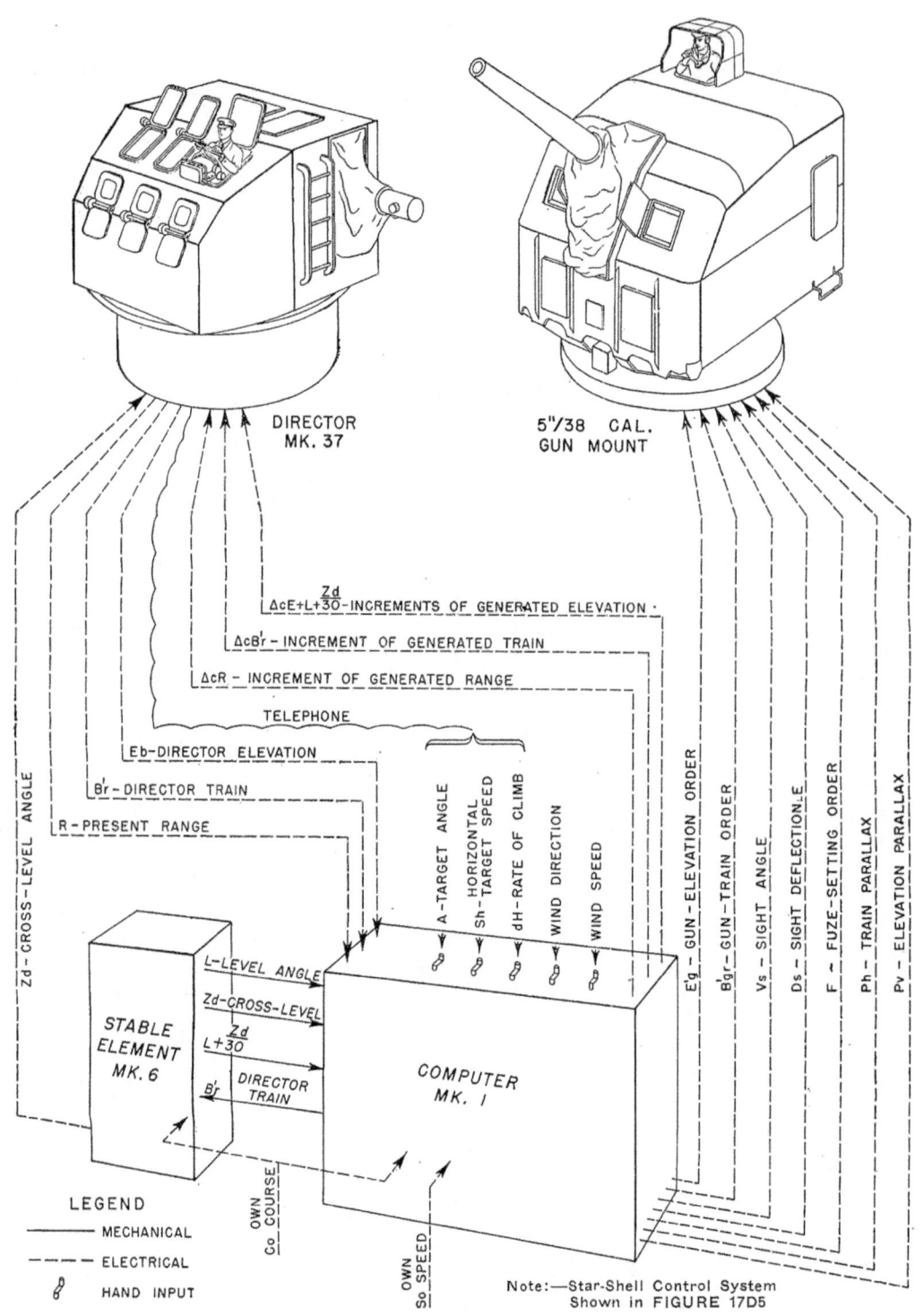

FIGURE 17A2. Mark 37 director system; diagrammatic.

CHAPTER 17

measuring wedges are positioned electrically. This arrangement provides a check on the computer target-position solution. When this solution is incorrect, the line of sight drifts off the target and the range finder wanders out of adjustment. The changes required in the set-up of A, Sh, and dH on the computer to correct its solution are computed and applied by a rate-control mechanism. This mechanism itself may be actuated automatically by the director personnel at the same time that they bring the line of sight back on the target and readjust the range finder, or by the computer personnel in response to indications from the director.

17A6. Illumination control. The director carries computing and transmitting equipment which supplies train and elevation orders to the searchlights for illumination purposes. Controls are provided to permit the addition of arbitrary corrections to the train and elevation orders to permit shifting of the light beams to the most desirable position. These searchlight orders clear through the fire-control switchboard, but are computed at the director and have no immediate connection to the plotting-room equipment.

· Star shells may be fired by the 5"/38 cal. guns to illuminate surface targets for their own or the main battery. A star-shell computer, attached to the Mark 1 computer in the plotting room, calculates star-shell gun train, elevation, and fuze-setting orders from basic information supplied mechanically by the Mark 1 computer and from spots received electrically from a star-shell spot transmitter in the director. It is thus possible for a director-computer-stable element group to control star-shell fire from some guns at the same time it is controlling service-projectile fire from other guns.

B. THE MARK 37 DIRECTOR

17B1. Primary function. The primary function of the Mark 37 director is to establish the target position by measuring these quantities:

SYMBOL	DEFINITION
B'r	*Director train.* The angle between the fore-and-aft axis and the vertical plane through the LOS, measured in the deck plane clockwise from the bow.
Eb	*Director elevation.* The elevation of the director LOS above the deck, measured in the vertical plane through the LOS. $Eb = E + L$.
R	*Present range.* The distance of the target from own ship measured along the LOS.

These three coordinates establish the target position with respect to the battery reference system, and are transmitted electrically to the computer. Their use at the computer is described more fully in Section D, but broadly, they provide:

1. The basis for gun-train and elevation orders.
2. Part of the initial computer set-up.
3. A check on the accuracy of the target position generated by the computer.

17B2. Secondary functions. The director has numerous secondary functions which are likely to overshadow the primary function unless the student is careful to realize that the determination of target position relative to own ship is the starting point for the fire-control problem calculations. The other functions may be summed up by simply stating that the director normally serves as the control center for the guns to which its director-computer-stable element group is connected by the switchboard set-up. This is a far-reaching statement, but in truth, when the system is operating according to design, the computer, stable element and guns operate entirely by remote control exercised at the *director*. Once the problem is set up correctly, assuming the target maintains course and speed, the only personnel actually contributing to the firing of service projectiles are the ammunition handlers who service the guns. All other operators are simply observing the operation, ready to step in when necessary. Even if changes occur in the problem conditions, the director personnel may introduce the necessary adjustments of the computer set-up by remote control.

RESTRICTED

DUAL-PURPOSE BATTERY FIRE-CONTROL SYSTEM

The functions associated with the director to provide such control include:
1. Initial estimate of target angle, target horizontal speed, and rate of climb.
2. Spotting service-projectile fire and transmission of the spots.
3. Computation and transmission of searchlight train and elevation orders.
4. Spotting star-shell fire and transmission of the spots.
5. Control of rate calculations made by the computer, and therefore, control of the primary settings of target angle, target horizontal speed, and rate of climb.

The director receives inputs that automatically keep its several lines of sight on the target and adjust the range-finder measuring wedges correctly, provided the computer set-up is accurate. This feature, of course, simplifies the tasks of the director pointer, trainer, and range-finder operator, but above all, it permits them to introduce rate corrections in the computer by remote control.

17B3. Inputs and outputs. All of the quantities transmitted to and from the director are sent normally by synchro transmission, with the exception of items 7, 8, and 9 below, which define target motion. In some cases target angle or its equivalent, target course, is transmitted by synchro. The quantities are:

Inputs:
1. Bearing correction ($\triangle cB'r$).
2. Elevation correction ($\triangle cE + L + Zd/30$).
3. Range correction ($\triangle cR$).
4. Train parallax for a 100 yard horizontal base (Ph) (in multi-director installations only).
5. Cross-level (Zd).

Outputs:
1. Director train ($B'r$).
2. Director elevation (Eb).
3. Observed present range (R).
4. Range spot (Rj).
5. Elevation spot (Vj).
6. Deflection spot (Dj).
7. Target angle (A).
8. Target horizontal speed (Sh).
9. Rate of climb (dH).
10. Battle and shell orders (in some installations).
11. Searchlight train order.
12. Searchlight elevation order.
13. Searchlight control orders.
14. Star-shell range spot (Rjn).
15. Star-shell elevation spot (Vjn).
16. Star-shell deflection spot (Djn).
17. On-target signals.

17B4. Component parts. Externally, the director appears as shown in figure 17B1. There are over 50 modifications of the Mark 37 director, the variations between them being relatively minor with regard to functions performed. This section does not apply to a specific modification, but deals primarily with the major features incorporated in all of them. The most important instruments, of course, are three director telescopes and a range finder, which permit measuring director train, director elevation, and present range.

The entire director is supported on a foundation built into the ship structure. This foundation is machined (as nearly as possible) parallel to the mean plane of the gun roller paths in order to reduce roller-path tilts to a minimum. The principal supporting elements of the director are the *base ring*, the *carriage*, and the *shield*. The shield (outer housing shown in the illustration) is attached to the carriage, and the carriage in turn is supported by roller bearings on the base ring. The base ring rests on the director foundation.

CHAPTER 17

The personnel (usually six or seven men) enter the director through the foundation and the base ring, passing through hatches in the floor of the carriage. The shield has five or six observation hatches in its top, three ports in the front plate for the telescopes, and a vertical slot on each side for the protruding range finder. One of the major sources of variation between director modifications is the shield thickness, which varies from the thin, weather-protective type on destroyers, to the heavy, splinter-proof, 1½ inch thick shields on battleships. Each telescope-port cover is operated

FIGURE 17B1. Mark 37 director.

by a handwheel, and on the heavier shields the observation hatches are similarly operated. In most of the director modifications, the shield and carriage support the following components, some of which are placed as shown in figure 17B2:

1. Three movable-prism telescopes (some modification of the Mark 60).
2. Slewing sight.
3. Stereoscopic range finder, 15 foot base (some modification of the Mark 42), and the necessary carriage for mounting.
4. Change of range receiver.
5. Cross-level gear with power drive only.
6. Training gear with hand and power drive.
7. Elevating gear with hand and power drive.
8. Director-train indicator.
9. Director-elevation indicator.
10. Either a battle-order transmitter or a spot transmitter.
11. Range-spot transmitter.
12. Searchlight trunnion-tilt corrector.
13. Searchlight control transmitter.
14. Star-shell spot transmitter.
15. Target-angle or target-course repeater.
16. Roller-path tilt corrector (multiple installations only).
17. Horizontal-parallax mechanism (multiple installations only).
18. Other mechanical and electrical elements not included in the preceding groups.

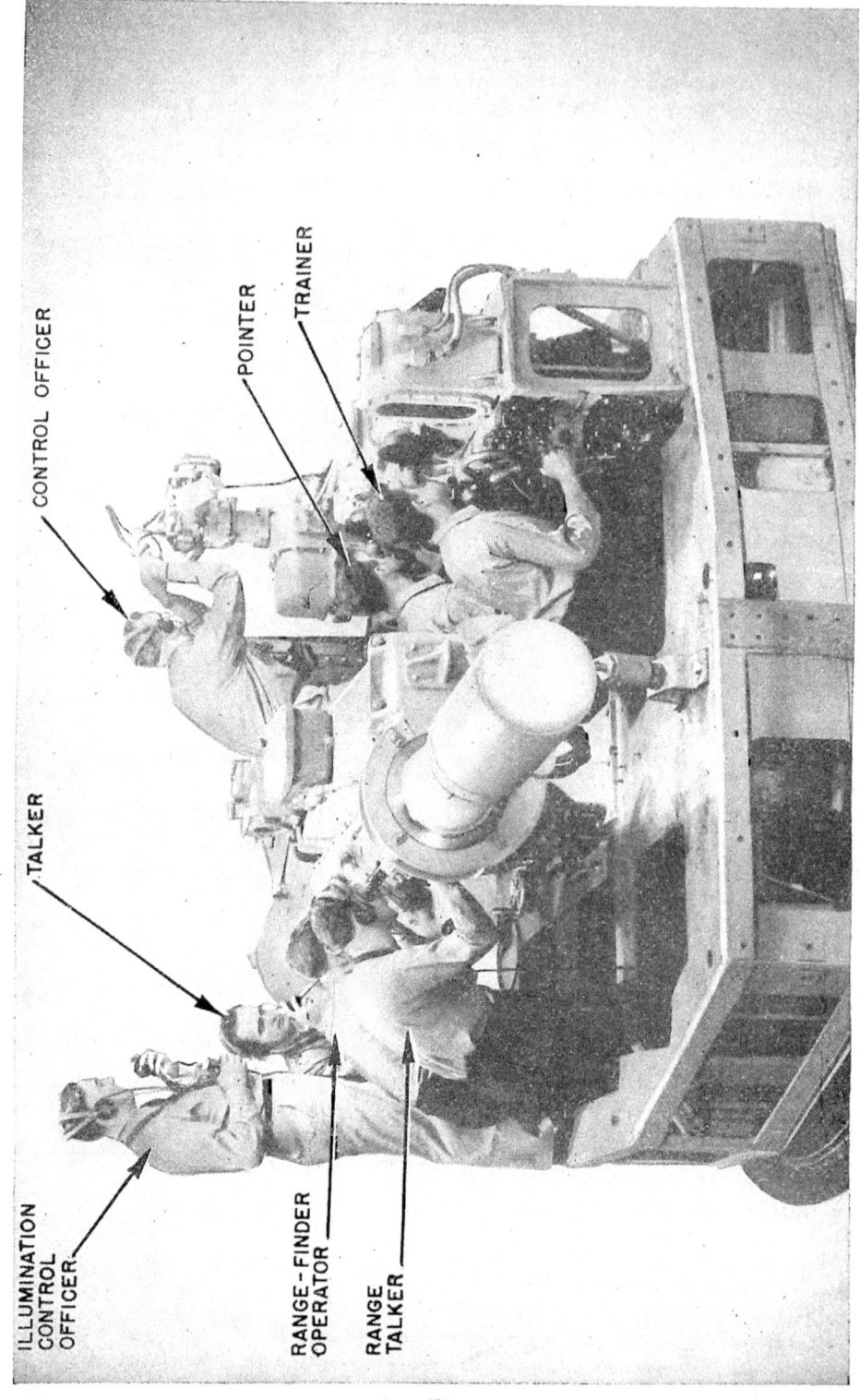

FIGURE 17B2. Disposition of personnel and equipment; Mark 37 director.

CHAPTER 17

17B5. General director operation. The three telescopes are mounted within housings called *optical boxes*. The three boxes are rigidly secured to a transverse frame, called the *optical box shelf*, at the front of the director. The longitudinal axes of the telescopes are in the sight plane established by the telescope lenses, prisms and crosshairs. Each telescope is supported in bearings so that it can rotate about its longitudinal axis within its optical box, and each box is installed so that its telescope axis is parallel to the director roller path. The line of sight is moved in elevation by rotation of a prism within the telescope. The prism axis of rotation is perpendicular to the telescope longitudinal axis, and parallel to the horizontal crosshair. (The telescope operating principle is shown in fig. 17B3.) The cross-level power drive is geared to all three telescopes and turns them about their longitudinal axes to keep the prism axes horizontal, thereby stabilizing the telescopes in cross-level. This arrangement therefore causes the prism rotation to occur in a vertical plane, or in other words, the sight plane is vertical.

The range finder is supported in a cradle called the *range-finder beam*. The beam is supported in a stand in such a manner that it can rotate about an axis in the sight plane and parallel to the roller path. The cross-level power drive is also geared to the range-finder beam, and through it holds the range-finder longitudinal axis horizontal.

The line of sight is kept on the target in train by turning the entire director, its total displacement from the fore-and-aft axis being a measure of director train (B′r). The line of sight is elevated by rotating the telescope prisms as explained above, and by rotating the range finder about its longitudinal axis. Since the elevation occurs in the vertical plane, the prism position with respect to the roller path is a measure of director elevation (Eb).

Train and elevation are controlled by similar mechanisms. Both mechanisms provide for (1) automatic, or remote control of the power drive; (2) local control of the power drive by means of handwheels; and (3) manual movement by direct gearing from the handwheels. The mechanisms are independent to the extent that train may be accomplished by one method while elevation is effected by another.

The cross-level drive, in addition to providing for measurements in the vertical plane, reduces the amount of training required to hold the line of sight on the target. It is considerably easier to move only the range finder and telescopes about their cross-level axes than to train the whole director to produce the same results. Cross-level is accomplished by automatic power-drive control only, for the director has no cross-level telescope or other means for indicating the value of cross-level. Further details and descriptions of other units at the director are mentioned in the operations described in succeeding articles.

17B6. Personnel. The disposition of the normal director complement is shown in figure 17B2. For most efficient operation, each of the men should be capable of handling all stations so that action can be continued in event of a casualty. Until a target is sighted the personnel act as lookouts through the roof hatches.

The *control officer* is in charge of the whole firing installation of which his director is a part. Thus he supervises the personnel of the director, the operators of the computer and stable element, and the crews of all guns and searchlights that are operating under the control of the director. He designates the target to his group, in some cases as designated to him by the captain or secondary battery officer, he estimates target angle (A), target horizontal speed (Sh), and rate of climb (dH), telephoning these quantities to the computer. The control officer also makes and transmits elevation and deflection spots, and issues battle orders. In some cases he may have control of the firing circuit by means of a portable firing key.

The *pointer* and *trainer* operate their respective handwheels as necessary to keep the director line of sight on the target. They can rate control in elevation and train. Each has a firing key which may be so connected that either man can control the firing, or that both keys must be closed. Pointers and trainers may act as talkers for the control officers.

The *range-finder operator* mans the range finder and makes range spots in AA fire. In surface fire he may spot in both range and deflection. He may also rate control in range.

The *range-finder talker* acts as the talker for the range-finder operator. In the event of casualty to the range synchro transmission he transmits the range-finder ranges and range spots by phone. He is also available as relief for the range-finder operator.

The *illumination officer* is in control of the searchlights and of such guns as are firing star shells. In the day time he may act as talker for the control officer or be a relief man for any position in the director.

The *talker* relays other control center messages to the control officer.

17B7. Train; manual control. Operation of the director in train is discussed in the following five articles. With few exceptions, which are explained later, the description applies to operation in elevation. The student should bear this in mind as these articles are read. As stated before, the trainer may control the director by any one of these methods: (1) *automatic,* (2) *local,* or (3) *manual.* In automatic or local the director is power driven; in manual it is driven by hand.

In front of the trainer between his handwheels is a selector lever having *automatic, local* and *manual* positions with which he chooses the type of control. When this lever is placed at *manual* the trainer's handwheels are directly geared to the training circle and the trainer supplies the motive power for training the director. This method is, of course, a standby used only in the event of power-drive failure.

17B8. Train; local control. The train power drive consists essentially of a control unit, which is called the *receiver-regulator,* and an electric motor. When the trainer shifts his selector lever to *local* the mechanical connection between the handwheels and training circle is broken and the handwheels are geared to the stator of a synchro in the receiver-regulator. Training is still controlled entirely by the trainer through the motion of his handwheels, but the power is supplied by the motor.

While the target is being picked up and as the computer is being set up for the fire-control problem the director is normally operated in *local.* When the computer has begun its solution and provides bearing correction ($\triangle cB'r$) the selector lever is shifted to *automatic* and that method becomes the normal type of control. Local control is then used only as a stand-by in the event of failure in the $\triangle cB'r$ transmission from the computer to the director.

17B9. Train; automatic control. When the selector lever is shifted from *local* to *automatic* it maintains the mechanical connection between the trainer's handwheels and the synchro in the receiver-regulator, and in addition, closes a switch which connects the $\triangle cB'r$ input to the synchro. The synchro then governs the response of the receiver-regular according to the algebraic sum of the handwheel movement and the $\triangle cB'r$ input. Therefore, in this type of control, the director's position in train is controlled both by trainer's handwheel movement and by the signal from the computer.

The signal $\triangle cB'r$ from the computer represents the calculated change in director train. This quantity takes into account changes in train due to relative target motion, changes in own-ship course, and the effect of deck tilt. The deck-tilt correction is required to account for the changes in train produced by ship roll and pitch.

The trainer's job is to make certain that the vertical crosshair of his telescope remains on the target. If the quantity $\triangle cB'r$ is sufficient to accomplish this, the trainer has no work to do. If, however, the crosshairs tend to drift off the target, the trainer must rotate his handwheels in order to keep the line of sight on the target.

17B10. Train; rate control. If the trainer has to turn his handwheels to keep on the target while the director is training in *automatic,* the computed bearing correction ($\triangle cB'r$) is in error. The error is due to incorrect setting of one or more of the primary quantities defining target motion; i.e., target angle, target horizontal speed, or rate of climb. Provided that the computer is set up for automatic rate control, the trainer can cause its rate-control mechanism to correct these settings as required to produce the correct value of $\triangle cB'r$. To do this, he simply closes a rate-control, or signal key on his left handwheel while he is turning the handwheels to keep on the target. This is called *rate-control,* and it continues until handwheel rotation is no longer necessary to keep the vertical wire on the target, which condition indicates that the bearing correction ($\triangle cB'r$) is then correct. In practice, the trainer puts his vertical wire on the target, holds his key closed, and turns his handwheels as necessary to keep the line of sight continuously on the target. Provided the pointer and range-finder operator rate control simultaneously, this brings the computer solution into agreement with actual conditions, and the handwheel motion decreases and ultimately ceases. When the target changes course or speed, handwheel motion will again be required until the solution has been corrected.

When the director is in *local*, rate controlling must be done at the computer by its operators. The director trainer keeps his key closed while on the target, but in this case it only actuates a signal which indicates to the computer operator that the line of sight is on in train.

17B11. Train; slewing. When the director is power driven (automatic or local control) the control officer may, if he desires, rapidly turn or *slew* the director in train and elevation. To do this he presses a key located on the grip of the slewing sight, and swings the sight to line up with the target the director is to follow. Mechanisms associated with the slewing sight then take control of the train and elevation power drives and train and elevate the director optics until the line of sight coincides with that of the slewing sight. The control officer uses this method to bring a target quickly into the telescope fields to help the pointer and trainer, thereby designating the target to the director personnel.

17B12. Director-train and elevation indicators. The *director-train indicator* is located on the optical box just to the right of the trainer's telescope. It receives bearing correction ($\triangle cB'r$) electrically from the computer and director train ($B'r$) mechanically. It has three dial groups arranged in a vertical line.

The upper group consists of a *dial* and *ring dial*. The inner dial bears a single index mark and is driven by a synchro which receives bearing correction ($\triangle cB'r$). The ring dial shows ten degrees of actual director train per revolution, being mechanically positioned by director train ($B'r$). This pair of dials is used when the director is in *local* or *manual* and the target is momentarily obscured by clouds. The trainer then keeps the zero calibration on the ring dial matched with the index mark on the inner dial, thus applying bearing correction ($\triangle cB'r$), until the target becomes visible once more.

The center group has a *ring dial* which shows coarse readings of director train against a fixed index from 0° to 360°. Used in conjunction with the ring dial of the upper group, it provides direct readings of $B'r$. In directors which receive target designations from transmitters located elsewhere on the ship, the coarse dial is a ring surrounding a central dial. This inner dial has a single index, and is driven by a synchro energized from the target designator or director-train transmitter of another director. The trainer can pick up a target as designated by training the director until the index on the coarse ring is opposite the index on the dial.

The lower group consists of one dial. It is calibrated to show up to 375° of actual director train on either side of zero, and is known as the *cable-twist dial*. Its purpose is to indicate to the trainer at a glance how far the director is trained in either direction from its zero train position. Since the director is limited to 375° train on either side of neutral, the dial provides an indication of how far the director has to go before it will hit its train stops.

The *director-elevation indicator* is located to the right of the pointer's telescope in the director. It receives own-ship course electrically from the ship's gyrocompass, elevation correction ($\triangle cE + L + Zd/30$) electrically from the computer, and director elevation (Eb) mechanically. Its three dial groups are set in a triangular arrangement. The top dial group consists of a single *dial*, driven by a synchro that receives own-ship course, which is indicated against a fixed index.

The dial group in the lower-right corner consists of a synchro-driven *dial* and a mechanically driven *ring dial*. The inner dial, bearing a single index is positioned by the synchro, which receives elevation correction ($\triangle cE + L + Zd/30$). The ring dial makes one revolution for every ten degrees of director elevation and is so calibrated. The pointer uses these dials in following an obscured target in the same way that the trainer uses the upper dial group on the director-train indicator.

In the lower-left corner of the face of this instrument is the *coarse-reading elevation dial*. This dial is calibrated in degrees to show 110° of elevation and 30° of depression. It is used in conjunction with the ring dial of the lower-right dial group for reading director elevation (Eb). In some modifications of this indicator, the coarse elevation dial is a ring surrounding a center dial having a single index that provides target designation. The pointer then uses the dials in elevating to a target designated from outside the director just as the trainer uses his designator dial group.

17B13. Elevation. It has already been indicated that the foregoing description of director train is also true for director elevation with but a few differences. Elevation and depression are accomplished by prism rotation in the telescopes and by rotating the range finder about its longitudinal axis. Thus in manual control the handwheels are directly geared to the telescope prisms and the elevating arc of the range finder. In *local or automatic*, the output shaft of the elevation motor is geared to the prisms and range finder. In *automatic*, the receiver-regulator is controlled by elevation correction ($\triangle cE + L + Zd/30$) plus any necessary handwheel motion.

It is evident from a study of previous chapters that the total elevation of the director line of sight, or director elevation, is the sum of target elevation and level angle; i.e., $Eb = E + L$. In order to hold the line of sight on the target in elevation, it is necessary to supply the changes in E, or generated elevation increments ($\triangle cE$), which are due solely to relative target movement. It is also obvious that changes in level must be supplied to hold the line of sight stationary in space as the ship rolls and pitches. With the deck plane horizontal, L is zero. Any movement from that position causes a change in level equal to the actual value of L. Hence $\triangle cE$ and L are elements of the elevation correction. In *automatic*, the sights are said to be stabilized in level, since the value of L included in the elevation correction holds the line of sight steady regardless of ship roll and pitch. It should be noted that automatic stabilization in level is not possible in *local* or *manual* since the director is not receiving signals from the computer. Instead, the pointer supplies elevation correction by handwheel movement.

17B14. Cross-level function. In the elevation correction, the third element, $Zd/30$, is termed *cross-level function*, and is supplied solely to counteract an effect produced by the telescope gearing. Figure 17B3 illustrates the operating principle of the *director telescopes*. Prism rotation is accomplished in the following manner. The input gear, driven by the elevation drive, turns the floating gear, which can rotate freely around the telescope but is not attached to it. The floating gear drives the gear at the end of the worm shaft. Rotation of the worm causes the prism to rotate about its axis, thus elevating or depressing the line of sight.

When the Zd input gear is turned by the cross-level drive, it drives the sector gear and turns the telescope. When the telescope is rotated, the pinion on the end of the worm shaft "walks" around the floating gear so long as the elevation input gear does not move, and causes rotation of the worm and the prism. A change in the line-of-sight elevation will thereby be introduced when Zd is entered. Because of the gear ratio used, the amount of elevation change due to this factor is equal to $Zd/30$. In order to prevent prism rotation caused by cross-leveling, the quantity $Zd/30$ is introduced through the elevation input pinion. This input rotates the floating gear in such a manner that the worm shaft is carried with the telescope to prevent its rotation with variations in Zd. In other words, it rotates the elevation input to take out $Zd/30$ at the same rate as the cross-level drive introduces it. The value of $L + Zd/30$ is developed in the stable element and transmitted mechanically to the computer. There $\triangle cE$ is added to obtain elevation correction ($\triangle cE + L + Zd/30$), and the total quantity is sent electrically to the elevation receiver-regulator.

17B15. Cross-level. As previously stated, the director optics are stabilized in cross-level by a *cross-level power drive*. Like the train and elevation drives, it consists of a receiver-regulator and a motor. The output shaft of this motor is connected through shafts and gearing to the director telescopes and range finder. By means of this drive the telescopes are rotated within their housings by an angle equal to Zd, and the whole range finder is rotated an equal amount about an axis in the sight plane. Figure 17B4 illustrates this stabilization.

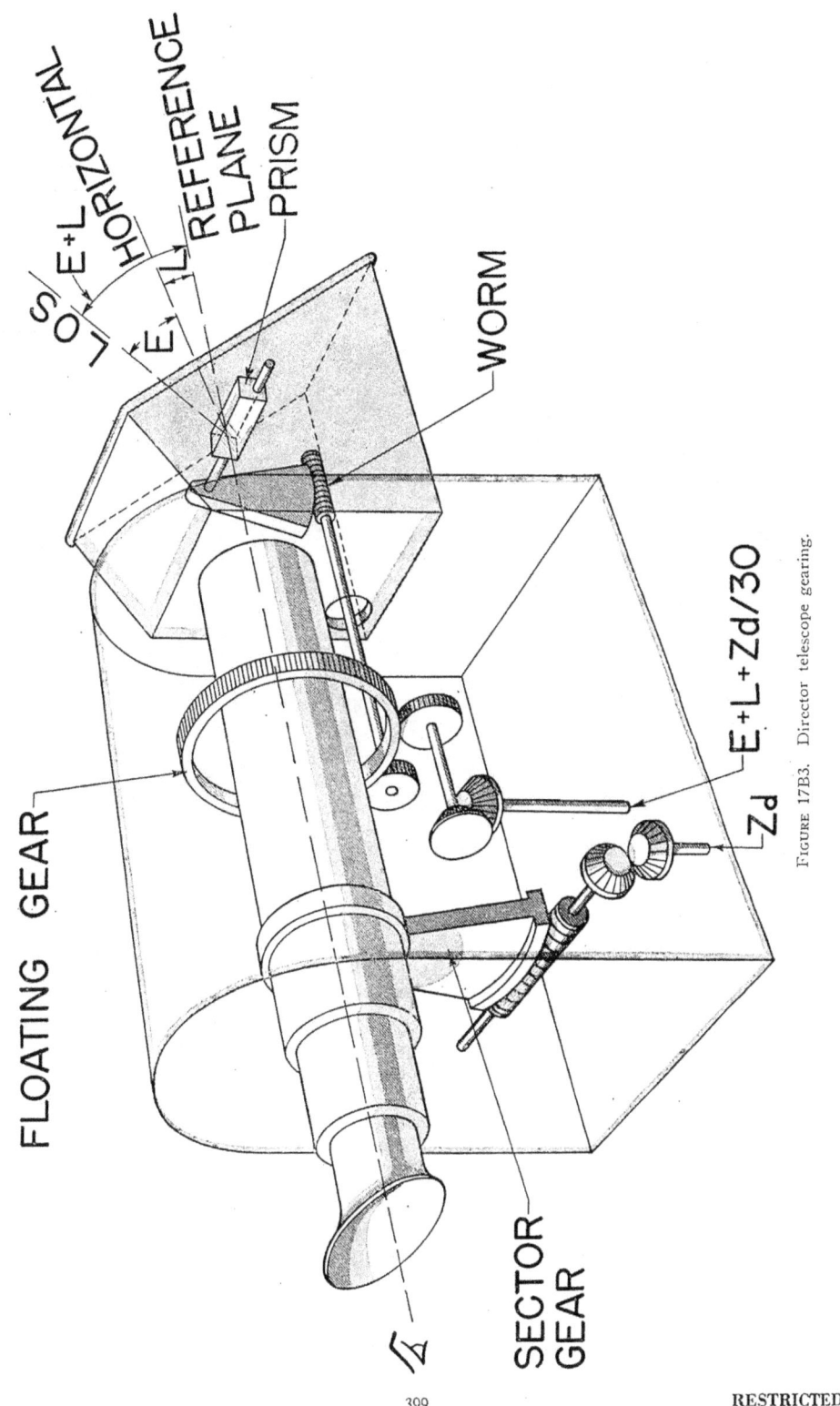

FIGURE 17B3. Director telescope gearing.

DUAL-PURPOSE BATTERY FIRE-CONTROL SYSTEM

The value of Zd is received continuously by the receiver-regulator directly from the stable element by synchro transmission. Stabilization in cross-level is entirely automatic and is independent of train and elevation control. No provision for manual stabilization in cross-level is made. In the event of casualty to the cross-level power drive the range finder is locked at the zero cross-level position and operation of the director is continued without stabilization in cross-level.

Limit switches in the cross-level receiver-regulator operate to cut the power to the electric motor when cross-level exceeds $17\frac{1}{2}°$ on either side of zero. When ship's motion brings Zd back to less than $17\frac{1}{2}°$, the limit switches automatically reclose the power circuit and stabilization in cross-level continues.

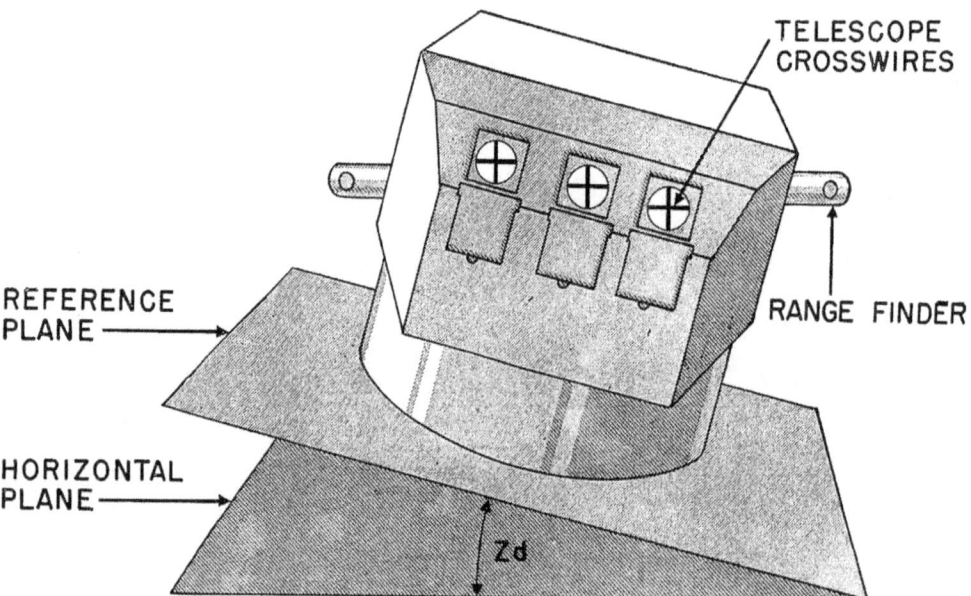

FIGURE 17B4. Director optics; cross-level stabilization.

17B16. Range. Located on the range-finder beam in front of the range finder is a *change-of-range receiver*, which contains a synchro and a servo. Generated range increments ($\triangle cR$), called *range correction* in this instance, is received from the computer by the synchro. The servo amplifies the quantity and drives it through a flexible shaft into a differential in the range finder. The range knob supplies the other input. The differential output, representing the sum of range corrections from both the computer and the range-finder operator, drives a mechanism that turns the range-finder measuring wedges. The differential output shaft also positions synchro generator rotors to transmit observed present range (R) to the computer.

It is the duty of the range-finder operator to see that the measuring wedges are so positioned that the target image and reticle marks are in fusion at all times. If the computer input ($\triangle cR$) is sufficient to do this the operator has nothing to do. If, however, this is not sufficient, the operator must use his range knob to keep the wedges in the proper position.

Deviation from the type of operation described above is made only when necessitated by casualty to the change-of-range receiver or to the $\triangle cR$ transmission system. In such an event the range-finder operator would be required to keep the measuring wedges continually set by hand. In this method of operation present range would still be transmitted automatically to the computer since the range synchro circuit is independent of the change-of-range receiver circuit. Should the present range circuit fail it would then become necessary for the range-finder talker to transmit present range to the computer by phone, such range to be set on the computer by hand.

CHAPTER 17

In the center of the range knob is a *rate-control*, or *signal button* corresponding to the rate-control keys on the pointer's and trainer's left handwheels. If the computer is set for automatic range-rate control, the range-finder operator presses the rate-control button as he turns the range knob to correct the position of the measuring wedges. If the range-rate control is to be handled at the computer, the range-finder operator presses the button to indicate when the transmitted range is correct.

17B17. Spotting. Bolted to the range-finder beam, and with its knob within easy reach of the range-finder operator, is the *range-spot transmitter*. Its handle is geared to synchros which transmit individual and total spots to the spot or battle-order transmitter (see below) and total spot to the computer. Two concentric *dials* are on the face of the instrument; the inner for total, the outer for individual spots. The outer dial and its corresponding synchro are so geared to the knob that they are returned to zero by a spring when the knob is pushed in. A detent restrains the knob at 50 yard increments, permitting the operator to enter a spot by the sense of touch alone.

Located on the control-officer's left is a *battle-order transmitter* on older installations, or a spot transmitter on newer installations. Both instruments are alike as far as spotting facilities are concerned. Each bears three sets of concentric *dials* like those on the range-spot transmitter. The outer ring in each group registers individual spots, the inner dials total spots. Deflection spots are shown on the lower set, elevation spots on the middle set. To the lower-left of each of these two groups is a corresponding spot knob, provided with a detent that clicks for each mil introduced. The upper dial group simply repeats, or indicates the total and individual spots transmitted from the range-spot transmitter for the control-officer's information, and has no knob. Thus, these instruments transmit total elevation and deflection spots and receive total and individual range spots.

At the lower end of the battle-order transmitter are two knobs which control annunciator circuits that actuate signal flags or drums on the fuze-setting indicator-regulators at the guns, and on some of the earlier computers. One knob sets the signals to read the battle orders: *fire, cease,* and *load hoist;* the other sends the shell orders; *illuminating, antiaircraft, common,* and *dive attack.*

The spot transmitter has, in place of the battle and shell-order knobs, three *solution indicators,* which are simply small dials having two white and two black quadrants. They are attached to the shafts of synchros that are actuated by signals from the computer representing the difference between observed and generated range, bearing, and elevation. As long as the computer set-up is correct these dials stand still, but they revolve when the computer solution is wrong.

17B18. Illumination control; searchlights. Searchlight mounts are trained and elevated like guns, and when controlled from a remote station, require searchlight train and elevation orders. These orders may be used to position the lights automatically by power drive, or to provide indications that can be matched manually by the searchlight operators. In surface fire, searchlight train and elevation are essentially the same as the train and elevation of the line of sight, except for the necessary parallax corrections which are introduced at each searchlight mount. This would be true in AA fire if the line of sight were not stabilized in cross-level, but since it is, the searchlight orders must include corrections for tilt of the searchlight trunnions. To afford control flexibility, provision must also be made for the introduction of arbitrary offsets which can be used to shift the light beams as desired.

Searchlight control is centered in the director. In early modifications, all the functions are performed by a single instrument located in front of the control officer. In later installations, there are two instruments: a *searchlight control transmitter* and a *searchlight trunnion-tilt corrector.* The corrector is mounted on the carriage under the control-officer's telescope and receives director train (B'r), director elevation (Eb), and cross-level (Zd) mechanically from their corresponding drives. This corrector computes train and elevation orders which include the trunnion-tilt corrections but not the arbitrary offsets, and transmits the orders electrically to the control transmitter which is mounted on the shield behind the illumination officer.

The illumination officer introduces the arbitrary train and elevation offsets through two knobs on the control transmitter, which adds the corrections to the train and elevation orders and then electrically transmits the final corrected orders to the searchlights. The arbitrary offsets are indicated by two dials actuated by the knobs. The control transmitter has a lever which actuates an annunciator system that transmits the searchlight orders *off close*, *on close*, and *on open* to the searchlight mounts, where they register on visual indicators. In still later directors, the searchlight trunnion-tilt corrector is omitted, and the searchlights are not used against air targets. However, the control transmitter is retained and receives its inputs electrically from transmitters in the director train and elevation receiver-regulators.

17B19. **Illumination control; star-shell fire.** Current installations have a *star-shell computer* attached to the main computer in the plotting room, as described in Section D. This instrument computes elevation, train, and fuze-setting orders for the guns firing star shells. On the director shield to the left of the illumination officer is a star-shell spot transmitter which electrically transmits star-shell range, elevation, and deflection spots to the star-shell computer. The illumination officer can thus control the star-shell fire from the director. The star-shell spot transmitter is similar in appearance to the spot transmitter next to the control officer. It has three pairs of dials for indicating total and individual spots, and three knobs for their introduction.

17B20. **Multiple installations.** On ships where there is but one director for the control of the dual-purpose battery, that director is the horizontal reference point, and its roller path is the reference plane. When there are two or more directors, they must transmit identical values of director train ($B'r$) and director elevation (Eb) for the same point of aim. To make this possible, each director in multiple installations has a *roller-path tilt corrector* and a *horizontal parallax corrector*.

The parallax corrector is in the train receiver-regulator, beneath the carriage. It receives electrically the unit parallax correction for a 100 yard base developed in the computer as explained in Section D. The electrical signal is changed to mechanical movement by a follow-up. The unit parallax correction is changed by a simple gear ratio to a correction applicable to the actual base length between the director and the reference point. This final correction is supplied to a mechanical differential, which also receives the actual director train. The output of this differential is actual director train corrected for horizontal parallax, or director train ($B'r$), and is used to actuate the synchro transmitters in the train receiver-regulator, and to drive the dials in the director-train indicator.

In the event that a casualty to the parallax transmission or to the follow-up mechanism prevents the automatic operation described above, manual standby is available for introducing the correction at the director. Two concentric *dials* are located in the face of the receiver-regulator, the outer one being graduated in terms of the parallax for a 100 yard base and the inner one bearing only an index arrow. The outer dial is driven by the follow-up mechanism, the inner by the parallax synchro receiver. Should the follow-up mechanism fail, a crank can be used to match the zero point on the outer dial against the index on the inner dial. If synchro transmission fails, unit parallax may be received by phone and set on the outer dial against a fixed index. In either case the crank delivers the mechanical motion that is normally provided by the follow-up.

The roller-path tilt corrector is a separate mechanism attached to the carriage. It computes the inclination of the roller path to the reference plane as measured in the sight plane, and delivers this computed correction to the elevation drive through a mechanical differential. The differential combines the roller-patch correction with the line-of-sight elevation above the roller path, and delivers director elevation (Eb) measured from the reference plane. The corrected value (Eb) drives the transmitters in the elevation receiver-regulator and the dials in the director-elevation indicator.

The roller-path inclination in the sight plane is a function of the angle of train from the high point. Consequently, director train ($B'r$) is supplied mechanically to the tilt corrector. The train of the high point and inclination at that point are set on the mechanism at installation. Changes in these settings are required only in the event of a slight shift in the roller path due to settling of the ship's structure.

CHAPTER 17

C. THE MARK 6 STABLE ELEMENT

17C1. Purpose. The instrument described in this section is the Mark 6 stable element, which appears in figure 17C1 and 17C2. Associated with it are a control and follow-up panel, and a motor generator, which are separately mounted in the plotting room. The primary purpose of the unit is to measure level and cross-level angles. It establishes a horizontal plane of reference, and measures the angles according to these definitions:

SYMBOL	DEFINITION
L	*Level angle.* The angle between the horizontal plane and the deck plane measured in the vertical plane containing the LOS.
Zd	*Cross-level angle.* The angle between the horizontal plane and the deck plane measured in a plane which is at right angles both to the vertical plane containing the LOS and to the deck.

The measured values of L and Zd are used in the system to perform the following functions:
1. Stabilize director optics (telescopes and range finder).
2. Compute target elevation (E) from director elevation (Eb).
3. Compute relative target bearing (Br) from director train (B'r).
4. Compute gun trunnion-tilt corrections (Vz and Dz), thereby providing for continuous fire.
5. Compute searchlight trunnion-tilt corrections.
6. Control the firing instant in selected level or selected cross-level fire, either by automatically closing the firing circuit or by indicating visually that the circuit should be closed.

The first five functions are performed at the director or in the computer, while the sixth is handled in the stable element. The electrical circuits pass through terminal tubes at the bottom of the case on the left side, while the mechanical connections to the computer are made through four shafts extending through the right side of the case at the bottom. The inputs and outputs required are:

Outputs:
1. Cross-level (Zd) electrically to director.
2. Cross level (Zd) or selected cross-level (Zdj) mechanically to computer.
3. Level (L) or selected level (Lj) mechanically to computer.
4. Level plus cross-level function (L + Zd/30) mechanically to computer.

Inputs:
1. Director train (B'r) mechanically from computer.
2. Own-ship course (Co) electrically from gyrocompass.
3. Own ship latitude manually before starting.
4. Selected level (Lj) or selected cross-level (Zdj) manually for corresponding type of fire.

17C2. General internal construction. This is a preliminary outline of the internal operation of the stable element. The student is advised to examine briefly figures 17C3, 4 and 5 before proceeding.

The heart of the instrument is a sensitive element, which is the unit that carries the gyroscope on a series of gimbal rings. The entire sensitive element is oriented within the instrument case by the director train input B'r so that level (L) and cross-level (Zd) are measured in and across the sight plane respectively.

The gyro spin axis remains vertical, thereby providing a true horizontal reference from which to measure the values of L and Zd. One of the devices used to keep the spin axis vertical is a latitude-correction weight carried on a horizontal screw attached to the case immediately surrounding the gyro. This weight is positioned on the screw by hand for the latitude of own-ship position. In addition, the screw is kept in the north-south vertical plane by a latitude-correction motor energized by the own-ship course input C.

RESTRICTED

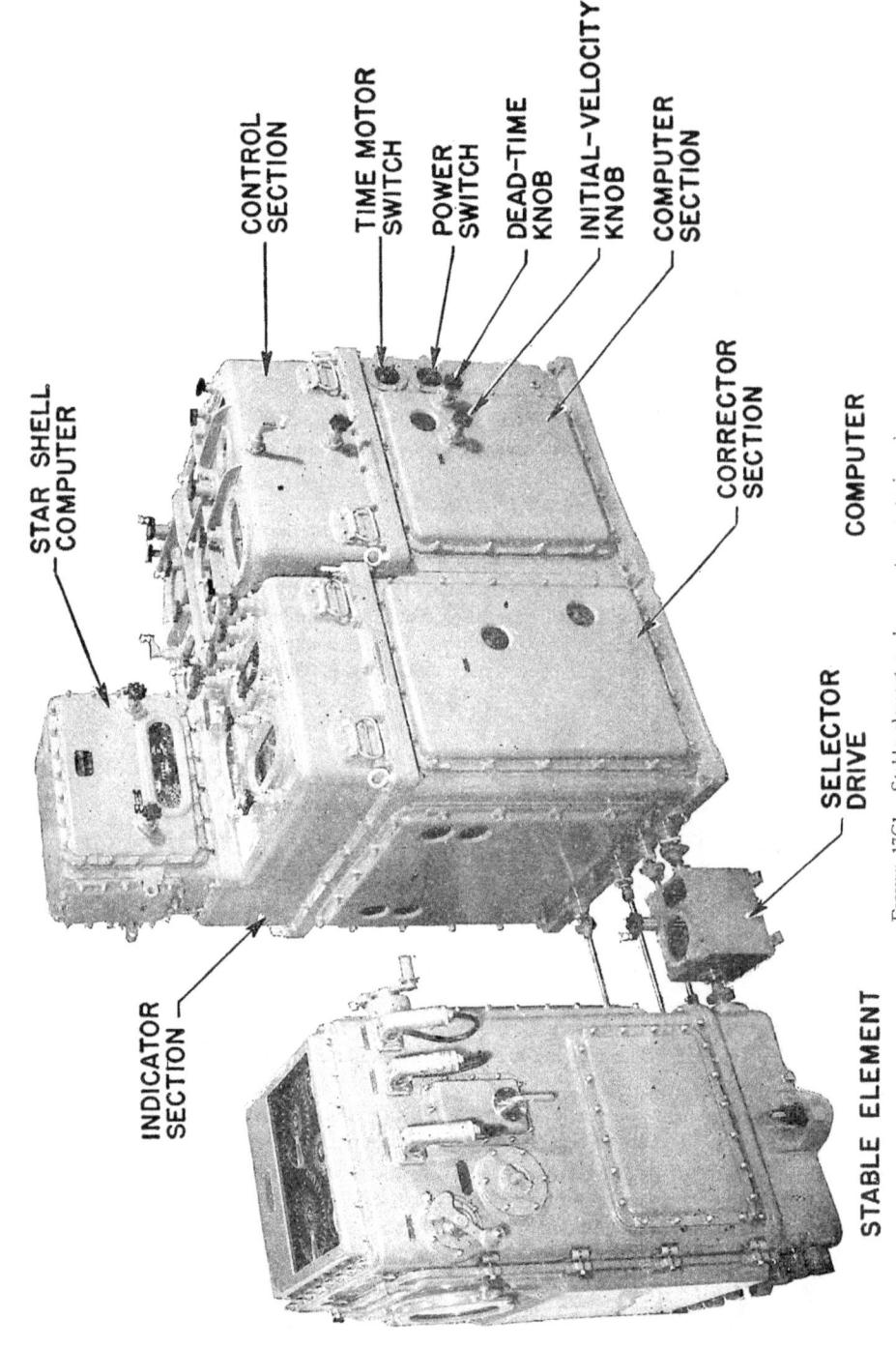

FIGURE 17C1. Stable element and computer; exterior view.

CHAPTER 17

Since the gyro cannot be used to drive transmitters without disturbing its stability, power devices, or follow-ups, are employed to generate the values of L and Zd, and to produce the stable-element outputs. There is a separate follow-up system for level, and another for cross-level, but both are alike in most respects. Each system is driven normally by a follow-up motor, but may be hand operated in case of motor failure.

17C3. Dials. The cranks, switches, keys, and dials are shown in figure 17C2. On top of the instrument are the indicators used by the operators in controlling the outputs.

The large dial group on the left indicates cross-level while the corresponding and identical group on the right registers level. The readings are made against fixed pointers, to the left of the dials, 2,000 minute readings representing zero values of level and cross-level. Each group consists of three dials: the outer ring, or *selected dial;* the intermediate, or *coarse-generated dial;* and the central, or *fine-generated dial.* The selected dial indicates the value transmitted to the computer, while the generated dials give the actual, or measured value. There is an index mark in the blank sector of both the intermediate and outer rings. When the transmitted values are the same as the generated values as in continuous fire, these indexes remain matched and their corresponding dials oscillate back and forth together.

Beneath each dial group is a set of automatic firing contacts. They are attached to the dial mounts which, of course, oscillate with the dials. When the selected and generated dials are matched, the contacts are closed. Thus, in continuous fire, the contacts are always closed, but in selected fire they are closed only when the selected angle equals the actual angle; i.e., when $L = Lj$, or $Zd = Zdj$, as the case may be. These contacts can be used to fire the guns automatically in selected fire, provided the switchboard firing-circuit set-up includes the automatic-firing key mounted on the instrument case. The contact-operating mechanism closes the contacts slightly before the dials arrive at their matched position to provide automatic firing-delay compensation. Adjustment is provided for varying the amount of lead within certain limits.

17C4. Controls. The left and back sides of the instrument have no operating attachments, but are fitted with inspection covers and windows through which the operators make necessary adjustments and observe the sensitive element operation. The cranks and switches are on the front and right sides.

The *level crank* (right side) and *cross-level crank* (front) are used (1) in manual follow-up to generate their respective angles, and (2) in selected fire to set the value of Lj or Zdj. Each is connected to its corresponding shafting by a clutch. These cranks cannot be used for manual follow-up and selected fire at the same time, since they must stand still to maintain a selected-angle setting, but must rotate continuously back and forth when used for follow-up. However, one crank may operate in manual while the other is being used in selected fire.

The *level follow-up switch* (right side) and *cross-level follow-up switch* (front) have two positions each: *automatic* and *manual.* When set on automatic, they energize their follow-up motors and disengage the crank clutches. On manual, they deenergize the motors, engage the cranks, and energize the galvanometers. One may be an automatic while the other is on manual.

The *selector switch,* the lowest lever on the front, has three positions: *level fire, continuous fire* and *cross-level fire.* It controls the crank clutches (this is in addition to the clutch control provided by the follow-up switches), and two additional clutches in the level and cross-level shafting between the generated and selected dials. Setting this switch for either selected method automatically engages the corresponding crank so that the latter may be used to select the angle. When level is generated by manual follow-up, this switch must be set for continuous or cross-level fire. Similarly, it must be set for continuous or level fire when manual follow-up is employed for cross-level.

17C5. Salvo signal and firing keys. There are three similar pistol-grip keys on the front of the instrument. Their circuits all clear through the fire-control switchboard where they may be cut in or out as necessary. The *salvo-signal key* is used by one of the stable-element operators to sound throughout the system a warning that the plotting room is ready, and that firing is about to begin.

DUAL-PURPOSE BATTERY FIRE-CONTROL SYSTEM

The *automatic-firing key* in the middle is used in either of the two selected-firing methods. Its circuit passes through the selector switch. Turning the switch to level fire connects the key in series with the contacts on the level dials, while the cross-level fire setting routes the circuit through

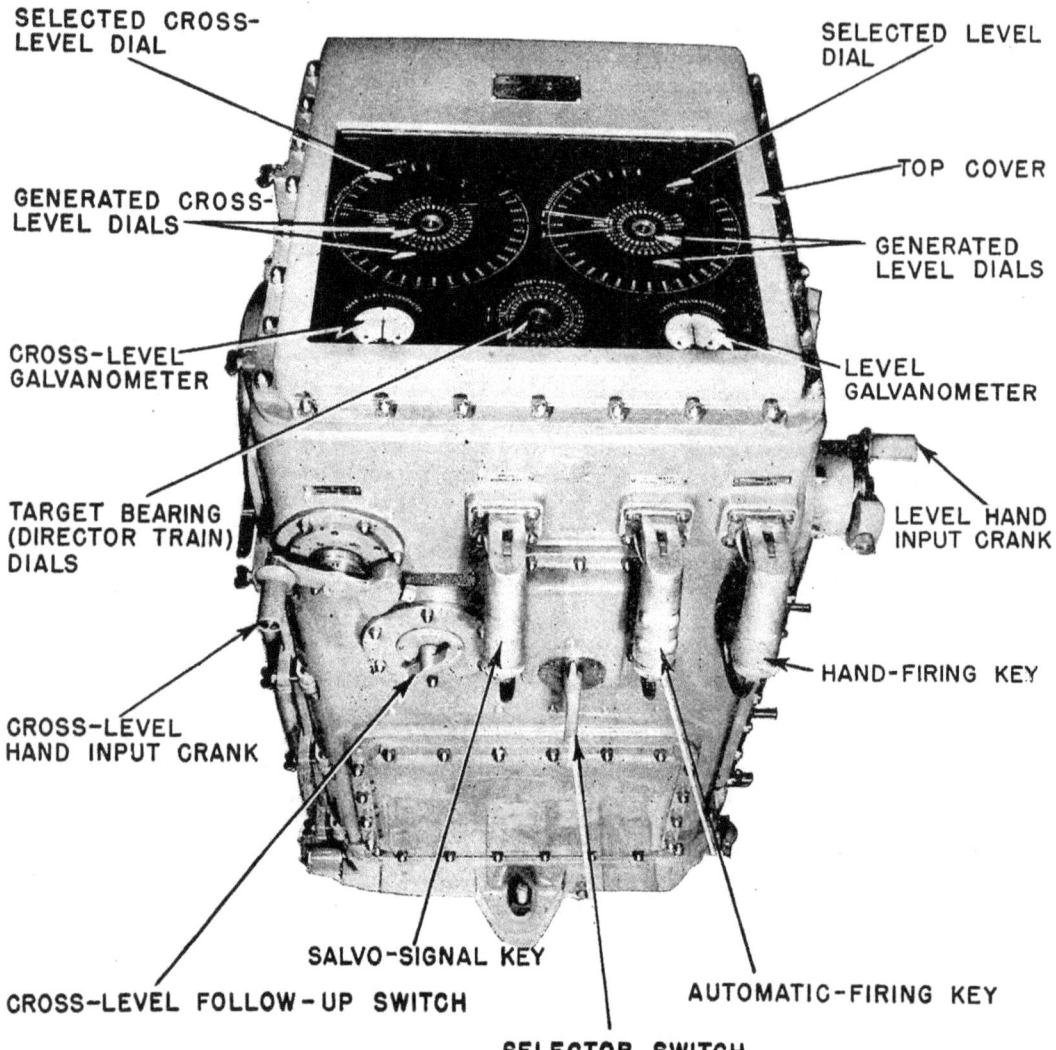

FIGURE 17C2. Stable element; top and front view.

the cross-level dial contacts. When the selector switch is set for continuous fire, the automatic key circuit is open and connected to neither set of dial contents. It is therefore necessary to cut out this key at the switchboard, or to close the hand-firing key at the stable element before the guns can fire.

The *hand-firing key* at the right has no electrical connections to the stable element. It is available for use in the selected firing methods if the automatic-firing key fails, and may be used as an additional break in the firing circuit in continuous fire if the necessary switchboard set-up is made.

CHAPTER 17

If this key is switched into the firing circuit in continuous fire, the stable-element operators can break the firing circuit in the event of plotting-room-equipment failure.

17C6. Operating the instrument. The normal assignment of personnel to the stable element is three men. One of them (not shown in fig. 17D4) attends the control and follow-up panel, where he supervises the control of gyro and follow-up power supplies, and checks on the proper receipt of electrical inputs. His principal activities are concerned with starting and securing, as outlined in a subsequent article. In general, the gyro must be started at least 10 minutes before it is to be used so that it will be running at normal speed and settled in the vertical. When the ship is in combat areas, the gyro is kept running at all times.

The other two men (see fig. 17D4) are the level and cross-level operators, stationed at their respective front corners of the instrument where they may conveniently reach the switches, keys and cranks. Their duties are outlined in the following discussions of follow-up and firing methods.

17C7. Follow-up methods. The normal and desired follow-up action is automatic. Each operator sets his follow-up switch on automatic, and the follow-up motors continuously generate values of level (L) and cross-level (Zd). If automatic follow-up fails in level, for example, the level operator turns his follow-up switch to manual. He then watches the level galvanometer, and turns the level crank back and forth as required to keep the galvanometer needle centered, or on the zero point. This operation produces the same effect in the system as automatic follow-up. The instrument outputs remain the same, unless motor failure occurs when selected-level fire is in use. In that case, the firing method must be changed to continuous or cross-level so that the level crank can be used for follow-up. Manual operation for cross-level is accomplished similarly by the cross-level operator.

Manual follow-up is a standby for follow-up motor failure (not loss of power), for it cannot be employed if no energy is available to actuate the galvanometers. Total power failure renders the stable element inoperative, since the gyro itself is electrically driven.

17C8. Continuous fire. The continuous-fire method is used against both air and surface targets. After making certain that the generated and selected dials of both groups are matched, one of the operators sets the selector switch on continuous fire, thereby locking each coarse-generated dial to its corresponding selected dial. The dial groups then oscillate in unison, and serve no purpose other than to inform the operators that instantaneous values of level (L) and cross-level (Zd) are entering the computer continuously.

The gun orders developed by the computer are therefore continuously correct, the guns are invariably properly positioned with respect to the line of sight and to the earth, and may be fired at any time. Firing keys throughout the system are pressed and generally locked, and each gun fires as its breech closes. As soon as the plotting room is ready, one of the stable-element operators sounds the salvo signal, and then closes the hand-firing key if it is included in the firing circuit. If automatic follow-up is in use for both level and cross-level, the operators have no further duties other than to observe and check the operation.

17C9. Selected Fire. The level-and cross-level firing methods are used only for surface fire and are identical as far as operation is concerned. The description here is for selected-level fire, but it applies to selected cross-level fire if cross-level is substituted for level, and vice versa, where appropriate. In this method, the selector switch is set on level fire. This releases the clutch between the generated and selected-level dials, and ties the level crank to the selected dial and to the level output shaft to the computer. The generated dials continue to indicate instantaneous values of L, and the cross-level dials operate as in continuous fire. The cross-level operator functions as in continuous fire, but the level operator turns the level crank to set the selected-level dial and drive the selected value Lj into the computer. His selection may *split the roll*, but not necessarily; i.e., it may be a value at about the half-way mark between the oscillation limits of the coarse-generated dial. Splitting the roll insures that the actual value will match the selected value twice on each roll cycle, even though the roll amplitude may vary.

When the operator is ready, he closes the automatic-firing key. Each time the actual or gener-

DUAL-PURPOSE BATTERY FIRE-CONTROL SYSTEM

ated value of L equals the selected value Lj, the level dial readings are the same, their small indexes are matched, and the automatic contacts come together, firing a salvo, provided all other breaks in the circuit are closed. If the automatic contacts fail, the level operator uses the hand-firing key, closing it an instant before the dial indexes meet to provide the firing-delay compensation.

In either level or cross-level fire, the computer continuously sends out gun orders, but they are computed on the basis of the selected angle (Lj or Zdj). The guns are therefore in the correct

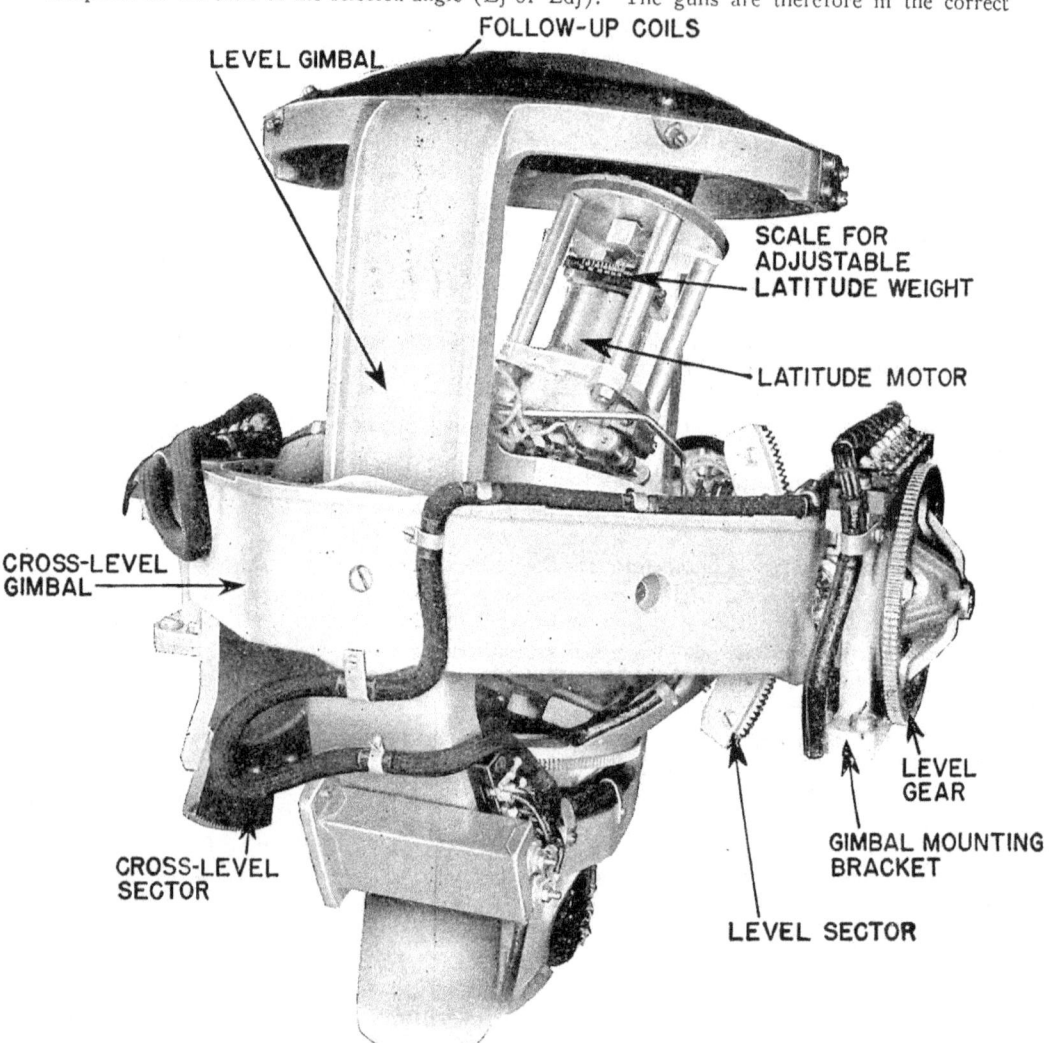

FIGURE 17C3. Sensitive element of a stable element.

firing position only when the ship roll and pitch make the actual angle (L or Zd) equal to the selected value. The director optics are stabilized in selected fire, just as in continuous fire, by the electrical Zd output and the mechanical $L + Zd/30$ output, of which neither is affected by the introduction of selected angles.

17C10. **Functional operation.** The preceding articles have outlined the purpose and methods of stable element operation. The internal mechanism that provides the instrument outputs is the subject of the following articles. The mechanical elements are relatively simple, but the electrical controls—especially those of the follow-up systems—are too complex for complete discussion in this text. The presentation is therefore limited in its technical aspects to a general functional description which should provide the student with a better understanding of gyro principles and their applications in stable elements.

In order to segregate the subject matter, the descriptions are broken into these five groups:
1. The sensitive element.
2. The mechanical transmissions.
3. The follow-up control systems.
4. The theory, construction, and control of the gyro proper.
5. The motor generator and control, and the follow-up panel.

17C11. Sensitive element. As considered in this text, the sensitive element, illustrated in figures 17C3 and 17C4, consists of seven principal parts: (1) *gyro*, (2) *gyro case*, (3) *gimbal ring*, (4) *rotating fork*, (5) *level ring*, (6) *cross-level ring*, and (7) *training gear*. Briefly the gyro spins within the gyro case, which is supported within the gimbal ring. The latter in turn is carried between the arms of the rotating fork, which revolves within the level ring. The cross-level ring, surrounding and supporting the level ring, is mounted on the training gear. The assembly of the first four parts is the truly sensitive, or vertical-indicating group. The last three parts provide the means for measuring level (L) and cross-level (Zd), and may be considered as the measuring group.

17C12. Vertical-indicating group. The gyro wheel is carried on an axle supported by ball bearings in the upper and lower ends of the gyro case. The wheel and case form the rotor and stator of a high-frequency, two-pole, squirrel-cage induction motor. The squirrel-cage winding, consisting of solid conducting bars, is in the gyro wheel, while the wire stator windings are in the gyro case. The theoretically greatest speed attainable by an induction motor operating on 60 cycle power is 3,600 r.p.m.; too low for efficient gyro operation. Consequently, a separate motor generator is provided to supply three phase, 70 volt, 146 cycle power which drives the gyro at about 8,000 r.p.m.

The gyro case pivots within the gimbal ring on the *case axis* (see fig. 17C4), which is perpendicular to the spin axis. The gimbal ring in turn is supported between the arms of the rotating fork on the *gimbal axis*, which is perpendicular to the case axis. This axis arrangement provides a universal mounting that gives the gyro three degrees of freedom and allows it to spin on a vertical axis even though the rotating fork may vary its position from time to time. The vertical position of the spin axis provides the equivalent of a horizontal reference from which to measure the deck inclination to the horizontal in and across the sight plane; i.e., level and cross-level.

The rotating fork rests on bearings in the lower end of the level ring. The *gimbal rotation motor*, attached to the level ring, drives the fork at about 18 r.p.m. through the ring gear visible near the lower end of the level ring. The fork, of course, turns the gimbal ring and gyro case with it. The axis of fork rotation is fixed with respect to the level ring, being vertical when the level ring is vertical. This rotation of the vertical-indicating group is part of the means for causing the gyro to seek the true vertical, as is explained in subsequent articles. The electrical connections to the vertical-indicating group are made through collector rings suspended from the fork, on the underside of the level ring.

17C13. Measuring group. The level ring, supporting the entire vertical-indicating group, is carried within the cross-level ring on the *level-axis*. The cross-level ring, pivoted on the *cross-level axis*, is supported by two *gimbal mounting brackets*. These brackets are secured to the training gear, which is a circular part with a ring gear on its periphery.

The training gear rests on roller bearings attached to the stable-element case, and is free to rotate in a plane parallel to the top of the instrument. When the stable element is installed, the case is leveled so that its top, and therefore the training gear are parallel to the deck or reference plane. This means that level and cross-level angles can be measured between the plane of training-gear rotation, or the gear itself, and the horizontal datum established by the gyro.

The training gear is oriented by the mechanical director train input (B′r) received from the computer. This turns the sensitive element so that the cross-level axis is in the sight plane, and the level axis is perpendicular to the sight plane.

When the deck plane is horizontal, the level and cross-level axes are also horizontal, and level (L) and cross-level (Zd) are both equal to zero. Under this condition, the center of the umbrella-shaped piece on the upper end of the level ring is in line with the gyro spin axis. When the deck plane tilts, the spin axis remains vertical, but the umbrella center is displaced from it by the angle L in the sight plane and the angle Zd across the sight plane.

The cross-level axis, being secured to the training gear, is parallel to the deck. The cross-level

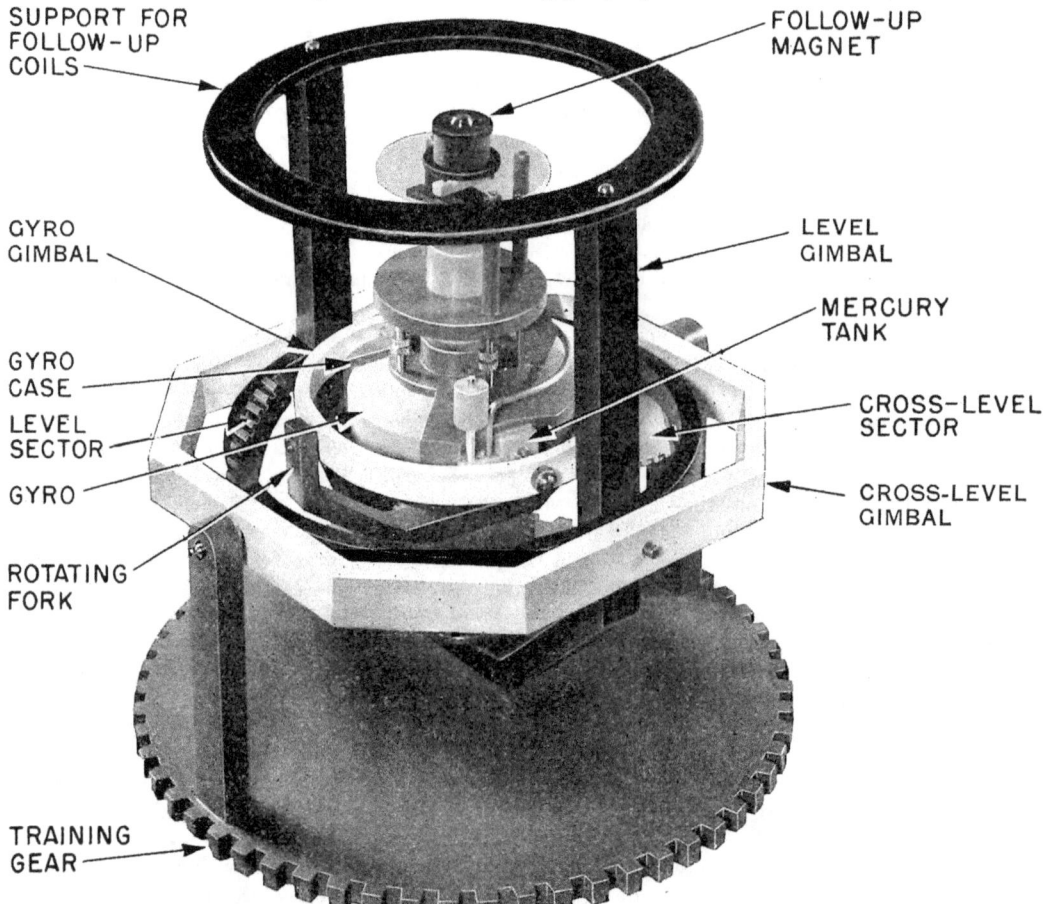

FIGURE 17C4. Gyro mounting; photo of a mockup.

follow-up system drives the cross-level ring about this axis until the umbrella is centered in cross-level; i.e., until its center is in the sight plane. This cross-level ring movement occurs in a plane perpendicular to the deck, and is therefore equal to Zd. When the umbrella is centered in cross-level, the level axis is horizontal. The level follow-up system drives the level ring about the level axis to center the umbrella in level; hence the level ring moves in the sight plane through the angle L.

Actually, both follow-up systems operate simultaneously as soon as any displacement begins. They do not wait for deck-plane motion to cease, but generate continuous values of L and Zd. In other words, at any instant the existing displacements of the level and cross-level rings from

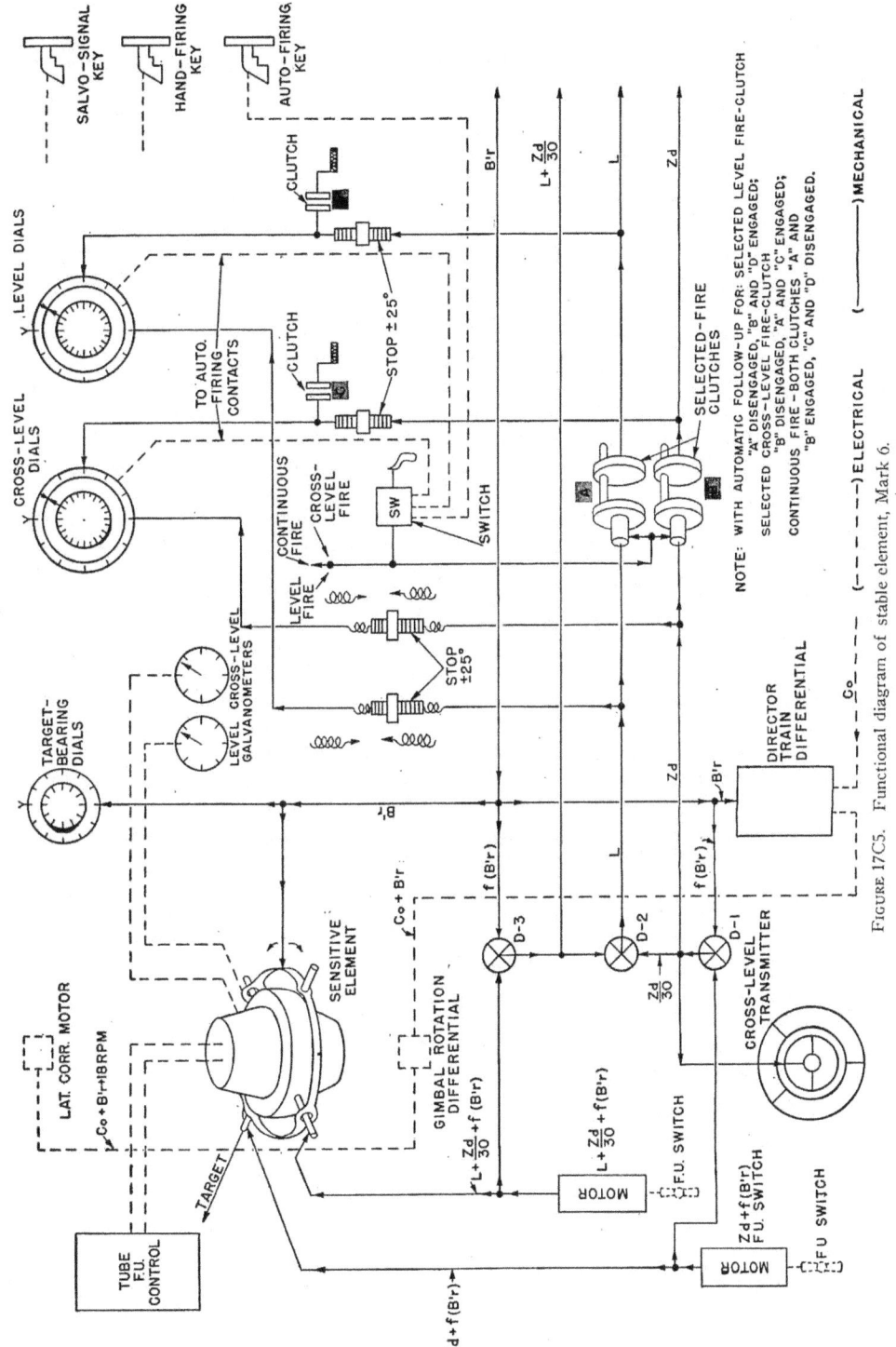

Figure 17C5. Functional diagram of stable element, Mark 6.

411

RESTRICTED

the positions they occupy with respect to the training gear when the deck plane is horizontal are the instantaneous values of L and Zd. It should be apparent, then, that the movement of the shafts driving these rings can provide mechanical measures of L and Zd.

17C14. Director-train transmission. The relations of the various mechanisms associated with the sensitive element are shown schematically in figure 17C5. The outer ring of the sensitive-element group represents the cross-level ring, its axis being held in the sight plane by director train (B'r). The housing surrounding the inner gyro mechanism represents the level ring, its axis being kept perpendicular to the sight plane by the cross-level ring. The quantity B'r also drives the target bearing dials on top of the instrument, and is used in determining cross-level (Zd), level plus cross-level function (L + Zd/30), and in holding the latitude-correction weight in the north-south vertical plane.

17C15. Cross-level transmission. The cross-level ring is connected to its follow-up motor by a gear train and shafting which may be driven either by the motor or by hand. The drive passes through the training gear, and when the latter turns, a pinion in the gear train "walks" on another gear in a manner similar to the walking of the pinion on the floating gear in the director telescopes. This walking is a function of director train, f(B'r), similar in nature to the cross-level function (Zd/30). The follow-up motor must therefore supply the quantity Zd + f(B'r) to the shaft driving the cross-level ring.

Director train (B'r) is changed by a gear ratio to f(B'r) and subtracted from Zd + f(B'r) by differential D-1 to obtain Zd. The differential output drives the cross-level transmitter which sends Zd to the director. The Zd line also goes to the generated cross-level dials, and through clutch B to the selected cross-level dial and to the computer. In continuous or level fire, clutch B is engaged, the generated and selected dials move in unison, and the instantaneous value of Zd is supplied to the computer. In cross-level fire, clutch B is disengaged and clutch C automatically ties the cross-level crank to the selected-dial shafting. Then the computer input becomes a selected value (Zdj), controlled by crank movement and indicated by the selected dial. However, the cross-level transmitter still sends the instantaneous value of Zd to the director. In manual cross-level follow-up, clutches B and C are both engaged, and the crank drives Zd into the line. This manual input feeds back through differential D-1 where it is added to f(B'r) to form the quantity Zd + f(B'r) which drives the cross-level ring shafting and also idles the follow-up motor.

17C16. Level transmission. The level gearing is similar in many respects to the cross-level gearing. However, in addition to passing through the training gear, it also passes through the cross-level ring. Consequently, cross-level ring motion causes walking of a pinion in the level drive, and shifts the level ring by a cross-level function which must be compensated by rotation of the level shafting. The gear ratio used makes this function equal to Zd/30 to match the function required at the telescopes. The level follow-up must therefore supply the quantity L + Zd/30 + f(B'r) to keep the level ring vertical.

Differential D-3 removes f(B'r) from this expression to obtain L + Zd/30. This is one of the required outputs, and is transmitted directly to the computer.

The function Zd/30 is obtained from Zd by a gear ratio and supplied to differential D-2 which subtracts it from L + Zd/30. The output L goes to the generated-level dials, and through clutch A to the selected-level dial and to the computer. In continuous or cross-level fire, clutch A is engaged, the generated and selected dials are locked together, and the instantaneous value of L goes to the computer. In level fire, clutch A is disengaged and clutch D automatically ties the level crank to the selected dial and computer. The computer input is then selected level (Lj). In manual level follow-up, clutches A and D are both engaged. The crank drives the value of L into the computer, and back through D-2 where it is added to Zd/30. Differential D-3 adds in f(B'r), and the quantity L + Zd/30 + f(B'r) drives the level ring shafting, idling the follow-up motor as in the case of cross-level.

In the shafting to the generated and selected dials of each system are limit stops which mechanically prevent the generation and transmission of angles exceeding 25°. In the stops on the generated

CHAPTER 17

side are electrical contacts which stop the follow-up motors shortly before the mechanical stops are reached. When ship roll and pitch produce level and cross-level angles great enough to cause the stops to operate, the stable element is shut down to prevent damage to the mechanism.

17C17. Follow-up signals. The primary function of the follow-up control system is to operate the motors automatically so that they will continuously hold the umbrella center on the spin axis. The secondary function is to actuate the galvanometers, thereby providing visual indication of the crank motion required to accomplish the same result. The control system centers about two follow-up coils and a follow-up magnet.

The level follow-up coil is imbedded in grooves on the upper side of the umbrella, while the cross-level follow-up coil is similarly carried on the underside. In effect, each coil is wound in the shape of a figure 8, the turns being clockwise in one loop and counterclockwise in the other. The line joining the loop centers of the level coil is at right angles to the level axis, which means that it lies in the sight plane when the training gear is oriented. The corresponding line of the cross-level coil is at right angles to the cross-level axis, and therefore lies across the sight plane. The cross bar or center of each figure 8 is on the umbrella center.

The follow-up magnet is mounted on top of the gyro case, as shown in figure 17C4, so that its axis coincides with the spin axis. The magnet is energized with alternating current which sets up an alternating magnet field. This field induces voltages in the follow-up coils. The actions of the level and cross-level follow-up controls are the same, hence the balance of this discussion is confined to the level system.

When the umbrella center is on the spin axis, the voltages induced in the two loops of the level coil are equal in magnitude. Since the loops are wound in opposite directions, their voltages at any instant have opposite polarity, and their resultant is zero. When a change in level angle occurs, the umbrella center moves off the spin axis in the sight plane. The magnet is then closer to one loop than to the other, and induces a higher voltage in the first loop than in the second. The resultant voltage across the coil is no longer zero, but has magnitude and direction that provide an indication of the follow-up action required to return the umbrella center to its neutral position on the spin axis.

17C18. Manual follow-up. The difference between automatic and manual follow-up lies in the methods of employing the follow-up coil voltage. When the level follow-up switch is set on manual, it connects the level coil directly to the level galvanometer. Any voltage applied to the galvanometer displaces its needle from zero, the direction of the displacement indicating the required direction of crank rotation. The operator then turns the level crank, thereby driving the level ring, until the umbrella center is back on the spin axis. This brings the net level-coil voltage back to zero, and centers the galvanometer needle.

17C19. Automatic follow-up. When the level follow-up switch is set on automatic, it connects the level coil to the level follow-up units on the control and follow-up panel. At the panel, the coil voltage is amplified and used to control rectifier tubes which supply direct current to the level follow-up motor. The direction of the current flow determines the direction of motor rotation, and is made such that the motor drives the level ring in the direction that will place the umbrella center back on the spin axis.

The response is made sufficiently sensitive to cause the umbrella center to *remain* very nearly in line with the magnet. In other words, the follow-up motor functions practically as soon as displacement begins, so that the generation or measurement of L is continuous. If the motor were permitted to drive until the signal voltage became zero, it would coast due to its inertia and actually drive the level ring slightly beyond the neutral position. Then the follow-up control would operate to run the motor in the reverse direction, and it would again coast beyond the desired position. To prevent this action, which is called *hunting,* an anti-hunt unit is included in the controls at the panel. It operates to shut off the power slightly before the neutral position is reached. A sensitivity control is provided to permit adjustment of this action to suit the requirements of individual ships.

17C20. Gyro principle. Two properties peculiar to a rapidly spinning mass such as the gyro wheel in the stable element are *gyroscopic inertia* and *precession*. These characteristics are not clearly evident unless the wheel is supported in a universal mounting which allows the wheel to revolve in any direction about its center of gravity.

Gyroscopic inertia is the tendency of the gyroscope to maintain its rotation plane. That is, the wheel resists any force tending to rotate it in a plane other than that in which it is spinning. Of course, this is equivalent to saying that a gyro tends to keep its spin axis pointing in the same direction in *space*.

Precession is the spin-axis movement occurring when an applied external force tends to alter the rotation plane. An effort to turn the spin axis about a second axis results in spin-axis rotation or precession about a third axis perpendicular to the second, and not about the second axis as one might suppose offhand. Stated another way, pressure at a point on the wheel rim will cause the point 90° away in the direction of spin to move in the direction of the pressure. The precession will continue until the applied force no longer tends to disturb the rotation plane.

As an illustration of this action, consider the stable element gyro, which spins clockwise (viewed from above) about a vertical axis. If one presses down on the north side of the gyro case, that side will not move, but the east side will go down. In this instance, it would appear that the pressure would cause the spin axis to turn about an east-west axis, whereas it actually precesses about a north-south axis.

17C21. Latitude correction. A vertical line, being fixed with respect to the earth, is continuously changing its direction in space. Such a line through the equator evidently revolves 360° in space about the earth's north-south axis every 24 hours, while a similar line through the poles is relatively fixed. It is evident that the rate at which the line changes its space direction depends upon its latitude. Since a free gyro keeps its spin axis fixed in space, it would not indicate true vertical for any extended period of time even if initially set in that position, but would appear to revolve about a north-south axis once in every 24 hours.

This effect of earth rotation is compensated by the latitude weight previously mentioned. The weight is simply a knurled nut. It is carried on a screw attached at right angles to the rotor shaft of the latitude motor shown in figure 17C3. The motor itself is a synchro unit, mounted on the upper end of the gyro case beneath the follow-up magnet so that the rotor axis coincides with the spin axis. Since the latter is vertical, the screw is horizontal and can be kept in a north-south direction by proper rotation of the latitude motor rotor.

The latitude weight applies a torque to the spin axis about an east-west axis; its effect is equivalent to a downward pressure on the north side of the gyro case. This torque causes the spin axis to precess about a north-south axis so that the east side of the gyro case moves down. This turns the spin axis in the same direction in space as earth rotation turns the true vertical. The value of the applied torque is adjusted by shifting the weight toward or away from the spin axis by hand according to a latitude scale alongside the screw. When the weight is properly set, it causes the spin axis to change its space direction at the same rate as a vertical line through the point occupied by the gyro. Theoretically, if the spin axis were initially set in the true vertical, the latitude weight would keep it there.

17C22. Latitude motor input. The latitude motor stator is attached to the gyro case. Consequently, the stator revolves within the level ring at 18 r.p.m. In addition, rotation of the training gear turns the stator. The electrical input to the motor positions the rotor with respect to the stator. Consequently, the electrical imput must be own-ship course (Co) corrected for training gear and gimbal rotation.

These corrections are made by two differential generators, shown schematically in figure 17C5. These units are similar in construction and appearance to a synchro generator. However, the rotors have windings similar to the stator windings, and have three leads. The differential stator receives an electrical input, and the rotor is turned by a mechanical input. The rotor output is an electrical signal representing the sum (or difference) of the electrical and mechanical inputs.

CHAPTER 17

The electrical input Co goes to the *director train differential*, mounted in the base of the instrument, while the mechanical input B'r drives the rotor. The electrical sum, Co + B'r goes to the stator of the *gimbal rotation differential* mounted on the level frame. The rotor of this differential is driven by the gimbal rotation motor at the same speed as the rotating fork, so that the 18 r.p.m. gimbal rotation is continuously added to Co + B'r. The output to the latitude motor stator, Co + B'r + 18 r.p.m., holds the latitude motor rotor stationary with the screw in the north-south direction.

17C23. Mercury tanks. A gyro that would behave as outlined in the preceding paragraphs would have to have absolutely frictionless bearings, perfect balance, and exact latitude correction. Furthermore, it would be necessary to place the spin axis in the true vertical by hand, an exceedingly difficult operation to perform accurately. Since there is friction in the bearings, ship motion transmits forces to the gyro tending to cause precession of an irregular character. In addition, any misadjustment of the latitude weight will cause precession about a north-south axis at a rate different from that required.

A mercury control system is provided to eliminate error from irregularities, and to cause the gyro to seek the true vertical automatically. Two mercury tanks, partially filled with mercury, are mounted on the gyro case diametrically opposite each other—one is shown in figure 17C4. They are connected at their lower ends by a mercury-flow tube and at their tops by an equalizing tube for air flow. At one end of the mercury-flow tube is a cut-out valve, actuated by an electromagnet.

When the spin axis is vertical, there are equal quantities of mercury in the two tanks and their weights balance each other. When the spin axis tilts, one of the tanks is higher than the other and mercury flows from the high one to the low. The difference in weight results in a force that tends to precess the spin axis. The pipes and orifices in the system are designed to retard the mercury flow so that the 18 r.p.m. gimbal rotation causes the effective mercury displacement to act at a point about 90° from the low point in the direction of gyro spin. Downward pressure at this point results in spin-axis movement toward the vertical. Actually, the mercury flow is gradual, and produces precession that has components tending to move the axis both away from and toward the vertical at the same time. The controlled lag in mercury flow results in righting components which have a greater net effect than the disturbing components, and the spin axis approaches the vertical in a spiralling motion. The righting forces are at work regardless of the cause of gyro displacement. Hence the spin axis will seek the vertical even though it is inclined at starting, and slight misadjustment of the latitude weight is compensated automatically.

Sharp turns and sudden speed changes will cause irregular flow of the mercury, tending to disturb the gyro rather than to right it. To prevent mercury flow during such changes in ship motion, the mercury cut-out valve closes, blocking the connecting tube. The valve is operated electrically by a cut-out control on the control and follow-up panel. During starting, the valve is made inoperative, since ship-motion changes are transitory and have no appreciable affect upon the settling of the gyro. After the gyro is settled, the valve is connected to the automatic control.

17C24. Motor generator. The motor generator is mounted in a convenient place in the plotting room. In external appearance it is similar to an elongated motor with no drive shaft extension. The motor and generator rotors are on a common shaft revolving within a single stator which has two sets of windings. The motor windings are connected to the regular 115 volt, 60 cycle ship supply. The generator windings deliver the 146 cycle power to the stable element through the control and follow-up panel. The size of the motor generator depends on the number of gyros to be driven in a given installation. Some are built to run one gyro, some four, and others six. Each control panel is wired to two motor generators so that the second may be used in case of failure of the first.

17C25. Control and follow-up panel. The control and follow-up panel supports the various parts of the follow-up control systems, the mercury cut-out control, pilot lights, switches, and fuzes required to operate and protect the instrument. The mercury cut-out control consists essentially of a pair of contacts controlled by the combined action of an own-ship speed motor and an own-ship course motor. These synchros close the contacts, thereby shutting the cut-out valve, whenever the rate of change of speed, the turning rate, or a combination of both exceed predetermined limits. Plotting rooms containing a single stable element have one of the modifications of the Mark 7 control and follow-up panel, which has one control section and two follow-up sections, one for level, the other for cross-level. Plotting rooms with two stable elements have one of the Mark 8 modifications which have a single control section and four follow-up sections, two for each stable element.

17C26. Starting the instrument. An indication of the devices on the control section is provided by the following sequence of starting operations:

1. Remove the sensitive element clamp and loosen the crank clamping screws. (When securing the instrument, after all circuits have been deenergized at the control panel, this step should be reversed. The cranks are clamped to prevent feed-back from the computer when the latter is in use and the stable element is not.)

2. On the control panel, turn the mercury control switch to *on*, which keeps the cut-out valve open. All other switches should be in their off positions.

3. Check that no fuses are blown.

4. Check that all the green power indicator lights are lit.

5. At the stable element, turn the selector switch to *continuous fire* and the follow-up switches to *automatic*.

6. Set the latitude weight.

7. Turn the gyro supply switch to the desired motor generator. (On Mark 8 panels this step is automatically accomplished with step 8.)

8. Turn the follow-up system switch to the filament-lighting position to heat tube filaments in follow-up circuits.

9. Using the gyro current switch and its associated ammeter, check the gyro-starting current in each phase from time to time until the gyro reaches normal speed (about 10 minutes). Keep the current on the meter only long enough to obtain a reading during starting to avoid overloading the meter while the needle is off scale. The normal operating current is 1.1 amperes in each phase. When the meter reads this value, it provides an indication that the gyro is up to speed.

10. See that the latitude motor is energized and that the screw heads north.

11. Turn the follow-up system switch to on, preferably after the gyro is up to speed, but not sooner than two minutes after the filaments are lighted.

12. Turn the mercury control switch to *automatic* after the gyro is up to speed. Until the gyro is very nearly in its settled condition, course and speed changes have a negligible effect on the mercury system operation.

D. THE MARK 1 COMPUTER

17D1. Function. This section presents descriptions of the Mark 1, Mod. 7 computer (serial numbers above 215) and the Mark 1 star-shell computer which sometimes is made an integral part of the main computer. When the word *computer* is used alone, it refers to the main computer illustrated in figure 17C1. The auxiliary unit is invariably designated *star-shell computer*, and is the small box-like element mounted on top of the main computer. It is described in subsequent articles. Computer functions are to:

1. Aid the director personnel in keeping the director line of sight on the target for AA or surface fire.

2. Permit the 5"/38 cal. guns to be pointed and trained, and fuzes and sights to be set continuously for AA or surface fire against a common target.

3. Permit, as an alternate to (2), the main-battery guns to be pointed and trained, and sights to be set for AA fire at a common target, using a selected value of fuze setting. (The method of employing the computer for the control of main-battery AA fire is not described here, but may be found in Bureau of Ordnance publications.)

CHAPTER 17

4. Permit (when equipped with the star-shall computer) guns of the 5"/38 cal. battery to be pointed and trained and fuzes to be set continuously for star-shell illumination of a surface target.

17D2. Outputs. In the interest of equipment standardization, the computer is designed to make nearly all the required calculations that can be computed centrally for 5"/38 cal. dual-purpose batteries, and may be used on any type of ship. The distribution of most of the outputs is represented graphically in figure 17D1. Provision is made for the computation and transmission of the following quantities, but not all of them are utilized in every installation:

Electrically to the guns:
1. Gun-train order (B'gr)—automatic.
2. Gun-train order (B'gr)—indicating.
3. Gun-elevation order (E'g)—automatic.
4. Gun-elevation order (E'g)—indicating.
5. Train parallax for a 100 yard horizontal base (Ph). (This quantity is also sent to directors in multi-director installations.)
6. Elevation parallax for a 100 yard horizontal base (Pv).
7. Fuze-setting order (F) for mechanical fuzes.
8. Sight deflection (Ds) for single mounts.
9. Sight deflection (Ds) for twin mounts.
10. Sight angle (Vs) for single mounts.
11. Sight angle (Vs) for twin mounts.

Electrically to the director:
1. Bearing correction ($\triangle$cB'r)—automatic.
2. Bearing correction ($\triangle$cB'r)—indicating.
3. Elevation correction ($\triangle$cE + L + Zd/30)—automatic.
4. Elevation correction ($\triangle$cE + L + Zd/30)—indicating.
5. Range correction ($\triangle$cR).
6. Target angle (A).
7. Rate-solution indications.

Mechanically to stable element:
1. Director train (B'r).

Mechanically to star-shell computer:
1. Gun-train order (B'gr).
2. Gun-elevation order (E'g).
3. Advance range (R2).
4. Star-shell deflection rate (WrD + KRdBs).

The two values in the pairs of quantities designated above as *automatic* and *indicating* are identical. They are sent individually by two separate sets of receivers at the mounts concerned. It is entirely possible to use the indicating signal to actuate the automatic receivers, and vice versa, provided suitable switching arrangements are available.

The computer uses Pv and Pe to represent parallax in elevation due to horizontal base and vertical base respectively. These correspond to Peh and Pev as used in Chapter 16. The correction for vertical base is included in sight angle and gun-elevation order.

At one time, each computer was supplied with two fuze-setting computing units, one for mechanical, the other for powder-train fuzes. Either unit could be installed as necessary. However, powder-fuze sections are no longer furnished, hence they are not considered in this section.

Sight angle and sight deflection are each received in the sight mechanisms by a single synchro (coarse) on single mounts, but by a pair of synchros (fine and coarse) on twin mounts. The speeds of the coarse receivers are not the same on the two types of mount, hence the computer has one set of transmitters used for single-mount control and another set used for twin-mount control.

RESTRICTED

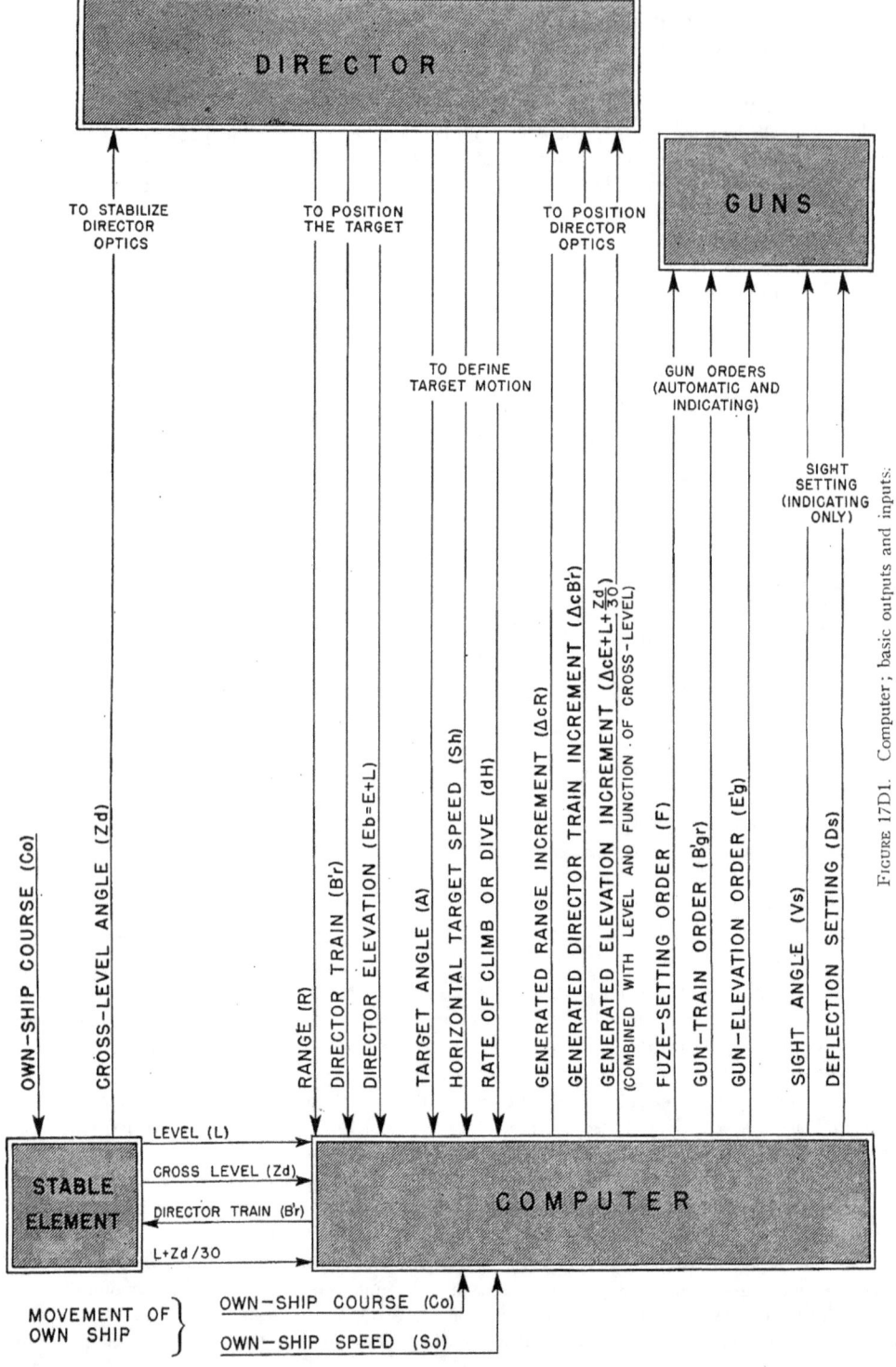

Figure 17D1. Computer; basic outputs and inputs.

CHAPTER 17

17D3. Inputs. The computer receives the following inputs from the sources and by the methods indicated:

Mechanically from the stable element:
 1. Level (L) or selected level (Lj).
 2. Cross-level (Zd) or selected cross-level (Zdj).
 3. Level plus cross-level function (L + Zd/30).

Electrically from director:
 1. Director elevation (Eb).
 2. Director train (B'r).

Electrically, or as alternative, manually:
 1. Normally from director:
   a. Observed present range (R).
   b. Target angle (A).
   c. Corrective settings to generated range, bearing, and elevation.
   d. Deflection spot (Dj).
   e. Elevation spot (Vj).
   f. Range spot (Rj).
 2. Normally from gyrocompass and pitometer log:
   a. Own-ship course (Co).
   b. Own-ship speed (So).

Manually:
 1. Horizontal target speed (Sh)—initial or corrective.
 2. Rate of climb (dH)—initial or corrective.
 3. Target diving speed (direct range rate) (dR).
 4. True direction true wind (Bw).
 5. True-wind speed (Sw).
 6. Initial relative target bearing (jBr).
 7. Initial range (jR).
 8. Initial target elevation (jE).
 9. Initial velocity.
 10. Dead time (Tg).
 11. Dip angle.

17D4. Target positioning summary. A brief description of the calculations made within the computer is provided in this and the next four articles. The presentation is by no means a complete discussion of the functional operation, but it will acquaint the student with the action within the instrument. The schematic diagram of the instrument is provided in figure 17D8.

The computer generates present range (cR), relative target bearing (cBr), and target elevation (cE) according to the equations of Article 15A8, and by the processes outlined in Section 15A, except that the effect of changes in own-ship course (Co) are taken into account in the case of cBr. In its calculations, the instrument uses generated present range (cR), but measured relative target bearing (Br) and measured target elevation (E) whenever the director is operating. Normally, the generated quantities cBr and cE are used only to drive comparison dials for rate control. Target elevation (E) is obtained by subtracting level (L) from director elevation (Eb). Relative target bearing (Br) is derived by adding the deck-tilt correction (jB'r) to director train (B'r).

17D5. Director positioning summary. Range correction ($\triangle$cR) is transmitted to the change-of-range receiver, which amplifies the quantity and drives the range-finder wedge-operating mechanism. This range correction is the same as generated range increments, but as transmitted to the director it is called range correction, for it corrects the range-finder wedge setting for the change in range as the problem progresses.

Bearing correction ($\triangle$cB'r) is transmitted to the director train power-drive control mechanism to cause the director line of sight to move in train according to the computed value of director

RESTRICTED

train. The value of $\triangle cB'r$ takes into account the deck-tilt correction, and the effect of own-ship course changes. In other words, $\triangle cB'r$ will hold the line of sight on the target in train while the ship rolls and pitches, and during changes in course, provided that the computer set-up is correct and that the director train drive is in *automatic*.

Elevation correction ($\triangle cE + L + Zd/30$), as indicated by its symbol, includes factors to account for changes in target elevation as the problem progresses, for the instantaneous value of level, and for the mechanical movement due to stabilization in cross-level. The computer generates $\triangle cE$ and combines it with $L + Zd/30$ received from the stable element. The total correction, sent to the elevation power drive, holds the line of sight on the target in elevation when the computer solution is correct and the elevation drive is in automatic.

The changes in director position produced by the bearing and elevation corrections show up in the form of changes in director train ($B'r$) and director elevation (Eb). As a result, they are fed back to the computer, since the director transmits $B'r$ and Eb continuously. In other words, the director and computer together form a regenerative group that generates director train and director elevation. It is therefore possible for the director to follow an obscured target when the computer is correctly set up and the target maintains its course and speed. This relationship can be employed in indirect fire against an invisible surface target when observations are available from an outside source. The computer is arranged so that it can use generated bearing in its calculation for surface fire, in which case it is regenerative by itself, and can control indirect fire without a director.

17D6. Rate control summary. The computer contains a rate-control mechanism which, in effect, is capable of computing target angle, target horizontal speed, and rate of climb when target speed is in excess of about 40 knots. This mechanism provides a means of comparing computed range, elevation, and bearing with the corresponding measured values, and may be operated by two methods: *automatic* and *manual*. Automatic is controlled from the director, manual by the computer personnel. The mechanism is subdivided so that bearing- and elevation-rate control can be accomplished in manual while range-rate control is automatic, and vice versa.

When the director is in automatic, the range, bearing, and elevation corrections transmitted to it are supposed to hold the line of sight on the target and keep the range finder in adjustment. Errors in the set-up of target angle, target horizontal speed, and rate of climb are evidenced at the director by the tendency of the line of sight and the range-finder measuring marks to wander off the target. The handwheel and range-knob motions required to keep the director instruments properly positioned are measures of the errors in the rates based on the target motion set-up. In automatic rate control, the computer analyzes these handwheel and knob motions, and converts them into corrections of target angle, target horizontal-speed, and rate-of-climb settings.

If the director is in local power-drive control, automatic rate control cannot be used, but the computer personnel can employ manual rate control. In this method, they compare observed and generated range, bearing and elevation by means of dials, and actuate the rate-control mechanism by hand to match the generated and observed quantities. The mechanism analyzes the hand inputs and corrects the target-motion set-up.

If the target speed is below about 40 knots, or if the rate-control mechanism fails, the computer is operated by a method known as *without rate control*. In this case, target angle, target horizontal-speed, and rate-of-climb settings are adjusted directly by hand as necessary to produce the correct generated quantities.

The target-angle setting within the instrument is transmitted to the target-angle repeater in the director automatically, so that the control officer knows its computed value. By the same token, he is aware of the changes he introduces in target angle settings by remote control.

17D7. Ballistics summary. The three linear rates and generated present range, plus other necessary quantities, are used to compute the ballistics which determine the required gun position with respect to the line of sight. The instrument computes advance range ($R2$) and predicted **target elevation ($E2$) by means of a number of approximations** which produce values that corre-

CHAPTER 17

spond closely to those developed in Section 15C. The ballistics computations, based on the advance target position, are essentially the same as those developed in Section 15C with the exception that no mechanism is provided for calculating air-density corrections.

Wind computations are made by means of empirical equations for a plane of fire that differs from that used in Chapter 15 because of the approximations employed. Sight deflection is computed for a slant plane that is perpendicular to the *sight plane* rather than a slant plane perpendicular to the vertical plane through the bore axis as described in Chapter 15. Also, the slant plane does not pass through the present target position, but through a point which may be above it or below it by a varying amount.

The data upon which the instrument design is based are curves similar to those in figure 15C6, except that they are for a 2,550 f.s. initial velocity. The calculations are automatically corrected for variations from this value between the 2,600 f.s. rated gun I.V. and a 2,350 f.s. lower limit. This results in more accurate computations throughout the gun life than would be obtained if the design were based on 2,600 f.s. The elements considered in the computation of sight angle, sight deflection, and fuze setting are:

Sight angle (Vs):
1. Relative movement of own ship and target during time of flight.
2. Superelevation.
3. Apparent range wind.
4. I.V. variations.
5. Complementary-error correction.
6. Elevation parallax for a 10-yard vertical base.
7. Elevation spot.

Sight deflection (Ds):
1. Relative movement of own ship and target during time of flight.
2. Drift.
3. Apparent cross wind.
4. Deflection spot.

Fuze setting (F):
1. Time of flight to the advance target position.
2. Correction of time of flight for relative movement of own ship and target during dead time.

17D8. Gun-positioning summary. The gun-train and gun-elevation orders are essentially the same as those developed in Section 16E. Gun-train order ($B'gr$) is director train ($B'r$) plus deck deflection (Dd). The latter quantity is composed of the deck-plane equivalent of sight deflection (Ds) and the trunnion-tilt correction in train. Gun-elevation order is director elevation (Eb) plus sight angle (Vs) minus the trunnion-tilt correction in elevation (Vz). These two orders are those that would correctly position a gun located at the horizontal reference point.

Train and elevation parallaxes are the deviations from the gun orders that are required to properly position a gun located 100 yards from the reference point. The parallaxes are converted at each mount by gear ratio to the values required for the actual distance between the mount and the reference point. The resulting corrections are used to offset the guns from their ordered positions by the amount necessary to converge their fire on the advance-target position. The elevation parallax order is used only in installations where the horizontal displacements are considerable, while the train-parallax order is universally employed.

17D9. Instrument arrangement. The actual instrument is built in four separable sections designated *control, indicator, computer* and *corrector*. The control and indicator sections are mounted on top of the computer and corrector sections respectively. The side of the assembled computer adjacent to the stable element is the *rear;* the opposite side is the *front*. The *right* and *left* sides are as seen by an operator standing at the front and facing toward the stable element.

The control section contains the rate-computing and rate-control mechanisms, and carries the ma-

DUAL-PURPOSE BATTERY FIRE-CONTROL SYSTEM

jority of the knobs, cranks and dials used in operating the computer. The computer section, beneath the control section, contains the mechanisms that calculate the ballistics. The indicator section indicates the final results of the ballistics calculations; i.e., sight angle, sight deflection, fuze setting, and advance range. The corrector section, beneath the indicator section, computes (and also indicates) the gun train and elevation orders, and the parallaxes.

The control section is illustrated in figure 17D2; the indicator section is shown in figure 17D3. The discussions which follow are concerned with the uses of the various dials and knobs shown in these two figures, to which frequent reference must be made by the student in order to aid in mentally fixing the location of the part referred to.

Many of the knobs and cranks have *in* and *out* positions, and have locking pins to hold them in either one. The locks are not specifically mentioned, for they are quite obvious on the computer itself. Also, most of the knobs have crank handles to facilitate rapid setting.

17D10. Target and ship dial group. The group of dials under the oval in the rear-center of the control section (see fig. 17D2) indicates the directions of wind, own-ship and target motion, and wind and own-ship speeds. The target dial at the rear has the outline of an airplane and a compass rose engraved upon it. The ship-dial group, at the front, consists of an inner ship dial, an intermediate compass ring, and an outer wind ring. The dial is engraved with a ship outline and a compass rose, the compass ring with a compass rose, and the wind ring with a single index pointing toward the compass ring. Between the dials, in line with their centers, is a pair of fixed indexes which represent the horizontal projection of the line of sight.

The target dial indicates only target angle (A), read against the fixed index. The ship dial indicates observed relative target bearing (Br) against the fixed index. On the compass ring, own-ship course (Co) is read opposite the ship bow, true target bearing (B) against the fixed index, and true direction true wind (Bw) against the pointer on the wind ring. The reading of the wind ring is the direction *from* which the true wind blows; i.e., for an east wind, it reads 90°.

Own-ship course (Co) is normally received by synchro from the gyrocompass, but it may be introduced manually through the ship-course crank (see fig. 17D2). The crank must be locked *out* for the electrical input, *in* for the manual input.

True direction true wind (Bw) is a manual input introduced through the *wind direction crank*. The instrument computes the wind angle between the wind direction and the line of sight, hence the wind setting need be changed only when wind direction changes.

Target angle (A) may be set initially by hand through the *target-angle crank* or by remote control. The actual knob (crank) position corresponds to target course (Ct); consequently, once the setting is correct, it requires no change as long as the target maintains its course. Inaccurate settings of target angle are modified either by the rate-control mechanism or by hand. On the target-angle crank is a two-position shift lever. For manual settings this lever must be set to *hand*. For remote control settings or changes from the rate-control mechanism, this lever and a similar lever on the target-speed crank must be set to *auto*. Near the target dial is a red signal light which is illuminated when a push button on the target angle repeater is depressed. When it is lighted, it indicates that target angle is being set by remote control, or that the knob shift-levers should be set to *auto* to permit such control.

True wind speed (Sw) is indicated in knots by the dial just to the left of the target and ship dials. The setting is made manually by means of the *wind-speed crank*.

Own-ship speed (So) is indicated in knots by the dial at the right of the target and ship dials (fig. 17D2). The setting is normally received by a synchro actuated from the pitometer log, or speed-measuring system. However, the input can be made manually through the *ship speed crank*. Like the ship course crank, the ship speed crank must be locked *out* for electrical reception, and *in* for manual input. When the computer is first energized, care should be taken that the own-ship speed dial indicates the correct speed, for it is possible that the receiver may be 40 knots out of agreement with actual speed. Since such variance would be quite evident, the operator can readily tell whether the reception is correct, and make the 40 knot change by hand if necessary.

CHAPTER 17

17D11. Target speed and time dial group. The indicators under the circular window on the right side of the control section (fig. 17D2) show target horizontal and vertical speeds, negative range rates (diving speeds), and elapsed time.

Rate of climb (dH), or target vertical speed, is shown on the dial in the right-rear part of the window. The setting is effected manually through the *rate-of-climb crank* or automatically through the rate-control mechanism. The crank must be pressed *in* for manual settings; locked *out* for automatic adjustments.

Target horizontal speed (Sh) is shown by the counters (left-front), and may be introduced or corrected by the *target-speed crank*, the *target-speed switch* (left-front corner of control section), or the rate-control mechanism. The target-speed crank has a two-position shift lever which must be set to *hand* for manual settings. Both this lever and the corresponding lever on the target-angle crank must be set to *auto* to permit speed settings by means of the target-speed switch or the rate-control mechanism.

The target-speed switch has three positions: *increase, normal,* and *dive attack.* Provided the shift levers are set on *auto*, the target horizontal speed setting automatically increases when the switch is on *increase,* and decreases when the switch is on *dive attack.* The switch is spring-returned from *increase,* but remains on *dive attack* until the operator returns it to *normal*. It thus assists the operator in making rapid approximate settings, and in the case of a dive attack, will run the horizontal speed setting down to zero.

The right-front dial in the target speed and time dial group (fig. 17D2) shows negative values of direct range rate (dR), but is called the *target diving-speed dial,* for it is used only in diving attacks against own ship. In such attacks, the target travels essentially along the line of sight, hence its total speed relative to own ship is equal to direct range rate. In a dive attack, the diving speed is set on this dial by means of the *range rate diving-speed crank* on the front right corner of the control section. This crank also has a *hand-auto shift lever,* which must be on *hand* for the manual setting, but on *auto* to permit the rate control mechanism to function.

Time is shown on the rear-left dial and ring. The ring is graduated in minutes, the dial in seconds. These dials, used for test purposes, are driven by the time motor, and may be set by means of the *time crank*. Pressing the crank *in* causes the minutes ring to return automatically to zero; in this position the crank can be turned to set the seconds dial to zero. With the time motor off, the crank can be pulled *out* and used to turn the time shafting throughout the instrument. The small dial in the center of the target speed and time group has a single index and revolves twice per second.

17D12. Control switches. On the right-front corner of the control section is a three-position *control switch* with *auto, semi-auto* and *local* settings. It governs the rate-control method employed for bearing and elevation, and establishes the set-up required for operation without the director in surface fire. When on *auto*, bearing and elevation rate corrections are introduced from the director whenever the pointer and trainer turn their handwheels with their rate-control keys closed. When on *semi-auto,* the same rate corrections are applied by the computer operators. When on *local,* the computer generates the equivalent of director train, and the instrument can be used for surface fire in *auxiliary operation;* i.e., without inputs from a director.

Just left of center on the front edge of the control section is the *range-rate control switch,* which has *auto* and *manual* positions. Set to *auto*, it permits range-rate control from the range finder in the director. Set to *manual*, it permits manual range rate control at the computer.

The control switch and the range-rate control switch are independent, but neither permits use of the rate-control mechanism unless the shift levers on the target angle, target speed, and range rate diving-speed cranks are all set to *auto*, and unless the rate of climb crank is *out.*

17D13. Bearing and elevation dial groups. The *dials* under the oval window at the right rear of the control section (fig. 17D2) form the *bearing group.* Observed relative target bearing (Br) from 0° to 360° is indicated by fine and coarse ring dials against fixed indexes. An inner dial within the fine ring has ten equally-spaced, unnumbered graduations, and is driven by gener-

RESTRICTED

ated relative target bearing (cBr). When the control switch is at *auto* or *local*, observed and generated bearing are matched automatically at all times, and the inner dial revolves at the same speed as the fine ring. When the control switch is at *semi-auto*, the inner dial rotates at a speed corresponding to the computed bearing rate. If the computed rate is correct, the dial will turn at the same speed as the ring; if the computed rate is incorrect, the difference in rotational speeds indicates to the operator that a bearing-rate correction is required.

Associated with these dials is the *generated bearing crank* at the right-rear corner of the control section. It has *in* and *out* positions. When turned while in either position, it applies a correction to generated relative-target bearing. In its *in* position it also introduces a bearing-rate correction to the rate-control mechanism.

Near the ring dials is an *indicator* which shows red when the trainer closes his rate-control key. A *solution-indicator dial*, with two black and two white quadrants, rotates when the trainer turns his handwheels in automatic rate control. When the trainer's signal shows red and the indicator is turning, bearing-rate corrections are entering the computer from the director, provided the several shift levers are set at *auto*, and the rate of climb crank is *out*.

Under the oval window at the left rear of the control section is the *elevation dial group*, with a pointer's signal and solution indicator. This group is identical to the bearing-dial group except for the ring-dial graduations and the quantities driving the dials. Observed target elevation (E) is indicated from 0° to 90°, while generated elevation (cE) drives the inner dial within the fine ring. The fine ring and dial are matched when the control switch is on *auto* or *local*, but the dials indicate zero target elevation in the latter case, since *local* is used only for surface fire without a director. *The generated elevation crank* on the left side of the case functions just like the generated bearing crank. The observed elevation indicated by the ring dials can be adjusted by means of the *synchronize elevation crank*, as explained later.

17D14. Range and height dial group. The window in the front-center of the control section (fig. 17D2) covers two sets of dials, one for range, the other for height. The *height dials* (the pair toward the rear) simply provide fine and coarse readings of computed target height against fixed indexes. Height is not used within the instrument, but it is indicated for use in selecting initial ballistic settings for special applications, such as the control of guns larger than 5 inch.

The *range dials* in the front part of the window consist of fine and coarse rings graduated in yards, and fine and coarse dials, each having a single index mark. The dials are driven by observed present range (R) received electrically from the range finder, while the outer rings are driven by generated present range (cR). When generated and observed range are equal, as they must always be for proper operation, the indexes at the zero graduations on the rings are opposite the indexes on the dials, and both observed and generated range can be read from the rings opposite the fixed indexes between them.

The ring and dial indexes may also be opposite each other when observed range is 36,000 or 72,000 yards greater than generated range. To the left and slightly to the rear of the range and height group is a lone dial which indicates coarse range values from 0 to 100,000 yards. This dial is electrically actuated from a ranging source other than the range finder in the director. It serves to inform the computer operator when the observed range is within the 35,000 yard tracking limit of the computer, and therefore shows when it is possible to match generated and observed range.

Like the bearing and elevation dials, the inner and outer dials of the range group must rotate at the same speed if the computed range rate equals the actual rate. However, the instrument bases its calculations on generated range, hence it is not sufficient to simply correct the rate by matching dial speeds. Generated range must be kept in agreement with observed range by the application of settings that correct generated range itself. When the range-rate control switch is on *auto*, and rate control in range is automatic, generated and observed range are kept matched automatically, and the required rate corrections are introduced by the range-finder operator. In

CHAPTER 17

manual range-rate control, the *range-rate control manual push button* and the *generated range crank* on the front face of the case are used to make the required settings. In its *out* position the crank corrects generated range and turns the ring dials, but does not alter the range rate. In the *in* position it corrects generated range, and if the push button is pressed, it also applies a range-rate correction.

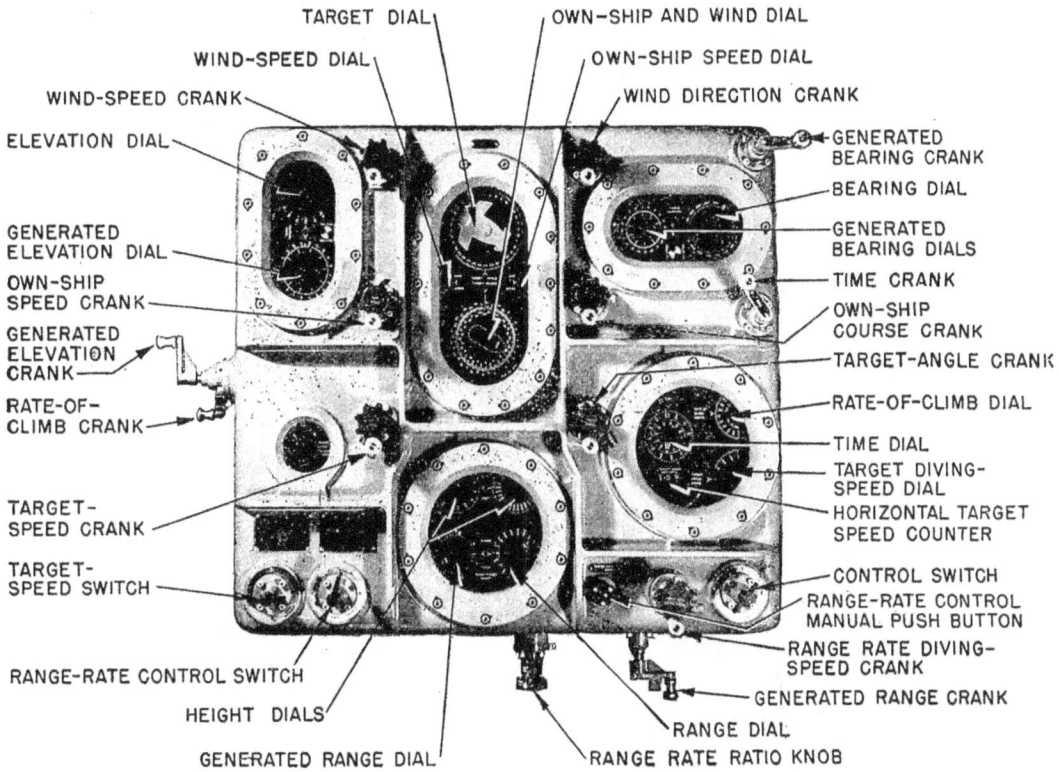

FIGURE 17D2. Control section, Mark 1 computer; top view.

In range-rate control, the correction applied to range rate is proportional to the correction applied to generated range. To provide a means for making rapid rate corrections when first starting to track a target, and yet assure the required smoothness of range-rate control as tracking progresses, a ratio mechanism is included in the instrument. This mechanism provides size adjustment of the range-rate correction applied for a given correction to generated range. The proportion is set by turning the *range rate ratio knob* mounted near the generated range crank. The knob carries a drum dial with five graduations numbered from 1 to 5, and must be *in* to effect a setting. For a given correction to generated range, the range correction is large when the drum is set at 1, and decreases as the drum setting is stepped up toward 5, as shown in the following table:

Range rate ratio	Change in range rate for a 100 yard change in range
1	88.3 knots
2	19.7 knots
3	11.1 knots
4	7.7 knots
5	5.9 knots

RESTRICTED

17D15. Synchronize elevation and dip dials. Director elevation is the sum of target elevation and level; i.e., $Eb = E + L$. Level (L), measured by the stable element, is subtracted from Eb in the computer to obtain E. The line of sight to the waterline of a surface target invariably dips below the horizontal, and when L is subtracted from the corresponding value of Eb, a negative value of E results. This negative value causes no appreciable error in the ballistics calculations, and is neglected from that standpoint. In continuous aim, the line of sight stays on the target and the computed value of E remains very small in surface fire.

However, when selected-level fire is controlled by the director pointer, the line of sight swings above and below the target, and a fixed value of Eb enters the computer. If the changing value of L is continuously subtracted from this constant value of Eb, fictitious values of E will be run into the computer, causing erroneous and unnecessary variations in sight angle. To prevent this, the target elevation shafting is locked at zero, and in effect, is disconnected from the Eb input. It is therefore necessary to synchronize the computed value of E with the actual value when restoring the set-up for continuous aim.

When selected-level fire is controlled by the stable element, Eb is no longer received from the director, but the Eb shafting is driven by level (L) plus the dip correction. The dip correction is simply a substitute for the negative value of E, that would be obtained if the line of sight were directed at the target. This correction depends upon the range to the target and the director telescope height above the waterline. Since this height is practically constant, range is the controlling value and the dip correction can be applied in terms of range.

Under the oval window near the center of the indicator section, shown in figure 17D3, are the *synchronize elevation dials* and the *dip dials*. Each group consists of a center dial (fine) and a concentric ring dial (coarse). Each of the dials of the synchronize elevation group has a single index mark and is used as a zero reader against the fixed index. The dip dial has a single index mark and is used as a zero reader against the fixed index, while the dip ring is calibrated in ranges from 500 yards to infinity and is read against the fixed index. The angular position of each range mark corresponds to the dip angle subtended at that range by the height of the director above the waterline.

Associated with these two dial groups is the *synchronize elevation crank,* which has *in, center,* and *out* positions. For continuous aim, this crank must be locked *in* and turned to match the synchronize elevation dial indexes at the fixed index; target elevation (E), as computed in the instrument, then corresponds to the actual value. For selected-level fire controlled by the director, the crank must be locked on *center* and turned until the target-elevation ring dials (on the control section) read zero. This locks the elevation shafting at a zero value of E. For selected-level fire controlled by the stable element, the target-elevation ring dials are set at zero with the crank on *center,* and then the crank is locked out and used to set present range on the dip ring, thereby introducing the required dip correction. To keep the proper correction, the setting should be adjusted as present range changes.

17D16. Spot indicators. The values of spots applied to the computer are indicated in the dial group under the oval window at the right side of the indicator section. *Elevation* and *deflection spots* are registered on *dials* calibrated in mils between the limits of -180 and $+180$. Range spots are indicated on two *counters,* one for spots up to 1,800 yards *out,* the other for spots up to 12,000 yards *in.* The large *in* limits are provided for control of guns larger than 5 inch. Spots are normally received and applied in the computer electrically, but *range, elevation* and *deflection spot cranks* are provided for manual setting when occasion demands. The cranks should be locked *out* for automatic reception, and *in* for manual use. When the range spot receiver is first energized, it may synchronize at a value differing from that transmitted by a multiple of 4,000 yards, and indicate an *in* spot exceeding 1,800 yards. Since the limits of electrical spot transmission are *in* and *out* 1,800 yards, false synchronization can be readily detected, and corrected by using the range spot crank.

17D17. Sight deflection, sight angle, fuze-setting order, and advance range. The remaining oval window at the left side of the indicator section encloses four sets of *counters* that show the computed values of *sight angle, sight deflection, fuze-setting order,* and *advance range.* The first three of these quantities are transmitted electrically to the guns, normally as computed by the instrument. However, *sight angle, sight deflection,* and *fuze cranks* are provided to permit manual setting of the transmitters so that arbitrary values may be sent to the guns. This feature is employed in connection with control of larger guns. Advance range is not transmitted, but is indicated for use in establishing star-shell firing data when a star-shell computer is not available.

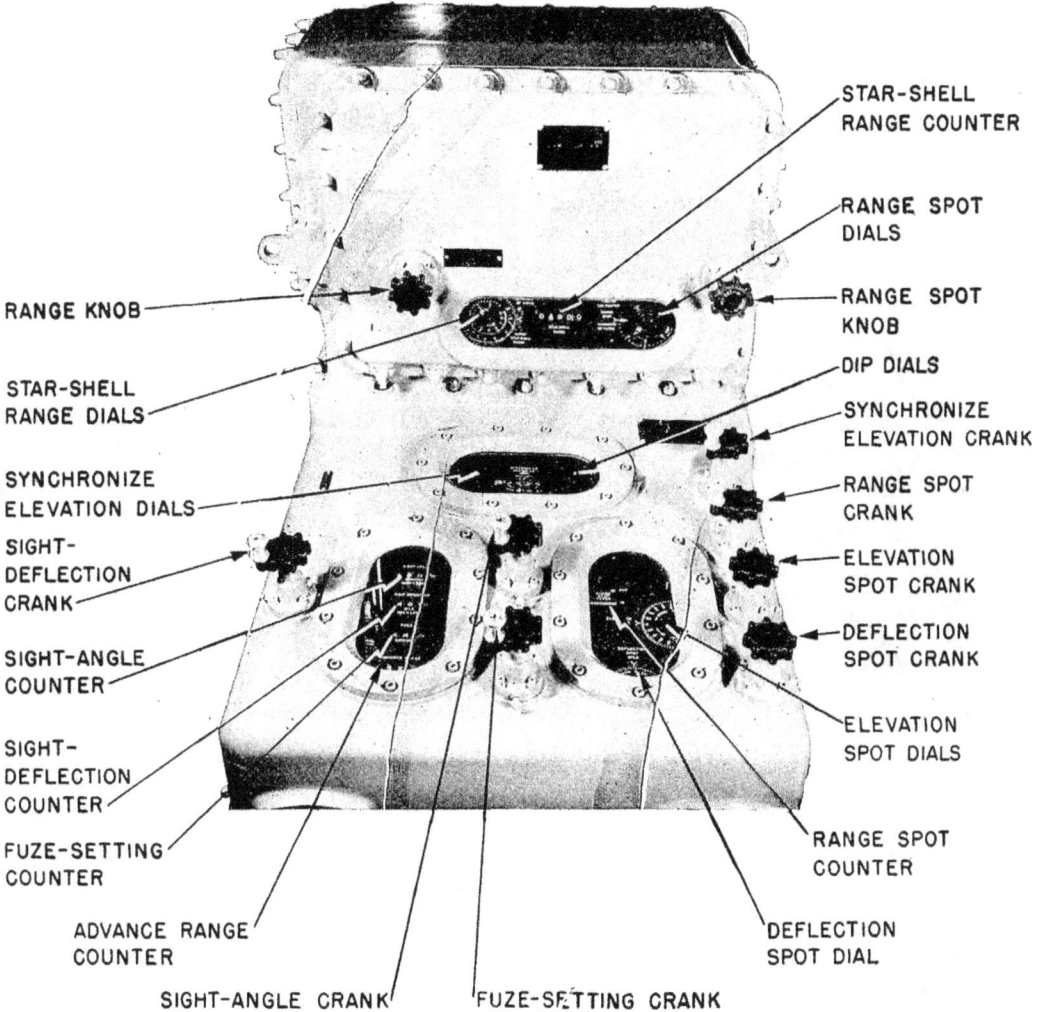

FIGURE 17D3. Indicator section, Mark 1 computer.

17D18. Other knobs, dials, and switches. On the left side of the instrument, (the computer section) are *initial-velocity* and *dead-time knobs,* and the *power* and *time motor switches* (see fig. 17C1). The settings introduced through these controls are semi-permanent in nature, hence their relatively inconvenient location is of no consequence.

The existing average initial velocity of the battery, between the limits of 2,350 and 2,600 f.s., is set on a dial by means of the initial-velocity knob in its *out* position. When not in use, the knob is locked *in.* The single setting takes into account loss due to erosion and variations due to powder temperature, and may even include a factor to partially counteract air-density variations. Dead time, between the limits of 0 and 6 seconds, is indicated on a dial, and set with a knob like the initial velocity knob. This setting is likely to be about 4 seconds, but varies with the installation.

The power switch controls all the servo motors which drive the various quantities into the mechanisms, including the time motor. The time-motor switch is provided to permit stopping the time motor even though the power switch is *on*. Circuits to the synchros are energized at the fire-control switchboard.

On the sides of the corrector section are numerous dials that indicate gun-train and elevation orders, train and elevation parallax, level, cross-level, trunnion-tilt elevation correction, and deck deflection; these dials are used for test purposes.

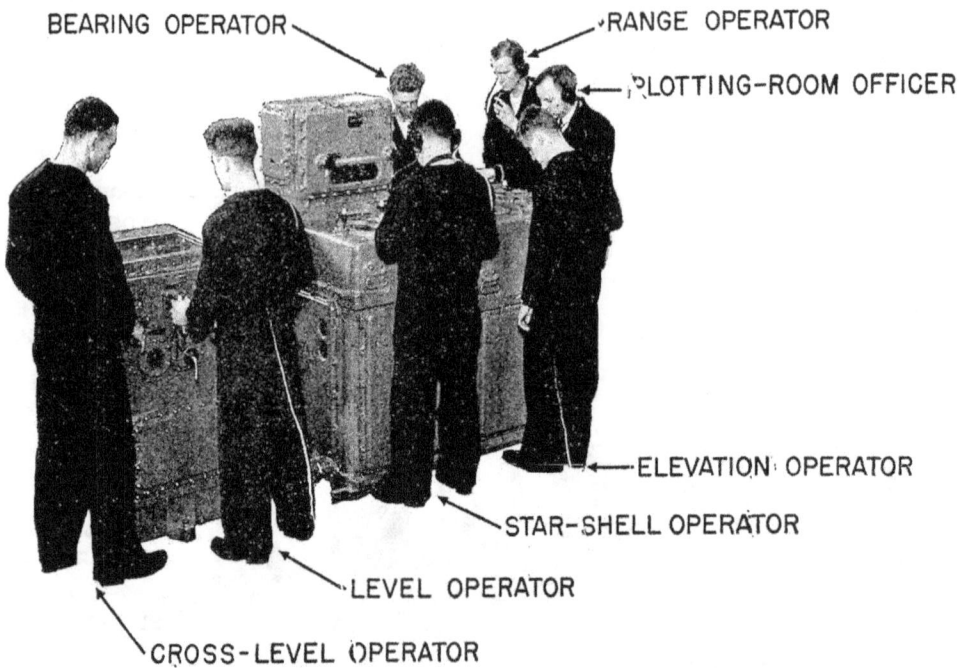

FIGURE 17D4. Disposition of personnel; stable element and computer.

17D19. Personnel. The operating personnel for the computer usually consists of four men under the immediate supervision of an officer, stationed as shown in figure 17D4 (note that two men in this figure are stationed at the stable element). One man stands near each of the three rate-control cranks, while the fourth stands at the indicator section. The three men at the rate-control cranks set up the problem on the control section. The fourth man at the indicator section operates the synchronize elevation crank as required, and stands by to make manual settings on the indicator section when necessary. He also operates the star-shell computer when it is in use. If the star-shell computer is in operation, and other knobs on the indicator section require attention, one of the other three men can assist, for they are not all needed at the control section in surface fire. To avoid confusion in subsequent articles, reference to the *computer operator* may be taken to mean any one of the four, as applicable. At least one other man is required to operate the fire-control switchboard, the actual number depending upon the complexity of the installation.

17D20. Operation. It should be quite evident from the discussions in the preceding articles that operation of the computer requires skilled personnel. The numerous alternatives and combinations of operational procedures make it essential that each operator be thoroughly indoc-

CHAPTER 17

trinated in his specific duties. Furthermore, each man should be acquainted with operation of the entire system in order that he may perform at highest efficiency. In the following articles, operating procedure is outlined, but this description is by no means all-inclusive. It serves to give a general picture of the part the computer plays in the control of the system as a whole.

Several conditions of operation are described, progressing from the secured condition to full engagement. There are certain set-ups recommended under secured and standby conditions which are designed to protect the computer and other parts of the system against unnecessary wear and consequent decrease in accuracy. Without full knowledge of the effects of changes, departure from the recommended settings is likely to be an invitation to trouble. Some of the settings, of course, are arbitrary, and serve largely to standardize procedure, and perhaps to aid in detecting unauthorized tampering. A tabular summary of operation under various conditions is provided in Article 17D26; it should be examined in connection with the descriptions.

17D21. Secured and stand-by conditions. If the computer and stable element are to remain unused for an appreciable period of time, they should be secured to reduce all wearing to a minimum. It is essential to have the computer power and time-motor switches and all switches on the fire-control switchboard turned *off*. Article 17D26 gives settings for other knobs.

The three types of stand-by listed in the table represent progressively-increasing degrees of preparedness with the stable-element gyro running. Stand-bys 1 and 2 permit setting the level and cross-level computer inputs at 2,000 minutes (representing zero values), and thereby cause the least wear in the computer with the gyro running. Selected level of 2,000 minutes is set at the stable element. Since the stable element cannot deliver selected level and selected cross-level at the same time, an auxiliary device, called a *selector drive* (see fig. 17C1) is provided in the cross-level shafting between the stable element and the computer. It permits disengaging the shaft in such a manner that the drive to the computer may be locked at 2,000 minutes, while the drive from the stable element may oscillate normally, and provides for synchronizing the stable element output with the computer input when the shafts are again engaged.

Stand-by 1, the condition of least readiness, is divided into forms a and b. Form a is limited to sea conditions in which neither roll nor pitch exceeds ±15°. Under more severe conditions of roll and pitch, the stable-element follow-ups and computer power circuits should be energized to protect the stable element as in form b. Stand-by 2 represents a condition of operation in which the director is stabilized and may be searching for a target. Some wear in the computer is unavoidable, but is held to a minimum by keeping level and cross-level inputs at 2,000 minutes and all speeds set at zero. Stand-by 3 is used when a target is expected, and settings are made as shown in the table (without regard for wear) to permit going into action rapidly.

17D22. Initial observation. Upon first detecting a target, the director pointer, trainer, and range-finder operator should close their rate-control keys while picking up the target. This actuates the signals at the computer and warns the computer personnel to prepare for tracking by setting up the instrument as shown in the table for initial observation. The computer operator, using the generated range crank in its *out* position, should match generated range with observed range as soon as the latter becomes less than 35,000 yards. Since the control switch is on *auto*, the bearing and elevation dials will synchronize automatically with their corresponding observed rings. When and if the target is visible, the settings of target angle, target horizontal speed, and rate of climb should be corrected to agree with the control-officer's estimates. During this operation, the rate-control mechanism should not function, and it is prevented from doing so by keeping the shift lever on the target speed crank at *hand* and the range-rate control switch at *manual*.

Some time may elapse between the initial observation and the moment the target comes within tracking range. During this interval, preparation should be made, insofar as possible, for tracking the target. If it appears that target speed is too low to permit the rate-control mechanism to function, provision should be made for operation without rate control. As soon as generated range is matched with observed range, and the bearing and elevation dials are synchronized, the plotting room should inform the control officer that it is ready to track.

RESTRICTED

DUAL-PURPOSE BATTERY FIRE-CONTROL SYSTEM

17D23. Automatic rate control. When the order is received to start tracking, the computer operator should turn the time motor switch *on* and immediately set the shift lever on the target-speed crank at *auto* and the range-rate control switch at *auto*. Rate corrections introduced at the director then act through the rate-control mechanism to correct automatically the set-up of target angle, target horizontal speed, and rate of climb. The operator must watch the bearing, elevation, and range dials, and the trainer's, pointer's, and range-finder operator's signals to determine whether or not operation is proceeding satisfactorily.

As previously stated, the shift levers on the target angle, target speed, and range rate diving-speed cranks must be at *auto*, and the rate-of-climb crank must be *out* to permit correction of the corresponding quantities by the rate-control mechanism. However, if the control officer announces a sudden change in target movement, corrections can sometimes be made more quickly if target angle, target horizontal speed, and rate of climb are slewed, or quickly changed manually, to agree with the new estimated values. To make such corrections, the shift lever for the quantity to be changed should be turned to *hand*, and restored to *auto* as soon as the change has been applied. Similarly, the rate-of-climb crank must be pushed *in* for the manual change, and then pulled *out*.

During problem solutions, a large, quick change in target angle may cause the rate-control mechanism to begin to compute an erroneous solution in which target angle is 180° from the correct value, and target speed is negative. This is evidenced by a sudden decrease in target speed to some fictitious value below 100 knots and then a cessation of rate corrections. This false solution can be overcome by holding the target-speed switch at *increase* until target speed reaches its former value.

17D24. Manual rate control; bearing and elevation. If the observed values of bearing and elevation do not progress smoothly and in the same direction, or if the signals from the pointer and trainer are intermittent, bearing and elevation-rate corrections should be made at the computer. The necessity for such operation may also be indicated by oscillation of target speed about its true value, since this oscillation may be caused by irregularities in observed bearing. For manual rate control in bearing and elevation, the control switch must be turned to *semi-auto*, and the generated bearing and elevation cranks must be locked *in* and turned by their respective operators so as to cause the inner, or generated dials to revolve at the average speed of the observed rings. The graduations on the inner dials serve as a guide in observing the speed of rotation; whether or not they are matched with the graduations on the outer rings is of no consequence. When it is possible to return to automatic rate control, the bearing and elevation cranks should be pulled *out*, the shift lever on the target-speed crank should be set at *hand*, and the control switch should be turned to *auto*. As soon as the generated dials have synchronized with the observed rings, the shift lever on the target-speed crank should be turned to *auto*, thereby restoring automatic rate control.

17D25. Manual rate control; range. If the observed value of range is not progressing smoothly and in the same direction, or if the signals from the range finder are intermittent, the range-rate control switch should be set at *manual* and range-rate corrections should be made manually at the computer. To do this, the operator locks the generated range crank *in*, depresses the range-rate control push button, and matches the generated range-ring indexes with the indexes on the observed range dials whenever the range-finder operator signals that the observed range is correct. Normally, the push button should be depressed while the crank is being turned, and released only when the setting has been completed. However, if the rate of divergence between generated and observed range is large, the operator should over-correct; that is, he should cause the index on the fine ring to overtake the arrow on the inner dial while the button is down. Then he should release the button and match the dial indexes.

In either automatic or manual range-rate control, the range-rate ratio knob should be set at 1 initially, or whenever there is a considerable change in target movement. Subsequently, as tracking progresses, the knob should be turned toward 5 until there is no oscillation in target speed. However, it is recommended that the ratio be kept as near to 1 as possible, to accelerate rate corrections.

17D26. Operation without rate control. The rate-control mechanism is not designed to function at target speeds lower than about 40 knots, the actual limiting speed depending upon the

KNOB, CRANK OR SWITCH	SECURED	STANDBY No. 1a Stable Element Running	STANDBY No. 1b Stable Element Running	STANDBY No. 2 Director Stabilized	STANDBY No. 3 Target Expected	Initial Observation	TRACKING Target Speed Above 40 Knots	TRACKING Target Speed Below 40 Knots
Generated Bearing	Out	→	Out-0° 00'	Out	→	Out-Automatic	Out-Automatic In-Manual	→
Generated Elevation	Out				→	Out-Automatic	Out-Automatic In-Manual	→
Generated Range	Out 30,000 Yards				Out-Expected	Out Match Observed	Out-Automatic In-Manual	Out Match Observed
Target Angle	Auto-0°						Auto-Computed	Hand-Estimated
Target Speed	Hand-0 Knots				Hand Expected	Hand Estimated	Auto-Computed	Hand-Estimated
Rate of Climb	Out-0 Knots					→	Out-Computed In-0 Knots (Diving Attack)	In-Estimated (0 Knots for surface target)
Range Rate—Diving Speed	Auto-0 Knots					→	Auto-Computed Hand Diving Speed	Auto-Computed
Wind Direction	0°			→	Determined			→
Wind Speed	0 Knots			→	Determined			→
Ship Course*	Out-0°			Out-Received				→
Ship Speed*	In-0 Knots			→	Out-Received			→
Synchronize Elevation	In-Synchronize						See Article 17D15	→
Initial Velocity	Determined							→
Dead Time	Determined							→
Range Rate Ratio	1					→	Set as required	→
Deflection Spot*	Out-0 Mils					→	Out-Received	→
Elevation Spot*	Out-0 Mils					→	Out-Received	→
Range Spot*	Out-0 Yards					→	Out-Received	→
Time	Center-0 Seconds						Indicated	
Sight Deflection	Out					→		→
Sight Angle	Out							→
Fuze	Out							→
Power Switch	Off		On			→	On	→
Time Motor Switch	Off	→						→
Control Switch	Local			Semi-Auto	Auto	→	Auto or Semi-Auto	Semi-Auto
Range Rate Control Switch	Manual					→	Auto or Manual	Manual
Target Speed Switch	Normal					→	Normal or Dive Attack	Normal
Level**	Selected-2,900'				Received	→		→
Cross-Level***	Selected-2,000'			→	Received	→		→
Fire Control Switchboard	All Circuits De-energized	All Circuits De-energized except Gyro in Stable Element.	Computer Power Circuit, and Stable Element Gyro and Follow-up Circuits, Energized.	All Circuits Energized except Director Elevation Transmission Circuit.	Director, Computer and Stable Element in full operation.			

*Normally Received Electrically. **Set at Stable Element. ***Set at Selector Drive.

problem conditions and the adjustment of the individual instrument. Therefore, at low surface or air-target speeds, or when the rate-control mechanism fails, it is necessary to operate without rate-control; i.e., to adjust manually the settings of target angle, target horizontal speed, and rate of climb.

For operation without rate control, the control switch must be at *semi-auto* and the range-rate control switch must be at *manual*. The shift levers on the target-angle and target-speed cranks must be set at *hand* and the rate-of-climb crank must be *in*. Target angle, target horizontal speed, and rate of climb should be adjusted so that the dials indicating generated bearing, elevation, and range turn at the same speeds as the corresponding observed dials. Since generated range is used as the basis for instrument computations, it must be kept in agreement with observed range by turning the generated range crank in the *out* position to match the inner and outer dial indexes whenever the range-finder signal is received. However, no bearing or elevation settings need be made.

The predominant element of target motion affecting generated bearing, elevation, or range depends upon target angle and target elevation, as shown in the table in Article 15B3. An operator should be able to visualize the actual problem conditions by inspection of the target and elevation dials, and then be able to adjust target angle, target horizontal speed, and rate of climb according to the rules outlined in that table. It is consoling to note that operation without rate control is normally required only for slow target speeds, and then usually only in connection with surface firing, in which case rate of climb should be set at zero, and the rules for adjusting target angle and speed are considerably simplified. Nevertheless, operators must be trained to perform the manual adjustments for fast-moving air targets if maximum benefit is to be obtained from the computer.

17D27. Special forms of attack. For practical purposes, a plane diving at own ship may be considered as traveling in the line of sight. The relative-motion problem is considerably simplified in such a case, and the computer has special provisions that are intended to accelerate the problem solution for a dive attack. The dive obviously must be against own ship, for a dive against another ship is a general problem. The control officer should notify the computer operator if a plane is observed beginning a dive against own ship. However, the computer operator can recognize such an attack from the instrument dials, for target angle becomes practically zero, range decreases rapidly, and target elevation remains very nearly constant at some value above 30°.

As soon as a dive attack is recognized, the computer operator must set the target speed switch at *dive attack*, set rate-of-climb at zero, and set the estimated diving speed on the diving-speed dial. Diving speed is set by means of the range rate diving-speed crank with its shift lever at *hand*, and is correct when the generated range decreases at the same rate as the observed range. Since target speed is brought to zero when the target-speed switch is turned to *dive attack*, the target-angle setting can be disregarded, but both target speed and rate of climb must be set at zero to prevent computation of a fictitious elevation rate. This procedure can be used at any target elevation, but must be employed when the target approaches at an elevation above approximately 30°, because the rate of climb setting is limited to —250 knots.

In a horizontal bombing attack, rate of climb is practically zero. If the rate-of-climb crank is pushed *in* and turned to set rate of climb at zero for such an attack, the rate-control mechanism will arrive at its solution in less time than if it were forced to reduce the rate-of-climb setting by computation. This type of operation may aid in establishing the range and elevation rates, but as soon as the setting has been made, the rate-of-climb crank should be pulled *out* to permit the rate-control mechanism to compute the true rate of climb.

17D28. Surface fire. As previously explained, the director and computer form a regenerative group. Bearing correction and elevation correction from the computer move the director line of sight in train and elevation, and the director in turn transmits revised values of director train and director elevation to the computer. Even though the target may be obscured, the director follows it provided the computer set-up is correct. This relationship is true for both surface and air targets. It is impracticable to maintain indirect fire against fast-moving air targets for an

CHAPTER 17

extended period of time, unless direct measurements of position are available. However, the regenerative relationship between the director and computer can be used in controlling indirect fire against moving or stationary surface targets if some observing station supplies range estimates and spots. The computer must be operated without rate control in such cases.

In the event that no director inputs are available, the computer may be used in *auxiliary operation*, in which case it is regenerative in bearing, and can be used to direct fire against visible or invisible surface targets. To set up the instrument for this type of operation, the operator turns the control switch to *local*, and has target elevation set to zero and dip introduced by means of the synchronize elevation crank. The computer then operates in conjunction with the stable element and transmits gun-train and elevation orders, and sight angle and deflection, firing being controlled at the stable element. Range, relative target-bearing estimates, and spots must be received by voice and introduced by hand.

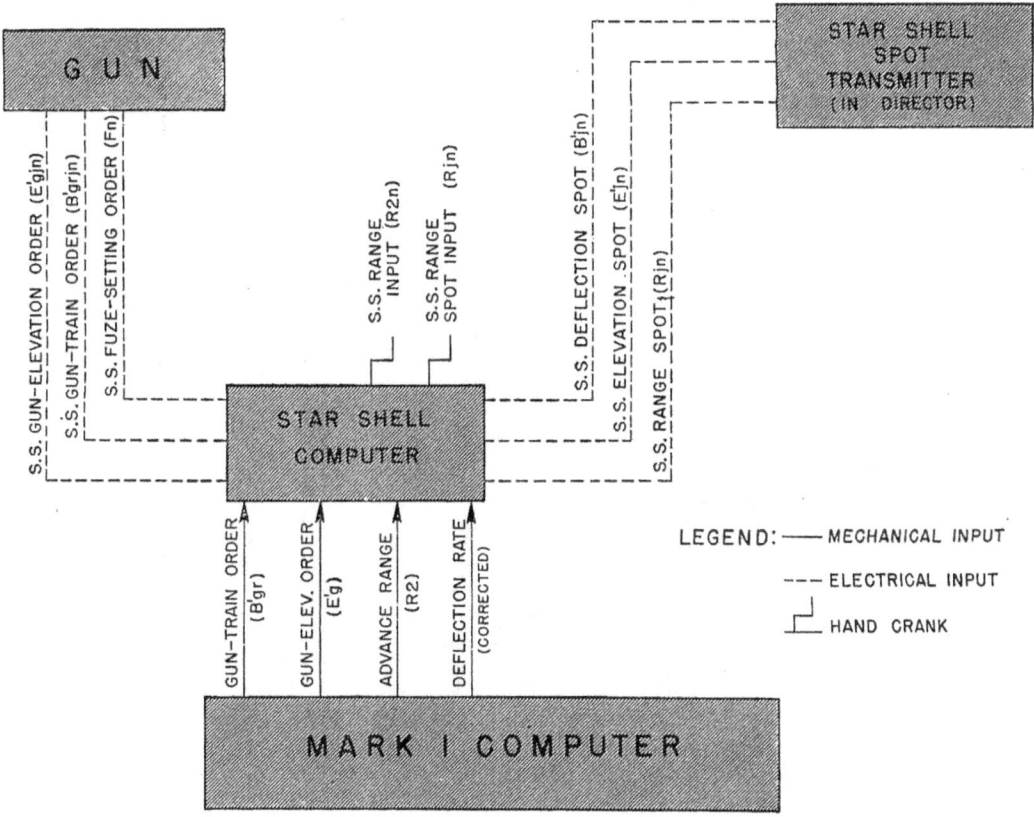

FIGURE 17D5. Star shell fire-control system; diagrammatic.

Both selected level and selected cross-level fire may be employed against surface targets, but never against air targets since the firing rate is too low. The computer personnel have no special duties in connection with cross-level fire, but in level fire they must operate the synchronize elevation crank as explained in Article 17D15.

17D29. Star-shell computer functions. The star-shell computer, Mark 1, forms an integral part of some Mark 1 computers. When supplied, it is mounted on the indicator section of the computer as shown in figure 17D3. Its functions are to compute and transmit the following data for controlling 5"/38 cal. guns firing star shells:

DUAL-PURPOSE BATTERY FIRE-CONTROL SYSTEM

1. Star shell gun-train order (B'grjn).
2. Star shell gun-elevation order (E'gjn).
3. Star shell fuze-setting order (Fn).

These outputs are transmitted over the regular circuits used for the gun-train and elevation orders and the fuze-setting order from the computer, their distribution being as shown in figure 17D5. The outputs are derived from the following inputs:

Mechanical from computer:

1. Gun-train order (B'gr).
2. Gun-elevation order (E'g).
3. Advance range (R2).
4. Star-shell deflection rate (WrD ÷ KRdBs).

Electrical from star-shell spot transmitter in director:

1. Star-shell deflection spot (B'jn).
2. Star-shell elevation spot (E'jn).
3. Star-shell range spot (Rjn).

Manual:

1. Star-shell range spot (Rjn).
2. Star-shell range (R2n).

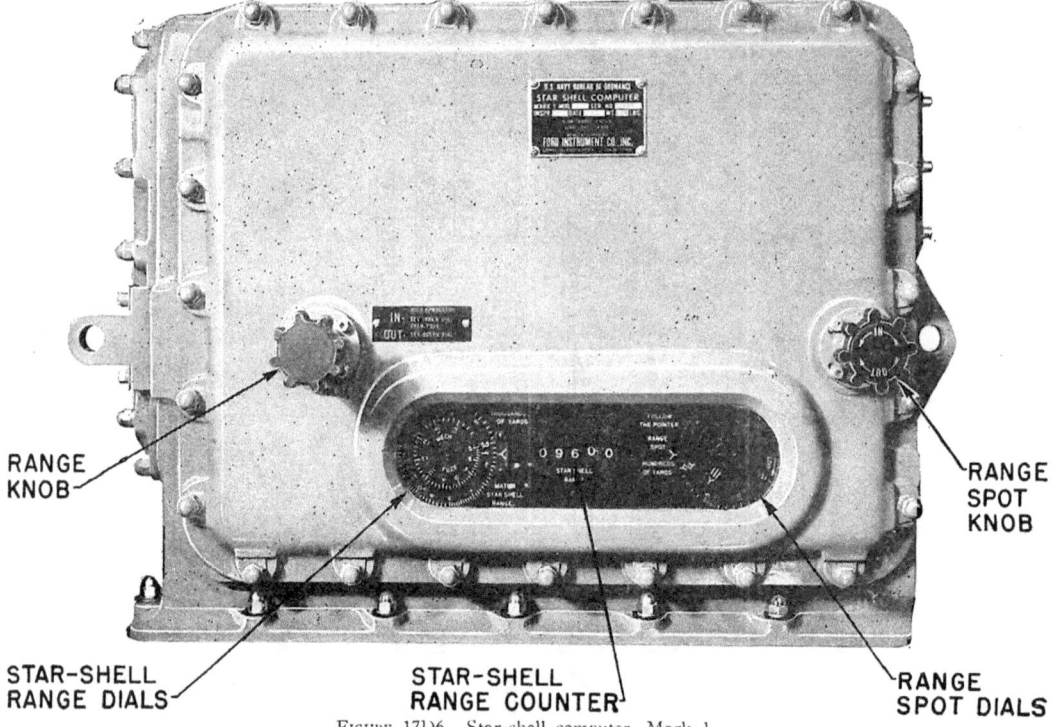

FIGURE 17D6. Star-shell computer, Mark 1.

The computed star shell gun-train order includes corrections for drift of the star due to wind, and for target and own-ship motion, so that at half the life of the star (about 30 seconds) the point of illumination is approximately in line with the target and own ship. The computed star shell gun-elevation and fuze-setting orders are such as to place the star 1,500 feet above and 1,000 yards beyond the target with spots at zero. The star-shell range spots are used to alter both the star shell gun-elevation and fuze-setting orders. Star-shell deflection and elevation spots are used to correct the train and elevation orders respectively.

CHAPTER 17

17D30. Star-shell computer; operation. The dials and knobs used in operating the star-shell computer are shown in figure 17D6. The star-shell range spot actuates a synchro motor that controls the center dial in the right-hand group. Surrounding this dial is a follow-the-pointer ring dial which is set by the knob at the right. The ring dial is graduated to read the spots applied to the instrument against the fixed index. The operator turns the knob to place the index on the ring dial opposite the index on the center dial, thereby introducing the range spot to the mechanism by manual follow-up.

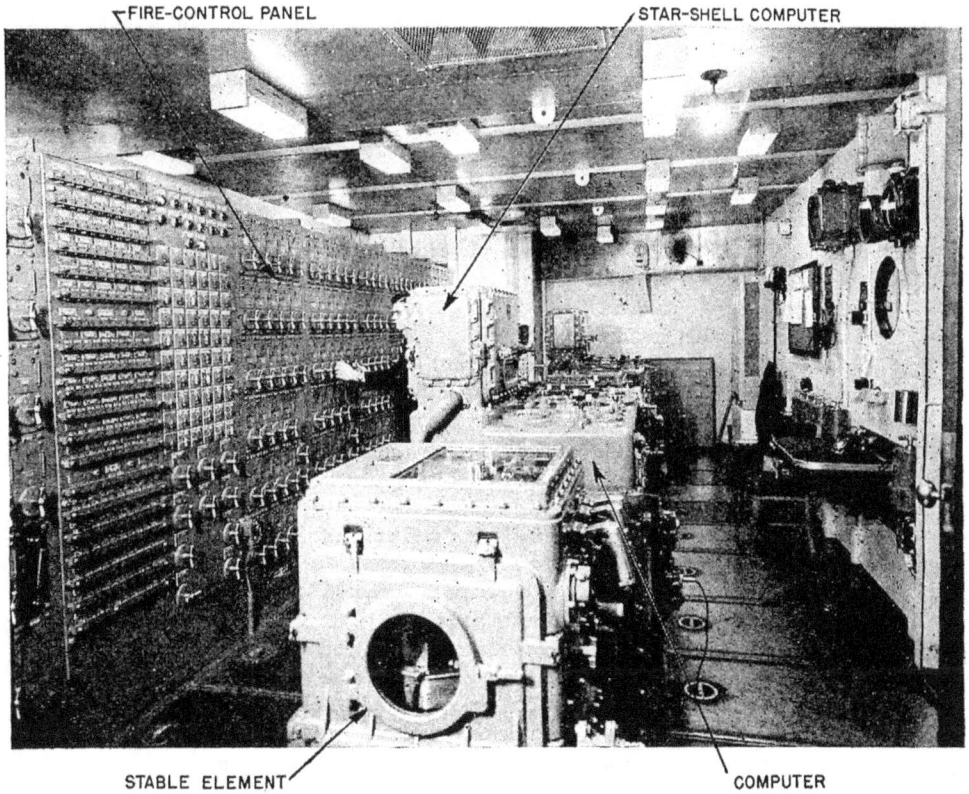

FIGURE 17D7. Secondary plot; 5"/38 cal. battery.

Star-shell range, which is the horizontal advance range measured to the point of star-shell burst, is indicated by the *counter* between the two dial groups. At the left of the counter are a *ring dial* and an *inner dial* which both indicate star-shell range against a fixed index. These dials are controlled by the *star-shell range knob* to the left. The operator presses this knob in and turns it to make the inner dial reading agree with that of the counter. He then pulls the knob out and turns it to make the outer-ring reading agree with the counter. These operations actuate the fuze-setting order transmitter and supply inputs to the mechanisms which compute the star shell gun-train and elevation orders.

To make operation of the star-shell computer possible, the main computer should be set up for control of surface fire against the same target as that to be illuminated by star-shell fire. The main computer can be used to control the guns firing service projectiles against the same target provided spots introduced for service projectiles are neutralized by opposite spots at the star-shell spot transmitter.

DUAL-PURPOSE BATTERY FIRE-CONTROL SYSTEM

Figure 17D7 shows a section of the plotting room for the secondary battery more or less typical of battleships, aircraft carriers and cruisers. Two Mark 1 computers with star-shell computers attached and two stable elements are shown, together with a section of the fire-control panel.

E. GUN-CONTROL EQUIPMENT

17E1. Introduction. The student will recall from the discussions in Section D of Chapter 6 that each 5"/38 cal. gun mount is equipped with an *elevation indicator-regular*, a *train-indicator-regulator*, a *fuze-setting indicator-regulator*, and a *sight-setting mechanism*. The subject of this section is the part these units play in applying the computer outputs to the control of the gun mounts. The three indicator-regulators have identical primary functions, implied by their names:

    1. Each regulates or controls the power drive with which it is associated, to make it respond either to the electrical signals from the computer or to motion of the handwheels or cranks used in local power-drive control.

    2. Each indicates the action needed to comply with the orders from the computer, and the response of the power drive concerned.

17E2. Train indicator-regulator. The train indicator-regulator, illustrated in figure 17E1, is at the trainer's station and receives:

    1. Gun-train order—automatic.
    2. Gun-train order—indicating.
    3. Train parallax for a 100 yard horizontal base.
    4. Train response (the angle in the deck plane between the fore-and-aft axis and the perpendicular plane containing the bore axis).

The instrument contains a fine and a coarse synchro motor for receiving automatic gun-train order, and a fine and a coarse synchro motor for receiving the indicating gun-train order. The automatic receivers govern the power drive in remote control, while the indicating receivers actuate some of the dials. Normally, both the indicating and automatic receivers are energized. The mechanism also contains a parallax synchro receiver, a parallax follow-up motor, and the necessary mechanism for converting the 100 yard parallax correction to that required for the actual mount horizontal displacement from the reference point.

The *train angle dial* and the *train angle ring dial* respectively provide fine and coarse readings of train response plus or minus the parallax correction applied to the mount. This means that when the gun is trained in accordance with the order, the dials read gun-train order, although the actual gun train is different from the ordered train by the parallax correction. These dials are used for checking purposes, and when occasion demands, may be employed by the trainer to place the gun at a designated position in train.

The *zero-reader dial* is attached to the coarse-indicating synchro shaft. It provides a coarse indication of the gun movement required to comply with the gun-train order. The synchro stator is driven by corrected train response; hence, when the zero reader is matched, the gun is close to its ordered position.

The *follow-the-pointer dial* is attached to the shaft of the fine-indicating synchro, whose stator is fixed. The *follow-the-pointer ring dial* is driven by corrected train response. When the indexes of the follow-the-pointer dial and ring are matched, the gun is trained as required, provided the zero reader is also matched.

The *parallax zero-reader dial* is attached to the parallax synchro shaft. The *parallax ring dial* and the parallax synchro stator are driven by the parallax follow-up motor, which also drives the 100 yard correction into the mechanism. The mechanism changes the correction to that applicable to the mount. When the follow-up motor is functioning, it continuously keeps the parallax zero-reader index matched against the fixed index and makes the parallax ring dial register the correction for a 100 yard base against the same index. If the follow-up motor fails, the trainer uses the *parallax crank* to introduce the parallax correction by turning it to keep the zero reader matched. If the parallax

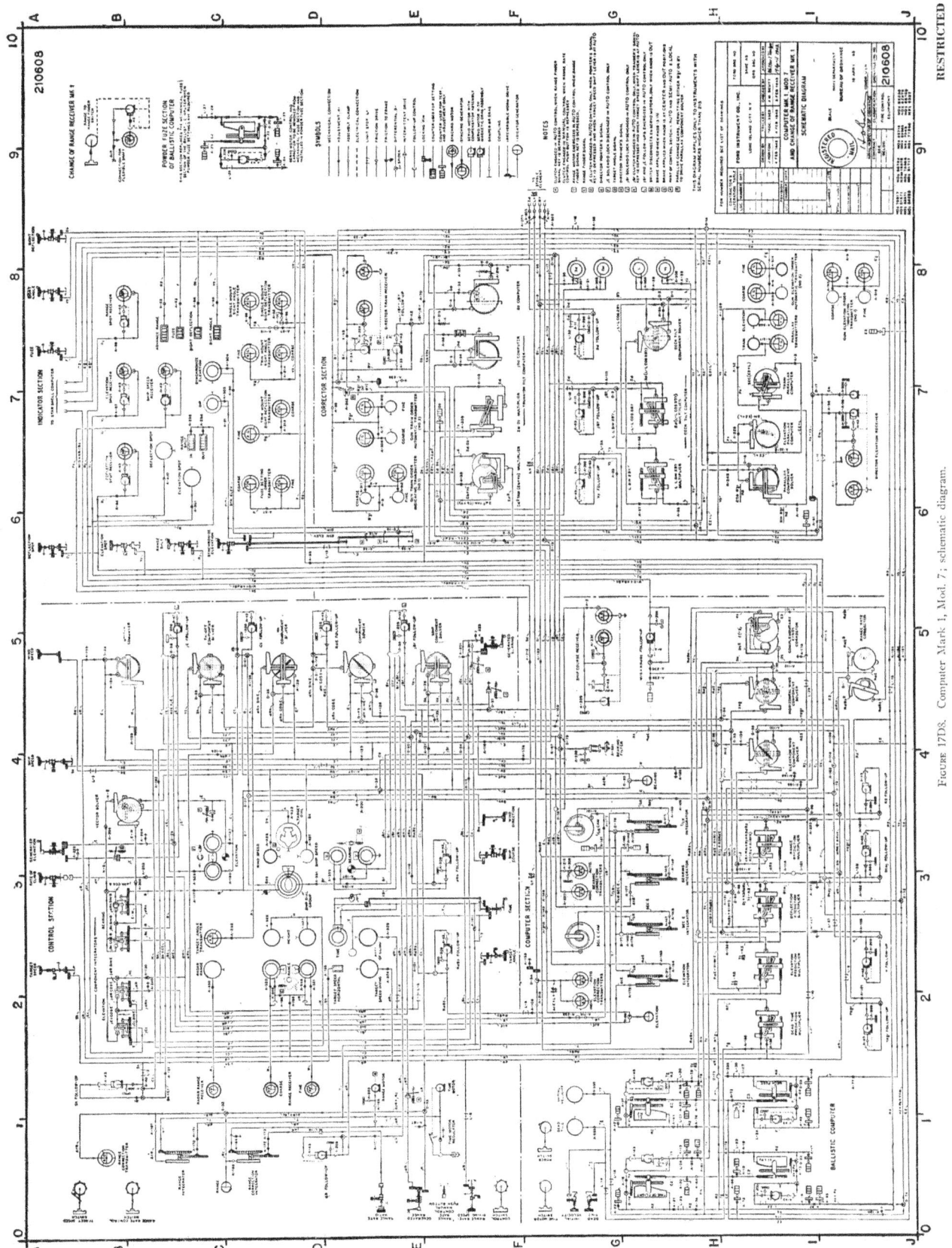

Figure 17D8. Computer Mark 1, Mod. 7; schematic diagram.

synchro fails, the zero reader and follow-up motor no longer function. The trainer may then set the value of the 100-yard correction on the parallax ring dial. The trainer also uses this crank to provide offsets between mounts when firing star-shell spreads.

The *star-shell signal indicator* is a small colored-glass window with lamps behind it. The lamps are lighted whenever the mount is ordered to fire illuminating projectiles, notifying the trainer that the gun-train order entering the indicator-regulator is for star-shell fire.

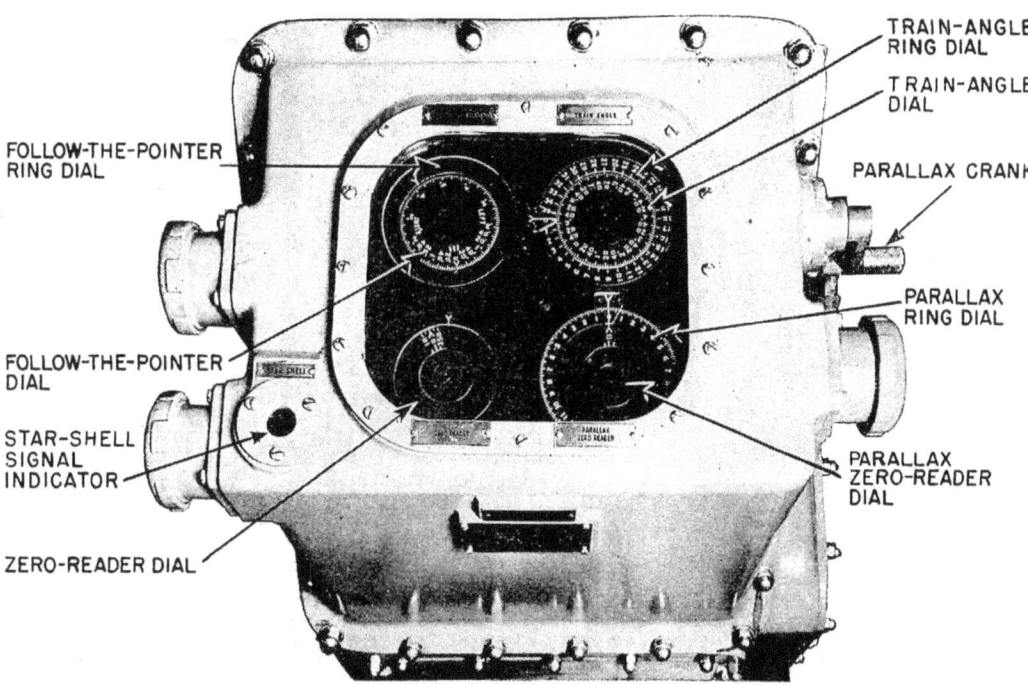

FIGURE 17E1. Train indicator-regulator; top view.

17E3. **Elevation indicator-regulator.** The elevation indicator-regulator, shown in figure 17E2, is at the pointer's station. It is quite similar to the train indicator-regulator, and receives:
1. Gun-elevation order—automatic.
2. Gun-elevation order—indicating.
3. Train response from the train drive assembly.
4. Elevation response (the perpendicular angle between the bore axis and the roller path).
5. Elevation parallax for a 100 yard horizontal base (in certain modifications only).

This instrument contains fine- and coarse-indicating synchros, and fine and coarse automatic synchros having functions like those of the corresponding units in the train indicator-regulator. A roller-path compensator is also provided to correct the response for roller-path tilt. In those indicator-regulators on guns corrected for elevation parallax due to the horizontal base, there is a parallax mechanism nearly identical to that in the train indicator-regulator. There is also a *star-shell signal indicator* provided for the pointer.

The train response received by this indicator-regulator is used to position part of the roller-path compensator. The pointer ordinarily has nothing to do with this device, as it requires only an initial setting made at installation, or subsequent alignment checks. However, it carries two scales which are visible through a port in the front of the instrument, and show the actual setting. The gun train from the roller-path high point is indicated by a circular scale, while the inclination at the high point is shown on a straight scale in the middle of the train-indicating dial.

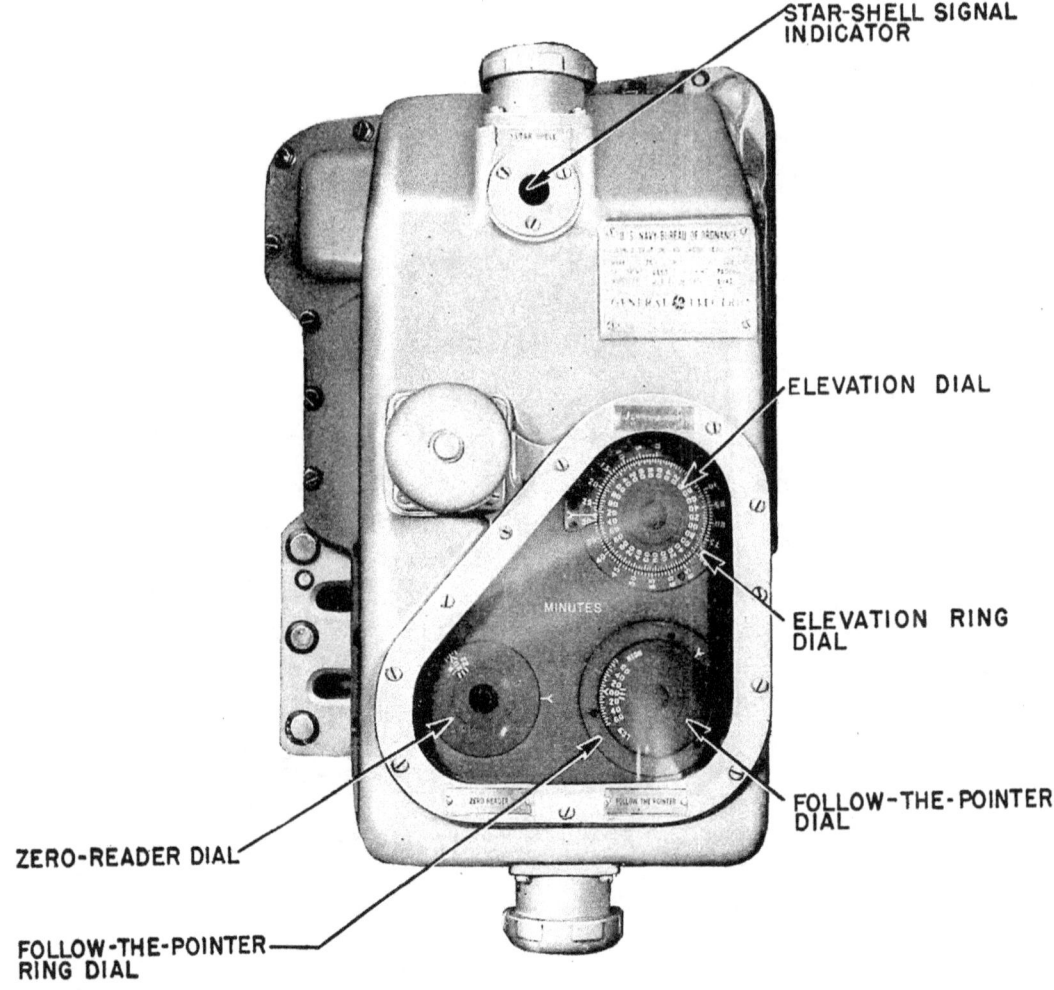

FIGURE 17E2. Elevation indicator-regular; top view.

The *elevation dial* and the *elevation ring dial* on the instrument top are fine and coarse reading respectively, and indicate corrected elevation response. This angle is equal to the gun response, or elevation above the roller path, plus or minus the roller-path tilt correction and the elevation-parallax correction when one is applied. Consequently, the elevation-dial group reading is equal to the gun elevation order when the gun is properly elevated. These dials are used by the pointer just as the trainer uses his train angle dials.

The *zero-reader dial* and the *follow-the-pointer dial* are attached to the indicating-synchro shafts and operate like the corresponding dials of the train indicator-regulator, except that they are actuated by gun-elevation order. The coarse indicating synchro stator and the *follow-the-pointer ring dial* are driven by corrected elevation response. Hence, when the zero-reader dial index is opposite the fixed pointer and the follow-the-pointer dial index is opposite the ring index, the gun complies with

CHAPTER 17

the order, and is elevated above the reference plane by gun elevation order plus or minus the parallax correction when one is applied.

17E4. Fuze-setting indicator-regulator. There are two principal types of fuze-setting equipment used with 5"/38 cal. guns. One, described in Chapter 6, is integral with the projectile hoist on mounts with handling rooms immediately beneath the gun. The other is a separate

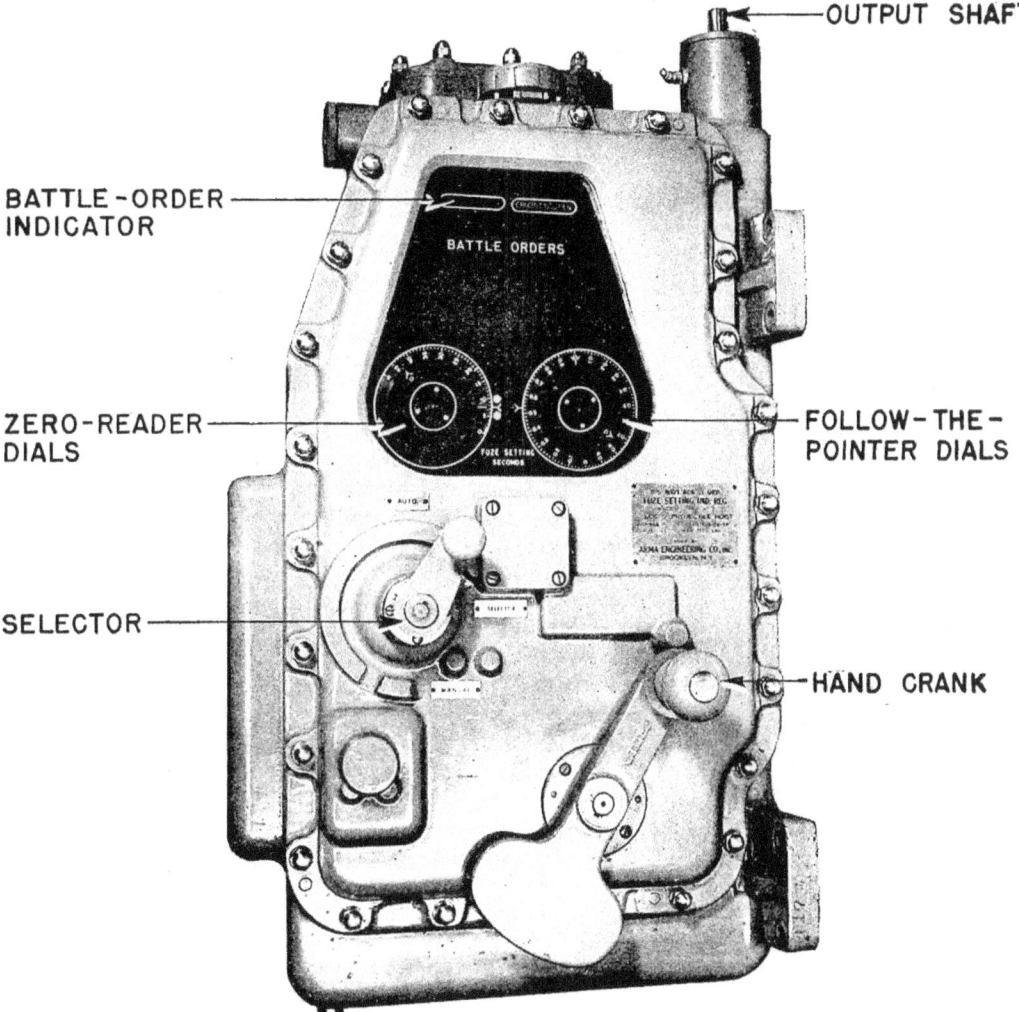

FIGURE 17E3. Fuze-setting indicator-regulator; front view.

unit containing three fuze-setting pots, mounted on the gun carriage behind the pointer on those installations served from hoists outside the working circle. Both types are controlled by fuze-setting indicator-regulators, of which there are several modifications. However, all these indicator-regulators are very much alike in their functions and operation.

Each fuze-setting indicator-regulator can provide automatic or manual fuze setting, the manual control being governed either by matching dials or by simply positioning the dials at the required fuze setting. The instrument contains a fine and a coarse synchro for receiving fuze-setting order from the computer, an electric power motor for driving the fuze-setting mechanism in automatic

RESTRICTED

control, and a pilot motor to control the power motor. The face of a typical unit is shown in figure 17E3.

The *zero-reader dials* (left group) and the *follow-the-pointer dials* (right group) are coarse and fine reading respectively. The inner dial of each group is attached to its corresponding synchro shaft, and bears a single index mark. Each ring-dial graduation represents one second on the zero-reader group and 1/50 second in the follow-the-pointer group. The actual fuze setting applied by the mechanism, or response, drives the ring dials and the zero-reader stator. The applied fuze setting can be read against the fixed pointers of the two groups. When the synchros are energized and follow-up is as ordered, the zero-reader-dial index is aligned with its fixed index, and the follow-the-pointer-dial index is opposite the zero index on its corresponding ring dial.

The *selector* sets up the indicator-regulator either for automatic or manual control. The *hand crank* is connected to the drive only when the selector is set for manual, and may then be used to match the zero-reader and follow-the-pointer dials, or to position the ring dials as required, thereby manually setting the fuzes.

In systems where the director has a battle-order transmitter, the fuze-setting indicator-regulator contains the battle-order indicators, as shown in the illustration. The upper indicator shows *fire, cease, load hoist,* and *circuit broken,* the latter signal indicating loss of power in the battle-order indicator system. The lower indicator shows *antiaircraft, illuminating, common,* and *dive attack.* Some fuze-setting indicator-regulators have a single star-shell indicator in place of the battle-order indicators.

17E5. **Sight-setting mechanism.** The line of sight for a 5"/38 cal. gun mount is established by movable-prism telescopes as explained in Chapter 6. Sight-angle and sight-deflection settings are obtained by moving elevation and deflection prisms respectively. The prism arrangement is such that the sight angle is set in the plane through the bore axis and perpendicular to the trunnion axis; i.e., in the vertical plane when the roller path is horizontal. Sight deflection is set in the slant plane which contains the line of sight and is parallel to the trunnion axis.

The sight-setting mechanism governs the telescope-prism settings through a system of shafts and gears whose movement or response registers on the mechanism dials shown in figure 17E4. There are three distinct dial groups: *sight angle, sight deflection,* and *range.* Sight setting, or follow-up, is accomplished manually in all methods of operation by means of the *sight-angle handwheel* and *sight-deflection handwheel.*

The *sight-deflection dials* consist of an inner dial and an outer ring. The dial has a single index mark, and is mounted on the shaft of a synchro which receives sight deflection from the computer. The ring is graduated in mils, the zero deflection marking being 500. The ring and the telescope deflection prisms are moved simultaneously by the sight-deflection handwheel. Normally, the synchro is actuated by the computer and the deflection setting is maintained by keeping the 500 mark opposite the dial index. However, the response, or setting, may be made by placing the required reading on the ring opposite the fixed index.

The *sight-angle dials* also consist of an inner dial and an outer ring, fundamentally differing from the sight-deflection dials only in the markings on the outer ring. The ring graduations are in minutes, 2,000 representing a zero sight angle. The synchro carrying the dial is driven by sight angle from the computer, while the ring and the telescope-elevation prisms are driven by the sight-angle handwheel. The response, or sight-angle setting, is normally maintained by keeping the 2,000 mark opposite the dial index, but may be made by placing the required value on the ring opposite the fixed index.

The *range dial,* geared to the sight-angle handwheel, is completely housed except for a small window. This dial indicates the range in yards that corresponds to the angle shown on the sight-angle ring dial, as shown in the range table for the gun. In other words, the agreement between the two dial groups applies only in surface fire. This range dial has particular usefulness in local control for surface fire where the personnel at the gun have to estimate range and set it directly on the sights.

CHAPTER 17

17E6. Gun-crew activities for director control. Among the personnel at the gun mount, those whose duties are most affected by the use of director control are the pointer, trainer, sight setter and fuze setter. It should be apparent that the numerous alternatives available to these crew members make it essential that each be thoroughly trained and capable of using his own judgment in

FIGURE 17E4. Sight-setting mechanism; front view.

changing from one operating method to another. The duties outlined in the following paragraphs provide a general picture of their activities in director firing methods.

Pointer and trainer. In automatic or remote control, the automatic synchros within the train and the elevation indicator-regulators have complete control of gun movement. The pointer and trainer have nothing to do but observe the operation to see that the equipment is functioning properly. The indicating synchros are operating at the same time, consequently the zero-reader and follow-the-pointer dials are continuously matched.

DUAL-PURPOSE BATTERY FIRE-CONTROL SYSTEM

Should the remote control fail in either elevation or train, the pointer or trainer, as required, changes to indicating control at once. To do this, he shifts his selector lever to high or low for local power-drive control and operates his handwheels to keep the indicating dials matched. If the power drive fails while the indicators still function, the operator can shift to manual and stay in indicating control.

If the indicator-regulators fail to provide either indications or remote control, the operator immediately changes to telescope control by simply looking through his telescope, for the sights are continuously set to provide for just such an occurrence. He then controls the gun movement with local power-drive control, or shifts to manual if necessary. It should be noted that this telescope control is still a form of director control, for the firing circuits and sight settings are controlled by the director system. Also, it is quite possible for the trainer to use one method while the pointer is using an entirely different procedure.

In any form of director firing, the pointer normally keeps his firing key closed. This presupposes that accurate training and pointing are provided continuously. The pointer can obviously break the circuit if for any reason the gun-elevation response is wrong temporarily, but the trainer has no such control of the circuit. Hence, if the trainer cannot maintain the correct train, he must so inform the pointer who will then break the circuit.

Fuze setter. Normally, fuze setting is fully automatic, and the fuze setter simply observes the operation. If the power motor fails to keep the indicator-regulator dials matched, the fuze setter shifts the selector lever to manual and matches the dial indexes by handcrank movement. If the synchro indications fail, he may set the ring dials in accordance with orders received from the plotting room by telephone. Also, in barrage fire with fixed-fuze settings, he may use these outer dials. In star-shell fire, used only against surface targets, the fuze-setting order entering the indicator-regulator is for star-shell fire, and the fuze setter operates as in service-projectile fire.

Sight setter. Primarily, the sight setter continuously operates the handwheels to keep the ring dial zero points (500 or 2,000) opposite the indexes on the central dials. This keeps the telescope prisms properly positioned so that either (or both) the pointer and trainer may turn to telescope control without delay in case of necessity. If the synchro transmission fails to actuate the inner dials, the sight setter turns the handwheels to make the ring dials read the required angles opposite the fixed pointers. The settings for this method of operation can be received by telephone from the computer if the director system is functioning. In local control, the sight setter may obtain the settings from tables, of which there are several varieties.

CHAPTER 18

MAIN BATTERY FIRE-CONTROL SYSTEM

A. GENERAL DESCRIPTION OF THE SYSTEM

18A1. Introduction. The main battery fire-control system described in this chapter is that used to control the 16-inch battery on newly constructed battleships. The main battery on these ships consists of nine guns (45 cal. on some, 50 cal. on others) mounted in three triple-gun, centerline turrets, two forward and one aft.

The fire-control equipment for the main battery is similar in principle to that employed for the control of the 5"/38 cal. battery. Some mechanisms necessary in the solution of the AA problem are not required for a main-battery control installation which is used solely against surface targets; hence, major caliber control units might be expected to be more simplified. However, accurate control of heavy guns for long-range fire requires a system just as elaborate as the Mark 37 director type.

The principal units employed in a major-caliber control system are comparable in function to those used in dual-purpose installations. They include directors, range keepers (computers), and stable verticals (stable elements). In effect, duplicate primary installations are provided for control flexibility. The system includes auxiliary arrangements which are available in the event of casualty to the primary units.

The main directors determine target bearing, range, level angle, and cross-level angle. Target angle and target speed are estimated by an observer at the director as in other control systems. Such measured and estimated quantities are transmitted to the plotting room units where they are used in computing sight angle, sight deflection, and gun-elevation and train orders. These computed quantities are transmitted electrically to the turrets where the sights are set by matching pointers, and the guns are positioned automatically or by matching pointers.

18A2. Components of the system. The principal units making up the main battery fire-control system are:

1. Main directors—two Mark 38.
2. Range keepers—two Mark 8 (Mods. 9 or 11).
3. Stable elements—two Mark 41 or 43 stable verticals (properly designated *gun directors*).
4. Range finders—one in each Mark 38 director and one in each turret.
5. Range receivers—two Mark 1 (with Mark 8 Mod. 9 range keeper only). The range receiver is incorporated in the graphic plotter of the Mark 8 Mod. 11 range keeper.
6. Auxiliary director—one Mark 40 director.
7. Auxiliary computers—one Mark 3 computer and one Mark 5 trunnion-tilt corrector in the fire-control tower, and one Mark 3 Mod. 1 or Mod. 2 computer in each turret-officer's booth.
8. Sight-setting indicators, sights, receiver-regulators, and other indicators in each turret.
9. Synchro transmitting, indicating, signaling and intercommunicating systems between the units in the system.

The main directors are mounted high in the superstructure to provide maximum visibility. The auxiliary director is at the main-battery control station in the fire-control tower, and usually is the main-battery reference point. Figure 18A1 shows the location of these directors and the guns.

The range keepers, stable elements, and range receivers are in the main-battery plotting room below decks. One set is shown in figure 18A2. The auxiliary computers in the fire-control tower and in each turret-officer's booth are used in auxiliary methods of control when the range keepers in the plotting room are inoperative.

18A3. Mark 38 director equipment. The director structure consists of a foundation ring secured to the ship's framework and a roller-bearing assembly which supports a cup-shaped plat-

RESTRICTED

FIGURE 18A1. Main-battery directors.

Figure 18A2. Main battery plotting-room equipment.

MAIN BATTERY FIRE-CONTROL SYSTEM

form covered by an armored shield. The director can rotate 198° right and left of the center position. An exterior view is shown in figure 18A3 and a plan view of the interior arrangements in figure 18A4. The inside working diameter and height are approximately 8 feet and 6½ feet respectively.

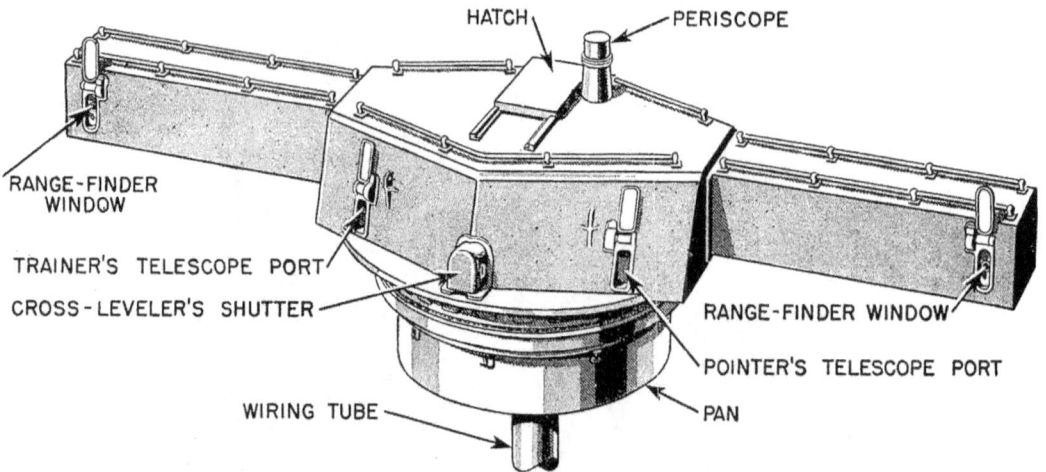

FIGURE 18A3. Mark 38 director; exterior view.

Equipment. The principal control units and equipment located in the director housing are:

1. *Training power drive*—an amplidyne (electric) power drive and receiver-regulator at the trainer's station. The drive is controlled by operation of the trainer's handwheels, by electrical inputs from the range keeper, or by a slewing control switch.

2. *Training gear*—includes trainer's telescope, handwheels, director-train indicator, director-train transmitters, and a horizontal parallax corrector.

3. *Leveler's telescope mount*—includes leveler's telescope, handwheels, roller-path tilt corrector, vertical parallax corrector, level-indicating dials, and level transmitters.

4. *Cross-leveler's telescope mount*—includes telescope, handwheels, cross-level indicating dials, and cross-level transmitters.

5. *Range finder*—26½-foot stereoscopic type with hand-operated range transmitter. Mount is not cross-leveled but may have an automatic level stabilizer.

6. *Spotter's periscope*—a 12 power periscope used for target and fall-of-shot observations, and for general lookout purposes. The line of sight can be inclined 15° above and below the deck (reference) plane.

7. *Battle-order indicator*—indicates range, deflection, and the battle orders transmitted from the range keepers in plot. It also contains the *plot ready* light operated from the range keepers.

8. *Cease firing contact maker*—sends cease firing signals to the guns.

9. *Time-of-flight buzzer*—operated by the time-of-flight mechanism in the range keeper. It sounds just before the projectiles reach the end of their flight and thus assists the spotter in identifying his ship's salvo.

18A4. General director functioning. The primary functions of the Mark 38 director are to obtain target bearing, level angle, and cross-level angle for the system. An additional function of measuring range is accomplished by the range finder mounted in the director. A functional diagram of the director is shown in figure 18A5.

The director is trained by any one of the following methods: (1) *manual*, in which the train handwheels are directly geared to the training rack, (2) *local*, in which rotation of the train hand-

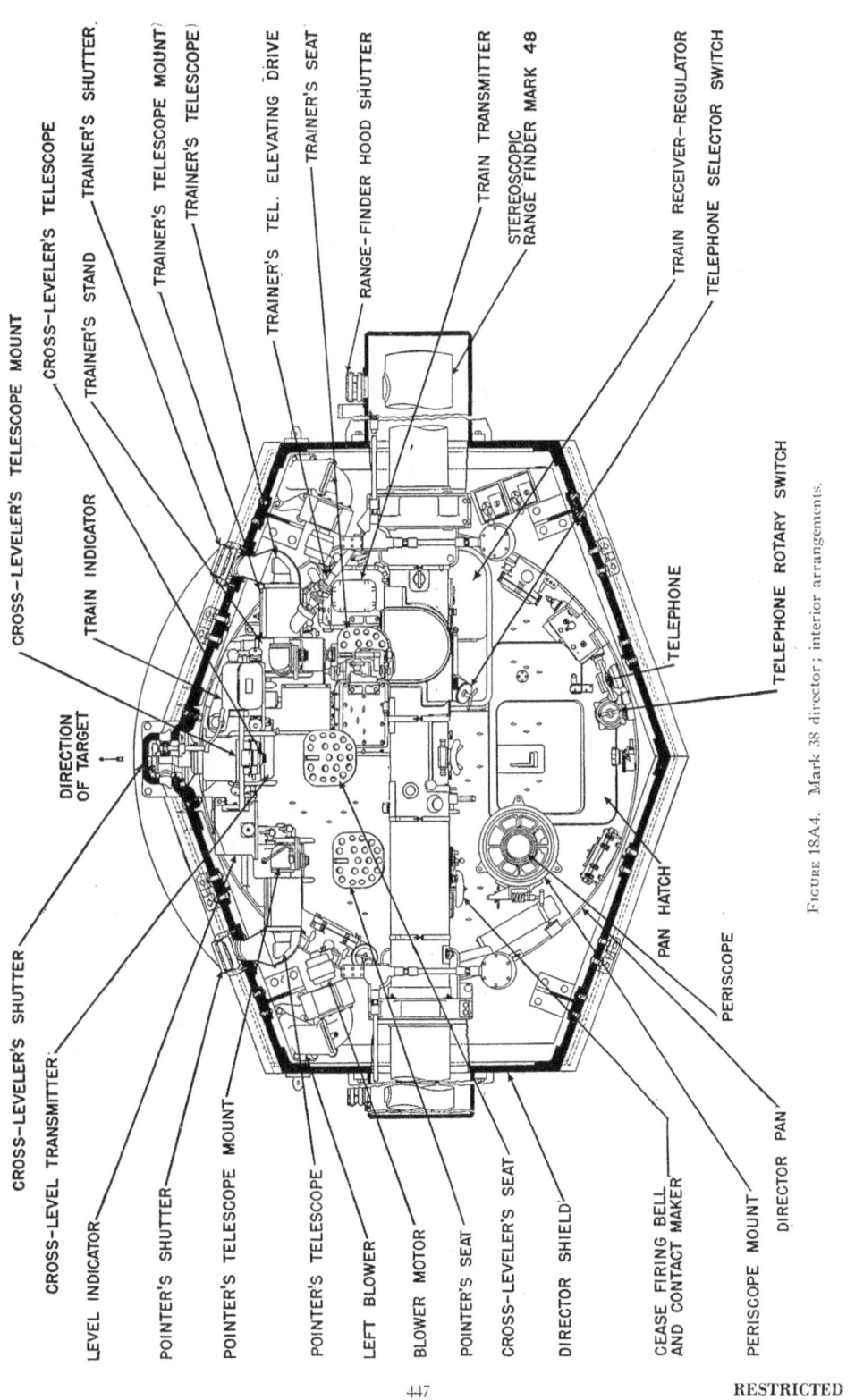

Figure 18A4. Mark 38 director; interior arrangements.

MAIN BATTERY FIRE-CONTROL SYSTEM

wheels controls the power drive through the receiver-regulator, (3) *automatic*, in which increments of director train generated in the range keeper control the power drive through the receiver-regulator, and (4) *slew*, in which a slewing switch, usually operated by the control officer, takes control of the power drive when in local or automatic.

Director train is corrected for horizontal parallax with respect to the main-battery reference point before it is transmitted to the range keepers. Gun-train order, sent to the turrets from the range keepers, is calculated for this reference point and corrected at the turrets for parallax due to their horizontal displacement from the reference point.

Level and cross-level angles are generated at the director by the pointer (leveler) and cross-leveler as they turn their handwheels to keep their crosswires on the horizon. If the horizon is not visible these values can still be generated by matching pointers on dials actuated by a stable element. Measured level angle is mechanically corrected for vertical parallax and director roller-path inclination before it is transmitted to the range keepers. The cross-level telescope is so mounted that it measures true cross-level; i.e., it is offset to eliminate vertical parallax due to horizon dip. The director telescopes are not stabilized in cross-level as they are in the Mark 37 director, hence level and cross-level angles are measured in planes perpendicular to the deck (reference) plane. The stable elements in the plotting room also generate level and cross-level angles in these same planes.

18A5. Director personnel. The personnel required for operating the director includes: (1) *spotter*, in immediate control of the director, (2) *trainer*, (3) *pointer* (leveler), (4) *cross-leveler*, (5) *range-finder operator*, (6) *parallax range setter and talker*, and (7) *stand-by range-finder operator and talker*.

The spotter estimates target angle, target speed, and wind conditions, and telephones these values to plot so that they may be entered into the range keeper.

Target designation transmitted from main-battery control in the fire-control tower is indicated on the train indicator at the trainer's station. The trainer matches pointers on target designation and director train dials and picks up the target in his telescope. Director train is automatically transmitted to the range keepers in plot. The trainer stays on the target by turning his handwheels until he is ordered to switch to automatic control. In automatic control increments of generated train transmitted from the range keeper control the training drive. If the range-keeper solution is correct, the trainer's crosswire will stay on the target. If the crosswire drifts off the target it may be brought back on by turning the handwheels. Generated and observed target bearings are indicated at the range keeper and a comparison of these values serves in rate control as a partial check on the range-keeper's solution of the problem. If firing circuits are controlled at the director, the firing key on either the level or cross-level handwheel is used, depending upon the firing method.

18A6. Stable element. The Mark 41 stable vertical (stable element), has a gyroscopic element very similar to that used in the Mark 6 stable element described for the Mark 37 system. The operating principle and the functions performed are the same for these two units. The Mark 41 level and cross-level gimbals are arranged so that level and cross-level angles are measured in planes perpendicular to the deck plane instead of in vertical planes.

Continuously generated values of level and cross-level are transmitted mechanically and electrically to the Mark 8 range keeper where they are used in determining sight angle, sight deflection, gun-train order and gun-elevation order. Selected values of level or cross-level are also transmitted electrically for certain types of firing described below. The stable element provides means for firing the battery automatically or by hand at a selected value of level or cross-level angle.

Inputs. The required inputs include:
1. Selected level or cross-level—set by hand.
2. Own-ship course—electrically from gyro repeater.
3. Director train—electrically from director transmitter.

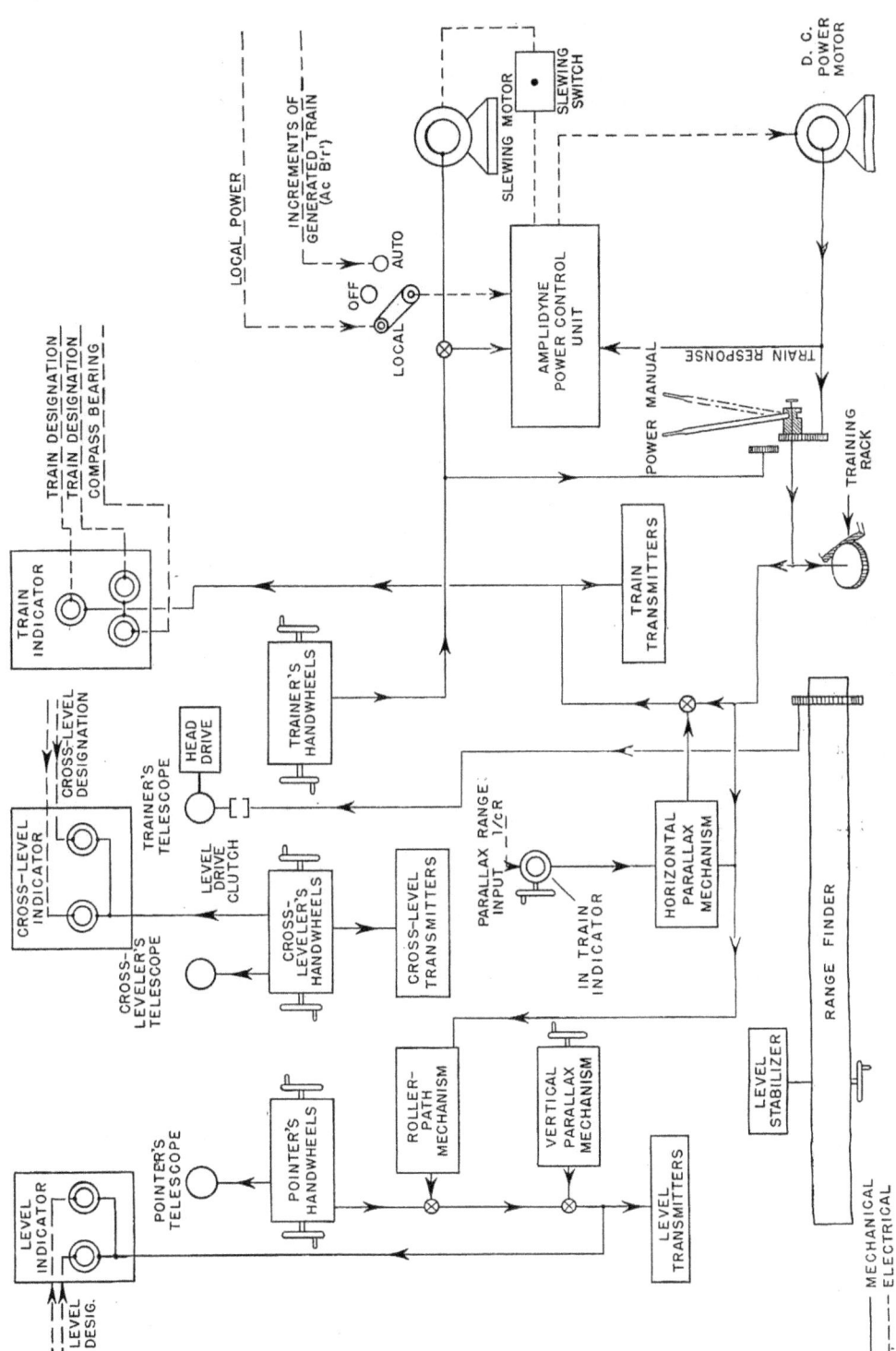

FIGURE 18.A5. Mark 38 director; functional diagram.

MAIN BATTERY FIRE-CONTROL SYSTEM

Outputs. The mechanical and electrical outputs are summarized as follows:
  1. Continuous level—mechanically to deck-tilt computer in range keeper.
  2. Continuous cross-level—mechanically to deck-tilt computer in range keeper.
  3. Continuous level—electrically to range keepers and directors.
  4. Continuous cross-level—electrically to range keepers and directors.
  5. Selected level (or cross-level)—electrically to range keepers and directors.

18A7. General range-keeper functioning. There are two Mark 8, mod. 9, range keepers in the main-battery plotting room. A range keeper solves the surface fire-control problem set up on its dials, indicates the range-keeper inputs and outputs, and transmits the values necessary to set sights, position the guns, and train the director. Battle orders, plot ready, and fall of salvo signals are also transmitted to the directors and turret-officer's booths. A top view of the range keeper is shown in figure 18A6.

Computations are based on initial velocities between 1,650 f.s. and 1,800 f.s. (target charges) or between 2,150 f.s. and 2,300 f.s. (service charges). Initial velocity may be set at any value within these limits by means of hand knobs.

The gun-elevation order calculated by the range keepers consists of level angle added to the sight angle corrected for the effect of trunnion tilt. This order may be for either a selected value of level angle or a continuously changing value which varies with the roll and pitch of the ship. Gun-train order transmitted from a range keeper is director train combined with the computed sight deflection corrected for the effect of trunnion tilt. The inputs and outputs of the range keeper are shown in figure 18A7.

No provision is made for automatic rate controlling; hence, observed range and bearing are checked with the values generated by the range keeper, and rate controlling is done manually by adjusting the hand inputs of target course and speed.

Graphic plotter. A graphic plotter attached to the range keeper automatically plots advance range, and generated present range and sight deflection. These values can be read from scales on the plotting board. A protractor with scale graduated for range rate in knots can be used to check generated range rate with that determined from the range-receiver plot. Time scales are automatically marked along each edge of the sheet at one minute intervals.

18A8. Range-keeper inputs and outputs. The inputs required to set up the problem and make the necessary corrections include:

Electrical
  1. Own-ship course.
  2. Own-ship speed.
  3. Director train (automatic and indicating).
  4. Level (selected or continuous).
  5. Cross-level (selected or continuous).

Mechanical (from stable vertical)
  1. Level (continuous).
  2. Cross-level (continuous).

Hand
  1. Target course.
  2. Target speed.
  3. True direction true wind.
  4. True-wind speed.
  5. Initial velocity.
  6. Powder charge.
  7. Range spots.
  8. Deflection spots.
  9. Selected train.
  10. Initial range setting.
  11. Initial bearing setting.

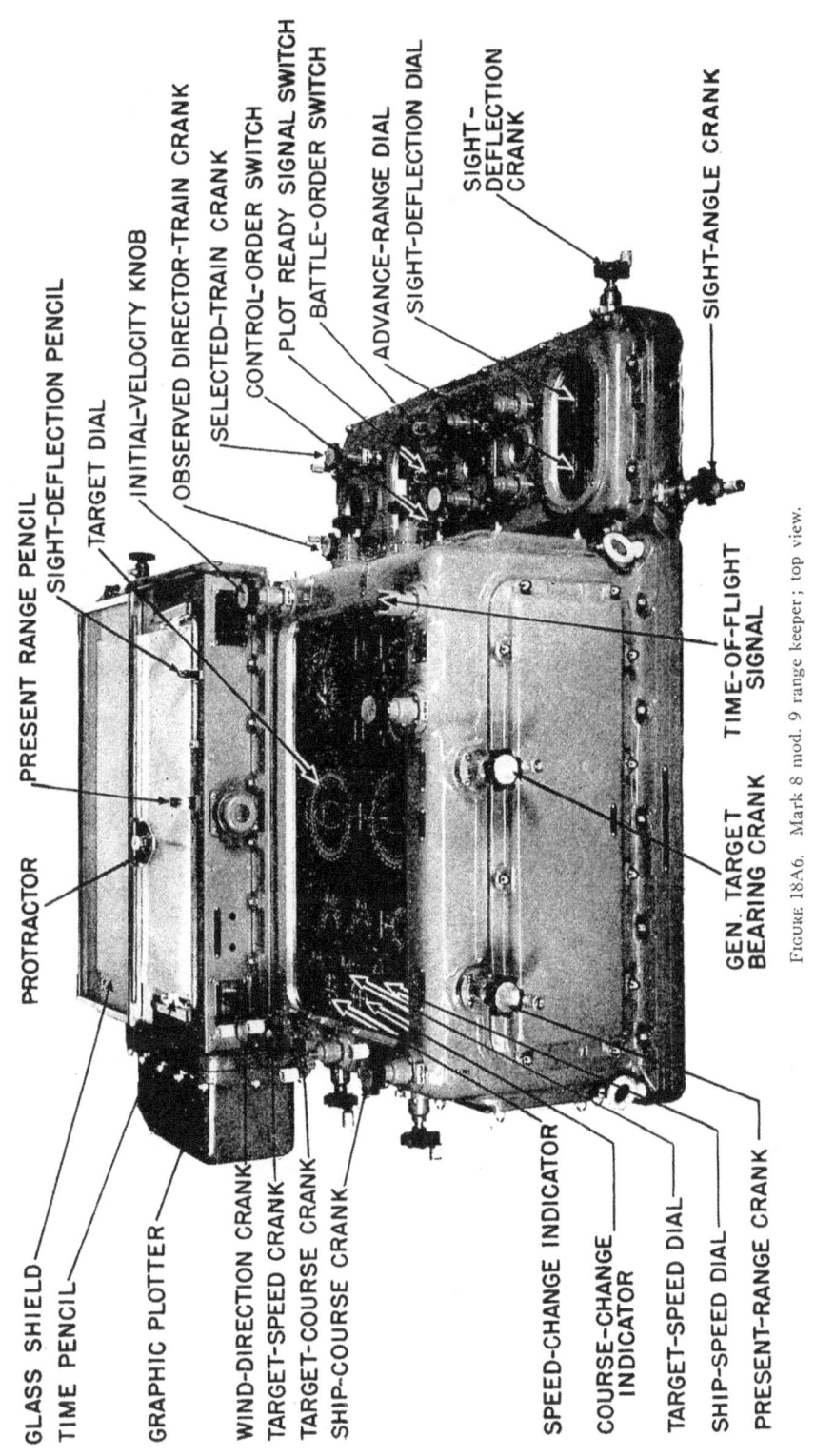

Figure 18A6. Mark 8 mod. 9 range keeper; top view.

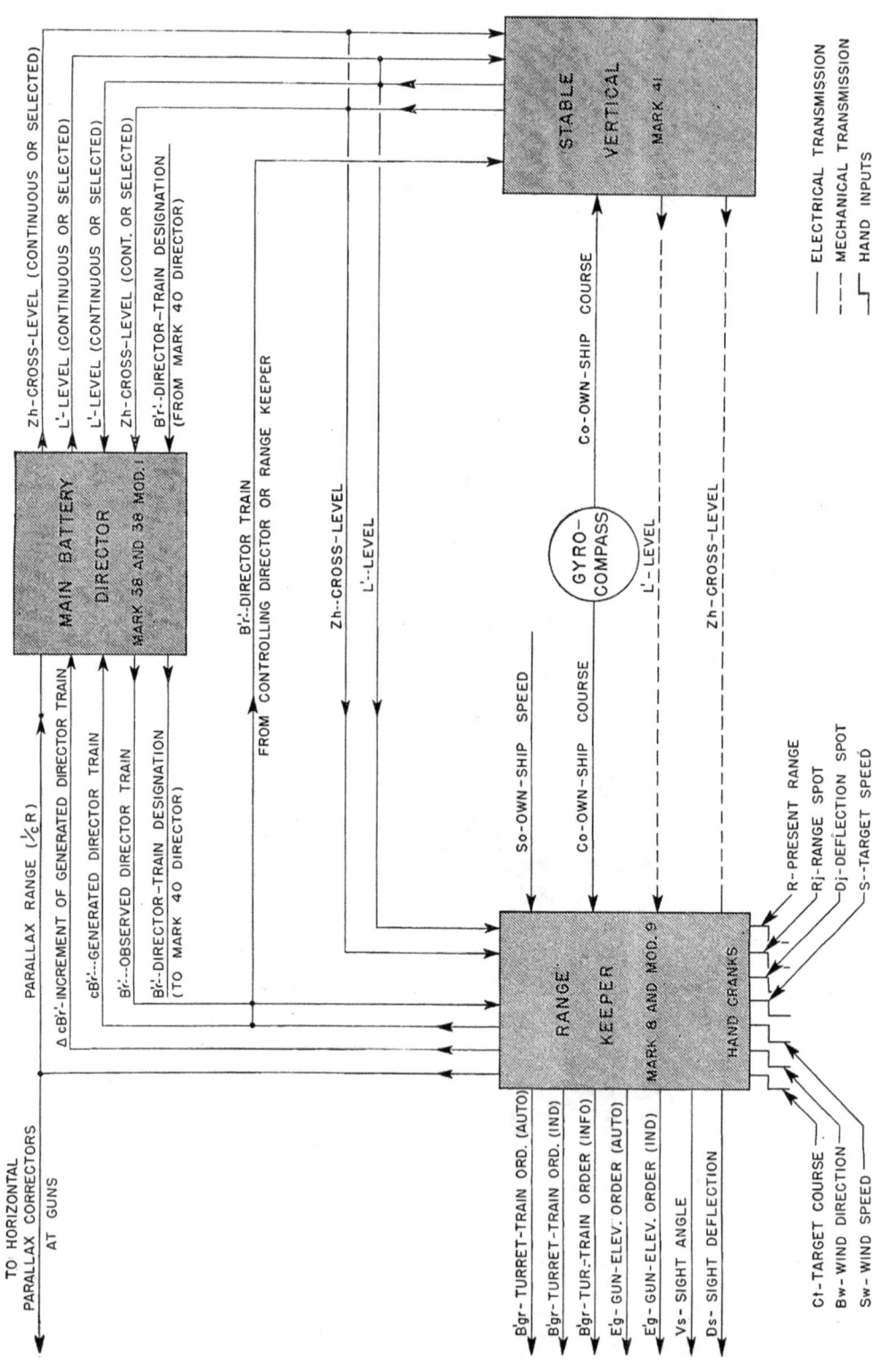

FIGURE 18A7. Mark 38 director system; inputs and outputs.

CHAPTER 18

Knobs are also provided for setting deflection, range, battle order, control order, and plot ready signal transmitters, and for setting sight deflection and sight angle which are normally handled electrically.

Outputs. The following outputs are all electrically transmitted:

1. Gun-train order—automatic.
2. Gun-train order—indicating.
3. Gun-train order—information.
4. Gun-elevation order—automatic.
5. Gun-elevation order—indicating.
6. Sight deflection.
7. Sight angle.
8. Generated director train.
9. Increments of generated director train.
10. Deflection.
11. Range.
12. Parallax range (1/cR transmitted to the convergence mechanisms of the various training units).

18A9. Range receiver. The range receiver is located near the range keeper. Ranges from the five main-battery range finders in the two directors and three turrets are received and recorded on a plotting sheet, one at a time. When a range-finder operator wishes to transmit a range he pushes a signal button at his station and a selector switch at the range receiver connects his transmitting circuit to the recording device. The recording device after being automatically positioned by the range transmitter at the range finder, depresses and prints a number on the plotting sheet, a number which identifies the range finder from which the range is received, and whose position on the paper indicates the range. All five ranges can be recorded in this manner in less than 10 seconds.

The range-receiver operator plots the mean range line and obtains the range rate from the slope of the line. He periodically gives the mean range and range rate to the range-keeper operator. The Mark 8, Mod. 11 range keeper incorporates the range receiver in the graphic plotter.

18A10. Auxiliary director system. The Mark 40 director is an auxiliary installation located in the fire-control tower (see fig. 18A8), and it includes the following units: (1) combined *spotting glass* and *trainer's periscope*, (2) two *periscopes* and *mounts* for leveler and cross-leveler, (3) Mark 3 *computer* and Mark 5 *trunnion-tilt corrector*, (4) *target designator*, (5) *target-bearing transmitter*, and (6) *level* and *cross-level transmitters*. The 15-foot base length spotting glass and the periscopes are mounted on top of the fire-control tower and can be simultaneously trained through an angle of approximately 300° by means of the trainer's handwheel drive. There are no power drives in the director.

The trainer establishes the LOS to the target or matches dials on the target designator to pick up a target designated by one of the other directors. The target-bearing transmitter sends director train to the Mark 38 directors and the range keepers. Since the fire-control tower is usually the reference point in train, director train need not be corrected for horizontal parallax.

The crosswires of the level and cross-level periscopes can be kept on the horizon so that either level or cross-level can be measured with each periscope. The angles are transmitted automatically to the range keepers and the trunnion-tilt corrector.

The Mark 3 computer and Mark 5 trunnion-tilt corrector, shown in figure 18A9, receive director train mechanically from the Mark 40 director, and solve the fire-control problem set up on the dials for sight angle and sight deflection and correct these values for the effects of trunnion tilt. Gun-elevation and train orders are then computed and sent to the turrets via the switchboard in turret 2.

Operation and personnel. The chief fire-control officer (gunnery officer) and 10 men are usually stationed at main-battery control to operate the Mark 40 director and the computing equipment. Their duties are:

1. The *chief fire-control officer* designates the target or targets and the kind of fire to be used.

MAIN BATTERY FIRE-CONTROL SYSTEM

2. The *trainer* keeps his vertical crosswire on the target and thus transmits target designation to the Mark 38 directors, director train to the range keepers and stable elements, and director train to the local computer.

3. The *spotter* estimates target course and speed, wind force and direction, and spots after firing begins.

4. The *leveler* generates level angle by keeping his crosswire on the horizon. When designated as firing pointer he controls firing of the battery by means of the firing key on his handwheels.

5. The *cross-leveler* generates cross-level angle by keeping his crosswire on the horizon in a direction 90° from the LOS to the target. He may be designated to fire the battery when the selected cross-level method is used.

6. The *range talker* receives range by telephone and gives this information to one of the local computer operators orally.

7. Five *computer operators* introduce hand inputs and operate follow-up cranks as necessary on the computer and trunnion-tilt corrector.

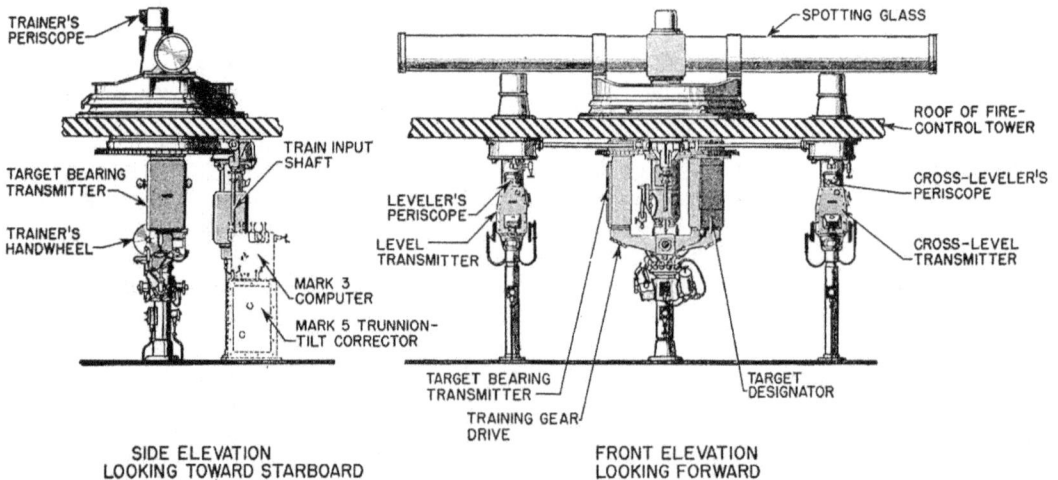

FIGURE 18A8. Mark 40 director.

18A11. Methods of firing. Salvo-type fire is used with the following methods of positioning the guns: (1) selected level, (2) selected cross-level, and (3) selected train. For the selected level method a value of level angle is set on the stable element or at the leveler's station in the controlling director, and this value is used by the range keeper in the calculation of gun-elevation and gun-train orders. Hence the guns are correctly positioned for firing only when the actual and selected-level angles are equal. The guns may be fired from the stable element either automatically or manually, or from the leveler's station in the controlling director.

The selected cross-level method is similar to the selected-level method. A selected value of cross-level is set on the stable element or at the cross-leveler's station in the director, and is used in the calculation of the gun-elevation and gun-train orders. The guns may be fired automatically or by hand from the stable element, or from the cross-leveler's station in the director.

In the selected-train method a fixed value of gun-train order is selected at the range keeper by means of the *selected train knob*. Rotation of the knob locks the indicating transmitters at the selected value of train, but the gun-train automatic transmitter continues to transmit the correct value of gun-train order. The selected-train dials on the range keeper indicate the point at which the computed value of gun-train order equals the selected value, and the guns may be fired at that time. This is done by means of a firing key at the range keeper.

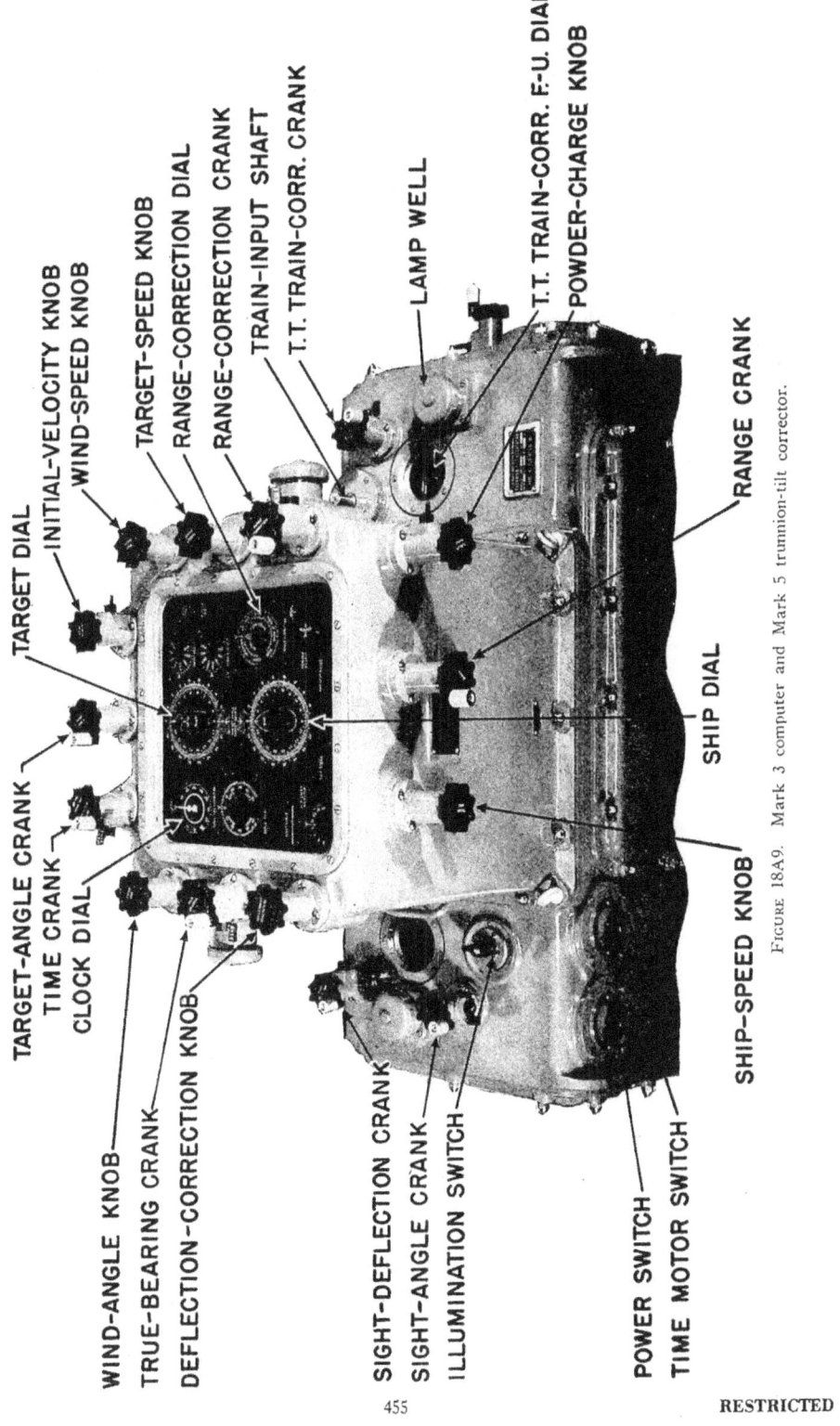

FIGURE 18A9. Mark 3 computer and Mark 5 trunnion-tilt corrector.

MAIN BATTERY FIRE-CONTROL SYSTEM

Either continuous or salvo fire may be used with continuous aim, for which the range keeper constantly transmits values of gun-elevation and train orders. The guns are positioned for firing in accordance with these gun orders, and are usually fired from the director or the stable element.

B. TURRET FIRE-CONTROL EQUIPMENT

18B1. Turret units. Each turret has the following fire-control equipment: (1) range finder, (2) sight-setting indicators, (3) telescope sights, (4) auxiliary Mark 3 computer, and (5) associated indicating and transmitting circuits. The general function and location of these units is described in relation to three turret control stations: (1) turret-officer's booth, (2) sight stations, and (3) gun-elevating and turret-training stations on the machinery floor.

18B2. Turret-officer's booth. The fire-control equipment in the turret-officer's booth includes: (1) a *range finder*, (2) a Mark 3 *computer*, (3) a *multiple turret-train indicator*, (4) a *battle-order indicator*, and (5) a *fire-control switchboard*. The range finder has a base length of 43 feet, and is designed for ranges from 4,500 yards to about 70,000 yards. Turret 1 has the coincidence type and the other two turrets the steroscopic type of range finder. Both are equipped with a range transmitter which sends electrically the observed ranges to the range receiver in the plotting room.

When the turret is in local telescope control the computer solves the ballistic problem and determines the values of sight angle and sight deflection required to set the sights. These values are indicated on counters at the computer, and are transmitted orally to the sight setters.

The multiple turret-train indicator has dials which show that the turret has been trained in accordance with the orders received from the controlling director. A direct-reading dial indicates turret train.

The battle-order indicator indicates range, deflection, and the following battle orders transmitted from one of the range keepers in plot:

1. *Fire* (commence firing).
2. *Cease* (cease firing).
3. *Auto* (director fire automatic).
4. *Ind.* (director fire indicating).
5. *Local* (local control).
6. *Circ. Brok.* (circuit broken).

18B3. Sight stations. The turrets have duplicate sight stations enclosed in flame-proof compartments right and left of the wing guns. Each sight station has a sight-pointer's telescope, sight-trainer's telescope, and a sight-setting indicator. Like elements of the two sight stations are interconnected. The arrangement is such that either station may be used for sight setting and sighting operations while the other serves as a stand-by.

18B4. Elevating and training stations. The gun-elevating and turret-training stations are located next to the hydraulic speed gear receiver-regulators on the machinery floor below the guns. Each gun has a separate elevating power drive and hence three elevation-control stations are necessary for a three-gun turret. The sight-pointer's and sight-trainer's handwheels are mechanically connected to indicating dials at the gun-elevating and turret-training stations, and turret train can be controlled directly from the sight-trainer's station.

18B5. Gun-laying and sighting personnel. Gun-elevating, turret-training, and sight-control stations require the following personnel:

1. Three *gun layers*, one for each gun-elevating-control station on the machinery floor.
2. One *gun-train operator* at the gun-train-control station on the machinery floor.
3. Two *sight trainers;* one in each sight station.
4. Two *sight pointers;* one in each sight station.
5. Two *sight setters;* one in each sight station.

18B6. Relation of sight control to turret control. The sight-setting indicator receives electrically and indicates sight angle, sight deflection, and battle orders transmitted from the plotting

CHAPTER 18

room. Sight-angle and deflection settings are made manually by matching pointers. The scales are marked 2,000 minutes for zero sight angle and 500 mils for zero deflection.

A gun-elevation transmitter at each sight-pointer's station transmits gun-elevation order generated by bringing the sights on the target to elevation indicators at the gun-layer's station. This gun-elevation order is equal to the depression of the pointer's sight, which in turn is equal to the sight angle set plus the level angle. Thus the pointer's sight serves as a local director transmitting gun-elevation order to the indicator at the gun-layer's station. Switching arrangements are provided so that the local gun-elevation order transmitter in turret 2 also can be connected to the indicators in the other two turrets.

The gun-elevation indicator at the gun-layer's station also shows gun-elevation order received electrically from the controlling director system. The indicator incorporates mechanisms which correct gun elevation for inclination of the turret roller path and for the difference in equivalent service rounds fired. The roller-path inclination compensator has two scales; one for setting inclination at any value from 0° to 1°, and the other for setting bearing of the high point. The correction for the difference in equivalent service rounds fired compensates the difference between the number used to obtain the I.V. setting on the range keeper, and the number actually fired by the guns. This correction is a function of both sight angle and difference in velocity loss. The sight angle is introduced mechanically from the sight-setting indicator and the other input is set by hand. The corrections are applied to the actual gun elevation before the latter is indicated on the elevation dials, and before it is transmitted to the follow-up mechanism in the receiver regulator.

The turret-train indicator and transmitter at the turret-training station shows the required train angle, the turret-train angle, and the turret train with horizontal parallax correction subtracted. The unit includes a horizontal parallax-correction mechanism which corrects the gun-train order received from the range keepers for the parallax error introduced by the horizontal displacement of the gun from the main-battery reference point. The required parallax range ($1/cR$) is received from the range keeper, and is introduced manually by matching pointers. The horizontal displacement is set for the particular gun position when the equipment is installed and need not be changed. Turret train with parallax correction subtracted is transmitted electrically to the multiple turret-train indicators in the turret-officer's booth. This value can also be sent to turrets 1 and 3 from the transmitter in turret 2.

18B7. Control arrangements. There are four methods available for controlling the elevating and training hydraulic-speed gear in the turrets:

1. *Primary control.* Director control from the forward or after main-battery director (Mark 38) through the plotting-room range keepers and the indicators at the gun-elevating and training stations in the turret. The operators at these stations match pointers on their respective indicators by turning handwheels, and thus control the hydraulic-power drives which elevate the guns and train the turret. Primary control may be by means of continuous transmission of corrected gun-elevation and train orders or by transmission of these orders calculated for selected values of level or cross-level. Turret sights are used only for transmitting sight angle to the erosion-correction mechanism in the gun-elevation indicators.

2. *Auxiliary control.* Director control by means of an alternate director (Mark 40) at the main-battery control station in the fire-control tower, or from the control equipment in turret 2. The Mark 40 director may be connected to the range keepers in the plotting room or it may use the auxiliary computer in the fire-control tower. In either case gun-elevation and gun-train orders are sent to the turret indicators. Gun-elevation order determined by the computer and sights in turret 2 may be transmitted to the other turrets through the switchboard in turret 2. The guns are elevated and the turret is trained from the same stations as in the case of primary control. Sights are used only to transmit the sight angle to the correction mechanism in the gun-elevation indicators.

3. *Local control.* Independent control of the guns in a turret by means of sights, range finder, and the auxiliary computer in the turret. Data from the plotting room instruments

RESTRICTED

MAIN BATTERY FIRE-CONTROL SYSTEM

or other range finders may also be used. Sight angle and sight deflection obtained from the computer are used to set the sights. The guns are elevated by the gun layers in accordance with mechanically or electrically transmitted gun-elevation orders from the sight-pointer's station. The turret is trained from the sight-trainer's station by means of a direct mechanical connection to the training hydraulic speed-gear control. The gun-train operator stands by in this method of control.

4. *Remote control.* Director control through the plotting room range keepers. Gun-elevation and gun-train orders from the range keepers actuate control mechanisms in the receiver-regulators. Thus the hydraulic speed gear is automatically controlled from the director and the guns are positioned without turning handwheels. However, inputs to the correction mechanisms are still manually introduced.

CHAPTER 19

MACHINE-GUN CONTROL

A. GENERAL

19A1. The fire-control problem. The problem of machine-gun antiaircraft defense differs from that solved by the Mark 37 director system in that for the shorter ranges involved the rate of change in target position is considerably greater, and the time afforded for firing is much shorter. Hence, to be effective, the machine-gun control system must solve the problem quickly, and permit a large volume of fire to be delivered in a short time.

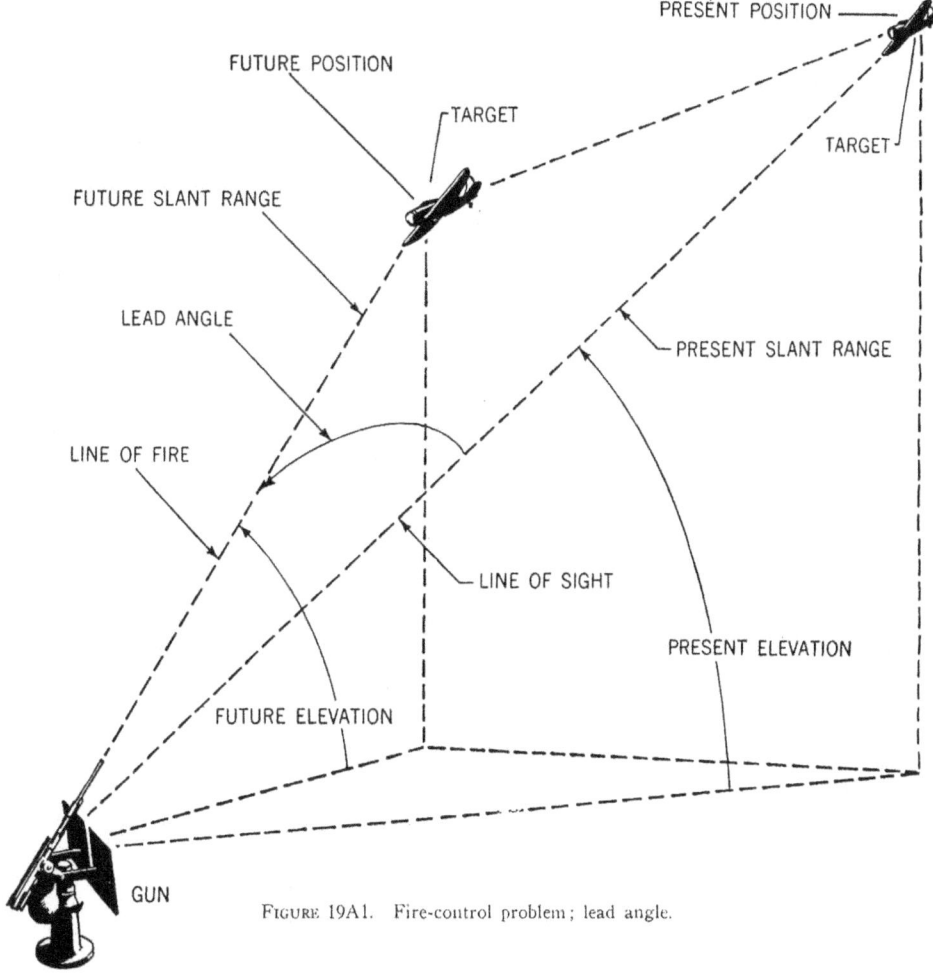

FIGURE 19A1. Fire-control problem; lead angle.

The caliber .50 and 20 mm. machine guns are free-swinging, and require local control by means of tracers and sights. Usually 1.1-inch and 40 mm. machine guns are controlled by directors which are specially designed to cope with the short-range problem. Such directors are greatly simplified to increase operating speed and decrease weight. The 1.1-inch and 40 mm. guns are also provided with ring sights for emergency use, and tracer ammunition to supplement director control.

19A2. Leading a moving target. It is well known that a gun must be pointed ahead of the LOS when firing at a moving target. This is due to the fact that the projectile requires a defi-

459

RESTRICTED

nite time to travel from the gun to the target, and during this time the target is continuously changing position. For example, if a target is moving at a relative speed of 270 knots or 150 yards per second across the LOS and is at such a range that it takes 5 seconds for the projectile to reach it, the target will have travelled 750 yards during the time of flight. Hence the gun should be pointed 750 yards ahead of the present target position. In other words, the gun must lead the target by 750 yards. The *lead angle* required is the angle between the LOS to the present target position and the line of fire required to hit the target at the end of the time of flight (see fig. 19A1).

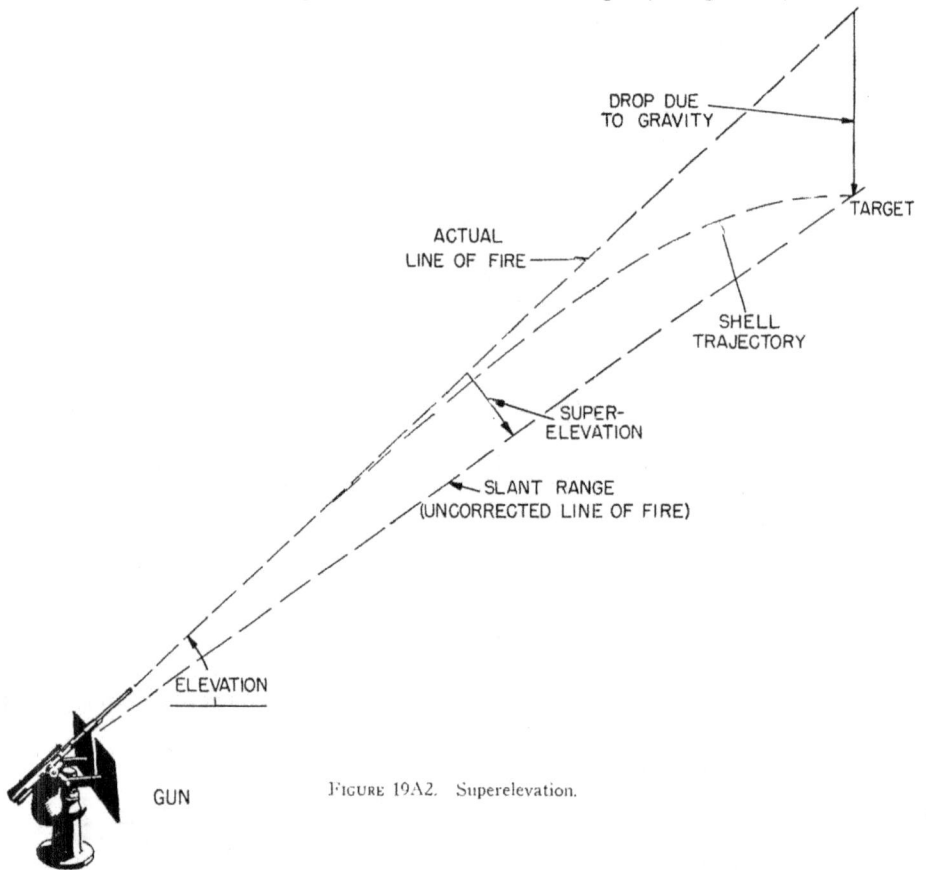

FIGURE 19A2. Superelevation.

19A3. Superelevation. The gun bore cannot be pointed directly at this future target position because gravity will cause the projectile to fall below the target. To compensate for this effect the gun elevation is increased by an additional angle called *superelevation*. This is shown in figure 19A2. The amount of superelevation required depends upon: (1) projectile I. V., (2) range, and (3) target elevation. For a given gun, superelevation is considered to be a function of target elevation and range.

19A4. Other ballistics. The numerous other elements which affect any trajectory such as wind, drift, variations in air density, and I.V., are present in machine-gun fire just as they are in main- or secondary-battery fire. However, the effects of these factors are usually small at short machine-gun ranges, and are neglected from the standpoint of making any computations for them. This is done in order to obtain a quick, practical solution to the problem. Any uncomputed errors are compensated by spotting, which is usually accomplished in machine-gun control by means of tracer observation.

CHAPTER 19

B. RING SIGHT AND TRACER CONTROL

19B1. Ring sights. Ring sights may be used in local control to establish the initial lead or aim-off before opening fire, after which tracer control may be employed. Sights may also be used when tracers are not visible as is the case when firing toward the sun.

Figure 4D6 shows a typical ring sight installation on a 20 mm. gun. The rear sight consists of two mutually perpendicular crosswires suspended in the eyepiece. The sighting axis established by the crosswire intersection and the center of the fore sight makes an angle of 24.5 minutes with the gun-bore axis. This is the gun elevation required to obtain a horizontal range of 750 yards with a 20 mm. gun. The *zero-range* point on the fore sight is used to establish a sighting axis parallel to the bore for close-range firing and check-up purposes. On certain of these sights the two axes for zero and 750 yards range are obtained by using the center of the fore sight and adjusting the height of the rear sight.

The fore-sight ring spacings give the correct aim-off for the across LOS velocities indicated. The target's speed and its direction of approach are observed, and from these the speed across LOS is estimated. The operator then moves the gun so that the ring corresponding to this estimated speed is placed on the target. The direction of aim-off is such that, should the gun be suddenly stopped, the target will pass through the center of the fore sight. For example, suppose that the target is at a range of 750 yards and is moving in such a manner that its across LOS velocity is 200 knots in a direction from right to left. The 200 knots ring on the right-hand side of center is placed on the target and kept there by tracking as long as the speed across the LOS remains the same. Thus the proper lead angle is obtained. Since the sighting axis is offset in the vertical direction also, superelevation is introduced. For *elevated* targets at 750 yards, the superelevation introduced is too large, since it is an elevation for a *horizontal* range of 750 yards. Thus the superelevation correction must be decreased with increased target elevation.

Recent designs provide an adjustable rear sight which takes into consideration both elevation and range variations in the determination of superelevation. The gun is elevated to the angle at which firing is expected to occur, and the elevation angle is read from a dial on the rear sight. The range is estimated, and a range-adjusting cylinder, which raises or lowers the rear sight, is turned until the estimated value of range on a range scale lines up with the elevation reading on an elevation scale. In this manner the sight is adjusted to give the correct superelevation for the range and elevation to be used. Initial-velocity loss due to erosion is also taken into consideration by providing three adjustable rear sights of different heights to be employed during the life of the barrel.

In general, proper use of the ring sight is difficult to learn because it involves rapid estimates of aim-off for various target speeds. The following suggestions are helpful when using a ring sight:

    1. In nearly every direction of attack the approach angle increases and hence the amount of aim-off must be increased as the target gets closer. For this reason the tendency is to shoot abaft the target.

    2. The initial lead should be either correct or too large as it is easier to drop back on the target than to catch up with it.

    3. The gun must be kept swinging in the direction of target motion as firing continues.

19B2. Tracer control. Tracers permit the gunner or director operator to observe the projectile flight; hence their use in controlling machine-gun fire provides several advantages, the most important of which are:

    1. No computations are necessary to determine the ballistic corrections. Total compensation is made visually when the gun is so moved that the tracers hit the target.

    2. There is no delay in applying corrections because the operator instinctively moves his gun to put the tracer stream on the target.

    3. Smoke and gun vibration interfere less with tracer observation than with the use of an open sight.

All machine guns use tracers either for primary control or to supplement other methods. At night they are visible to the burn-out point, but in daylight the maximum range at which they can be seen under favorable conditions is reduced to approximately 1,000 yards in the case of projectiles fitted with small-size tracers. Under certain other conditions they may not be visible at all. The chief disadvantages of tracer control are the following:

1. It is difficult to estimate distances at the target with the unaided eye.

2. Since the tracer and the aircraft target are both moving at high speeds, an optical illusion is created, causing the observer to see a decided curve in the tracer stream. This optical illusion increases the difficulty of estimating the tracer position relative to the target.

3. The gunner has a tendency to direct his fire in accordance with the bright portion of the tracer stream. This results in insufficient lead angle and the projectiles fall behind the plane.

4. Tracers are indistinct against a bright sky or when firing into the sun.

19B3. Hints on tracer control. The following rules should be observed when using tracer control:

1. Watch the tracers at the target. This prevents lining up the target with the bright portion of the tracer stream and thus obtaining an incorrect lead angle.

2. Let the target "suck in" the tracers. When hitting, the tracer stream at the target looks as though it is parallel to the fuselage and appears to be drawn into the plane. Hits are not obtained when the normal curve of the tracer stream appears to cross the target.

3. Keep the tracer stream ahead of the target. If the stream falls behind the target, it is difficult to catch up again. Also, it is better to concentrate on the forward end of the target rather than on the rear, because the misses ahead tend to disconcert the pilot.

4. On opening fire use a large lead angle. It is a common fault to open fire with lead angle too small. If at first the stream is ahead of the target, the gunner should not stop tracking or move the gun back. The correct procedure is merely to decrease the tracking rate and let the target catch up with the tracer stream.

5. Do not sight along the gun barrel. Sighting in such a manner makes it impossible to watch the tracers at the target, puts the gunner in an awkward position for tracking, and increases the possibility of having vision obscured by smoke.

6. Keep in mind that the angle of lead increases as the target gets closer. This is true for nearly every direction of attack.

C. THE MARK 14 SIGHT AND MARK 51 DIRECTOR

19C1. General. The Mark 14 sight, shown in figure 19C1, is designed primarily for the control of AA machine guns (20 mm. in particular) against rapidly moving targets at the short ranges for which these guns are effective. The correct offset (for relative target motion and superelevation) between the LOS and the bore axis is automatically established by the motion of the sight as the target is tracked. The gunner tracks the target by moving the mount in train and elevation to keep the luminous cross of the sight mechanism constantly centered on the target. The computing mechanism is actuated by this movement and provides the proper LOS offset. The only manual operations required are range setting, smooth tracking to keep the cross on the target, and the introduction of spots. To a large extent this sight eliminates the guess work involved in leading the target by means of telescopic and ring sights, and therefore hits are obtained sooner. Quick action is very important because the usual effective firing time for air targets is only a few seconds at best.

For controlling the 1.1-inch and 40 mm. guns modifications of the Mark 14 sight are incorporated in Mark 51 directors which transmit elevation and train orders to the guns. A Mark 51 director is shown in figure 19C2. This type of instrument is placed close to the guns which it controls, thereby obviating horizontal and vertical parallax corrections.

19C2. Basic functions of Mark 14 sight. The Mark 14 sight incorporates two gyroscopically operated mechanisms which automatically compute lead angle and superelevation. In other words, the sight makes corrections for relative target motion and gravity effect on the projectile. Lead-angle computations are always correct for any value of trunnion tilt, and superelevation is

CHAPTER 19

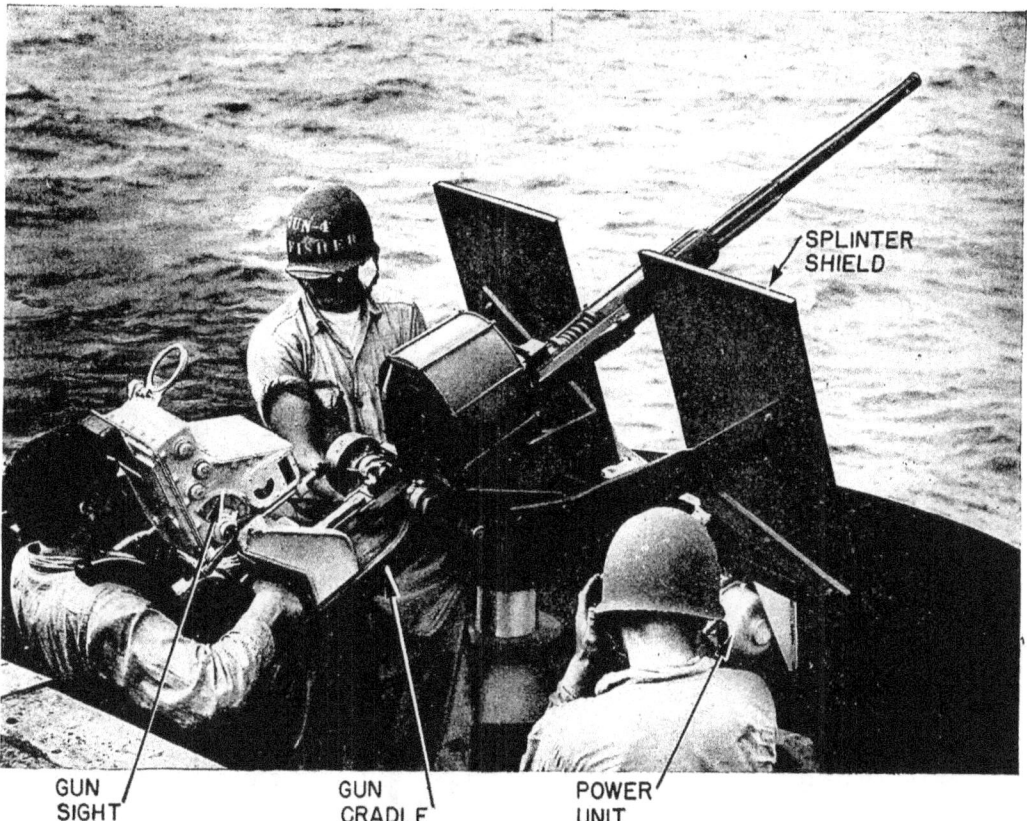

FIGURE 19C1.—Mark 14 sight; 20 mm. installation.

computed and applied in the vertical plane. Hence, the computing mechanism construction inherently obviates any trunnion-tilt corrections. Other ballistics are not computed, but are accounted for by spot adjustments as necessary.

The luminous cross which establishes the sighting axis is the reflected image of an illuminated reticle. Sight movement during tracking causes the two gyroscopes to precess, and thereby to actuate two mirrors which shift the position of the cross. The offset of the cross is such that the sighting axis is displaced from the bore axis by required lead angle corrected for superelevation.

19C3. Tracking rate and lead angle. In tracking a target, the gunner simply follows it by keeping his sight on continuously. In general, tracking a target requires moving the sight in both elevation and train, and the resultant angular velocity of the sight is called the *tracking rate*. The Mark 14 sight uses this rate in computing the lead angle instead of range and bearing rates. Calculations of tracking rate are illustrated in the three cases below.

Case I. Suppose a target is moving in such a manner that, in order to track it, the operator must continually train the sight to the left at a rate of 2° per second but need make no change in elevation. In this case, the tracking rate is 2° per second. If the time of flight of the projectile is 3 seconds, the target will be 6° to the left of its present position at the end of the time of flight. Hence, to hit the target, the gun must be aimed 6° to the left of the present target position. In other words, the lead angle is 6° left. It should be noted that the lead angle is obtained by multiplying the tracking rate by time of flight.

Case II. Suppose an airplane is approaching on a heading which requires no training but does require elevating the sight at a rate of $\frac{1}{2}$° per second. If the time of flight is 3 seconds, the target will be on a line $1\frac{1}{2}$° above the present line of sight when the projectile meets it. Hence, in order

RESTRICTED

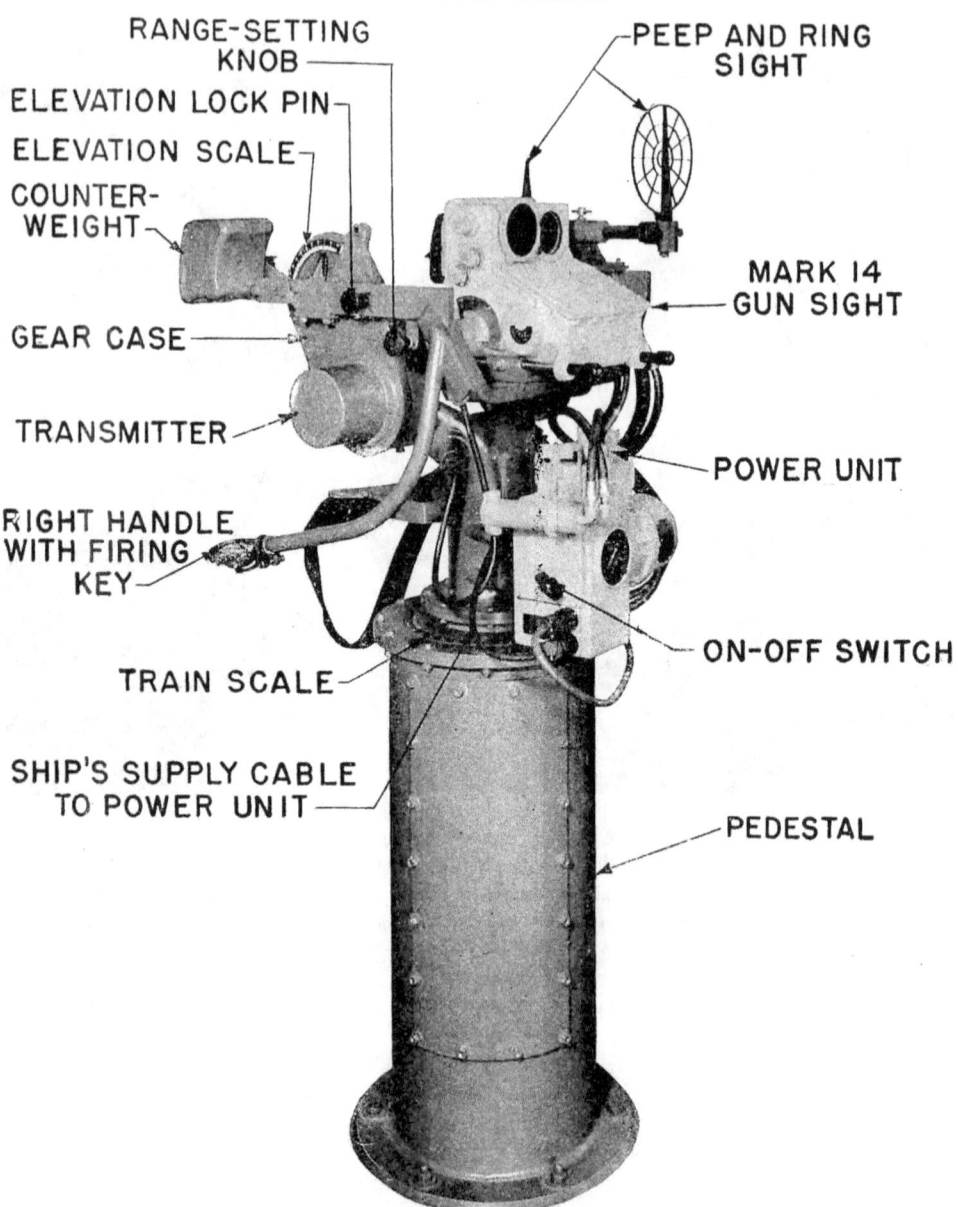

FIGURE 19C2. Mark 51 director.

to hit the target, the gun must be aimed 1½° above the present line of sight and the lead angle is 1½° upward.

Case III. Generally, the operator has to train and elevate the sight simultaneously to track a target. Suppose a target is moving so that the sight must be trained to the left at a rate of 2° per second and elevated at a rate of ½° per second. Assume a time of flight of 3 seconds. In this case the tracking rate is a combination of the rates described in Cases I and II. In order to hit the target, the gun has to be aimed ahead in both train and elevation. As in Case I, the gun has to lead by 6° in train and as in Case II, by 1½° in elevation. The *lead angle* is the resultant of the separate train and elevation lead angles.

From these examples it is seen that the lead angle is the product of the tracking rate and time of flight. In practice the tracking rate is automatically established by keeping the LOS on the target, and the time of flight, considered to be directly proportional to range for a given gun, is introduced through the range setting.

19C4. Gyroscopic precession. Gyroscopic effects exist whenever an attempt is made to deflect a spinning wheel from its plane of rotation. These effects may be observed if the spinning wheel, called the rotor, is mounted in a gimbal system which permits the spin axis to swing in any direction. A gyroscope so mounted is shown in figure 19C3. The gyroscopic effect called *precession* is the tendency of the gyro rotor to swing its spin axis at right angles to the direction of a force applied to the gyro mount. This principle is utilized in the Mark 14 sight.

In figure 19C3a the *applied force* A rotates the gyro unit about a vertical axis clockwise as viewed from above. Rotation in this direction causes the gyro to precess, and the spin axis moves in the direction shown. If the force A is sustained, the spin axis will become parallel to the torque axis, which is the axis about which the gyro unit is rotated. A general rule for determining the direction of precession is that the gyro rotor tends to move so as to make the spin and torque axes parallel and the directions of rotor spin and rotation about the torque axis the same.

When a torque is applied to the spring-restrained gyro shown in figure 19C3b, the rotor spin axis tilts until the precessional force is balanced by the spring tension. The precessional force, and hence the angle of tilt, is proportional to the speed at which the gyro is turned about the torque axis. A spring-restrained gyro can be used to measure tracking rate, and the amount of rotor tilt for a given tracking rate can be controlled by varying the spring tension. This is the basic unit for the Mark 14 sight.

19C5. Description of the Mark 14 sight. Figure 19C5 is a schematic representation of the lead-angle computing mechanism in the Mark 14 sight. This mechanism is placed in the sight case as shown in figure 19C4. The separate systems for train and elevation are similar. Each of them consists essentially of a pivoted mirror actuated by an air-driven, spring-restrained gyro unit.

19C6. Gyroscope units. The *gyro rotors* (see fig. 19C5) are mounted in ball bearings and the *gimbal frames* are supported by frictionless *spring suspensions*. The maximum precessional tilt of the rotors is limited to a small angle. The suspensions are not designed to absorb the heavy shocks of gunfire, and a damping device must be used. This consists of *damping disks* which rotate within housings containing a damping fluid at each end of the gimbal shafts. The damping action also retards the gyro response and makes it possible for the operator to smoothly track a rapidly moving target.

The gyros are oriented so that both spin axes are perpendicular to the gun bore. When the train axis of the sight is vertical, the spin axis of the train gyro unit is horizontal and that of the elevation gyro unit is vertical.

The train gyro controls the motion of the *train mirror* by means of the *train gyro lever* which is rigidly attached to the gimbal shaft. The mechanical linkage for controlling the *elevation mirror* consists of the *elevation gyro lever* and the *bell crank* which permits rotation of the mirror about an axis perpendicular to the elevation gyro axis. The mechanical linkages move the mirrors in such a manner that the lead angle indicated by the luminous cross is proportional to the resultant of the train gyro precession and the elevation gyro precession, and is independent of trunnion tilt.

19C7. Range springs. The *range springs* (see fig. 19C5) serve as the restraint for the gyros, and their tension can be varied by means of the *range adjustment handle*. Thus, the lead-angle components computed by the gyros for a certain tracking rate can be varied in accordance with the range. For short ranges (short time of flight), the two spring supports for each gyro unit are moved toward each other and the spring action is stiff. Then the lead angle computed is less than that for the high range settings, for which the spring supports are moved apart and the spring tension is less. Range setting is the method used to adjust the lead angle for time-of-flight differences.

19C8. Superelevation. Superelevation is introduced by a *superelevation weight* attached to to each gimbal shaft as shown in figure 19C5. The action of these weights is to turn their respective gimbal shafts, thereby offsetting the mirrors, in order to compensate superelevation. When both

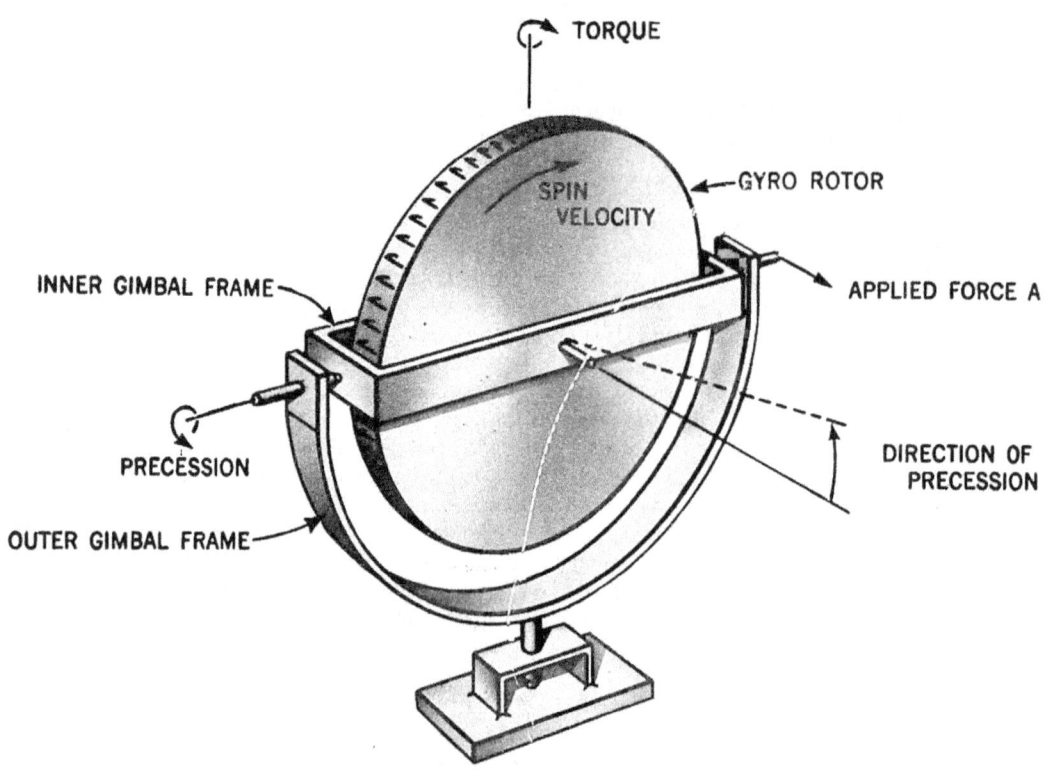

a. ESSENTIAL ELEMENTS

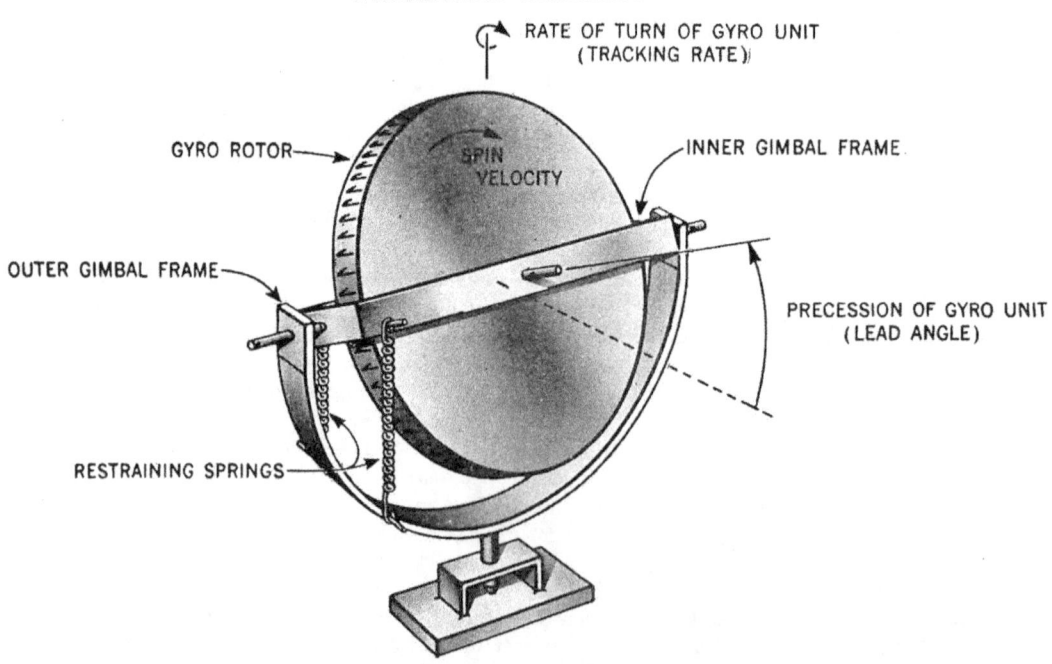

b. SPRING-RESTRAINED GYROSCOPE

FIGURE 19C3. Gyroscope.

CHAPTER 19

FIGURE 19C4. Mark 14 sight; sectional view.

bore and trunnion axes are horizontal, the train weight has no effect, while the elevation weight has maximum effect. This is apparent from the fact that the train weight is on a vertical line through its gimbal shaft, and therefore has no leverage and cannot apply a turning effort or torque. At the same time, the elevation weight has its maximum leverage, and thus applies its greatest torque.

When the trunnion axis tilts, leverage is acquired by the train weight and lost by the elevation weight, and each gyro unit computes part of the superelevation. The sight construction is such that the resultant mirror tilts caused by the superelevation weights displace the luminous cross only in a vertical plane, regardless of the trunnion-axis inclination.

As the bore elevates above the horizontal, both weights lose leverage and their effects decrease, becoming zero when the bore axis is vertical. The range springs resist the torque applied by the

MACHINE-GUN CONTROL

weights just as they resist the precession due to tracking. Since the springs are stiffest when adjusted for short ranges, the gimbal-shaft rotation caused by the weights varies in the same manner as superelevation, namely: (1) maximum for horizontal firing and decreasing as target elevation increases; (2) maximum for long ranges and minimum for short ranges. Thus, with the weights properly designed and adjusted for a particular gun, superelevation is automatically introduced at any elevation and range, and the correction is independent of trunnion tilt. Since the lead angle computation is also correct for any value of trunnion tilt, it is evident that no separate trunnion-tilt corrections are needed.

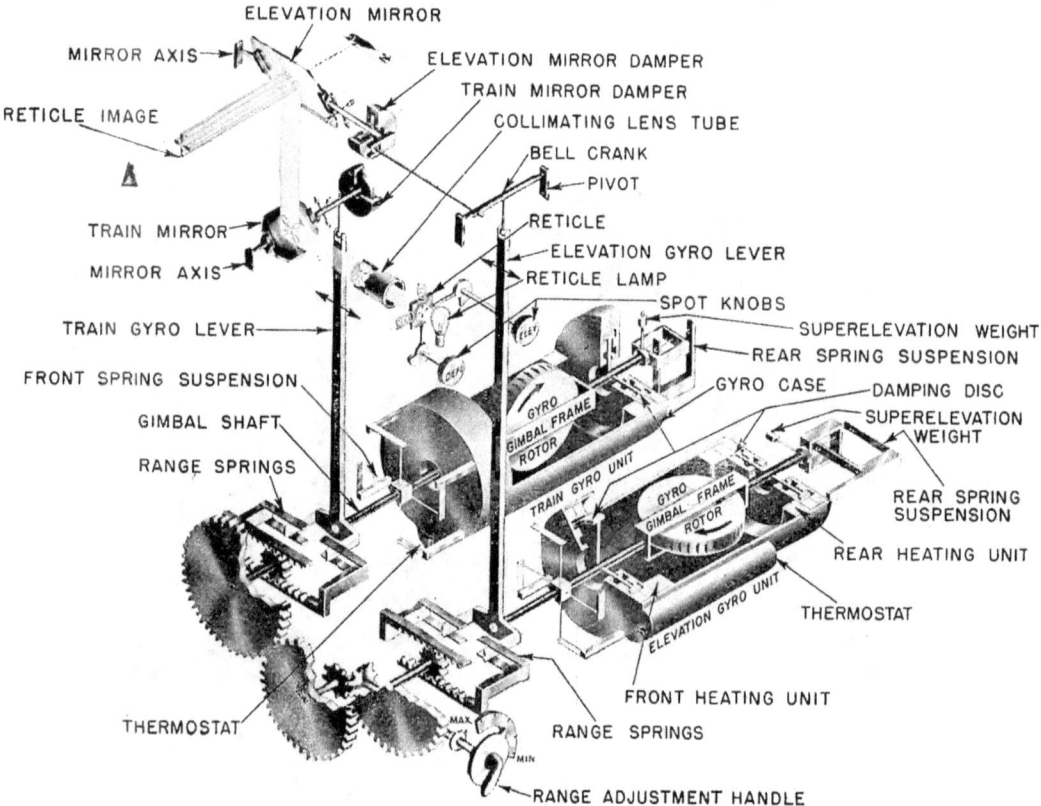

FIGURE 19C5. Mark 14 sight; schematic.

19C9. Optical system. The luminous cross seen by the gunner as he looks through the sight is the image of the cross in the *reticle* (see fig. 19C5) which is illuminated by the *reticle lamp*. As previously stated, movement of either the train or elevation mirror shifts the image position. In addition, the reticle is so mounted that it can be moved sidewise or up and down by means of the *spot knobs* to cause image shifts in elevation and deflection. The spot knobs thus provide means for introducing spots as necessary.

The *collimating lens* (see figs. 19C4 and 19C5) focuses the luminous cross an infinite distance away. This means that when the gunner focuses his eye on the target, regardless of its distance, the cross is always sharp and distinct, appearing superimposed on the target. It also means that the cross will not shift with respect to the target if the gunner moves his head slightly in any direction. In other words, there is no parallax in the sight. Etched on the front and rear windows are a circle and cross which are used to help the gunner locate the luminous cross and are not employed when the target is being tracked. A light filter is so pivoted on the front of the case that it can be placed over the window to reduce glare or be swung out of the way when not needed.

CHAPTER 19

The *train mirror* (see figs. 19C4 and 19C5) has a metallic surface and functions as an ordinary mirror. The *elevation mirror* is clear glass which reflects the reticle image into the operator's eyes and also permits him to view the target directly. The train and elevation mirrors rotate about axes which are respectively parallel and perpendicular to the gun-bore axis. When the gun is stationary, both mirrors make angles of 45° with the light from the reticle and the reticle image is reflected at right angles to the elevation mirror axis. Angular displacement of the train mirror causes the luminous cross to move parallel to the elevation mirror axis and angular displacement of the elevation mirror causes the cross to move perpendicular to this axis. Simultaneous angular displacement of both mirrors moves the cross to a position which results from motions in both of these directions. Any displacement of either mirror changes the angle of reflection from the elevation mirror, and hence shifts the sighting axis.

The *elevation* and *train mirror dampers* (see fig. 19C5) are installed to minimize the effects of mechanical vibration. These dampers are similar to those used on the gyros.

19C10. Optical system operation. As an example of sight operation, suppose that the sight is being trained in the horizontal plane clockwise as seen from above and that the gyros are spinning in the directions shown by the arrows. Since there is no motion in elevation, the elevation gyro is not affected, but the train gyro precesses. Appling the general rule for direction of precession, the train gyro rotor tilts counterclockwise as viewed from the left end of the gimbal shaft. This motion, transmitted through the train gyro lever, turns the train mirror clockwise into a more-nearly vertical position. Hence, the point where the light rays strike the elevation mirror shifts to the right, and the reflection to the operator's eye is along a line which points to the left of the bore axis. This line is displaced horizontally from the bore axis by a lead angle proportional to the angle through which the train mirror has turned. The sight is so designed that this displacement is equal to the lead angle required for the particular training rate and range setting used. To put the cross on the target, the operator must move the sight case, and hence the gun bore, to the right (ahead) of the target an amount equal to the lead angle computed. Similarly, the elevation gyro rotates its mirror when the sight is moved in elevation and the sighting axis is displaced from the bore axis by the required elevation lead angle.

19C11. Air and heat control. The correct lead angle is computed only when the gyro rotors spin at constant speed. This requires an air supply at unvarying pressure and temperature. The air-supply system functions to clean, heat and regulate the pressure of the air supplied from the power unit described below. Air enters the system at about 7 pounds p.s.i. pressure and passes through a filter within the sight case where dirt and moisture are removed. Next, the air passes over a heater unit which brings the temperature to approximately that of the gyro units. The air then enters a pressure regulator which eliminates supply pressure variations. From the regulator the heated air passes through the thermostat unit which controls the air heater, and then goes directly to the two gyro units. Needle valves in the air nozzles at the gyros permit rotor speed adjustment. A gauge on the right side of the case indicates air pressure at the needle valves.

An intake valve in the base of the case admits air whenever the vacuum within the case becomes greater than normal. This may happen when first starting the unit, since the air may be exhausted from the case faster than it can be returned through the air-supply system.

The gyro unit accuracy depends partly upon the uniformity of the damping action, which in turn depends upon a constant temperature of the damping fluid. Thermostatically controlled *heating units* attached to each gyro case maintain the required constant temperature. Normally the sight should be turned on 30 minutes before using so that the parts will be at the correct operating temperature. However, in emergencies, the sight can be used sooner with reduced accuracy.

19C12. Power unit. The power unit, shown in figure 19C6, functions to supply clean and dry compressed air to drive the gyros. The diaphragm-type *air pump* is driven by an electric motor which receives power from the ship's 115 volt, single phase, alternating-current supply. The pump delivers air at a pressure between 6 and 8 pounds p.s.i. and this is indicated on the *pressure gauge*. The air from the pump passes through a drier and filter before being delivered into the pressure hose connecting the pump to the air-supply system in the sight. After the air has passed through the gyros it is exhausted into the case and then returned to the pump through a second hose.

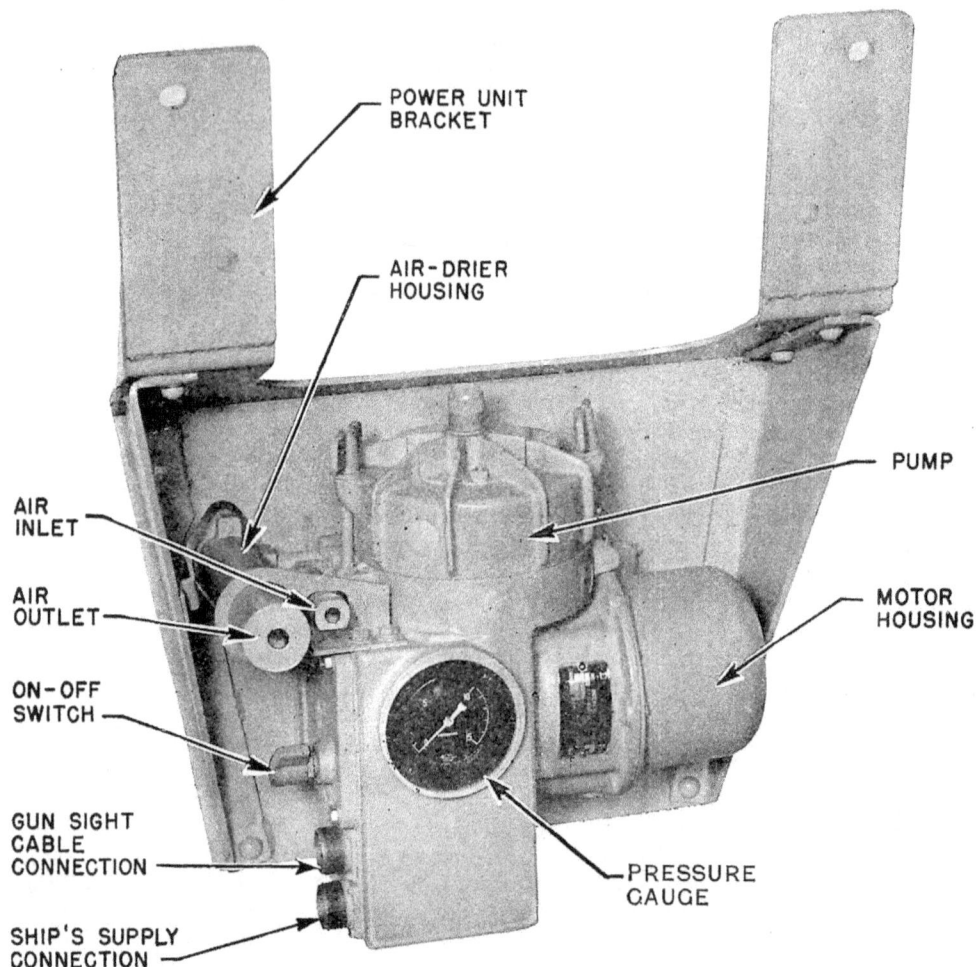

FIGURE 19C6.—Power unit; 20 mm. installation.

19C13. **Mark 51 director.** The Mark 51 is a simple, light-weight director which has proved to be very effective for controlling 40 mm. and 1.1-inch guns. It is placed a short distance away from the guns, where it is relatively free from gun vibration and interference from smoke.

The director, shown in figure 19C2, consists of three main parts: (1) the *pedestal,* (2) the *head,* and (3) a modification of the Mark 14 sight. The pedestal is bolted to the deck and supports the head on ball bearings. The *train transmitters,* and a *train damper* on some modifications, are housed within the pedestal and geared to the head. The train damper functions to smooth out irregularities in motion as the head is trained, and thus aids the operator in tracking smoothly. It consists of an impeller which rotates within a housing partly filled with oil. The gear ratio is such that the impeller is driven at a relatively high speed when the head is turned. A slip clutch is provided to enable the operator to slew the director in train without having to drive the impeller at an excessive rate. The train angle is limited by mechanical stops to 375° in either direction from the center position.

The head supports a *carriage* which has a platform for mounting the gun sight. The *handle bars* attached to the carriage are used by the operator to move the carriage in elevation and the head in train. The right handle contains a *firing* key which is employed when the guns are fired from the director. The sight *power unit* is attached to the head. The carriage is pivoted on two short shafts which project into the *gear cases* attached to the head extensions, or arms. As the carriage is ele-

FIGURE 19C7. 40 mm. quad with Mark 51 director.

vated, gears within the cases drive the elevation transmitters. An *elevation damping unit*, similar to the train damper, is installed in the right-hand case. The counterweights are provided to balance the weight of the gun sight.

The *ring sight* is used when the Mark 14 sight is inoperative. When the ring and peep are swung down into position the ring can be viewed through the gun-sight windows.

The two modifications of the sight used with the director differ only in their ballistic computations. The Mod. 3 is designed for 1.1-inch guns and has a maximum range setting of 2,800 yards. The Mod. 4 is used with 40 mm. guns and has a maximum range of 3,200 yards.

The director is oriented so that the 0°–180° line of the train scale on the pedestal is parallel to the fore-and-aft axis of the ship, and the training axis is parallel to that of the gun mount. It is important that roller paths be accurately paralleled upon installation since there is no provision for compensating any misalignment between guns and director. Sight alignment between director and guns is accomplished by boresighting the director with the guns, and then the sight with the director. The director is aligned with the guns first so that when, for any reason, the Mark 14 sight is changed it is only necessary to align the sight with the director.

19C14. Operation and personnel. The Mark 14 sight used on the 20 mm. gun requires two operators: gunner and range setter. The Mark 51 director requires a pointer and a range setter.

The gunner's or pointer's sole duty in connection with the sight is to keep the luminous cross accurately on the target by tracking smoothly. The range setter stands on the right side of the sight and performs the following functions:

1. Makes the range setting by means of the range handle.
2. Observes the tracers as they pass the target to determine the corrections needed.
3. Makes the appropriate corrections by means of the range handle and the spot knobs to obtain hits.

In performing these functions the range setter must keep in mind that misses may be due to either or both of the two following causes: (1) incorrect lead angle, and (2) ballistic factors such as wind, drift, or superelevation not correctly accounted for. When the lead angle is incorrect the tracers will either pass ahead of or behind the target. The Mark 14 sight computes the lead angle by multiplying the tracking rate by the time of flight (range), and since the tracking rate is correct as long as the target is tracked smoothly, any error in the lead angle is due to an incorrect range input. Hence, the range setting should be changed to correct for errors in lead. Lead angles that are too large can be reduced by lowering the range setting, and those that are too small can be corrected by increasing the range setting. These changes can be made either continuously or in steps of 400 yards each.

When a target is approaching the firing ship and the tracers are passing behind it, the lead is obviously too small. The range setter could correct this by increasing the range setting. However, since the target is approaching, the lead angle is decreasing and it will not be long before the lead angle computed will be equal to that required for hitting the target. Accordingly, it is common practice to make changes in range only for large errors in lead angle when the tracers trail an approaching target. However, if the tracers pass ahead of an approaching target, the range setting must be modified, since the lead angle is too large and will become larger as the range decreases.

Even though the lead angle is correct, the target may still be missed due to errors introduced by drift, wind and other ballistic factors. These errors must be compensated by means of spots in elevation and deflection.

For example, suppose the target is moving at right angles to the LOS and that the tracer stream, while remaining on in train, is passing above the target. The lead is correct, but a *down* spot must be applied through the elevation spot knob. This moves the luminous cross upward, compelling the pointer or gunner to decrease the gun elevation in order to place the cross on the target. If the tracer stream passes to the right or left of an *approaching* target, deflection spots must be applied. This moves the cross to the right or left, compelling the gunner to move the gun in the opposite direction. Thus it is seen that corrections other than those for incorrect lead are introduced by means of the spot knobs.

CHAPTER 20

TORPEDO CONTROL

A. PRINCIPLES OF TORPEDO FIRE

20A1. Introduction. Torpedoes and their release gear are described in Chapter 9, Section B, and their general operating characteristics should be understood before this chapter is studied.

The method used for firing torpedoes varies with the different classes of ships which use them as an offensive weapon and with the various installations on the same class of ship. The torpedo-control problem for surface vessels is quite different from that for submarines. A destroyer attack will usually be seen by the target in the early stages and therefore the torpedoes must be fired at comparatively long ranges. Thus the target has time to maneuver and avoid being hit. For this reason destroyers usually fire their torpedoes in rapid succession with the individual torpedoes on slightly different courses so as to produce a spread. This procedure greatly increases the possibility of obtaining a hit. A division of destroyers in battle usually fires its torpedoes as a unit with each ship firing a spread. This combination of spreads produces a pattern which is almost certain to give at least one hit. With spread firing it is possible to neglect certain minor errors which would be considered if single torpedoes were fired to hit a given point of aim.

Underwater attack is usually made by a single submerged submarine at comparatively low speed against a target which has little opportunity to maneuver after the torpedo is fired. Submarines endeavor to come in close to their targets and at these short ranges each torpedo is fired to hit. Spreads are used only to allow for changes in target motion in the interval between launching successive torpedoes.

20A2. Control of firing. The two kinds of control used for torpedo fire are: (1) bridge control, and (2) local control. Bridge control is remote control from directors located on the signal bridges of destroyers or in the conning towers or control rooms of submarines. This is the primary method of control on modern ships. The torpedo-control problem is solved at the director, and the torpedo course and torpedo gyro settings are transmitted to indicators at the tube mount so that the tube can be properly trained by matching pointers.

Local control is accomplished at the tube mount, and is used primarily when the director control system is inoperative. In this method the torpedo-control problem is usually solved with the aid of a portable angle solver. The tube sight is offset properly, and then the tube is trained until the sight is on the point of aim. Spreads may be fired by setting spread gyro angles on the torpedo gyroscope or by training the mount.

20A3. Methods of firing. The firing methods described below are those which are ordinarily used. Obviously, the method employed will depend upon the type of installation and the requirements of the torpedo problem. The methods described are those for both movable and fixed tubes.

Movable tubes. Most modern installations have movable tubes with which two firing methods are ordinarily used: (1) straight fire with spread gyro angles, and (2) curved fire with spread gyro angles.

In straight fire with spread gyro angles, shown in figure 20A1a, the torpedo-control problem is solved by the director and the tube is trained by matching pointers. The gyro-angle settings on the torpedoes differ from zero by only that amount necessary to produce the desired spread. The mean torpedo course will be parallel to the tube axis.

Curved fire usually is controlled from the director also. A basic gyro angle is employed to produce curved fire, and in addition the basic gyro angle for each torpedo may be modified to produce a spread as shown in figure 20A1b. The torpedoes turn through the set gyro angles after they are in the water and straighten out on courses which differ from the direction of the tube axis by the amount of the gyro angles.

RESTRICTED

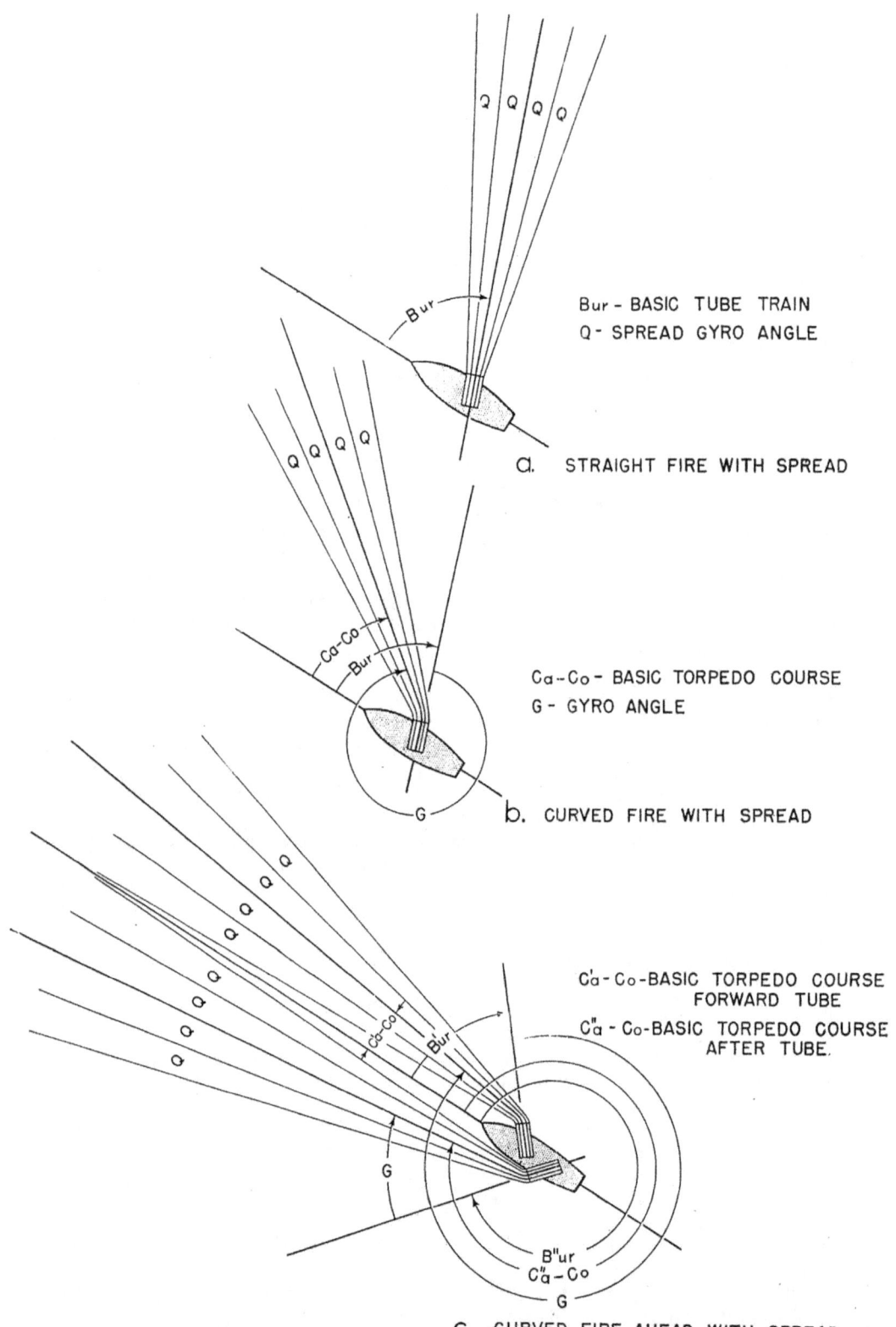

FIGURE 20A1. Torpedo fire.

CHAPTER 20

If the director transmission system fails, firing is shifted from either of the above described methods to the firing-course method described below. If the communication system fails also, or the director installations are shot away, local control must be used.

Fixed tubes. Fixed tubes are still used on some ships and on all submarines. Under certain conditions movable tubes may be clamped in place and used as fixed tubes. Two methods are commonly used by ships for firing from fixed tubes: (1) firing-course method, and (2) curved fire ahead.

In the firing-course method the tubes are clamped in some designated position. The torpedo problem is solved to determine the correct torpedo course and the ship is swung until the tube is correctly positioned for firing. A spread may be obtained by spread gyro settings or by swinging the ship, although the latter method is rarely used.

Curved fire ahead (see fig. 20A1c) is a method which was formerly used when it was desired to fire both port and starboard batteries at the same target. The tubes were clamped in their firing position and the gyro settings were such that the mean torpedo track was along own-ship heading at the moment of firing. A spread was produced by varying the gyro settings (see fig. 20A1b and c). This method was considered to be effective for long-range surface attacks. Curved fire is the standard method used by submarines, although straight fire (see fig. 20A1a) may be employed if the target bearing is not changing too rapidly.

20A4. Standard symbols and definitions for torpedo control.

SYMBOL	DEFINITION
B_{ur}	*Basic tube train.* The computed angle between the fore-and-aft axis of own ship and the axis of the tube mount measured from the ship's bow (0°-360°) clockwise.
B'_{ur}	*Actual tube train, forward tubes.* Measured as above.
B''_{ur}	*Actual tube train, aft tubes.* Measured as above.
C_a	*Torpedo course.* The angle between the north and south line and the final track of the actual torpedo, measured clockwise from the north (0°-360°).
C'_a	*Torpedo course, forward tube.* Measured as above.
C''_a	*Torpedo course, after tube.* Measured as above.
(C_a-C_o)	*Basic torpedo course.* The angle between the ship's bow and torpedo track (0°-360°).
D	*Sight angle.* The computed angle between the line of sight and the final course of the torpedo.
D'	*Corrected sight angle.* Sight angle with latitude and intercept offset corrections applied.
G	*Gyro angle.* The angle between the axis of the torpedo tube and the final track of the actual torpedo, measured clockwise from 0°-360°. Gyro-angle order is the gyro angle transmitted to the torpedo tube.
G_m	*Latitude correction.* The correction required to compensate the error due to proving (balancing) a torpedo gyro in one latitude and firing it in another.
H	*Target run.* Distance covered by target during the time of torpedo run.
I	*Track angle.* The angle between the target course and the reverse course of the torpedo; measured clockwise from the target bow (0°-360°).
L_t	*Intercept offset.* An arbitrary change in sight angle right or left to produce an offset angle, thus changing the point of intercept.
L_u	*Tube offset.* The angle between the tube mount axis and the basic tube train measured right or left of the basic tube train.
Q	*Spread angle.* The angular difference between the final course of two adjacent torpedoes from a tube mount exclusive of any change in basic gyro angle of the torpedoes.
S_a	*Torpedo speed.* The average speed in knots of the actual torpedo from the muzzle of the tube to the point of intercept.

TORPEDO CONTROL

SYMBOL	DEFINITION
Ta	*Time of torpedo run.* The time between the instant of firing the torpedo and the instant it crosses the target track when torpedo speed correction is zero.
Tt	*Time of target run.* The time between the instant of firing the torpedo and the instant the target crosses the torpedo track.
U	*Torpedo run.* The distance in yards travelled by the torpedo measured from the muzzle of the tube to the point of intercept. The distance the torpedo runs at its designated speed beyond the target ship's track is called *over run*.

Firing point. The point where the torpedo begins its run; usually considered to be the position of the firing ship when the torpedo is launched.

Point of aim. That point on the target which it is desired to hit with the torpedo.

Torpedo range. The normal distance in yards which the torpedo will run at its designated speed. Any additional distance which the torpedo may cover is called *over range*.

Firing range. The distance to the target in yards at the instant the torpedo is fired.

Effective range. The maximum distance at which the target can be located at the instant of firing and still permit the torpedo to reach the point of intercept at the designated speed. Effective range varies with the course and speed of the target as well as with torpedo range.

Range allowance. The margin allowed to insure that the torpedo run will be less than the torpedo range. It is equal to effective range minus firing range.

B. TORPEDO CONTROL PROBLEM

20B1. Graphical solution. The problem of determining the basic torpedo collision course can be solved graphically with sufficient accuracy for illustrative purposes. A mooring board is convenient for this purpose but the simple layout can be quickly drawn up on any sheet with the aid of a scale, a protractor, and a compass.

The complete solution of the problem also includes the calculation of the torpedo run and the time required for this run. For this complete solution two triangles are required: (1) a speed triangle, and (2) a distance triangle (see fig. 20B1). The angle between the torpedo course and the LOS, called the sight angle, is determined from the speed triangle. The sight angle is then used in the construction of the distance triangle from which are determined the torpedo run and the time required for this run.

Speed triangle. The speed triangle, shown in figure 20B1b, is similar to the line-of-fire diagrams used in the surface fire-control problem for gun fire. The LOS and target course are drawn in their proper relative positions and a convenient scale is selected for plotting the speeds of target and torpedo on the diagram. The target speed (S), is laid off along target course and appears as WG in the figure. With G as a center and a radius equal to the torpedo speed (Sa), an arc is drawn through the LOS to establish point K. The line KG completes the speed triangle WKG and D is the sight angle required. A torpedo set for a speed of Sa knots and launched on a course which makes an angle D with the line of sight will intercept the target if the latter does not change its course and speed during the torpedo run. It is to be noted that own-ship course and speed do not enter into the solution of this problem.

Distance triangle. The speed and distance triangles are similar and may be superimposed. It is important, however, for the student to recognize that these two triangles are constructed with different scales, and therefore must be used independently in the two parts of the problem.

The distance between own ship and the target is laid off along the LOS to some convenient distance scale as shown in figure 20B1c. The line K'G' is laid off at the angle D determined from the speed triangle. W'G' is laid off along the target course. The triangle formed by the intersections of these three lines is the required distance triangle. The side W'G', on the same scale used for the range, represents the target run (H), and the side K'G' the torpedo run (U).

CHAPTER 20

The time of the torpedo run in minutes (Ta) is determined analytically by dividing the length of the torpedo run in yards by the torpedo speed in yards per minute. Similarly, the time for the target run (Tt) can be calculated by using the target run and the target speed. Obviously the values of Ta and Tt are equal if a hit is obtained.

20B2. Analytical solution. The required values can also be obtained by an analytical solution of the speed and distance triangles. The known parts of these triangles make it possible to use the simple law of sines for the solutions. This law expressed in equation form for the general triangle is illustrated in figure 20B1a.

Speed triangle. For the speed triangle WGK of figure 20B1b, three parts are known: (1) target angle (A), (2) target speed (WG), and (3) torpedo speed (Sa). Using the law of sines:

$$\frac{Sa \text{ (torpedo speed)}}{\sin A \text{ (target angle)}} = \frac{S \text{ (target speed)}}{\sin D \text{ (sight angle)}}$$

$$\text{or } \sin D = \frac{S \sin A}{Sa}$$

Distance triangle. For the distance triangle of figure 20B1c three parts are known after the sight angle (D) has been determined from the speed triangle: (1) target angle (A), (2) sight angle (D), and (3) range (R). Again using the law of sines:

$$\frac{U \text{ (torpedo run)}}{\sin A \text{ (target angle)}} = \frac{R \text{ (range)}}{\sin \theta} = \frac{H \text{ (target run)}}{\sin D}$$

$$\theta = 180 - (A + D)$$

$$U = \frac{R \sin A}{\sin \theta}$$

$$H = \frac{R \sin D}{\sin \theta}$$

The time in minutes required for the torpedo run can be calculated by dividing the torpedo run in yards by the torpedo speed in yards per minute.

Example. Given the following data, calculate sight angle, torpedo run, target run, and time required for torpedo run:

$$Co = 030°$$
$$Ct = 210°$$
$$Br = 60°$$
$$S = 12 \text{ knots}$$
$$Sa = 27 \text{ knots}$$
$$R = 6{,}000 \text{ yards}$$

These data were used to construct the torpedo triangles in figure 20B1.

Solution. Target angle (A) required in the solution, must be determined first:

$$A = 180 + (Co + Br) - Ct$$
$$A = 60°$$

In many cases the target angle will be given instead of the target course.

For the speed triangle:

$$\sin D = \frac{S \sin A}{Sa} = \frac{12 \times .866}{27}$$
$$\sin D = .385$$
$$D = 22° \, 39'$$
$$I \text{ (track angle)} = A + D$$
$$I = 82° \, 39'$$

RESTRICTED

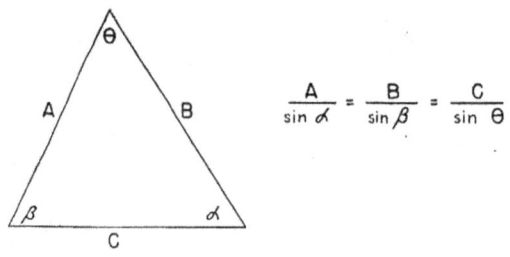

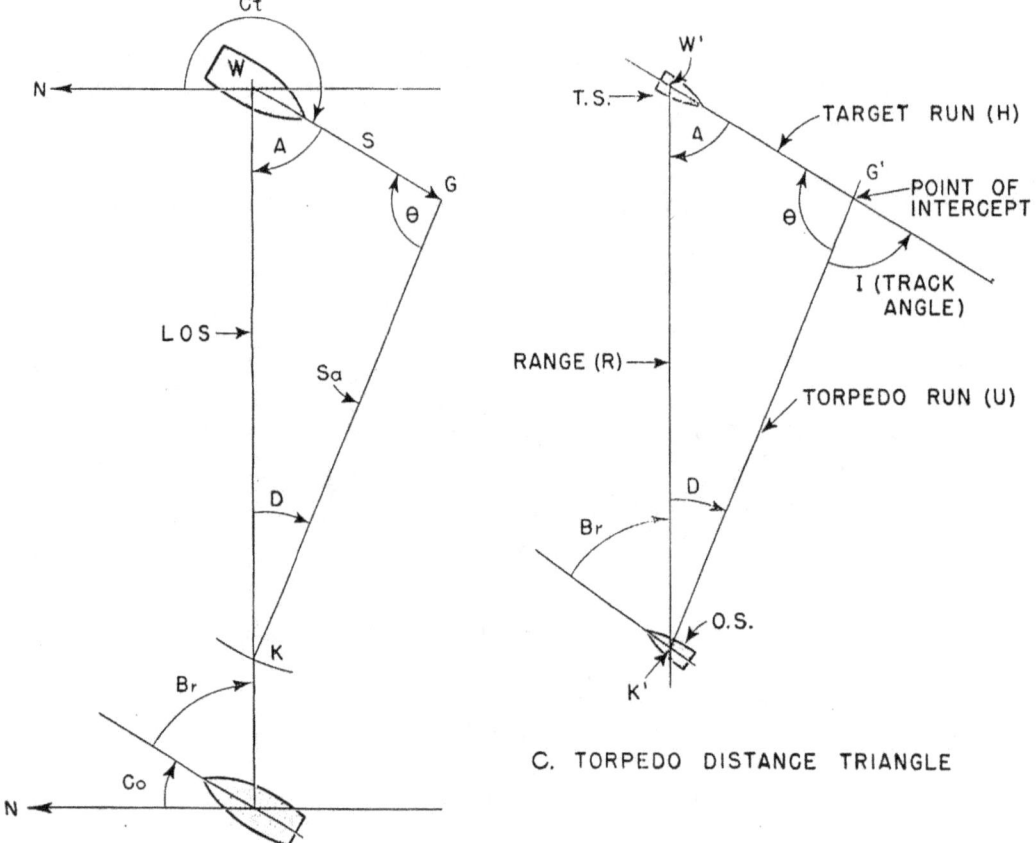

FIGURE 20B1. Torpedo fire triangles.

CHAPTER 20

For the distance triangle:

$$U = \frac{R \sin A}{\sin \theta}$$

$$\theta = 180 - (A + D) = 97° \; 21'$$

$$U = \frac{6,000 \times .866}{.992} = 5,238 \text{ yards}$$

$$H = \frac{R \sin D}{\sin \theta} = \frac{6,000 \times .385}{.992}$$

$$H = 2,328 \text{ yards}$$

$$Ta \text{ (min.)} = \frac{U \text{ (yards)}}{\text{Torpedo speed (yards per minute)}}$$

$$Ta = \frac{5,238 \times 60}{27 \times 2,027} = 5.7 \text{ minutes}$$

$$Tt = \frac{2,328 \times 60}{12 \times 2,027} = 5.7 \text{ minutes}$$

C. TORPEDO DIRECTOR SYSTEM

20C1. General. The torpedo director system includes a *director,* a *torpedo-course indicator* at the tubes, and the associated transmission and communication systems. The systems used for above-water tubes are basically the same for all ships but individual installations may vary considerably in mechanical arrangements. The discussion in this section is confined to the equipment and arrangement found on modern destroyers. This may consist of two quintuple tube mounts located on the ship's centerline, and either one or two Mark 27 torpedo-director systems. On ships having only one director system the director is mounted on the centerline of the signal bridge forward of the Mark 37 gun director. When two directors are used they are mounted well forward in the wings of the signal bridge, one on each side.

20C2. General operation. The Mark 27 torpedo director mechanically computes the sight angle from the torpedo-speed triangle, corrects this sight angle for torpedo creep, combines the corrected sight angle with relative target bearing to obtain basic torpedo course, and transmits basic torpedo course and gyro angle to the torpedo-course indicator at the tube mount. The sight angle can be modified at the director to change the point of intercept without changing the problem set-up. The basic torpedo course can also be modified to produce tube offset. This provides a spread angle between tubes and should not be confused with spread gyro angle set on the torpedoes to produce a spread between torpedoes of a given tube mount. Dials at the director show the problem graphically and indicate the values transmitted to the tubes.

The gyro angle and basic torpedo course transmitted to the torpedo-course indicator actuate zero-reader dials. The proper gyro setting and tube train are obtained by turning handcranks until the actual gyro setting and torpedo course match with those ordered by the director.

20C3. Mark 27 director. The Mark 27 director is shown in figures 20C1 and 20C2. Its principal parts are: (1) *stand,* (2) *case,* and (3) *telescope unit.* The stand is bolted to the deck and supports the case on ball bearings. A training circle secured to the stand meshes with a gear driven by the *training handwheel* mounted in the case. Thus, rotation of the training handwheel turns the case with respect to the stand. This rotation is limited to about 390°.

The case houses the units used to solve the torpedo problem, transmit basic torpedo course and gyro angle to the tubes, and receive own-ship course from the gyro repeater. The knobs for hand inputs and all dials are attached to the case. All inputs are manually introduced except own-ship course. The *own-ship course crank* may be used to introduce this value manually if the auto-

TORPEDO CONTROL

matic system fails. This is done by pushing own-ship course crank to the *in* position and then either matching pointers on zero-reader dials or setting the actual course on an indicating dial. The own-ship-course electrical system is disconnected when the crank is *in*.

The telescope is normally positioned by a small electric follow-up motor within the case. This can also be done with the *sight-angle crank*. For manual operation this crank must be pushed to the *in* position and zero-reader dials matched, or the computed sight angle set on the *sight-angle dial*. The telescope-train motor is disconnected when the crank is *in*.

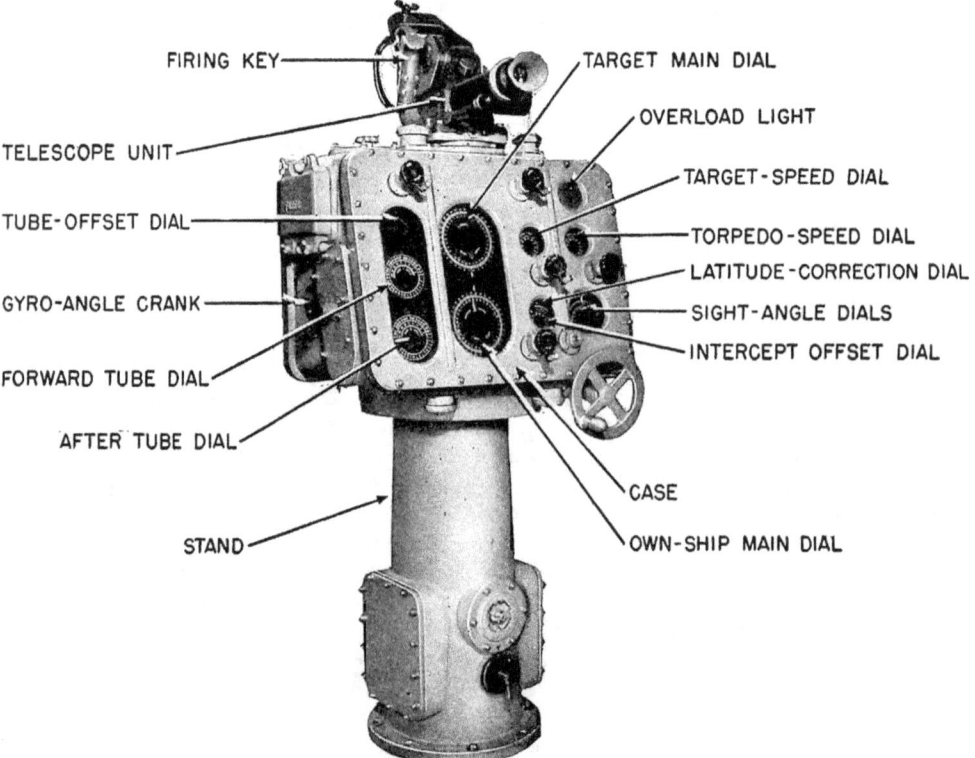

FIGURE 20C1. Mark 27 director; left rear view.

The telescope unit consists of a tiltable mirror, and fixed-prism telescope mounted on a telescope pivot. The tiltable mirror makes it possible to hold the LOS on the target in elevation as the ship rolls and pitches. The pivot is supported on a ball-bearing assembly attached to the top of the case. The pivot is geared to the computing mechanism within the case, and is automatically turned through the calculated sight angle with respect to the case. A scale on the pivot is graduated from 295° through 0° to 65°, and may be used to read the approximate sight angle. However, this quantity can be read more accurately from the *sight-angle dials*.

20C4. Director solution of the problem. The director solution of the torpedo-fire triangle for sight angle is illustrated in figure 20C3. The triangle shown is similar to the speed triangles discussed in Article 20B1. From the figure it is evident that the torpedo track will intersect the target course at point M only when Xt is equal to Xa. These values are given by the following equations:

$$Xt = S \sin A$$
$$Xa = Sa \sin D$$

Hence to secure a hit, $S \sin A$ must equal $Sa \sin D$. The target speed, torpedo speed, and target

CHAPTER 20

angle are introduced by hand, and the computer determines the sight angle necessary to make these two values equal by means of mechanical component solvers. The director telescope is turned automatically through an angle equal in magnitude to the sight angle, but opposite to its direction with respect to the LOS. Thus, when the telescope is on the point of aim, the case has been trained through an angle equal to the sum of relative target bearing and sight angle. This total angle of train is the basic torpedo course and is the value transmitted to the torpedo-course indicator if there is no tube offset.

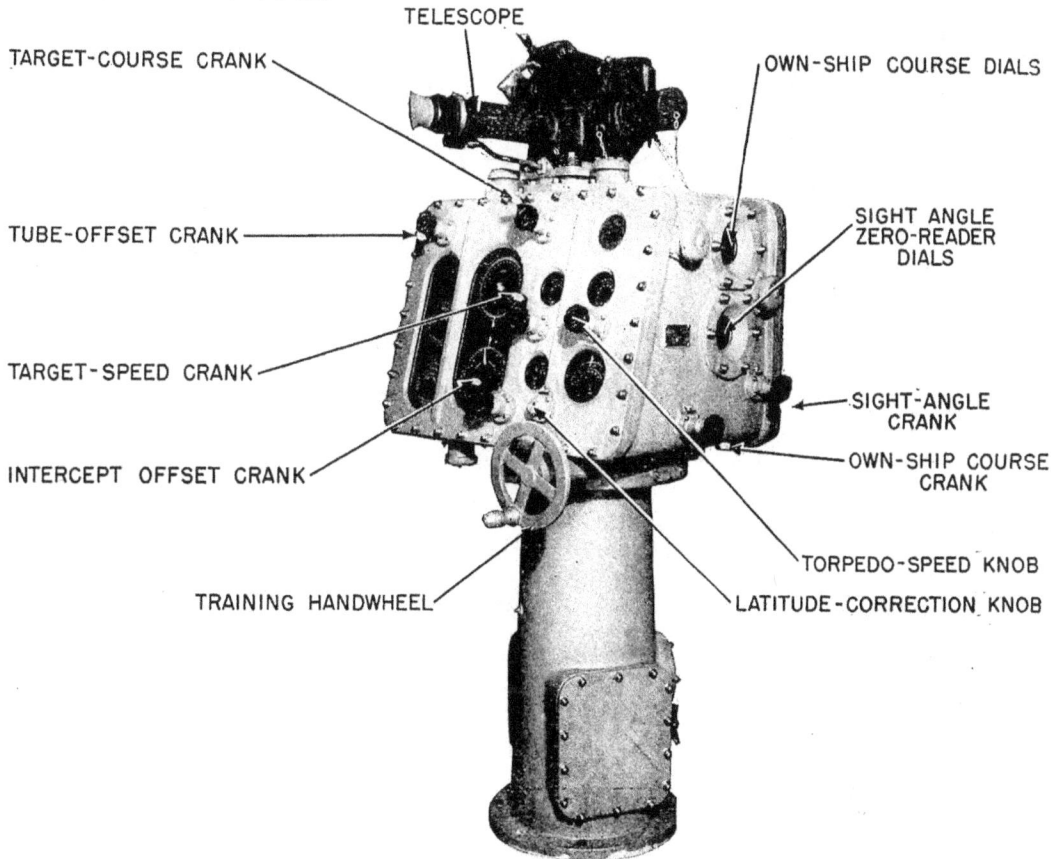

FIGURE 20C2. Mark 27 director; right rear view.

The director solves continuously for sight angle as long as the operator keeps the telescope on the target and the target speed and course remain constant. Changes in own-ship course are automatically entered by the own-ship course receiver.

Director inputs. The inputs and outputs for the director and torpedo-course indicator are shown schematically in figure 20C5. The following director inputs are required to determine the sight angle and hence the basic torpedo course:

    1. *Own-ship course* (C_o). This is received automatically and continuously from a gyro repeater by means of electrical transmission, or it may be introduced manually.

    2. *Target speed* (S), introduced manually.

    3. *Torpedo speed* (S_a), introduced manually.

    4. *Target course* (C_t), introduced manually.

    5. Training crank hand input to keep the telescope on the target.

TORPEDO CONTROL

The sight angle calculated from the above data is further corrected by:

6. *Latitude correction* (Gm). This correction compensates for the inherent tendency of the torpedo gyro to creep to the right in north and left in south latitudes due to earth rotation. The magnitude of the correction depends upon the length of torpedo run and the latitude and is set on the *latitude-correction dial*.

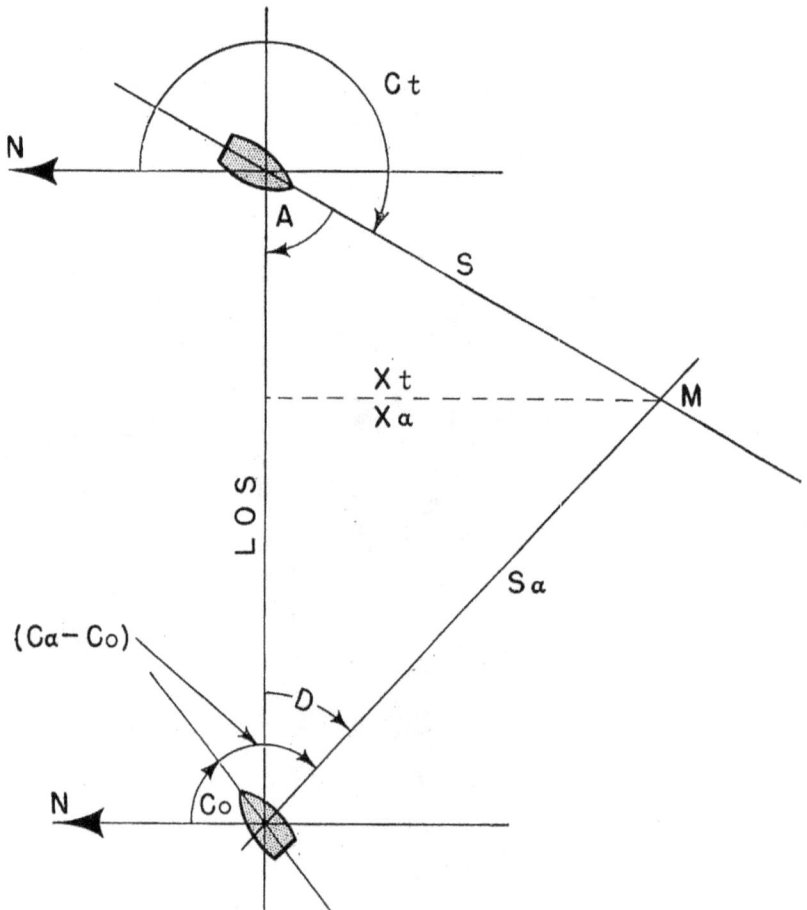

FIGURE 20C3. Director solution of torpedo triangle.

7. *Intercept offset* (Lt). This is an arbitrary correction which is introduced when it is desired to change the sight angle by some predetermined value to take care of unexpected target maneuvers. The changes are made in the sight angle directly without altering the problem set-up.

The following inputs modify the basic torpedo course determined by using the corrected sight angle:

8. *Gyro angle* (G), introduced manually.
9. *Tube offset* (Lu), introduced manually.

The gyro angle is transmitted directly to each torpedo course indicator and does not enter into the director calculations. The value of the basic torpedo course transmitted will be the same for each tube unless a tube offset is used. A tube offset modifies the torpedo-course order so that the

CHAPTER 20

forward tube is turned forward and the after tube aft, with respect to the basic torpedo course, through an angle equal to the tube offset angle.

Director outputs. The following outputs are obtained from the director:

1. *Corrected sight angle* (D'). This is the sight angle (D), computed from the torpedo-fire triangle, modified by latitude and intercept-offset corrections. It is used to automatically offset the director telescope.

2. *Basic torpedo course* ($Ca-Co$). When modified by tube-offset angle, the basic torpedo courses for the two tubes are different and are designated by ($Ca'-Co$) and ($Ca''-Co$) for the forward and after tubes respectively. These values are transmitted to the torpedo-course indicators.

3. *Gyro angle* (G). This value is transmitted directly to the torpedo-course indicators.

FIGURE 20C4. Torpedo-course indicator.

20C5. Torpedo-course indicator. The Mark 1 torpedo-course indicator shown in figure 20C4 indicates the following quantities on its dials: (1) basic torpedo-course order ($Ca-Co$), (2) gyro-angle order (G), (3) actual torpedo course used, (4) actual gyro angle used, and (5) actual tube train. The mechanical inputs of gyro angle and tube train introduced at the tube mount are added in a differential to obtain basic torpedo course. These mechanical inputs actuate the ring dials of the torpedo-course and gyro-angle dial groups shown schematically in figure 20C5. The electrical inputs from the director actuate the center dials of these same dial groups. The gyro angles can be set and the tube properly trained by matching pointers on these dials. Transmission of basic torpedo course instead of tube train makes it possible to change the gyro setting at the tube and still obtain the correct torpedo course by matching pointers on the torpedo-course dials. As the gyro setting is changed, the tube train required is also modified by an equal amount

RESTRICTED

in the opposite direction. The common practice is to transmit basic torpedo course, and the gyro setter at the tube sets only enough gyro angle to keep the tube trained to safe bearings. This method permits using small gyro angles which is always desirable.

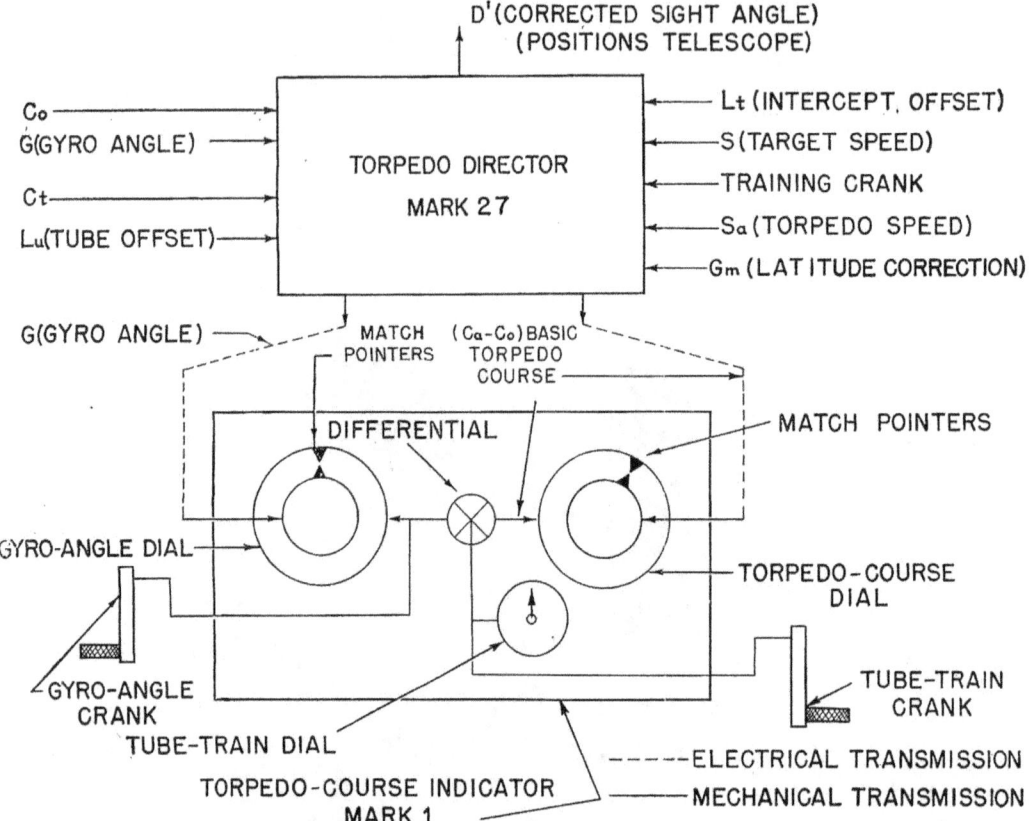

FIGURE 20C5. Torpedo director system; schematic.

20C6. Gyro-setting mechanism. This mechanism, shown in figure 20C6, permits setting both basic and spread gyro angles while the torpedoes are in the barrels of the tube. Thus the gyros can be set at any time up to just before firing. When the *spindle-engaging lever* is moved to the *in* position spindles extend upward through the bottom of the barrels and engage the torpedo gyro-setting mechanisms. The gyro spread-setting handcrank is then rotated until the desired spread angle is indicated on the *gyro-angle dial*. This operation sets the spread gyro angle on all torpedoes simultaneously. A spread setting of 3° indicated on the dial gives a spread of 3° between adjacent torpedoes. The gearing is so arranged that the angles set on the torpedo gyros differ by the amount necessary to obtain the desired spread. For example, if a quintuple tube mount is used and a 3° spread is desired, the actual gyro angles set on the torpedoes will be 0° for the center one, 3° for those adjacent to the center, and 6° for those on the outside.

The basic gyro angle is set on all torpedoes simultaneously by means of the basic gyro-setting handcrank with spindle-engaging lever in. The gyro spindle-engaging lever must be in the *out* position when the gyros are not being set, since any movement of the torpedo within the tube with the spindle engaged may damage the mechanisms.

20C7. Local control. A sight at the tube mount serves as a standby for local control in the event of director-system casualty. The sight is mounted on a circular base attached to the *train-*

CHAPTER 20

ing-handcrank stand shown in figure 20C6. The sight angle is determined by means of a portable angle solver and the sight is offset an amount equal to the algebraic sum of the sight angle and the basic gyro angle set on the torpedoes. Hence, when the line of sight is on the point of aim, the tube will be correctly trained for firing the torpedoes on the calculated collision course.

FIGURE 20C6. Gyro-setting mechanism.

20C8. Example of director control. A destroyer equipped with two quintuple tube mounts and a single Mark 27 Director is on course 040° when it sights an enemy cruiser at an estimated range of 6,000 yards. The estimated target course and speed are 210° and 30 knots respectively. The torpedo-control officer orders both tube mounts to stand by for curved fire with a spread gyro angle of 4°. He decides to use a 30° gyro angle and a 10° tube offset and these values are cranked into the director along with the latitude correction.

The torpedo problem set-up is shown in figure 20C7a. The fixed straight line joining the centers of the *main dial groups* represents the line of sight. Each main dial group consists of three concentric dials. The zero graduations of both outer ring dials indicate true north and these ring dials are automatically positioned by electrical inputs from a gyro repeater. The outlines of own ship and target ship are on the inner ring dials. The arrow on the center dials indicates the direction of the torpedo track.

Basic torpedo course ($C_a - C_o$), can be read approximately from the own-ship inner ring dial against the arrow on the inner dial. Corrected sight angle (D') is indicated by the displacement of the arrow from the line of sight. However, both of these values can be read more accurately on other director dials. Track angle (I) is read directly from the target center ring dial since it is the angle between the arrow on the center dial and the target heading. This angle is not indicated elsewhere. Other values obtained from the main dial groups are noted in the figure.

The dials shown in figure 20C7b indicate tube offset, basic torpedo-course order and gyro-angle order transmitted from the director to the torpedo-course indicators. Each tube dial group has three dials with the one at the center fixed. The values read from these dials are indicated in the figure.

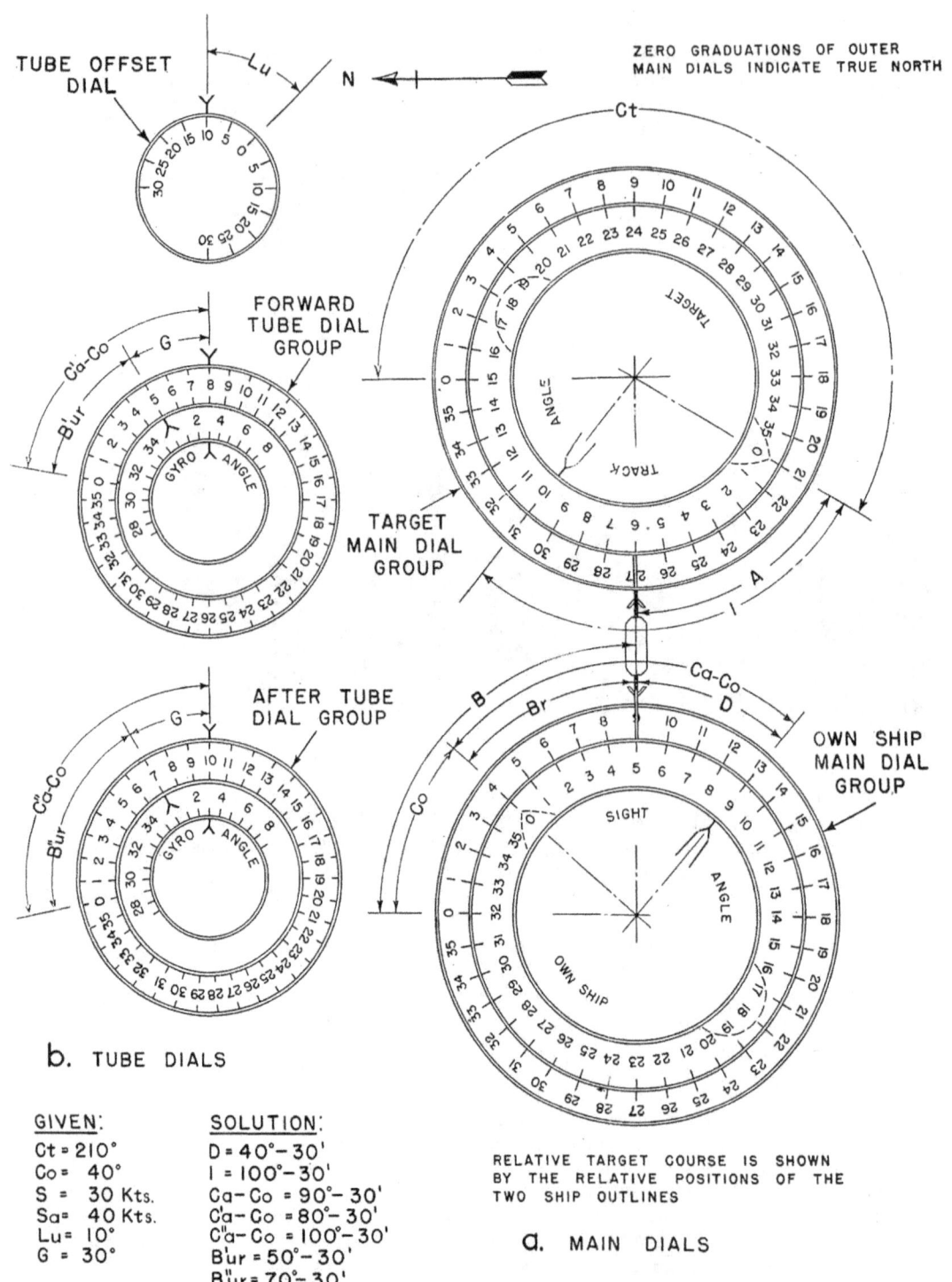

FIGURE 20C7. Director dials.

CHAPTER 20

The computer mechanism generates a corrected sight angle of 40° 30' and when the director trainer is on the target the generated basic torpedo course is 90° 30'. Since the firing is to be done from the starboard side of own ship the tube offset of 10° will make the basic torpedo course for the forward tube (Ca' — Co) equal to 80° 30' and that for the after tube (Ca''' — Co) 100° 30'. These latter values are transmitted to the tubes along with the gyro angle of 30°. After the spread gyro angles have been set and the basic gyro angle introduced by matching pointers on the gyro dial, the pointers on the torpedo-course dials can be matched by training the forward tube 50° 30' and the after tube 70° 30'. These angles are designated as B'ur and B''ur on the tube dials at the directors and are also shown on the tube train dials at the torpedo-course indicators.

After the tubes are positioned the torpedoes are fired in succession on orders from the torpedo-control officer. The firing set-up is illustrated schematically in figure 20C8.

20C9. Torpedo control organization. A typical torpedo control organization for a destroyer is given in the following outline:

TITLE	STATION	DUTIES
Captain	Bridge	Designates target and method of fire to be used. Conns ship, if necessary, to bring the torpedo battery in position for firing. Estimates target angle and target speed if there is sufficient time.
Torpedo officer	Torpedo-control station	Controls torpedo battery, makes set-up on the director using Captain's estimate if available, or own estimates. When more than one ship is firing, makes set-up ordered by the Division Commander, and fires in accordance with standard doctrine.
Torpedo control talker	Torpedo-control station	Serves as torpedo officer's talker and operates firing switches.
Director trainer	Torpedo director	Maintains correct line of sight by training director. Operates director firing key.
Hand-input operator	Torpedo director	Operates hand inputs for own-ship course or sight angle in case of casualty to automatic source for either.
Torpedo-battery captain	Tube mounts	In general charge of both tube mounts. Sees that orders are properly carried out at the mounts and handles casualties at the mounts.
Mount captain	One at each tube mount	In charge of mount and wears head phones.
Mount trainer	One at each tube mount	Trains mount to keep the zero readers of the torpedo-course dial matched, or trains the tube as ordered.
Gyro Setter	One at each tube mount	Sets gyro angles either by keeping zero readers of the gyro-angle dial matched, or as ordered. Also sets the depth-setting mechanism on the torpedoes.

RESTRICTED

TORPEDO CONTROL

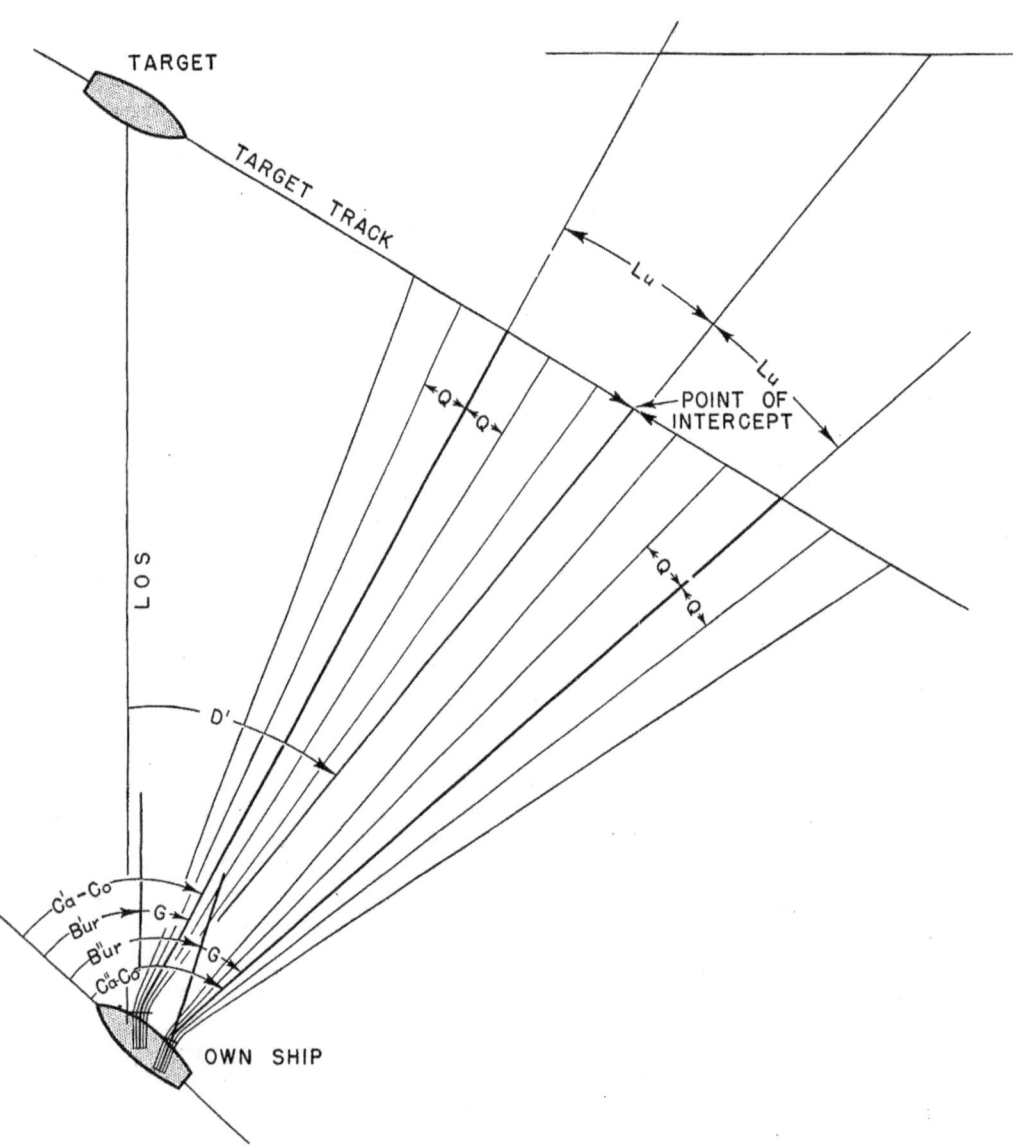

FIGURE 20C8. Torpedo firing problem.

CHAPTER 21

SPOTTING

A. SALVO ANALYSIS AND RELATION TO SPOTTING

21A1. Definitions. It is the job of the spotter to note where shots fall so that necessary corrections can be made. The spotter, therefore, must have at least an elementary knowledge of salvo analysis. The following definitions relate to terms used in connection with such analysis.

Salvo. A *salvo* consists of two or more shots fired at the same target at the same time, or within a short interval of time such as the duration of a firing signal.

M.P.I. The *mean point of impact* (M.P.I.) of a salvo is the point which is at the *geometrical* center of the points of impact of the several shots of a salvo, *excluding wild shots.*

Wild shot. A *wild shot* is a shot whose dispersion is abnormal in comparison with the dispersions of other shots fired in the same salvo. A shot may be wild either in range or in deflection.

Pattern. The *pattern in range* of a salvo is the distance, measured parallel to the line of fire, between the point of impact of the shot falling farthest beyond the M.P.I. and the point of impact of the shot falling farthest short of the M.P.I., *excluding wild shots.* The *pattern in deflection* of a salvo is the distance, measured at right angles to the line of fire, between the point of impact of the shot falling farthest to the right of the M.P.I. and the point of impact of the shot falling farthest to the left of the M.P.I., *excluding wild shots.*

Dispersion. The dispersion of a shot is the distance between its point of impact and the M.P.I. of the salvo. The *dispersion in range* is the component of dispersion measured parallel to the line of fire, and the *dispersion in deflection* is the component of dispersion measured at right angles to the line of fire. Dispersion in *range* is positive when the shot falls beyond the M.P.I. Dispersion in *deflection* is positive when the shots fall to the right of the M.P.I. The algebraic sum of the dispersions in range (or deflection) of the several shots of a salvo must equal zero.

Apparent mean dispersion. The apparent mean dispersion in range (or deflection) of a salvo is the *arithmetical* mean of the dispersions in range (or deflection) of the several shots of that salvo, *excluding wild shots.*

True mean dispersion. The true mean dispersion is the mean dispersion in range (or deflection) of an infinite number of shots, all assumed to have been fired under conditions as nearly similar as possible, and excluding wild shots.

Hitting space. The hitting space in range is the distance, measured parallel to the line of fire, between the point of fall at the surface of the water of a shot that just pierces the top of the target and of one that just pierces the bottom of the target (i.e., the waterline on the engaged side of the target). It includes the projection of the target's vertical height upon the plane of the water and the target's horizontal dimension in the line of fire (or depth). It may also include a distance in front of the target within which impacts are likely to produce underwater or ricochet hits on the target. The hitting space in deflection is the material width of the target. The term "hitting space," as generally used, refers to the hitting space in range.

Error of the M.P.I. The error of the M.P.I. is the distance of the M.P.I. of a salvo from the *center of the hitting space,* measured parallel to the line of fire in range and at right angles to the line of fire in deflection.

21A2. Accidental errors causing dispersion. It might be thought that if a battery of guns is fired at the same instant with the same sight angle and sight deflection set all the projectiles would hit at the same point. The fact that such is not the case is due to *accidental errors* causing dispersion.

If the battery of guns were stationary and rigidly fixed in elevation and train, variations in range and deflection would be caused by *gun errors,* which are:

SPOTTING

1. Differences in weight and temperature among individual powder charges.
2. Differences in projectile weights.
3. Variations in angles of projection. That is, the axes of projectiles diverge, in varying amounts, from the continuation of the bore axis as they leave the various guns.
4. Differences in projectile seating causing variations in density of loading and initial velocity.
5. Differences in erosion among the several guns with corrections not precisely made.
6. Differences in droop among similar guns, and variation in droop with temperature changes.
7. Variations in the amount the various gun mounts yield, and irregularities in the action of the recoil mechanisms.
8. Differences in the alignment of guns.

The above variations, no matter how slight, justify acceptance of the fact that dispersion in the points of fall of projectiles from several guns fired in salvo can be expected even under the ideal conditions specified.

21A3. Accidental errors due to personnel. Further accidental errors are introduced by movement of the ship when the guns are positioned by pointers and trainers. The ship's motion can cause misalignment of the sights on the target when the guns are fired. Also, one pointer may be bringing his line of sight on the target by elevating while another may be doing so by depressing, introducing opposite, although slight, vertical components of velocity due to the motions of the guns about their trunnions. The same effect may result from failure to fire simultaneously, in which case different guns fire at slightly different velocities of own-ship motion. Personnel errors of this type are practically eliminated by automatic positioning of the guns from a central control station.

21A4. Effect of accidental errors on pattern sizes. Accidental errors cause normal dispersion yielding a certain pattern size at a particular range. The pattern sizes at various ranges for a battery of a ship are observed and used as standards for that battery in spotting the M.P.I. of a salvo to the target. By definition, a shot whose dispersion is abnormal in comparison with the dispersion of other shots fired in the same salvo, is a wild shot. A wild shot is considered to result from a *mistake* rather than from accidental errors. For example, a powder bag loaded with its ignition pad forward is a mistake and will probably cause excessive dispersion. Also, if a sight setter were to set a range of 15,000 yards, instead of the correct setting of 10,500 yards, it would be classed as a mistake and the result would be a wild shot.

21A5. Law of probability applied to dispersion. Having enumerated the accidental errors which cause dispersion, it is desirable to see how these errors affect the chances of hitting a target, and how the control of a battery of guns may be influenced by a knowledge of the errors. The answers may be obtained by application of the laws of probability, which deal with the prediction of future occurrences on the basis of information gained from past experiences.

The law which pertains to symmetrically distributed accidental errors is known as the *normal probability law*. The accidental errors of gunfire follow a fairly symmetrical distribution and this law is applied to them. It is based on the following considerations:

1. Positive and negative errors of the same size are equally probable and hence will occur with equal frequencies.
2. Small errors are more probable than large errors, and hence will occur with greater frequency than large errors.
3. Very large errors will not occur, or, more specifically, such errors are due to mistakes and are not classifiable as accidental errors.

21A6. Equal frequency of errors. The first consideration is interpreted as follows: assuming the M.P.I. of a six-shot salvo is at the desired point of impact, one-half of the shots may be expected to fall short of the desired point of impact, and the other half over. Also, there is equal frequency of right and left errors in deflection. Therefore, if the M.P.I. is maintained at the center of the target (or hitting space) in both range and deflection, an equal frequency of shorts

CHAPTER 21

and overs, and rights and lefts in any salvo can be expected. Figure 21A1 illustrates the equal frequency of shot dispersions from the M.P.I. as caused by the accidental errors of gunfire previously described.

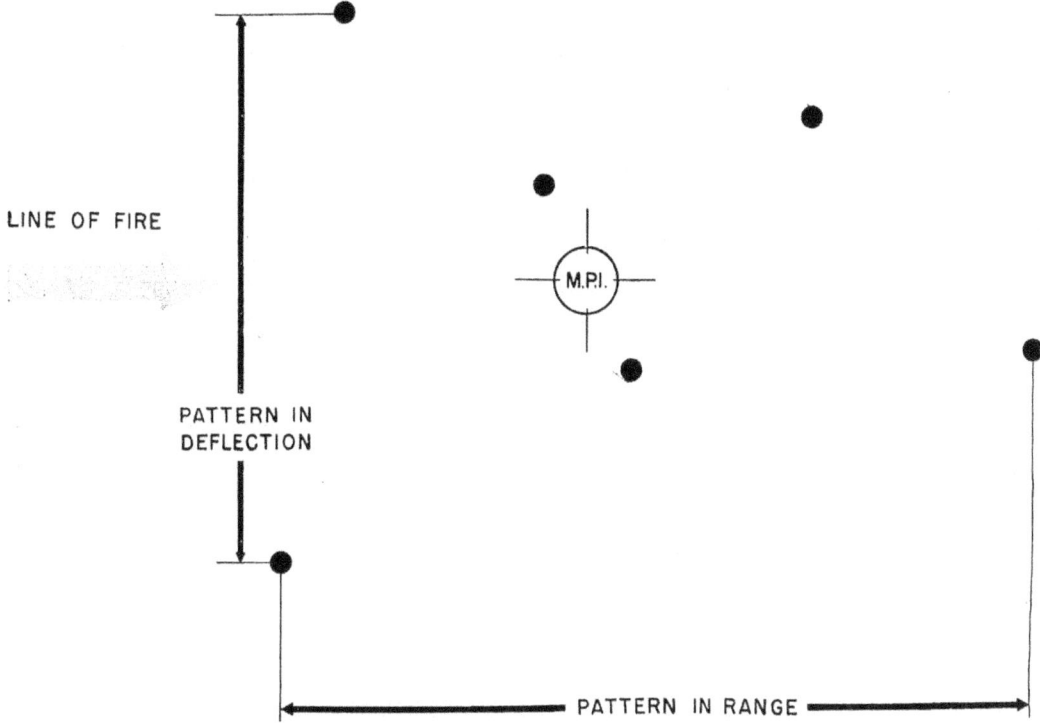

FIGURE 21A1. Pattern of a six-shot salvo.

21A7. Frequency of small errors. There is also indicated in figure 21A1 the pattern in range and deflection. It is evident that if the M.P.I. is located at the point of aim and the dispersion is not excessive, some or all of the shots will hit the target (or hitting space) depending on the size of the latter.

Assuming that the M.P.I. is at the point of aim (or center of the hitting space) and that the hitting space is smaller in range than the range pattern of the salvo, some of the shots will hit over the target and some will hit short of the target. However, according to the second consideration of normal probability, small errors are more probable than large errors and hence more shots will be clustered about the M.P.I. It therefore follows that, with normal dispersion and with the M.P.I. centered at the desired point of impact, more hits will be obtained than misses (see fig. 21A2).

On the other hand, if the M.P.I. is not at the desired point of impact, as shown in figure 21A3, the majority of shots miss the target. Hits may still be obtained at the extreme pattern edges but the *probability* of a number of hits is greatly reduced depending on *pattern size* and the *error of M.P.I.* The same laws of probability apply to deflection hits as well as to range hits.

21A8. Desirable pattern sizes. Assuming that the M.P.I. is at the desired point of impact, a pattern of small dispersion is expected to yield the maximum number of hits. However, where excessive accidental errors are allowed to creep in, excessive dispersion results. In such a case, the pattern size is so spread that few shots are grouped around the desired point of impact, and therefore, a minimum number of hits are probable. Excessive dispersion should not be confused with wild shots, as such dispersion still follows the laws concerning symmetrical distribution.

RESTRICTED

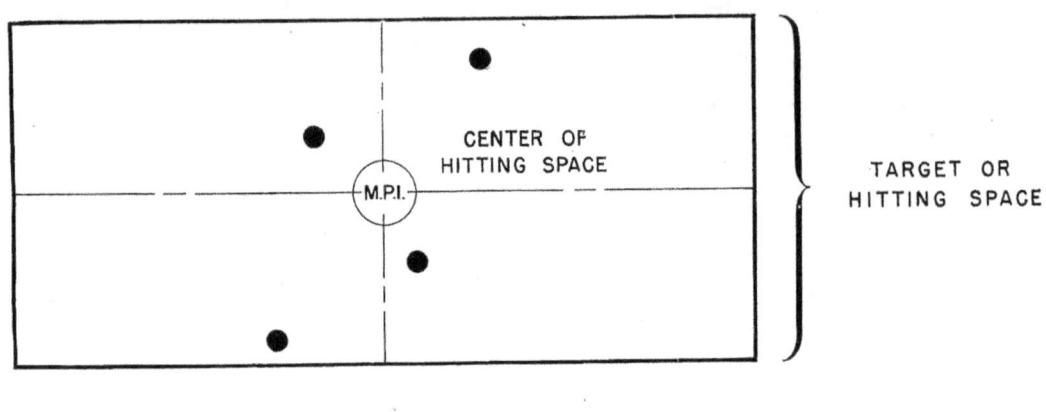

FIGURE 21A2. Clustering of shots about the M.P.I.

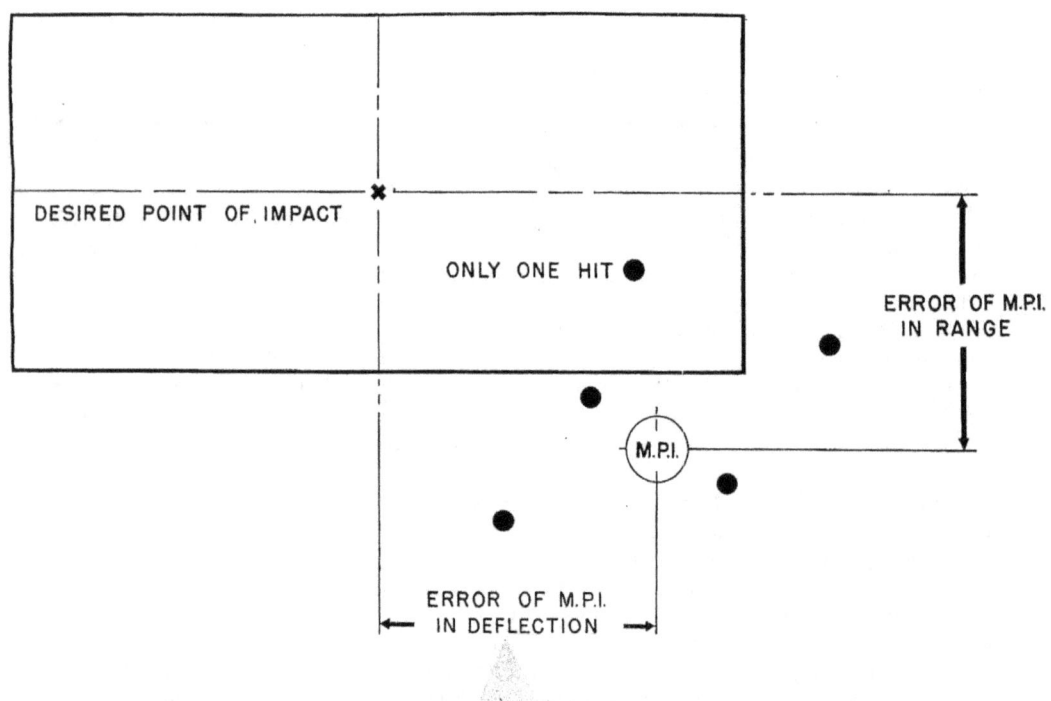

FIGURE 21A3. Effect of error of M.P.I.

492

CHAPTER 21

From the standpoint of a maximum number of hits, it would seem desirable to eliminate accidental errors entirely so as to have a zero pattern in range and deflection (that is, have all the shots grouped about the M.P.I.), and in truth such a small pattern would be desirable if the M.P.I. were always at the desired point of impact. However, since it is difficult in actual firing to place the M.P.I. at the desired point of impact, and it is better to hit with some shots than to miss with all, a fairly large pattern gives a certain advantage. In practice, a compromise is effected; the pattern is not as small as it might be, nor is it too large for real effectiveness.

21A9. Accidental errors causing shift of M.P.I. Since dispersion of individual shots is caused by accidental errors, it is logical to assume that no two salvos will be exactly alike and hence the geometrical center of the shots in a given salvo (or M.P.I.) can be expected to differ in location from the M.P.I.'s of other salvos. Thus there is a dispersion of individual M.P.I.'s from the aggregate M.P.I. of several salvos, which can be determined only by analyzing target practice results.

The average shift between successive M.P.I.'s is usually so small as to be difficult for the spotter to estimate. Therefore, the spotter should not be too quick to spot when only one or two salvos seem to wander off the target during a string that is, in general, satisfactory.

21A10. Error of M.P.I. The only accidental errors so far considered have been those which the spotter must recognize as causing dispersion of individual shots in a salvo and dispersion of the M.P.I. for several salvos. Recognition of these accidental gun errors helps the spotter to distinguish them from another class of errors which more decidedly affect the M.P.I. of a salvo. This latter class of errors is known as *control errors* or *control inaccuracies*.

Before the first salvo is fired at a target, as many effects causing error of the M.P.I. as possible are accounted for. If this salvo has an error of M.P.I., the causes for the error are the result of control inaccuracies. There are four general control inaccuracies which may cause error of the M.P.I. These are listed in the order of the probable frequency of their occurrence and the magnitude of their effect:

1. Range keeper set up with incorrect range, courses, and speeds.
2. Ballistic corrections based on conditions not existing at time of firing (initial velocity, wind, air density, and others).
3. Battery not properly aligned with director.
4. Indeterminate errors (Class B and personnel errors).

21A11. Incorrect range-keeper set-up. The present range to the target is correct only to the extent that measurement by eye estimate, or range finder is precise. An error in this basic range is directly registered as an error in M.P.I.

Measurements of own-ship course and speed usually are reasonably accurate, but any inaccuracies result in error of M.P.I.

The determination of target's course and speed are made directly from the spotter's estimate of target angle and speed, or by rate controlling to establish the values. These two variables are the most difficult to determine, are the chief cause of an incorrect range-keeper set-up, and, hence, the chief source of M.P.I. error.

21A12. Incorrect ballistics. It is shown in preceding chapters that the computer determines corrections necessary to compensate for any variations from standard conditions. If the determination of these corrections or ballistics is based on incorrect values of ballistic wind, initial velocity, air density, and others, the total ballistic will be in error and result in a corresponding error in M.P.I.

21A13. Improper battery alignment. The alignment referred to here is not intended to mean alignment between the guns of a battery. Such misalignment results in greater dispersion and larger pattern sizes but does not materially affect the error of the M.P.I. of a salvo.

The improper alignment referred to here means, rather, that which may exist between the controlling director and the battery as a whole. This type of error is generally caused by failure to director-check the battery and *all* directors. Thus, a battery which is aligned with one director is not necessarily aligned with another which may be in control. Frequent director checks can assist in elimination of this cause of error in M.P.I.

21A14. Indeterminate errors. One of the two classes of indeterminate errors results from the fact that some of the computations by fire-control instruments are mechanical approximations. These approximations result in *class B errors*, which are small in magnitude for normal ranges and therefore usually acceptable as causing minimum error in M.P.I. However, at extremely short or long ranges the errors may become large, depending upon the instrument concerned, and thus may seriously affect the M.P.I. When class B errors are known to be large they are not admissible as accidental, and steps must be taken to make correction for them.

The other class of indeterminate errors is assignable to control personnel. An example would be the director pointer or trainer being off the point of aim when the salvo is fired. Only training and experience can prevent the occurrence or reduce the magnitude of such mischance. Small errors of this type merely cause a slight shift or dispersion of M.P.I. and should not be corrected for by the spotter. The director operator should inform the spotter of large discrepancies in the point of aim in order that the spotter may differentiate this error from others.

21A15. Summation of errors. During actual firing a spotter cannot sit in his aloft station and carefully analyze each fall of shot to determine the exact geometrical M.P.I. of a salvo. Splashes last only a matter of seconds and the firing situation is probably changing even during that short interval. The spotter must have previously analyzed all probable errors and the reasons for them. This knowledge, coupled with practical training, will enable him to make instant and intelligent decision as to the proper correction or spot necessary to hit the target.

In summation, the general errors attending shipboard gunfire are:

    1. Accidental gun errors causing dispersion of shots. These errors are only compensated to the extent of achieving desired pattern sizes. Most of the errors are eliminated by careful design, normal upkeep of the battery, and the training of gun crews.

    2. Accidental gun errors causing a shift in the M.P.I. of successive salvos. The shift is usually small in magnitude and should only be corrected when excessive.

    3. Control inaccuracies causing error of M.P.I. It is the primary duty of the spotter to "spot" the corrections in range and deflection (also fuze range in AA fire) necessary to bring the M.P.I. of a salvo to the desired point of impact. He should recognize the first two classes of errors in order to judiciously spot the error of M.P.I. caused by control inaccuracies.

B. THE SPOTTER AND METHODS OF SPOTTING

21B1. General function of the spotter. The primary function of the spotter is the correction of actual range and deflection errors of the M.P.I. to bring the shots on the target, or in other words, to establish the hitting gun range and deflection. The basis of these corrections is his own observation. Failure to promptly and accurately correct initial errors may conceivably be the deciding factor in a naval engagement, which fact makes apparent the importance of the spotter in relation to the accurate control of gunfire.

There are other functions in relation to fire control which the spotter must perform before and during the firing. These functions, in general are:

Preliminary.

    1. Search for and locate the enemy.

    2. Describe the enemy forces (general bearing line, number of ships, and deployment).

    3. Estimate all values necessary for initial positioning and tracking of the prescribed target (R, E, A, dH, and Sh).

During firing.

    4. In addition to the primary function of spotting the fall of shot, report changes in target angle and speed. If necessary, report changes in range.

    5. Keep control informed of the tactical situation, if required to do so.

21B2. Other functions and duties. The spotter is required to perform other duties in connection with the control of gunfire. The exact nature thereof varies in different ships, de-

CHAPTER 21

pending upon the particular fire-control organization in effect. The spotter's detailed duties and spotting procedures are prescribed in the ship's battle bill and in Fleet doctrinal publications. However, it is not the purpose of this text to enumerate specific duties or to lay down detailed instruction for their accomplishment. Rather, discussion here is limited to the spotter's most important functions; that is, the estimation of target course and speed, and the observation and correction of the error of M.P.I. by standard methods of spotting.

21B3. Estimation of target course and speed. The spotter who can estimate target angle (used in the determination of target course) and target speed with fair accuracy will greatly assist the range-keeper operator in arriving at an early solution of the problem, and also aid him in checking the range-keeper set-up. On the other hand, if the spotter's estimates are inaccurate the range-keeper operator is required to make many changes, particularly in the case of a surface problem where automatic rate controlling cannot be used.

To estimate target angle the spotter must know the structural details of all possible targets. Silhouettes and usually scale models of all probable targets are furnished each ship. The spotter should study the details of these visual aids, not only for the purpose of recognizing the enemy, but also for the purpose of estimating target angle. In estimating target angle, the spotter should make use of prominent objects such as bridges, breaks in the deck, stacks, masts, and other features. By observing the opening and closing of the distance between such details the spotter can estimate the angle the enemy ship makes with the line of sight.

Target speed at best can be but roughly estimated. Here again, knowledge of enemy ships is valuable, and particularly so in the case of maximum speeds. If in line of battle, target speed may be estimated as about one or two knots less than the maximum speed of the slowest ship in the formation, considering any damage reports which may have been received.

The best speed estimations can be made only by men who have had extensive training and experience. The aids used by a spotter in a direct estimation of target speed are smoke from the stacks, bow wave, and stern wake.

21B4. Spotting error of M.P.I. in deflection. Correcting the deflection is extremely important, not only because salvos must be on to hit, but also because, in long-range surface fire, it is very difficult to spot on in range if the deflection is excessive. If other means for spotting in range are not available, such as aircraft observation or shipboard measuring devices, a salvo which is widely off a single target should first be spotted in deflection and the succeeding salvo spotted in range. With a number of targets in formation it should be possible to spot at once in both range and deflection. In estimating deflection spots, the estimated target width in mils should be used as a guide. Also, telescopes, binoculars, or other optical devices fitted with mil scales furnish a suitable means of estimating deflection errors.

With a high-speed target, the spotter should bear in mind that the apparent M.P.I. in deflection should be held abaft the point of aim in order to allow for the travel of the target while the splashes are forming. In other words, it must not be assumed that full splashes form instantaneously at the impact of a salvo. The time lag is only a few seconds at most, varying with the caliber of projectiles, but is sufficient to allow considerable movement of a high-speed target.

21B5. Spotting error of M.P.I. in range. With good visibility and a spotting height of 100 feet or more, the error of M.P.I. can usually be estimated with reasonable accuracy at ranges up to about 15,000 yards. This is accomplished by observing the position of the bases or slicks of the splashes relative to that of the waterline of the target. Experience in determining this range error can be gained by actual firings at target practice and by practice on a spotting board which simulates the actual conditions of splashes relative to the target as viewed from the spotter's station. *Spotting diagrams* are used in conjunction with such training to indicate the relative displacement of a salvo from the target. Figure 21B1 shows a typical example for a spotting height of 100 feet and target heights of 10 feet, 20 feet, and 30 feet. Such a diagram is constructed by means of dip curves for the particular height of a spotting station. Range lines are drawn at intervals of 500 or 1,000 yards from minimum range to that of the horizon. The distances between these range lines rep-

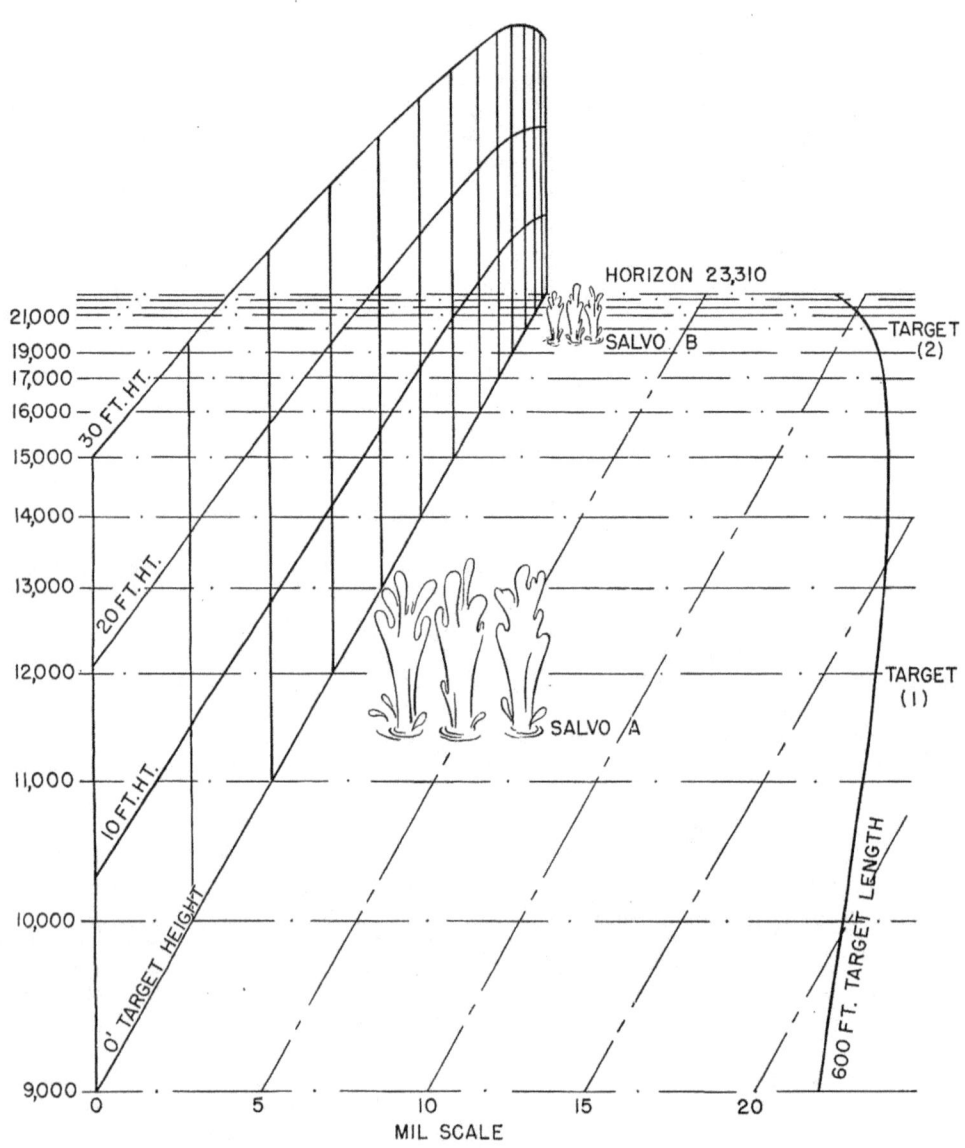

NOTE: DIAGRAM FOR SPOTTING HEIGHT OF 100 FT.
RANGES BEYOND HORIZON

HT.(FT)	RANGE (YARDS)	HT.(FT)	RANGE (YARDS)	HT.(FT)	RANGE (YARDS)
10	30,700	50	39,700	90	45,400
20	33,600	60	41,300	100	46,600
30	36,100	70	42,800	110	47,600
40	37,900	80	44,200	120	48,900

FIGURE 21B1. A typical spotting diagram.

CHAPTER 21

resent the apparent range differences as viewed by an observer at the height in question. The method of construction and the dip curves are available in Fleet training publications aboard ship, and are not described here.

Study of figure 21B1 makes it evident that salvo A is short of the imaginary extension of the waterline of a target at 12,000 yards by about 500 yards. However, the error of a salvo fired at a target at 20,000 yards is not as apparent. The 500 yard error of such a salvo (indicated on the diagram as salvo B), appears as little or no error in apparent distance from the extended waterline of the target. Thus, it can be seen that, for shipboard spotting at such a range, the splashes must be in line with some portion of the target before the spotter can reasonably tell whether the salvo is over or short, to say nothing of estimating the *amount* of the M.P.I. error.

21B6. Methods of spotting. It is evident from the preceding discussion that certain doctrinal aids, or methods of spotting, must be employed by the shipboard spotter to assist him in this difficult estimate of the M.P.I. error in range.

When describing the various methods of spotting it must be borne in mind that these methods are but general guides and are in no way intended to replace or abrogate actual spotting procedures which are laid down in detail in Fleet and Force doctrines. Such doctrines must be perfectly known and rigidly followed by the spotter under all conditions covered therein. This demand might seem restrictive, but it is purposely so because only proved and sound principles should be employed. Individual untried ideas, which are likely to produce negative results, are excluded by the ruling.

Subject to the foregoing qualifications, there are five methods of spotting, the use of any one of which is determined by the type of battery firing, type of target, range, and visibility. The methods are as follows:

1. Direct method.
2. Bracket and halving method.
3. Ladder method.
4. Barrage method.
5. Tracer method.

21B7. Direct method. Spotting by the *direct method* is, as its name implies, the spotting of salvos (splashes) direct to the target. This is the most desirable procedure, but its use is very limited. It requires a spotting height of 120 feet for ranges up to 15,000 yards. It is also necessary that the splash be relatively close to the target, and that the range-keeper set-up be fairly accurate. These considerations limit this method to main-battery fire at relatively short ranges.

A thoughtful analysis of the problem with reference to the spotting diagram in figure 21B1 reveals that the greatest limitation of the direct method is the matter of range. Deflection spots can be made with equal accuracy at any visible distance. If, then, air spots are available, and the plane spots in range with the ship spotting in deflection, the direct method can be used by the main battery at any range at which a portion of the splash is visible to the shipboard spotter. Air spotters cannot spot accurately in deflection unless they have a line of sight containing the firing ship and the target. Such a line of sight is, of course, a practical impossibility.

The direct method of spotting in range is also used when mechanical means are available for measurement of the M.P.I. error. Such means include stereo range finders, and other devices which cannot be described here because of their confidential nature.

21B8. Bracket or halving method. The *bracket* or *halving method* is used at long ranges by the main battery when no air spot is available. It has as its basis the fact that at great distances it will be impossible to tell if a splash is short of or over a target, unless the two are in line. If the splash and target are not in line, or if the splash does not line up with a ship in the formation whose range is approximately that of the target, the first spot is made in deflection only. When target and splashes line up, succeeding spots are made in range in increments of not less than pattern size until the target has definitely been crossed. The next spot is in the opposite direction, and half the size of the initial spot. This should produce a straddle. If it fails to do so, the

497 **RESTRICTED**

spotter should recognize the probability that the range-keeper set-up is badly in error. Once a straddle is obtained, *centering* deflection spots only should be made, as long as the spotter is sure he has both short and over splashes. If, however, over splashes are obscured by the target or by shorts, every third or fourth salvo should be fired with an up-shot of one pattern size so as to lift the entire salvo over the target. Only thus can the spotter know his straddles were not all shorts in reality.

Normally, when using this method, *slow fire* is employed until at least one straddle is obtained. In slow fire the firing of a salvo is held up until the previous salvo has landed, the spot has been made, and corrections applied for the succeeding salvo. When a straddle occurs *rapid fire* is usually ordered by the controlling officer. In rapid fire, continuous salvo fire is maintained without waiting for each salvo to land and for spots to be made and applied to each succeeding salvo. In such a case a spot is made whenever possible and applied to the gun range of the salvo to which the spot was applicable. Certain tactical conditions might make it necessary for a battery to *open* with rapid fire.

21B9. Ladder method. The *ladder method* is one designed to find quickly the hitting gun range, given an estimated opening range and a fast-moving target. The basic technique is to fire three ranging salvos. The first salvo is with range-keeper range less a given value (for example, a pattern size, or 500 yards or 1,000 yards), the second with range-keeper range, and the third salvo with range-keeper range plus the same increment used in the first salvo. Fire is checked until all three salvos land. A spot is then applied to the second salvo, which should give hitting gun range. Using this range as a basis, each succeeding salvo is fired in increments of 100 yards, or any other pre-selected value, so that if the range-keeper data is correct, succeeding salvos will land in such order as: on the target, over 100 yards, over 200 yards, over 100 yards, on, short 100 yards, short 200 yards, short 100 yards, etc. Spotting in this manner consists of keeping the ladder oscillating about the target as a center, and watching for changes in target course and speed. The use of this procedure or variations of the basic method is a common practice in the case of destroyer main batteries, and the secondary batteries of battleships, particularly if ranging accuracy is doubtful.

The three ranging salvos may be omitted in one variation of the foregoing method. In this case, fire is opened deliberately short or over, and succeeding salvos are fired to approach the target in steps not less than a pattern size. As soon as the target has been crossed the steps are reversed and halved. When the target has again been crossed the steps are again reversed and halved. This procedure is continued, thus keeping the M.P.I. of the salvos moving back and forth (rocking) across the target. Steps are never reduced below 100 yards.

21B10. Barrage method. The *barrage method* is more a feature of control than a type of spotting. It consists of placing a zone of fire short or over the target, according to whether the target is closing or opening range. The gun range is kept constant until the target passes through the zone and is then lowered or raised as necessary to force a second passage through the barrage, continuing the movement of the barrage as may be necessary. This system is used by smaller caliber, fast-firing batteries against light, fast-moving targets. It is the spotter's function to determine when the barrage should be shifted, and in which direction and how much it should be moved.

21B11. Tracer method. The tracer method of control consists of spotting shots by noting the error as the tracer passes the target. It is used chiefly in spotting machine-gun fire as described in Chapter 19. This method may also be used in addition to others in spotting larger caliber projectiles (i.e., 5"/38 cal. or larger), particularly at short ranges. At longer ranges, tracers are of doubtful assistance to the spotter because of the difficulty in keeping them in sight or in judging when they pass the target. Under certain light condition the tracers cannot be seen clearly, and should not be relied upon except for short-range firing.

21B12. Short-range spotting. Spotting the fall of shot at very short ranges differs from other spotting problems in that range errors are not difficult to judge. However, in determining

CHAPTER 21

deflection errors at short ranges, consideration must be given to the travel of the target and the spotter's position relative to the line of projectile flight. For example, with the firing ship and target on opposite courses, target to starboard, a shot fired with correct deflection but long in range will appear to the spotter to be in error to the left of the target. Special short-range splash diagrams are constructed as an aid to the spotter in this type of firing; instructions for their use are available in publications carried by all ships.

CHAPTER 22

ORGANIZATION AND INTERIOR COMMUNICATIONS

A. ORGANIZATION

22A1. Introduction. In order that a ship's armament and fire control can be utilized to perform its primary function—that of destroying the enemy—an efficient organization of personnel and matériel, and excellent interior communication must be provided. On practically all ship types, and particularly the larger ones, the fire-control stations are widely separated yet each is dependent upon the others for the gathering of information, computation of data to be used by the guns, and transmission of that data to the guns. Therefore, one of the important phases of shipboard gunnery is to effectively organize the ship's personnel and matériel, and to insure rapid and accurate communication and application of fire-control data.

22A2. Action bill. In all U. S. Navy combatant ships both large and small an *action bill* is drawn up for the guidance and coordination of the four main subdivisions of ship organization during battle. Figure 22A1 shows such an action bill in diagrammatic form. Only the fire-control subdivision (representative of battleship installations), however, is fully outlined; the complete development of this organization being known as the *battle bill*. Each main subdivision is administered by a separate organization and functions under the Commanding Officer, through a chain of command as illustrated in the case of fire control.

The purpose of this discussion is to deal with the fire-control organization. It should be realized, however, that all four major subdivisions (see fig. 22A1) are closely related in fighting the ship, and that a thorough understanding of each is essential to the coordination of the various activities.

22A3. Organization variations. Efforts are being made continuously aboard each ship to arrive at the most efficient organization. During the test of battle, certain flaws in a seemingly-perfect system may sometimes disrupt the desired coordination of effort and lead to loss of the ship. Naturally, in efforts to perfect the finer details of an efficient organization some mistakes may be made. It is none the less desirable to develop a general pattern which represents the best results in achieving optimum coordination.

One of the first things an officer assigned to gunnery should do after reporting aboard for duty is to study the action bill thoroughly. This bill includes the organization plan of the ship, and all subdivisions are explained in detail. The duties of the personnel in each of the fire-control stations are outlined, and the relationship of each station to the entire battle organization is explained. The new officer should study the existing system with great care before he assumes that flaws exist. Often the reasons for a particular plan may not be obvious in daily drills under relatively peaceful circumstances.

22A4. Organization similarities. It must not be understood that the details shown in figure 22A1 are typical of all ships because variation will be found, even among ships in the same class. Organizations necessarily are dissimilar because the requirements of ships differ. Naturally, however, every effort is made to standardize the organization for each type of ship; e.g., a battleship's action bill closely follows the typical chart shown in figure 22A1. It is even desirable that the organizations for other ship classes closely parallel this master pattern. To the extent that this is possible it enables all members of the naval service to be generally familiar with a typical set-up; thus an interchange of personnel does not present a major training problem, and new hands may be worked into the existing organization aboard any ship with facility.

Hence, though the organization for various ships may differ in detail, the one shown in figure 22A1 embraces the essential features common to all combatant types. For example, a heavy cruiser organization is almost identical with that of a battleship. Also, a destroyer main-battery organization is similar to that of a battleship's secondary battery. Antiaircraft organizations for practically all ship types are patterned after the one shown in figure 22A1.

CHAPTER 22

22A5. Command. In figure 22A1 the various subdivisions and stations are connected by lines which show both the *chain of command* and the *flow of information*. Opposed arrows on the lines indicate the transmission of data and resulting orders for proper use of the information.

It is to be noted that all lines emanate from command because the Captain, from his battle station, exercises control over all ship activities. In particular, he is responsible for the supervision of his ship's course and speed, the use of armament, the preservation of watertight integrity, the rectification of casualties to personnel and matériel, and the receipt and dissemination of information. The task of coordinating the many activities, so as to obtain a smoothly functioning organization, is a tremendous one. It is obviously necessary that some agency be available to assist the Captain in performing his functions.

22A6. Combat information center. The function of the combat information center (C.I.C.) is to assist command in planning a correct course of action, and to assist command and fire control in the execution of that plan. The C.I.C. is, briefly, an agency for the collection, evaluation and distribution of combat information, and for facilitating the use of that information. It is not something strange and complex, nor is it merely another plotting room under a new name. It clarifies and simplifies the work of command in preparing for or carrying out combat operations.

Rapid progress in facilitating an early solution of the fire-control problem was made possible by improvements in equipment and methods, and by development of techniques for obtaining accurate target information before the target could be seen. The coordination of intricate gunnery problems with considerations of tactics, ship damage, and exterior communications is the Commanding Officer's primary function. During an engagement the Captain is beseiged with a vast number of informational items, each of which must be weighed individually to determine whether it should be used, discarded, or mentally filed for future reference. The C.I.C. assists the Captain in carrying out his responsibilities by filtering and evaluating nearly all incoming information. The Captain is given the required data as he needs it, and thus is free to concentrate on his decisions and carry the burden of command. C.I.C. may also include an organization to which the Captain delegates secondary decisions and control duties as occasion may require.

The C.I.C. may be arbitrarily divided into two sections, *evaluation* and *control,* whose functions are illustrated schematically in figure 22A2. The control activity with respect to own weapons is of primary interest in this discussion. With its facilities for detecting, tracking, and communication, C.I.C. can be of great assistance to fire control, particularly when visual observation is poor. Some of the functions it can perform are:

1. Designate or suggest suitable targets.
2. Coach gun control on targets.
3. Provide an initial solution for computing mechanisms in plot.
4. Control, designate, or warn automatic weapons in fire against planes and torpedo boats.
5. Spot fall-of-shot with respect to target, if requested to do so by gun-control stations.

22A7. Chief fire-control officer. The fire-control organization functions under the *chief fire-control officer,* known as the *gunnery officer.* It has three principal subdivisions: main battery, secondary battery, and AA battery. Although the gunnery officer exercises command over each subdivision, he normally performs only supervisory functions in connection with all but the main battery which he directly controls. During battle, the gunnery officer is in immediate communication with the Captain and with the officers in charge of the main-battery subdivisions such as plotting, turret, and spotting stations, as well as with those officers in control of the secondary and AA batteries.

In general, the broad duties of the gunnery officer may be outlined as follows:

1. He informs the Captain as to the general gunnery situation and the practicability of opening fire.

2. Prior to opening fire with the main battery he designates targets at which to fire, the fire distribution to be employed, the gun directors to be used, the firing method, and the controlling spotters.

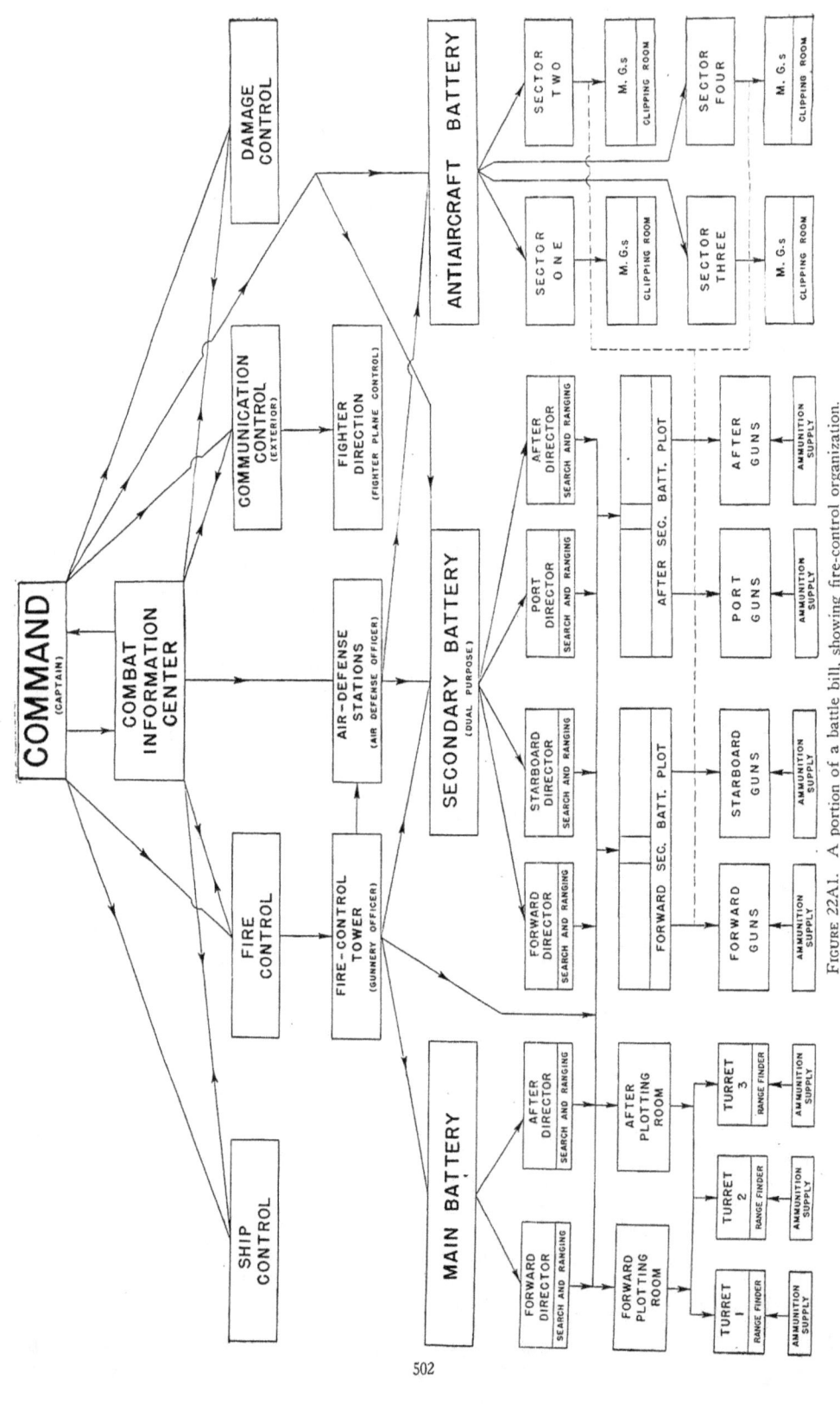

FIGURE 22A1. A portion of a battle bill, showing fire-control organization.

CHAPTER 22

3. When directed by the Captain, he gives the commands "commence firing" and "cease firing".

4. He supervises the salvo signals, and determines the rate of fire.

5. He keeps in touch with the various armament subdivisions and their activities. He is advised as to casualties, and directs such redistribution as may be necessary to maintain the maximum efficiency of the armament as a whole.

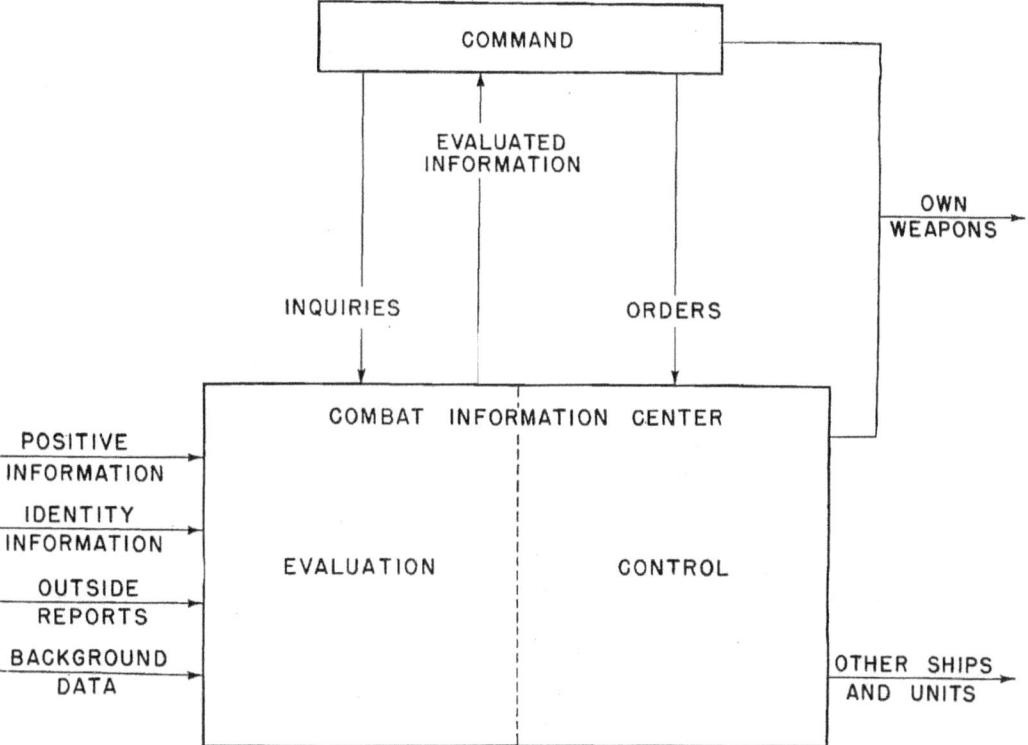

FIGURE 22A2. Combat information center.

As noted in item (2) above, one of the gunnery officer's chief functions is target designation for all the armament subdivisions, which is made in accordance with the Captain's instructions. In the case of a main-battery target, it is probable that there will be a specific assignment for all ships of the battle line, in the form of a fire-distribution signal. Thus the designation of a main-battery target may actually be originated from without the ship, and the order transmitted by the Captain to the gunnery officer. If more than one target is assigned, the Captain specifies the manner of dividing fire. Targets for secondary and AA batteries are less likely to be designated from without the ship, and these decisions may be made by the Captain. On the other hand, targets may be selected and fire opened by officers controlling the respective batteries without direct orders to do so, if this is in accordance with standard instructions.

22A8. Air defense. The defense against air attack has so complicated the handling of a ship's armament that a special officer is detailed to deal with this phase of the gunnery problem; such an officer is called the *air-defense officer*. His primary function is to coordinate the ship's defenses against aircraft in order that the gunnery officer may devote his entire thought and energies to the use of the ship's offensive armament against the primary objective. The air-de-

ORGANIZATION AND INTERIOR COMMUNICATIONS

fense officer may be delegated to open fire on hostile planes on his own initiative, without direct orders from the Captain, either in accordance with standard doctrine or upon the Captain's standing orders.

The air-defense officer may not necessarily control the firing groups directly, but may act in a supervisory capacity over the various gun groups through their respective control officers. He may designate targets, cause a shift in the guns firing at certain planes, and generally direct the ship's AA defenses. It is usual for an air attack to emanate from many sectors about the ship, and it is vitally necessary that all enemy aircraft be under fire. It is the function of the air-defense officer to see that the ship's defensive effort be efficiently coordinated, and that an effective fire be maintained so that the ship's defenses will not be penetrated.

22A9. Battery control. Each main subdivision of any ship's armament (main battery, secondary battery, AA battery, torpedoes, mines, and depth charges) is controlled by a *battery-control officer* assigned to it, each control officer being directly responsible to and in contact with the gunnery officer and the Captain. The gunnery officer is always the *main-battery control officer* since he directly commands main-battery operation. The secondary battery may be directed by a *secondary-battery control officer* (sometimes the assistant gunnery officer), but it may also be supervised by the air-defense officer or the gunnery officer, depending upon the particular ship's organization. The AA battery (which may include both light and heavy machine-gun batteries) is usually under the supervision of a specially designated *AA control officer*. It has already been observed that the Captain can exercise direct supervision over the main battery through the gunnery officer, but it is also to be noted that the Captain can exercise direct control over the secondary and AA batteries through their control officers (see fig. 22A1). This provision is necessary since it might become desirable to subordinate battery control to tactical considerations.

Each battery can be subdivided into groups containing two or more guns (or turrets). Such a group may be controlled by an officer designated as a *group control officer*. This subdivision of a battery is justified because it affords greater flexibility, particularly when divided fire is necessary. In the case of AA machine-gun groups it is desirable to achieve still greater flexibility of control. This is because aircraft targets may approach from many directions, and in their runs may cross over into areas covered by other gun groups. The result from the standpoint of ship's defenses might be a heavy barrage in one area while another area was left open to attack. The problem is handled by the assignment of certain guns to cover definite sectors; the control of fire thus achieved is called *sector control*, and is governed by standard instructions and procedure.

22A10. Sector control. Sector assignments for a battery (such as a 20 mm. or 40 mm. battery) are made in accordance with the total number of guns in that battery. This determines how many may be assigned to any one sector, and that sector's effectiveness. At long range the sector is covered by the larger battery, and at short range by the smaller guns. The effect is to achieve maximum sector coverage from the longest possible firing range down to the bomb-release point.

In the following discussion an imaginary example of sector assignments for 20 mm. guns is used so as not to prejudice the security of existing installations. Let it be assumed that there are 36 guns. The guns can be divided into four groups, and the sectors assigned to each group are numbered correspondingly (1, 2, 3, 4). Each sector can be further divided into three sub-sectors which overlap one another to a certain extent. Guns in each group are assigned to the sectors and sub-sectors as follows (see fig. 22A3):

Group No.	Sector No.	Sub-sector No.	Guns
1	1 (0°– 90°)	A (350°– 45°)	1, 3, 5
		B (45°–110°)	11, 15, 17
		C (350°–110°)	7, 9, 13
2	2 (90°–180°)	A (70°–135°)	25, 27, 29
		B (135°–180°)	23, 33, 35
		C (70°–190°)	19, 21, 31

CHAPTER 22

Group No.	Sector No.	Sub-sector No.	Guns.
3	3 (180°–270°)	A (170°–290°)	20, 22, 32
		B (180°–225°)	24, 34, 36
		C (225°–290°)	26, 28, 30
4	4 (270°– 0°)	A (250°– 10°)	8, 10, 14
		B (250°–315°)	12, 16, 18
		C (315°– 10°)	2, 4, 6

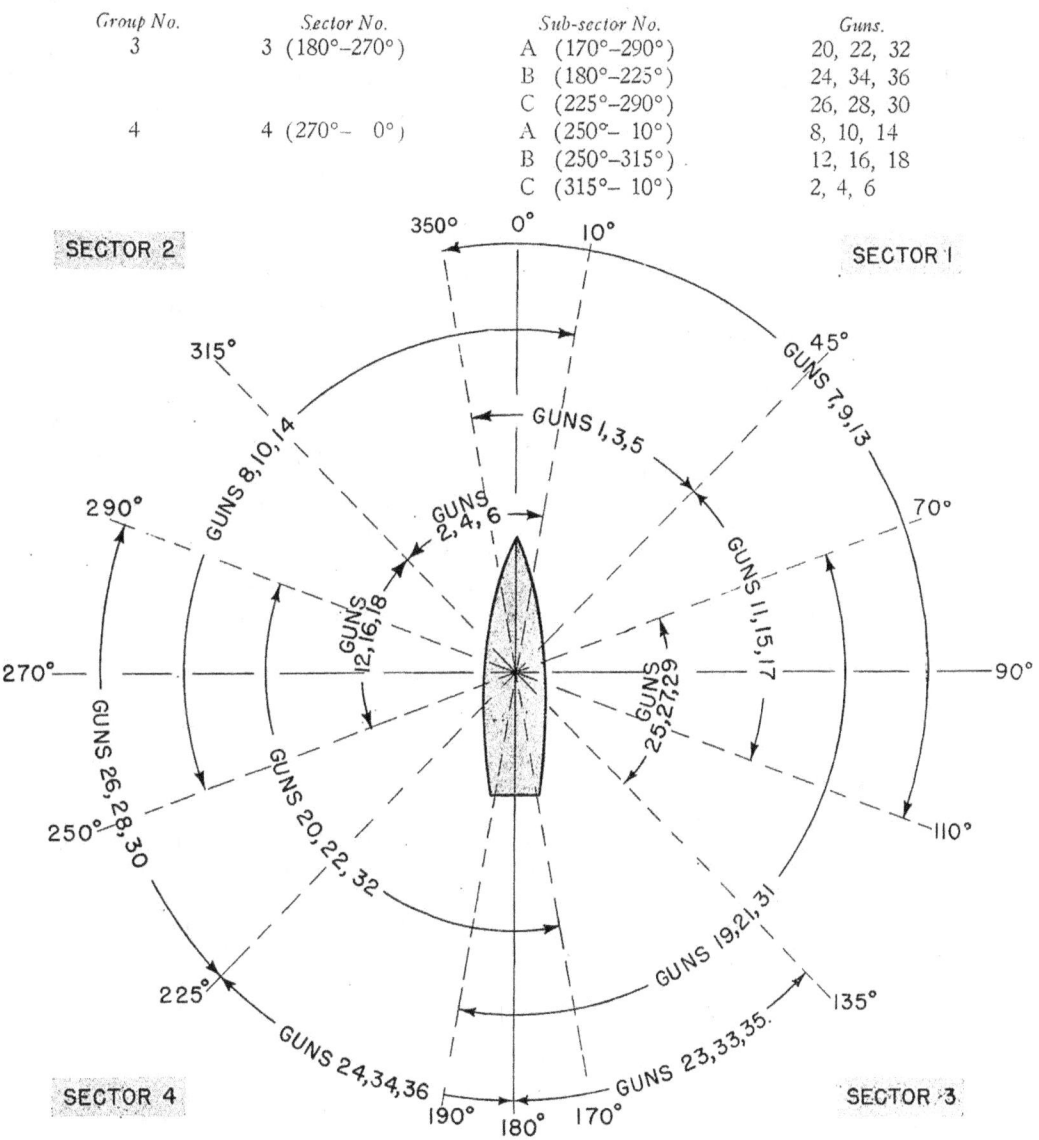

FIGURE 22A3. Sector assignment of 20 mm. guns.

A group control officer can direct the activities of each group. He is responsible for the particular sector assigned to that group, including fire doctrine, and the designation of targets necessary to keep the planes which are attacking the ship in his sector under fire. There also may be an *assistant group control officer* for each sub-sector, who may be an officer or petty officer regularly assigned to a gun of the sub-group.

B. COMMUNICATIONS

22B1. General requirements. An organization can only be effective in proportion to the efficiency of communications between the various subdivisions. The breakdown of interior communications during battle causes even the most efficient organization to lose its ability to fight the ship effectively.

The development of efficient communications is most important in the training of a fire-control organization. A good organization requires, first of all, a satisfactory system of telephones and other transmitting instruments, and secondly, properly trained personnel. The first necessity is met by the sound-powered battle telephone system. The second requirement, however, is not one that can be satisfied by mechanical perfection of instruments or by prescribing hard and fast rules for personnel. It demands careful indoctrination and training so that every individual is familiar with the problem at least insofar as it concerns his immediate station and those with whom he is in communication, and is prepared to act intelligently in any contingency.

Many communication difficulties arise from the fact that the communicating parties cannot see each other. For example, failure to receive acknowledgment from a station to which a message is addressed may mean that the station is out of commission, that its talker is momentarily so occupied that he cannot reply, or that he is asleep. It may be impossible to determine immediately which of these conditions exists. Judgment must be exercised in such situations so that communications may be quickly reestablished, with minimum disruption of other activities.

Efficient fire-control communication, upon which success in action depends, is obtained by following standardized procedure, and employing intelligent talkers who have undergone intensive training and are completely familiar with any type of situation which may be encountered.

22B2. Requirements for brevity. Fire-control telephone circuits are party lines providing communication between several stations. For instance, on the Captain's battle circuit there may be as many as ten parties at the same time. When any talker pushes down his button and speaks, all others on the circuit receive his message. With many communications to be transmitted, each must be of the utmost brevity consistent with clarity. In action, or at any other time, conversation in the usual sense has no place on a fire-control telephone circuit.

The demand for brevity has resulted in the development of numerous stereotyped phrases and modes of expression. They constitute what may be termed the "fire-control language". Familiarity with this language can be acquired only through actual experience, but certain elements must be learned as a preliminary phase of such training.

22B3. Fire-control language. All numerals indicating measurements of distance, range, spot, director angle, deflection bearing, etc., are transmitted over the telephone circuit by speaking each digit separately, with the exception that when the last two digits are zeros they are transmitted "double 0". Zero is generally expressed as "Oh" except when it occurs in an expression of bearing; in such a case the word "zero" is used. Deflection is always called "scale".

Examples:

Message	Telephone transmission
Range 13,350	"Range one-three-three-five-O"
Range 12,500	"Range one-two-five-double-O"
Range 12,000	"Range one-two-O-double-O"
Deflection 540	"Scale five-four-O"
Spot up 100 yards	"Up one-double-O"
Spot up 5 mils (AA)	"Up five"
Spot left 5 mils	"Left five"
Spot right 5 mils	"Right O-five" (note use of O for a right spot)
Bearing 105 degrees	"Bearing one-zero-five"

22B4. Call and acknowledgment procedure. An important feature of the fire-control language is the manner of calling a station, and of replying, or acknowledging. Ordinary practice in a telephone conversation is to call a person by name, and to await acknowledgment before proceeding with the message. This is logical when the party calling has no assurance that the party called is listening. On a fire-control circuit such a question does not exist. The first step in manning battle stations is to establish communications; once established they must be maintained as long as communication lines remain intact. If one station has information to transmit to another, it is

CHAPTER 22

unnecessary to call that station and receive an acknowledgment before giving the message. It is presumed that all stations are on the circuit and listening continuously.

Where several stations are on the same circuit and it is desired to address a message to one of them, the correct procedure is to give the call of that station and follow it immediately with the message. The station addressed then acknowledges by repeating his own call followed by "aye, aye". For example air defense forward wishes director 3 to train to 125° relative. He says: "Sky three train one two five", and sky 3 answers, "Sky three aye, aye".

If the station transmitting the message does not receive an acknowledgment immediately, it repeats the call and message. If after several repetitions no acknowledgment is received, the calling station should establish communication by some other means and determine the casualty.

If a message is heard by the station addressed, but not understood, that station should say at once "Repeat", and the transmitting station should then repeat both the call and the message.

When a message is addressed to several stations simultaneously, the procedure is the same as for a single station except that a collective call (or the individual calls of the several stations) precedes the message. Acknowledgments are made in numerical sequence, as briefly as possible. In such cases the "aye, aye" may even be omitted. For example, if air defense gave a message to the four AA directors, the acknowledgments should be "One", "Two", "Three", "Four". If a station fails to answer in its proper turn, the next following should, after a short pause, take up the sequence. The station that was late may then answer at the end.

During standby intervals, and sometimes during active periods, it may be desirable to communicate with a station for the purpose of testing the circuit. If a station is addressed solely for test, the procedure is to speak the call as usual, followed immediately by the call of the calling station and the word "testing". The acknowledgment is made in the usual way.

22B5. Multiple circuits. In all the fire-control stations there are numerous outlets from the various battle-telephone circuits. There usually is a principal circuit over which the officer-in-charge of the station listens, but in most cases it is necessary for him to shift to various others. In order that all necessary circuits may be guarded continuously, he has a talker on each of them. A selective switch controlling the connection of his telephone permits the officer-in-charge to make a shift as the situation may demand, but even when these selective switches are provided, he will have to communicate over several circuits through his talkers who are usually non-rated enlisted men.

22B6. Rules for talkers. Talkers hold very important positions, yet they cannot be expected to possess the experience and judgment of the officer-in-charge. In order that there may be no possibility of having the substance of a message altered due to misinterpretation by a talker, the responsibility for framing a message rests with the officer originating it. Once delivered to a talker, it must be transmitted in *exactly the same words,* and the talker receiving the message at the other end must repeat it to the officer concerned *in the exact form in which it is heard.* In brief, there are two principles governing fire-control communication by talker:

1. The person originating the message must state it in the exact words in which it is to be transmitted.

2. The talker transmitting a message must repeat the exact words of the originator, and the talker hearing the message must repeat word for word what he heard.

Acknowledgments belong in the same category as messages. An acknowledgment signifies that the message has been heard and *understood.* It is the responsibility of the person addressed to understand the message, and it is he who must originate the acknowledgment. An exception to this rule applies in the case of the transmission of data, in which case the acknowledgment consists of a repetition of the message. For example, sky 1 transmits the target angle to plot. The correct message is "Target angle two seven zero", and the correct acknowledgment is an exact repetition of the message.

Another exception applies when a message is not heard distinctly, or not understood. If a talker does not hear a message distinctly, he must call for "repeat". If the person addressed does not understand the message, he must also call for "repeat". It is to be noted that the responsibility

ORGANIZATION AND INTERIOR COMMUNICATIONS

for understanding a message does not rest with the talker, and as long as the talker hears a message distinctly, he must repeat it exactly as heard, whether or not he comprehends its significance. It is the duty of the person addressed to determine whether or not the message is intelligible. Much confusion and loss of time result from talkers asking for a repeat on messages which they do not understand, but which would probably be clear to the officers addressed. Of course it is highly desirable that talkers do understand all messages that they are likely to transmit and receive. The handling of messages thereupon becomes much more accurate and rapid.

22B7. Battle-telephone circuits. It is impossible to even partially list all the battle-telephone circuits and outlets of a battleship, for there may be over 50 circuits and more than 500 outlets. In addition, these installations are not the same on different types of ships, and there may even be minor variations among ships of the same class. This discussion, therefore, is limited to presentation of a few general rules concerning circuit designations, and a listing of designating symbols for some of the more important battle circuits. The complete description of telephone circuits can be found aboard ship and should be studied with care.

A ship's battle-telephone circuits are designated by letters and numbers; the characteristic letter indicating a battle circuit is J (there are also a few J-circuits belonging to ship-, damage-, and communication-control organizations). Other letters and numbers indicate the individual circuits (see accompanying table). As a general rule, the numbers preceding the designating letters distinguish between circuits which serve the subdivisions of an activity (occasionally, circuits for *different* purposes under the same activity). Numbering usually is according to standard shipboard procedure; odd numbers indicating starboard or forward circuits, even numbers port or after circuits.

PARTIAL LISTING OF BATTLE-TELEPHONE CIRCUITS

GENERAL DESIGNATIONS	PRINCIPAL USE
GENERAL CIRCUITS	
JA	Captain's battle circuit.
1JV	Maneuvering circuit.
1JZ	Emergency circuit.
MAIN BATTERY CIRCUITS	
1JD and 1JE	Forward and after sight-setter's circuits.
2JD and 2JE	Forward and after main-battery control circuits (sometimes called the turret-officers' circuits).
1JB	Spotting circuit, Spot 1.
JC	Spotting circuit, Spot 2.
1JW and 2JW	Range-finder circuits (forward and after).
SECONDARY AND AA CIRCUITS	
(1, 2, 3, or 4) JP	Secondary-battery group control circuits and corresponding units.
5JP	Secondary-battery control (all stations).
JK	Secondary-battery fuze setters.
JQ	Secondary-battery sight setters.
JN or JNT	Illumination control (for both main and secondary batteries).
JW	Range finders and spotters.
4JW	Navigational ranging and bearing.
JL	Lookouts.
JCX	Control circuit for ranging and bearing.
JS	Search, bearing, and ranging.
JY	AA machine guns and control circuits.
X5J	Magazine sprinkling control (X is used as a prefix for all auxiliary circuits).

CHAPTER 23

SUGGESTIONS TO JUNIOR GUNNERY OFFICERS

A. THE JUNIOR OFFICER AND THE GUNNERY DEPARTMENT

23A1. Function of the gunnery department. The primary function of a fighting ship is to damage or destroy the enemy. It is the special task of the gunnery department to provide efficient battle organization and to maintain matériel in fighting condition so that the primary function can be fulfilled.

23A2. Assignment of junior officers to duty. Under present conditions, it is difficult to predict the exact type of duty a junior officer may be assigned aboard ship. On a battleship, aircraft carrier, or cruiser, a junior officer without previous sea experience may be detailed as assistant navigator, to an engineering division, to communication duties, or to the gunnery department as junior officer of a gunnery division. Aboard a smaller ship he may have one or a combination of these duties, or special assignment as needs aboard that ship may dictate.

During war time the need for keeping a stable shipboard organization precludes rotation of assignments which would round out the experience of the young officer and fit him for command. Thus, a junior officer who is assigned gunnery duties upon reporting aboard probably will remain in this work and have the opportunity to become more and more proficient in one particular job. The new junior officer is expected to assume some degree of responsibility practically at once, and there is no opportunity for the peace-time "shake down" period during which instruction in various duties is regulated. In time of war the individual is expected to *take the initiative* in the matter of instruction, and must be prepared to make practical application of what he knows and learns at any time.

There is always plenty to learn in the gunnery department. The recent past has witnessed continual development and improvement of older gunnery installations, and the evolution of much new ordnance and fire-control equipment. Learning about the mechanisms already in existence, and keeping up with new developments on one's own and other ships requires a high degree of concentrated effort. Improvements and new methods have been suggested, and in some cases developed, by alert junior gunnery officers.

23A3. Gunnery assignments. In the following discussion, the battleship is used as an example, but the discussion is sufficiently broad to apply, in modified form, to any other type of ship. The possible assignments a junior officer may expect in the gunnery department of a battleship are the following: (1) junior turret officer, or turret handling-room officer, (2) battery officer of a secondary battery group of guns, (3) junior division officer of the fire-control division, (4) range-finder officer, (5) group control officer of the secondary or dual-purpose battery, or (6) battery officer or group control officer of an AA machine gun group.

A junior officer attached to a deck or gunnery division will of course have other duties in addition to those connected with gunnery. He will find that he has deck watches to stand, normal administrative and upkeep divisional functions, and various extra items from time to time. This being true, the prospective junior gunnery officer should be prepared to make that part of his time available for gunnery duties count for the most. A detailed description of procedures for carrying out individually the gunnery duties listed above cannot be attempted. However, hints regarding the general accomplishment of gunnery assignments by junior gunnery officers may prove useful.

23A4. Study for a gunnery assignment. Regardless of the type of gunnery duty assigned, a great amount of study on the part of any officer is necessary. For example, the study of gunnery principles and their practical application is necessarily limited in a book such as this, which includes a wide scope of subject matter dealing with matériel and its use. Figuratively

RESTRICTED

speaking, the surface has only been scratched here, and further study is vitally necessary. The junior officer must apply himself until he knows his equipment better than the enlisted men under him, and knows his own duties and those of his men. He must then learn the functions of other officers in the gunnery organization, and particularly the officer whom he might be called upon to relieve. A junior officer should not be surprised or caught unprepared if he is ordered to perform more responsible work at any time.

There are a number of ways in which a junior officer aboard ship may learn his gunnery department duties. In general, there are four steps toward the mastery of a gunnery assignment and toward a thorough understanding of both the equipment and how it is employed:

1. Study all available literature pertaining to the ordnance equipment with which the officer is working. The importance of making frequent reference to the appropriate publications, such as Ordnance Pamphlets and Ordnance Data, cannot be overemphasized. The various types of literature available are listed in Section B of this Chapter.

2. Study the gunnery (or torpedo) doctrine applicable to the ship in which he is serving.

3. Talk with and ask questions of experienced officers and rated enlisted men.

4. While observing and participating in daily drills or actual firings, analyze the experiences gained for the vast wealth of practical knowledge which may be derived from actual use of the equipment.

23A5. Shipboard gunnery duties. Though it is vitally necessary that the junior officer prepare himself for proficient performance in his present and future assignments, it is of first importance that he shall fit into the gunnery team as a useful member the moment he reports aboard. Shipboard training is the heart of efficient gunnery performance. Paradoxically, however, training often must be subordinated to immediate wartime demand for a gunnery team fighting in the actual game rather than a team that will be ready to fight when all of its members have learned their jobs.

The junior officer has much of his time occupied by the upkeep and repair of equipment as well as by frequent operation of the equipment at daily drills. Upkeep and repair of ordnance and fire-control equipment is an assignment whose importance cannot be overemphasized. Only with thorough, painstaking care can this equipment be expected to perform at its maximum efficiency during battle action. Therefore, the junior gunnery officer must accelerate his preparatory studies, intelligently correlating his training with the actual performance of duty.

23A6. Upkeep. The routine upkeep to be performed is outlined, in general, in the appropriate OP, supplemented in some cases by an OD, OTI and OCL (see table of literature, Section 23B). Check-off lists should be used to facilitate the carrying out of routine tests and upkeep procedures, and to avoid inadvertent omission of any item.

The value of proper lubrication, for example, cannot be overemphasized. Many of the faults which develop into serious items of repair are initially caused by failure to observe existing instructions. Here again, the OP for a particular item of equipment outlines the lubrication of that equipment in detail and should be closely followed.

23A7. Repair. Repairs are divided into three classes according to the agency making them: ship's force, tender, and navy yard.

The repairs that can be performed by the ship's force are necessarily limited by existing facilities and lack of personnel qualified to do the work. This class of repairs normally is confined to the replacement of worn parts, and to minor adjustments. Naturally, in cases of emergency, any repair can be attempted by the ship's force if it is deemed necessary. As a major portion of ship's force repair work is confined to the replacement of parts, it is the duty of the officer concerned to see that a complete set of spare parts is kept on hand, and that a follow-up system is maintained to insure replacement of items which have been used.

Ships are assigned periodical tender and navy yard overhaul periods. During these intervals, such repair work as cannot be performed by the ship's force is undertaken. Shipboard records must be kept of the outstanding repairs to be accomplished during these periods.

CHAPTER 23

Aboard tenders are personnel who are specially trained in such intricate and delicate repair work as that concerned with range keepers, range finders, and other similar mechanisms. Also, there are facilities aboard a tender to perform repairs upon a relatively broad scope.

During navy yard overhaul periods all repairs which cannot be accomplished by ship's force or the tender are undertaken. These include such work as the lifting of guns and carriages, replacement of worn-out guns, lifting of directors, rewiring of worn-out electrical gear, and other extensive mechanical and electrical changes which require a considerable period of time.

The officer concerned with the equipment is responsible for making necessary repair requests and for seeing to it that the job is done properly no matter what agency is at work. Repairs which are not checked may result in failure *after* the special facilities are no longer available.

23A8. Alterations. The word *repair* is defined in *Navy Regulations* as "such work as may be necessary to restore the ship or article under consideration to serviceable condition without any alteration in design, without the addition of any article or parts, and without the removal of any articles or parts that are not to be replaced".

Quoting from the same reference, "the word *alterations* shall be construed to mean all work not included in the above definitions of the word *repair,* including all changes in design that may be deemed advisable in making repairs. The term alterations includes all additions to any articles or parts and the removal of any articles or parts that are not replaced by similar ones. It shall include all changes in the character of the material of which any article or part is made. No alterations shall be made to any vessels (or articles aboard), either by ship or navy yard force until specifically authorized by the Bureau concerned."

It is necessary that an officer give much thought to both repairs and alterations. Thus, when he writes a request for an ordnance alteration (Ordalt), the necessity for such can be clearly shown, as well as the service and material required. Such effort will assist the gunnery officer in originating a request for the alteration, and assist the Commanding Officer, who must sign the request and submit it to higher authority.

B. SOURCES OF INFORMATION

23B1. Ordnance publications. Listed in the table at the end of this section are the current types of publications made available by the Bureau of Ordnance. The most important from the standpoint of a junior officer's studies of ordnance and fire-control equipment are Ordnance Pamphlets, Ordnance Technical Instructions, Ordnance Data, and Ordnance Circular Letters. Files of the appropriate publications are maintained aboard ship and the officer should avail himself of every opportunity to consult them.

On board ship, stowage and filing of all listed literature which has a security classification of restricted or non-classified is under the cognizance of the gunnery officer. All secret or confidential information is under the cognizance of the communications officer and may be obtained on custody receipt from him.

23B2. Procurement of ordnance publications. The Bureau of Ordnance distributes all Bureau of Ordnance publications through Bureau of Ordnance Publications Distribution Centers. Commissioning allowances of ordnance publications are automatically furnished to applicable ships. New ordnance publications are forwarded as available to interested activities without request. The *Navy Department Bulletin,* issued semimonthly, and distributed to all ships and stations, lists new Bureau of Ordnance publications.

Requests for nonregistered ordnance publications, not automatically received, should be directed to the nearest Ordnance Publications Distribution Center: Washington, D. C.; Mare Island, California; Adak, Alaska; Pearl Harbor, Hawaii; Brisbane, Australia; Espiritu Santo, New Hebrides; and Exeter, England. Distribution-center mailing addresses may be obtained from the *Standard Navy Distribution List.* Aviation ordnance publications are also available at many naval air stations. Registered ordnance publications are obtained in accordance with instructions set forth in *Navy Regulations* and the *Registered Publications Manual.* The U. S. Naval Gun Factory, Washington 25, D. C. is the central depository for all Bureau of Ordnance drawings, and furnishes commissioning allowances of drawings to applicable ships. Emergency requests for drawings may also be directed to navy yards at Boston, Mare Island, New York, Norfolk, Pearl Harbor, Philadelphia, and Puget Sound, which receive vandykes of all Bureau of Ordnance drawings from the Bureau of Ordnance Distribution Center at the U. S. Naval Gun Factory, Washington 25, D. C.

23B3. Fleet publications. In addition to the ordnance publications listed in the table, certain Fleet Training Publications (FTP) are prepared by the Bureau of Ordnance and issued to the service through the Vice Chief of Naval Operations, Division of Naval Communications, Registered Publication Section. The two most important are Orders for Gunnery Exercises (OGE) and Gunnery Instructions (GI). These may be drawn from the communications officer on board ship. Orders for Gunnery Exercises consist of several sections; the first is devoted to gunnery exercises in general, and the remainder are applicable to individual types of ships. Gunnery Instructions contains general information with which all officers connected with gunnery should be familiar. They may be drawn from the communications officer.

23B4. Doctrinal publications. Within each Fleet or Force there are established gunnery and torpedo doctrines; such doctrines being further divided as to types of ships. Thus, there are separate gunnery doctrines for battleships, heavy cruisers, light cruisers and destroyers. Torpedo doctrines are established for light cruisers and destroyers. Doctrines serve two purposes: first, they prescribe the manner in which guns and torpedoes are to be used tactically, and second, they prescribe the methods of control and procedures to be used for guns and torpedoes under varying conditions and against different types of targets. It can easily be seen that they are essential to standardized gunnery and torpedo control on the same types of ships, and for purposes of coordinating the ships in a Fleet or Force. They are promulgated as confidential bulletins or confidential registered publications, and may be drawn from the communications officer.

23B5. Information bulletins. Since the start of the present war various series of information bulletins have been distributed to the service. The two of principal interest to officers connected with gunnery are Information Bulletins, Commander-in-Chief, U. S. Fleet, and Bureau of Ordnance Information Bulletins, both of which are confidential and may be drawn from the communications officer.

Information Bulletins, CominCh, are published about once a month. While these bulletins are not confined to matters connected with gunnery alone, practically all items in them touch upon gunnery in some form, and should be studied.

Bureau of Ordnance Information Bulletins are published quarterly. They are devoted to technical developments in ordnance, and the status of production. They frequently contain valuable information concerning the control of guns, and should be studied by all officers connected with the gunnery department.

CHAPTER 23

TABLE OF ORDNANCE PUBLICATIONS

Type of Literature	Contents	Purposes	Remarks	Classification
Bureau of Ordnance Manual	1. A combination of administrative and such technical instructions relating to matters coming under the cognizance of the Bureau which are deemed necessary for a clear understanding of the requirements and work of the Bureau.	1. To provide uniform procedure and insure the correct and expeditious performance of work under the Bureau.	This publication is issued under the authority of the Secretary of the Navy and has the force and effect of Navy regulations.	Unclassified
Ordnance Circular Letters Symbol: NavOrd OCL	1. General policy matters of the Bureau of Ordnance and other non-technical information. 2. Information regarding the operating capabilities of ordnance equipment. 3. Changes to the *Bureau of Ordnance Manual*.	1. To disseminate information regarding Bureau policy. 2. To provide information, confidential or otherwise, regarding the capabilities and limitations of ordnance equipment in order that its tactical use may be directed intelligently. 3. To disseminate advance notice of changes to the *Bureau of Ordnance Manual*.	Issued in Eight Classes A—Ammunition and pyrotechnics. C—Chemical warfare equipment. G—Guns and Mounts. F—Fire control and optical equipment. M—Mines and depth charges. N—Nets and booms. T—Torpedoes. V—Aviation ordnance. Numbered serially within each class each year. Example: NavOrd OCL A2-44.	Secret Confidential Restricted Unclassified
Ordnance Pamphlets Symbol: OP Registered secret or Registered confidential Symbol: ORD	1. Information regarding the description, operation, and maintenance of ordnance equipment. (This includes information now contained in Range Tables, Nomenclature Lists, certain Ordnance Data, and certain manufacturers' booklets.)	1. To furnish officers and enlisted men with detailed information which will enable them to use the equipment. 2. To be used for instructional purposes in schools.	Serially numbered.	Secret Confidential Restricted Unclassified (All classes may be registered.)
Ordnance Data Symbol: NavOrd OD	1. Information regarding inspection and test data of ordnance equipment. 2. Ordnance equipment lists.	1. To furnish material for inspectors and research personnel. 2. To provide planning information for BuOrd and BuShips.	Serially numbered.	Secret Confidential Restricted Unclassified

513

RESTRICTED

TABLE OF ORDNANCE PUBLICATIONS, CONT'D.

Type of Literature	Contents	Purposes	Remarks	Classification
Ordnance Technical Instructions Symbol: NavOrd OTI	1. Up-to-the-minute instructions amplifying those contained in Ordnance Pamphlets. OTI's are not to be a substitute for Ordnance Pamphlets. 2. The rapid dissemination of condensed instructions, pending completion of a preliminary Ordnance Pamphlet, and of modifications of a non-permanent nature.	1. To provide for the easy reference of officers and enlisted men in loose-leaf notebook form, the instructions here defined, plus reference to other publications giving instructions on each class of ordnance equipment.	Classified and numbered the same as NavOrd OCL's.	Restricted or Unclassified so that they are readily available to enlisted personnel.
Ordnance Handling Instructions Symbol: NavOrd OHI	1. Condensed information on installation, use, and care in handling ordnance equipment.	1. To provide, when necessary, a leaflet or tag attached to ordnance equipment, which will give personnel, particularly enlisted ratings, the pertinent information regarding installation, use, and care in handling.	Classified and numbered the same as NavOrd OCL's.	Unclassified
Ordnance Alterations Symbol: NavOrd ORDALT	1. Instructions, plans and drawings for making changes in ordnance equipment already installed or in use.	1. To provide instructions for alterations to installed ordnance equipment, or other ordnance equipment in use.	Serially numbered.	Secret Confidential Restricted Unclassified (Usually Restricted)
Ordnance Specifications Symbol: NavOrd OS	1. Bureau of Ordnance requirements to be met in the manufacture and inspection of ordnance equipment.	1. To give specifications for manufacturers producing ordnance equipment.	Serially numbered.	Secret Confidential Restricted Unclassified
Ordnance Standards Symbol: NavOrd OSTD	1. Bureau of Ordnance methods and practices which have become standardized. (Markings, sizes, formulas, etc.)	1. To provide a guide for the standardization of Bureau of Ordnance methods of drafting and manufacture.	Serially numbered.	Confidential Restricted Unclassified
Army Publications Symbols: varied	1. Information applicable to ordnance equipment procured from the Army.	1. To furnish to the Naval Service applicable Army publications.	Identified by Army numbers.	Secret Confidential Restricted Unclassified

CHAPTER 23

TABLE OF ORDNANCE PUBLICATIONS, CONT'D.

Type of Literature	Contents	Purposes	Remarks	Classification
Ordnance Modification Instructions Symbol: NavOrd OMI	1. Instructions for modifying ordnance equipment.	1. To provide instructions for modifying ordnance equipment, when modification is relatively simple. NavOrd ORDALT used if modification is relatively involved and a record of accomplishment of the modification is desired.	Classified and numbered same as NavOrd OCL's. (To be bound in NavOrd OTI binder.)	Restricted Unclassified
Ordnance Charts Symbol: NavOrd CHART	1. Enlarged views (usually colored) which outline the operation and component parts of ordnance equipment.	1. To furnish officers and enlisted men with instructions to enable them to understand and use ordnance equipment. 2. To be used for instructional purposes in schools.	Classified and numbered the same as NavOrd OCL's.	Confidential Restricted Unclassified
Bureau of Ordnance Drawings	1. Officially signed drawings of plans of ordnance work and ordnance material.	1. To furnish drawings for the manufacture and use of ordnance equipment.	Serially numbered.	Secret Confidential Restricted Unclassified
Bureau of Ordnance Sketches	1. Line sketches—drawings showing preliminary or experimental work. 2. List sketches—lists of drawings.	1. Line sketches—to provide an official record of preliminary work and information for the manufacture of experimental devices. 2. List sketches—to list all drawings pertaining to one unit, assembly, or type of ordnance material. To list all drawings pertaining to a ship or class of ships.	Serially numbered.	Secret Confidential Restricted Unclassified

515

SUGGESTIONS TO JUNIOR GUNNERY OFFICERS

C. REPORTS AND REQUISITIONS

23C1. Purpose of reports. Certain reports are required by the Bureau of Ordnance for the following purposes:
 1. To keep an accurate record of all matériel issued.
 2. To keep an accurate record of all matériel on hand, in store, and in reserve.
 3. To keep fully informed of the progress of manufacture and all other work under the cognizance of the Bureau, or plans and specifications for such matériel.
 4. To insure compliance with war plans.
 5. In some cases, to collect data required by the Navy Department or other Bureaus.

The *BuOrd Manual* lists the reports required of units ashore and afloat, and also lists the forms used in making them. These forms incorporate complete printed instructions. All are numbered and dated, the date being that of the last edition of the form. Form N. Ord. 1, in addition to its use for requesting blank forms from the Bureau of Ordnance, provides a complete list of forms used afloat, showing the number of form, date, title, when forms should be submitted, and other information necessary for proper compliance. Form N. Ord. 1-a includes the same information for shore stations.

23C2. Responsibility for reports. All officers and petty officers in the gunnery department are responsible for, and should thoroughly acquaint themselves with the filling in of required data on these forms. Data used is gathered from an actual inventory of matériel, and from check of the marks, modifications, and serial numbers. Instructions in the *BuOrd Manual* should be referred to along with those on the forms themselves. The reports, signed by the Commanding Officer, are submitted to the Bureau at periodic intervals, and when changes occur in the number of items aboard, such as after expenditure, transfer, or receipt of any material or equipment.

23C3. Purpose of requisitions. After the building of a new ship has been ordered, a list of all ordnance equipment required is made out. This list is known as the *allowance list* for that vessel, and shows the exact quantity of each item that should be aboard. The allowance list includes all the Title A, B, and C articles which are supplied by the Bureau of Ordnance or which will be required for the proper installation and maintenance of the ship's battery. Spare parts are also indicated thereon.

Each officer should acquaint himself with his ship's allowance list, which may be found in the gunnery office. Needless to say, it should be determined that the full number of items listed, including spare parts, is aboard. If not, it is the duty of the officer concerned to requisition the required equipment through appropriate channels.

23C4. Requisitions defined. Requisitions for matériel are of two general types: *not-in-excess* requisitions, and *in-excess* requisitions. A not-in-excess requisition is for items which are shown on the allowance list but are not on shipboard, or are present in less than specified quantity. An in-excess requisition is for some item which is not shown on the allowance list, or would cause the number of items on board to exceed the assigned number. Matériel in-excess of allowance must not be placed on the same requisition with not-in-excess matériel.

Requisition forms are provided for all items except ammunition, for which requisition shall be made by letter. The officer should acquaint himself with the means for requisitioning ordnance and fire-control equipment by studying the requirements set forth in *BuOrd Manual,* as well as the actual forms necessary.

In general, all not-in-excess requisitions are forwarded to the nearest source of supply, with the exception of optical matériel and ammunition, which are requisitioned from the Bureau of Ordnance, or in accordance with the exceptions listed in the *BuOrd Manual.* All in-excess requisitions must be forwarded to the Bureau of Ordnance.

The exact nomenclature given in the allowance list must be used. This is especially important in the case of articles having no drawing numbers. If an item is not properly identified, the requisition probably will be disapproved because necessary information is lacking.

23C5. Shipboard gunnery files. The junior officer should consult the files of the gunnery department aboard his ship for copies of the various types of reports, letters, and requisitions. In these and other pertinent files the officer can see examples of Ordalts and various other types of work requests, routine and special ordnance reports, and requisitions, as well as reports on gunnery exercises and data concerning the performance of the various batteries.

Frequent reference to the gunnery files will keep the officer posted concerning past gunnery performance as well as current doctrines and procedures. No officer should ever report to a ship feeling that his training is complete, or his knowledge perfect concerning installations and routines. He should be receptive to continued training, and be on the alert to learn not only standard, but also improved practices.

23C6. General remarks. Finally, the junior officer should never feel that the complications of modern ordnance and fire control are beyond his abilities of comprehension, or that necessary information for the use and upkeep of equipment is lacking. Every item has *something* written about it to assist those who must understand its operating principles, use, upkeep, and repair. The publication for a particular item of equipment is probably aboard. If not, it is the duty of the officer concerned to see that the publication is requested, not only for his own requirements, but also for those of enlisted personnel who must be informed. The officer should never forget that his men are anxious to better their training for the battles tomorrow.

APPENDIX A

STANDARD FIRE-CONTROL SYMBOLS

A1. Introduction. The symbols contained in this appendix are quoted from *Ordnance Data* 3447,

Standard fire-control and torpedo-control symbols. These deal with quantities used in surface and antiaircraft gunnery problems; torpedo-control symbols are provided in Chapter 20. Standard symbols are divided into two classes as follows:

    Class I—Those quantities which are used frequently enough to justify their being rigidly defined and given an associated symbol.

    Class II—All quantities not listed as Class I. (No quantities in this class are listed here. They are defined in the chapters requiring their use.)

All quantities may be broken down into basic quantities and modifying quantities, as in these examples:

QUANTITY	BASIC QUANTITY	MODIFYING QUANTITIES
Target bearing	Bearing	Target-relative to north
Relative target bearing	Bearing	Target-relative to own ship
Speed of target	Speed	Target
Speed of own ship	Speed	Own ship
Speed of wind	Speed	Wind

It follows that if symbols are assigned to all the common basic and modifying quantities it should be possible by proper combination of these symbols to pick a logical symbol for any quantity. This is what is done. Capital letters are assigned to the various common basic quantities; small letters to the various common modifying quantities. It will be noted that various letters are not assigned any meaning.

As stated before, all quantities that are used frequently and are susceptible to rigid definition should be listed as Class I. In making such a list one difficulty is soon apparent; that is, while such quantities as relative target bearing, level, cross-level, etc., are measured in more than one way (thus more than one value is possible), in common usage the general terms are used regardless of the manner in which the quantities are measured. In order not to needlessly complicate the problem of definitions (by assigning a different name to each way a quantity can be measured), and still permit such quantities to be rigidly defined the following compromise is adopted: The quantity in question is given its usual name regardless of how it is measured but assigned a different symbol for each method of measuring it. This gives a rigid definition and symbol for each Class I quantity and still provides a general name for quantities which differ only in the mechanical method of measuring them.

    No list of Class II quantities is given for obvious reasons. In the event it is desired to assign a symbol to a Class II quantity, the basic principle of combining a defined capital letter with none, one, or two defined small letters should be followed. After selecting such a symbol it should be checked to make certain that it meets the following requirements: First, the same symbol is not used for a Class I quantity; and second, that the letters chosen to form the symbol convey in general the rigid definition of the quantity. In the event this is not possible, form a symbol at random from the letters not assigned a meaning. Should it be desired to define a quantity in a plane different from that already provided for, it is recommended that double prime (quotation mark on the typewriter) be used. For example, bearings of a machine gun, which is trained in a slant plane, could be represented by B'' if it were not convenient to use B_s.

    The word "bearing" should be used to indicate angles in a horizontal plane, while "train" should designate angles in the deck plane. Thus "director train" is the correct term, instead of "director bearing", when the angular displacement of the director in the plane of the deck is meant.

APPENDIX A

A2. Basic symbols.

SYMBOL	DEFINITION
A	Target angle.
B	Bearing (of target, measured in horizontal plane; unless modified).
B'	Same as B but measured in plane of deck.
C	Course—measured in horizontal plane.
D	Lateral deflection (angular measure).
E	Elevation (of target, unless modified)—measured in vertical plane.
E'	Same as E, but measured in plane perpendicular to the deck.
F	Fuze setting.
G	
H	Height of target (normally in feet).
I	Angle of climb or dive (inclination).
J	
K	K1, K2, etc. Constants.
L	Level angle—measured in vertical plane (positive if, when facing target, the deck is tilted down).
M	Roll.
N	Pitch.
O	
P	Parallax.
Q	
R	Range.
S	Speed (of target unless modified).
T	Time.
U	
V	Vertical (elevation) prediction (angular measure).
W	
X	Horizontal component of velocity perpendicular to vertical plane through the line of sight.
Y	Horizontal component of velocity in vertical plane through the line of sight.
Z	Cross-level angle measured about an axis in the vertical plane through the line of sight (positive if, when facing the target, the *left* trunnion is down).
Z'	Cross-level angle measured about an axis in the plane, perpendicular to the deck, through the line of sight (positive if, when facing the target, the *left* trunnion is down).
Note:	1. In general, prime after a basic symbol indicates that the quantity is measured in the plane of the deck of own ship, or in a plane perpendicular to the deck of own ship.
	2. Double prime (quotation mark on the typewriter) can be used to modify a basic quantity when it is desired to define a quantity which is measured in a plane not otherwise provided for.

A3. Modifying symbols.

SYMBOL	DEFINITION
a	Relative to air.
b	Director, of.
c	Before a quantity means the value of that quantity as computed (generated) by the mechanism, as opposed to the observed value of the same quantity. After a quantity means relative to rate control.
d	Before a quantity means a time rate of change or differential of that quantity,

RESTRICTED

STANDARD FIRE-CONTROL SYMBOLS

	equivalent to $\frac{d}{dt}$. After a quantity means in or relative to deck plane, or plane perpendicular to the deck.
e	Elevation.
f	Due to standard trajectory.
g	Gun, of.
h	Horizontal, projection of.
i	Searchlight (illumination).
j	Before a quantity means a correction or partial correction to that quantity, usually generated by the mechanism; after a quantity means an arbitrary correction (spot) to that quantity.
k	
l	
m	Loss of initial velocity.
n	
o	Own ship; of, or due to.
p	Parallax; of, or due to.
q	
r	Relative to own ship.
s	Relative to line of sight to target or in a slant plane (because several slant planes may be involved, each definition should definitely specify the plane used).
t	Target; of, or due to.
u	
v	Vertical projection of.
w	Wind; of, or due to.
x	
y	
z	Cross-level; of, or due to.
f( )	f1( ), f2( ), f3( ), etc. means function of the quantity in parenthesis.
Δ	Before a quantity means change in that quantity during some specific time, usually time of flight; increment of a quantity.
$\int$	Before a quantity means the integral of that quantity.
Number 0	After a quantity indicates that it is the predicted value of that quantity for dead time seconds prior to the present time.
Number 2	After a quantity indicates that it is the predicted value of that quantity for advance position; i.e., for the instant a projectile, which is fired at the present time, hits (bursts in antiaircraft fire).
Number 3	After a quantity indicates that it is the predicted value of that quantity for fuze position; i.e., for the instant a projectile, fired dead time seconds from the present time, hits (bursts in antiaircraft fire).
Note:	1. All quantities are relative to earth, or true (relative to North), unless modified by a, r, or s.
	2. A dot between two quantities signifies their product.
	3. In definitions when the term "line of sight" is used "line of sight to the target" is understood.
	4. In definitions when the term "deck plane" or "plane of the deck" is used, the standard reference plane of the ship is meant.

A4. Class I quantities.

SYMBOL	DEFINITION
A	*Target angle.* The angle between the vertical plane through the fore-and-aft axis of the target and the vertical plane through the line of sight, measured in a horizontal plane clockwise from the bow of the target.

APPENDIX A

B
: *True target bearing.* The angle between a north and south vertical plane and the vertical plane through the line of sight (from the point upon which the range keeper or data computer bases its solution), measured in a *horizontal plane* clockwise from the north.

Bg
: *True gun bearing.* The angle between the north and south vertical plane and the vertical plane through the gun axis (plus parallax correction, when such a correction is made), measured in a horizontal plane clockwise from the north.

Bgr
: *Relative gun bearing.* The angle between the vertical plane through the fore-and-aft axis of own ship and the vertical plane through the gun axis (plus parallax correction, when such a correction is made), measured in a horizontal plane clockwise from the bow.

B'gr
: *Gun train.* The angle between the fore-and-aft axis of own ship and the plane through the gun perpendicular to the deck (plus parallax correction, when such a correction is made), measured in the deck plane clockwise from the bow. (Gun-train order is the ordered gun train as transmitted to the gun).

Br
: *Relative target bearing.* The angle between the vertical plane through the fore-and-aft axis of own ship and the vertical plane through the line of sight (from the point upon which the range keeper or data computer bases its solution), measured in a *horizontal plane* clockwise from the bow.

B'r
: *Director train (stabilized sight).* The angle between the fore-and-aft axis of own ship and the vertical plane through the line of sight (from the point upon which the range keeper or data computer bases its solution), measured in the deck plane clockwise from the bow.

B'r'
: *Director train (sight not stabilized).* The angle between the fore-and-aft axis of own ship and the plane through the line of sight (from the point upon which the range keeper or data computer bases its solution) perpendicular to the deck, measured in the deck plane clockwise from the bow.

Bw
: *True direction true wind.* The angle between north and the direction *from* which the true wind is blowing, measured in a horizontal plane clockwise from north.

Bwa
: *True direction apparent wind.* The angle between north and the direction *from* which the apparent wind is blowing, measured in a horizontal plane clockwise from north.

Bwr
: *Relative direction true wind.* The angle between the vertical plane through the fore-and-aft axis of own ship and the direction *from* which the true wind is blowing, measured in a horizontal plane clockwise from the bow.

Bwra
: *Relative direction apparent wind.* The angle between the vertical plane through the fore-and-aft axis of own ship and the direction *from* which the apparent wind is blowing, measured in a horizontal plane clockwise from the bow.

Ca
: *Target course relative air (target reading).* The angle between the north and south vertical plane and the vertical plane through the fore-and-aft axis of the target, measured in a horizontal plane clockwise from north.

Co
: *Own-ship course.* The angle between the north and south vertical plane and the vertical plane through the fore-and-aft axis of own ship, measured in a horizontal plane clockwise from north.

Ct
: *Target course.* The angle between the north and south vertical plane and the vertical plane through the direction of motion of the target with respect to the earth, measured in a horizontal plane clockwise from north.

Cw
: Bw ± 180° ⎫
Cwa
: Bwa ± 180° ⎪ Definition same as for corresponding quantities, except substitute
Cwr
: Bwr ± 180° ⎬ *toward* for *from.*
Cwra
: Bwra ± 180° ⎭

Cws
: *Wind angle.* The angle between the direction *toward* which the wind is blowing

STANDARD FIRE-CONTROL SYMBOLS

and the vertical plane through the line of sight, measured in a horizontal plane clockwise from the direction *toward* which the wind is blowing.

Dd   *Deck Deflection.* Gun-train order minus director train (also minus parallax correction when the parallax correction is applied to gun-train order in the computer), when director train is B'r. (Stabilized sight).

Dd'   *Deck deflection.* Gun-train order minus director train (also minus parallax correction when the parallax correction is applied to gun-train order in the computer), when director train is B'r'. (Director sight not stabilized).

Df   *Drift.* Lateral deflection angle due to drift of the projectile, measured in the horizontal plane.

Dfs   *Drift.* Lateral deflection angle due to drift of the projectile, measured in the slant plane through the advance position of the target.

Dh   *Horizontal deflection angle.* The angle between the vertical plane through the line of sight and the vertical plane through the gun, measured in a horizontal plane from the line of sight; positive if the gun is to the right of the line of sight, and negative if the gun is to the left.

Ds   *Sight deflection.* The angle between the line of sight and the plane through the gun axis perpendicular to the trunnions, measured in the plane through the line of sight containing the gun trunnions; positive if the gun is to the right of the line of sight, negative if the gun is to the left.

dB   *True angular-bearing rate.* The time rate of change of true target bearing.

dBr   *Relative angular-bearing rate.* The time rate of change of horizontal relative target bearing.

dBs   *Angular-bearing rate in slant plane.* The time rate of change of true target bearing measured in the slant plane through the line of sight.

dE   *Angular elevation rate.* The time rate of change of target elevation.

dH   *Rate of climb.* The time rate of change of height.

dR   *Range rate.* The time rate of change of range.

dRh   *Horizontal range rate.* The time rate of change of horizontal range.

E   *Target elevation.* The elevation above the horizontal of the line of sight (from the point upon which the range keeper or data computer bases its solution), measured in the vertical plane through the line of sight.

E'   *Perpendicular position angle.* The elevation above the horizontal of the line of sight (from the point upon which the range keeper or data computer bases its solution), measured in the plane perpendicular to the deck, through the line of sight.

Eb   *Director elevation (stabilized sight).* The elevation of the director line of sight above the deck (plus parallax correction, when such a correction is made) measured in the vertical plane through the line of sight.

E'b   *Director elevation (sight not stabilized).* The elevation of the director line of sight above the deck (plus parallax correction, when such a correction is made), measured in a plane perpendicular to the deck, through the line of sight.

Eg   *Vertical gun elevation.* The elevation of the gun above the horizontal, measured in a vertical plane through the gun axis.

E'g   *Gun elevation.* The elevation of the gun above the deck (plus parallax correction, when such a correction is made) measured in the plane through the gun perpendicular to the deck. (Gun-elevation order is the gun elevation transmitted to the gun).

F   *Fuze setting.* Fuze setting in seconds.

H   *Height.* The vertical distance of the target above the horizontal plane through the point upon which the range keeper or data computer bases its solution.

APPENDIX A

I	*Angle of climb or dive.* The vertical angle which the target speed vector makes with the horizontal; positive for ascending target.
I.V.	*Initial velocity.* The range-table initial velocity (muzzle velocity) of the projectile. Since this quantity is not used alone in computations the standard "I.V." will be retained.*
L	*Level angle.* Level angle measured about an axis in the horizontal plane; the angle between the horizontal plane and the deck plane, measured in the vertical plane through the line of sight. (Positive when the deck toward the target is below the horizontal plane. Applies to stable elements Mark 2 and Mods.)
L'	*Level angle.* Level angle measured about an axis in the deck; the angle between the horizontal plane and the deck plane measured in the plane perpendicular to the deck through the line of sight. (Positive when the deck toward the target is below the horizontal plane.) (Applies to stable element Mark 3, and stable verticals Mark 29, Mod. 1, and Mark 30 and Mods.)
Lj	*Selected level angle.* Any arbitrary value of level for directors which measure level as L.
L'j	*Selected level angle.* Any arbitrary value of level for directors which measure level as L'.
M	*Roll.* Angle of roll measured about an axis in the deck; the angle, measured in the athwartship plane perpendicular to deck, between its intersections with the horizontal plane and with the deck plane. (Applies to stable element, Mark 1). (Positive when starboard side of ship is up).
m	*Loss of initial velocity.* The difference between range-table initial velocity and actual initial velocity of the projectile.
N	*Pitch.* Angle of pitch measured about an axis in the horizontal; the angle, measured about the intersection of the horizontal plane with the athwartship plane perpendicular to the deck, between the vertical plane and a plane perpendicular to the deck through this axis. (Applies to stable element, Mark 1. Positive when bow of the ship is up).
R	*Range.* The distance from the point upon which the range keeper or data computer bases its solution to the target (measured along the line of sight).
RdBs	*Linear deflection rate.* The component of target motion perpendicular to the vertical plane through the line of sight (horizontal bearing rate times horizontal range or slant bearing rate times slant range).
RdE	*Linear elevation rate.* The component of target motion perpendicular to the line of sight in the vertical plane through the line of sight (elevation rate times range).
Rh	*Horizontal range.* The projection of the range on a horizontal plane.
S	*Surface target speed.* The speed of surface target.
S	*Air-target speed relative to ground.* The speed of air target along a line of flight with respect to the earth. (May or may not be horizontal flight.)
Sa	*Air-target air speed.* The speed of air target along the line of flight with respect to the air. (May or may not be horizontal flight.)
Sh	*Air-target horizontal ground speed.* The horizontal component of air target speed with respect to the earth.
Sha	*Air-target horizontal air speed.* The horizontal component of air-target speed with respect to the air.
So	*Own-ship speed.* The speed of own ship relative to the earth.
Sw	*True wind speed.* The horizontal velocity of wind with respect to the earth.
Swr	*Apparent wind speed.* The horizontal velocity of wind with respect to own ship. (Velocity of apparent wind.)

---

* I.V. is an abbreviation, not a symbol.

STANDARD FIRE-CONTROL SYMBOLS

T	*Time.* (Clock.)
Tg	*Dead time.* The time in seconds between cutting a fuze and firing the projectile.
Vf	*Superelevation.* The difference between the elevation of the gun above the horizontal (uncorrected for wind) and the advance target elevation. It is the angle the gun must be elevated above the advance target to allow for curvature of the trajectory in a vertical plane.
Vs	*Sight angle.* The angle between the gun axis and the plane through the gun trunnion parallel to the line of sight, measured in the plane through the gun axis perpendicular to the trunnions. (Positive if axis of gun is above the line of sight; negative if below.)
VV	*Vertical deflection angle.* The difference between vertical gun elevation and target elevation (E).
Xo	*Horizontal component of ship velocity.* The horizontal component of own-ship velocity perpendicular to the vertical plane through the line of sight.
Xt	*Horizontal component of target velocity.* The horizontal component of target velocity perpendicular to the vertical plane through the line of sight.
Xw	*Cross wind.* The component of *true* wind perpendicular to the vertical plane through the line of sight. (Positive when cross wind is from left to right.)
Xwr	*Apparent cross wind.* The component of *apparent* wind perpendicular to the vertical plane through the line of sight.
Yo	*Horizontal component of ship velocity.* The horizontal component of own-ship velocity in the vertical plane through the line of sight.
Yt	*Horizontal component of target velocity.* The horizontal component of target velocity in the vertical plane through the line of sight.
Yw	*Range wind.* The horizontal component of *true* wind in the vertical plane through the line of sight. (Positive when along the line of fire.)
Ywr	*Apparent range wind.* The horizontal component of *apparent* wind in the vertical plane through the line of sight.
Zd	*Cross-level angle.* Cross-level angle measured about an axis in the deck; the angle, measured about the intersection of the plane of the deck with the vertical plane through the line of sight, between the vertical plane and a plane perpendicular to the deck through this axis. (Positive if when you face the target the right hand side of the ship is up.)
Z'd	*Cross-level angle.* Cross-level angle measured about an axis in the deck; the angle, measured about the intersection of the plane of the deck with the plane through the line of sight perpendicular to the deck, between the vertical plane and a plane perpendicular to the deck through this axis. (Positive if when you face the target the right hand side of the ship is up).
Zdj (Z'dj) (Zhj) (Z'hj)	*Selected cross-level angle.* Any arbitrary value of cross-level when cross-level is measured as indicated in the main symbol.
Zh	*Cross-level angle.* Cross-level angle measured about an axis in the horizontal; the angle, measured about the intersection of the horizontal plane with the vertical plane through the line of sight, between the vertical plane and a plane perpendicular to the deck through this axis. (Positive if when you face the target the right hand side of the ship is up.)
Z'h	*Cross-level angle.* Cross-level angle measured about an axis in the horizontal; the angle measured about the intersection of the horizontal plane with the plane perpendicular to the deck through the line of sight, between the vertical plane and a plane perpendicular to the deck through this axis. (Positive if, when you face the target, the right hand side of the ship is up.) (Applies to stable verticals, Mark 30, Mods. 2 and 3, and stable element, Mark 3.)

APPENDIX B

DEFINITIONS

B1. Purpose. The definitions of the various terms in this chapter are provided to establish a basis for a common language and a common understanding throughout the Navy. The definitions are to be used in connection with the analysis and discussion of gunnery exercises. A thorough knowledge of the definitions is essential in order that misunderstandings may be avoided.

B2. General.

1. *Gunnery exercises.* The term "gunnery exercises" includes the various practices and official rehearsals, as prescribed for all forms of armament.

2. *Base course* is the selected reference course from which the firing vessel, target (tow) and observing vessels are to maneuver.

3. *General bearing line* (G.B.L.) designates the true bearing of the target of the flagship of the O.T.C. from that vessel.

4. The *horizontal* is the imaginary plane, tangent to the earth's surface, at the point instantaneously occupied by a surface ship. The ship is considered to move only in the horizontal plane. The *horizontal* is used as a reference plane from which to measure the relative movements of elevated (aerial) targets.

5. A *vertical* is a plane, perpendicular to a reference plane, passing through a designated line, point or points in the reference plane. If no reference plane is specified, it is assumed to be the horizontal, and the "vertical" is then the true vertical plane.

6. The *reference plane* for a battery is an arbitrarily chosen, imaginary plane within the ship, from which the vertical angles of all guns and fire-control instruments are measured. It is usually the plane whose inclination to the horizontal is the mean of the inclinations of the turret or mount roller-path planes to the horizontal, or a selected director roller path, or the horizontal (as determined while checking in dry dock).

7. *Level angle* is the angle between the reference plane and the horizontal, measured in the plane perpendicular to the reference plane through the line of sight.

8. *Cross-level angle* is the angle between the reference plane and the horizontal, measured in the plane perpendicular to the reference plane and normal to the vertical plane through the line of sight. Cross-level angle is positive when the right trunnions of guns or directors are depressed.

9. *Trunnion tilt* is the instantaneous inclination of the axis of the trunnions to the horizontal.

10. *Roll* is the instantaneous value of the angle between the reference plane and the horizontal, measured in an athwartship plane normal to the reference plane.*

11. *Pitch* is the instantaneous value of the angle between the reference plane and the horizontal, measured in a fore-and-aft plane normal to the reference plane.*

12. *Line of sight* is the straight line joining the sight and the point of aim.

13. *Target angle* is the relative bearing of own ship from the target, measured in the horizontal from the bow of the target clockwise from 0° to 360°.

14. *Position angle* is the vertical angle between the line of sight to an elevated target and the horizontal.

15. *Relative target bearing* is the bearing of the target from the firing ship measured in the horizontal from the bow of own ship clockwise from 0° to 360°.

16. *True target bearing* is the true compass bearing of the target from the firing ship.

17. *Generated target bearing* (relative or true) is the relative or true bearing of the target as determined mechanically or graphically from previous positions of own ship and target and established or estimated rates of change of bearing.

---

* NOTE: For purposes of naval gunnery both roll and pitch are measured and recorded as rates; that is, in terms of amplitude per unit time, ordinarily as total degrees of roll or pitch per minute.

DEFINITIONS

18. *Wind direction* is the direction by compass *from* which the wind is blowing.

19. *Wind angle* is the angle between the direction toward which the wind is blowing and the line of sight measured clockwise to the projection of the line of sight in the horizontal plane.

20. *True wind* is the actual wind blowing unaffected by the motion of the observing station, indicated by true direction and true speed (velocity).

21. *Apparent wind* is the wind apparent to the observing station and is the resultant of the true wind and motion of the observing station, indicated by apparent direction and apparent speed (velocity).

22. *Altitude* is the height of an elevated target above the horizontal plane; it is ordinarily measured in feet.

23. *Vertical parallax* is the vertical angle formed at the target by a line of sight from each of two stations (such as two guns, two directors, or a gun and a director) mounted at different heights on a ship.

24. *The vertical parallax correction* is the correction required to compensate the vertical parallax.

25. *Horizontal parallax* (convergence) is the horizontal angle at the target formed by a line of sight from each of two stations (such as two guns, two directors, or a gun and a director) on a ship.

26. *The horizontal parallax correction* is the correction required to compensate the horizontal parallax of the director, turret or gun from the director or reference point used as the origin in train for the system.

27. *Director correction* is the inclination of the line of sight of a gun director to the selected reference plane. Director correction is expressed in minutes and is positive when the line of sight is elevated above the reference plane and is negative when the line of sight is depressed below the reference plane. This correction is applied to the angle of elevation to obtain the gun angle, and compensates the inclination of the reference plane from the horizontal at the instant of firing.

28. *Bearing rate* is the rate of change of bearing of the target from own ship caused by the relative motion of own ship and target at right angles to the line of sight. It may be expressed in knots or in degrees per minute.

29. *Range rate* is the rate of change of range in yards per minute caused by relative motion of own ship and target along the line of sight. It is designated as the horizontal-range rate for surface targets and as the slant-range rate for aerial targets.

30. *Elevation rate* is the rate of change of position angle in degrees per minute. *Angle of climb or dive* is the angle between the horizontal and the actual direction of motion of the target measured in degrees at the target.

31. *Rate of climb* is the rate of change of altitude measured in feet per minute.

32. *Fuze setting* is the instant measured on the fuze scale (zero of scale is the instant of discharge) at which the fuze is to function. In mechanical fuzes this is the time of flight of the projectile. In powder fuzes, the burning rate of the powder in the fuze train ordinarily requires a "setting" differing from the time of flight. Fuze *setting* is expressed in seconds, and fifth-second fraction of time.

33. *Dead time* is the sum of the time lags that accumulate during the solution of a particular fire-control problem, the transmission of that solution to the gun in terms of elevation, train, and fuze setting, and the loading and firing of a projectile from the gun correctly laid to that elevation, and train, and with the corresponding fuze setting. In the fast-moving antiaircraft problem an accurate allowance for the average dead time of a particular battery is of extreme importance. The average dead time is made up of:

    a. *Observational dead time.* The range-keeper solution is obtained by repeated observations of the target. There is an appreciable time delay between the instant that an observation registers in the mind of the observer and the instant that the results of that observation are entered into the solution. The solution obtained is thus late by the amount of the observational dead time; the latter being a function of target speed and the speed of observer reaction.

APPENDIX B

b. *Transmission dead time.* There may be a dead time between the solution of the particular problem in the range keeper and the receipt of the outputs of that solution at the gun or battery. With full automatic transmission the transmission dead time will be relatively small; with manual transmission and manual operation of the guns the transmission dead time may amount to several seconds.

c. *Loading dead time.* The time between the instant a projectile is removed from the fuze pot and the instant that the projectile starts its travel toward the target is the loading dead time for that shot. The loading dead time may vary slightly among successive shots, may vary among the different guns of a battery due to the unequal abilities of the individual gun crews, or may change with change in angle of gun elevation due to differences in the ease with which the gun may be loaded. For practical purposes the average loading dead time of the battery is that dead time for which correction must be made. Any departure from this average loading dead time in the firing of a single shot will affect the apparent fuze dispersion of that shot. For a battery, the variation of the guns of the battery from the average loading dead time will have a resultant effect upon the apparent fuze pattern.

34. *Preparation time* (used in barrage fire) is that time elapsing between the making of the necessary set of observations of course, speed, range, and position angle of the target and the opening of the barrage fire with the information obtained set on the guns in terms of elevation, deflection, and fuze setting. It includes:

    a. The selection of the barrage.

    b. The setting of sights and fuzes.

    c. The loading and firing of the first shot.

35. *Sight angle* is the vertical angle between the line of sight of a gun and the axis of the gun bore.

36. *Director sight angle* is the vertical angle between the line of sight of a gun director and the axis of the bore of the gun(s) controlled by the director.

37. *Gun angle* is the angle between the reference plane and the axis of the gun bore.

38. A *mil* is the unit of deflection, and is the angle at the sight (gun or director) which subtends a horizontal distance at the target at right angles to the line of sight, equal to one one-thousandth of the target distance. One mil is equivalent to $3'.44$ or $3'26''$ of arc.

39. *Trunnion-tilt correction in train* is the correction to deflection to compensate the lateral errors due to the tilt of the trunnions at the instant of firing.

40. *Trunnion-tilt correction in elevation* (vertical correction for trunnion tilt) is the correction to the gun elevation to compensate the errors in range due to tilt of the trunnions at the instant of firing.

41. *Gun-train order* is the signal transmitted by the director, or from the director to associated instruments and as corrected by them to the guns, indicating to the guns the correct position in azimuth in the reference plane.

42. *Gun-elevation order* is the signal transmitted by the director, or from the director to associated instruments and as corrected by them to the guns, indicating the correct gun angle.

B3. Range.

1. *Distance* is the linear measurement between two points.

2. *Range* is the same as distance, and in ordinary use is considered to mean the distance as determined by range finder, or other instrumental means, corrected for known errors. In analysis of gunnery performances the word "range" should not be used without a proper qualifying adjective or phrase.

3. *Range-finder range* is the linear distance to a target as read from the range-finder scale.

4. *Range-finder correction* is the correction to be applied to the range-finder range to correct for known (or estimated) errors.

5. *Corrected range-finder range* is the algebraic sum of the range-finder range and the range-finder correction.

DEFINITIONS

6. *Mean range-finder range* is the mean of the range-finder ranges taken simultaneously or approximately so, from two or more range finders.

7. *Mean range-finder correction* is the correction to be applied to the mean range-finder range to correct known or estimated errors of the range finders in use.

8. *Corrected mean range-finder range* is the mean of the corrected range-finder ranges of two or more range finders taken simultaneously or approximately so, or the mean range-finder range plus (or minus) the mean range-finder correction.

9. *Horizontal range* is the horizontal distance to a point directly beneath an elevated target and is equal to the target distance multiplied by the cosine of the angle of position.

10. *Present range* is the best available estimate of the target distance. It is usually based on the corrected mean range-finder range at the time of opening fire, but it may be based on the spotter's estimate of the target distance. It is continuously determined by the range keeper or by plotting.

11. *Advance range* is the value of sight-bar range computed by the range-keeper or by plotting, combining with the present range, the corrections and predictions necessary to compensate own-ship and target motion during dead time and time of flight, plus ballistic corrections and spots. In some installations the trunnion-tilt correction in elevation is combined automatically; in others the advance range must be further corrected for trunnion tilt. The advance range, corrected for trunnion tilt, if necessary, is transmitted to the guns as the gun-elevation order.

12. *Navigational range* is the range or distance to the target from the firing vessel. It is determined after the firing from consideration of all available data.

13. *Gun range* is the range in yards (shown in the latest range table in effect) corresponding to the angle of elevation above the horizontal for which a gun is laid when fired. It is equal to the sight-bar range less the control ballistic, plus the gun erosion correction when this correction is introduced by erosion correctors. When sight scales and range converters are graduated to the proper range table, when the point of aim is the desired point of impact, when there is no uncorrected vertical parallax between gun and director, and when erosion correctors are not used, the gun range is equal to the sight-bar range.

14. *Sight-bar range* is the range in yards shown on the sight scale of the gun, the range converter or director at the instant of firing.

15. *Hitting gun range* is the gun range that, if used, would place the mean point of impact at the center of the target. It is equal to the gun range corrected for the error of the mean point of impact.

16. *Maximum gun range* is the gun range corresponding to the maximum gun angle at which all guns of the firing battery may be fired without structural interference.

17. *Impact range* is the distance from the firing point to the point of impact of the projectile. This can be accurately determined only at calibration practices.

B4. Ballistics. Ballistics are computed corrections to be applied to the measured or estimated values of range and to the midpoint of the deflection scale to compensate known or predicted errors and variations from selected standard conditions.

1. The *gun ballistic* consists of the corrections to compensate the following variations from standard conditions: (1) initial velocity errors due to temperature of the powder, gun erosion, and variation of projectile weight; (2) errors due to atmospheric conditions of pressure, temperature and wind; (3) errors produced by relative movements of gun and target; and (4) drift.

2. *Arbitrary ballistic* is a correction to compensate the indeterminate errors in the total ballistic used on previous firings, obtained by analysis of those firings. When used, it is combined with the gun ballistic in determining the gun range and deflection. A special arbitrary ballistic known as "the cold-gun correction" is often used for an opening salvo.

3. The *control ballistic* consists of corrections to the gun range and deflection to obtain the sight-bar range and deflection to compensate the following:

    a. In pointer fire:
        Point-of-aim correction.
        Sight-scale range-table correction.

APPENDIX B

    b. In director fire:
        Point-of-aim correction.
        Vertical parallax correction.
        Range-converter range-table correction.

4. *Total ballistic* is the correction determined prior to firing to be applied to the present range, and in deflection to the midpoint of the deflection scale to obtain the sight-bar range and deflection, and in the case of antiaircraft fire, the fuze setting. It includes the gun ballistic, arbitrary ballistic and control ballistic.

5. *Ballistic used* is the correction in range actually applied to present range, and in deflection to the midpoint of the deflection scale; in range it is equal to the difference between present range and advance range. For the first salvo the ballistic used should be the same as total ballistic if there are no errors in application.

6. *Hitting present range ballistic* is the correction in range which, if applied to the present range, will give the sight-bar range necessary to place the mean point of impact in range at the center of the target.

7. *Hitting navigational range ballistic* is the correction in range which, if applied to the navigational range, will give the sight-bar range necessary to place the mean point of impact in range at the center of the target. It can only be determined by post-firing analysis. If this analysis shows that the present range was the same as the navigational range, then the hitting present range ballistic was the hitting navigational range ballistic.

8. *Hitting deflection ballistic* is the correction in deflection which, if applied to the midpoint of the deflection scale, would place the mean point of impact (in deflection) at the center of the target. For purposes of analysis the mil used in computing the hitting deflection ballistic is one one-thousandth of the navigational range.

9. *Ballistic wind* is an effective wind, determined by computation, of such force and direction that its action on the projectile during the time of flight will be the same as the combined actions produced by the various true winds acting in the strata through which the projectile will pass as it moves along its trajectory.

10. *The point-of-aim correction* is the correction applied to gun range and deflection to compensate the vertical and horizontal angles between the line of sight to the point of aim and the line from the gun to the desired point of impact.

11. *The sight-bar range-table correction* is the correction applied to the gun range to correct for errors caused by sight scales being graduated for an obsolete range table or for a type of ammunition of different initial velocity or different type of projectile than that of the ammunition in use. Using the same sight angle in column 2 of both range tables, it is the difference between the ranges given in column 1 of the two tables.

12. *Range-converter range-table correction* is the correction applied to the gun range to correct for errors caused by the range converter being graduated for an obsolete range table or for a type of ammunition of different initial velocity or different type of projectile than that of the ammunition in use. Using the same sight angle in column 2 of both range tables, it is the difference between the ranges given in column 1 of the two tables.

13. *Bearing prediction* is the angular lateral deviation due to motion of target and own ship across the line of sight, including effect of wind due to own-ship motion upon a projectile during the time of flight.

14. *Wind deflection prediction* is the angular lateral deviation due to the effect of the true wind upon the projectile during the time of flight.*

15. *Drift angle* at any range is the angular lateral deviation due to rotation of a projectile in flight.

---

* In practice the effect of the wind due to own-ship motion is most frequently not included in the bearing or range prediction but is compensated in the wind deflection and wind range predictions by using apparent wind.

DEFINITIONS

16. *Range prediction* is the predicted change in range due to the target's motion relative to own ship, including effect of wind due to own-ship motion during the time of flight. (Includes effect of wind on target in antiaircraft firing.)

17. *Wind range prediction* is the predicted change in range due to the effect of true wind upon the projectile during the time of flight.

18. *Elevation prediction* is the angular vertical deviation, or change in position angle, due to motion of the target and own ship during the time of flight. It includes the effect of the true wind on the target.

19. *Wind elevation prediction* is the angular vertical deviation due to the apparent wind's effect on the projectile during the time of flight.

20. *Superelevation* is the angle the gun must be elevated above the line of sight when firing at elevated targets, to compensate curvature of the trajectory.

21. *Deflection* is the lateral angular correction applied to the target bearing to obtain gun-train order and thus offset the gun from the line of sight the proper amount to compensate the total effects of lateral motion of target and own ship, wind effect and drift. Sights and fire-control instruments are designed for, and deflection is computed for the application of deflection in the horizontal plane. However, unless corrected for trunnion tilt, the deflection is actually applied in the reference plane.

22. *Corrected sight deflection.* This term is frequently employed when a separate instrument is used to correct for the effect of trunnion tilt, and the value so designated indicates that it is the sight deflection corrected for the effect of trunnion tilt in train.

23. *Sight depression* is the vertical angle the gun or director sights must be set below the gun bore (or gun pilot in the case of a director) to secure the gun elevation corresponding to the computed advance range, when the gun and director are elevated to bring the line of sight on the point of aim.

24. *Corrected sight depression.* This term is frequently employed when a separate instrument is used to correct for the effect of trunnion tilt and the value so designated indicates that it is the sight depression corrected for the effect of trunnion tilt in elevation.

B5. Armament.

1. *A capital ship* is an armored vessel of war, not an aircraft carrier, mounting a battery of a caliber greater than 8 inches.

2. *Armament* of a ship includes all offensive weapons.

3. *Turret guns* are all guns of 6 inches in caliber and larger mounted in multiple in a turret or in a similar structure designated as a mount.

4. *Broadside guns* are guns of less than 6 inches in caliber, except antiaircraft guns and guns used for saluting purposes only.

5. *Antiaircraft guns* are intermediate-caliber guns designed for use against aircraft.

6. *Dual-purpose guns* are guns designed for use against both aircraft and surface targets.

7. *Antiaircraft machine guns* are minor-caliber fully-automatic guns designed for use against aircraft.

8. *Main battery.* The term main battery includes those guns of the largest caliber on board.

9. *Secondary battery.* Only turret ships are considered to have a secondary battery; the secondary battery includes all except the turret guns and those guns specifically designated for use against aircraft. In ships having no broadside guns other than dual-purpose guns this battery may be designated as the secondary battery.

10. *Antiaircraft battery* includes all antiaircraft guns and antiaircraft machine guns.

B6. Salvo analysis.

1. *A salvo* consists of two or more shots fired either simultaneously by means of a master key as in director or master-key firing, or on the same firing signal as in pointer fire. In gunnery exercises, a salvo is considered to have been fired when it is clearly the intent to fire two or more shots, even though as a result of casualty, only one shot may actually be fired.

APPENDIX B

2. The *pattern* of a salvo in range is the distance measured along the line of fire between the shot of the salvo falling or bursting at the greatest distance from the firing point and the shot falling or bursting at the shortest distance, excluding wild shots. In deflection it is the distance measured at right angles to the line of fire from the shot falling or bursting at the greatest distance to the right to the shot falling or bursting at the greatest distance to the left of the line of fire, excluding wild shots.

3. The *mean point of impact* is the point which is at the geometrical center of all the points of impact of the several shots of a salvo, excluding wild shots. When firing time-fuzed projectiles the point of detonation of such a projectile is considered to be the point of impact.

4. *The dispersion* of a shot is the distance of the point of impact of that shot from the mean point of impact of the salvo. Dispersion in range is measured along the line of fire and in deflection at right angles to the line of fire.

5. *The apparent mean dispersion* of a salvo in range (or deflection) is the average of the dispersions in range (or deflection) of the several shots of the salvo, excluding wild shots.

6. *The true mean dispersion* is obtained when an infinite number of shots are fired. The true mean dispersion is greater than the apparent mean dispersion of a small number of shots. In analyses of gun practices, to obtain the true dispersion of a salvo the apparent mean dispersion of the salvo is multiplied by a factor depending upon the number of shots in the salvo, excluding wild shots, representing the probable ratio of the apparent mean dispersion of that number of shots to the true mean dispersion of an infinite number of shots.

7. *Error of the mean point of impact* is the distance of the mean point of impact from the target or other reference point measured parallel to the line of fire for range, and at right angles to the line of fire for deflection.

8. A *straddle* is obtained from a salvo in range (or deflection) when, excluding wild shots, a portion of the shots of that salvo fall or detonate short and other shots of the salvo beyond the target (right and left of the target for deflection).

9. A *wild shot* is a shot whose point of impact is abnormal when compared with the remaining shot impacts of the same salvo.

10. *The danger space* is such distance measured along the line of fire in front of the target that, if the target were moved toward the firing point, a shot striking the base of the target in its original position would strike the top of the target in its new position.

11. *The hitting space* for a material target is the distance behind the target that a shot striking the top of the target will strike the horizontal plane through the base of the target. Hitting space must be computed from column 19 of the range tables or directly from the angle of fall of the shot. Hitting space is the basis for computing the limits of a constructive target when such is used for scoring hits in certain forms of gunnery exercises. Allowance is added for depth of the assumed target in range, and in some cases, for an arbitrary distance short of the point of aim.

12. *Line of fire* as used in analyses of gunfire is the true bearing of the target from the firing ship at the instant a shot (salvo) is fired.

B7. Control stations.

1. *Control* is that station from which the armament of the ship is controlled. Control is, as a rule, in actual control of the main battery only.

2. *Secondary control* is that station from which the broadside battery is controlled.

3. *Air defense* is that station from which the antiaircraft battery is controlled.

B8. Types of control.

1. *Primary.*

    a. *Director control* is that method of control in which the guns are trained and elevated in accordance with signals transmitted by a director and associated instruments, and as corrected by them. It includes *indirect control* which is that method of director control employed when the target is obscured to the director, the line of sight being stabilized by use of the stable element and oriented by using the bearing, elevation and range generated by the computer.

DEFINITIONS

    b. *Barrage control.* Any method of control in which, due to near proximity of a diving air target, or the size and speed of a surface target, or the fact that the battery is not director controlled, it is desirable to resort to one of the methods of barrage fire.

  2. *Secondary.* Any secondary method of gun control resorted to when the primary method has failed. It includes:

    a. *Stand-by control,* wherein a ship having two plotting rooms employs the stand-by plotting room only for control of fire.

    b. *Auxiliary control,* wherein a ship having but one plotting room employs an auxiliary range keeper usually located on or near the director and associated instruments for control of fire.

    c. *Emergency control,* wherein the control of fire of a group of guns or turrets is maintained from a local station at one of the guns or turrets. The values of range, deflection, and (for antiaircraft fire) fuze setting are determined at the local station.

    d. *Local control* is that method of control in which the guns are pointed and trained by telescope, and fired locally, the values of range, deflection, and fuze setting being determined at the gun or an adjacent local control station.

    e. *Indicating control* is that method employed when, the remote control system having failed, it is still possible to follow the signals transmitted from a distant station by manually matching indicators in the elevation, train, and fuze-setting indicators.

    f. *Telescope control* is that method of control in which the guns are pointed and/or trained locally by telescope, but the values of range, deflection and fuze setting are transmitted from a control station. It does not necessarily mean that the guns are fired locally.

B9. Types of fire.

1. *Salvo fire* is that type of firing in which a number of guns that are aimed at the same target and ready to fire are fired simultaneously by means of a master key at the control station, by an automatic contactor, or on the same salvo signal. Note: Salvo interval may be controlled by time (stop watch), point of roll, or time of flight; it should be adjusted so that the guns do not fire at the same instant or just prior to a salvo landing or detonating in order to prevent blinding of spotters and director operators.

2. *Split salvo.* When less than the full number of guns in a multiple gun mount or mounts is ordered to fire on one salvo signal, the salvo is a split salvo. Split salvos may be used to reduce dispersion, conserve ammunition, or reduce salvo interval.

3. *Partial salvo.* When less than the full number of guns (in single gun mount batteries) or less than the full number of mounts or turrets (in multiple gun mount batteries) is ordered to fire on one salvo signal, the salvo is a partial salvo. Partial salvos may be used for the firing of ranging shots or ladders, to conserve ammunition, or to reduce salvo interval.

4. *Continuous fire* is that type of fire in which each gun is fired individually and when ready without regard for the condition of readiness of other guns.

5. *Continuous turret fire* is that type of fire in which each turret is fired as soon as all of the guns in that turret are ready to fire without regard for other turrets.

6. *Slow fire.* That type of fire wherein the fire is deliberately delayed to allow for the application of corrections or spots, or to conserve ammunition.

7. *Rapid fire* is that type of fire in which no check fire for purposes of applying corrections to fire control is used.

B10. Methods of fire.

1. *Pointer (trainer) key* is that method of firing in which the gun pointer (trainer) fires his gun, either in response to a salvo signal, or when ordered to do so by the gun captain.

2. *Master key* is that method of firing which is controlled by a firing key at the director or control station. The master key may be operated manually or automatically. Master key includes the methods of firing in items of 3-7 below.

APPENDIX B

3. *Continuous aim* consists of keeping the director pointer's, trainer's, leveller's, and cross-leveller's's wires on the target (horizon) at all times. The master key may then be closed at any time.

4. *Selected level.* The pointer's (leveller's) wire is laid in elevation; when it rolls across the target (horizon) the master key is closed. This manner of firing may be advantageously employed when the pointer (leveller) has difficulty keeping on the target (horizon) or when advance range is such that motion in level would cause the guns to strike the elevation stops.

5. *Selected cross-level.* The cross-leveller's wire is laid in elevation, the pointer's (leveller's) wires are kept on the target (horizon). When the cross-leveller's wire rolls across the horizon the master key is closed. This manner of firing may be advantageously employed when the motion of the guns in train due to cross-roll is so great that the trainers have difficulty keeping zero readers (pointers) matched.

6. *Selected train.* The method of firing in which the guns are trained in accordance with the signals from a hand-operated transmitter and are fired when corrected gun-train order is in agreement with the gun-train order selected as shown by comparison dials.

7. *Cranking.* In the cranking method the pointer cranks his line of sight above (below) the target, and then cranks rapidly and at a uniform speed down (up) across the target. The stabilized sight firing key may be used, in which case it is closed as the pointer starts cranking. If the independent key is used it is closed as the pointer's horizontal wire crosses the target.

B11. Methods of ranging.

1. *Instrument.* The primary means, either by stereo, coincidence, sound, or other instruments of obtaining the range to a distant gun target.

2. *Horizon method.* The range to the target is obtained by observing that point of the target in line with the horizon, and comparing the observation with a diagram based on dip curves.

3. *Estimate.* The range may be estimated by eye or by glasses in the field of which a mil scale has been scribed. During periods of reduced visibility the distance a target can be seen under the existing conditions of weather, light, and sea may be estimated and will then become the *range of the day/night*.

B12. Ranging shots.

1. The *range ladder* is a method of fixing the hitting gun range by opening fire with a gun range assuredly over or short of the target with the application of successive range corrections between successive salvos or groups of shots toward the target, and continued in that direction until target is crossed. A specialized form of range ladder is the *initial ladder,* consisting of a definite number of salvos (greater than two) fired as a rule with large steps or increments, and calculated to locate the target within the limits of the ladder.

2. A *deflection ladder* may be used to fix the hitting deflection by arbitrarily varying the deflection to locate the target within deflection limits.

3. In an *up ladder* the first salvo is fired with a gun range short of the best range, the second salvo with an increased gun range, the third salvo still higher, etc., until the target is located or crossed.

4. In a *down ladder* the first salvo is fired with a gun range over the best range, the second salvo with a decrease in gun range, the third salvo still further decreased, etc., until the target is located or crossed.

5. In a *bracket* two salvos are fired with gun ranges or deflection set either side of the best range or deflection, the object being to include the target between the two sets of splashes.

6. *Barrage* is a method used against a fast-opening or closing target, whereby a gun range or fuze setting is used which will place the initial shots ahead of the target in the direction of the target's anticipated advance. The gun range or fuze setting is unchanged until the target passes through the zone of fire, at which time the gun range or fuze setting is changed to a new advanced position.

B13. Methods of spotting.

1. *Direct* is that method in which the spotter's correction is based on an actual estimate of the error of the mean point of impact from the target in range and deflection.

DEFINITIONS

2. *Halving.* When the first salvo is all over or short a spot is made large enough to cause the succeeding salvo to cross the target. When the target has been crossed the succeeding spots are half the last one used, applied in the opposite direction, until the target has been brought within the limits of the pattern.

3. In a *continuous ladder* the salvos are moved back and forth across the target.

4. In an *out ladder* the range is increased in small uniform increments after each salvo or at regular intervals until the target is crossed. The ladder may or may not be reversed or repeated.

5. In an *in ladder* the range is decreased in small uniform increments after each salvo or at regular intervals until the target is crossed. The ladder may or may not be reversed or repeated.

6. In a *rocking ladder* the range is varied within narrowly restricted limits by alternating laddering in and out for the purpose of increasing the normal pattern of the battery.

7. *Barrage.* In barrage spotting, as used with barrage fire, the spotter waits until the target enters the barrage zone; then, counting on the shots in the air to produce hits, fire is shifted and the spotter spots to the next zone.

B14. Methods of range keeping.

1. *Direct,* wherein the fire-control problem is solved continuously and every shot fired is directed to hit. Changing quantities of bearing, range, position, angle, level and cross-level bearing continuously observed are plotted or continuously introduced into a range keeper to predict gun range (fuze setting), deflection, and, if required, the superelevation.

    a. *Automatic rate control.* Employed in range keepers and directors for the solution of the fire-control problem. Manual correction of generated values of range, bearing, and elevation to make them compare with observed values of the same quantities automatically corrects the target course, speed, and angle of climb or dive (for aircraft) to maintain the generated and observed values identical.

    b. *Manual rate control.* Employed in range keepers and directors for the solution of the fire-control problem. Target course, speed, and angle of climb or dive (for aircraft) are manually corrected to maintain the generated values of range, bearing, and elevation coincident with the observed values of range, bearing and position angle.

2. *Auxiliary.* Any method of determining gun range (fuze setting) deflection, and elevation, in which range tables, graphic plotters, mechanical computers, nomograms, tables, or estimates are used to replace the director and range-keeper equipment for control of fire. Auxiliary or secondary methods include:

    a. *One-minute barrage.* A modified creeping-zone barrage used as a rule against aircraft, wherein fire is concentrated in that period of one minute just prior to the bomber target reaching the bomb-release point.

    b. *Line of sight barrage.* An antiaircraft barrage fired with fixed fuzes and varying sight angle, with each shot so directed that the bursts will occur on the instantaneous line of sight to the plane.

    c. *Fixed zone barrage.* A barrage fired with fixed sight angles and a fixed fuze setting. Each zone is selected in an effort to place initial bursts or shots ahead of the target in an area through which it is anticipated that the target will pass. Once the target has passed through the zone of fire a second zone may be selected and a new barrage fired.

    d. *Creeping zone barrage.* A barrage in which the gun range or the fuze settings are varied within a zone limit in such a way as to advance along the target track at a rate slower than that of the actual target advance.

B15. Methods of illumination.

1. *Star-shell* illumination consists of illuminating by projectiles fired by own or other ships, so placed as to disclose to the control party the target and fall of shot in its vicinity.

    a. *Star gun.* A gun (mount or turret) designated to fire illuminating projectiles. The remainder of the same battery may or may not be designated to fire at the target.

    b. *Star spread.* A star-shell spread is made up of single stars fired from a battery, each

APPENDIX B

gun of which is offset in deflection from the director line of sight by amounts necessary to accomplish the desired spread.

    c. *Star search.* A star search consists of several star spreads fired by one or more ships on different bearings in order to search out a designated arc or sector.

2. *Searchlight.* Illumination of the target by the searchlights provided. It may be accomplished by director or by local control. Usually a secondary or relatively short range method of illumination.

B16. Fire distribution.

1. *Divided fire.* One ship firing at two or more targets.
2. *Concentration.*
    a. *Single fire.* One ship firing alone at a single target.
    b. *Double concentration.* Two ships firing at the same target.
    c. *Triple concentration.* Three ships firing at the same target.
    d. *Quadruple concentration.* Four ships firing at the same target.

APPENDIX C

STANDARD COMMANDS

C1. General. In the interest of uniformity the following commands are laid down and defined, covering all standard methods of primary and secondary control used in recognized doctrines of gun fire. The standard commands should be used as a drill manual, all members of control teams and gun batteries being familiar with the commands and the interpretation of each.

1. *"All Stations Report When Manned and Ready"*. Command given by control immediately upon control circuit being manned.

2. *"Ready"*. Each gun (mount or turret) report "ready" to control as soon as there are sufficient personnel on station to fire a gun at a reduced rate by local methods. Control report "ready" to conn as soon as one gun can be fired.

3. *"Manned and Ready"*. Each station report "manned and ready" to control when the station has been placed in the fully ready condition, with sufficient personnel present to service the station using primary methods. Control report "Manned and Ready" to conn when all stations have reported.

4. *"Test Circuits"*. This command is interpreted to mean: test telephones, voice tubes, salvo signals, cease firing, and firing circuits where applicable. Guns report "circuits OK" if all circuits test properly; otherwise report only those circuits which fail.

5. *"Test Transmission"*. Directs the members of the control party responsible for signals transmitted from the director to the battery to test the accuracy of transmission thereof, including: elevation (auto and indicating), train (auto and indicating), sight setting, fuze setting, battle order, shell order, shutter order.

6. *"Sight Angle............"*. Set sight angle in minutes.

7. *"Range............"*. Set range in yards.

8. *"Scale............"*. Set deflection in mils.

9. *"Dip and Convergence............"*. Set dip and convergence in yards.

10. *"Parallax range............"*. Set parallax range in yards.

11. *"Stations"*.
    a. Ammunition parties prepare to send up or break out ammunition.
    b. Gun crews take their designated stations. Prepare to service the battery.
    c. Start motors.

12. *"Air (Surface) (Land) Target"*. Directs preliminary dispositions for type target designated.

13. *Common (High Explosive) (Illuminating) Projectile*. Directs preparations to fire the type of ammunition indicated.

14. *"Action Starboard (Port)"*. This is interpreted as a preliminary command by conn or control, used to position director and guns on the side designated, with ammunition distributed and ammunition train formed for rapid supply to the gun on that side. On this command cut in all stable elements and follow-ups, all guns (lights) match or shift to automatic remote control.

15. *"Target is Destroyer (Aircraft) (Submarine) Bearing"*. Conn must designate accurately by type and bearing the target it is desired to take under fire. In the same manner control designates target to guns so that in case of necessity guns may be able to go intelligently to local control. Director trainer and pointer report to control when on target, using the expression *"On Target"*.

16. *"Start Tracking"*. The control party starts tracking the designated target. After this command the entire battery is in condition to open fire at a moment's notice using primary methods of control.

C2. Commands for various types of control and fire.

1. *"Director (One) (Two) Control"*. Places the designated director in control.

2. *"Indirect Control"*. Directs control as defined in Appendix B.

APPENDIX C

3. *"Telescope Control"*. Directs control as defined in Appendix B. Given by the gun captain in case casualty makes is necessary.

4. *"Local Control"*. Directs control as defined in Appendix B. Given by the gun captain in case casualty makes it necessary.

5. *"Divide Control"*. Places each director in control of its own battery of guns.

6. *"Fire Salvo, Interval .... Seconds"*. Directs salvo fire as defined in Appendix B.

7. *"Split Salvo"*. Directs that in accordance with ship doctrine, part of the battery will fire on the first salvo followed by part of the battery on the succeeding salvo, the groups to alternate salvos until commanded otherwise.

8. *"Fire Continuous"*. Directs the type of fire defined in Appendix B.

9. *"Pointer (Trainer) Key"*. Directs the methods of fire defined in Appendix B.

10. *"Master Key"*. A command to produce the methods of fire defined in Appendix B. On this command gun pointers close and lock firing keys, and set firing circuit transfer switches on "Motor Generator".

11. *"Continuous Aim"*. Directs the method of master key fire defined in Appendix B.

12. *"Selected Level"*. Directs the method of master key fire defined in Appendix B.

13. *"Selected Cross-Level"*. Directs the method of master key fire defined in Appendix B.

14. *"Selected Train"*. Directs the method of fire defined in Appendix B.

15. *"Down (Up) Ladder .... Hundred Yard Steps"*. A command to produce a ladder in accordance with Article B12, Appendix B.

16. *"Bracket .... Hundred Yard Steps"*. A command to use ranging shots by bracket method defined in Appendix B.

17. *"Barrage"*. A command to control fire by barrage method defined in Appendix B.

18. *"Spot Direct"*. Directs controlling spotter to spot by method defined in Appendix B.

19. *"Spot Halving"*. Directs controlling spotter to spot by method defined in Appendix B.

20. *"Use Continuous Ladder"*. Directs controlling spotter and control party to use continuous ladder defined in Appendix B.

21. *"Ladder Out .... Yard Steps"*. A command from controlling spotter to control party to run a continuous *out* ladder.

22. *"Ladder in .... Yard Steps"*. A command from controlling spotter to control party to run a continuous *in* ladder.

23. *"Rocking Ladder............Yard Spread............Yard Steps"*. A command from controlling spotter to control party to alternately ladder *in* and *out*.

24. *"Prepare to Illuminate Bearing..........by Searchlight and (or) Star shell"*. Command given by conn to control preparatory to illumination.

25. *"Lay for Star-shell Spread to Starboard (Port)"*. A command from control to guns when a star-shell spread is to be fired. Guns load fuze setters with star shells, elevate, and offset train dials in accordance with doctrine.

26. *"Illuminate"*. A command by conn to control to illuminate an arc or bearing. In the case of star-shell illumination, *"Illuminate"* is a command for all or designated guns to load star shells, and director pointer to close firing key. In the case of searchlight illumination, "Illuminate" is a command to open searchlight shutters.

27. *"Match Parallax"*. A command from control to guns to match parallax zero readers.

28. *"Match Pointers (zero readers) in Train (elevation)"*. A command from control to guns (searchlights) to match coarse and fine dials on indicators specified.

29. *"Load"*. A command from control to trip salvo latches, open breech and load with ammunition previously specified. This command may be omitted, in which case the action is taken at the command "Commence Firing".

30. *"Commence Firing"*. A command from control to load and open fire, given after trainer reports "On Target".

31. *"Check Fire"*. A command to discontinue fire, probably temporarily. All other functions of control party and gun crew continue. On this command sound cease firing gong (howler).

RESTRICTED

STANDARD COMMANDS

32. *"Resume Fire"*. A command to continue fire after check fire has been given.

33. *"Shoot"*. A command to fire a salvo or split salvo, used when the control officer desires to control personally the rate of fire.

34. *"Cease Firing"*. A command to stop the firing preparatory to shifting the method of fire or the target fired upon. Cease firing is not a command signifying the end of action. On this command sound cease firing gong (howler). Gun captains report to control rounds of ammunition fired and condition of guns.

35. *"Cease Tracking"*. A command to resume the fully ready condition. All stations remain alert.

36. *"At Ease"*. A command given when it is desired to allow the men at stations to relax at their stations.

37. *"Secure from General Quarters (Set Condition II/III Watch 1/2/3)"*. A command from conn repeated by control to secure or to set condition of readiness designated.

38. *"Strike Arcs"*. A command by control to the searchlights to establish the arc in the lights with shutters closed for the purpose of quick illumination or for warming up and testing.

39. *"Break Arc"*. A command to extinguish searchlight arcs.

40. *"Open (close) Shutters"*. A command by control when it is desired to open or close the searchlight shutter.

41. *"Take Cover"*. A command to gun crews to proceed to gun-crew shelter on the double. Given when angle of elevation is such as to permit shell fragments to fall on deck, or to avoid a spray gas attack.

42. *"Repel Dive Attack"*. A command to discontinue all other action, guns to go to designated sectors, and to open fire automatically in accordance with doctrine on any plane sighted in the assigned sector.

43. *"Drop Down"*. At this command all hands on deck drop down on their faces to avoid bomb fragments. The command ordinarily will only be given after aircraft have been observed to reach a bombing position, to have released their bombs, and just before the bombs reach the surface.

44. *"Replace Ammunition"*. A command to replace ammunition expended from ready racks, etc., and report completion.

45. *"Report Casualties"*. During firing, reports shall be made at once of any casualty that puts a gun out of action for an appreciable time, or of any serious damage to or near the gun requiring immediate assistance for the safety of the ship. Detailed reports should be made as opportunity permits but should not overload communication channels.

C3. Casualty reports and commands.

1. *"Silence"*. A command used in case of serious casualty or in case of doubt as to the seriousness of a casualty. On the command "Silence" every member of the gun crew, director crew, ammunition party and all in the vicinity freeze in their tracks and remain there without noise or confusion until further orders are given or they hear the command "Carry On".

2. *"Carry On"*. On this command normal service of the battery is continued from the point interrupted by the command of "Silence".

3. Follow-up side reports *"Time Motor Stopped"*. Control orders *"Use Hand Crank"*.

4. Follow-up side reports *"Sight Angle (deflection) (fuze range) Out"*. Control orders *"Follow By Hand"*.

5. Range-keeper operator reports *"Range-keeper Power Off"*. Control orders *"Telescope Control, Master Key (pointer key)"*.

6. Gun pointer (trainer) reports *"Power Out"*, gun captain orders: *"Shift to manual, match zero readers"*.

7. Gun pointer (trainer) reports *"Indicating Out"*, gun captain orders: *Point (train) By Telescope"*.

8. Gun pointer reports *"Misfire"*. Gun captain order *"Kick It Out"*. In case of repeated misfire by motor generator shift to local battery for remaining salvos.

APPENDIX C

9. Rammer man reports *"Rammer Out"*. Gun captain orders *"Load By Hand"*.

C4. Standard commands for the 40 mm. battery.

1. *"Stations"*.
    a. Pointer.
        1. Set control-selector switch in *manual*.
        2. Start elevating-power motor.
        3. Start firing-circuit motors.
        4. Place control-selector switch on *automatic* (after approximately matching with director in *manual*, or after director matches with mount).
    b. Trainer.
        1. Set control-selector switch in *manual*.
        2. Start train-power motor.
        3. Set control-selector switch on *automatic* (after approximately matching with the director in *manual*, or after director matches with mount).
    c. First loaders.
        1. Start cooling-pump motors.
        2. Check to see top cover plate closed and latched.
        3. See rear firing-selector switch on "safe" (stop fire).
        4. See feedway firing stop lever on *red arrow*.
        5. Check recoil indicator set at 6.

2. *"Load"*.
    a. First Loaders.
        1. Retract hand operating lever and secure in retract position.
        2. Place first clip in feedway, forcing down until empty clip is ejected at side.
        3. Place second clip in feedway.
        4. Take third clip from second loader.

3. *"Standby"*.
    a. First loader.
        1. Place operating lever in forward position and latch.
        2. Set firing-selector switch on *automatic* (or on *single shot* if directed).
    b. Mount captain.
        1. Inspect to see that one round is in loading tray of each barrel.

4. *"Track"*.
    a. Director pointer.
        1. Track designated target and establish rates.

5. *"Commence Firing"*.
    a. Director pointer.
        1. When rates are established on target, close director firing key.
    b. Gun pointer.
        1. Press down on foot-firing pedal, and hold depressed as long as it is safe to fire.
    c. First and second loaders.
        1. Keep feed ways filled as necessary. Use ammunition from clipping room as a primary source, but not to decrease the rate of fire.

C5. Standard commands for the 20 mm. battery.

1. *"Stations"*. All crew members take designated stations.
    a. Gunner—In straps. Test train and elevation.
    b. First loader—Take up tension on one magazine.
    c. Second loader—Open ready service box and standby to take up tension on remaining magazines on further orders.
    d. Column operator—Cast gun loose. Adjust column.
    e. Mark 14 sight operator—Set range at 2,000; set spots at zero unless otherwise directed. See oil at proper temperature and air at right pressure. Test gyros for precession.

STANDARD COMMANDS

    f. Clip-room crews—Standby for further orders.
2. *"Load and Lock"*.
    a. Gunner—Cock, and put trigger-safety catch on *safe*.
    b. First loader—Inspect bore from muzzle end. Ship first magazine, pick up second.
    c. Second loader—Take up tension on remaining magazines.
    d. Clip-room crew—Take up tension on all magazines.
3. *"Track"*.
    a. Gunner—Lay gun on designated target and track.
    b. Column operator—Keep column adjusted.
    c. Mark 14 sight operator—Keep range set as directed.
4. *"Standby"*.
    a. First loader—Set trigger safety catch to *"Fire"*.
5. *"Commence Firing"*.
    a. Gunner—Press trigger. When magazine is expended, place trigger-safety catch on *safe* and order *change*. Cock magazine catch.
    b. First loader—Remove expended magazine and replace.
    c. All—Continue fire on designated target when ready.
6. *"Cease Firing"*.
    a. Gunner—Release trigger and put trigger safety catch on *"Safe"*.
    b. First loader—Remove magazine and replace with full magazine.
    c. All—Stand by to resume action instantly.
7. *"Easy"*.
    a. Gunner— ⎱ Remove magazine, inspect bore for "bore clear" and uncock.
    b. First loader ⎰
    c. Second loader—Remove tension from all magazines and restow them in ready-service box.
    d. Column operator—Lower column. Insert elevation and train stops when gunner directs.
    e. Clip-room crew—Remove tension from all magazines.

Note: The above commands are all applicable to *any* firing that the 20 mm. battery may do. However, all of these commands will not necessarily be given. For example, the only two orders that may be given are *"stations"* and *"commence firing"*, and crews must be prepared to perform the necessary operations without further orders.

C6. Commands for caliber .50 battery.
1. *"Place Ammunition"*.
    a. Gunner—Brings gun to horizontal pointing in safe direction.
    b. Loader—Places ammunition box on gun.
2. *"Stand By"*.
    a. Gunner—Continues to point gun in safe direction.
    b. Loader—Brings cartridge belt into position.
3. *"Half Load"*.
    a. Loader—Closes cover. Grasps train lever to prevent gun from whipping around.
    b. Gunner—When cover is closed pulls operating handle to rear once and releases it.
4. *"Load"*.
    a. Loader—Continues to hold train lever.
    b. Gunner—Pulls operating handle to rear the second time and releases it.
5. *"Track"*.
    a. Gunner—Brings sights on target, or points gun at estimated position to bring tracers on target.
6. *"Commence Firing"*.
    a. Gunner—Fires in accordance with doctrine.
7. *"Cease Firing"*.
    a. Gunner—Points gun in safe direction. Opens feed cover.

APPENDIX D

5-INCH RANGE TABLE FROM O.P. 551

2,600 f.s. initial velocity to 18,200 yards.
weight of projectile—54 pounds
length of projectile—4.15 calibers
radius of ogive—5.25 calibers

A.A. COMMON PROJECTILE

Explanatory Notes

1. Columns 1 to 8 give the elements of the standard trajectories and need no explanation.

2. Columns 10 to 19 give the differential effects due to variations from standard conditions. In these columns the effect is to be taken as proportional to the cause. For example, column 13 is tabulated for a 10-knot wind; in case of a 20-knot wind column 13 is to be multiplied by 2.

3. Column 10, although computed for a velocity loss, may be used to determine the change in range due to either plus or minus variations in initial velocity. It is listed as a positive effect in order to avoid minus tabulations.

4. Column 11 gives the effect on the range for a change of $\pm$ one pound in weight of projectile, the charge remaining the same.

5. In computing the range table, it is assumed that the density aloft varies with altitude according to an exponential law, and that the surface density is 1.2034 kg./m^3 which corresponds to 59°F. and 29".53 barometer.

6. Column 12 gives the mean effect on the range of ± 10 per cent variation in the density of the air. Ballistic density should be used in entering column 12. A ballistic density is a single fictitious air density constant in magnitude which would have the same total effect on the projectile during flight as the actual densities at the various altitudes. A method of making up a ballistic density is given in Bulletin of Ordnance Information, O.P. No. 561-I of May, 1923.

7. When ballistic density is not available and surface conditions only are known use Plate 1, which gives results corresponding to an AVERAGE ballistic density.

8. Column 12 also gives the mean effect on the range of ± 10 per cent variation in the ballistic coefficient.

9. Columns 13 and 16 give the effects of a 10-knot ballistic wind, which is the combined weighted winds in the several vertical zones of the trajectory. When, however, no measurements of upper air winds can be obtained, the surface true wind must necessarily be used in entering these columns. Methods of making up ballistic wind are given in Bulletin of Ordnance Information, O.P. No. 561-I of May, 1923, and in Method of Computing Range Tables, O.P. No. 500 of April, 1929.

10. Column 19 shows how much the point of impact is raised or lowered on a vertical screen by raising or lowering the sight bar 100 yards, the actual range remaining fixed.

11. The change in range due to a variation of ± 1 minute in the angle of elevation may be deduced from column 2b.

12. The powder is assumed to give normal velocity at 90°F. For each degree increase in temperature the initial velocity is increased approximately 2 f.s.; a decrease in temperature causes a corresponding decrease in velocity.

13. The firings upon which this range table is based are summarized below:

5-inch 38 cal. Gun Mark XII
Rifling, Uniform Twist 1/30
AA Common Projectile Mark XXXI
45 sec. combination—long point fuze

5-INCH RANGE TABLE FROM O.P. 551

Number of Rounds	Elevation	Mean Velocity	Fall of Shot (Actual)	
			Range	Mean Error
		Ranging Sheet 683, 28 Nov., 1932.		
5	15°	2650 ± 3	11827	± 26
5	15°	2596 ± 3	11585	± 26
5	15°	2602 ± 3	11615	± 15
		Ranging Sheet 684, 2 Dec., 1932		
5	45°	-----	18589	± 43
5	45°	-----	18434	± 47
4	45°	-----	18418	± 62
		Ranging Sheet 695, 16 Jan., 1933.		
5	5°	2604 ± 2	6880	± 14
5	10°	2597 ± 4	9858	± 35
5	30°	-----	16645	± 77
5	45°	-----	18192	±178
		Ranging Sheet 696, 23 Jan., 1933.		
5	10°	2595 ± 3	10126	± 35
4	20°	-----	14218	± 24
5	45°	-----	19045	± 72
		Ranging Sheet 708, 31 Mar., 1933.		
5	15°	2593 ± 3	12226	± 25
5	15°	2587 ± 4	12113	± 32

Sine	0°	5°	10°	15°	20°	25°	30°	35°	40°	45°	50°	55°	60°	65°	70°	75°	80°	85°	90°
Cos	90°	85°	80°	75°	70°	65°	60°	55°	50°	45°	40°	35°	30°	25°	20°	15°	10°	5°	0°
	.000	.087	.174	.259	.342	.423	.500	.574	.643	.707	.766	.819	.866	.906	.940	.966	.985	.996	1.000

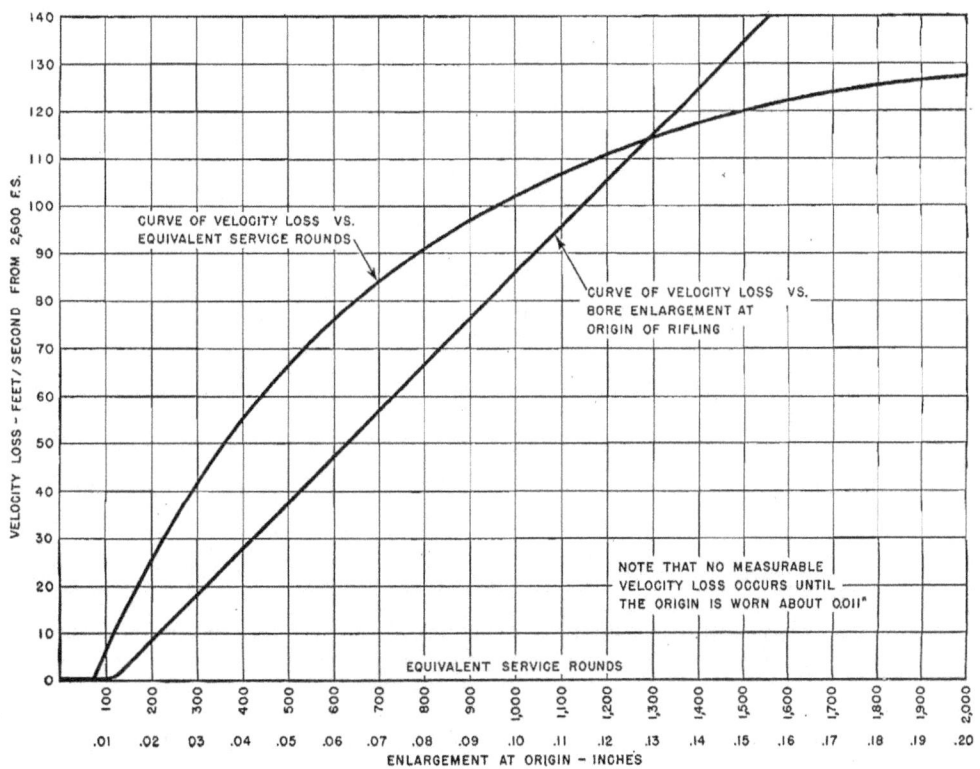

5"/38 CALIBER EROSION DATA
GUNS MK. 12 AND MODS.
VELOCITY LOSS IN FEET PER SECOND
SERVICE CHARGE WITH AA COM. PROJECTILES
MK. 31 AND MODS.

5-INCH RANGE TABLE FROM O.P. 551

PLATE 1

INSTRUCTIONS

Align T and B to get point on support D.
Align D with R to get M, multiplier for column 12.
Align M and R_1 to get error in yards due to change in density of air.

EXAMPLE:

  Given—Barometer ....inches = 30.3
     Temperature ....degrees F = 40°
     Range ....yards = 10,000
  Result—Error ....yards = —310

For unusual conditions, when the result is beyond the Error Scale, complete the solution by using the multiplier for column 12. The product of this multiplier and column 12, with the proper sign, is the error in yards.

NOTE: The best recent estimate of ballistic density to different altitudes is in very close, but not in exact agreement with standard density when surface conditions are standard. When surface density is not standard, the disagreement is usually greater and is a function of surface density and maximum ordinate. In O.P. 561–I of May, 1923, pp. 98-114, corrections for density are discussed. The use of Plate 1 will not give agreement with results obtained from column 12 and surface observations only, but should be a more accurate figure, in that it takes into account the ratio between mean measured and standard density for the actual maximum ordinate obtained.

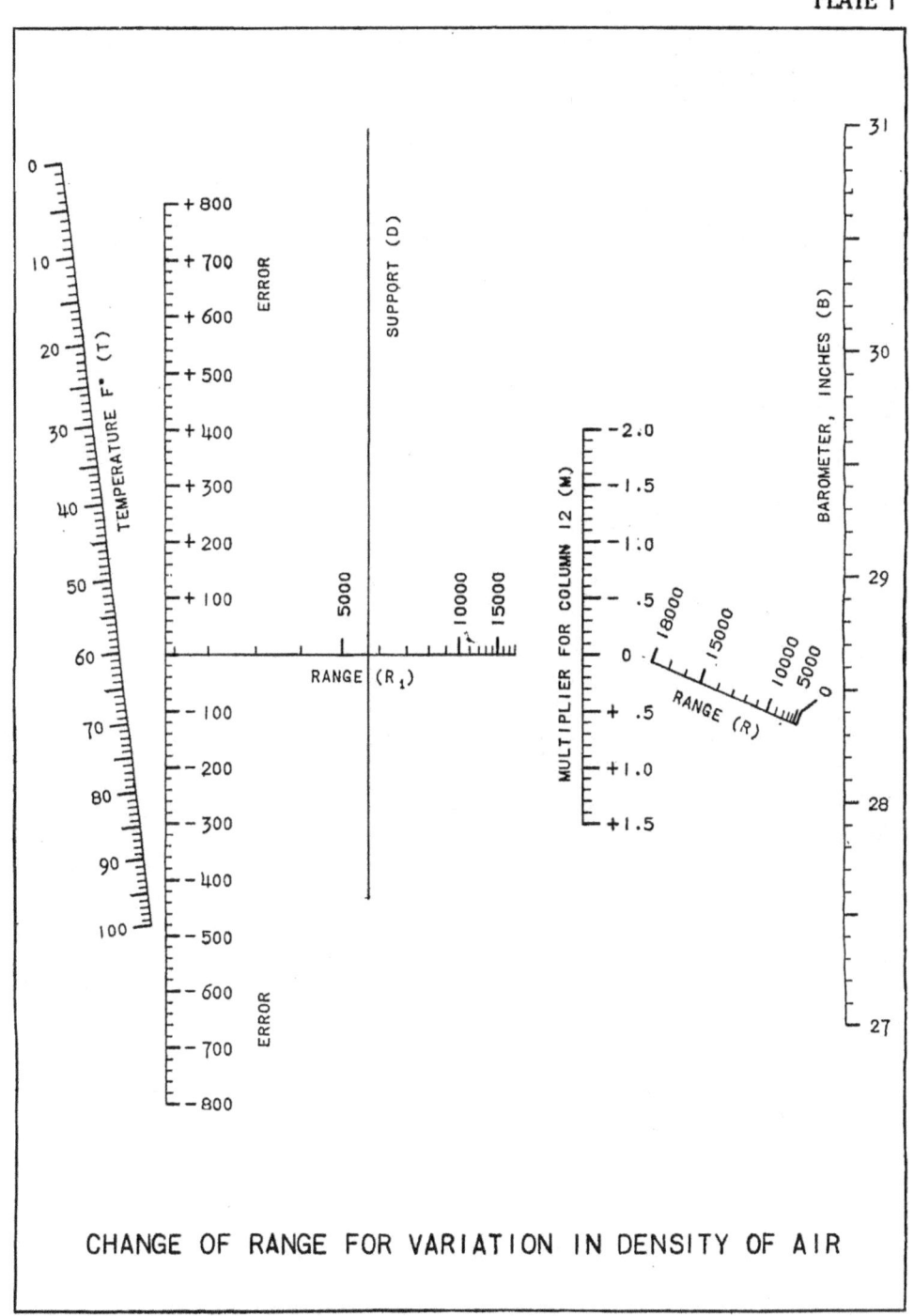

PLATE 1

CHANGE OF RANGE FOR VARIATION IN DENSITY OF AIR

RANGE TABLE FOR 5-INCH GUN

INITIAL VELOCITY = 2600 F.S. WEIGHT OF PROJECTILE = 54 POUNDS. LENGTH OF PROJECTILE = 4.15 CALIBERS. RADIUS OF OGIVE = 5.25 CALIBERS

1	2	2a	2b	3	4	5	6	7	8	10	11	12	13	14	15	16	17	18	19
Range	Angle of elevation		Increase in angle of elevation for 100 yards increase in range	Angle of fall	Time of flight	Striking velocity	Drift	Danger space for a target 20 feet high	Maximum ordinate	Change of range for variation of +10 feet per second initial velocity	Change of range for variation of −1 pound in weight of projectile	Change of range for variation in density of air of −10 per cent	Change of range for wind component in plane of fire of 10 knots	Change of range for motion of gun in plane of fire of 10 knots	Change of range for motion of target in plane of fire of 10 knots	Deviation for lateral wind component of 10 knots	Deviation for lateral motion of gun perpendicular to line of fire, speed of 10 knots	Deviation for lateral motion of target perpendicular to line of fire, speed of 10 knots	Change in height of impact for variation of 100 yards in sight bar
Yards	° ′	Minutes	Min.	° ′	Seconds	F.S.	Yards	Yards	Feet	Yards	Yards	Yards	Yards	Yards	Yards	Yards	Yards	Yards	Feet
1000	26.2	26.2	2.8	28	1.21	2365	0.2	1000	6	8	11	6	—	7	7	0.2	6.5	6.8	2.4
1100	29.0	29.0	2.8	31	1.34	2342	0.3	1100	7	9	12	7	—	7	8	0.3	7.2	7.5	2.7
1200	31.8	31.8	2.9	34	1.47	2319	0.3	1200	9	10	13	8	—	8	8	0.4	7.8	8.2	3.0
1300	34.7	34.7	2.9	37	1.60	2296	0.4	1300	11	10	13	10	—	9	9	0.5	8.4	8.9	3.2
1400	37.6	37.6	3.0	40	1.73	2273	0.4	1400	13	11	14	12	—	9	10	0.6	9.1	9.7	3.5
1500	40.6	40.6	3.0	44	1.86	2251	0.5	1500	15	12	15	14	—	10	10	0.7	9.7	10.4	3.8
1600	43.6	43.6	3.0	48	1.99	2229	0.6	1600	17	12	16	16	—	10	11	0.8	10.4	11.2	4.2
1700	46.6	46.6	3.1	52	2.13	2207	0.7	1700	19	13	16	18	—	11	12	0.9	11.1	11.7	4.5
1800	49.7	49.7	3.1	56	2.27	2185	0.8	717	21	14	17	20	—	12	13	1.1	11.7	12.8	4.8
1900	52.8	52.8	3.2	1 00	2.41	2163	0.9	569	23	15	17	22	—	12	14	1.2	12.4	13.6	5.2
2000	56.0	56.0	3.2	1 04	2.55	2141	1.0	491	26	15	18	24	1	13	14	1.3	13.1	14.4	5.5
2100	1 59.2	59.2	3.3	1 08	2.69	2119	1.1	436	29	16	18	26	1	14	15	1.4	13.8	15.2	5.9
2200	1 02.5	62.5	3.3	1 12	2.83	2097	1.2	397	32	16	19	29	1	14	16	1.6	14.4	16.0	6.2
2300	1 05.8	65.8	3.4	1 16	2.98	2075	1.4	364	35	17	19	32	2	15	17	1.8	15.0	16.8	6.6
2400	1 09.2	69.2	3.4	1 20	3.13	2054	1.5	335	39	18	20	35	2	15	18	2.0	15.6	17.6	7.0
2500	1 12.6	72.6	3.5	1 25	3.28	2033	1.7	310	43	18	20	38	2	16	18	2.1	16.3	18.5	7.4
2600	1 16.1	76.1	3.5	1 30	3.43	2012	1.8	289	47	19	20	41	3	16	19	2.3	17.0	19.3	7.8
2700	1 19.6	79.6	3.6	1 35	3.58	1991	2.0	270	51	19	20	44	3	17	20	2.5	17.7	20.2	8.3
2800	1 23.2	83.2	3.6	1 40	3.73	1970	2.2	253	55	20	21	47	3	18	21	2.7	18.3	21.0	8.7
2900	1 26.8	86.8	3.7	1 45	3.89	1949	2.4	238	60	20	21	50	4	18	22	2.9	19.0	21.9	9.1
3000	1 30.5	90.5	3.7	1 50	4.05	1928	2.6	225	65	21	21	53	4	19	23	3.2	19.6	22.8	9.6

RANGE TABLE FOR 5-INCH GUN

INITIAL VELOCITY = 2600 F.S. WEIGHT OF PROJECTILE = 54 POUNDS. LENGTH OF PROJECTILE = 4.15 CALIBERS. RADIUS OF OGIVE = 5.25 CALIBERS

1	2	2a	2b	3	4	5	6	7	8	10	11	12	13	14	15	16	17	18	19
Range	Angle of elevation		Increase in angle of elevation for 100 yards increase in range	Angle of fall	Time of flight	Striking velocity	Drift	Danger space for a target 20 feet high	Maximum ordinate	Change of range for variation of +10 feet per second initial velocity	Change of range for variation of −1 pound in weight of projectile	Change of range for variation in density of air of −10 per cent	Change of range for wind component in plane of fire of 10 knots	Change of range for motion of gun in plane of fire of 10 knots	Change of range for motion of target in plane of fire of 10 knots	Deviation for lateral wind component of 10 knots	Deviation for lateral motion of gun perpendicular to line of fire, speed of 10 knots	Deviation for lateral motion of target, perpendicular to line of fire, speed of 10 knots	Change in height of impact for variation of 100 yards in sight bar
Yards	° '	Minutes	Min.	° '	Seconds	F.S.	Yards	Yards	Feet	Yards	Yards	Yards	Yards	Yards	Yards	Yards	Yards	Yards	Feet
3000	1 30.5	90.5	3.7	1 50	4.05	1928	2.6	225	65	21	21	53	4	19	23	3	20	23	10
3100	1 34.2	94.2	3.8	1 55	4.21	1907	2.8	213	70	22	21	57	4	19	24	3	21	24	10
3200	1 38.0	98.0	3.9	2 01	4.37	1886	3.0	202	76	22	21	61	5	20	25	4	21	25	11
3300	1 41.9	101.9	3.9	2 07	4.53	1865	3.2	192	82	23	21	65	5	20	26	4	22	26	11
3400	1 45.8	105.8	4.0	2 13	4.69	1845	3.4	183	88	23	21	69	5	21	27	4	22	27	12
3500	1 49.8	109.8	4.0	2 19	4.86	1825	3.7	175	94	24	21	73	6	21	27	4	23	27	12
3600	1 53.8	113.8	4.1	2 25	5.03	1805	3.9	167	100	24	21	77	6	22	28	5	23	28	13
3700	1 57.9	117.9	4.2	2 31	5.20	1785	4.2	160	107	25	21	81	6	22	29	5	24	29	13
3800	2 02.1	122.1	4.3	2 37	5.37	1765	4.5	153	114	25	21	85	7	23	30	5	25	30	14
3900	2 06.3	126.3	4.3	2 44	5.54	1745	4.8	146	122	26	21	90	7	23	31	6	25	31	14
4000	2 10.6	130.6	4.4	2 51	5.71	1725	5.1	140	130	26	21	95	8	24	32	6	26	32	15
4100	2 15.0	135.0	4.4	2 58	5.89	1706	5.4	134	138	26	21	100	8	24	33	6	27	33	16
4200	2 19.4	139.4	4.5	3 05	6.07	1687	5.7	128	147	27	21	105	9	25	34	7	27	34	16
4300	2 23.9	143.9	4.6	3 13	6.25	1668	6.1	122	156	27	20	110	10	25	35	7	28	35	17
4400	2 28.5	148.5	4.6	3 21	6.43	1649	6.5	117	165	28	20	115	10	26	36	7	29	36	17
4500	2 33.1	153.1	4.7	3 29	6.61	1630	6.9	112	175	28	20	120	11	26	37	8	29	37	18
4600	2 37.8	157.8	4.8	3 37	6.80	1611	7.3	108	185	28	20	125	11	27	39	8	30	39	19
4700	2 42.6	162.6	4.9	3 45	6.99	1593	7.7	104	195	29	19	130	12	27	40	9	31	40	20
4800	2 47.5	167.5	4.9	3 54	7.18	1575	8.1	100	206	29	19	135	13	28	41	9	32	41	20
4900	2 52.5	172.5	5.0	4 03	7.37	1557	8.5	96	218	30	18	140	13	28	42	9	32	42	21
5000	2 57.5	177.5	5.1	4 12	7.56	1539	9.0	92	230	30	18	146	14	29	43	10	33	43	22

547

RESTRICTED

RANGE TABLE FOR 5-INCH GUN

INITIAL VELOCITY = 2600 F.S. WEIGHT OF PROJECTILE = 54 POUNDS. LENGTH OF PROJECTILE = 4.15 CALIBERS. RADIUS OF OGIVE = 5.25 CALIBERS

1	2	2a	2b	3	4	5	6	7	8	10	11	12	13	14	15	16	17	18	19
Range	Angle of elevation	Angle of elevation	Increase in angle of elevation for 100 yards increase in range	Angle of fall	Time of flight	Striking velocity	Drift	Danger space for a target 20 feet high	Maximum ordinate	Change of range for variation of +10 feet per second initial velocity	Change of range for variation of −1 pound in weight of projectile	Change of range for variation in density of air of −10 per cent	Change of range for wind component in plane of fire of 10 knots	Change of range for motion of gun in plane of fire of 10 knots	Change of range for motion of target in plane of fire of 10 knots	Deviation for lateral wind component of 10 knots	Deviation for lateral motion of gun perpendicular to line of fire, speed of 10 knots	Deviation for lateral motion of target perpendicular to line of fire, speed of 10 knots	Change in height of impact for variation of 100 yards in sight bar
Yards	° '	Minutes	Min.	° '	Seconds	F.S.	Yards	Yards	Feet	Yards	Yards	Yards	Yards	Yards	Yards	Yards	Yards	Yards	Feet
5000	2 57.5	177.5	5.1	4 12	7.56	1539	9.0	92	230	30	18	146	14	29	43	10	33	43	22
5100	3 02.6	182.6	5.2	4 21	7.76	1521	9.5	88	243	30	18	152	15	29	44	10	34	44	23
5200	3 07.8	187.8	5.3	4 31	7.96	1504	10.0	85	256	31	17	158	15	30	45	11	34	45	24
5300	3 13.1	193.1	5.4	4 41	8.16	1487	10.5	82	269	31	17	164	16	30	46	11	35	46	24
5400	3 18.5	198.5	5.5	4 51	8.37	1470	11.0	79	283	32	16	170	17	31	47	12	35	47	25
5500	3 24.0	204.0	5.7	5 01	8.58	1453	11.6	76	297	32	16	176	17	31	48	12	36	48	26
5600	3 29.7	209.7	5.8	5 12	8.79	1436	12.2	74	311	33	15	183	18	32	49	13	36	49	27
5700	3 35.5	215.5	5.8	5 23	9.01	1420	12.8	71	326	33	15	189	19	32	50	13	37	50	28
5800	3 41.3	221.3	5.9	5 34	9.23	1404	13.4	69	342	33	14	195	20	33	52	14	38	52	29
5900	3 47.2	227.2	6.0	5 46	9.45	1388	14.0	66	359	34	14	202	21	33	53	15	38	53	30
6000	3 53.2	233.2	6.1	5 58	9.67	1372	14.7	64	377	34	13	208	22	33	54	15	39	54	31
6100	3 59.3	239.3	6.3	6 10	9.90	1357	15.4	62	395	34	12	214	22	34	55	16	39	55	32
6200	4 05.6	245.6	6.4	6 23	10.13	1342	16.1	60	414	35	12	221	23	34	56	16	40	56	33
6300	4 12.0	252.0	6.5	6 36	10.36	1327	16.8	58	434	35	11	227	24	35	58	17	41	58	34
6400	4 18.5	258.5	6.6	6 49	10.59	1312	17.6	56	454	35	10	234	25	35	59	18	41	59	35
6500	4 25.1	265.1	6.8	7 02	10.83	1298	18.4	54	475	36	9	241	26	35	61	19	42	61	36
6600	4 31.9	271.9	6.9	7 16	11.07	1284	19.2	52	496	36	8	248	27	36	62	19	43	62	37
6700	4 38.8	278.8	7.0	7 30	11.31	1270	20.0	51	518	36	8	254	28	36	64	20	44	64	39
6800	4 45.8	285.8	7.1	7 44	11.56	1257	20.9	49	541	36	7	261	29	36	65	21	44	65	40
6900	4 52.9	292.9	7.2	7 59	11.81	1244	21.8	48	565	37	6	268	30	37	67	21	45	67	41
7000	5 00.1	300.1	7.4	8 14	12.06	1232	22.8	46	590	37	5	275	31	37	68	22	46	68	43

RANGE TABLE FOR 5-INCH GUN

INITIAL VELOCITY = 2600 F.S. WEIGHT OF PROJECTILE = 54 POUNDS. LENGTH OF PROJECTILE = 4.15 CALIBERS. RADIUS OF OGIVE = 5.25 CALIBERS.

1	2	2a	2b	3	4	5	6	7	8	10	11	12	13	14	15	16	17	18	19
Range	Angle of elevation		Increase in angle of elevation for 100 yards increase in range	Angle of fall	Time of flight	Striking velocity	Drift	Danger space for a target 20 feet high	Maximum ordinate	Change of range for variation of +10 feet per second initial velocity	Change of range for variation of −1 pound in weight of projectile	Change of range for variation in density of air of −10 per cent	Change of range for wind component in plane of fire of 10 knots	Change of range for motion of gun in plane of fire of 10 knots	Change of range for motion of target in plane of fire of 10 knots	Deviation for lateral wind component of 10 knots	Deviation for lateral motion of gun perpendicular to line of fire, speed of 10 knots	Deviation for lateral motion of target, perpendicular to line of fire, speed of 10 knots	Change in height of impact for variation of 100 yards in sight bar
Yards	°	Minutes	Min.	°	'	Seconds	F.S.	Yards	Yards	Feet	Yards	Yards	Yards	Yards	Yards	Yards	Yards	Yards	Yards	Feet
7000	5	00.1	300.1	7.4	8	14	12.06	1232	22.8	46	590	37	5	275	31	37	68	22	46	46	43
7100	5	07.5	307.5	7.5	8	30	12.31	1220	23.8	45	615	37	5	282	32	37	69	23	46	46	44
7200	5	15.0	315.0	7.6	8	46	12.57	1208	24.8	43	640	38	4	289	33	37	71	24	47	47	46
7300	5	22.6	322.6	7.7	9	02	12.83	1197	25.8	42	670	38	3	296	34	38	72	24	48	48	47
7400	5	30.4	330.4	7.8	9	19	13.09	1186	26.9	41	700	38	2	303	35	38	74	25	49	49	49
7500	5	38.4	338.4	8.1	9	36	13.36	1175	28.0	40	730	39	1	310	36	38	75	26	49	49	50
7600	5	46.5	346.5	8.2	9	53	13.63	1165	29.1	39	760	39	0	317	38	38	77	27	50	50	52
7700	5	54.7	354.7	8.4	10	11	13.90	1155	30.3	37	790	39	−1	324	39	38	78	27	51	51	54
7800	6	03.1	363.1	8.5	10	29	14.17	1146	31.5	36	820	40	−2	331	40	39	80	28	52	52	55
7900	6	11.6	371.6	8.6	10	47	14.45	1137	32.7	35	855	40	−3	338	42	39	81	29	52	52	57
8000	6	20.2	380.2	8.8	11	06	14.73	1128	34.0	34	890	40	−4	345	43	39	83	30	53	53	59
8100	6	29.0	389.0	9.0	11	25	15.01	1119	35.3	33	925	40	−5	352	44	39	85	31	54	54	61
8200	6	38.0	398.0	9.2	11	44	15.29	1111	36.6	32	960	40	−6	359	46	39	86	32	54	54	63
8300	6	47.2	407.2	9.3	12	04	15.58	1103	38.0	31	1000	41	−7	366	47	39	88	33	55	55	64
8400	6	56.5	416.5	9.5	12	24	15.87	1095	39.4	30	1040	41	−8	373	48	40	89	34	55	55	66
8500	7	06.0	426.0	9.6	12	44	16.16	1088	40.8	30	1080	41	−8	380	50	40	91	35	56	56	68
8600	7	15.6	435.6	9.8	13	05	16.45	1081	42.3	29	1120	41	−9	386	51	40	92	36	56	56	70
8700	7	25.4	445.4	10.0	13	26	16.75	1074	43.8	28	1160	42	−10	393	53	40	94	37	57	57	72
8800	7	35.4	455.4	10.1	13	47	17.05	1067	45.2	27	1205	42	−11	400	54	40	96	38	58	58	74
8900	7	45.5	465.5	10.3	14	09	17.35	1060	47.0	26	1250	42	−12	406	56	40	97	39	58	58	76
9000	7	55.8	475.8	10.4	14	31	17.65	1054	48.6	26	1295	42	−13	413	58	40	99	40	59	59	78

549   RESTRICTED

RANGE TABLE FOR 5-INCH GUN

INITIAL VELOCITY = 2600 F.S. WEIGHT OF PROJECTILE = 54 POUNDS. LENGTH OF PROJECTILE = 4.15 CALIBERS. RADIUS OF OGIVE = 5.25 CALIBERS.

1	2	2a	2b	3	4	5	6	7	8	10	11	12	13	14	15	16	17	18	19
Range	Angle of elevation		Increase in angle of elevation for 100 yards increase in range	Angle of fall	Time of flight	Striking velocity	Drift	Danger space for a target 20 feet high	Maximum ordinate	Change of range for variation of +10 feet per second initial velocity	Change of range for variation of −1 pound in weight of projectile	Change of range for variation in density of air of −10 per cent	Change of range for wind component in plane of fire of 10 knots	Change of range for motion of gun in plane of fire of 10 knots	Change of range for motion of target in plane of fire of 10 knots	Deviation for lateral wind component of 10 knots	Deviation for lateral motion of gun perpendicular to line of fire, speed of 10 knots	Deviation for lateral motion of target perpendicular to line of fire, speed of 10 knots	Change in height of impact for variation of 100 yards in sight bar
Yards	° ′	Minutes	Min.	° ′	Seconds	F. S.	Yards	Yards	Feet	Yards	Yards	Yards	Yards	Yards	Yards	Yards	Yards	Yards	Feet
9000	7 55.8	475.8	10.4	14 31	17.65	1054	48.6	26	1295	42	−13	413	58	40	99	40	59	99	78
9100	8 06.2	486.2	10.6	14 53	17.96	1048	50.3	25	1340	42	−14	420	60	40	101	41	60	101	80
9200	8 16.8	496.8	10.8	15 15	18.27	1042	52.0	24	1390	42	−15	426	61	40	102	42	60	102	82
9300	8 27.6	507.6	10.9	15 37	18.58	1036	53.8	24	1440	42	−16	433	63	40	104	43	61	104	84
9400	8 38.5	518.5	11.1	16 00	18.89	1031	55.6	23	1490	42	−17	440	65	40	106	44	62	106	86
9500	8 49.6	529.6	11.2	16 23	19.20	1026	57.5	22	1540	43	−18	446	67	40	108	45	63	108	88
9600	9 00.8	540.8	11.4	16 46	19.52	1021	59.4	22	1595	43	−19	453	69	40	110	46	64	110	90
9700	9 12.2	552.2	11.6	17 09	19.84	1016	61.3	21	1650	43	−20	459	70	40	111	47	64	111	92
9800	9 23.8	563.8	11.7	17 32	20.16	1011	63.3	21	1705	43	−21	465	72	40	113	48	65	113	95
9900	9 35.5	575.5	11.9	17 56	20.48	1007	65.3	20	1765	43	−22	472	74	40	115	50	65	115	97
10000	9 47.4	587.4	12.0	18 20	20.81	1003	67.4	20	1825	43	−23	478	76	40	117	51	66	117	99
10100	9 59.4	599.4	12.2	18 44	21.14	999	69.5	20	1885	43	−24	484	78	40	119	52	67	119	101
10200	10 11.6	611.6	12.4	19 08	21.47	995	71.7	19	1950	43	−25	491	80	40	121	53	68	121	103
10300	10 24.0	624.0	12.5	19 32	21.80	991	73.9	19	2015	43	−26	497	82	40	123	54	69	123	106
10400	10 36.5	636.5	12.7	19 56	22.13	987	76.2	18	2080	44	−27	503	83	40	124	55	69	124	108
10500	10 49.2	649.2	12.9	20 20	22.47	984	78.5	18	2145	44	−28	510	85	40	126	56	70	126	110
10600	11 02.1	662.1	13.0	20 44	22.81	981	80.8	18	2210	44	−29	516	87	40	128	58	70	128	113
10700	11 15.1	675.1	13.2	21 09	23.15	978	83.2	17	2280	44	−30	522	89	40	130	59	71	130	115
10800	11 28.3	688.3	13.3	21 34	23.49	975	85.6	17	2350	45	−31	529	91	40	132	60	72	132	118
10900	11 41.6	701.6	13.5	21 59	23.83	972	88.1	17	2425	45	−32	535	93	40	134	61	73	134	120
11000	11 55.1	715.1	13.7	22 24	24.18	969	90.7	16	2500	45	−33	541	95	40	136	62	74	136	123

550

RANGE TABLE FOR 5-INCH GUN

INITIAL VELOCITY = 2600 F.S. WEIGHT OF PROJECTILE = 54 POUNDS. LENGTH OF PROJECTILE = 4.15 CALIBERS. RADIUS OF OGIVE = 5.25 CALIBERS.

1	2		2a	2b	3		4	5	6	7	8	10	11	12	13	14	15	16	17	18	19
Range	Angle of elevation			Increase in angle of elevation for 100 yards increase in range	Angle of fall		Time of flight	Striking velocity	Drift	Danger space for a target 20 feet high	Maximum ordinate	Change of range for variation of + 10 feet per second initial velocity	Change of range for variation of – 1 pound in weight of projectile	Change of range for variation in density of air of –10 per cent	Change of range for wind component in plane of fire of 10 knots	Change of range for motion of gun in plane of fire of 10 knots	Change of range for motion of target in plane of fire of 10 knots	Deviation for lateral wind component of 10 knots	Deviation for lateral motion of gun perpendicular to line of fire, speed of 10 knots	Deviation for lateral motion of target, perpendicular to line of fire, speed of 10 knots	Change in height of impact for variation of 100 yards in sight bar
Yards	°	′	Minutes	Min.	°	′	Seconds	F. S.	Yards	Yards	Feet	Yards	Yards	Yards	Yards	Yards	Yards	Yards	Yards	Yards	Feet
11000	11	55.1	715.1	13.7	22	24	24.18	969	90.7	16	2500	45	–33	541	95	40	136	62	74	136	123
11100	12	08.8	728.8	13.8	22	49	24.53	967	93.3	16	2575	45	–34	547	97	40	138	64	74	138	126
11200	12	22.6	742.6	14.0	23	14	24.88	965	96.0	15	2655	45	–35	553	99	40	140	65	75	140	128
11300	12	36.6	756.6	14.2	23	39	25.23	963	98.7	15	2735	45	–36	560	101	40	142	66	76	142	131
11400	12	50.8	770.8	14.4	24	04	25.59	961	101.5	15	2815	46	–37	566	103	40	144	68	76	144	133
11500	13	05.2	785.2	14.5	24	29	25.95	959	104.3	15	2900	46	–38	572	105	40	146	69	77	146	136
11600	13	19.7	799.7	14.7	24	54	26.31	957	107.2	14	2985	46	–39	578	107	40	148	70	78	148	138
11700	13	34.4	814.4	14.9	25	20	26.67	955	110.1	14	3075	46	–40	584	109	40	150	72	78	150	141
11800	13	49.3	829.3	15.0	25	46	27.04	953	113.1	14	3165	46	–41	591	112	40	152	73	79	152	144
11900	14	04.3	844.3	15.2	26	12	27.41	951	116.2	13	3255	46	–42	597	114	40	154	74	80	154	147
12000	14	19.5	859.5	15.4	26	38	27.78	950	119.3	13	3350	46	–43	603	116	40	156	76	80	156	150
12100	14	34.9	874.9	15.6	27	04	28.15	949	122.5	13	3445	46	–44	609	118	40	158	77	81	158	153
12200	14	50.5	890.5	15.7	27	30	28.53	948	125.8	13	3545	46	–45	615	120	40	160	78	82	160	156
12300	15	06.2	906.2	15.9	27	56	28.91	947	129.1	12	3645	46	–46	622	123	40	162	80	82	162	159
12400	15	22.1	922.1	16.1	28	22	29.29	946	132.5	12	3745	46	–47	628	125	40	164	81	83	164	162
12500	15	38.2	938.2	16.3	28	48	29.68	946	135.9	12	3850	46	–47	634	127	40	166	82	84	166	164
12600	15	54.5	954.5	16.5	29	14	30.07	945	139.4	12	3955	46	–48	640	130	40	169	84	85	169	167
12700	16	11.0	971.0	16.7	29	40	30.46	944	143.0	11	4060	46	–49	646	132	40	171	85	86	171	171
12800	16	27.7	987.7	16.9	30	06	30.85	943	146.7	11	4170	47	–50	653	134	40	173	86	87	173	174
12900	16	44.5	1004.5	17.1	30	32	31.25	942	150.4	11	4280	47	–51	659	137	40	176	88	88	176	177
13000	17	01.6	1021.6	17.3	30	58	31.65	942	154.2	11	4395	47	–52	665	139	40	178	89	89	178	180

RANGE TABLE FOR 5-INCH GUN

INITIAL VELOCITY = 2600 F.S. WEIGHT OF PROJECTILE = 54 POUNDS. LENGTH OF PROJECTILE = 4.15 CALIBERS. RADIUS OF OGIVE = 5.25 CALIBERS.

1	2	2a	2b	3	4	5	6	7	8	10	11	12	13	14	15	16	17	18	19
Range	Angle of elevation		Increase in angle of elevation for 100 yards increase in range	Angle of fall	Time of flight	Striking velocity	Drift	Danger space for a target 20 feet high	Maximum ordinate	Change of range for variation of + 10 feet per second initial velocity	Change of range for variation of − 1 pound in weight of projectile	Change of range for variation in density of air of − 10 per cent	Change of range for wind component in plane of fire of 10 knots	Change of range for motion of gun in plane of fire of 10 knots	Change of range for motion of target in plane of fire of 10 knots	Deviation for lateral wind component of 10 knots	Deviation for lateral motion of gun perpendicular to line of fire, speed of 10 knots	Deviation for lateral motion of target perpendicular to line of fire, speed of 10 knots	Change in height of impact for variation of 100 yards in sight bar
Yards	° ′	Minutes	Min.	° ′	Seconds	F. S.	Yards	Yards	Feet	Yards	Yards	Yards	Yards	Yards	Yards	Yards	Yards	Yards	Feet
13000	17 01.6	1021.6	17.3	30 58	31.65	942	154.2	11	4395	47	−52	665	139	40	178	89	89	178	180
13100	17 18.9	1038.9	17.5	31 24	32.05	942	158.1	11	4510	47	−53	671	141	40	180	90	90	180	183
13200	17 36.4	1056.4	17.7	31 50	32.46	941	162.0	11	4625	47	−54	677	144	40	183	92	91	183	186
13300	17 54.1	1074.1	17.7	32 16	32.87	941	166.0	10	4745	47	−55	684	146	40	185	93	92	185	189
13400	18 12.0	1092.0	18.1	32 42	33.28	941	170.1	10	4865	48	−56	690	148	40	187	95	93	187	193
13500	18 30.1	1110.1	18.3	33 08	33.69	941	174.3	10	4990	48	−57	696	151	40	190	96	94	190	196
13600	18 48.4	1128.4	18.5	33 35	34.11	941	178.6	10	5115	48	−58	703	153	40	192	98	95	192	199
13700	19 06.9	1146.9	18.7	34 02	34.53	941	183.2	9	5245	48	−59	709	156	40	195	99	96	195	203
13800	19 25.6	1165.6	19.0	34 29	34.95	941	187.4	9	5375	49	−60	715	158	40	197	101	97	197	206
13900	19 44.6	1184.6	19.2	34 56	35.38	942	191.9	9	5510	49	−61	722	161	40	200	102	98	200	209
14000	20 03.8	1203.8	19.4	35 23	35.81	942	196.5	9	5645	49	−62	728	163	40	202	104	98	202	213
14100	20 23.2	1223.2	19.7	35 50	36.24	942	201.2	9	5785	49	−63	734	165	40	204	105	99	204	217
14200	20 42.9	1242.9	20.0	36 17	36.68	943	206.0	9	5925	49	−64	741	168	40	207	107	100	207	220
14300	21 02.9	1262.9	20.2	36 44	37.12	943	210.9	9	6070	49	−65	747	170	40	210	108	101	210	223
14400	21 23.1	1283.1	20.5	37 11	37.57	944	215.9	8	6215	49	−66	753	173	40	212	110	102	212	227
14500	21 43.6	1303.6	20.8	37 38	38.02	945	221.0	8	6365	50	−66	760	175	40	214	112	102	214	231
14600	22 04.4	1324.4	21.0	38 05	38.48	946	226.2	8	6515	50	−67	766	177	40	216	113	103	216	234
14700	22 25.4	1345.4	21.3	38 32	38.94	947	231.5	8	6670	50	−68	772	180	40	219	115	104	219	238
14800	22 46.7	1366.7	21.5	38 59	39.40	948	236.9	8	6830	50	−69	779	183	40	222	116	105	222	242
14900	23 08.2	1388.2	21.8	39 26	39.87	949	242.4	8	6995	50	−70	785	185	41	224	118	106	224	246
15000	23 30.0	1410.0	22.0	39 53	40.34	950	248.0	8	7165	50	−71	792	188	41	227	120	107	227	250

RANGE TABLE FOR 5-INCH GUN

INITIAL VELOCITY = 2600 F.S. WEIGHT OF PROJECTILE = 54 POUNDS. LENGTH OF PROJECTILE = 4.15 CALIBERS. RADIUS OF OGIVE = 5.25 CALIBERS.

1	2		2a	2b	3		4	5	6	7	8	10	11	12	13	14	15	16	17	18	19
Range	Angle of elevation			Increase in angle of elevation for 100 yards increase in range	Angle of fall		Time of flight	Striking velocity	Drift	Danger space for a target 20 feet high	Maximum ordinate	Change of range for variation of +10 feet per second initial velocity	Change of range for variation of −1 pound in weight of projectile	Change of range for variation in density of air of −10 per cent	Change of range for wind component in plane of fire of 10 knots	Change of range for motion of gun in plane of fire of 10 knots	Change of range for motion of target in plane of fire of 10 knots	Deviation for lateral wind component of 10 knots	Deviation for lateral motion of gun perpendicular to line of fire, speed of 10 knots	Deviation for lateral motion of target perpendicular to line of fire, speed of 10 knots	Change in height of impact for variation of 100 yards in sight bar
Yards	°	′	Minutes	Min.	°	′	Seconds	F. S.	Yards	Yards	Feet	Yards	Yards	Yards	Yards	Yards	Yards	Yards	Yards	Yards	Feet
15000	23	30	1410	22	39	53	40.34	950	248	8	7165	50	−71	792	188	41	227	120	107	227	250
15100	23	52	1432	22	40	21	40.82	951	254	8	7340	50	−72	799	190	41	230	121	108	230	254
15200	24	14	1454	23	40	49	41.31	952	260	8	7515	50	−73	805	192	41	232	123	109	232	258
15300	24	37	1477	23	41	17	41.80	953	266	8	7695	51	−74	812	195	41	235	125	110	235	263
15400	25	00	1500	23	41	45	42.30	954	272	8	7880	51	−75	819	198	41	238	127	111	238	267
15500	25	23	1523	24	42	13	42.80	955	278	8	8070	51	−76	826	201	41	241	129	112	241	271
15600	25	47	1547	24	42	41	43.31	956	285	7	8260	51	−77	833	203	41	244	130	114	244	275
15700	26	11	1571	25	43	09	43.83	957	292	7	8455	51	−78	840	206	41	247	132	115	247	280
15800	26	36	1596	26	43	38	44.36	959	299	7	8655	52	−79	847	208	41	250	134	116	250	285
15900	27	02	1622	26	44	07	44.90	961	306	7	8860	52	−80	854	211	41	253	136	117	253	290
16000	27	28	1648	27	44	36	45.44	963	313	7	9070	52	−81	861	214	42	256	138	118	256	295
16100	27	55	1675	27	45	05	45.99	965	321	7	9285	52	−82	868	217	42	259	140	119	259	300
16200	28	22	1702	28	45	35	46.55	967	329	7	9505	52	−83	876	220	42	262	142	120	262	306
16300	28	50	1730	29	46	05	47.12	969	337	7	9735	52	−84	883	223	42	265	144	121	265	311
16400	29	19	1759	29	46	35	47.71	971	345	7	9975	53	−85	891	226	42	268	146	122	268	317
16500	29	48	1788	30	47	05	48.31	973	353	7	10220	53	−86	898	229	42	272	148	124	272	322
16600	30	18	1818	31	47	35	48.92	975	362	6	10475	53	−87	906	232	43	275	150	125	275	328
16700	30	49	1849	31	48	06	49.54	977	371	6	10740	53	−88	914	235	43	278	152	126	278	334
16800	31	20	1880	32	48	37	50.18	979	380	6	11015	54	−89	922	239	43	282	154	128	282	340
16900	31	52	1912	33	49	09	50.83	981	390	6	11300	54	−91	930	242	43	286	156	130	286	346
17000	32	25	1945	35	49	41	51.50	984	400	6	11595	54	−92	938	246	44	290	159	131	290	353

553   RESTRICTED

RANGE TABLE FOR 5-INCH GUN

INITIAL VELOCITY = 2600 F.S. WEIGHT OF PROJECTILE = 54 POUNDS. LENGTH OF PROJECTILE = 4.15 CALIBERS. RADIUS OF OGIVE = 5.25 CALIBERS

1	2		2a	2b	3		4	5	6	7	8	10	11	12	13	14	15	16	17	18	19
Range	Angle of elevation			Increase in angle of elevation for 100 yards increase in range	Angle of fall		Time of flight	Striking velocity	Drift	Danger space for a target 20 feet high	Maximum ordinate	Change of range for variation of + 10 feet per second initial velocity	Change of range for variation of − 1 pound in weight of projectile	Change of range for variation in density of air of − 10 per cent	Change of range for wind component in plane of fire of 10 knots	Change of range for motion of gun in plane of fire of 10 knots	Change of range for motion of target in plane of fire of 10 knots	Deviation for lateral wind component of 10 knots	Deviation for lateral motion of gun perpendicular to line of fire, speed of 10 knots	Deviation for lateral motion of target, perpendicular to line of fire, speed of 10 knots	Change in height of impact for variation of 100 yards in sight bar
Yards	°	′	Minutes	Min.	°	′	Seconds	F. S.	Yards	Yards	Feet	Yards	Yards	Yards	Yards	Yards	Yards	Yards	Yards	Yards	Feet
17000	32	25	1945	35	49	41	51.50	984	400	6	11595	54	92	938	246	44	290	159	131	290	353
17100	33	00	1980	36	50	13	52.19	987	410	6	11900	54	93	946	250	44	293	161	132	293	360
17200	33	36	2016	38	50	46	52.90	990	421	6	12220	54	95	955	253	45	298	163	134	298	367
17300	34	14	2054	40	51	20	53.64	993	432	6	12555	55	96	964	257	45	302	166	136	302	374
17400	34	54	2094	42	51	55	54.42	996	444	6	12905	55	98	974	261	46	306	168	138	306	382
17500	35	36	2136	44	52	32	55.25	999	456	6	13275	55	100	984	265	46	311	171	140	311	391
17600	36	20	2180	47	53	11	56.12	1002	469	5	13665	56	102	995	269	47	316	174	142	316	400
17700	37	07	2227	51	53	52	57.04	1006	484	5	14080	56	103	1006	273	48	321	177	144	321	410
17800	37	58	2278	58	54	36	58.03	1010	501	5	14540	56	105	1019	278	49	327	180	147	327	422
17900	38	56	2336	69	55	25	59.13	1014	520	5	15075	56	107	1033	283	51	333	183	150	333	435
18000	40	05	2405	78	56	22	60.40	1019	542	5	15715	57	109	1048	288	53	340	187	153	340	451
18100	41	23	2483	115	57	34	61.95	1024	569	5	16520	57	112	1066	294	56	350	192	158	350	472
18200	43	18	2598		59	22	64.40	1030	613	5	17750	57	116	1088	300	62	363	199	164	363	507

554

RANGE TABLE FOR 5-INCH GUN

INITIAL VELOCITY = 2600 F.S. WEIGHT OF PROJECTILE = 54 POUNDS. LENGTH OF PROJECTILE = 4.15 CALIBERS. RADIUS OF OGIVE = 5.25 CALIBERS

1	2	2a	2b	3	4	5	6	7	8	10	11	12	13	14	15	16	17	18	19
Range	Angle of elevation		Increase in angle of elevation for 100 yards increase in range	Angle of fall	Time of flight	Striking velocity	Drift	Danger space for a target 20 feet high	Maximum ordinate	Change of range for variation of +10 feet per second initial velocity	Change of range for variation of −1 pound in weight of projectile	Change of range for variation in density of air of −10 per cent	Change of range for wind component in plane of fire of 10 knots	Change of range for motion of gun in plane of fire of 10 knots	Change of range for motion of target in plane of fire of 10 knots	Deviation for lateral wind component of 10 knots	Deviation for lateral motion of gun perpendicular to line of fire, speed of 10 knots	Deviation for lateral motion of target perpendicular to line of fire, speed of 10 knots	Change in height of impact for variation of 100 yards in sight bar
Yards	°	′	Minutes	Min.	°	′	Seconds	F.S.	Yards	Yards	Feet	Yards	Yards	Yards	Yards	Yards	Yards	Yards	Yards	Feet
18235	45	10	2710		60	25	66.3	1035	648	4	18780	58	−117	1097	307	67	373	205	168	373	528
18200	47	28	2848	−125	63	05	68.8	1042	695	4	20120	59	−117	1101	314	73	387	213	174	387	536
18100	49	33	2973	−125	63	31	71.0	1047	738	3	21270	59	−116	1103	319	80	399	219	180	399	602
18000	50	52	3052	−79	64	25	72.4	1051	765	3	22050	59	−116	1103	321	86	407	222	185	407	627
17900	51	53	3113	−61	65	07	73.4	1054	786	3	22650	59	−115	1102	323	90	413	225	188	413	647
17800	52	43	3163	−50	65	42	74.3	1056	804	3	23160	59	−115	1100	324	94	418	227	191	418	664
17700	53	28	3208	−45	66	12	75.0	1058	820	3	23600	59	−115	1098	325	97	422	229	193	422	680
17600	54	08	3248	−40	66	38	75.7	1059	834	3	24000	59	−114	1095	325	101	426	231	195	426	694
17500	54	45	3285	−37	67	03	76.3	1061	846	3	24360	59	−114	1092	325	104	429	233	196	429	708
17400	55	20	3320	−35	67	25	76.8	1062	858	3	24700	58	−114	1089	325	107	432	234	198	432	721
17300	55	53	3353	−33	67	47	77.4	1064	869	3	25030	58	−113	1086	326	110	436	236	200	436	735
17200	56	24	3384	−31	68	08	77.9	1065	879	3	25340	58	−113	1082	326	113	439	237	202	439	748
17100	56	54	3414	−30	68	27	78.3	1066	889	3	25630	58	−113	1078	326	115	441	238	203	441	760
17000	57	23	3443	−29	68	46	78.8	1067	899	3	25910	58	−112	1074	326	118	444	239	205	444	772
16900	57	51	3471	−28	69	04	79.2	1069	908	3	26170	57	−112	1070	326	120	446	240	206	446	784
16800	58	18	3498	−27	69	22	79.7	1070	918	3	26420	57	−112	1066	326	123	449	241	208	449	797
16700	58	43	3523	−25	69	39	80.1	1071	927	2	26660	57	−111	1061	326	125	451	242	209	451	809
16600	59	08	3548	−25	69	55	80.4	1072	935	2	26880	57	−111	1057	326	127	453	243	210	453	821
16500	59	32	3572	−24	70	11	80.8	1073	943	2	27100	56	−111	1052	326	129	455	243	212	455	833
16400	59	55	3595	−23	70	26	81.1	1074	950	2	27320	56	−110	1047	327	130	457	244	213	457	844
16300	60	18	3618	−23	70	40	81.5	1075	958	2	27530	56	−110	1042	327	132	459	244	215	459	855

RESTRICTED

TABLE OF NATURAL TRIGONOMETRIC FUNCTIONS

(Note: When entering tables with an angle larger than 45 select such angle from right hand side and obtain values in column corresponding to the function at *bottom* of page.)

Angle	Sin	Cos	Tan	Cot	Sec	Csc	
0°	.0000	1.0000	.0000	∞	1.000	∞	90°
1	.0174	.9998	.0175	57.29	1.000	57.30	89
2	.0349	.9994	.0349	28.64	1.001	28.65	88
3	.0523	.9986	.0524	19.08	1.001	19.11	87
4	.0698	.9976	.0699	14.30	1.002	14.34	86
5	.0872	.9962	.0875	11.43	1.004	11.47	85
6	.1045	.9945	.1051	9.514	1.006	9.567	84
7	.1219	.9925	.1228	8.144	1.008	8.206	83
8	.1392	.9903	.1405	7.115	1.010	7.185	82
9	.1564	.9877	.1584	6.314	1.012	6.392	81
10	.1736	.9848	.1763	5.671	1.015	5.759	80
11	.1908	.9816	.1944	5.145	1.019	5.241	79
12	.2079	.9781	.2126	4.705	1.022	4.810	78
13	.2250	.9744	.2309	4.331	1.026	4.445	77
14	.2419	.9703	.2493	4.011	1.031	4.134	76
15	.2588	.9659	.2679	3.732	1.035	3.864	75
16	.2756	.9613	.2867	3.487	1.040	3.628	74
17	.2924	.9563	.3057	3.271	1.046	3.420	73
18	.3090	.9511	.3249	3.078	1.051	3.236	72
19	.3256	.9455	.3443	2.904	1.058	3.072	71
20	.3420	.9397	.3640	2.747	1.064	2.924	70
21	.3584	.9336	.3839	2.605	1.071	2.790	69
22	.3746	.9272	.4040	2.475	1.079	2.669	68
23	.3907	.9205	.4245	2.356	1.086	2.559	67
24	.4067	.9135	.4452	2.246	1.095	2.459	66
25	.4226	.9063	.4663	2.145	1.103	2.366	65
26	.4384	.8988	.4877	2.050	1.113	2.281	64
27	.4540	.8910	.5095	1.963	1.122	2.203	63
28	.4695	.8829	.5317	1.881	1.133	2.130	62
29	.4848	.8746	.5543	1.804	1.143	2.063	61
30	.5000	.8660	.5774	1.732	1.155	2.000	60
31	.5150	.8572	.6009	1.664	1.167	1.942	59
32	.5299	.8480	.6249	1.600	1.179	1.887	58
33	.5446	.8387	.6494	1.540	1.192	1.836	57
34	.5592	.8290	.6745	1.483	1.206	1.788	56
35	.5736	.8192	.7002	1.428	1.221	1.743	55
36	.5878	.8090	.7265	1.376	1.236	1.701	54
37	.6018	.7986	.7536	1.327	1.252	1.662	53
38	.6157	.7880	.7813	1.280	1.269	1.624	52
39	.6293	.7771	.8098	1.235	1.287	1.589	51
40	.6428	.7660	.8391	1.192	1.305	1.556	50
41	.6561	.7547	.8693	1.150	1.325	1.524	49
42	.6691	.7431	.9004	1.111	1.346	1.494	48
43	.6820	.7314	.9325	1.072	1.367	1.466	47
44	.6947	.7193	.9657	1.036	1.390	1.440	46
45	.7071	.7071	1.0000	1.000	1.414	1.414	45
	Cos	Sin	Cot	Tan	Csc	Sec	Angle

INDEX

Note: The following items are indexed according to the chapters, sections and articles in which they occur. For example, 14A3 indicates Chapter 14, Section A, and Article 3.

Action bill, 22A2
Adjustment, range finders, 11A16
Advance range, indication of, 17D17
Advance range section, range keeper, 14F10
Advance time, definition of, 15D4
Air control, in gyro units, 19C11
Air defense, 22A8
Air density, variations in, 13B8
Ammunition, bag, 2C3; blank, 2C12; bomb-type, 2C7; case, 3C2; chemical, 2C9; classification of, 2C2; containers, 3C3; definitions of, 2C1; dummy drill, 2C13; fixed, 2C5; general types of, 3A; gun, 3A; handling of, 5E; of caliber .50 machine gun, 7D2; impulse, 2C11; pyrotechnic, 2C8; safety precautions, 2C15; semi-fixed, 2C4; trench-warfare, 2C10; of 20 mm gun, 7C13
Angles, of Mark 6 stable element, 17C1
Application of spots, 14B10
Astigmatizers, 11A16
Attack, special forms of, 17D27
Automatic follow-up, Mark 6 stable element, 17C19
Automatic rate control, Mark 1 computer, 17D23
Auxiliary director system, main battery, 18A10
Azides, characteristics of, 2B14; uses of, 2B14

Ballistics, arbitrary, 13C4; automatic corrections, 14F2; basic computing mechanisms and, 14E1; computations, 15C1, 17D7; control, 13C1; data, 15C4; definitions of, 1-8, 13B2; definitions and symbols used in, 15C2; density, 13B8; of the gun, 13B3, summary of, 13B17; incorrect, 21A12; initial, 13C7, 14F3; on machine-gun fire, 19A4; problem of, 13B1; sample computation of, 13B17; statement of the problem, 15C3; total, 13C5
Ballistic table, preparation of, 12A9
Barrel assembly, 40 mm gun, 7B3
Batteries, antiaircraft, 10B1; main, 10B1; secondary, 10B1
Battery alignment, as cause of error, 21A13; checking of, 16E12
Bearing dial group of Mark 1 computer, 17D13
Bearing rate, definition of, 13A8; use of, 13A9
Black powder, characteristics of, 2B3; composition of, 2B3; uses of, 2B3

Bombs, 1-7
Bomb-type ammunition, explosives used in, 9A3; in general, 9A1
Booster extender, operation in depth charge, 9D6
Boosters, characteristics of, 2B7
Boresighting, of gun, 12D4; preparation for, 12D3
Breech assemblies, in general, 4B1; mechanisms of, 4B2; operating mechanism, 4B8; safety precautions regarding, 4B23
Breech mechanism, 5B2; closure, 6B1; operating shaft, 6B2; of 5"/38 caliber gun, 6D6; of 40 mm gun, 7B4; of 3"/50 caliber gun, 6C3; of 20 mm gun, 7C7

Caliber .50 machine gun, ammunition of, 7D2; automatic firing of, 7D10; cocking of, 7D9; description of, 7D3; extracting and ejecting, 7D12; feeding of, 7D11; firing of, 7D6; functioning of, 7D4, 7D5; in general, 7D1; recoiling cycle of, 7D7
Caliber .45 automatic pistol, description of, 8C1; general data on, 8C3; operation of, 8C2
Caliber .30 Springfield rifle, description of, 8B1; general data on, 8B3; operation of, 8B2
Cams, 14E3; plate, 14E3, 6B3; positive motion, 14E3; reducing displacement of, 14E4
Carbine, U. S., caliber .30, 8E
Carriage, of 5"/38 caliber gun, 6D17
Carrier, description of, 5B2
Check levers, 40 mm gun, 7B8
Chemicals, 1-7
Circuits, battle telephone, 22B7; multiple, 22B5
Cocking, of caliber .50 machine gun, 7D7; of 20 mm gun, 7C17
Combat information center, object of, 22A6
Command, during battle, 22A5
Commands (see Appendix C)
Communications, brevity of, 22B2; general requirements of, 22B1; language of, 22B3
Compensation, for erosion, 16D11; for firing delay, 16E11; for horizontal displacement, 16D10; for horizontal parallax, 16D4
Compensator wedges, 11A4

557 **RESTRICTED**

INDEX

Component section, range keeper, 14F8
Component solvers, 14C5; basic, 14C6; schematic representation of, 14C6
Computations, positioning, 15A
Computer, elementary, 14F
Continuous fire, operation of, 17C8; in stable element operation, 17C8
Control and follow-up panel, of stable elements, 17C25
Control arrangements, need for, 16C4; of turret fire control, 18B7
Control switches, of Mark 1 computer, 17D12
Conversion of energy, of torpedo, 9B12
Corrections, computation of, 13B10; multipliers for, 14E5
Corrections, point-of-aim, 13C3; range-table sight scale, 13C2
Counterrecoil mechanism, characteristics of, 4C16
Cradle, 5E3
Cross-level, elevation correction, 17B14; and power drive, 17B15
Cross-level angle, 16E3
Cross-level transmission of Mark 6 stable element, 17C13
Counterrecoiling of caliber .50 machine gun, 7D8
Cyclonite (RDX or hexogen), 2B9
Cylinders, differential, 4C19; main, 4C18

Dead time, 13D3, 14B8
Deck-tilt correction, 16E8
Definitions (see Appendix B), of advance range 14B9; of advance time, 15D4; of ballistics, 14D5; of bearing rate, 13A8; of cams, 14E3; of control ballistic, 13C1; of component solvers, 14C5; of dead time, 13D3; of differentials, 14D5; of fire control, 10A1; used in fire control, 10A1; of firing mechanism, 4B12; of a generator, 4B20; of gunnery, 1-3; of horizontal parallax, 16D3; of integrator, 14D8; of magnification, 11A13; of mean range line, 14B4; of ordnance, 1-2; of partial deck deflection, 16E5; of perpendicular angle, 16A7; of primer, 3B1; of projectile, 3D1; of range finding, 11A1; of range keeping, 13A2; of range rate, 13A7; used in rate control, 15B2; of refraction, 11A3; of relative target motion, 13A3; of reference plane, 16A7; of requisitions, 23C4; of salvo analysis, 21A1; of sight angle, 13B17; of semi-automatic gun, 6A1; of symbols used in director control, 16E1; and symbols of Mark 6 stable element angles, 17C1; used in target positioning, 15A2; of tracer, 2E7; of TNT, 2B8; for torpedo control, 20A4; of torpex, 2B10; of vertical angle, 16A7; of vertical parallax, 16D5
Deflection section of range keeper, 14F11
Deflection setting, 4D11
Depth charges, 1-7; booster extender operation of, 9D6; in general, 9D1; operation of, 9D4; projectors, 9D3; pistol-operation, 9D5; release gear of, 9D2; safety precautions regarding, 9D7
Depth mechanism, operation of, 9B14; of the torpedo, 9B8
Depth perception, 11A11
Detonation, of explosives, 2A6; of fulminate of mercury, 2B13
Detonators, characteristics of, 2B7
Dials, of Mark 1 computer, 17D18
Differential gears, 14C4; schematic representation of, 14C4
Director, function of, 16C2
Director check, 16E13
Director control, example of torpedo, 20C8; gun-crew activities for, 17E6; with computed gun orders, 16E1; with measured gun orders, 16D1
Director elevation indicator of Mark 37 director, 17B12
Director firing, development of, 10A5; Mark 37 system, 17A
Director positioning, with Mark 1 computer, 17D
Director system, auxiliary, 18A10; elementary, 16C1; need for, 16C4
Director train indicator of Mark 37 director, 17B12
Director-train transmission of sensitive element, 17C14
Dispersion, caused by accidental errors, 21A2; law of probability applied to, 21A5
Disrupters, characteristics of, 2B7
Double loading stop of 20 mm gun, 7C11
Drift, 13B9
Dual-purpose battery fire-control system, corrections in, 17A4; description of, 17A2; flow of information, 17A5; functions of, 17A1; illumination control, 17A6; installations, 17A3

Electrical circuits, care of, 4B21
Electric element, of primer, 3B3

INDEX

Elementary computer, functions of, 14F1
Elevating gear, major caliber gun, 5D1
Elevating and training gear, 4C5; of 5"/38 caliber gun, 6D18
Elevating and training stations of turret, 18B4
Elevation, Mark 37 director, 17B13
Elevation dial group, of Mark 1 computer, 17D13
Erosion, cause and effect of, 4A13; compensation for, 4A15, 16D11; control of, 4A14; curves, Appendix D; error due to, 13B7
Error, and calibration, 11A17; causing dispersion, 21A2; causes of, 13B5, 13B10, 13D4; causing shift of M.P.I., 21A9; complementary, 15C16; due to personnel, 21A3; effect on pattern sizes, 21A4; frequency of, 21A6, 21A7; in fire control, 21A14, 21A15; in general range, 14D9; of M.P.I., 21A10; in rates, 14D9; range keeper, 14F14; unit of, 11A14
Exploder, operation of the torpedo, 9B11
Explosive D, description of, 2B11; uses of, 2B11
Explosives, characteristics of, 2A5; classification of, 2A2, 2B1; definitions of, 2A1; explosive D, 2B11; high, 2A4; low, 2A3; reactions of, 2A; service uses of, 2B1; propellent, 2B2
Extractors, of 3"/50 caliber gun, 6B5

Feeding caliber .50 machine gun, 7D11
Fire, buzzer, 16A3; director, 16A6; indirect, 16E9; master key, 16A4; methods of, 16E10; pointer key, 16A2; types of, 15D1
Fire-control language, call and acknowledgment, 22B4; in general, 22B3
Fire control, definition of, 1-9, 10A1; development of, 10A4; equipment for, 1-10; flexibility of, 10A9; history of, 10A2; language of, 22B3; modern armament, 10B; multiple circuits, 22B5; in general, 10A1; symbols and definitions of, 10B7
Fire-control organization, battery control officer, 22A9; chief fire-control officer, 22A7; sector control, 22A10
Fire-control problem, in general, 10B5; presentation of, 10B6; sight settings, applying and firing, 10B5; correcting sight settings, 10B5; determination of sight settings, 10B5; maintaining sight settings, 10B5
Fire-control stations, interior communications of, 22A1; organization of, 22A1

Fire-control systems, of battleship, 10B3; of cruiser and destroyer, 10B4; dual-purpose battery, 17; in general, 10B2; main battery, 18A
Firing, methods of, 18A11
Firing attachments, electrical, 4B17; of 5"/38 caliber gun, 6D10; in general, 4B15; percussion, 4B16
Firing circuits, 4B18
Firing-control mechanism, 7B12
Firing delay, compensation for, 16E11
Firing mechanisms, bag-gun, 4B13; case-gun, 4B14; definition of, 4B12; description of, 5B3; in general, 6C5; of 5"/38 caliber gun, 6D9; of 40 mm gun, 7B5; of 20 mm gun, 7C8
5"/38 caliber gun, ammunition of, 6D3; classes of, 6D1; description of, 6D1, 6D4; firing mechanism of, 6D9; firing attachments of, 6D10; hand operating lever of, 6D7; housing of, 6D5; operation of, 6D1
Flow of information in dual-purpose battery fire-control system, 17A5
Follow-up methods for Mark 6 stable element, 17C7
Follow-up signals, functions of, 17C17
40 mm gun, ammunition of, 7B1; description of, 7B1; distinctive features of, 7B1; summary of operation, 7B14; mount of, 7B15; personnel for, 7B16
Fulminate of mercury, characteristics of, 2B13; detonation of, 2B13; substitute for, 2B14; uses of, 2B13
Functional operation, range keeper, 14F7
Fuze settings, analysis of, 15
Fuzes, and tracers, 3E; arming of, 3E4; classification of, 2E1; design of, 3E2; safety precautions regarding, 3E9; terms, 3E3; time, 3E6
Fuze-setting order, computation of, 17D17
Fuze settings, 15C18; mechanism of, 6D20
Fuze-setting zones, 15D3

Gas checks, 4B5
Gas ejector, 6D11
Gas-ejector system, 5B4
Gas-expelling apparatus, 4B22
Gases of explosives, 2A5
Gears (use as gear constants), 14E2
Generator, definition of, 4B20; sources of power, 4B20
Graphic range keeping, description of, 14B2;

INDEX

limitations of, 14B12; personnel, 14B3; purpose of, 14B1

Graphic solution, AA problem, 15D2

Graphic tracking, definition of, 14A3; description of, 14A2; purpose of, 14A1

Gun ammunition, description of, 3A2; designation of, 3A1; safety precautions for, 3A3

Gun assembly, 4A1, 5B1

Gun ballistic, in general, 13B3; steps in computing, 13B18; summary of, 13B17

Gun barrels, 4A

Gun-control equipment, primary functions of, 17E1

Gun crew of 20 mm gun, 7C16

Gun mounts, elevating and training gear, 4C5; free swinging, 4C6; in general, 4C1; manual drive, 4C7; pedestal mounts, 4C3; turret mounts, 4C4; types of mounts, 4C2

Gun orders, computation of, 16E7

Gun-position, angles of, 16A8; summary of, 15C15; computation of, 17D8

Gun range line, 14B7

Gun sights, development of, 10A3; functions of, 4D1; in general, 12C1; open sight, 4D3; principle of, 12C2; sighting axis, 4D2; small gun sights, 4D13; of 20 mm gun, 7C15

Gun-sight alignment in general, 12D1

Gunnery, definition of, 1-3

Gunnery department, functions of, 1-1

Junior gunnery officers, assignment to duty, 23A2, 23A3; shipboard duties of, 23A5

Guns, 1-7, caliber .50 machine gun, 7D; caliber .45 automatic pistol, 8C; caliber .30 Springfield gun, 8B; classification by use, 4A3; construction of, 4A5; combination of, 4A8; built-up, 4A6; designation by size, 4A2; gun assembly, 4A1; safety precautions, 4A17; radially expanded, 4A7; in service, 4A4; 16"/45 cal., 5B, 5C, 5D; 3"/50 cal., 6C; 5"/38 cal., 6D; 40 mm., 7B; 20 mm., 7C; Thompson sub-machine gun, 8D; U. S. carbine, 8E

Gyro principle, in stable element, 17C20

Gyro setting mechanism, torpedo, 20C6

Gyroscope, initial turn, 9B19; operation of, 9B16

Gyroscopic precession in Mark 14 sight, 19C4

Gyroscope units, use in machine-gun fire, 19C6

Hand operating lever, 6B7; of 5"/38 cal. gun, 6D7

Hand operating mechanism, 40 mm gun, 7B13

Hangfires, 4B10

Heat in explosives, 2A5, 2A6

Heat control in gyro units, 19C11

Height dial group of Mark 1 computer, 17D14

Hexogen (RDX or cyclonite), 2B9

Horizontal displacement, 16D9; compensating for, 16D10

Horizontal parallax, compensating for, 16D4; definition of, 16D3

Housing of, 5"/38 caliber gun, 6D5

Hydraulic recoil system, 6D12

Ignition tubes, 3B4

Illuminating circuits, 4B19

Illumination control, in dual-purpose battery fire-control system, 17A6; in Mark 37 director, 17B18; star-shell fire, 17B19

Indicator-regulators, elevation, 17E3; fuze-setting, 17E4; train, 17E2

Indirect fire, problem of, 13B6

Influence, in explosives, 2A6

Information, sources of, 23B1

Initial-velocity, loss of, 15C13

Initiation, of explosives, 2A6

Inputs of range keepers and computers, 14D3, 14F4, 17D3

Integrators, 14C7; schematic representation of, 14C7

Interior communication of fire-control stations, 22A1

Interrupted-screw plug, 4B4

Jams, of 20 mm gun, 7C18

Knobs, of Mark 1 computer, 17D18

Latitude correction in Mark 6 stable element, 17E21

Latitude motor input of Mark 6 stable element, 17C22

Level angle, 16E3

Level transmission, of Mark 6 stable element, 17C16

Line of fire, diagram of, 13B12

Line of sight, local, 16A1; remote, 16A5; use of, 12A5

Loader assembly, 40 mm gun, 7B6

Loading equipment, 4B9

Loading operation of 5"/38 caliber gun, 6D16

Lost motion, 12D5

Machine-gun control, firing at moving target, 19A2; in general, 19A1; superelevation in, 19A3

Machine guns, definition of, 7A1; operation of, 19C14; personnel for, 19C14; types of, 7A1; uses of, 7A1

INDEX

Magazine of 20 mm gun, 7C14
Magazine interlock gear of 20 mm gun, 7C10
Magnification in range finders, 11A13
Main-battery fire-control system, components of, 18A2; description of, 18A; director equipment of, 18A3; range keepers for, 18A7; range receiver, 18A9
Manual follow-up of Mark 6 stable element, 17C18
Manual rate control, of Mark 1 computer, 17D24; operation of, 17D24; range of, 17D25
Mark 1 computer, automatic rate control of, 17D23; bearing dial group of, 17D13; dip dials of, 17D15; elevation dial group of, 17D13; function of, 17D1; height dial group, 17D14; initial observation operations of, 17D22; inputs of, 17D3; instrument arrangement of, 17D9; operation of, 17D20; operation of without rate control, 17D26; outputs of, 17D2; range dial group of, 17D14; secured conditions of, 17D21; ship-dial group of, 17D10; spot indicators, 17D16; synchronize elevation dials of, 17D15; target dial of, 17D10
Mark 51 director, 19C13
Mark 41 stable element, 18A6
Mark 14 sight, basic functions of, 19C2; description of, 19C5; in general, 19C1
Mark 6 stable element, controls of, 17C4; definitions and symbols, 17C1; dials of, 17C3; follow-up methods, 17C7; functional operation of, 17C10; internal construction of, 17C2; operating personnel, 17C6; purpose of, 17C1; starting operations of, 17C26
Mark 38 director, description of, 18A3; equipment in, 18A3; functions of, 18A4; personnel for, 18A5
Mark 37 director, component parts of, 17B4; general operation of, 17B5; inputs and outputs of, 17B3; operation in train, 17B7; personnel for, 17B6; primary function of, 17B1; secondary functions of, 17B2
Mark 27 director, description of, 20C3
Matériel, care of, 1-12; knowledge of, 1-11
Mean range line, 14B4
Measured quantities, sources for, 13C8
Measuring group of Mark 6 stable element, 17C13
Mechanical representation of quantities, 14C2
Mercury tanks in Mark 6 stable element, 17C23
Mine assembly, operation of, 9C5

Mines, 1-7; classification of, 9C1; explosion of, 9C7; firing mechanisms of, 9C2; in general, 9C1; laying and sweeping, 9C3; moored contact, 9C4; safety devices for, 9C6
Misfires and hangfires, 4B10
Motion, of ship, 13B16; of target, 13B15
Motor generator of stable element, 17C24
Mounts, 4C; parallel motion, 4D12; sight, 4D8; yoke sight, 4D9
Multiple installations of Mark 37 director, 17B20
Multiplication by a constant, 14C3
Multipliers, 14E5; division with, 14E5; screw-type, 14E5; sector-type, 14E5
Naval weapons in general, 1-7
Ogive, definition of, 3D2; description of, 3D2
Operating cycle, 3"/50 cal. gun, 6B8
Operating mechanism of breech closure, 4B8
Operating spring, 3"/50 cal. gun, 6B4
Optical systems, adjuster scale reading, 11A16; internal adjuster, 11A16; main-scale reading, 11A16; operation of, 19C10; use of in machine-gun fire, 19C9
Ordnance, definition of, 1-2
Ordnance publications, doctrinal, 23B4; Fleet, 23B3; in general, 23B1; information bulletins, 23B5; procurement of, 23B2; table of, 23B5
Organization during battle, action bill for, 22A2; command of, 22A5; in general, 22A1; information center, 22A6; similarities in, 22A4; variations in, 22A3
Outputs of Mark 1 computer, 17D2; of Mark 37 director, 17B3
Pallet mechanism of torpedo, 9B17
Pallets, 6B6
Parallax, corrections for fire-control system, 17A4; horizontal, 17A4; vertical, 17A4
Parallelism, 12D6; conditions for, 16E14
Partial deck deflection, definition of, 16E5
Pattern sizes, determination of, 21A8
Percussion of explosives, 2A6
Percussion element of primer, 3B2
Personnel, disposition of, 6D24; of 5"/38 caliber gun, 6D24; of Mark 1 computer, 17D19; of Mark 37 director, 17B6
Pistol operation in depth charge, 9D5
Plotting board, 14B
Plotting-room systems, 16E2
Positioning computations, 15A1; definitions used in, 15A2; relative, 15A3

INDEX

Powder, black, 2B3; smokeless, 2B4
Powder bags, description of, 3C1
Powder hoist, 5E5
Powder scuttle, description of, 6D21
Powder temperatures, changes in, 13B6; errors due to, 13B6
Power, control of, 4C12; transmission of, 4C11
Power drive, automatic control, 4C9; equipment for, 4C10; local control, 4C8; principle of, 4C10
Power transmission, 4C11
Power unit, function in gyro operation, 19C12
Prediction, of deflection, 15C10; of elevation, 15C9; of range, 15C8
Present range section, range keeper, 14F9
Pressure, of explosives, 2A5
Primers, classes of, 3B1; definition of, 3B1; electric element of, 3B3; percussion element of, 3B2
Primers and igniters, 2B6
Prisms, double-reflecting, 11A7; single-reflecting, 11A6
Projectile, after end of, 3D3; body and exterior finish of, 3D4; characteristics of, 12A8; classification of, 3D6; definition of, 3D1; forces acting upon, 12A8; forward end of, 3D2; high capacity, 3D9; markings of, 3D11; rotating band of, 3D5; special-purpose, 3D10; thick-walled armor-piercing, 3D7; variations in weights of, 13D4
Projectile hoist, in general, 5E2; of 5"/38 caliber gun, 6D19
Projectile stowage, 5E1; of depth charges, 9D3
Propellant, function of, 2B2
Propelling charges, 3C
Propelling mechanism of torpedo, 9B7, 9B13

Rammer, 5E4; of 40 mm gun, 7B7
Ramming cycle, 6D15, 7B9
Rammer mechanism, 6D14
Range, advance, 14B9; increase, 14B9; measure of, 17B16; prediction of, 15C8; rate, 15A7
Range dial group of Mark 1 computer, 17D14
Range finder, adjustment of, 11A16; base length of, 11A13; coincidence, 11A10; definition of, 11A1; principle of, 11A8; range triangle, 11A2; stereoscopic, 11A12; types of, 11A9
Range increments, 14C7
Range keeper, description of, 14D2; elementary, 14D1; errors, 14F14; inputs and outputs, 18A8; for main-battery system, 18A7; mechanical, 14C; mechanisms, 14D8; operation of, 14F5; symbols and definitions, 14D6
Range keeping, basic mechanisms, 14C1; definitions of, 13A2; equation of, 14D7; example of, 13A10; graphic, 14B1; in general, 13A1
Range measurement, 11
Range rate, definition of, 13A7; errors in, 13A11; measurement of, 14B6; use of, 13A9
Range setting, 4D10
Range springs, use in machine-gun fire, 19C7
Range tables, Appendix D; compilation of data for, 12B5; corrections for, 12B6; preparation of, 12A10, 12B2; standard conditions for, 12B1; use of, 12A11, 13B4
Ranging, experimental, 12B3; interruptions in, 14B5
Rate control, 14F6, 15B1; on the Baby Ford, 14D9; corrections for, 15B3; definitions used in, 15B2; practical correction, 15B6
Rate-control mechanism of Mark 1 computer, 17D6
Rate correction, 15B6; interdependence in, 15B7; practical, 15B6
Rate errors, detecting, 15B4
Rates, bearing, 15A7; control, 15B1; elevation, 15A7; horizontal, 15A6; in and about the LOS, 15A7; range, 15A7; vertical, 15A6
Ray filters, 11A16
RDX (cyclonite, hexogen), compositions using, 2B9; production of, 2B9
Receivers, elevation, 16C3; functions of, 16C3; train, 16C3
Recoil brake, grooved-wall type, 4C15; throttling-rod type, 4C14
Recoil-counterrecoil system, 40 mm gun, 7B10; counterrecoiling, caliber .50 gun, 7D8
Recoil mechanism, of caliber .50 machine gun, 7D7; characteristics of, 4C13; description of, 5C2; safety precautions for, 4C21
Recuperator, 5C3, 6D13
Reference marks, 11A15
Reference plane, 16A7; inclination of, 16E3
Reference point, 16D2
Reference system, 16A7
Reflection, 11A5
Refraction, explanation of, 11A3
Relative target motion, definition of, 13A3; evaluation of, 13A4; problem of, 13A5
Relative target movement, 15C6
Release gear, depth charge, 9D2

INDEX

Reports, purpose of, 23C1; responsibility for, 23C2

Representation, of differential gears, 14C4; of component solvers, 14C6

Requisitions, definition of, 23C4; purpose of, 23C3

Reticles, 11A15

Rifling, classes of, 4A12; purpose of, 4A12

Ring sights, 19B1; use in machine-gun fire, 19B2

Roller-path inclination, 16D7; compensating for, 16D8; correction for, 17A4

Rotating band, description of, 3D5; purpose of, 3D5

Safety devices, of mines, 9C6; gun slide securing, 4C20

Safety precautions, in firing, 4A17, 5F; in gun loading, 4D14, 5F; in handling ammunition, 2C15; in handling depth charges, 9D7; in handling gun ammunition, 3A3; in handling small arms, 9B20; in handling torpedoes, 9B20; in handling projectiles, 3E9; in magazine or handling room, 5F; from Navy regulations, 1-14, 2B15, 2C15, 3A3, 3C4, 3D12, 4A17, 4C21, 4D14; in opening breech of gun, 4B23

Salvo analysis and spotting, definitions used in, 21A1

Salvo latches, 4B11; of 5"/38 cal. gun, 6D8; of 3"/50 caliber gun, 6C4; safety precautions regarding, 4B23

Sealing, by compression, 4B7; by expansion, 4B6

Searchlights, illumination control, 17B18

Sector control, 22A10

Selected fire, in stable element operation, 17C9; operation of, 17C9

Semi-automatic guns, definition of, 6A1; description of, 6A1; uses of, 6A1

Semple centrifugal plunger, construction of, 3E5

Sensitive element, principal parts of, 17C11

Servos, 16B

Set-up (range keeper), maintenance of, 14D4; summary of, 14D5

Shield, description of, 6D23

Shipboard ordnance, alterations of, 23A8; repairs of, 23A7; upkeep of, 23A6

Ship dial group of Mark 1 computer, 17D10

Sight angle, computation of, 17D17

Sight control, relation to turret control, 18B6

Sight deflection, 13B17; computation of, 17D17

Sight scales, 12C4

Sight settings, 15D5; analysis of, 15; applying and firing, 10B5; correcting, 10B5; determination of, 10B5; maintaining, 10B5; mechanism for, 17E5

Sight stations, of turret, 18B3

Sighting gear, 5D3

Sights, 4D, 6C8; of 5"/38 caliber gun, 6D22; 16"/45 cal. gun, 5B, 5C, 5D

Slide of 5"/38 caliber gun, 6D17

Slide assembly, in general, 5C1; of 40 mm gun, 7B2

Sliding-wedge block (vertical-sliding wedge), 4B3

Small arms, cartridges, 2C6; description of, 8A4; in general, 8A1; operating principles of, 8A3; safety precautions regarding, 8A5; types of, 8A2

Small-arms cartridges, classification of, 2C6

Smokeless powder, development of, 2B5

Spot indicators of Mark 1 computer, 17D16

Spotter, functions and duties of, 21B1, 21B2

Spotting, barrage method, 21B10; bracket or halving method of, 21B8; direct method of, 21B7; error of M.P.I. in deflection, 21B4; error of M.P.I. in range, 21B5; in general, 17B17; ladder method, 21B9; methods of, 21B6; short range, 21B12; tracer method, 21B11

Spotting and salvo analysis, 21A1

Stable element, Mark 6, 17C; Mark 41, 18A6

Stabilized director sights, 16E4

Stand, of 5"/38 caliber gun, 6D17

Star-shell computer, functions of, 17D29; operation of, 17D30

Star-shell illumination, 15D7, 17B19

Steering mechanism of the torpedo, 9B9

Study hints, 1-6

Surface problem, 13, 14

Superelevation, in machine-gun control, 19A3; in machine-gun fire, 19C8

Surface fire, control of, 17D28

Switches, of Mark 1 computer, 17D18

Symbols (See Appendix A.)

Synchronize elevation dials of Mark 1 computer, 17D15

Synchros and Servos, 16B1, 16B2, 16B3

Systems, variations in director, 16E15

Talkers, rules for, 22B6

Target, motion of, 13A3, 13B15

INDEX

Target course, estimation of, 21B3
Target dial of Mark 1 computer, 17D10
Target motion, corrections, 15B5; horizontal rates, 15A6; relative, 15A5, 15C5; vertical rates, 15A6
Target position, advance, 15C7; data used in, 15B1; generated, 15A8; relative, 15A3; symbols and definitions in, 15A2
Target speed, computation of, 17D11; estimation of, 21B3
Telescope, fixed-prism, 4D6; movable-prism, 4D7; sight, 4D4; straight, 4D5; trainer and finder, 11A16
Tetryl, description of, 2B12; uses of, 2B12
Thompson submachine gun, description of, 8D1; general data on, 8D3; operation of, 8D2
3"/50 caliber gun, breech mechanism of, 6C3; description of, 6C1; types of, 6C2
Time and data records, 14B11
Time of flight, 15C17
Time section, range keeper, 14F12
TNT, definition of, 2B8; grades of, 2B8; uses of, 2B8
Torpedo control, control of firing, 20A2; director, 20C8; in general, 20A1; local, 20C7; organization for destroyer, 20C9; symbols and definitions for, 20A4
Torpedo-course indicator, 20C5
Torpedo director system, in general, 20C1; general operation of, 20C2
Torpedo control problem, analytical solution of, 20B2; director solution of, 20C4; graphical solution of, 20B1
Torpedo firing, control of, 20A2; methods of, 20A3
Torpedoes, 1-7; above-water tubes, 9B2; conversion of energy supply of, 9B12; depth mechanism of, 9B14; director system of, 20C1; energy supply of, 9B6; explosive charge of, 9B5; exploder-operation of, 9B11; functional essentials of, 9B4; in general, 9B1; major divisions of, 9B3; operation of, 9B10; pallet mechanism of, 9B13; propelling mechanism of, 9B7, 9B13; safety precautions for, 9B20; steering engine and valve, 9B18; steering mechanism, 9B9, 9B15
Torpedo tubes, above-water type, 9B2
Torpex, definition of, 2B10
Tracer, definition of, 2E7; description of, 3E8
Tracer control, hints on, 19B3; of machine-gun fire, 19B2

Tracking board, 14A
Tracking rate, in machine-gun fire, 19C3
Tracking sheet, 15D6
Train, automatic control, 17B9; local control, 17B8; operation of Mark 37 director in, 17B7; rate control, 17B10; slewing, 17B11
Training gear, 5D2
Train indicator-regulator, 17E2
Trajectory, analysis of, 12A1; characteristics of, 12A4; definitions of, 12A3; elements of, 12A2; elements affecting, 13D4; principle of rigidity, 12A6; symbols of, 12A3
Trajectory curvature, 15C11
Transmitter section, range keeper, 14F13
Transmission systems, units of, 16B1
Trigger mechanism, 40 mm gun, 7B11
Trunnion bearings, 5C4
Trunnion tilt, 12C3, 13D5
Turret fire-control equipment, 18B; elevating and training stations, 18B4; control arrangements, 18B7; personnel for, 18B5; sight control, 18B6; sight stations of, 18B3; turret-officer's booth, 18B2; turret units of, 18B1
Turret installations, main battery, 5A2; major caliber, 5A1; structure of, 5A2
20 mm gun, ammunition of, 7C13; barrel of, 7C4; breechblock of, 7C5; breech casing of, 7C4; breech mechanism of, 7C7; cocking of, 7C17; double-loading stop, 7C11; firing mechanism of, 7C8; in general, 7C1, 7C19; gun crew of, 7C16; gun sights of, 7C15; jams and breakages of, 7C18; magazine of, 7C14; magazine interlock gear of, 7C10; mounts of, 7C2; recoil and counterrecoil of, 7C6; recoiling parts of, 7C5; summary of operation of, 7C12; trigger mechanism of, 7C9
Twin mounts, differences from single, 6D25

Underwater explosions, 9A2
U. S. carbine, caliber .30, description of, 8E1; general data on, 8E3; operation of, 8E2

Velocity, of explosives, 2A5
Vertical-indicating group of Mark 6 stable element, 17C12
Vertical parallax, compensating for, 16D6; definition of, 16D5

Wind, apparent, 13D1, 13D2, 15C12; ballistic, 13B14; effects of, 13B14

X and Y components, determination of, 13A6

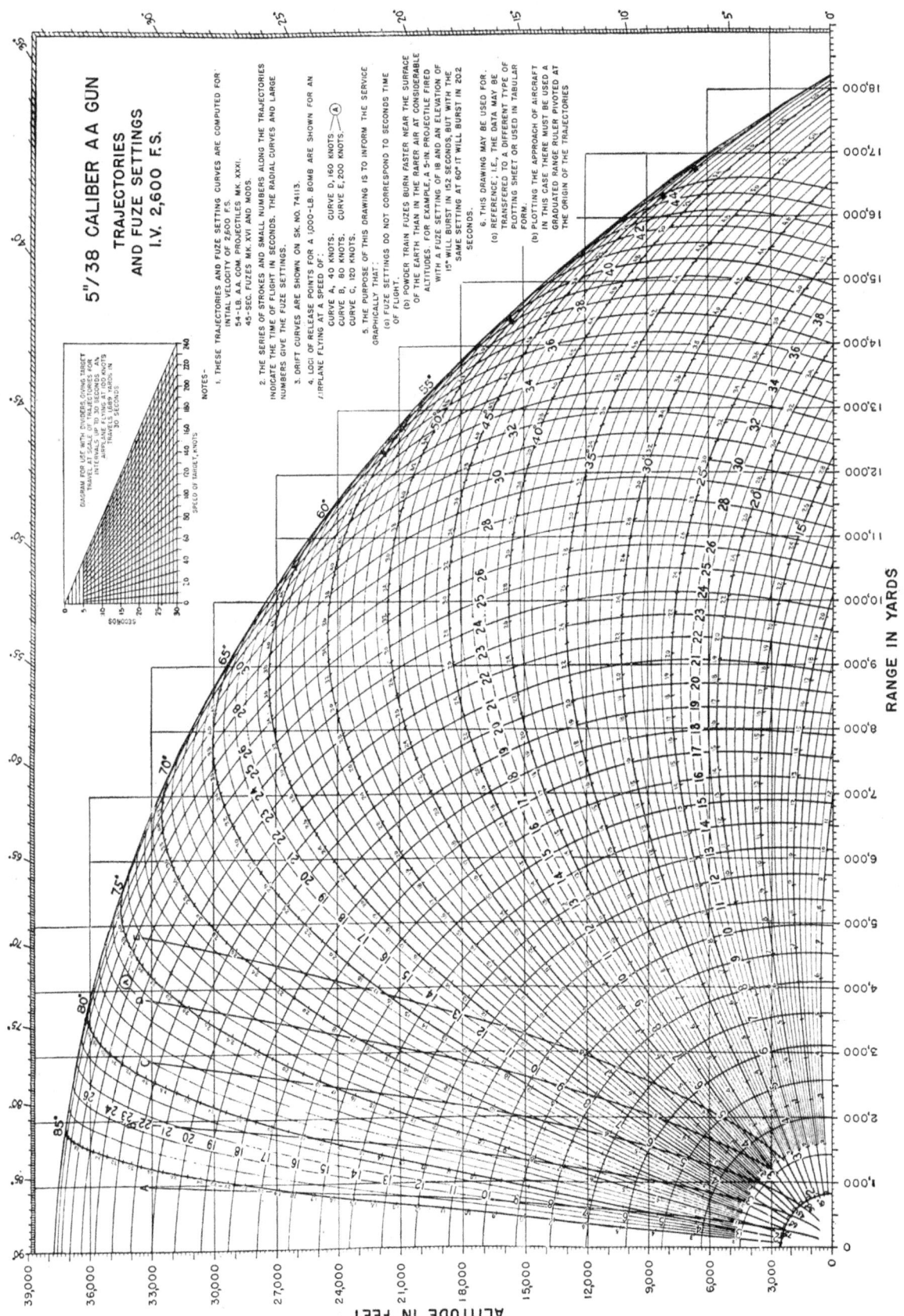

©2013 Periscope Film LLC
All Rights Reserved
ISBN#978-1-937684-22-8
www.PeriscopeFilm.com

www.ingramcontent.com/pod-product-compliance
Lightning Source LLC
Chambersburg PA
CBHW080719300426
44114CB00019B/2424